Advanced Metallization
Conference 1999
(AMC 1999)

Advanced Metallization Conference 1999 (AMC 1999)

Proceedings of the Conference held September 28–30, 1999, in Orlando, Florida, sponsored by Continuing Education in Engineering, University Extension, University of California at Berkeley, California, U.S.A. An Asian session of the Conference was held October 14–15, 1999, in Tokyo, Japan.

EDITORS:

Mihal E. Gross
Bell Labs, Lucent Technologies
Murray Hill, New Jersey, U.S.A.

Thomas Gessner
Chemnitz University of Technology
Chemnitz, Germany

N. Kobayashi
Hitachi, Ltd.
Tokyo, Japan

Y. Yasuda
Nagoya University
Nagoya, Japan

Materials Research Society
Warrendale, Pennsylvania

Single article reprints from this publication are available through
University Microfilms Inc., 300 North Zeeb Road, Ann Arbor, Michigan 48106

CODEN: MRSPDH

Published by:

Materials Research Society
506 Keystone Drive
Warrendale, PA 15086
Telephone (724) 779-3003
Fax (724) 779-8313
Web site: http://www.mrs.org/

Library of Congress Cataloging-in-Publication Data

Advanced Metallization Conference 1999 (AMC 1999) : proceedings of the conference held
    September 28–30, 1999, in Orlando, Florida; Asian session was held October 14–15, 1999
    in Tokyo, Japan / editors, Mihal E. Gross, Thomas Gessner, N. Kobayashi, Y. Yasuda
        p.cm.—(Materials Research Society conference proceedings)
        "Sponsored by Continuing Education in Engineering, University Extension, University
        of California at Berkeley, California, U.S.A."
        ISBN 1-55899-539-0
        ISSN: 1048-0854
        I. Gross, Mihal E. II. Gessner, Thomas III. Kobayashi, N. IV. Yasuda, Y. V. Series:
    Materials Research Society conference proceedings

Manufactured in the United States of America

CONTENTS

Preface xix

Acknowledgments xxi

Materials Research Society Conference Proceedings xxiii

KEYNOTE PRESENTATION

The MARCO/DARPA Interconnect Focus Center 3
   James D. Meindl

INTEGRATION OF DAMASCENE
ARCHITECTURES

*Critical Reliability Issues for sub-0.25-micron Generation Copper Wiring 9
   A.K. Stamper, W.A. Klaasen, and R.A. Wachnik

Comparison of Al and Cu Metallizations for Dual Damascene Applications 17
   L.A. Clevenger, S.G. Malhotra, A.H. Simon, R. Iggulden,
   and S.J. Weber

Microstructure of Aluminum Damascene Interconnects Inlaid by Chemical
Vapor Deposition 27
   Akira Furuya and Yoshihiro Hayashi

Integration of the 3MS Low-k CVD Material in a 0.18μm Cu Single
Damascene Process 33
   T. Gao, W.D. Gray, M. Van Hove, H. Struyf, H. Meynen,
   S. Vanhaelemeersch, and K. Maex

High-Performance Copper Dual Damascene Interconnects Using an
Organic Low-k Dielectric 41
   Ikuhiro Yamamura, Kazuhiko Tokunaga, Koichi Ikeda, Toshiaki Hasegawa,
   Koji Miyata, Tetsuya Tatsumi, Mitsuru Taguchi, Keiichi Maeda,
   Naoki Komai, Masanaga Fukasawa, Hideyuki Kito, and Shingo Kadomura

*Invited Paper

v

Comparison of Via Partial-Etch Approach and Via Full-Etch Approach in
Dual Damascene Process 47
  L.C. Chao, M.H. Lui, J.C. Liu, C.C. Chen, C.S. Tsai,
  and M.S. Liang

COPPER: DEPOSITION PROCESSES
AND PROPERTIES

*Factors Influencing Fill of IC Features Using Electroplated Copper 53
  J. Reid and S. Mayer

Effects of Plating Current Density and Solution Additive on the
Microstructure and Recrystallization Rate of Electroplated Copper Films 63
  Haebum Lee, Sergey D. Lopatin, Dae-Yong Kim, Valery M. Dubin,
  and S. Simon Wong

A Novel Electrolyte for Wafer Plating 69
  A. Thies, H. Meyer, J. Helneder, M. Schwerd, and
  T. Gebhart

Copper Room Temperature Resistance Transients as a Function of
Electroplating Parameters 77
  S.G. Malhotra, P.S. Locke, A.H. Simon, J. Fluegel, C. Parks,
  P. DeHaven, D.G. Hemmes, R. Jackson, and E. Patton

Mechanistic Studies of the Room Temperature Recrystallization of
Electroplated Damascene Copper and Sputter-Deposited Copper 85
  M.E. Gross, R. Drese, D. Golovin, W.L. Brown, C. Lingk,
  S. Merchant, and M. Oh

Microstructure of Electroplated Cu Films 93
  H. Okabayashi, K. Ueno, N. Ito, S. Saitoh, and E. Nomura

Pattern-Dependent Surface Profile Evolution of Electrochemically
Deposited Copper 101
  Tom Ritzdorf, Dakin Fulton, and Linlin Chen

The Influence of Additives on the Room-Temperature Recrystallization
of Electrodeposited Copper 109
  C.R. Stafford, M.D. Vaudin, T.P. Moffat, N. Armstrong, and
  D.R. Kelley

*Invited Paper

Use of On-Line Chemical Analysis for Copper Electrodeposition 117
R.J. Contolini, J.D. Reid, S.T. Mayer, E.K. Broadbent, and
R.L. Jackson

**Characterization of Electroplated and MOCVD Copper for Trench
Fill In Damascene Architecture** 125
K. Weiss, S. Riedel, S.E. Schulz, H. Helneder, M. Schwerd,
and T. Gessner

Copper Electroplating: Processing and Integration 133
C.S. Hsiung, K. Hsieh, W.Y. Hsieh, and W. Lur

Crystal Texture of Electroplated Damascene Cu Interconnects 137
A.H. Fischer, A. von Glasow, A. Huot, and R.A. Schwarzer

**Effect of Plating Current Density on ECD Cu Gap Fill and
Film Properties** 143
M.H. Tsai, W.J. Tsai, S.L. Shue, C.H. Yu, and M.S. Liang

**Roles of Additives During Filling Process of Damascene Structures
With Electrochemical Deposited Copper** 149
E. Richard, I. Vervoort, S.H. Brongersma, H. Bender, G. Beyer,
R. Palmans, S. Lagrange, and K. Maex

**A New Post-ECD Annealing Technique for Cu Interconnects
Using a High-Pressure Gas Ambient** 155
Kohei Suzuki, Takuya Masui, Takao Fujikawa, Yoji Taguchi,
and Tomoyasu Kondo

Forcefill Applied to ECD Copper 161
G.C.A.M. Janssen and T. Ohba

Rapid Thermal Annealing of Electro Chemically Plated Cu Films 167
G.P. Beyer, P. Kitabjian, S.H. Brongersma, J. Proost, H. Bender,
E. Richard, I. Vervoort, P. Hey, P. Zhang, and K. Maex

**Formaldehyde-Free Electroless Copper Deposition for
Ultra-Large-Scale-Integration (ULSI) Metallization** 173
Y. Shacham-Diamand, N. Petrov, E. Sverdlov, R. Cheung,
and L. Castro

Electroless Growth of 10-100 nm Copper Films 181
S. Lopatin

New Copper Interconnect Technology Using High-Pressure
Anneal Process 187
    T. Fujikawa, K. Suzuki, T. Masui, and T. Ohnishi

Adhesion Study on Copper Films Deposited by MOCVD 195
    S. Riedel, K. Weiss, S.E. Schulz, and T. Gessner

Alcohol and Amine Adducts of Cu(hfac)$_2$ for CVD Copper 201
    A.W. Maverick, H. Fan, M.S. Bufaroosha, Z.T. Cygan,
    A.M. James, F.R. Fronczek, G.L. Griffin, and J.Y. Boey

Effects of Process Variables on Cu(TMVS)(hfac) Sourced Copper
CVD Films 207
    D. Yang, J. Hong, and T.S. Cale

New Copper Precursor for Chemical Vapor Deposition of Pure
Copper Thin Films 213
    Weiwei Zhuang, Lawrence J. Charneski, David R. Evans,
    and Sheng Teng Hsu

Material Issues in Silver Metallization 219
    E. Kondoh and T. Asano

BARRIERS FOR COPPER

Improvement of Diffusion Barrier Performance by a Thin Al Interlayer
Deposited Between Barrier and Copper 227
    Kyoung-Ho Kim, Se-Joon Im, Soo-Hyun Kim, and Ki-Bum Kim

Electrical properties and adhesion of Cu/Zr/TaN Layer for
Cu Metallization 233
    C.J. Uchibori, N. Shimizu, and T. Nakamura

Process Development and Film Characterization of CVD Tantalum
Nitride as a Diffusion Barrier for Copper Metallization 239
    Se-Joon Im, Soo-Hyun Kim, Ki-Chul Park, Sung-Lae Cho,
    and Ki-Bum Kim

Integration of Vacuum Arc Plasma Deposited Diffusion Barrier
Films with Electrochemical Cu Deposition Process 245
    O.R. Monteiro, I.G. Brown, I.C. Ivanov, D. Papapanayioutou,
    and C. Ting

Preclean Barrier and Seed Layers for Dual Damascene Copper
Metallization 251
    J.M. Gilet, J. Torres, M. Swaanen, and R. Gonella

Modification of Ta-Based Thin-Film Barriers by Ion Implantation
of Nitrogen and Oxygen 257
    E. Wieser, J. Schreiber, C. Wenzel, J.W. Bartha, B. Bendjus,
    V. Melov, M. Peikert, W. Matz, B. Adolphi, and D. Fischer

Effects of Copper Diffusion Barrier on Physical/Electrical
Barrier Properties and Copper Preferred Orientation 265
    C.S. Liu, S.L. Shue, C.H. Yu, and M.S. Liang

Improvement of <111> Texture and Suppression of Agglomeration
of Cu Film by Ti/CVD-TiN/Ti Underlayer 271
    Mitsuru Sekiguchi, Haruhiko Sato, Takeshi Harada, and
    Shinichi Domae

Ideal Microstructure of Thermally Stable TiN Diffusion Barriers
for Cu Interconnects 277
    M. Moriyama, T. Kawazoe, M. Tanaka, and M. Murakami

Low Resistivity TiN Films at Low Deposition Temperature Using
Flow Modulation Chemical Vapor Deposition (FMCVD) 283
    H. Hamamura, R. Yamamoto, H. Komiyama, and
    Y. Shimogaki

X-ray Photoelectron Spectroscopic Study of Post-Treatment on
Chlorine in CVD-TiN Films 289
    Yoko Uchida, Hidekazu Goto, Tsuyoshi Tamaru, and
    Yoshitaka Nakamura

Conformal MOCVD-TiN Films Deposited From TDEAT and
Ammonia 295
    S.Q. Xiao, R. Tobe, K. Suzuki, X.B. Xu, A. Sekiguchi,
    H. Doi, O. Okada, and N. Hosokawa

Electrical Integrity Monitoring of Electroless Barriers for
Copper Interconnects 301
    Yosi Shacham-Diamand, Barak Israel, and Yelena Sverdlov

SIMS Study of Copper Quantification and Diffusion in Silicon,
Silicon Dioxide, and Silicon Nitride 307
    K.K. Harris, F.A. Stevie, J.M. McKinley, S.M. Merchant,
    and M. Oh

Diffusion Properties of Cu in Nanocrystalline Diamond Thin Films 313
  D. Zhou, F.A. Stevie, E. Anoshkina, H. Francois-Saint-Cyr,
  K. Richardson, A. Hussain, and L. Chow

Studies of Particle Formation Phenomenon Observed Between Silicon
Nitride Capping Layer and Copper 319
  Kia Seng Low, Werner Pamler, Markus Schwerd, Heinrich Koerner,
  Hans-Joachim Barth, and Anthony O'Neill

LOW-k DIELECTRICS

*Sacrificial Macromolecular Porogens: A Route to Porous
Organosilicates for On-Chip Insulator Applications 327
  R.D. Miller, W. Volksen, J.L. Hedrick, C.J. Hawker,
  J.F. Remenar, P. Furuta, C.V. Nguyen, D. Yoon, M. Toney,
  D.P. Rice, and J. Hay

Optimization of Interconnection Systems Including Aerogels by
Thermal and Electrical Simulation 335
  R. Streiter, H. Wolf, U. Weiss, X. Xiao, and T. Gessner

Characterization of Nanoporous Low-k CVD Film for ULSI
Interconnection 343
  Choon Kun Ryu, Si-Bum Kim, Sam-Dong Kim, and
  Chung-Tae Kim

Evaluation of Nanoglass in a Single Damascene Structure 349
  C.B. Case, M. Buonanno, G. Forsythe, H. Maynard, J. Miner,
  W.W. Tai, and J.J. Yang

SiLK H for Damascene and Gap Fill Application 357
  G. Passemard, J.C. Maisonobe, C. Maddalon, A. Achen,
  M. Assous, C. Lacour, N. Lardon, R. Blanc, and
  O. Demolliens

A Novel Integration Approach to Organic Low-k Dual Damascene
Processing 365
  K. Miyata, M. Fukasawa, T. Tatsumi, M. Taguchi, T. Hasegawa,
  K. Ikeda, and S. Kadomura

*Invited Paper

*Applications for Organosilicon Gases in PECVD Processes for
Low-k Dielectrics 371
    M.J. Loboda

Black Diamond™ A Low-k Dielectric for Cu Damascene Applications 379
    Wai-Fan Yau, Yung-Cheng Lu, Kuowei Liu, Nasreen Chopra,
    Tze Poon, Ralf Willecke, Ju-Hyung Lee, Paul Matthews,
    Tzufang Huang, Robert Mandal, Peter Lee, Chi-I Lang,
    Dian Sugiarto, I-Shing Lou, Jim Ma, Ben Pang, Mehul Naik,
    Dennis Yost, and David Cheung

Effect of Thermal Conductivity and Fluorine of Low-k HDP-SiOF
Dielectric on EM Behavior of Al-Cu Line 387
    Dong-Chul Kwon, Yong-Jin Wee, Yun-Ho Park, Hyeon-Deok Lee,
    Ho-Kyu Kang, and Sang-In Lee

Low-k (≈2.0) Plasma CF Polymer Films Modified by *In Situ* Deposited
Carbon Rich Adhesion Layers 395
    M. Uhlig, A. Bertz, M. Rennau, S.E. Schulz, T. Werner, and
    T. Gessner

Integration of Cu/FSG Dual Damascene Technology for
Sub-01.8μm ULSI Circuits 403
    C.C. Liu, C.Y. Tsai, Y.M. Huang, J.Y. Wu, and W. Lur

A Modified Capacitance/Voltage Technique to Characterize
Copper Drift Diffusion in Organic Low-k Dielectrics 409
    F. Lanckmans, L. Geenen, W. Vandervorst, and
    Karen Maex

Oxazole Dielectric (OxD) as a Potential Low-k Dielectric:
Properties and Preliminary Integration Results With a Cu
Damascene Architecture 417
    Manfred Engelhardt, Hans Helneder, Guenter Schmid,
    Michael Schrenk, Markus Schwerd, Uwe Seidel,
    Heinrich Körner, and Recai Sezi

Plasma CVD of Low-k a-C:F Films Using Substitutional PFC 425
    T. Shirafuji, Y. Hayashi, and S. Nishino

The Use of Fluorinated Oxide (FSG) for Al-Cu and Cu Interconnects 431
    W. Chang, T.I. Bao, C.L. Chang, S.M. Jang, C.H. Yu, and M.S. Liang

*Invited Paper

BLOк™ —A Low-k Dielectric Barrier/Etch Stop Film for Copper
Damascene Applications 437
    P. Xu, K. Huang, A. Patel, S. Rathi, J. Ferguson, B. Tang,
    J. Huang, C. Ngai, and M. Loboda

Low Dielectric Constant Mechanism of Amorphous FluoroCarbon
(a-C:F) Film 443
    E.G. Loh, F.R. Hutagalung, H. Komiyama, and Y. Shimogaki

Extreme Low Dielectric Interlayers—A Nanoporous Polymers Approach 449
    Thomas J. Markley, Xiaoping Gao, Michael Langsam,
    Lloyd M. Robeson, Mark L. O'Neill, Paul R. Sierocki,
    Shahrnaz Motakef, and David A. Roberts

Process Optimization of Hydrogen Silsesquioxane (HSQ) Via Etch
for 0.18μm Technology and Beyond 455
    S.C. Yang, M.H. Huang, Y.H. Chiu, B.R. Young, H.J. Tao,
    C.S. Tsai, Y.C. Chao, and M.S. Liang

O_2 Plasma Treatment of Low-k Organic Silsesquioxane for Novel
Intermetal Dielectric Application 461
    T. Yoshie, S.C. Chen, and J. Kanamori

Vapor Deposition Polymerization of Parylene Integral Foam Thin Films 467
    James Erjavec, John Sikita, Stephen P. Beaudoin,
    and Gregory B. Raupp

Optical Metrology for Monitoring the Cure of SiLK Low-k Dielectric
Thin Films 473
    F. Yang, W.A. McGahan, C.E. Mohler, and L.M. Booms

An Integrated Low-k HDP-FSG for 0.15μm Copper Interconnects 479
    Hichem M'saad, Manoj Vellaikal, Wen Ma, and Kent Rossman

ALUMINUM, TUNGSTEN, AND
DRAM METALLIZATION

High-Temperature Stability of Conducting Ir-Ta-O Film in
Oxygen Ambient 487
    Fengyan Zhang, Sheng Teng Hsu, Jer-shen Maa, Shigeo Ohnishi,
    and Norito Fujiwara

A Study on the Metallorganic Chemical Vapor Deposition (MOCVD)
of Ru and RuO2 Using Ruthenocene Precursor and Oxygen Gas 495
    Hyun-Mi Kim, Sung-Eon Park, and Ki-Bum Kim

Optimization of Pre-Metal Deposition Sputter Cleans for
1-Gigabit DRAMS 501
    R.C. Iggulden, L.A. Clevenger, J. Gambino, R.F. Schnabel,
    and S.J. Weber

Dependence of Metal Contact Resistance on Contact Size Down
to 0.14μm and on Aspect Ratio Up To 12 507
    Hyunchul Sohn, Chang-Young Kim, In-Haeng Lee,
    Choon-Hwan Kim, Hyung-Soon Park, Kyung-Bok Lee,
    Woo-Hyun Kim, Soo-Jin Kim, Inn-Cheol Ryu, Hyug-Jin Kwon,
    Heung-Lak Park, and Dong-Joon Ahn

Tungsten Gate Structure Formation by Reduced Temperature
Conversion of Tungsten Nitride 513
    C.J. Galewski, A.R. Londergan, C.A. Sans, T.E. Seidel, D. Clarke,
    and Q. Zhu

Reliable Pretreatment Technology for Hot Process During W Polymetal
Gate Formation 521
    Yasushi Akasaka, Kiyotaka Miyano, Kouji Matsuo, Kazuaki Nakajima,
    and Kyoichi Suguro

Thermally Stable W Bit Line Process for 256M bit DRAM and Beyond 527
    Jeongeui Hong, Youngjun Lee, Won-Hwa Jin, Kyu-Hyun Kim,
    Tae-Seok Kwon, Pilsung Kim, Won-Jun Lee, Hong-Seok Kim,
    and Sa-Kyun Rha

Mechanisms of Tungsten Plug Corrosion in Borderless Aluminum
Interconnects 533
    S. Muranaka, I. Kanno, and H. Sasai

Bias Sputtered Tungsten as a Diffusion Barrier and Nucleation Film
for Tungsten CVD in Oxide Vias 539
    S.B. Herner, H-M. Zhang, B. Sun, Y. Tanaka, K.A. Littau,
    and G. Dixit

Surface Reaction Model of Al Growth Using DMAH: Elementary
Reaction Simulation and Surface Analysis 545
    M. Sugiyama, H. Ogawa, T. Nakajima, T. Tanaka, H. Itoh,
    J. Aoyama, Y. Egashira, K. Yamashita, Y. Horiike, H. Komiyama,
    and Y. Shimogaki

Theoretical Study on the Reactivity of Oxidized Aluminum Surfaces: Effects of Adsorbed Metallic Atoms (Au, Cu, Ti, V) 551
T. Tanaka, T. Nakajima, and K. Yamashita

Compositional Defect Analysis on a 300 mm Wafer by Using Auger Based Defect Review Tool 557
R. Oiwa, T. Ohba, T. Morohashi, K.D. Childs, D.F. Paul, and S.P. Clough

SILICIDES

Thermally Stable SALICIDE Technology Using Epitaxial CoSi$_2$ With Low Junction Leakage and Lower Ideal Factor 565
Tatsuo Sugiyama, Ryuji Etoh, Masato Kanazawa, Kikuko Tsutsumi, and Shinichi Ogawa

Effects of Nitrogen and Ge Ion Implant on Cobalt Silicide Formation 571
M.Y. Wang, S.M. Jeng, S.L. Shue, C.H. Yu, and M.S. Liang

Thermal Desorption Issues Related to Silicidation and Back-End Metallization 577
Hua Li, G. Vereecke, M. Schaekers, M.R. Baklanov, E. Sleeckx, K. Maex, and L. Froyen

Retardation of CoSi$_2$ Formation on Sub 0.18μm Gate Poly Due to Oxide Resputtering by *In Situ* RF Sputter Etch Precleaning 583
Ju-Hyuk Chung, Jang-Eun Lee, Ja-Hum Ku, Eung-Joon Lee, Jong-Wang Park, Young-Hyun Lee, Chul-Sung Kim, Suh-Hu Park, U-In Chung, Geung-Won Kang, and Moon-Yong Lee

ICP-Ar/H$_2$ Precleaning and Plasma Damage-Free Ti-PECVD for Sub-Quarter Micron Contact of Logic With Embedded DRAM 589
T. Taguwa, K. Urabe, T. Yamamoto, and H. Gomi

CMP/CLEANING/ETCHING

*Copper and Tantalum Dissolution and Planarization in H$_2$O$_2$-Based Slurries 597
M. Hariharaputhiran, S. Ramarajan, Y. Li, and S.V. Babu

*Invited Paper

**Effect of Plasma Treatment on the CMP Behavior of Organic
Low-k Material** 605
    S.J. Hong, H.D. Chung, B.U. Yoon, S.R. Hah, J.T. Moon,
    and S.I. Lee

**Oxygen Enhanced Dry Etching of Low-ϵ Organic SOG (ϵ=2.9) for
Dual-Damascene Interconnects** 609
    Shuntaro Machida, Atsushi Maekawa, Takao Kumihashi,
    Takeshi Furusawa, Kazutami Tago, Takashi Yunogami,
    Takafumi Tokunaga, and Kazuo Nojiri

**Characterization of Plasma Etch Related Residues Formed on Top
of ECD Cu Films** 615
    M.R. Baklanov, T. Conard, F. Lankmans, S. Vanhaelemeersch,
    D. Holmes, and K. Maex

**Effects of Physical/Reactive Clean, Dry/Wet Clean and Barrier Metal
Thickness on Cu Dual Damascene Process** 621
    C.S. Liu, S.L. Shue, C.H. Yu, and M.S. Liang

**Application of High-Pressure Argon Sputter Etch "Contouring" for the
Integration of High Reliability Low Cost VLSI and ULSI Multilevel
Metallization** 627
    G.M. Grivna

Optical Emission Diagnostic to Avoid Contact Etch Stop 633
    S.C. McNevin, K.V. Guinn, M. Cerullo, J. Ashley Taylor,
    and K. Tokashiki

**Characterization of a Novel Method of Cleaning Wafer Back Sides
and Effecting a Bevel Etch in a Single Processing Module** 637
    C. Dundas, T. Ritzdorf, Gary Curtis, and Steve Peace

**Impact of Post Window Etch Cleans Process on Reliability of
0.25µm Vintage Windows** 643
    Y.S. Obeng, J.S. Huang, S.H. Kang, X. Lin, and A.S. Oates

R&D Program for Reduction of PFC Emission at ASET 649
    Hironori Matsunaga, Takuya Fukuda, Nobuo Aoi, and Hitoshi Itoh

PROCESS MODELING

*Equipment Simulation for Copper Metallization Technology 655
Christoph Werner and Alfred Kersch

Simulation of Ionized Magnetron Sputtering of Copper 663
M.O. Bloomfield, D.F. Richards, and T.S. Cale

Full 3D Microstructural Simulation of Refractory Films Deposited
by PVD and CVD 669
T. Smy, R.V. Joshi, and S.K. Dew

Modeling of Feature-Scale Planarization in Cu CMP Using MESA™ 677
T. Laursen, S.R. Runnels, S. Basak, M. Grief, and K. Murella

A Modified Removal Rate Model for Both *In Situ* and *Ex Situ* Pad
Conditioning CMP 683
Eric Tseng, Michael Meng, and S.C. Peng

RELIABILITY

*Electromigration Reliability Study of Submicron Cu Interconnections 691
C-K. Hu, R. Rosenberg, W. Klaasen, and A.K. Stamper

*Reliability Improvement of Aluminum Dual-Damascene Interconnects
by Low-k Organic SOG Passivation Dielectrics 699
H. Kaneko, T. Usui, T. Watanabe, S. Ito, M. Kawai, and
M. Hasunuma

Reliability Enhancement of Copper Interconnects Using Wafer-Level
Electromigration Testing 705
V.V.S. Rana, J. Educato, S. Parikh, M. Naik, T. Pan, P. Hey,
D. Yost, and D. Pierce

3-D Electrical Resistance Model of Via and Contact Reliability
in Integrated Circuit Metallization 713
S.H. Kang, A.S. Oates, and Y.S. Obeng

Influence of Test Structure Shape and Test Conditions in High
Accelerated Cu-Electromigration Tests 721
J. Ullmann, T. Kötter, W. Hasse, and S.E. Schulz

*Invited Paper

**Electromigration Induced Edge Drift Velocity Measurement
by Blech Pattern With Multiple Voltage Probes** 727
    S. Shingubara, T. Osaka, H. Sakaue, T. Takahagi, and
    A.H. Verbruggen

**A Novel Interconnect Stack for Improved Resistance to
Electromigration and Stress-Induced Voiding** 735
    R. Jaiswal, T.S. Lee, K.K. Looi, I. Lim, A. Velaga,
    S.M. Merchant, J.S. Huang, J. Li, S. Karthikeyan,
    J. Zhang, S. Lai, and G. Yao

**Contact Resistance Modelling as a Function of Silicide Type
for ULSI Devices** 741
    G.K. Reeves, A.S. Holland, and P.W. Leech

**The Growth of Extrusions at W-Terminated AlSiCu Lines
and an Approach for an Extrapolation to Use Conditions** 747
    F. Ungar and A.V. Glasow

Author Index 753

Subject Index 759

PREFACE

This volume contains the proceedings of the "Advanced Metallization Conference (AMC) 1999" and the "Advanced Metallization Conference-Asia (ADMETA) 1999," affiliated conferences that were held in Orlando, Florida, September 28-30, 1999, and in Tokyo, Japan, October 14-15, 1999, respectively. This year's meetings, the sixteenth in the series, reflected the rapid advances and changes taking place in the field of metallization systems for integrated circuit (IC) applications. The aim of the AMC and ADMETA conferences is to provide forums within the IC metallization community, across industrial, academic and government institutions, for presentation and discussion of leading edge research and development, and technology. The revolution in materials and processes for IC metallization involving Cu and low-k dielectrics provides exciting challenges moving forward.

This year's AMC conference featured papers on Cu and low-k dielectrics spanning materials properties, processing, integration, and reliability. A majority of the 11 invited talks, 37 contributed talks, and 67 poster presentations were related to these subject areas. The keynote presentation by Prof. James D. Meindl (Georgia Institute of Technology) provided an overview of the MARCO/DARPA Interconnect Focus Center, a cooperative research program involving several top universities in the U.S.A. whose mission it is to explore new interconnect strategies for the future. One of several highlights of the conference was a panel discussion on "Optimal Scaling and Materials for Interconnects as a Function of Different Chip Products." Leading experts in the industry, including Alina Deutsch (IBM), Bob Havemann (SEMATECH, Texas Instruments), Avi Liebermensch (Sun Microsystems), Jon Reid (Novellus) and Douglas Yu (TSMC), took questions from the panel moderator Daniel Edelstein (IBM) and the audience. One consistent message was that optimization of interconnect performance will require increasing communications between metallization technologists, integrators, circuit designers, and reliability engineers. Over 250 attendees contributed to the success of the conference, now informally dubbed the "Copper and Low-k Workshop" in keeping with the origin of the conference as the "Tungsten Workshop" in the early 1980's.

The ADMETA conference was held October 14-15 in Tokyo and was chaired by Dr. N. Kobayashi (Chair/Hitachi) and Prof. Y. Yasuda (Vice Chair/Nagoya University). The keynote speech "Low Cost and High Performance DRAM Technology" was given by Dr. Gurtej Sandhu (Micron Technology). In total, 6 invited papers and 18 contributed papers, as well as 11 poster papers, were presented, focusing on leading edge work in the area of DRAM, Al, and high-k dielectric materials, processes, and technologies, in addition to Cu and low-k dielectrics. Over 190 attendees contributed to the success of the conference. Highlight papers selected by the Asian Executive Committee are included in this book.

In keeping with the conference goals of information exchange, highlights of papers from the ADMETA conference were presented in Orlando by Prof. Y. Yasuda (Nagoya University), and highlights of papers from the AMC conference were presented in Tokyo by Dr. Mihal E. Gross (Bell Labs, Lucent Technologies).

The AMC 2000 conference will be held October 3-5, 2000, in San Diego, California, and will be chaired by Dr. Girish Dixit (Applied Materials) and Dr. Daniel Edelstein (IBM). The ADMETA 2000 conference will be held October 18-20, 2000, at the University of Tokyo, Tokyo, Japan and will be chaired by Prof. Y. Yasuda (Nagoya University) and Dr. T. Ohba (Selete/Fujitsu). For further information, please contact Linda Reid, University of California at Berkeley, by phone 510-642-4151, fax 510-642-6027, or e-mail at course@unx.berkeley.edu, or visit the web site at http://www.berkeley.edu/unex/eng/metal. For ADMETA, please contact H. Doshida by phone 81-3-5821-7210, fax 81-3-5821-7439, or e-mail at QYZ05607@nifty.ne.jp. We encourage you to participate in these conferences and look forward to meeting you there in 2000.

Mihal E. Gross
Thomas Gessner
N. Kobayashi
Y. Yasuda

January 2000

ACKNOWLEDGMENTS

The continued success of the "Advanced Metallization Conference" relies on the efforts of many people. It is primarily based on the strong support and contributions from its speakers, authors, panelists, and attendees, and from the semiconductor metallization community in general. The editors would like to thank the members of the Executive Committees for the United States and Asia Conferences, listed below, who actively participated throughout the year, served as session chairs, and reviewed the manuscripts. The editors also thank Linda Reid and Jenny Black Deer of the University of California at Berkeley for their organizational and logistical support, and Kristin Richter (Chemnitz University of Technology) for her assistance in preparing these proceedings.

1999 Conference Executive Committee - United States

Thomas Gessner, Chemnitz University of Technology (Co-Chair)
Mihal Gross, Bell Labs, Lucent Technologies (Co-Chair)

Robert S. Blewer, Sandia National Laboratories
Timothy S. Cale, Rensselaer Polytechnic Institute
Robin Cheung, Applied Materials
Girish Dixit, Applied Materials
Daniel Edelstein, IBM T.J. Watson Research Center
Russell Ellwanger, Applied Materials
Guido C.A.M. Janssen, Delft University of Technology
Rajiv Joshi, IBM T.J. Watson Research Center
Ki-Bum Kim, Seoul National University
Jeff Klein, Motorola, Inc.
Heinrich Körner, INFINEON Technologies AG
Mong-Song Liang, TSMC
Karen Maex, IMEC
Andrew McKerrow, Texas Instruments Inc.
Peter Moon, Intel Corporation
Takayuki Ohba, Selete/Fujitsu
Gary W. Ray, Novellus
Linda Reid, University of California, Berkeley
Gurtej Sandhu, Micron Technology, Inc.
Yosi Shacham-Diamand, Tel-Aviv University
Tom Smy, Carleton University
Shih-Wei Sun, United Microelectronics Corporation
Joaquin Torres, STMicroelectronics
Shi-Qing Wang, Allied Signal, Inc.
Chris Yu, Cabot Corporation

MATERIALS RESEARCH SOCIETY CONFERENCE PROCEEDINGS

Tungsten and Other Refractory Metals for VLSI Applications, Robert S. Blewer, 1986; ISSN 0886-7860; ISBN 0-931837-32-4

Tungsten and Other Refractory Metals for VLSI Applications II, Eliot K. Broadbent, 1987; ISSN 0886-7860; ISBN 0-931837-66-9

Tungsten and Other Refractory Metals for VLSI Applications III, Victor A. Wells, 1988; ISSN 0886-7860; ISBN 0-931837-84-7

Tungsten and Other Refractory Metals for VLSI Applications IV, Robert S. Blewer, Carol M. McConica, 1989; ISSN 0886-7860; ISBN 0-931837-98-7

Tungsten and Other Advanced Metals for VLSI/ULSI Applications V, S. Simon Wong, Seijiro Furukawa, 1990; ISSN 1048-0854; ISBN 1-55899-068-2

Tungsten and Other Advanced Metals for ULSI Applications in 1990, Gregory C. Smith, Roc Blumenthal, 1991; ISSN 1048-0854; ISBN 1-55899-112-3

Advanced Metallization for ULSI Applications, Virendra V.S. Rana, Rajiv V. Joshi, Iwao Ohdomari, 1992; ISSN 1048-0854; ISBN 1-55899-152-2

Advanced Metallization for ULSI Applications 1992, Timothy S. Cale, Fabio S. Pintchovski, 1993; ISSN 1048-0854; ISBN 1-55899-192-1

Advanced Metallization for ULSI Applications in 1993, David P. Favreau, Yosi Shacham-Diamand, Yasuhiro Horiike, 1994; ISSN 1048-0854; ISBN 1-55899-235-9

Advanced Metallization for ULSI Applications in 1994, Roc Blumenthal, Guido Janssen, 1995; ISSN 1048-0854; ISBN 1-55899-279-0

Advanced Metallization and Interconnect Systems for ULSI Applications in 1995, Russell C. Ellwanger, Shi-Qing Wang 1996; ISSN 1048-0854; ISBN 1-55899-341-X

Advanced Metallization and Interconnect Systems for ULSI Applications in 1996, Robert Havemann, John Schmitz, Hiroshi Komiyama, Kazuo Tsubouchi, 1997; ISSN 1048-0854; ISBN 1-55899-385-1

Advanced Metallization and Interconnect Systems for ULSI Applications in 1997, Robin Cheung, Jeffrey Klein, Kazuo Tsubouchi, Masanori Murakami, Nobuyoshi Kobayashi, 1998; ISSN 1048-0854; ISBN 1-55899-412-2

Advanced Metallization Conference in 1998 (AMC 98), Gurtej S. Sandhu, Heinrich Koerner, M. Murakami, Y. Yasuda, N. Kobayashi, 1999; ISSN 1048-0854; ISBN 1-55899-484-X

Advanced Metallization Conference 1999 (AMC 1999), Mihal E. Gross, Thomas Gessner, N. Kobayashi, Y. Yasuda, 1999; ISSN 1048-0854; ISBN 1-55899-539-0

Keynote Presentation

THE MARCO/DARPA INTERCONNECT FOCUS CENTER

JAMES D. MEINDL
Pettit Professor of Microelectronics, School of Electrical & Computer Engineering,
Director, Microelectronics Research Center
Director, MARCO/DARPA *Interconnect Focus Center*
Georgia Institute of Technology
Atlanta, GA, 30332 USA

Introduction

The U.S. semiconductor industry, largely through the Semiconductor Industry Association (SIA), has led the nation for more than 15 years in organizing precompetitive research consortia. Sematech, located in Austin, Texas and founded in 1987 is an industrial research consortium that emphasizes applied research and development with a one-to-three year time horizon. The Sematech agenda deals primarily with manufacturing equipment, materials and processes used throughout the U.S. and more recently the world-wide semiconductor industry.   The Semiconductor Research Corporation (SRC)   located in Research Triangle Park in North Carolina was founded in 1982 by SIA.   SRC funds generic semiconductor research in universities with a typical three-to-eight year time period to impact products.

The latest SIA sponsored research consortium is the Focus Center Research Program (FCRP) initiated in 1998. The FCRP is a cooperative effort. Participants are member companies of SIA and SEMI/SEMATECH (SEMI is the Semiconductor Equipment Manufacturing Institute, an association of suppliers of the semiconductor industry.) and the U.S. Department of Defense (DOD). The FCRP is managed jointly by the Microelectronics Advanced Research Corporation or MARCO, a wholly owned subsidiary of SRC, and the Defense Advanced Research Projects Agency (DARPA). The first two focus centers, selected in 1998 are: 1) the Design and Test Focus Center and 2) the Interconnect Focus Center (IFC).   All research conducted by these two centers is pre-competitive and uses cross-disciplinary, broad-based resources to optimize the generation of new technology.   The centers target long-range research results needed by industry in the eight-to-fifteen year time frame.   Professors, research engineers and scientists, graduate students and industry assignees conduct research using university facilities and equipment. The program will be reviewed every two years.   This article describes salient features of the organization and research program of the *IFC*.

The *Interconnect Focus Center (IFC)*

Organization

The overall leadership of the *IFC* is the responsibility of the director who is immediately supported by an assistant director concerned with technical issues and a principal administrative coordinator who deals with administrative and financial matters.   The initial six universities that participate in the research of the *IFC* are MIT, Stanford University, Rensselaer Polytechnic Institute (RPI), State University of New York at Albany (U@A), Cornell University and Georgia Institute of Technology (GIT), the prime contracting university.   A salient feature of the organization of the *IFC* is a *leadership council* currently consisting of five members and chaired

3

by the director. Principal functions of leadership council members are to support the director in overall leadership of the center and to *vertically* integrate the various research efforts at their individual institutions. As illustrated in Figure 1, the research program of the *IFC* is organized as a two-dimensional matrix. Each of the rows of the matrix represents one of the seven principal tasks or research thrusts of the center; each of the columns is identified with one of the six universities active in the center. A clear matrix element indicates active effort within a particular task at a particular site; a shaded matrix element denotes that a specific site is not active in the research within a given task. A *task leader or co-task leaders* are responsible for *horizontal* integration of the efforts at all sites participating in a particular task. In addition, *site task leaders* at each active site support the integration efforts of their task leaders. A key objective of the organizational structure of the *IFC* is to achieve a *whole that is greater than the sum of its individual parts*.

Research Program

The primary objective of *Task I: System Architecture and Circuit Innovation* of the *IFC* is to invent new communication or interconnect—centric microarchitectures that are intended to complement and replace the conventional arithmetic or transistor—centric microarchitectures that have been a dominant feature of microprocessors since their inception. Conventional microarchitectures use global concentrations of register files for immediate memory, instruction unit hardware for control and arithmetic-logic unit hardware for computation. This global lumping of similar resources necessitates the use of long global interconnects in critical paths that limit the clock frequency and therefore the performance of a microprocessor. Communication centric microarchitectures use distributed clusters of register files, instruction units and arithmetic-logic units that are capable of operating at relatively large clock frequencies due to the short length of the local interconnects within a cluster. Global communication among clusters is accomplished by a switching network designed to minimize critical path dependencies. Novel interconnect circuits for such purposes as local clock synchronization are a second important objective of Task I. Currently, principal investigators at Stanford, MIT and GIT participate in Task I research.

Task II: Physical Design Tools is directed toward creating a library of generic three-dimensional interconnect microcell models. These models are selected as needed from the library and then concatenated to synthesize a macromodel of a complex multilevel meandering interconnect network for prediction of its response time and crosstalk. In essence, Maxwell's equations are solved to describe the electric and magnetic fields that determine the values of the self and mutual conductances, capacitances and inductances of a generic interconnect microcell assuming a given geometry and material set. These passive element values characterize the equivalent circuit of the microcell that is deposited in the library. Investigators from Stanford, MIT and RPI participate in this task.

The prime topic of investigation of *Task III: Novel Communications Mechanisms* is optical interconnects; wireless RF interconnect networks are also of interest. Photonic interconnections offer the prospect of improved latency, bandwidth, power dissipation and crosstalk compared with conventional electrical wiring. Several complementary approaches are under investigation. These include: 1) a hybrid approach in which compound semiconductor chips containing high performance photon emitters and detectors are bonded to separately fabricated silicon CMOS substrates; 2) a silicon monolithic approach in which single or poly-crystal silicon waveguides surrounded by silicon dioxide are used in conjunction with off-chip light sources; and 3) a heteroepitaxial approach involving a continuous multi-layer epitaxial deposition of silicon,

4

followed by silicon-germanium, followed by gallium arsenide. The net result of the heteroepitaxial depositions is a set of islands of high quality single crystal islands of gallium arsenide surrounded by a sea of silicon. The purpose of the GaAs islands is fabrication of high efficiency vertical cavity lasers and photodetectors for input/output as well as on-chip global interconnects. Investigators from Stanford, MIT, U@A, RPI and Cornell participate in this task.

Task IV: Chip-to-Module Interconnects is focused on novel low cost wafer-level-batch-packaging (WLBP) that completes all final external lead attachment and packaging operations prior to dividing a 200 or 300 mm wafer into individual chips. An additional objective is to enable complete electrical testing of all chips prior to wafer dicing. Plans call for the use of a compliant polymer interposer for the chip package and flexible copper leads with solder bumps to attach *and* detach, if necessary, the packaged chip from a printed wiring board.

There are two principal objectives of *Task V: Materials and Processing*. The first is new high conductivity metal alloys, low permeability liners and low permittivity dielectrics for sub-70 nm multi-level interconnect networks; the second is novel three dimensional structures for gigascale integration that use multiple levels of transistors in addition to multiple levels of wiring as currently used. As copper film thickness approaches the value of the mean free path of electrons in copper, a rapid increase in resistivity is observed due to increased surface and grain boundary scattering. New surface treatments are needed to reduce this effect. To reduce substantially the lengths of the longest interconnects on a chip, and hence increase its clock frequency, new bonded and deposited three-dimensional structures are under investigation.

The prime target of *Task VI: Process Modeling, Simulation and Technology Assessment* is creation of new interconnect manufacturing process and equipment models that are physically based for prediction of interconnect topography, chemical composition, microstructure, doping concentration, resistivity, dielectric constant and thermal and mechanical properties across a wafer. For example, the intent is prediction of metal film properties at the bottom, on the sides and on the top surround of an array of 50 nm diameter vias with 10:1 aspect ratios distributed across a 300 mm wafer.

Task VII: Reliability and Characterization is aimed at development of new generic methodologies and physical models for circuit level assessment of multi-level interconnect network reliability considering factors such as electromigration and stress voiding as well as delamination. To help achieve this aim novel in-situ material characterization techniques including microspectroscopy, X-ray microtomography and ultrasonic microscopy are under development.

Conclusion

In summary, future opportunities for gigascale integration will be governed by a hierarchy of limits imposed by interconnects. The agenda of the *IFC* clearly demonstrates that critical barriers at all levels of this hierarchy including fundamental, material, device, circuit and system limits are currently under investigation. An unusual degree of autonomy has been granted to the *IFC* by its managing agencies, MARCO and DARPA in defining and pursuing this agenda. This feature of the *IFC* is in itself an experiment in research management aimed at encouraging the exploration of high risk, high payoff new frontiers in interconnect science and technology. The investigators of the *IFC* are strongly motivated to create *a multi-campus virtual corporate laboratory* that will provide the return on investment that fulfills the high expectations of the sponsors of the Focus Center Research Program.

Figure 1. Interconnect Focus Center Research Program Organization

Task Areas, Leaders and Institutions	Stanford	MIT	New York			Georgia Tech
			Albany	Rennsselaer	Cornell	
I. System Architecture and Circuit Innovations			■	■	■	
II. Physical Design Tools			■		■	■
III. Novel Communications						■
IV. Chip-to-Module Interconnects	■				■	
V. Materials and Processing	■				■	■
VI. Process Modeling, Simulation and Technology Assessment			■		■	■
VII. Reliability and Characterization					■	■

Integration of Damascene
Architectures

Critical Reliability Issues for sub-0.25-micron Generation Copper Wiring

A. K. STAMPER, W. A. KLAASEN, and R. A. WACHNIK*, IBM Microelectronics, Essex Junction, Vermont, * Hopewell Junction, New York

ABSTRACT

Second order parameters, such as via size, via slope, via overlay, damascene wire trench aspect ratio, and electron flow direction have major effects on damascene-copper electromigration reliability. Wiring reliability degraded as the the vias became more vertical and with decreasing via size, due to the increased via aspect ratio and reduced via contact area. In addition, the step coverage of the PVD copper seed layer, as opposed to the copper volume in the wire, was the dominant factor in electromigration lifetime. Controlling these geometry-based parameters and optimizing the step coverage of the PVD copper seed layer are critically important in obtaining consistently reliable dual-damascene wiring.

INTRODUCTION

Primarily because it has 40% lower bulk resistivity, copper is beginning to replace aluminum in high-performance CMOS BEOL wiring [1]. Most of the previously published reliability data on copper wiring has focused on the bulk electromigration (EM) or stress migration properties of plated, CVD, or PVD copper, and relatively little information is available on geometry-dependent effects with integrated copper wires and vias [1-3]. In this paper, we have investigated the effects of via dimension, via-to-underlying metal overlay, via slope, wire aspect ratio, and the ratio of copper to PVD barrier metal in the wire.

EXPERIMENT

Multi-level copper wiring, passivated with Si_3N_4/SiO_2 dielectric, was fabricated using additive single- and dual-damascene processes in which damascene trenches and vias are etched into an insulator, standard non-ionized PVD conductive liner and copper seed layers were deposited, copper was electroplated onto the wafer, and CMP was used to remove excess copper and conductive liner material [1]. The first level of copper wiring (M1) was fabricated using single-damascene processing and was contacted to damascene tungsten local interconnects (M0) and CMOS devices on the silicon substrate using damascene tungsten studs (C1). The next and subsequent copper wiring (M2) and via (V1) levels were fabricated using dual-damascene processing. Representative SEM cross-sections are shown in Figure 1. The via-to-underlying metal wire overlay was varied by fabricating aligned and misaligned test sites on the same mask set and the via size and slope were controlled by a combination of lithographic critical dimension control and via RIE chemistry. Electromigration was performed at 250°C and 295°C using standard minimum-dimension single via-to-wire test sites. Stress migration was performed at 225°C with no applied current on via-chain test sites with dimensions similar to the electromigration test-sites. The electromigration stressing was performed with electron flow from either the wire into the via or from the via/contact into the wire (Figure 2). Because the electromigration median fail time was typically a factor of ~10-30X higher when the electron flow was from the wire into the via, the discussion in this paper is limited to the worst-case electromigration electron flow direction, which is from the via or contact into the copper wire (Figure 2a). Failure analysis of electromigration fails was performed by dry etching through the dielectric to expose the test-site and taking SEM top- and tilt-angle view micrographs of the fails.

RESULTS

Electromigration Resistance versus Time Behavior

Figure 3 shows a typical single via-to-wire test-site resistance versus time profile during electromigration stressing (electron flow into the wire). The resistance was initially approximately constant, jumped 5-40% (R_{VOID}) after ~10-1000 hours of stressing (T_{VOID}), and then increased approximately linearly (R_{SLOPE}) for the next few hundred hours of stressing. We defined an electromigration fail as a 20% increase in resistance (T_{FAIL}) and fitted the data to a lognormal distribution, with corresponding median fail time (T_{50}) and sigma values. The jump in resistance at T_{VOID}, which corresponds to the formation of a void across the full width of the copper wire, and the slope of resistance versus time after T_{VOID} are caused by the increasing void dimensions.

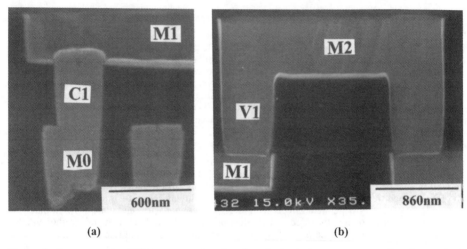

(a)   (b)

Figure 1: Representative SEM cross-section showing single-damascene tungsten local interconnect (M0), single-damascene tungsten stud (C1), single-damascene first level copper wiring (M1), and dual-damascene copper wires (M2) and vias (V1).

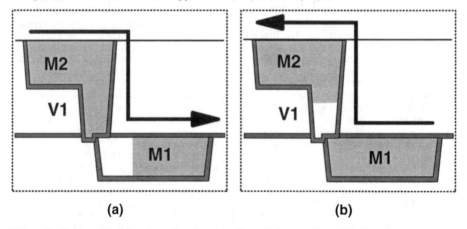

(a)   **(b)**

Figure 2: Cross-section drawings showing the effect of electron flow direction on electromigration-induced voiding from (a) the via or contact into the copper wire M1, and (b) the copper wire M1 into the dual damascene copper via V1 (voids are shown in white).

<u>Damascene Wires</u>

As the PVD conductive liner or PVD copper seed thicknesses in the trench increase, the effective aspect ratio of the trench also increases because the PVD film step coverage on the trench sidewalls and bottom are about 35% and 50%, respectively (Figure 4). In addition, as the

PVD liner or seed thicknesses increase, the cross-sectional area ratio of copper to total wire cross-sectional area decreases. Calculations, using Figure 4, of the aspect ratios prior to PVD copper seed (AR_{SEED}) and electroplated copper fill (AR_{PLATE}) as well as the ratio of copper to total wire cross section ($RATIO_{COPPER}$) are shown below:

$$AR_{SEED} = (h1 + h2 - h4) / (w1 - 2*w2) \qquad (1)$$

$$AR_{PLATE} = (h1 + h2 + h3 - h4 - h5) / (w1 - 2*w2 - 2*w3) \qquad (2)$$

$$RATIO_{COPPER} = (h6 - h4) * (w1 - 2*w2) / (h6 * w1) \qquad (3)$$

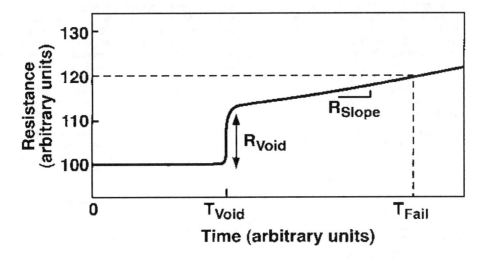

Figure 3: Resistance versus time data for typical electromigration fails.

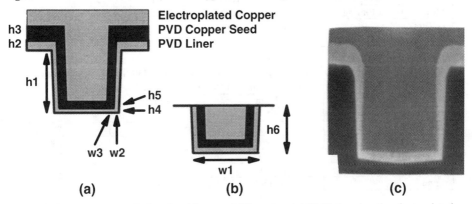

Figure 4: Damascene trench drawing (a) pre and (b) post metal CMP showing the electroplated copper, PVD copper seed, and PVD conductive liner deposition profiles; (c) SEM pre-metal CMP.

To determine the effect of the aspect ratio prior to the PVD copper seed and electroplated copper depositions as well as the percentage of copper in the wire on electromigration lifetime, we varied nominal PVD conductive liner thickness from 50nm to 80nm. Although the single via test-site electromigration lifetimes for 410nm-wide wires were unaffected by this change, (Table 1), the T_{50} of 280nm-wide wires was increased by a factor of ~7X as the nominal PVD liner thickness was decreased from 80nm to 50nm (Tables 1 and 2; Figure 5). Since AR_{PLATE} for the V1-M1 and V2-M2 test-sites were similar, we conclude that the reduced V1-M1 and CA-M1 T_{50} was not caused by increasing AR_{PLATE}. For current densities under 25mA/μm^2, it has been reported that the current density dependence of the electromigration lifetime is given by $\tau = \tau_0 j^{-n}$ with n = 1.1 +/- 0.2 [2]. Because the current density has an approximately linear effect on T_{50}, we conclude that the relatively small changes in $RATIO_{COPPER}$ as the conductive liner was increased from 50nm to 80nm could not have caused the large changes observed in the M1 test-site T_{50} data. As AR_{SEED} increases, the PVD copper seed step coverage in the wiring trench sidewall and bottom decreases, which reduces the copper available as a seed layer for the plated copper. We believe that this degradation in PVD copper seed step coverage increases either the as-deposited porosity or voiding of the plated copper, which degrades the electromigration lifetimes. From these data, we conclude that the increase in AR_{SEED}, which was about 2.3:1 for the 80nm conductive liner M1 test-sites, induced the degradation in M1 electromigration lifetime. The need for high aspect-ratio PVD copper gapfill, which has been previously reported [4], will necessitate the use of ionized PVD processes for damascene-copper wires as the aspect-ratio requirements exceed ~2:1.

Electromigration lifetimes were dramatically improved by adding a second contact or via (Tables I and II; Figure 5). We believe that the T_{50} improvement associated with the addition of a second contact or via occurs due to a combination of a reduction of the effective current density at the via/wire interface and because the fail does not occur until the electromigration-induced void extends over both contacts or vias (Figure 6).

Table I: 250°C 25mA/μm^2 EM summary for V1-M1 280nm and V2-M2 410nm wide wires (one to three wafers per cell).

	V1-M1			V2-M2	
Nominal Liner	80nm	50nm	80nm	50nm	80nm
Number of vias	2	1	1	1	1
% Cu in Wire	64%	77%	64%	83%	73%
AR prior to PVD Cu seed	2.3:1	1.9:1	2.3:1	1.4:1	1.6:1
AR prior to Cu Plating	3.5:1	2.7:1	3.5:1	2.8:1	3.4:1
T_{50} (hrs)	1,300	650	90	440	430
sigma	0.4	0.4	0.4	0.3	0.4

Table II: 250°C 25mA/μm^2 electromigration log normal median fail time (T_{50}) and sigma summary for C1-M1 280nm wide wires (one to three wafers per cell).

Nominal Liner	50nm	80nm	50nm
Number of contacts	1	1	2
% Cu in Wire	77%	64%	77%
AR prior to PVD Cu seed	1.9:1	2.3:1	1.9:1
AR prior to Cu Plating	2.7:1	3.5:1	2.7:1
T_{50} (hrs)	210	39	2,000
sigma	0.5	0.6	0.6

Dual Damascene Vias

The C1-M1 and V1-M1 electromigration test-sites were located in the chip dicing channel and had nominally identical M1 dimensions and metallization. Despite this, the C1-M1 test-sites had about a factor of 3 lower T_{50} and a higher log-normal sigma as compared to V1-M1 test-sites

(Tables I and II; Figure 5). In addition, the M1-V1 test-sites with electron flow into the via (not shown) had a factor of ~10-30X higher T_{50} as compared to the C1-M1 test-sites. During the single-damascene M1 RIE, the M1 is overetched beyond the top of the C1 stud to achieve robust M1-C1 contact yields and resistance (Figure 1). Becuase of the M1 overetch, the M1 copper thickness above the C1 stud is locally reduced and we attribute the lifetime degradation in C1-M1 test-sites due to both an increase in the effective current density and to the creation of a irregular M1-C1 interface. From these data, we conclude that dual-damascene wiring has significantly improved electromigration reliability as compared to single-damascene wiring

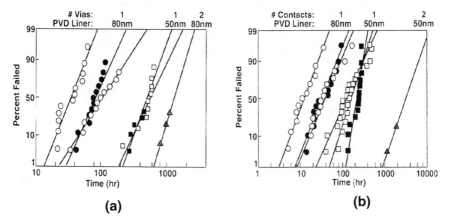

(a)   **(b)**

Figure 5: 250°C log normal 25mA/μm^2 electromigration test-site data taken from 280nm wide M1 single-damascene copper wires for (a) dual-damascene copper V1-M1 and (b) single-damascene tungsten stud C1-M1 test-sites with no intentional V1 or C1 misalignment.

Table III summarizes stress migration results versus via size and slope for misaligned and aligned M2-V1-M1 chain test-sites. These via chain test-sites, which consisted of about 10,000 vias, had similar wire and via cross-sections as compared to the single-via electromigration test-sites discussed above. Although all structures had 100% yield to a 20% resistance increase threshold after 1000 hours of stressing, the structures with intentional via misalignment had lower yields to 5% resistance increase thresholds. In addition, the resistance increase in these structures increased as the via size decreased, particularly in the misaligned via structures with 90° slopes. As via size decreases or slope increases, the effective via aspect ratio increases, which degrades the via sidewall step coverage of the PVD copper seed deposition. Although the resistance increases with these test sites were relatively small, we believe that they are due to a combination of void condensation acceleration with decreasing via contact area and an increase in the as-deposited porosity or voiding in the electroplated copper caused by increased via aspect ratios.

Table III: 225°C stress migration yields (no applied current) to a 5% resistance increase specification after 1000 hours of stressing for M2-V1-M1 via chains with 10,000 vias (two wafers per cell). The nominal via alignment was about +/- 90nm and the misaligned via chains were intentionally misaligned by 200nm.

Via Test Site	Via Bottom Size (Slope)			
	500nm (90°)	460nm (90°)	430nm (87°)	390nm (87°)
Aligned	100%	100%	100%	100%
Misaligned	72%	62%	96%	77%

The aligned electromigration V2-M2 test-site T_{50} systematically decreased by more than 50% and the log-normal sigmas increased as the via size decreased from 500nm to 390nm; and, for both 90° and 87° vias, a 40nm reduction in via size decreased the T_{50} by about 35% (Table IV). In addition, the T_{50} of misaligned vias was more than 50% lower for all via sizes and slopes as compared to nominally aligned vias. Although we defined an electromigration fail as a 20% increase in resistance, we observed that the resistance versus time behavior prior to T_{FAIL} was a strong function of the via size and slope angle. In particular, the resistance jump, R_{VOID}, which occurred when the copper void extended across the copper wire, and the slope after the resistance jump, R_{SLOPE}, both increased with decreasing via size, increasing via misalignment, and increasing via slope. Since the electron flow was into the wire, the electromigration-induced copper voiding occured in the copper wire and we believe that the R_{VOID} and R_{SLOPE} increases are due to an increase in the rate of void condensation because of reduced via contact area. From these data trends, which are similar to the stress migration results discussed above, we conclude that void creation is accelerated by decreasing via-to-metal contact area for this technology node.

Table IV: 295°C 25mA/μm^2 electromigration summary for aligned and misaligned single via V2-M2 test-sites (one wafer per cell). The nominal via alignment was about +/- 90nm and the misaligned via chains were intentionally misaligned by 140nm.

Via Bottom Size (Slope)	Aligned			Misaligned		
	T_{50} [hrs] (sigma)	R_{Void} (%)	R_{Slope} (%/hour)	T_{50} [hrs] (sigma)	R_{Void} (%)	R_{Slope} (%/hour)
500nm (90°)	136 (0.4)	5	0.14	52 (0.5)	11	0.25
460nm (90°)	91 (0.5)	5	0.22	33 (0.7)	12	0.28
430nm (87°)	94 (0.7)	32	0.19	42 (0.5)	15	0.23
390nm (87°)	62 (0.9)	26	0.25	44 (0.7)	20	0.32

Failure Analysis of Electromigration Fails

Figures 6 and 7 show top view SEM micrographs of C1-M1 and V1-M1 electromigration fails. Note that, in Figure 6, the dry etch delayering method removed the PVD conductive liner. The SEM micrographs show homogeneous copper voids, in the direction of electron flow, for both C1-M1 and V1-M1 test-sites. In addition, no evidence of copper extrusions to adjacent lines was observed in the SEM's or in the electrical data.

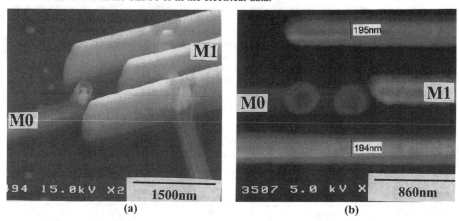

Figure 6: Top view SEM micrographs of electromigration fails from (a) single C1-M1 and (b) double C1-M1 test-sites after delayering.

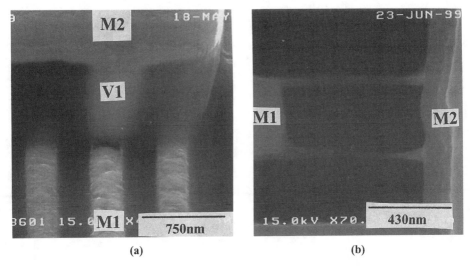

(a)   (b)

Figure 7: Top view SEM micrographs of V1-M1 electromigration fails.

SUMMARY

Second order parameters, such as via size, via slope, via overlay, trench aspect ratio, type of damascene process, and electron flow direction have been shown to have major effects on damascene copper electromigration and stress voiding reliability. In particular, dual-damascene wiring had improved wiring reliability as compared to single-damascene wiring; increased aspect ratios prior to the PVD copper seed deposition degraded wiring reliability; and increased via contact area and slope improved wiring reliability. Controlling these geometry-based parameters and optimizing the gapfill of the PVD copper seed layer copper are critically important in obtaining consistently reliable electroplated copper dual damascene wiring.

ACKNOWLEDGEMENTS

The authors wish to acknowledge the numerous engineers at IBM who contributed to the damascene copper program. Regarding this paper, we gratefully acknowledge the assistance of Brian Lefebvre and Robert Cooke for the SEM analysis; and Edward Barth, Peter Biolsi, William Cote, Edward Cooney, Daniel Edelstein, Robert Geffken, Ronald Goldblatt, Stephen Greco, James Harper, John Heidenreich, C.K. Hu, Stephen Luce, Tom McDevitt, Vincent McGahay, Paul McLaughlin, Darryl Restaino, Andrew Simon, and Jill Slattery for useful discussions.

REFERENCES

1. Tech. Dig. IEEE Int. Elec. Dev. Mtg., Wash. D.C., Dec. 7-10, 1997: H. Aoki, et al., pp777-80; D. Edelstein et al., pp773-6; S.C. Sun, pp765-8; S. Venkatesan, et al., pp769-72.

2. C.K. Hu, R. Rosenberg, H. S. Rathore, D. B. Nguyen, and B. Agarwala, Proc. 1999 Int. Interconnect Tech. Conf., May 24-26, 1999, San Francisco, CA pp. 267-269.

3. C.-K. Hu, R. Rosenberg, and K.L. Lee, Appl. Phys. Letter, 74, 2945 (1999).

4. C. A. Nichols, S. M. Rossnagel, and S. Hamaguchi, J. Vac. Sci. Tech.B, vol. 14, 5 pp3270-5 (1996);   P.F. Cheng, S.M. Rossnagel, D.N. Ruzic, J. Vac. Sci. Tech. B, vol.13, 2 pp203-8 (1995).

Comparison of Al and Cu Metallizations for Dual Damascene Applications

L.A. Clevenger*, S.G. Malhotra**, A.H. Simon**, R. Iggulden*, and S.J. Weber***
* IBM Microelectronics, DRAM Development Alliance, Hopewell Junction, NY 12533
** IBM Microelectronics Division, Hopewell Junction, NY 12533
*** Infineon Technologies, DRAM Development Alliance, Hopewell Junction, NY 12533

Abstract

SEM and electrical analysis for Al and Cu filling of high aspect ratio (4.2 to 1 aspect ratio, 0.2 μm via size) structures used in 1Gbit DRAM and high performance logic applications was investigated. With Al metallization, high quality filling of these structures is obtained with the use of a CVD TiN liner and a CVD Al seed layer where the thickness of these layers can be varied to obtain a high yielding and efficient process flow. The CVD TiN also provides an extra benefit in preventing the formation of high resistance $TiAl_3$ during the later stages of the Al metallization process. For Cu metallization, it will be shown that high quality filling of these structures is related to the Cu seed layer step coverage and continuity. Cu seed layer enhancements, such as the use of a CVD Cu seed, offer the possibility of extending Cu technology to even smaller ground rules. It is demonstrated that a high yielding dual damascene integration scheme for 1Gbit DRAM and advanced logic applications can be obtained using either advanced Al or Cu metallization approaches.

Introduction

Until recently, most interconnections used for VLSI BEOL wiring applications were created by the reactive ion etching of Al.[1] As BEOL wiring is scaled down to sub-0.30 μm dimensions, Al RIE has become increasing difficult due to etching of tall/narrow Al lines, oxide gap fill, smaller overlay tolerances and the need for interconnections with better electrical resistance. Dual damascene[2] offer advantages over Al RIE of simple oxide etches, CMP planarization after each wiring level and the ability to use either Al or Cu metallization. Al dual damascene is compatible with previous Al RIE technologies, cost effective and has limited contamination issues.[3-6] Cu dual damascene integration schemes produce interconnections which have a much lower resistance, higher allowable current density and increased scaliablity compared to Al damascene or RIE interconnections.[7-9] In this work, we have investigated the SEM and electrical performance of Al and Cu filling of high aspect ratio (4.2 to 1 aspect ratio, 0.2 μm via size) dual damascene structures used in 1Gbit DRAM and high performance logic applications. We demonstrate that a high yielding dual damascene integration scheme for advanced BEOL applications can be obtained by using advanced Al or Cu metallization approaches. For Al processes, CVD liners and seeds must be used to obtain reliable filling. The exact thickness of these CVD layers can be optimized to achieve a desired process complexity. For Cu processes, the filling of these structures with electroplated Cu requires a seed layer which has a thickness above a minimum value. We show that Cu seed enhancement processes,[10-14] such as CVD Cu, can be used to extend Cu technology to smaller ground rules.

* email - lacleven@us.ibm.com

Experimental

The dual damascene structures used in this work were built by first etching trenches in an oxide dielectric (M1), then filling them with CVD W, and then planarizing with conventional W CMP approaches. Next, the subsequent dual damascene level was formed by depositing a 930 nm thick layer of SiO_2, followed by the via etch (V2), and then the line etch (M2). A typical example of the structure formed is shown below in Fig. 1 where the M2 lines are interspersed with the V2 vias. The Al or Cu metallization sequence used to fill these type of structures consisted of the deposition of a thin liner followed by a seed layer which was then followed by a thick Al or Cu fill layer. The details behind the compositions of the liner, seed and fill layers are listed below in Table 1. The quality of the fill for either the Al or Cu metallizations after deposition and metal CMP was characterized by a combination of SEM and electrical analysis of 100K via chain, sheet resistance and contact resistance structures.

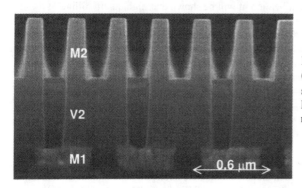

Figure 1: A 4.5 to 1 aspect ratio, 0.2 µm via, dual damascene structure after the V2 via and M2 line etches and before Al or Cu metallization.

Table I: Al and Cu filling Processing conditions

	Al Metallization	Cu Metallization
Liner	IPVD Ti (<30 nm)/CVD TiN (15 nm)	IPVD Ta (< 80nm)
Seed	CVD Al (<30 nm)	IPVD Cu (<200 nm) OR CVD Cu (<80 nm)
Fill layer	PVD Al(Cu) 1 µm (<410Oc)	Plated Cu

Results and Discussion

For Al dual damascene, good filling of 4.5 to 1 aspect ratio, 0.2 µm structures was obtained when a CVD TiN liner and CVD Al seed layer was used. Figure 2a demonstrates the Al fill of the structures of Fig.1 using an IPVD Ti/CVD TiN liner, followed by a thin CVD Al seed followed by a thick Al layer deposited at approximately 400°C. Both the high aspect ratio vias and the closely adjacent metal line structures have void free fill. The key parameters for this filling are the thin, highly conformal CVD TiN liner which acts as a diffusion barrier and wetting

18

layer for the thin CVD Al seed layer and the thin CVD Al seed layer which acts as a wetting layer for the final PVD Al reflow layer (which provides most of the Al for filling).

If the CVD TiN liner and/or CVD Al seed layers are left out of the process flow, incomplete fill of dual damascene structures will result. Figures 2b-2d demonstrate the filling for dual damascene structures for the cases of no CVD TiN liner (Fig. 2b), replacing the CVD TiN liner with PVD TiN (Fig. 2c) and replacing the CVD Al seed with PVD Al seed (Fig. 2d). Figure 2b demonstrates that when the CVD TiN is absent, voids are seen at the bottom of via and incomplete filling of the damascene structures is obtained. This emphasizes the key roll that the CVD TiN liner plays as a wetting layer for the Al seed layer. Figures 2c and 2d demonstrate that if either the CVD TiN liner and/or the CVD Al seed layer are replaced with PVD films of equilivant thicknesses (Fig. 2c, CVD TiN replaced with PVD TiN or Fig. 2d, CVD Al replaced with PVD Al), voids are formed in the bottom of via structures, indicating that the conformality and wetting characteristics of the PVD seed and liners are inferior to their CVD counterparts.

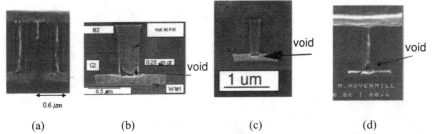

    (a)   (b)   (c)   (d)

Figure 2: Cross-sectional SEM filling of 0.175 to 0.25 μm dual damascene via structures with a) IPVDPTi/CVD TiN/CVDAl/Warm Al, b) IPVD Ti/PVD Al/Warm Al, c) IPVD Ti/IPVDTiN/CVDAl/Warm Al and d) IPVD Ti/CVDTiN/PVD Al/WarmAl.

The use of a CVD TiN liner is also essential to forming a low resistance Al conductor by preventing TiAl$_3$ formation in the Al lines during the warm Al deposition.[15] As shown in Fig. 3 samples without any TiN liner or with a PVD TiN liner have large amounts of TiAl$_3$ formation after the final high temperature Al metallization step and samples with the CVD TiN liner have no TiAl$_3$ formation. The TiAl$_3$ formation in Al lines heated above 400°C without any TiN liner is a known phenomenon. The fact that PVD TiN is not as effective as CVD TiN in preventing TiAl$_3$ formation suggests that these films have different microstructures and that the CVD TiN is more effective in stopping Ti-Al reactions.[15]

While a CVD TiN liner and CVD Al seed are required for high quality Al dual damascene filling, there are process simplifications that can be used to optimize the process flow described in Table I. Hoinkis et al.[16] demonstrated that the CVD Al seed layer can be thinned to being discontinuous and still allow for to reliable filling by the PVD Al layer. SEM and electrical analysis presented in Fig. 4 below demonstrate that the IPVD Ti portion of the IPVD Ti/CVD TiN liner step listed in Table I can be removed without degrading the quality of the Al dual damascene fill. From all these observations, we conclude that as long as the CVD TiN is conformal and continuous and the CVD Al seed layer is deposited as a thin wetting layer, even to the point of being discontinuous, then high quality filling of Al dual damascene structures can be expected.

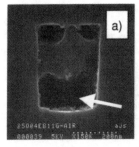  

Figure 3: SEM cross-section
for filling of M2 line structures with Al metallizations without a CVD TiN liner (a), with a PVD TiN liner instead of CVD TiN (b), and for a standard CVD TiN liner (c) process as detailed in Table I. The TiAl₃ phase (marked by arrows) has been chemically removed.

(a)   (b)

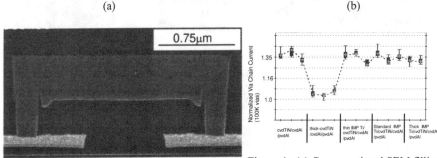

Figure 4: (a) Cross-sectional SEM filling
of 0.175 μm dual damascene via structures using a CVD TiN liner, CVD Al seed and Warm PVD Al fill layer, and b) normalized 100K via chain current for various Al dual damascene metallization techniques with various IPVD Ti/CVD TiN or CVD TiN liner combinations. All the via chain current values are approximately equal to one another.

For the 4.2 to 4.5 to 1 aspect ratio, 0.2 μm via size dual damascene structures used in this work, similar SEM filling characteristics were observed using either the Al or Cu metallization approaches listed in Table I. Figure 5 is a SEM micrograph of a via structure similar to the one shown in Fig. 1 filled with Cu. The metallization process used for this sample was an IPVD Ta liner, followed by an IPVD Cu seed and plated Cu fill. As is demonstrated by this figure, the Cu filling of these challenging structures is void free, indicating a high quality liner and seed combination which allows for void free Cu electroplating.

The success of the Cu metallization integration scheme detailed in Table I in filling aggressive dual damascene structures is dependent on the Cu seed layer step coverage and uniformity. Figure 6 below presents normalized via chain current (100K vias) for 4.2 to 1 aspect ratio, 0.2 μm vias metallized with different Cu seed metallization schemes. In this figure, the basic Cu metallization process was as detailed in Table I - IPVD Ta liner followed by IPVD Cu seed followed by Cu electroplating. The processes that were varied for this experiment were the IPVD Cu seed thickness and deposition pressure. As is shown in Fig.6, a large spread in the via chain current, which is an indication of poor quality Cu filling of vias, occurs only for the split with the thinnest IPVD Cu liner and does not depend on the deposition pressure of the IPVD seed. This suggests that at the thinnest IPVD Cu seed thickness, the Cu seed in the via structure becomes partially discontinuous and does not allow the complete fill of the via structures by Cu plating to occur. It should be noted that for the thinnest IPVD Cu seed sample, random SEM cross-sections of numerous via chain macros showed complete Cu filling similar to the standard IPVD Cu liner thickness sample shown in Fig. 5. This suggests that for the thinnest IPVD Cu seed samples, the discontinuous seed layer that leads to incomplete filling of vias does not occur on a large scale. Rather, the voiding of Cu in a via is a statistically small occurrence which most likely depends on the exact structure of each via that is being filled.

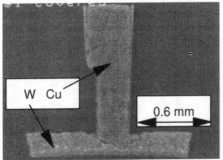

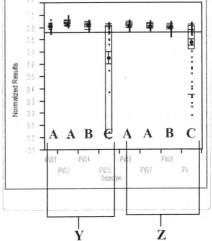

Figure 5: Cross-sectional SEM filling of 0.175 μm dual damascene via structures using a IPVD Ta liner, IPVD Cu seed and Electoplated Cu fill layer.

Figure 6: normalized 100K via chain current for dual damascene structures filled by IPVD Ta/IPVD Cu/Plated Cu metallization. For this experiment, the "A" represents a standard IPVD Cu seed thickness, "B" thinner and "C" thinnest seed thickness. "Y" represents the standard IPVD Cu deposition pressure and "Z" is a higher IPVD Cu deposition pressure.

While decreasing the IPVD Cu seed thickness increases the spread in 100K via chain current (as shown in Fig. 6), which is characteristic of random isolated voids in the 100K via chain, it has little effect on the single via contact resistance or line sheet resistance. Instead, the decrease in the deposition pressure for the Cu seed layer more strongly effects these two parameters. Figure 7 below presents normalized 0.175 μm line width sheet resistance (Fig 7a.) and single via contact resistance (Fig. 7b) for different IPVD Cu seed thickness and deposition pressures. Figure 7a demonstrates that the Cu films with Cu seeds deposited at a lower pressure have a slightly lower sheet resistance than films with seeds deposited at a higher deposition pressure. One possible explanation for this result would be that the lower deposition pressure of the IPVD Cu seed results in a microstructure formed in the deposited seed layer which promotes the plating of larger grain Cu films, leading to a lower sheet resistance.

The normalized single via contact resistance shown in Fig. 7b demonstrates a similar trend to sheet resistance data from Fig, 7a, - Cu films with Cu seeds deposited at a lower pressure have a slightly lower contact resistance. This is expected since the single via contact resistance structure used for this work is sensitive to the sheet resistance of the deposited film. It must be noted however, that for all thickness of IPVD Cu liner, there is little spread in the single via contact resistance indicating good Cu filling of these single via structures. This confirms the result of Fig. 5 which demonstrated complete filling by SEM analysis of a random via structure. The results of Figs. 5, 6 and 7b taken together reinforce the observation that while single via structures can be filled with the thinnest IPVD Cu liner, large via chains have isolated voids which cause a large decrease in the 100K via chain current.

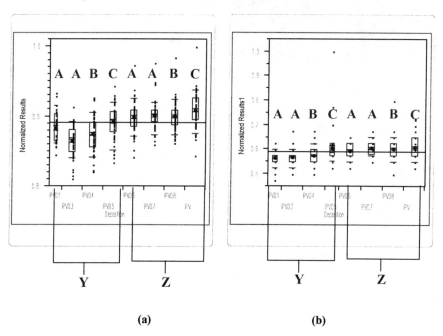

(a)   (b)

Figure 7: Normalized 0.175 μm wide sheet resistance (a) and single via contact resistance (b) for dual damascene structures filled by IPVD Ta/IPVD Cu/Plated Cu metallization. For this experiment, the "A" represents a standard IPVD Cu seed thickness, "B" thinner and "C" thinnest seed thickness. "Y" represents the standard IPVD Cu deposition pressure and "Z" is a higher IPVD Cu deposition pressure.

To improve the Cu plating of aggressive aspect ratio structures which require the use of very thin Cu seed layers, alternative "seed friendly" processes such as CVD Cu seed layers,[12] electrolytic Cu seed repair[10,11,13] and direct plating on liners[14] are being investigated across the IC industry. These approaches offer advantages over conventional IPVD Cu seed layers of the ability to deposit thinner Cu layers that are conformal and continuous (CVD Cu), minimizing and repairing the etching of Cu seed layers during the initial stages of the electrodeposition process (electrolytic seed repair) or allowing for the direct plating of Cu on different types of liner materials. All these approaches offer solutions to normal IPVD Cu seed layer limitations in the areas of continuity and conformality for very aggressive via structures.

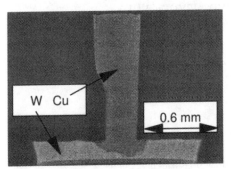 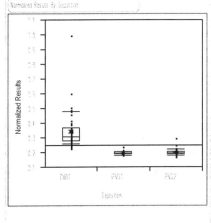

Figure 8: Cross-sectional SEM filling of 0.175 μm dual damascene via structures using a IPVD Ta liner, CVD Cu seed and Electoplated Cu fill layer.
Figure 9: normalized single via chain contact resistance for dual damascene structure filled by Cu metallization as described in Table I using either a CVD Cu or IPVD Cu seed layer. In this figure, the thickness of the seeds layer was IPVD2>IPVD1>CVD1.

An example of filling of the dual damascene structures discussed in this paper using a CVD Cu seed layer instead of an IPVD Cu seed is shown in Fig. 8 above. This figure shows a SEM micrograph of a via structure similar to the one shown in Fig. 5 except that a very thin CVD Cu instead of an IPVD Cu was used for the Cu liner. In this case, the CVD Cu thickness was less than half the thinnest IPVD Cu liner used in Figs. 6 and 7. As is demonstrated by fig. 8, the Cu filling of this aggressive via structure using this thin CVD Cu seed is void free, indicating a continuous CVD seed layer which allows for void free Cu electroplating. In addition, the relatively similar single via contact resistance using either a CVD Cu seed layer or IPVD Cu seed layers shown in Fig. 9 above provide further conformation of similar single via filling between using a very thin CVD Cu or a thicker IPVD Cu seed layer. The slightly higher average 0.175 μm line sheet resistance (Fig. 10) and contact resistance (Fig. 9) is most likely due to a higher level of impurities in the CVD Cu liner film.

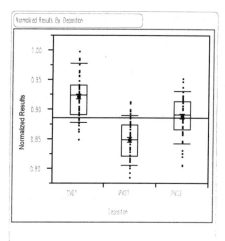

Figure 10: Normalized 0.175 μm sheet resistance for dual damascene structures filled by Cu metallization as described in Table I using either a CVD Cu or IPVD Cu seed layer. In this figure, the thickness of the seeds layer was IPVD2>IPVD1>CVD1

Conclusions

We have shown that both Al and Cu metallization techniques can be used to fill 0.2 μm, 4.2 to 1 aspect ratio vias used for advanced DRAM and logic dual damascene structures. Filling these structures with Al is enabled by the use of a CVD TiN liner and a CVD Al seed layer. The CVD TiN liner is also necessary to prevent the formation of high resistance $TiAl_3$ during the PVD Al fill step. The exact thickness of the CVD TiN liner and the CVD Al seed can be optimized to achieve a desired process complexity. Filling similar structures with an IPVD Cu seed and electroplated Cu requires an IPVD seed layer which has a thickness above a minimum value in order to provide a continuous and conformal seed layer in a via structure to allow electroplating. Alternative "seed friendly" processes, such as a CVD Cu seed, offer the possibility to deposit a thinner seed layer which is conformal and continuous thus extending Cu metallization to very small ground rule structures.

References

1) R. Ravikumar, H. Cichy, R.G. Filippi, E.W. Kiewra, D.L. Rath, and G. Stojkovic, 1999 Internation Interconnect Techology Confrence, IEEE, p 146-148, 1999

2) R. F. Schnabel, D. Dobuzinsky, J. Gambino, K.P. Muller, R. Wang, D.C. Peng, and H. Palm, Microelectronic Engineering, 37/38, 59 (1997)

3) T. Licata, M. Okazaki, M. Ronay, W. Landers, T. Ohiwa, H. Poetzlberger, H. Aochi, D. Dobuzinsky, R. Filippi, D. Knorr, and J. Ryan, Proceedings on the VLSI Multilevel Interconnect Technology Conference, p. 596 (1995)

4) L.A. Clevenger, G. Costrini, D. Dobuzinsky, R. Filippi, J. Gambino, M. Hoinkis, L. Gignac, J.L. Hurd, R.C. Iggulden, C. Lin, R. Longo, G.Z. Lu, J. Ning, J.F. Nueztel, R. Ploessl, K. Rodbell, M. Ronay, R.F. Schnabel, D. Tobben, S. J. Weber, L. Chen, S. Chiang, T. Guo, R. Mosley, S. Voss, and L. Yang, 1999 International Interconnect Technology Conference, IEEE, p. 127 (1998)

5) J. Proost, K. Maex and L. Delaey, Appl. Phys. Lett, 73, 2748, (1998)

6) R. Iggulden, L.A. Clevenger, G. Costrini, D. Dobuzinsky, R. Filippi, J. Gambino, L. Gignac, C. Lin, R. Longo, G.Lu, J. Ning, K. Rodbell, M. Ronay, F. Schnabel, J. Stephens, D. Tobben and S. Weber, Proceedings of the VLSI Multilevel Interconnect Technology Conference, p. 19, (1998)

7) D. Edelstein, J. Heidenreich, R. Goldblatt, W. Cote, C. Uzoh, N. Lastig, P. Roper, T. Mcdevitt, W. Motsiff, A. Simon, J. Dukovic, R. Wachnik, H. Rathore, R. Schulz, L. Su, S. Luce and J. Slattery, Proceedings of the IEEE International Electron Devices Meeting (IEDM), p. 773, (1997)

8) P. Singer, Semiconductor International, p. 91, June 1998

9) S. Venkatesan, Proceedings of the IEEE International Electron Devices Meeting (IEDM), p. 769, (1997)

10) U. Landau, J. D'Urso, A. Lipin, Y. Dordi, A. Malik, M. Chen and P. Heg, Proc. Electrochem. Soc. Mtg. (ECS), 263 (1999)

11) L. Chen and T. Ritzdorf, Proc. Electrochem. Soc. Mtg. (ECS), 263 (1999)

12) R. Kroger, M. Eizenberg, D. Cong, N. Yoshida, L.Y. Chen and L. Chen, Materials Research Society Porceedsing, vol. 564, (1999), to be published

13) Y. Shacham-Diamand, B. Israel and Y. Sverdlov, Advanced Metallization Conference, (1999), to be published

14) A.R. Ivanova, C.J. Galewski, C.A. Sans, T.E. Seidel, S. Grunow, K. Kumar, and A.E. Kaloyeras, Materials Research Society Proceedings, vol. 564, (1999), to be published

15) L.M. Gignac, K.P. Rodbell, L.A. Clevenger, R.C. Iggulden, R.F. Schnabel, S.J. Weber,C. Lavoie, C. Cabral, Jr., P.W. Dehaven, Y.-Y. Wang and S.H. Boettcher, in Advanced Metallization and Interconnect Systems for VLSI Applications, (1998), p. 79

16) M. Hoinkis, L.A. Clevenger, R. Iggulden, and S.J. Weber, Advnaced Metallization Conference, (1998)

Microstructure of Aluminum Damascene Interconnects Inlaid by Chemical Vapor Deposition

Akira Furuya[†] and Yoshihiro Hayashi
Silicon Systems Research Laboratories, NEC Corporation, 1120 Shimokuzawa, Sagamihara, Kanagawa 229-1198, Japan
[†]E-mail: furuya@mel.cl.nec.co.jp

Microstructure of ultra fine Al damascene interconnects, whose minimum pitch was 0.32μm, were successfully fabricated by chemical vapor deposition. Microstructure of these lines, investigated by transmission electron diffraction, has revealed that the dominant texture was (random)$\|$/(111)$\perp$ and increases with narrowing trench width. For Al-reflow sputtering, (111)$\|$/(022)$\perp$ decreases with narrowing trench width. Here, the (hkl) texture to the substrate and (hkl) texture from the trench are denoted as (hkl)$\|$ and (hkl)$\perp$, respectively. These differences in microstructure is discussed by considering the differences in the film growth characteristics and the influence of the interfacial energy.

INTRODUCTION

Signal delay due to interconnect resistance and interdielectric capacitance is one of the most important problems for future ULSI devices. In order to reduce the resistance and the capacitance, studies of low electrical resistivity or low dielectric constant materials have been undertaken. It has been reported that the signal delay is dominated by the capacitance component at local interconnects near transistors since the interconnect resistance is smaller than the CMOS output-resistance.[1] The materials with low electrical resistivity are not necessary for local interconnects, while it is strongly desired for the global interconnects. Therefore, we think that Al interconnects are the strongest candidates for the local interconnects in future devices. Protection of the transistor from metal contamination due to Cu is a severe problem.

The damascene method is a candidate for fabricating fine interconnects since a) it reduces charging damage in the transistors,[2] b) the dual-damascene method reduces the number of process steps[3] and c) the self-aligned structure reduces failure related to misalignment between interconnects and vias.[4] In contrast, low migration resistance has been reported for damascene interconnects due to its poor microstructure.[5]

In our previous article, the microstructure of ultra fine Al-damascene interconnects (0.80-0.32 μm pitch), fabricated by electron beam (EB) lithography, reflow sputtering (RS), and chemical mechanical polishing (CMP) techniques, were investigated by transmission electron diffraction (TED). In this article, the microstructure of ultra fine Al-damascene interconnects, inlayed by using chemical vapor deposition (CVD) techniques, is investigated and differences due to the deposition method will be discussed.

EXPERIMENTAL

Multilayered SiO_2(500nm)/SiON(100nm)/SiO_2(100nm) films were deposited on Si substrates by plasma enhanced CVD. The top SiO_2 film was patterned using EB lithography and reactive ion etching (RIE). The depth was controlled at 5000 Å by using the SiON film as the etch stop. The trench was designed with line widths W = 0.10 - 0.50μm for single straight line and line pitches P = 0.32-0.80μm and both the line space S and the line width W = P/2 for open short check pattern.

Titanium(10nm) film was deposited on the SiO_2 substrate as a liner by long throw sputtering. Subsequently, Al-CVD(800nm) films were deposited using dimethyl aluminum hydrate (DMAH) with H_2 carrier gas. The substrate was heated to 160°C. After annealing the sample in N_2/H_2(50%) ambient at 400°C for 30min, Al damascene interconnects were fabricated by CMP with silica slurry.[6]

Microstructure of the interconnects was observed by TED. TED samples were fabricated by thinning from the substrate side (Fig. 1) by dimple grinding, followed by Ar ion milling. TED photographs from the interconnect surface were observed at 20 locations for each of 0.80 and 0.32 µm pitched lines. In this article, the (hkl) texture to the substrate and (hkl) texture from the trench wall will be denoted as (hkl)∥ and (hkl)⊥, respectively.

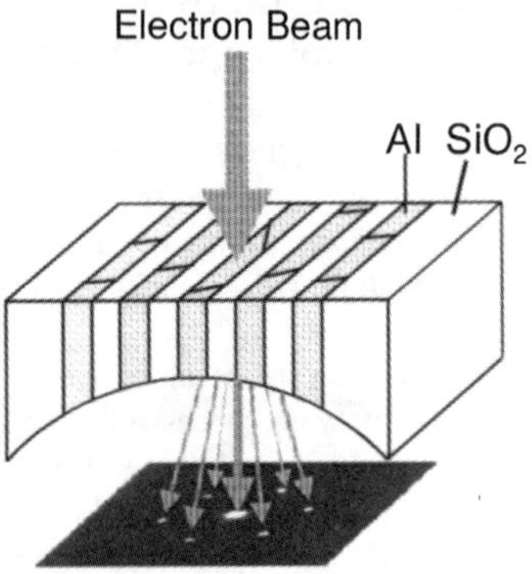

RESULTS

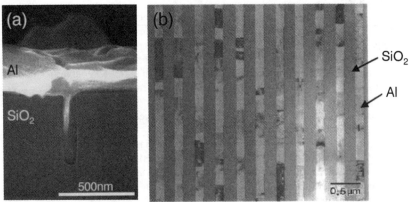

Fig. 1 Schematic diagram of TED observation.

Fig. 2 (a) Cross sectional SEM photograph of single straight line of 0.10µm width. (b) Plane view of TEM photograph of snake line of 0.32µm pitch.

Both single straight lines of 0.10µm width and serpentine lines of 0.32µm pitch were successfully filled by Al-CVD (Fig. 2 (a) and (b)). The interconnect has a bamboo structure, in which small Al grains span line width. The textures parallel to the substrate that are observed by TED are listed in Table I. The (111)∥ texture of RS is also listed dor a reference. Regarding the CVD samples, there is no specific (hkl)∥ texture for either wide or narrow interconnects. On the other hand, in the case of Al-RS filling, reduction of (111)∥ was observed in the narrower lines.[5, 7]

In the next, texture from the trench wall will be presented. TED analysis enables us to get the exact (hkl)⊥. However, the exact plane is usually a high index plane which is not useful to reveal the behavior of microstructure in this experiment. Therefore, arrangement of the trench wall and the typical low index planes, (111), (022), and (002) have been investigated. Angles between the

Table I   Relationship between texture to substrate and line pitch.

Deposition method	Texture to substrate	Percentages of grains(%)	
		0.80µm pitch	0.32µm pitch
CVD	(111)‖	10	30
	(100)‖	15	10
	(233)‖	35	10
	(211)‖	5	5
	(011)‖	30	15
	(310)‖	5	10
	(others)‖	0	25
RS	(111)‖	75	41

direction normal to the trench wall and the <111>, <022>, and <002> directions were calculated using the TED patterns. Then the plane which has the minimum angle in the three planes is defined as the crystal plane parallel to the trench wall. For example, if the angle between (111) plane and the trench wall is the minimum, we define (111) as the parallel plane to the trench wall. Distribution of the parallel plane to the trench wall is shown in Fig. 3. The legend in the graph indicates the texture to the parallel plane to the substrate. There is no dominant texture from the trench wall in the (111)⊥, (022)⊥, and (002)⊥ whose plane to the substrate is (random)‖ at the wide line. On the other hand, (111)⊥ is dominant and it has (random)‖ in the narrow lines.

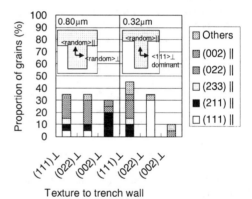

Texture to trench wall

Fig. 3 Proportion of grains which have (111)⊥, (022)⊥ and (002)⊥ texture for wide (0.80µm) and narrow (0.32µm) pitched lines. Cross' sectional schematic diagrams of the trench are also shown. Arrows in these diagrams indicate the typical direction of crystal planes.

DISCUSSION

It is found that the texture of the Al damascene lines depends on both trench width and deposition method. The effect of the trench wall is probably due to the lowering of interfacial energy. To investigate the differences with deposition method, the texture from the trench wall for RS-derived interconnects is shown in Fig. 4. For wide lines, the texture is strongly (111)‖/(022)⊥, while for narrow lines, the texture is weakly (111)‖/(022)⊥. Therefore, it is revealed that the texture of Al-CVD damascene interconnects is (random)‖/(111)⊥ and increases with narrowing trench width. For Al-RS, (111)‖/(022)⊥ decreases with narrowing trench width.

These differences between CVD and RS textures are probably due to a difference of filling modes between these two depositions and the influence of interfacial energy. The CVD achieves a conformal growth while RS achieves a bottom up growth. Therefore, most grains derived by CVD in the trench should consist of grains that grow from the side wall. Therefore, it should strongly reflect the grain orientation of initial film formation at the side wall especially in narrow lines. On the other hand, the RS-film should consist mostly of grains that grow from the bottom. Therefore, it should strongly reflect the grain orientation of initial film formation at the bottom especially for wider lines. Grains in a thin film orient to reduce their surface energy since the surface energy is usually much larger than the interfacial energy. The surface energy of fcc metals is ordered generally $\gamma(111)<\gamma(100)<\gamma(110)$;[8] therefore, the texture of an Al thin film on a flat substrate is usually $(111)\|/(random)\perp$. In the case of the grains in the trench, it should

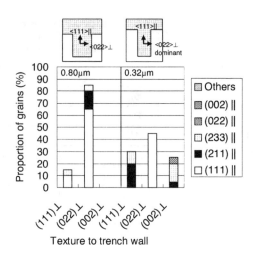

Fig. 4 Proportion of grains which have $(111)\perp$, $(022)\perp$ and $(002)\perp$ texture for wide $(0.80\mu m)$ and narrow $(0.32\mu m)$ pitched line of RS derived interconnects. Cross sectional schematic diagrams of trench are also shown. Arrows in these diagrams indicate the typical direction of crystal planes.

be $(random)\|/(111)\perp$ at the side wall and $(111)\|/(random)\perp$ at the bottom at the initial stages of film growth (Fig. 5 (a)). Therefore, grains deposited by CVD and RS should basically reflect the $(random)\|/(111)\perp$ texture and $(111)\|/(random)\perp$, respectively.

The interfacial influence is believed to appear much stronger in the narrower lines, since the fraction of the interfacial area between the grains and the side-wall increases while the fraction of the surface area decreases. Moreover, the smaller grains size in the narrower trench result in easier reorientation. Therefore, the grains in the narrower trench would be affected by the interfacial energy, since the proportion of the interfacial area increases and less energy is necessary for changing their orientation in the narrower trench.

In the case of the microstructure of the CVD film, the grains in the wide trench reflect not only the texture of the side wall but also that of the bottom (Fig. 5 (b)). Therefore, the texture of CVD in the wide lines becomes $(random)\|/(random)\perp$ (Fig. 5 (f)), because it is affected by both initial texture of $(111)\|/(random)\perp$ at the bottom and $(random)\|/(111)\perp$ at the side wall. The texture of CVD in the narrow line becomes $(random)\|/(111)\perp$ dominant (Fig. 5 (g)), since the grain of CVD dominantly reflects the initial texture of the side wall (Fig. 5 (c)).

The effect of the interfacial energy also increases the $(111)\perp$ texture in the narrow line. The (111) plane, which is the closest packed planes of Al atoms, has the lowest surface energy and thus forms the most stable surface. Therefore, Al (111) seems to form the most stable interface which have the lowest interfacial energy. According to this assumption, the grains in the narrow line, which are affected by the interface strongly, should change from $(random)\perp$ to $(111)\perp$ in order to reduce the total surface energy of grains by reducing the interfacial energy. This interfacial effect could enhance the $(111)\perp$ texture at the narrow line.

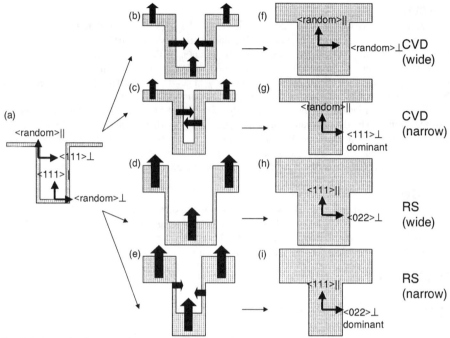

Fig. 5 Schematic diagram of texture behavior during film formation for CVD and RS.

In the case of the microstructure of the RS film, the grains in the wide trench dominantly reflect the $(111)\|/(random)\perp$ texture of the initial film formation at the bottom (Fig. 5 (d)). The interface with a high index plane of $(random)\perp$ seems to have high interfacial energy compare to the low index plane, because a high index plane is an unstable surface with high surface energy. According to this assumption, the grains reduce their interfacial energy during their filling by changing its texture from the $(random)\perp$ to the $(022)\perp$ texture which is the low index plane normal to $(111)\|$. This interfacial effect results in the $(111)\|/(022)\perp$ texture for the wide line deposited by RS (Fig. 5 (h)). The grains in the narrow trench are influenced by the interface more strongly than in the wide line. Therefore, some of the grains rotate from $(022)\perp$ to $(111)\perp$ and $(002)\perp$, which probably have lower interfacial energy. Moreover, the texture reflects not only the texture of the bottom but also that of the side wall (Fig. 5 (e)). The growth from the side wall also increase proportion of the grains which have the $(111)\perp$ and $(022)\perp$. Therefore, for the narrow line, the $(111)\|/(022)\perp$ texture is still dominant, however, there appears another texture (Fig. 5 (i)).

It is revealed that the texture in the narrow line is strongly affected by the trench wall and that in the wide line is strongly affected by the substrate. This difference in the growth characteristics results in a $(111)\perp$ texture for the line deposited by CVD and $(111)\|$ texture for the line deposited by RS. Therefore, to obtain highly (111) oriented interconnects, filling by RS is suitable for the wide lines with low aspect ratios and filling by CVD is suitable for the narrow lines with high aspect ratios.

CONCLUSION

Microstructure of fine Al interconnects inlaid by CVD was investigated by TED observation. It is revealed that the microstructure of Al-CVD lines is strongly $(111)\|/(022)\perp$ in the wide lines and weakly $(111)\|/(022)\perp$ in the narrow lines. This is different from the microstructure of Al-RS lines of $(random)\|/(random)\perp$ at the wide line and $(random)\|/(111)\perp$ at the narrow line. The microstructure difference is due to (a) the effect of interfacial energy: $\gamma(111)<\gamma(100)<\gamma(110)<\gamma$(high index plane) and (b) the difference in the film growth characteristics: bottom up growth of RS and conformal growth of CVD. The texture in the narrow line is strongly affected by the trench wall and that in the wide line is strongly affected by the substrate. This difference in the growth characteristics results in a $(111)\perp$ texture for the line deposited by CVD and a $(111)\|$ texture for the line deposited by RS. Therefore, to obtain highly (111) oriented interconnects, filling by RS is suitable for wide lines with low aspect ratio and filling by CVD is suitable for narrow lines with high aspect ratios.

ACKNOWLEDGMENT

The authors thank Drs. Hiroyuki Abe, Masao Fukuma and Takemitsu Kunio for their encouragement. They are also grateful to Mr. Ken'ichi Tokunaga, Mr. Ken Nakajima, Mr. Shin'ya Yamazaki, Mr. Toshiyuki Takewaki, Dr. Kuniko Kikuta, Dr. Kazutoshi Shiba, Mr. Shinobu Saito, and Mr. Takayuki Onodera for their help in conducting the experiments.

REFERENCES

1. S. Takahashi, M. Edahiro and Y. Hayashi, Technical Digest of International Electron Devices Meeting (1998) 833.

2. K. Shiba and Y. Hayashi, to be presented at International Electron Devices Meeting 1999.

3. K. Shiba, H. Wakabayashi, T. Takewaki, K. Kikuta, A. Kubo, S. Yamasaki and Y. Hayashi, 1998 International Conference on Solid State Devices and Materials (1998) 280.

4. K. Ueno, K. Ohto, K. Tsunenari, K. Kajiyana, K. Kikuta and T. Kikkawa, Technical Digest of International Electron Devices Meeting (1992) 305.

5. P. R. Besser, J. E. Sanchez, Jr, D. P. Field, S. Pramanick and K. Sahota, Mat. Res. Soc. Symp. Proc. **473** (1997) 217.

6. Y. Hayashi, T. Onodera, T. Nakajima, K. Kikuta, Y. Tsuchiya, J. Kawahara, S. Takahashi, K. Ueno and S. Chikaki, 1996 Symposium on VLSI Technology, Digest of technical paper (1996) 88.

7. A. Furuya, T. Takewaki, K. Kikuta and Y. Hayashi, "*Advanced Metallization Conference in 1998*", edited by G. S. Sandhu, H. Koerner, M. Murakami, Y. Yasuda, and N. Kobayashi, Mater. Res. Soc., Pittsuburgh, PA (1999), 691.

8. S. M. Foilies, M. I. Baskes and M. S. Daw, Phys. Rev. B **33** (1986) 7983.

Integration of the 3MS Low-k CVD Material

in a 0.18μm Cu Single Damascene Process

T. Gao[1][2], W.D. Gray[3], M. Van Hove[1], H. Struyf[1], H. Meynen[4], S. Vanhaelemeersch[1]
K. Maex[1][2]

[1] IMEC, Kapeldreef 75, B-3001 Heverlee, Belgium.

[2] E. E. Dept., K.U. Leuven, Leuven, Belgium

[3] Affiliate researcher at IMEC for Dow Corning

[4] Dow Corning S.A., Parc Industriel, B-7180 Seneffe, Belgium

Tel: 32-16-281778, Fax: 32-16-281214, E-mail: gao.teng@imec.be

ABSTRACT

A low-k CVD material is applied in a sub 0.18μm technology utilizing the Cu-Damascene approach. Been measured with a dedicated test vehicle, the dielectric constant of the material between two nearby trenches achieves 2.5 after the process.

INTRODUCTION

With the scaling down of the device technology, low-k material becomes more and more important for high-density interconnection for reducing interconnect RC delay. From an integration point of view, the implementation of low-k material for interconnect systems is a difficult task due to the process-induced changes of the material properties for most low-k films.

Generally, the lowering of the dielectric constant of films requires either an addition of organic components or an increase of the porosity formation. These kind of materials are very sensitive to most commonly used pattern definition processes including O_2 plasma etching, ashing, and wet strip for remaining polymer residues. Some low k films are also unstable during high thermal processing (>450°C).

Conference Proceedings ULSI XV © 2000 Materials Research Society

In order to integrate low-k materials successfully, several characteristics need to be present for device integration such as electrical properties, thermal stability, adhesion to other materials, mechanical strength, and chemical stability.

The integration of low-k material into classical interconnect systems has been studied intensively. Both organic [1] and inorganic [2] materials have been integrated successfully into sub 0.25μm subtractive technologies.

However, the integration of low-k materials into more advanced technologies utilizing damascene architecture and Cu metallization is much more difficult compared to the classical architecture. The damascene approach, which increases the exposed area of low-k material, requires low-k materials with much better chemical stability. Without these characteristics, the material between the trenches will no longer show low-k behavior after processing [3].

In this paper we present results of a low-k CVD SiOC film deposited in a regular PECVD chamber from a Trimethylsilane (3MS) precursor manufactured by Dow Corning Corporation [4]. The 3MS Low k film showed superior characteristics for process integration and has been integrated into a 0.18μm single damascene process with Cu metallization. With a dedicated test vehicle, the dielectric constant of the material between two trenches is measured as 2.5 after finishing the complete metallization process.

EXPERIMENT

Trenches were etched in 500nm PECVD 3MS low-k film with a 200nm PECVD oxide as cap layer. Reference splits containing only PECVD oxide were processed together with the 3MS low-k/oxide for comparison. A medium density plasma RIE etch tool was used for the trench etching using C-F chemistry. Initial experiments with O_2 plasma (with low pressure, medium density plasma, and low power strip process) used for resist and polymer strip resulted in a bowing problem (Fig. 1 left). After optimization of the strip process (with low pressure, high density plasma and higher power strip process), good via profiles were achieved (Fig. 1 right). A wet strip step was not used after the O_2 plasma strip in this experiment.

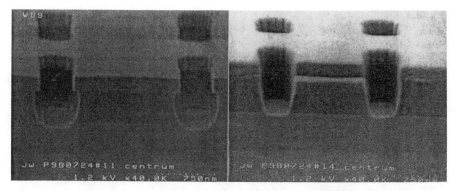

Figure 1. Via profile after non-opitimized dry strip (left) and after optimized dry strip (right)

30nm PVD TaN was deposited as a Cu diffusion barrier followed by a 150nm PVD Cu seed layer. Both layers were deposited in Endura IMP chambers. A thick Cu layer was then deposited using a Semitool electroplating process. Cu CMP and TaN CMP with two kinds of alumina-based slurries were applied to remove the Cu and the TaN layers above the oxide cap layer.

FIB cross-section showed excellent trench profiles with good Cu filling for $0.35\mu m$ width trenches after the metallization process (Fig. 2). No delamination of Cu lines or dielectric films was observed.

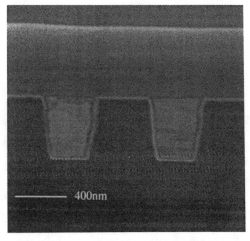

400nm

Fig. 2.Cross-section of the capacitor structure

RESULTS

The measured interconnect resistance of 0.3μm and 0.35μm width Cu lines is similar for the 3MS low-k and the oxide reference wafers (Fig. 3). Equally high metal yield (>90%) was measured on very large (0.25cm²) meander-fork structures (0.35μm width) for both splits (Fig. 4).

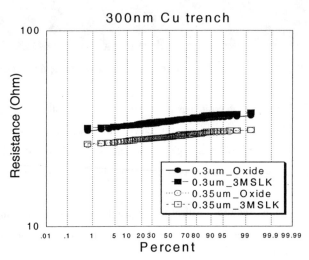

Figure 3. Line resistance of 0.3μm and 0.35μm trenches

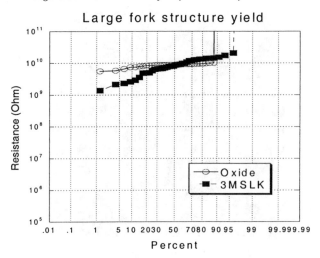

Figure 4. Yield of 0.25cm² fork structure

Special meander intraline capacitors were designed with three different spacings. The spacings of 0.4, 0.7, and 1.0μm were patterned as meander structures having effective lengths of 9706μm each. The intraline capacitance was measured immediately after Cu CMP and again immediately after annealing at 450°C for 90 minutes in N_2 (Fig. 5). Electrical measurements were taken at 100kHz. The capacitance data shows that the 3MS low-k material exhibits slightly lower dielectric constant after an anneal step. This confirms that 3MS low-k has very good thermal stability.

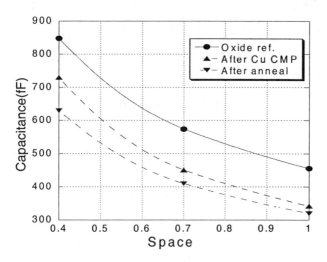

Figure 5. Measured intraline capacitance

Taking into account the trench profile, the 3MS low-k dielectric constant between two closely spaced trenches was extracted using the simulation software Raphael® TMA. Without the anneal step, the dielectric constant of the 3MS low-k film is clearly increased compared to the bulk value (k=2.6). The increase of the dielectric constant can be due to two factors: 1) moisture absorption into the sidewall which might be altered during etch and strip processes, and 2) interface charges induced by the plasma processes. This kind of process-induced sidewall "damage" can not be avoided and can create serious problems for advanced technology generations using the low-k damascene approach.

Due to the sidewall modification (absorbed moisture and induced charges), the effective dielectric constant in the spacing can be increased, especially for the smaller spacings. The simulation results indeed indicate that for 1.0μm, 0.7μm, and 0.4μm spacings the effective k-value is degraded respectively to 2.8, 2.9 and 3.2 (Fig. 6).

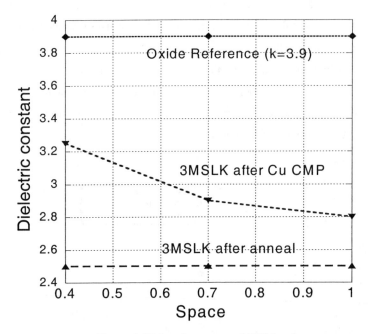

Figure 6. Dielectric constant of 3MS low-k

Nevertheless, this kind of sidewall modification on 3MS low-k material can to be recovered by an annealing step. The resulting effective dielectric constant of 3MS low-k after annealing is 2.5, which is close to the bulk value (2.6), and the dielectric constant of the material is independent of the trench spacing. This result suggests that the low-k property of the 3MS low-k material has not been destroyed during etch and strip processing. The deposited porous film most probably absorb moisture during process, which can be removed easily by a thermal anneal. This annealing step can also release the charges induced by the plasma processes.

It is also possible that the annealing has caused some rearrangements in the amorphous random covalent network of the as-deposited 3MS low-k film [5]. The fact that the dielectric constant is slightly lower than the bulk value after annealing might indicate that such a rearrangement results in the enhancement of pre-existing porosity in the films.

CONCLUSIONS

The 3MS low-k dielectric film, which is deposited using a standard PECVD process, can be successfully integrated into a 0.18μm Cu-damascene technology. After etch, strip, Cu metallization, and CMP processing, the dielectric constant of the material is increased probably due to absorbed moisture and induced charges. The material can be easily recovered by a thermal anneal step to remove the moisture and the charges. The annealing might also enhance the porosity of the film.

ACKNOWLEDGMENTS

The authors would like to thank H. Bender, C. Vrancken, B. Deweerdt, M. Van Dievel, and the IMEC Pilot-line for support. The authors would also like to thank P. Dembowski and M. J. Loboda for helpful discussions. K. Maex is a research director of the Fund for Scientific Research – Flanders (FWO).

REFERENCES

[1] M. Miyamoto, T. Furusawa, A. Fukami, Y. Homma and T. Nagano, VLSI Multilevel Interconnection Conference 1997, p. 37.

[2] S. W. Chung, J. W. Kim, D. W. Suh, N. H. Park, J. W. Park, and J. J. Kim, Dielectrics for ULSI Multilevel Interconnection Conference 1998, p. 151.

[3] T. Gao, W. D. Gray, M. Van Hove, E. Rosseel, H. Struyf. H. Meynen, S. Vanhaelemeersch, and K. Maex, International Interconnect Technology Conference 1999, p. 53.

[4] M.J. Loboda, European Workshop Materials for Advanced Metallization 1999, p. 19.

[5] A. Grill and V. Patel, J. Appl. Phys., Vol. 85, No. 6, p. 3314 (1999).

HIGH PERFORMANCE COPPER DUAL DAMASCENE INTERCONNECTS USING AN ORGANIC LOW-K DIELECTRICS

Ikuhiro YAMAMURA, Kazuhiko TOKUNAGA, Koichi IKEDA, Toshiaki HASEGAWA, Koji MIYATA, Tetsuya TATSUMI, Mitsuru TAGUCHI, Keiichi MAEDA, Naoki KOMAI, Masanaga FUKASAWA, Hideyuki KITO, and Shingo KADOMURA

LSI Business & Technology Development Group, CNC, Sony Corporation
4-14-1 Asahi-cho, Atsugi-shi, Kanagawa, 243-01 Japan

ABSTRACT

We developed a copper dual damascene process using an organic low-k dielectric and evaluated the electrical characteristics of test devices by using this process. Using a "dual hard mask" process, we fabricated good 0.4 μm pitch dual damascene structures without damaging the organic low-k material. It is demonstrated that the stage delay time of a ring oscillator using this interconnect system is 20% less than that of one made of silicon dioxide.

INTRODUCTION

As the dimensions of ULSI devices continue to shrink, the RC delay of their interconnects will limit the device performance. Copper interconnects have therefore started to be used to lower the resistance, and the use of low-k dielectrics has been explored as a means of decreasing the capacitance of interconnects. One of the most promising solutions to the problem of RC delay is thus thought to be a combination copper and low-k material interconnect system.

In this paper, we describe the integration of a copper dual damascene (D.D.) structure with an organic low-k dielectric: FLARE™ (Allied Signal, Inc.; K=2.8). We also demonstrate the effect of this interconnects on the delay time of a CMOS circuit.

EXPERIMENTAL

A cross section of our test device is shown schematically in Fig. 1. We evaluated the effects of the organic low-K dielectrics with a 0.18 μm gate transistor. The gate oxide thickness was 3.5 nm. We used a full cobalt salicide and a dual gate process. The tungsten electrodes were formed by CVD-W method. The first metal layer is a single damascene

Conference Proceedings ULSI XV © 2000 Materials Research Society

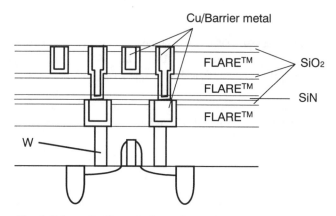

Cu/Barrier metal

FLARE™   SiO₂

FLARE™   SiN

FLARE™

W

Fig. 1 Schematic diagram of sample structure.

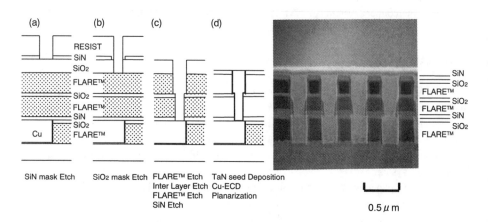

(a)   (b)   (c)   (d)

RESIST
SiN
SiO₂
FLARE™
SiO₂
FLARE™
SiN
SiO₂
Cu   FLARE™

SiN
SiO₂
FLARE™
SiO₂
FLARE™
SiN
SiO₂
FLARE™

0.5 μ m

SiN mask Etch   SiO₂ mask Etch   FLARE™ Etch   TaN seed Deposition
                                   Inter Layer Etch Cu-ECD
                                   FLARE™ Etch   Planarization
                                   SiN Etch

Fig.2 Organic low-k / Cu D.D. process flow.   Fig. 3 Cross-sectional SEM photograph.
                                               0.4 μ m pitch line & space.

formation. The via electrodes between the first and second metal and the second metal wiring were fabricated by using the D.D. method.

From the many materials that have been proposed for use in low-k applications, we selected the organic film, FLARE™, because it meets the basic requirements for dielectric application in the copper D.D. process, such as small outgassing, good adhesion, and resistance to chemical reagents. and process availability. FLARE™ (based on a polyarylether) has a K value of 2.8, and glass transition (Tg) and decomposition temperature (Td) are above 400℃.

To fabricate the copper D.D. structure, we developed a SiNx/SiO2 stacked hard mask process we call the "dual hard mask process" [1]. The process flow for the fabrication of a copper D.D. structure is outlined in Fig. 2. A SiNx/SiO2/FLARE™/SiO2/FLARE™/SiNx sandwich was formed on the first copper metal wiring. The FLARE™ films were formed by a spin-on method, and each was cured at 400℃. After using the SiNx layer as a hard mask for etching the trench pattern (Fig. 2(a)), the SiO2 layer was patterned for via formation (Fig. 2(b)). The photoresist was removed spontaneously when the upper FLARE™ film was patterned. The final D.D. structure was completed as shown in Fig. 2(d). A SEM photograph of a cross section of the two-layer interconnects is shown in Fig. 3. The FLARE™ film adhesion to the SiO2 and SiNx films was good enough for use in this process.

The dual hard mask process was constituted based on two concepts: (1) the organic polymer should not be exposed in oxygen plasma, and (2) the processing should be easily broken down into discrete steps (dielectric-stacked-film formation, patterning using reactive ion etching (RIE), and metallization). A SiNx/SiO2 stacked hard mask is therefore applied, and ammonia chemistry is used for etching the organic polymer [2].

RESULTS and DISCUSSIONS

The wiring resistance of the FLARE™ interconnect system was the same as that of an SiO2 interconnect system (Fig. 4), and the Cu resistivity was uniformly 2.0 $\mu \Omega$cm. This showed that effective dual damascene interconnects can be fabricated by using our dual hard mask process.

The dependence of the leakage current on the voltage across wiring lines separated by 0.2 μm is shown in Fig. 5. No unusual current was observed, and there was no breakdown at electric field up to 0.5 MV/cm. This showed that the use of FLARE™ as the intermetal dielectric (IMD) in an interconnect system does not cause any degradation or damage. The dependence of IMD capacitance on space width is shown in Fig. 6, where it can be seen that using of FLARE™ instead of silicon dioxide reduced the capacitance by 30% when the space width was 0.3 μm.

The via resistances measured for various size vias is shown in Fig. 7. The average resistance for a 0.2 μm via was 0.5 Ω, and good uniformity was obtained across the entire wafer. The contact resistance of the copper D. D. electrode was less than one tenth that of the

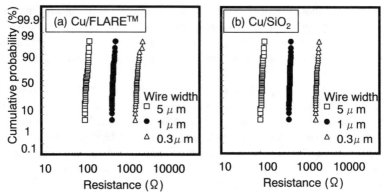

Fig. 4 Wiring resistance of (a) Cu/FLARE™ and (b) Cu/SiO₂ interconnections. The wire length was 10 mm.

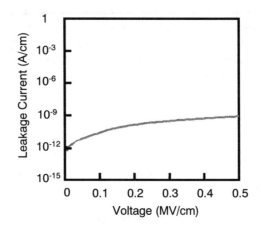

Fig. 5 Leakage current as a function of the supplied voltage measured with a comb structure. The sample space width was 0.2 μ m, and the wire height was 0.4 μ m.

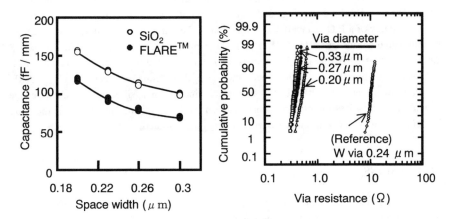

Fig. 6 IMD capacitance vs. space width.

Fig.7 Via resistance measured with a Kelvin test pattern. Reference sample was W via electrode with Al wire [3].

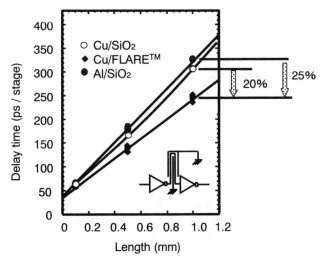

Fig. 8 Ring oscillator stage delay time vs. wiring load length.

conventional tungsten via electrode. These results indicate that a reliable copper D.D. interconnects was formed in this study.

We evaluated the effect of using a low-K dielectric on the performance of a CMOS circuit. We tested a 101-stage ring oscillator with second-level load wiring. The measured ring oscillator delay of a F/O=1 inverter as a function of the wiring load length is shown in Fig. 8. We fabricated a transistor using 0.18 μ m node technology. The gate oxide thickness was 3.5 nm, the V_{dd} was 1.8 V, the gate length was 0.18 μ m, and the wiring space was 0.4 μ m. For an inverter with a 1 mm wiring load, the stage delay time obtained with a FLARE^{TM} interconnect system was 20% less than that obtained with a silicon dioxide interconnect system, and 25 % less than that obtained with an Al/SiO_2 system. The effects of the copper interconnects with a low-K dielectric were clearly demonstrated.

CONCLUSIONS

A copper dual damascene interconnect system using FLARE^{TM} has been successfully demonstrated by using a novel dual hard mask process. The FLARE^{TM} reduced the delay time of an inverter with a 1 mm wiring load length by 20%. This copper dual damascene structure with an organic low-k dielectric is a promising one for use in LSIs with node sizes of 0.13 μ m and beyond.

ACKNOWLEDGEMENTS

We thank Mr. Shinichi Hishikawa for his continuous encouragement. We are also grateful to Tokyo Electron Ltd. and to Electroplating Engineers of Japan Ltd. for their cooperation in developing the fabrication process.

REFERENCES

[1] The Dual Hard Mask Process is reported in detail by K. Miyata et al. in this proceedings after Advanced Metallization Conf. Assian Session 1999 pp.23-24.
[2] M. Fukasawa et al., Proc. on Dry Process Symp. P.175 (1998).
[3] K. Tokunaga, K. Ikeda, T. Miyamoto, T. Hasegawa, M. Fukasawa, H. Kito, and S. Kadomura, 1999 Int. Conf. on SOLID STATE DEVICES AND MATERIALS (SSDM '99) pp. 498-499 (1999).

Comparison of Via Partial-Etch Approach And Via Full-Etch Approach in Dual Damascene Process

L.C. Chao, M.H. Lui, J.C. Liu, C.C. Chen, C.S. Tsai, M.S. Liang

Advanced Module Technology Division, R&D, Taiwan Semiconductor Manufacturing Company, Ltd., 9, Creation Rd. I, Science-Based Industrial Park, Hsin-Chu, Taiwan

Abstract

A promising via partial-etch dual damascene process is reported in this work. This new approach can reduce both the trench and via etching stop layer thickness without any via fence issue and concern of protection coating residue. The integration challenges like pattern sensitivity and scaling capability along with via Rc and trench to via misalignment window of via partial-etch approach are investigated and compared to that of via full-etch approach. In addition, extension of this new via partial-etch approach to application of none trench etching stop layer will also be discussed.

Introduction

Conventional self-align dual damascene process requires a thick trench etching stop layer like SiN for trench depth control during via etching which has disadvantage of increasing interconnect capacitance by the high k value of Si_3N_4. Recently, the via full-etch approach (so-called counter bore approach) has been extensively studied to reduce the trench etching stop layer thickness. However, the via full-etch approach will require a thick via etching stop layer as well due to the initial patterned via being over-etched during trench etch. Thus, coating of organic protection materials inside the via hole has been tested to reduce via etching stop layer thickness. Unfortunately, via fence issue as shown in Figure 1, resulting from the etching selectivity between dielectric film and the organic material, as well as post-etching removal of the protection coating are always concerns for via Rc stability and interconnect reliability[1,2]. Although the via fence can be eliminated by using a lower-selectivity trench etch recipe, the process window is limited since via top facet tends to be a trade-off as indicated by Figure 2.

Via partial etch approach seems to be a promising alternative to minimize the trade-offs between via fence and facet with reduced thickness of high-k etching stop layer. Since it does not employ any organic protection layer in the via, it should be free of via fence and thus gives more leeway to minimize via top facet. It is therefore the purpose of this work to evaluate the process- and production-worthiness of USG via partial etch process in terms of both physical and electrical data.

Experiment

In this via partial etch approach, Si_3N_4 as the etching stop layer and USG as IMD of via and trench is deposited and then patterned with via photo. The via is partially etched to different depth and ashed to remove photoresist. A thin via hole liner is then deposited and followed by trench photo and trench etching. Figure 3 shows the sketch of this via partial-etch approach before trench etch. By using via partial-etch approach, organic protection coating of via full-etch approach can be eliminated which also solves the associated fence issue. The thin via hole liner is deposited to suppress via facet as demonstrated in Figure 4 and thereby reduce the required trench etching stop layer thickness.

The dual damascene etching is followed by a barrier and seed deposition. Copper is then electroplated on the wafers followed by Cu CMP.

Results

Figure 5 shows the via Rc distribution of via partial-etch approach with liner deposition and via full-etch approach with organic protection coating. The average via Rc and Rc distribution of via partial-etch approach with splits of via oxide remain 5KA and 6KA is better than that of via full-etch approach due to no via fence and no residue of protection coating for via partial-etch approach. It has to be noted that the Rc distribution of via partial-etch approach start to broaden for via oxide remain depth less than 4KA which results from punching through of via etching stop layer during trench etch.

One important integration implication of via partial etch process is the misalignment between via and trench lithography process. As misalignment occurs, residual photoresist in the via holes will block the etching of the remaining USG and results in the shrinkage of the via hole bottom size. This effect is demonstrated in Figure 6, where via Rc for both vai full etch and partial etch with different extent of misalignment is compared. It can be seen that the via partial-etch approach has a smaller window for trench to via misalignment due to potential via under-etch in higher extent of misalignment. Nonetheless, the process window can be improved by proper modification of the trench etching recipe. Compared with Figure 6, data in Figure 7 shows that the new trench etching recipe does improve both Rc and its distribution. Figure 7 also indicates that as oxide remain is decreased to 2000A°, the Rc performance is almost as good as that of no misalignment case even with 0.12um misalignment. Based on these data, therefore, although misalignment is a potential issue for the via partial approach, a reasonable process window(misalignment within 0.12um) can be obtained by suitable combination of oxide remain and trench etch recipe.

The feasibility of doing without the intermediate etch stop layer in via partial approach is also investigated. Without this stop layer, control of trench profile and etching thickness, and thus Rs, will be a challenge regardless of the damascene approach. For example, a trench profile with microtrench is normally obtained. But this phenomena can be eliminated by moving the etching recipe to a high pressure and low power region, as shown in Figure 8, or via full etch, since an organic protection layer is required for this approach, the impact is even more obvious, in addition to the via fence and facet issues. Due to its higher flowability near via, the organic material is thinner around the holes, The resultant surface topography will be directly transferred to the etching front, which would be leveled off if there were a etch stop layer. On the other hand, for via partial etch approach, there is no concern of this kind.

Without the intermediate trench etch stop layer in via partial approach using the above-mentioned microtrench-free recipe, Rc and its distribution has been shown to have a similar dependence on oxide remain, as shown in Figure 9. It is even more encouraging to find from Figure 10 that the trench Rs without trench etch stop layer is comparable to the trench Rs with trench etch stop layer, indicating good control of the trench thickness even without the stop layer.

Conclusions

The physical and electrical data of both full via etch and partial via etch approach have been compared in this work. The latter process can get around the via fence issue and shows better Rc and Rc distribution than the traditional full vai etch approach. However, the control of the oxide remain thickness after partial via etch is crucial for obtaining good Rc performance. Moreover, even that misalignment is a potential issue for the via partial approach, a reasonable process window can be obtained by suitable combination of oxide remain and trench etch recipe. It has also been successfully shown that a non-intermediate stop layer structure can be implemented into this partial vai etch approach.

References

1. A. Blosse, U. Raghuram, S. Thekdi, B. Koutny, G. Lau, S.L. Koh, and C. Goodenough, Proceedings of the International Interconnect Technology Conference, 1999, pp. 215-217.
2. S. Kordic, J.Torres, L. Liauzu, C. Verove, P. Vannier, R. Gonella, Ph. Gayet, E. Van der Vergt, and J.M. Gilet, Proceedings of VLSI Multilevel Interconnection Conference, 1999, pp. 53-62.

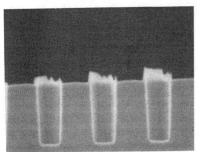

Figure 1. via fence formation due to use of organic protection layer in via

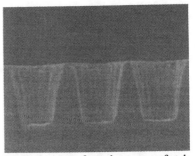

Figure 2 via top facet due to use of etch recipe with low oxide/protection layer sel..

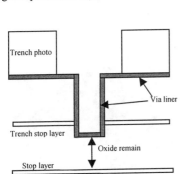

Figure 3. sketch of partial via approach

Figure 4. No facet with via liner compared with Figure 2.

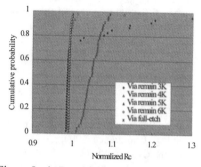

Figure 5. via Rc of full and partial via etch approach.

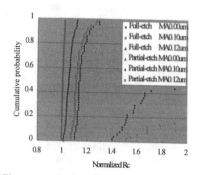

Figure 6. Effects of misalignment on via Rc.

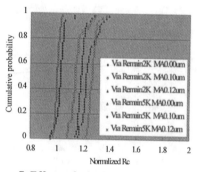

Figure 7. Effects of misalignment on Rc with a modified trench etch recipe.

Figure 8. Via facet improved by a high pressure and low power recipe.

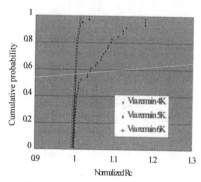

Figure 9. via Rc of non-stop layer structure with partial via etch approach

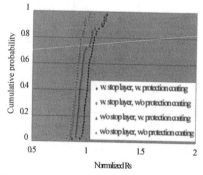

Figure 10. Trench Rs of w/i and w/o stop layer with partial via etch approach

Copper: Deposition Processes and Properties

Factors Influencing Fill of IC Features Using Electroplated Copper

J. Reid and S. Mayer

Novellus Systems, Portland Technology Center, Wilsonville, OR 97070

Abstract

The demonstrated capability of copper electroplating processes to produce void-free fill of high aspect ratio (AR) damascene features has made it the process of choice for copper interconnect formation. To achieve this filling performance, as well as to extend it to future device geometries, key electroplating process parameters and incoming wafer characteristics must be optimized to encourage acceleration of deposition near the base of damascene features (bottom-up fill). This paper describes mechanisms of bottom-up fill performance based on process chemistry, and reviews the effects of feature geometry, seed layer characteristics, and applied current waveform.

Introduction

The transition from Al to Cu interconnects is underway or in development at most major semiconductor manufacturers. As evidenced by numerous press announcements and technical papers [1-3], interest in copper interconnects is now widespread.   Use of a dual-damascene process flow [4] allows construction of the multilevel Cu stack with potentially fewer steps and reduced costs as compared to conventional Al-based interconnects. A new process technology for damascene interconnect fabrication is the electrodeposition of Cu for feature fill.   This process is unique among copper deposition methods because it allows filling of high AR features from the bottom up, avoiding the introduction of a center seam or void.

Existing process chemistries developed for printed circuit board (PCB) applications are providing sufficient capability to fill sub-0.25 μm generation IC features. Many manufacturers are, however, implementing copper at smaller device dimensions where such process chemistries do not provide void-free filling. The extendibility requirement for filling smaller geometries has driven extensive development of the copper plating and seed layer deposition processes. As a result, improved seed layer integrity, modifications of plating current waveforms, and new plating process chemistries have led to filling improvements. The success of copper plating to fill high AR features is built upon the achievement of rapidly accelerated Cu deposition at the feature base relative to the field deposition rate. To date, the bottom-up fill process has been described in the literature as being driven primarily by the establishment of a diffusion gradient of suppressing polymers [5]. The effect of leveling additives which may reduce the copper growth rate at the entrance to features has also been widely discussed [6,7]. In this paper, these fill mechanisms, as well as others suggested by the effect of process parameters on fill performance, are discussed.

Conference Proceedings ULSI XV © 2000 Materials Research Society

Damascene Feature Profile

For a given bottom-up fill rate capability of an electroplating process, feature geometry can play a key role in determining the maximum aspect ratio of a feature which can be filled. This is a reflection of the fact that a feature with a narrow and necked opening will tend to close sooner than a feature having a generous opening (but otherwise similar internal dimensions) owing to a small but significant rate of deposition at the neck. Typical examples of combined feature and seed profiles that can produce either voids or complete filling are shown in Figures 1a and 1b, respectively.

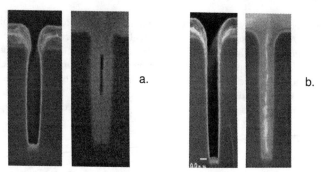

a.   b.

Figure 1 - Seed profile of a trench feature showing (a) pinch off near the opening and fill result following DC plating, and (b) a wide opening and fill result following DC plating.

Seed Layer Coverage

All electroplating processes require an electrically conductive surface film (seed) to promote nucleation and growth. The seed surface must also be largely free of oxides for efficient charge transfer to begin. PVD copper seed films with thin (10-40 Angstrom), acid soluble oxide layers easily meet this criteria for damascene plating. Existing PVD processes can yield a variety of coverage profiles inside high AR features (Figs. 2 and 3), some of which are discontinuous or exhibit significant agglomeration of Cu material. With a continuous seed layer, plating readily initiates on all surfaces, thereby allowing subsequent bottom-up fill as shown in Figure 2. In contrast, plating onto agglomerated or discontinuous seeds can result in voids near the base which develop when nucleation and plated growth fails on the lower sidewalls or base of the vias. For the seed shown in Figure 3, both a failure to initiate plating and evidence of seed dissolution near the via base are observed. Propagation of the initial agglomerated structure has started higher in the via.

One possible cause for this observation is electrical discontinuity between the wafer surface and the via base, leading to a lower applied voltage at the via base. If a 200 Å Ta barrier is intact on the sidewalls of a 0.2 μm wide x 1.0 μm deep via, the resistance drop from the wafer surface to the base of the via is about 20 Ohms. During plating at a nominal current of 30 mA/cm^2, the current within the via will be approximately 2×10^{-6} amps and the voltage drop from the top to the base of the via is on the order of 0.05 mV. This very small voltage drop suggests that only in the case of discontinuous barriers could resistance play a role in the formation of bottom voids. Other factors which could impede filling on agglomerated seed material include a low fraction of metallic Cu surface coverage, loss of Cu metal due to increased oxidation of a higher surface area, difficulty in wetting highly irregular or oxidized surfaces, galvanic cells established by the bath chemistry which dissolve seed at the via base, and Ostwald-ripening type processes causing growth of larger agglomerates and dissolution of smaller agglomerates [8].

Figure 2 - Cross sectional via images showing (left to right) smooth seed coverage only, metal profile following partial fill by electroplating, and complete fill following electroplating.

Figure 3 - Cross sectional via images showing (left to right) agglomerated seed coverage, metal profile following partial fill by electroplating, and final fill result following electroplating.

Electroplating methods have been proposed [3,9] which yield some degree of initially conformal deposition to join together isolated agglomerations of seed material, thus allowing subsequent bottom-up fill on a continuous conductive surface. The success of these methods suggests that an initial lack of metallic copper coverage near the via base contributes directly to fill initiation problems. Methods such as these facilitate initial plating on poor seed layers at the expense of increasing the aspect ratio (through conformal deposition) of the feature which must eventually be filled.

Process Chemistry

Additives present in acid copper plating solutions are responsible for the rapid plating rate within high AR features compared to the plating rate on the adjacent field. Such additives include polymers such as polyethylene glycol which suppress current at a given voltage, levelers such as cationic surfactants and dyes which suppress current at locations to which their mass transfer rate is most rapid, mercapto-containing molecules (brighteners) which accelerate current (relative to the suppressors) where they adsorb, and chloride ion which is required for the suppression caused by polymers. To illustrate the effect of these additives, polarization curves measured at a Pt rotating disc electrode for a PCB-type plating solution are shown in Figure 4. In Figure 4a, behavior of the copper electrolyte including sulfuric acid, copper sulfate, and chloride ion is shown as applied potential is swept from 0 to −280 mV. In this simple system, current increases in an approximately exponential manner as potential is increased, until mass transfer of copper to the interface begins to limit overall current (to a degree dependent on the mass transfer associated with electrode rotation). During filling, plating should be carried out at current and mass transfer conditions which avoid copper depletion at the wafer surface because this effect will be exaggerated slightly [10] in small features and can impact fill efficiency.

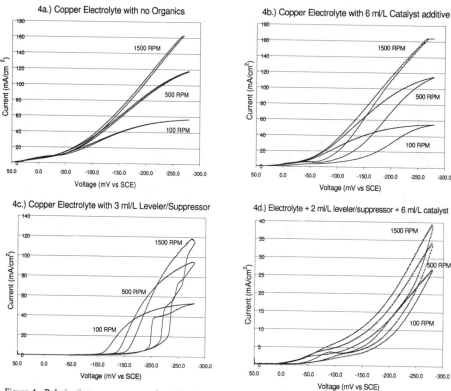

Figure 4 - Polarization curves measured at 100, 500, and 1500 RPM in (a) copper plating solution containing copper sulfate, sulfuric acid, and chloride ion, (b) with addition of 6 ml/L of a catalytic additive to (a), (c) with addition of 3 ml/L of a suppressor/leveler additive to (a), and (d) with addition of 6 ml/L catalytic additive and 2 ml/L suppressor/leveler additive to (a). (All anodic and cathodic scans were taken at 2 mv/sec.)

Figure 4b shows polarization behavior in the presence of a more current-accelerating (catalytic) additive mixture. With this additive component, it can be seen that current is not increased relative to the base electrolyte system. The catalytic effect is only observed when this component is used with more suppressing components. Mass transfer behavior in the presence of the catalytic component is similar to the base electrolyte (i.e. current proportionately increases with RPM/flow). However, for the case of the catalyst additive, hysteresis behavior at lower levels of mass transfer is noted between the forward and reverse scans of the plot. When hysteresis is observed the current at a given voltage is lower when scanning toward larger applied voltages than the current when scanning toward zero applied voltage. Hysteresis reflects maintenance of additive adsorption levels existing at a proceeding time, and thus a tendency to maintain an existing current level until surface adsorption re-equilibrates at a mass transfer dependent rate. This hysteresis behavior is a potentially desirable aspect of a process chemistry intended to deliver bottom-up fill since it implies that once an accelerated rate of growth is achieved within a feature it will tend to propagate itself.

Figure 4c shows polarization behavior in the presence of the more current-suppressing additive mixture. In the presence of this additive, the current at a given voltage is higher at a lower mass transfer rate, except at high currents where mass transfer of copper limits overall current. This behavior reflects the presence of leveling agents which suppress current to a degree dependent on their mass transfer rate to the surface. The suppression of growth achieved by levelers (at areas of high mass transfer such as the wafer field) could contribute to relatively accelerated growth in vias if proper concentration profiles were achieved. Alternatively, levelers which diffuse into damascene features may act to disrupt fill where it eventually adsorbs and suppresses current.

Figure 4d shows polarization behavior in the combined presence of accelerating and suppressing additive components. Starting with this system, further addition of the accelerating or suppressing additive components will increase or decrease, respectively, the current at a given mass transfer and applied potential condition. At the component concentrations shown, the current at a given potential is between the values observed in the presence of suppressing and accelerating species alone only at low voltages. At higher voltages, suppression of current is greater than observed in the presence of the suppressing component alone. This suggests a synergistic suppression effect under certain conditions as has been recently noted [7]. Polarization behavior of the overall additive system will represent current-voltage characteristics on the wafer surface. Maximum differentiation between currents at the wafer surface and within the via can clearly be achieved using a system which causes an overall strong suppression behavior, but which fails to manifest that behavior at the via base.

Bottom-up fill takes place when the additive adsorption conditions at the base of the via result in a much higher current than the additive adsorption condition on the field adjacent to the via. (Recall that voltage applied at the base of the via is virtually identical to the voltage on adjacent field.) As illustrated in Figure 5, the current at the base of the feature can easily be 5-10X the value on the field if additive adsorption differences exhibiting desired polarization behavior are established at each geometric position. The case shown compares the 500 RPM polarization behavior using the complete additive system to represent field polarization, and polarization behavior in the presence of the more catalytic additive to represent behavior at the via base. Many other combinations are possible, with the most extreme being no additives of any type at the feature base combined with suppressors present at the surface. The ability to achieve this adsorption combination could lead to feature base acceleration rates approaching 100X that of the field (based on the polarization data).

Filling Trends and Observations

Any proposed mechanisms for bottom-up fill should explain several measured trends in filling performance as a function of process chemistry conditions. These trends relate strongly to elimination of center voids of the type shown in Figure 1b, and somewhat less to the ability to eliminate bottom voids of the type shown in Figure 3. The following outline of points can be made in summarizing our experience and observations in working with Cu plating bath additive systems.

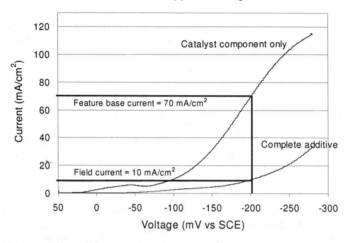

Figure 5 - Polarization behavior of copper electrolyte with catalytic additive only and with complete additive system. Plot shows possible current density differences between wafer areas (feature base vs. field) due to differences in additive adsorption.

1. Strongly accelerated fill has not been achieved using suppressing polymers alone or suppressors with chloride ion. Growth in several systems tested appears to be largely conformal as shown in Figure 6.

2. Suppressing polymer concentration can be increased to at least 4X beyond an optimal (plateau) level without impacting fill. These concentrations can extend upwards of 1000 mg/L, well beyond a level at which concentration gradients within features are likely to develop.

3. Bottom-up filling can be achieved in the absence of levelers. A typical time evolution profile using an additive system containing only a polymer suppressor, chloride, and an accelerator is shown in Figure 7. It is seen that an accelerated bottom-up growth component is operative.

4. The addition of levelers to bottom-up filling chemistries often results in top center voids as shown in Figure 8.

5. As chloride ion concentration is increased from zero, the filling acceleration in vias increases to a maximum value then decreases as chloride concentration is further increased. Conformal fill has been noted at very high chloride levels.

6. As accelerator concentration is increased, bottom-up fill increases from near zero to a maximum rate and then diminishes to conformal behavior. This is similar to the trend for chloride ion (point 5).

7. For a given additive system, too low a copper concentration diminishes bottom-up fill capability in high aspect ratio features.

8. In the absence of a leveler-like species, accelerated copper growth continues over a damascene feature following bottom-up fill. Figure 9 shows the metal thickness profile over a set of trenches with and without leveler added to a bath which exhibits bottom-up fill. The nearly 2X thickness increase of copper over dense features reflects a continuation of the accelerated copper growth beyond the time of filling completion. The addition of a leveler component serves to suppress current on the rapidly growing surface once it protrudes above the field, thereby leading to a relatively uniform deposit thickness. A lack of suppressing polymer on the rapidly growing surface (over features) could also contribute to continued rapid growth. To test this in a leveler-free solution, trenches were plated under normal filling conditions until they were approximately flush with the adjacent field. At this time, current was then turned off for 30 seconds. This off-time should allow for complete equilibration of polymer adsorption, a behavior which requires less than two seconds. Following this equilibration period, plating current was resumed using the normal process. The profiles observed were identical to those without the zero current equilibration, indicating accelerated growth following fill results from a strongly adsorbed excess of accelerator species which had accumulated on the surface during bottom-up filling.

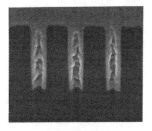 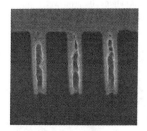 

| PEG only | PEG / Cl only | Ultrafill suppressor / Cl only |

Figure 6 - Via filling results in the presence of suppressor molecules. (PEG = polyethylene glycol)

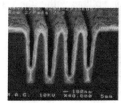 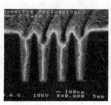 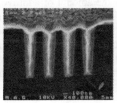 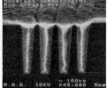

Figure 7 - Fill evolution profile in trenches following 10, 15, 20, and 30 seconds at a current of 10 mA/cm^2.

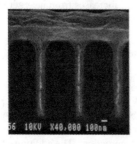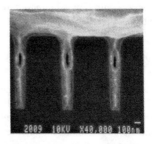

Figure 8 - Trench fill capability of a two-component (polymer suppressor / mercapto accelerator) organic additive plating solution without (left) and with (right) leveler added.

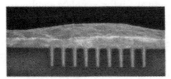 a.  b.

Figure 9 – Metal thickness field profile over dense features following (a) accelerated bottom-up filling, acceleration continues after filling (no leveler present), and (b) following bottom-up filling, acceleration stopped following fill (leveler present).

Filling Mechanisms

Based on these observations, several mechanisms and factors appear to contribute to bottom-up fill while other possible mechanisms may be only marginally important. The following scenario for the establishment and propagation of bottom-up fill is proposed:

When a wafer is first immersed in a plating solution, a concentration gradient of suppressing and accelerating species may exist on the wafer surface between the via base and the field. This will happen when the quantity of additive species required to form an adsorbed layer on the surface within the via exceeds the amount of additive contained in the solution volume within the via. This effect could account for less suppression at the via base by slow diffusing polymer species when relatively low suppressing polymer concentrations (<100 mg/L) are present in solution. However, a simple calculation indicates that unless a species is consumed rapidly, this concentration gradient will last a second or less for suppressing polymer species which are present at up to several hundred mg/L in solution, and should not exist at all when polymer concentrations exceed 1000 mg/L (where bottom-up fill is still observed). It is well known that most polymer species used as suppressors are not readily consumed or decomposed at the cathode and (as noted above) increasing the polymer level beyond that required for optimal fill does not degrade fill as would be expected for a fill mechanism driven by a polymer diffusion gradient. Beyond this, it is observed that bottom-up fill will take place on a well seeded surface both if a wafer is immersed in a plating solution with the current on and if the wafer is placed in the bath and allowed to equilibrate with the additives in solution for at least 10 seconds. For these reasons, it appears that diffusion-limited adsorption of the polymer is not critical to bottom-up fill in at least some commercially available additive systems.

When current flow begins we will assume that all additive species have reached an equilibrium level on all surfaces of the wafer. This should be expected to lead to initial currents on all surfaces which are approximately equivalent, an effect which agrees with the observed relatively small amount of bottom-up growth seen in the first 5-10 seconds of a filling process (see Figure 7) and the essentially conformal initial growth shown in Figure 2 (center).

After some period of time which depends on current density and chemical concentrations, two phenomena may begin to contribute to fill. First, the accumulation of accelerating mercapto species (or their more accelerating breakdown products) on the surfaces within the features takes place. Accumulation may simply result from the strong mercapto adsorption strength and the high relative surface area within the feature relative to a diffusion path out of the feature, or it may be the result of decreasing surface area within the via onto which adsorbed mercapto species are concentrated. Too much accelerator in a bath probably disrupts fill because the accumulation of accelerating species also begins to take place on the wafer surface and differentiation from the via base is lost. The accumulation of catalytic species on the growing surfaces within features is strongly supported by the continued rapid Cu growth above features in the absence of leveler. Such behavior is not interrupted by discontinuing current flow to allow polymer re-equilibration. It is, however, disrupted by reversal of interfacial potential to a value causing oxidation or desorption of the adsorbed catalytic material or by addition of a leveling additive which suppresses current at protruding geometries. It has been shown using rotating disc electrodes that the same potential reversals which disrupt accelerated growth over features cause desorption or oxidation of catalytic species adsorbed on electrodes [11].

To explain the observed effect of chloride on fill, a second phenomena which contributes to fill initiation may involve the co-deposition of chloride in the Cu film resulting in a depleted chloride concentration within features, or competitive displacement of Cl on the copper surface by the accumulating mercapto species. Low chloride concentration leads to weak or negligible polymer adsorption and higher current for a given potential, as has been well documented in the literature [12]. This effect results in poor suppression of the current at the wafer surface and therefor a lack of possible relative acceleration in the feature. As Cl concentration is increased, good surface suppression is obtained but the levels of Cl within the via may not be adequate maintain polymer adsorption in the presence of the accumulating accelerator mercapto species and good bottom-up fill is obtained. Further increase in Cl may disrupt fill because Cl is able to compete successfully with the mercapto species for adsorption sites and subsequently attract suppressing polymer adsorption within the via.

Finally, once bottom-up fill begins in a bath showing hysteresis behavior between the cathodic and anodic sweeps of a polarization curve (Figure 4b for example), the high current which is initially established will tend to continue.

While most levelers have been found to impede fill, the benefits of a molecule with ideal properties could be realized both with respect to fill and surface uniformity.

Current Waveform Effects

Bipolar pulse plating has been applied successfully to improve fill performance of commercial additive systems [3]. Such plating methods are believed to accelerate dissolution of copper at the opening of high aspect ratio features or to cause the adsorption behavior of additives to vary in a way that enhances filling. Beyond these changes, pulse plating using additives has been found to respond in generally the same way as DC plating to process variable changes. This suggests that the basic fill mechanisms remain the same.

References

1.) Takashi, K., Proc. of IITC 1999, pp. 99-284, May 24-26, 1999.

2.) Andricacos, P., Uzoh, C., Dukovich, J., Horkans, J., Deligiani, H., IBM J. Res. Dev., **42**, 567 (1998).

3.) Reid, J., Bhaskaran, V., Contonlini, R., Patton, E., Jackson, B., Broadbent, E., Walsh, T., Mayer, S., Martin, J., Morrissey, D., Schetty, R. Menard, S., Proc. of IITC 1999, pp. 284 - 286, May 24-26, 1999.

4.) Singer, P., Semiconductor International, pp. 80, August, 1997.

5.) Deligianni, H., Dukovic, J.O., Andricacos, P.C., Walton, E.G., Electrochemical Soc. Proc., no. 267, May 2-6, 1999.

6.) Chen, L., Ritzdorf, T., Electrochem. Soc. Proc., no. 275, May 2-6, 1999.

7.) Kelly, J., Tian, C., West, A., J. Electrochem. Soc., **146**, 2540(1999).

8.) Mayer, S., Contonlini,R., Bhaskaran, V., Jackson, B., Reid, J., Martin, J., Morrissey, D., Schetty, R., Electrochemical Society Proceedings, no. 732, Oct. 17-22, 1999.

9.) Mirkova, L., Rashkov, S., Nanev, C., Surface technology, **15**, 181, 1982.

10.) K. Takahashi, M. Gross, J. Electrochem. Soc., In press.

11.) Mayer, S., Reid, J., To be published

12.) Yokio, M., Hayashi,T., Konishi,S., Denki Kagaku (Journal), **52**, 218, 1984.

EFFECTS OF PLATING CURRENT DENSITY AND SOLUTION ADDITIVE ON THE MICROSTRUCTURE AND RECRYSTALLIZATION RATE OF ELECTROPLATED COPPER FILMS

HAEBUM LEE, SERGEY D. LOPATIN*, DAE-YONG KIM, VALERY M. DUBIN[†] and S. SIMON WONG
Center for Integrated Systems, Stanford University, CIS 017, MC 4070, Stanford, CA 94305-4070, haebum@leland.stanford.edu
*Advanced Micro Devices, Sunnyvale, CA 94088-3453
[†]Intel Corp., 5200 N.E. Elam Young Parkway, Hillsboro, OR 97124-6497

ABSTRACT

The effects of plating current density and additive inclusion on the microstructure evolution of electroplated Cu films have been investigated. Cu plating was done with and without additives at current densities raging from 3.5 to 60 mA/cm^2. Films were characterized up to 60 days after the deposition. Plan view TEM image and X-ray analysis reveal that the microstructure of plated films show strong dependence on plating current density. Grain growth during recrystallization process is retarded for films plated with additives. (111) x-ray peak intensity decreases while (200) intensity increases as the plating current density is increased independent of the inclusion or exclusion of additives. Study of microstructure evolution with time shows that the recrystallization rate increases drastically as the plating current density is increased. No significant change in microstructure is observed for plating current density $\leq$ 5 mA/cm^2. Continuous decrease of (111) peaks, increase of (200) peaks, and development of tensile stress as a function of time is observed in all films plated at higher current densities. Furthermore the (111) texture degrades faster, and the biaxial stress develops faster during recrystallization process as the plating current density is increased.

INTRODUCTION

The electromigration failure of polycrystalline Al and Cu metal interconnects has been known to be highly dependent on the microstructure of the film[1,2]. Therefore, a study of microstructural variation with the deposition condition is critical for optimizing the metallization condition and integration scheme for maximum reliability. The transition from Al to Cu for advanced metallization technology involves changes in deposition technique as well as the processing architecture, which will influence the microstructure and texture of the metal. In particular, electroplating has emerged as the most promising deposition technique for Cu metallization. This choice is based on advantages including low capital cost, high throughput, high quality films, good via/trench filling capability, and compatibility with low-K materials. A remarkable feature of electroplated Cu is the recrystallization at room temperature[3-7]. Therefore, the manufacturing process of Cu interconnect must account for the continual change in physical properties of Cu due to the microstructure evolution at room temperature. There have been studies on the factors that affect the microstructure of electroplated Cu films[8-10]. However, thorough study of microstructure that correlates the processing parameters and the recrystallization process is still lacking. This paper describes the microstructure variation of Cu films electroplated with different current densities and

63

solution contents, and studies the texture evolution associated with the recrystallization of the films.

EXPERIMENT

Copper films with 1μm thickness were electroplated at room temperature on a sputtered Cu seed layer (1000Å) with sputtered Ta diffusion barrier (250Å). Thermally grown, blanket and via patterned SiO_2 were used as substrates. Plating current distribution has been optimized across the wafer to obtain better than 3% variation in electrical resistivity of electroplated films. In order to determine the effect of plating current density on the microstructure of the plated film, current density was varied from 3.5 to 60 mA/cm². For some films, commercially available standard additives consisted of brightener and leveler were included in the plating solutions to study the effects. Films were characterized as a function of time up to 60 days after the deposition. The film texture was determined with X-ray diffraction using both the conventional Bragg-Brentano 2θ scan and the Schulz reflection method. The grain size distributions in the Cu films were obtained with the plan-view transmission electron microscopy (TEM) technique. During the TEM sample preparation, the samples were carefully maintained at room temperature to avoid thermally induced grain growth. The grain diameter was calculated assuming circular grain shape. About 1000 grains per sample were used to estimate the distributions.

RESULTS AND DISCUSSION

Microstructure variation as a function of plating current density was studied for Cu films plated without additives. All the analyses were performed 10 days after deposition for this series of samples. Plan view TEM image of a film plated at 5mA/cm² in Fig.1(a) reveals that the grains are mostly small (median grain size is 0.177μm) and with polyhedral shape with rather curved boundaries. However, a few giant grains with straight - lined grain boundaries as shown in Fig.1(b) are found to be sporadically located among regions of fine grains. The bimodal grain size distribution is an indication that abnormal grain growth, or recrystallization has occurred after the electrodeposition. As the plating current density increases, however, the microstructure of the film changes drastically. Figure 2 (a)-(f) represent the structures of films plated at

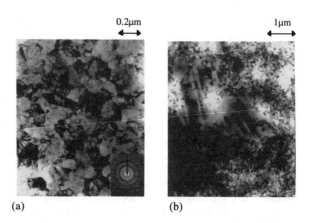

(a)   (b)

Fig.1. Plan view TEM images of (a) small grains and (b) a giant grain located among regions of fine grains in Cu film electroplated at 5mA/cm² (10 days after deposition)

(a) 10mA/cm²   (b) 20mA/cm²   (c) 30mA/cm²

(d) 40mA/cm²   (e) 50mA/cm²   (f) 60mA/cm²

Fig.2. Plan view TEM images of Cu films electroplated at different plating current densities (10 days after deposition)

10, 20, 30, 40, 50, and 60mA/cm², respectively. These pictures clearly show that the films contain mostly the giant grains which are very different in sizes (>1μm) even when the plating current density is raised slightly to 10mA/cm². The grains are even larger as the current density is further increased. It is also noted in the X-ray texture analysis that the (111) peak intensity decreases rapidly while (200) signal increases as the plating current density is increased (Fig.3). The other Cu peaks show negligible changes. Fig.4 compares the dependence of grain size on plating current density for the films plated with and without additives. Additive inclusion seems to lead to the retardation of the grain growth during recrystallization process in the plated films.

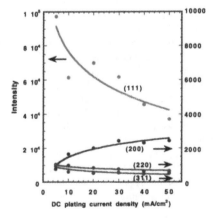

Fig.3. Texture variation with plating current density

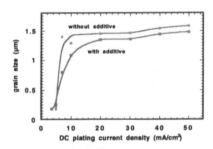

Fig.4. Dependence of grain sizes on plating current density for films plated with and without additives (10 days after deposition)

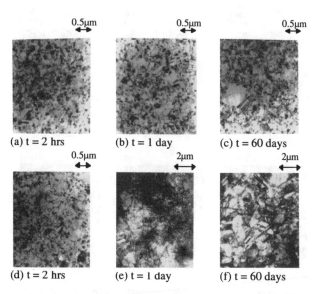

0.5μm	0.5μm	0.5μm
(a) t = 2 hrs	(b) t = 1 day	(c) t = 60 days
0.5μm	2μm	2μm
(d) t = 2 hrs	(e) t = 1 day	(f) t = 60 days

Fig.5. Plan view TEM images as a function of time for Cu films plated with additives; (a), (b), (c): at 5mA/cm^2 ; (d), (e), (f): at 10mA/cm^2

The evolution of microstructure with time was also studied on the films plated with additives at different current density. As shown in Fig.5 and Fig.6, the grain sizes are initially small in all films regardless of the plating current density. However, the recrystallization rate changes drastically with the plating current density. No significant grain growth is observed up to 60 days for plating current density ≤ 5 mA/cm^2, whereas significant recrystallization occurs in the films plated at higher current densities. Films recrystallize faster as the plating current density is increased. This behavior is believed to be due to the higher energy state of the resultant films plated at higher current density. These films have higher intrinsic stress introduced by higher defect density, which may act as the driving force for faster recrystallization[3-5]. Texture evolution associated with recrystallization as a function of plating current density was also studied and shown in Fig.7. Decrease of (111) peaks and increase of (200) peaks with time is observed as reported elsewhere,[5-7] in all films plated with current densities ≥ 7.5 mA/cm^2. Furthermore, the degradation of (111) and the development of (200) texture is enhanced at higher plating

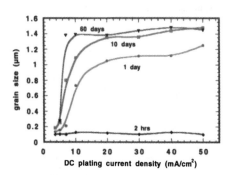

Fig.6. Grain size evolution for Cu films plated with additives at different plating current density

current density. No significant texture variation is observed up to 10 days for the films plated at 3.5 and 5 mA/cm². The evolution of biaxial stress in the films as a function of plating current density are shown in Fig.8. Stress develops to the tensile direction in films during recrystallization, and the tensile stress builds up to a higher magnitude with increasing plating current density.

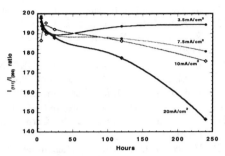

Fig.7. Evolution of texture at room temperature for films plated at different current density and with additives

CONCLUSIONS

The microstructure evolution of electroplated Cu films show strong dependence on the plating conditions. Regardless of additive inclusion, (111) x-ray peak intensity decreases while (200) intensity increases as the plating current density is increased. The recrystallization rate changes drastically with varying plating current density, i.e. recrystallization process is accelerated as the plating current density is increased. This behavior is believed to be due to the higher energy state of the films as a result of electroplating at higher current density. However, the driving force for the recrystallization process is yet to be identified. No significant change in microstructure during recrystallization is observed for films plated at current density ≤ 5 mA/cm². For higher plating current densities, the grain size grows, (111) texture degrades, (200) texture enhances,

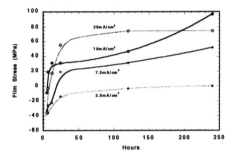

Fig.8. Evolution of (a) (111) intensity and (b) biaxial stress at room temperature for films plated with different current density and with additives

and tensile stress develops as a function of time. The grain growth during the recrystallization process induces the stress development to the tensile direction in the film. In order to reduce the strain energy, (200) texture, which has a lower strain energy than (111) texture, begins to develop[11,12]. Higher plating current density leads to larger development of tensile stress and faster degradation of texture in the films during recrystallization process. The inclusion of additive appears to retard the grain growth during recrystallization process.

ACKNOWLEDGMENTS

We are grateful to Takeshi Nogami for useful technical discussions. This work was partially supported by the Semiconductor Research Corporation (SRC).

REFERENCES

[1] S. Vaidya and A. K. Sinha, Thin Solid Films, vol. 75, pp. 253-259 (1981).

[2] C. Ryu, K. Kwon, A. L. S. Loke, H. Lee, T. Nogami, V. M. Dubin, R. A. Kavari, G. W. Ray, and S. S. Wong, IEEE Transactions on Electron Devices, vol. 46, no.6, pp. 1113-1120, June (1999).

[3] B. M Hogan, AESF Conference - SUR/FIN '84.

[4] C. Lingk and M. E. Gross, J. Appl. Phys. 84, 5547 (1998).

[5] J. W. Patten, E. D. Mc Clanahan, and J. W. Johnston, J. Appl. Phys. 42, 4371 (1971).

[6] I. V. Tomov, D. S. Stoychev, and I. B. Vitanova, J. Appl. Electrochem. 15, 887 (1985).

[7] Q.-T. Jiang, R. Mikkola, B. Carpenter, and M. E. Thomas, Advanced Metallization Conference, (1998).

[8] H. Lee, S. D. Lopatin, T. Nogami, and S. S Wong, MRS Fall Meeting Symposium A1.9, Boston, MA, December (1998).

[9] S. S. Wong, H. Lee, C. Ryu, A. Loke, K. Kwon, Advanced Metallization Conference, pp. 53-54, Tokyo, Japan, September (1998).

[10] T. Ritzdorf, L. Graham, S. Jin, C. Mu, and D. Fraser, Proceedings of the IEEE Int. Interconnect Technology Conf., 166 (1998).

[11] E. M. Zielinski, R. P. Vinci, and J. C. Bravman, J. Appl. Phys. 76, 4516 (1994).

[12] C. V. Thompson and R. Carel, J. Mat. Sci. Eng. 32B, 211 (1995).

A NOVEL ELECTROLYTE FOR WAFER PLATING

A.Thies[1] and H. Meyer, ATOTECH Germany, Berlin
J. Helneder and M. Schwerd, INFINEON Technologies AG, Munich
T. Gebhart, SEMITOOL Germany, Piding

ABSTRACT

A novel electrolyte for wafer plating with inert anodes is presented. In addition to sulfuric acid and a copper salt, it contains a redox mediator system that prevents water decomposition at the anode and keeps the copper concentration constant with an inherent replenishment mechanism. The additives are shown to be exceptionally stable with this electrolyte, eliminating the necessity for additive replenishment for about 2000 wafers. The electrolyte is shown to have strong bottom up filling behaviour for trenches down to 0.22 µm width and 1.1 µm heights. Physical data was found to be fully within spec.

INTRODUCTION

Today's standard electrolyte for the galvanic deposition of copper on wafers consists of a sulfuric acid solution of copper with some additives and soluble anodes often containing phosphorus. The counter reaction of the copper deposition is copper dissolution. These electrolytes have the advantage that the copper concentration is constant during the deposition. However, they need intensive care, as the anode has to be prepared very carefully to obtain best uniformity and the common additive systems are not very stable in the presence of a copper anode. Particles in the plating chamber due to anode replacement or film preparation can become a problem. Copper can also be deposited from electrolytes with inert anodes. Particle count and anode preparation are no longer a problem and the distance between anode and wafer keeps constant. The counter reaction is now water decomposition, accompanied by evolution of oxygen bubbles. In the usual wafer upside-anode downside configuration, gas bubbles that hit the wafer can become a problem, as they reduce the local deposition rate. For configurations with inert anodes , the copper and the additives have to be replenished with an external replenishment equipment. The investigations clearly show that the combination of the Atotech electrolyte with inert anode is clearly favored amongst the others due to the highly stable additives and the maintenance free inert anode. This combination makes an easy "Push Button" operation without any expensive dosing equipment and in house analytic for the additives possible. Also, system availability is increasing with the leave out of any anode preparation and conditioning after long (>10h) system idle states.

THEORY

The new electrolyte contains copper sulfate, sulfuric acid and an additional redox mediator system that is stable in two oxidation stages. As redox mediator system, iron is used. The inherent replenishment system consists of very high purity metallic copper pieces in a cartridge ("copper module ") that is built into the system (Fig.1). At the anode the mediator is oxidized (Eq. 1a), at the cathode, copper deposition takes place (Eq. 1b). The replenishment takes place as a pure chemical reaction between copper and iron (Eq 1c):

anodic reaction : $\quad 2Fe^{2+} -> 2Fe^{3+} + 2e^-$ $\qquad E^0 = 0.77$ V $\qquad$ *Eq. 1a*

cathodic reaction : $\quad Cu^{2+} + 2e^- -> Cu^0$ $\qquad E^0 = 0.345$ V $\qquad$ *Eq. 1b*

replenishment : $\quad 2Fe^{3+} + Cu -> 2Fe^{2+} + Cu^{2+}$ $\qquad\qquad$ *Eq. 1c*

---

[1] A. Thies, ATOTECH Deutschland GmbH, Erasmusstr. 20-24, D-10553 Berlin, email : andreas.thies@atotech.de

Conference Proceedings ULSI XV © 2000 Materials Research Society

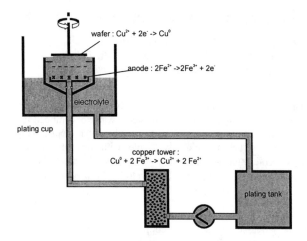

wafer : $Cu^{2+} + 2e^- \rightarrow Cu^0$

anode : $2Fe^{2+} \rightarrow 2Fe^{3+} + 2e^-$

electrolyte

plating cup

copper tower :
$Cu^0 + 2\,Fe^{3+} \rightarrow Cu^{2+} + 2\,Fe^{2+}$

plating tank

Fig.1 Principle working scheme of the electrolyte. The counter reaction for the copper deposition is the oxidation of the redox mediator at the anode. The copper concentration is kept constant by a chemical reaction between the oxidized mediator and the metallic copper in the copper tower. No gas evolution occurs at the anode.

As the standard electrode potential of the Fe^{2+}/Fe^{3+} redox couple is higher than the standard electrode potential for the Cu/Cu^{2+} redox pair by about 0.4 V, copper is easily and quickly dissolved. The overall reaction is therefore stochiometrically balanced and the copper concentration stays constant. With this principle, the increase of the anion concentration is avoided. Iron deposition does not occur, as the iron reduction is thermodynamically unfavored (Eq. 2):

$$Fe^{2+} + 2e^- -> Fe \qquad\qquad E^0 = -0.44\ V \qquad\qquad Eq.\ 2$$

As long as copper is available, the copper deposition is favored by 0.345-(-0.44) V = 0.8 V. In no case ever any iron incorporation in the copper deposit was found (see results section). Compared to water decomposition as anodic reaction, the use of the redox mediator reduces the total cell voltage necessary for the deposition. It was found to be 1.35 V for a current density of 2.5 Adm^{-2} and dropped during the deposition to 1.23 V. This decrease is very reliable and can be used to monitor the deposition and the bath. The lack of oxygen evolution strongly increases the life time of the additives. As with conventional baths, two different principal organic components are used : a leveller and a brightener. Both can be analytically determined by cyclic voltammetry stripping (CVS) and cyclic pulsed voltammetry stripping (CPVS).

EXPERIMENTAL

The experiments were conducted in a SEMITOOL Equinox plating tool in the Material Innovation Laboratory of INFINEON Technologies in Munich on six inch wafers. The tool was equipped with a copper replenishment tower and an ATOTECH inert anode that could easily be fitted in. Bath volume was 20 l, bath temperature 25 °C and the flow rate was 7 liter per minute. The wafer was rotated with 20 RPM. During approx. 1 month, a total of 16000 Amin and 2000 wafers were plated. The charge per wafer was 7.55 Amin at 2.5 Adm^{-2} for a normal six inch test wafer. Two patterned test wafers were daily deposited with only 1 Amin to obtain partial filling that was checked by SEM/FIB. The wafers were plated in runs of 25 pieces. Every 2nd and every 2nd last wafer of a lot was checked for uniformity. During the whole marathon the bath was not replenished. To check inorganic and organic compounds, a bath sample was drawn and analyzed daily.

RESULTS

Stability of the bath

The concentration of copper, iron and sulfuric acid during the marathon is plotted in Fig.2a. The copper concentration increased from 35 gl^{-1} to 42 gl^{-1}. This is due to the direct and indirect copper corrosion caused by oxygen during the long span of operation. Copper in the replenishment tower can be corroded directly via

$$2Cu + O_2 + 4H^+ \xrightarrow{\ Cu\ } 2Cu^{2+} + 2H_2O \qquad\qquad Eq.\ 3a$$

This is an inhomogeneous electrochemical reaction that takes place only at the interface metallic copper/electrolyte in the copper tower. However, a homogenous reaction, the oxidation of the iron, can additionally take place :

$$4Fe^{2+} + O_2 + 4H^+ \xrightarrow{\hspace{1cm}} 4Fe^{3+} + 2H_2O \qquad\qquad Eq.\ 3b$$

The iron(III) formed by Eq. *3b* can now react via equation *1c* to form additional copper ions. If the time span of the experiment is diminished and oxygen contact is reduced as far as possible, the increase in copper can be reduced close to zero, as has been experimentally confirmed later [unpublished results]. As found experimentally (Fig. 2a) and seen from equations *3* , the increase in copper is accompanied by an increase of the pH value.

Leveller and brightener were examined by CVS and CPVS, respectively. These methods do not measure the additive concentration, but rather the effect of the additives on the copper deposition. The result is the effectiveness of the additives, expressed in a formal concentration. During the bath life, the concentration of both additives decreases due to consumption and drag out. This normal decrease is found for the leveller during the first 10 days and for the brightener for the first 20 days. The steep decrease in leveller concentration after 10 days and in the brightener concentration after 20 days was unexpected and cannot be explained in the moment. Surprisingly, the leveller decrease is not reflected in total organic content (TOC) of the bath, whereas the decrease in brightener is seen in TOC value, too. This means that the decomposition products gas out. However, the bath was not replenished and the trench filling properties of the bath even for small structures did not change.

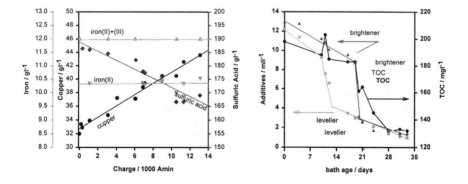

Fig 2 Bath analysis during the marathon run. a) Development of the inorganic concentrations b) Development of the organic additives.

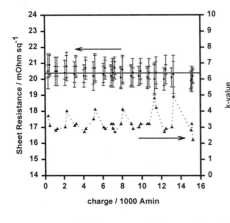

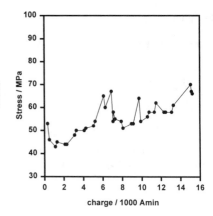

Fig. 3 Resistance and k-value during the course of the experiment

Fig. 4 Stress by wafer during the course of the experiment.

Resistance and Stress

The sheet resistance of unannealed samples is plotted in Fig. 3. The values were obtained by a four point probe measurement with 49 data points and an edge exclusion of only 3 mm to the seed layer edge. The mean value is 20.5 mΩ sq^{-1}, and decreases slightly during the marathon. The k-value (uniformity) is around 3 %; we attribute the small peaks to the idle time over the weekends or holidays. The stress as determined by wafer bow and shown in Fig. 4 increases slightly during the experiment.

Purity of the deposit

Purity of the deposit as well as contamination of the wafer on the copper plated surface and the back side was examined by SIMS and VPD AAS. As SIMS measurements need to be calibrated carefully with a standard that was not available in this very sensitive range and element composition, all data is shown together with a sputtered copper layer. SIMS depth profiles obtained by sputtering through a 1μm copper layer are shown in Fig 5 with ECD layers on the left and PVD layers on the right side. In both layers phosphorous, iron, manganese and nickel are below the detection limit of this method. The sulfur content in both layers is the same. However, in the ECD layer, an increase in the chloride concentration at the diffusion barrier is found. The sodium concentration in the PVD layer is somewhat higher (4x) than in the ECD layer. The surface concentration of iron was cross checked by TOF-SIMS. With this method, the detection limit of iron in silicon is about 2E9 atoms·cm^{-2}. In Fig. 6, the intensity is shown as a function of the mass. Again, no difference between a PVD and an ECD layer was detected. As additional reference, a sample with an iron concentration of 1.1E12 atoms/cm^2 was tested and a clear iron peak was found. The back side contamination during the marathon was tested on silicon wafers using VPD AAS and was found to be below a contamination level of 1E11 atoms/cm^2.

The Xray pattern is shown in Fig. 7. It was obtained from of a copper layer of 1.5 μm. The ratio of the intensities I(111)/I(200) was found to be 4.06. From the Cu(311) peak, the internal stress was determined by Xray measurements to be 40 +/- 15 MPa, in excellent agreement with the result from wafer bow data. Pole figures show a [111] fiber structure.

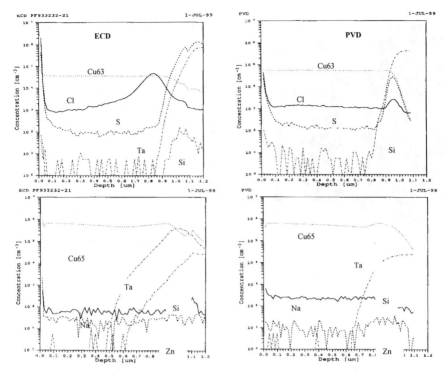

Fig. 5 SIMS depth profile measurements of ATOTECH ECD copper layers. the measurements are referenced to a PVD sputtered copper layer. ECD results on the left , PVD results on the right side.

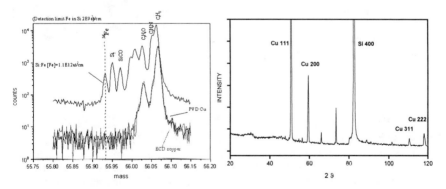

Fig. 6 TOFSIMS measure-
ments for the backside
contamination of Iron on
Wafers

Fig. 7 Xray pattern of a 1.5 μm copper
layer. I(111)/I(200) = 4.06 :1

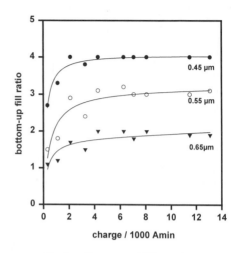

0,35μm   0,28μm   0,23μm   charge / 1000 Amin

Fig. 8   Cross sections of voidless filled   Fig. 9   Bottom-up filling ratio as a
trenches after 510, 7079 and 13090 Amin   function of bath age and trench size.

Fig.8 shows cross sections of trenches obtained by FIB-analysis as a function of the bath age after plating 1 Amin per wafer. Between the beginning of the marathon (510 Amin) and close to the end (13090 Amin), no change in the filling behavior is seen . The trenches between 0.23 and 0.35 μm width are filled voidlessly with copper grains that have the size of the trench. The bottom up filling ratio for structures larger than above 0.35 μm that are only partially filled with 1 Amin is shown in Fig. 9. The bottom-up filling ratio – defined as the ratio of the thickness at the bottom of the trench to the nominal layer thickness at the surface – depends on the trench size. The smaller the trench, the higher the ratio.

Via filling (Fig. 10) was observed using a repeated cut and analysis to observe the center of the via after the CMP process. No seams or voids there detected and the grain size in the range of the via size indicates defect free filling.

Fig. 10 Cross section of a via
(1,0μm depth, ∅ 0,4μm)

Electrical Results

The resistivity of unpatterned copper layers of around 1 μm was determined to 1.75 μOhms cm. To investigate the electrical properties further, patterned wafers were plated and polished by cmp. Tantalum (50nm) was used as a diffusion barrier. The structures were not passivated. The results of the electrical tests for lines with a length of 1000 μm and different cross sections are shown in Fig. 11. None of the prepared and tested lines failed. The cross sections were determined separately by FIB. However, for wider trenches, dishing occurs which falsifies the bulk conductivity that was calculated.

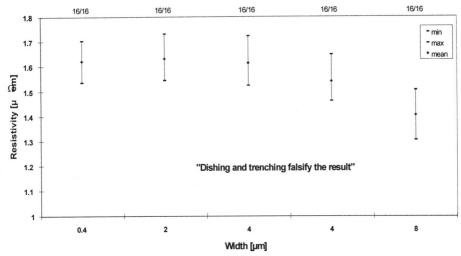

Fig. 11 Resistivity of lines of different width after cmp. Line length 1000 μm, line depth 0.9 μm. The cross section of the lines was determined by FIB at selected points.

CONCLUSIONS

A marathon run with a novel electrolyte for inert anode and inherent copper replenishment mechanism was conducted. During 4 weeks, 16000 Amin, equivalent to about 2000 six inch wafers with 1 μm copper, were plated. During the run, the properties of the deposited copper as well as the condition of the bath were monitored.

During the whole marathon, trenches down to .23 μm with an aspect ratio of 4.5 were filled without voids, indicating a strong bottom –up filling mechanism that did not change. The uniformity of the deposit did not vary during the run and the mean copper thickness stayed constant. The deposit shows [111] orientation, the internal stress of unannealed layers increased moderately from 45 to 70 MPa. Within the detection limits of the analytical methods used, no iron in the bulk or at the surface was detected ($< 1E9 Atom/cm^2$).

During the time span of four weeks, the copper concentration increased slightly from 35 to 42 gl^{-1}, paralleled by a small decrease in the acid concentration. The iron concentration was constant. The bath operates with two organic additives (leveller and brightener) that were determined separately from each other. Both decreased during the marathon without changing the filling of trenches down to 0.22 μm (AR 4.5) or the bottom-up fill ratio. This indicates a very wide process window for the ECD process.

Both, the wide process window and the long term stability of the electrolyte with its additives in combination with a maintenance free anode, enjoys all the advantages necessary for an easy "push button" operation which is relevant for cost-effective processing.

COPPER ROOM TEMPERATURE RESISTANCE TRANSIENTS AS A FUNCTION OF ELECTROPLATING PARAMETERS

S.G. MALHOTRA*, P.S. LOCKE*, A.H. SIMON*, J. FLUEGEL*, C. PARKS*,
P. DEHAVEN*, D.G. HEMMES**, R. JACKSON***, E. PATTON***
*IBM Microelectronics Division, Hopewell Junction, NY 12533
**Novellus Systems, Inc., Hopewell Junction, NY, 12533
***Novellus Systems, Inc., Wilsonville, OR 97070

ABSTRACT

Cu resistance transients were investigated as a function of electroplating parameters rotation speed, flow rate, and current. It is shown that the plating parameters strongly affect the resistance transient time. Specifically, high plating currents induce a faster transient. Also, high agitation (rotation speed and flow) causes the center of the wafer to transform more slowly than the rest of the wafer. The difference in transient times for the plating conditions is shown to be primarily a function of plating impurity incorporation. By extrapolation, the transient time non-uniformity within each wafer is also most likely a function of a spatial distribution of plating impurities.

INTRODUCTION

The transition in metallization from Al to Cu for sub-0.25 μm interconnects became necessary to improve circuit speed and required a shift in deposition technique from physical vapor deposition (PVD) to electroplating [1]. A key difference between PVD Al alloys and electroplated Cu is the existence of a resistance transient, in which the as-plated Cu has an initial resistivity of 2 to 3 μΩcm, and over a period of time at room temperature, decreases by 20-25% [2-4]. The analysis of this phenomenon to date has focused on quantifying the mean resistance transient across entire blanket wafers with a 4-point probe measurement [2] and relating it to microstructure and stress evolution [3-5]. Also, the influence of substrate topography on the kinetics of recrystallization was reported, where the initial recrystallization at the corners of patterned features was related to the higher stress and/or dislocation densities at the upper corners [6]. The objective of the present work is to expand upon the understanding of the mean transient. Specifically, the correlation between plating condition (rotation speed, flow rate, current) and spatial distribution of resistance transients is investigated using 1 μm blanket Cu films electroplated with the Novellus Sabre™ Plating System.

EXPERIMENT

Nominally 1 μm blanket Cu films were electroplated with the Novellus Sabre™ Plating System onto 200 mm Si wafers coated with a Cu seed layer and Ta barrier deposited with the Novellus Hollow Cathode Magnetron System. The four plating conditions that were used for this study are as follows:

1) 150 rotations per minute (RPM) / 8 amperes current/ 16 liters per minute flow (LPM)
2) 50 RPM/ 3 amps/ 4 LPM
3) 150 RPM/ 3 amps/ 16 LPM
4) 50 RPM/ 8 amps/ 4 LPM.

Conference Proceedings ULSI XV © 2000 Materials Research Society

The resistance transient was monitored with a conventional 4-point probe sheet resistance measurement tool (Prometrix RS50). The measurement pattern consisted of a center-point measurement as well as a series of circular measurements at radii R/3, 2R/3 and R from the center, where R was the pre-programmed maximum measurement radius of 94 mm. Secondary Ion Mass Spectrometry (SIMS) was done with a Cameca ims-5f tool using cesium primaries and negative secondary ions. Mass resolution was sufficient to separate sulfur from the oxygen dimer and quantification was achieved using ion implant references in copper. Grain growth was monitored via x-ray diffraction by measuring the change in the peak width of the Cu <222> reflection as a function of time, the assumption being made that changes in the peak width were exclusively due to changes in the average grain size of the film. Scans of the Cu <222> reflection were made with a Siemens D500 diffractometer using monochromatic Fe radiation from a sealed tube source operating at 35 kV and 35 mA. Scan range was 132 to 142 degrees 2θ with a stepsize of 0.04° and a counting time per step of 2 seconds. Slits used were I-III=1°, IV=0.015mm. Once mounted on the diffractometer, the sample was not removed until completion of the experiment, so that the same region of the sample was measured at all times.

RESULTS

Figures 1 - 4 show the Rs as a function of time for each plating condition at the wafer center and at radii of 32, 63, and 94 mm. The mean is also included. It can be seen that the resistance transient time and variation from center to edge are strong functions of the electroplating conditions. The difference in transient time both across a wafer and between different wafers can be due to the variation of the following parameters, which will all be addressed: plated film thickness, grain size, and/or impurity incorporation [3-5]. The plated Cu thickness was determined at the center and edge of each wafer via scanning electron microscopy (SEM), and the results are shown in Table 1. In order to determine the grain size before and after the transients occurred, the Cu <111> and <222> diffraction peaks were collected as a function of time at a radial value of 50 mm. Figures 5 and 6 show the Cu <222> reflection as a function of time for the 150 RPM/ 8 amps/ 16 LPM and 150 (RPM) / 3 amps/ 16 (LPM) wafer, respectively. The full width half maximum (FWHM) of the Cu <222> was determined as a function of time from the raw data in Figures 5 and 6, and is shown in Figure 7. Table 2 shows the amount of C, O, S, and Cl impurities incorporated in the films at a nominal radial position of 70 mm from the center using secondary ion mass spectrometry (SIMS) analysis. Figures 8 - 11 show the completion of the resistance transient as a function of the impurities at that same radial position (nominally 70 mm from the center).

The data in Figures 1 - 4 indicates that the transient time is a strong function of current, where a high current induces a quicker transient (Figs. 1 and 4), and that the transient time nonuniformity across the wafer is a strong function of agitation (rotation and/or flow), where high agitation increases the transient time at the center of the wafer (Figs. 1 and 3). The differential in transient times within each wafer and between the different wafers in Figs 1 through 4 is most likely not a plated film thickness effect. As shown in Table 1, the thicknesses at the center and edge of each wafer are very similar. It has been shown that thicker films transform faster [3], and the plating profile in the wafers used for this study was designed to be concave, or thicker at the edge. However, the transient time in 1 μm films was not shown to differ drastically than that of 2 μm films in a previous study, so a thickness difference of hundreds of nm is not enough to account for the much slower transients in the wafer center [3].

78

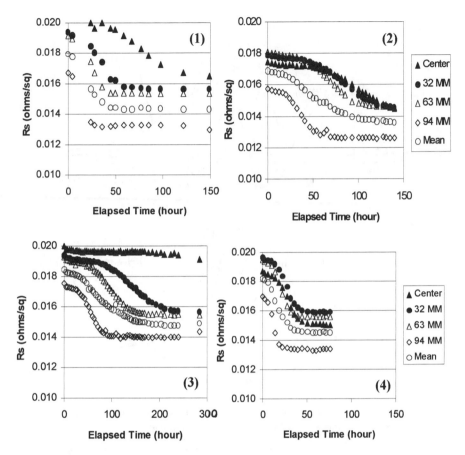

Figures 1-4: Resistance transients as a function of radial position for
(1) 150 RPM/8 amp/16 LPM, (2) 50 RPM/3 amp/4 LPM, (3) 150 RPM/3 amp/16 LPM,
and (4) 50 RPM/8 amp/4 LPM

Sample	Ctr Thickness (μm)	Edge Thickness (μm)
150 RPM/8 amp/16 LPM	1.25	1.30
50 RPM/3 amp/4 LPM	1.40	1.50
150 RPM/3 amp/16 LPM	1.10	1.30
50 RPM/8 amp/4 LPM	1.30	1.30

Table 1: Center and edge Cu film thickness for each plating condition determined by SEM
cross-section analysis. .

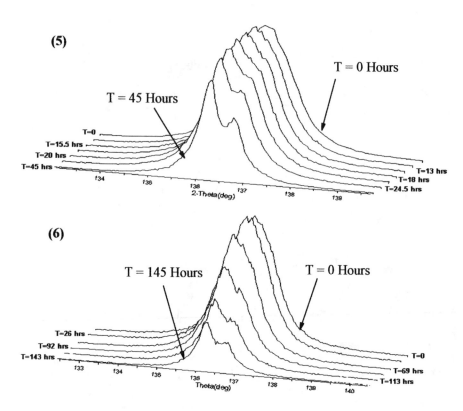

(5)

T = 0 Hours

T = 45 Hours

T=0
T=15.5 hrs
T=20 hrs
T=45 hrs
134   135   136   137   138   139
2-Theta(deg)
T=13 hrs
T=18 hrs
T=24.5 hrs

(6)

T = 145 Hours

T = 0 Hours

T=26 hrs
T=92 hrs
T=143 hrs
133   134   135   136   137   138   139   140
Theta(deg)
T=0
T=69 hrs
T=113 hrs

Figures 5-6: Cu <222> diffraction peaks as a function of time at room temperature for (5) 150 RPM/ 8amp/16 LPM and (6) 150 RPM/ 3amp/16 LPM wafer.

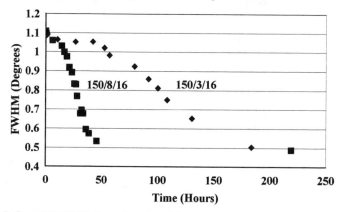

150/8/16   150/3/16

FWHM (Degrees)

Time (Hours)

Figure 7: Cu <222> FWHM as a function of time at room temperature for 150 RPM/ 8amp/16 LPM wafer and 150 RPM/ 3amp/16 LPM wafer .

Sample	C (atoms/cm^3)	O (atoms/cm^3)	S (atoms/cm^3)	Cl (atoms/cm^3)
150 RPM/ 8 amp/ 16 LPM	2.9E18	1.8E18	8.4E17	1.7E18
50 RPM/ 3 amp/ 4 LPM	4.2E18	2.7E18	1.9E18	2.3E18
150 RPM/ 3 amp/ 16 LPM	7.6E18	5.3E18	2.8E18	4.6E18
50 RPM/ 8 amp/ 4 LPM	3.2E18	2.5E18	6.4E17	1.2E18

Table 2: Impurity concentrations (atoms/cm3) in each plated film as determined by SIMS.

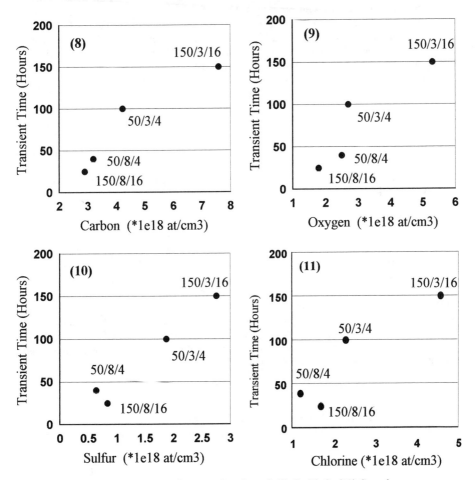

Figures 8 - 11: Resistance transient time as a function of (8) C, (9) O, (10) S, and (11) Cl concentration at a nominal R = 70 mm.

One other possibility for the different transient times is related to initial grain size. One would expect that smaller, more defective grains would drive a faster recrystallization and resistance transient. The grain growth process can be seen for the 2 wafers with high agitation in Figures 5 and 6, where the broad Cu <222> diffraction peaks narrow and split into the Fe Kα1 and Kα2 components. The Cu <222> FWHM data in Figure 7 indicates that both the initial and final grain sizes for the wafer with a short (150 RPM/8 amp/16 LPM) and long (150 RPM/3 amp/16 LPM) transient time are the same. Debye-Scherrer analysis of the FWHM data indicated that the initial grain size was 70 - 100 nm, but could not be applied reliably to determine the final grain size. Previous work on plated Cu indicated that the final grain size is on the order of 1 μm [4].

The remaining mechanism that could affect the resistance transients is plating impurity incorporation, as one would expect a high impurity incorporation to increase the transient time because the grains would be heavily pinned [5]. The C, O, S, and Cl impurities arise from the copper sulfate, levelers, and brighteners that constitute the plating bath. The data in Table 2 indicates that the amount of incorporated impurities is a strong function of the plating current, so that for a given agitation, the lower current condition results in higher incorporated impurities. The higher impurity incorporation then increases the resistance transient time, as shown in Figures 8 - 11. In each figure, for a given agitation, the lower current condition yields the highest impurities and longest resistance transient. The qualitative explanation for this trend is that the impurity incorporation is controlled by a mass transport mechanism. The lower current, and hence slower deposition rate, allows sufficient time for the plating impurities to be incorporated in the film. The condition with the high flow and low current (150 RPM/3 amp/16 LPM) had the highest impurity level of all because the increased flow provided a rapid supply of impurities that could be incorporated into the film. By the same argument, the transient time non-uniformity across each wafer can also be attributed to a spatial distribution of impurities, although confirmation of this hypothesis is required.

CONCLUSIONS

The following trends were observed: (1) by decreasing the electroplating current for a given agitation (flow and rotation) the impurity incorporation in plated Cu increases, and (2) the resistance transient time at a given wafer location increases as the plating impurity concentration (C, O, S, and Cl) increases. The longest resistance transient occurred for the high agitation/ low current condition (150 RPM/ 3 amps/ 16 LPM). This trend is qualitatively explained as being a mass transport phenomenon. A high flow rate and low deposition rate from a low plating current allows sufficient time for the plating impurities to be incorporated in the film. In addition, the resistance transient times within each wafer are a function of radial position, which is most likely a result of a spatial distribution of the plating impurities.

ACKNOWLEDGMENTS

The authors would like to thank Steven Boettcher for the SEM analysis. The authors also appreciate the insight and helpful discussions provided by James Harper and Cyril Cabral.

REFERENCES

1. D. Edelstein, et al., IEDM Technical Digest, p. 773 (1997).
2. T. Ritzdorf, et al., Proc. IEEE Int. Interconnect Tech. Conf., p. 166 (1998).

3. C. Cabral, et al., Proc. Adv. Metall. Conf. (1998) in press.
4. L.M. Gignac,et al., MRS Proc. (1999) in press.
5. J.M.E. Harper, et al., J. Appl. Physics, 86, 2516 (1999).
6. C. Lingk and M.E. Gross, J. Appl. Physics, 84, 5547 (1998).

MECHANISTIC STUDIES OF THE ROOM TEMPERATURE RECRYSTALLIZATION OF ELECTROPLATED DAMASCENE COPPER AND SPUTTER-DEPOSITED COPPER

M.E. GROSS[1*], R. DRESE[#], D. GOLOVIN[##], W.L. BROWN[1], C. LINGK[###], S. MERCHANT[2], M. OH[2]

[1]Bell Labs, Lucent Technologies, Murray Hill, NJ 07974
[2]Bell Labs, Lucent Technologies, Orlando, FL 32819

ABSTRACT

The room temperature recrystallization of electroplated Cu films is clearly related to the influence of organic and inorganic additives in the plating bath, however, the underlying mechanisms are still not well understood. In this paper we present preliminary results on (1) a radial variation in recrystallization rate that suggests a non-uniformity during plating related to the additives; (2) a volume influence on the recrystallization rate of Cu plated over wide trenches; and (3) the recrystallization of a thick (1 μm) sputtered Cu film *at room temperature* driven by recrystallization of the overlying EP Cu film. It is notable that we observe high levels of C, N, and O in the sputtered Cu layer after recrystallization, indicating a high mobility of the additive-derived inclusions in the film. The concentration profiles of Cl and S are significantly different, decreasing rapidly close to the interface between the two Cu layers.

I. INTRODUCTION

Electroplated (EP) Cu provides a technology-enabling bottom-up fill of sub-0.25μm damascene (inlaid) structures that has greatly facilitated the introduction of Cu as a replacement for Al in advanced integrated circuits (IC).[1] Bottom-up filling is achieved through the use of a combination of organic and inorganic additives that modify the charge current density at the wafer/cathode surface. The resulting films are slightly compressive with equiaxed grains of about 0.1-0.2μm. Given the strong surface interactions of the additives, it is not surprising to find that they are incorporated into the Cu. These inclusions may temporarily stabilize the small-grained microstructure through impurity pinning of grain boundaries;[2] however, directly or indirectly, they ultimately have a destabilizing effect that turns out to be advantageous for IC interconnect performance.[3,4]

A striking consequence of the use of additives in the Cu plating is the secondary recrystallization or grain growth that occurs *at room temperature* only in films deposited from additive-containing baths.[5-8] This is the strongest indicator that the influence of the additives extends beyond plating, whether directly, as inclusions in the matrix, or indirectly, through an increased density of defects, dislocations, and twins in the films.[2,5,7,9] The small as-plated grains recrystallize over a period of hours to days after plating to a final grain size of several microns. The decrease in grain boundary area is accompanied by a 20% drop in sheet resistance and provides superior electromigration resistance.[10,4] The influence of additives on the mechanical

---

* e-mail contact: mihal@lucent.com
# now at RWTH-Aachen, Germany; ##Cornell University, Ithaca, NY; ###Ludwig-Maximilians-Universität-München, Germany

properties, grain size, orientation, conformality, and grain growth have been reported prior to the recent interest in IC applications;[5,11,12] however, a detailed mechanistic understanding is still lacking.

There is a paradox presented by the expectation that inclusions will pin grain boundaries,[2,13] yet a dynamic process is observed in the EP Cu films *at room temperature*. While the direct effect of the additive inclusions may be to pin grain boundaries, indirectly, they are responsible for higher densities of defects, twins, and/or strain whose stored energy is sufficient to overcome the pinning. The lack of analytical tools with both the sensitivity and the spatial resolution to map the inclusions and defects, without introducing additional damage during sample preparation, has led us to seek clues to the mechanisms driving the recrystallization by indirect means. Changes in hardness and sheet resistance with time after plating have previously been correlated with grain growth in electroplated (EP) Cu films.

We recently reported that the nucleation of grain growth is localized at step edges when Cu is plated into damascene trenches or over isolated steps patterned in SiO_2-covered Si wafers.[7] The inhibitory effect of the additives is expected to be greatest at these sites during plating, which may result in their having the highest concentration of defects, twins, or strain, that is, the highest stored energy to initiate recrystallization. In the case of EP Cu deposited over 2 and 5 μm wide trenches, we observed that once nucleated, the grain growth extended laterally first *over* the trenches and subsequently between the trenches. This suggested a thickness dependence of the recrystallization rate. Another observation from cross-section transmission electron micrographs (TEMs) that the 1,000Å PVD (physical vapor deposition) Cu seed layer, after recrystallization of the overlying EP Cu layer, is indistinguishable from that layer, led us to examine recrystallization of the PVD layer more closely.

In this paper we present preliminary results on the thickness dependence of the recrystallization rate of Cu on damascene-patterned wafers and in blanket films, and evidence that the EP Cu layer can be used to drive the recrystallization of an equally thick PVD Cu layer. These observations provide further clues as to the role of additives in the recrystallization process.

II. EXPERIMENT

200 mm Si wafers with blanket PETEOS SiO_2 films or with arrays of 5 μm/5 μm trench/space gratings, 1 μm deep etched in oxide. Prior to plating, a 500Å Ta diffusion barrier layer was sputter deposited, followed immediately by a 1,000Å sputtered Cu conduction layer. Electroplated Cu was deposited to a coulometric equivalent of 1 μm using a commercial fountain plater with a consumable Cu anode and commercial plating bath (Enthone OMI CuBath SC; MLO 70/30 and MD additives). Sheet resistance measurements (121 points) were recorded on a KLA-Tencor Prometrix RS75 system.

Recrystallization rates were determined by analysis of plan-view focused Ga[+] ion beam (FIB) secondary electron images recorded at successive time intervals, as described elsewhere.[7] Damascene samples were stored on dry ice to prevent recrystallization after plating prior to measurement. Each image was recorded in a different area of the same sample to avoid complications from implanted Ga in the previously imaged area. Cross-section images were generated by first milling a rectangular trench with the Ga[+] beam at high dose, then rotating the sample 45° and recording the secondary electron image at low dose.

III. RESULTS

A. Radial Variation in the Recrystallization of Electroplated Cu Films

Sheet resistance measurements have typically been used to monitor the changes in the EP Cu films after plating The drop in sheet resistance can be correlated with the decrease in grain boundary volume as the 0.1 μm equiaxed grains grow to several microns in width. A closer examination of the data, measured at 121 points across the wafer in five equally spaced concentric rings and at the center, reveals a variation in rate across the wafer. Recrystallization is more rapid at the edge of the wafer and decreases towards the center, as shown in Figure 1a. By comparing the initial sheet resistance values for all the rings, as well as the final ones, we can eliminate thickness variation as a factor in the radial variation. This leaves us with the perplexing possibility that there is some variation in additive influence and/or incorporation in the film related to the wafer rotation, process conditions, and configuration of the fountain plater.

We do see evidence for thickness variation contributing to a difference in recrystallization rate in a duplicate wafer that was plated at the same time. Figure 1b shows a similar trend in recrystallization from the edge inward. However, the center of this film is transforming significantly more slowly. The higher sheet resistance at the center of the wafer is consistent with a film that is about 0.1 μm thinner, assuming a constant resistivity. It is unlikely that this difference in thickness alone accounts for the difference in recrystallization rate. Rather, the same factors that are causing variations in the center thickness, process conditions that are affecting the interaction of the additives, would also affect the recrystallization rate. Analyses of impurity concentrations between center and edge by secondary ion mass spectrometry (SIMS) have been ambiguous and have not shown any clear trends.

To the extent that the radial variations in recrystallization rate are somehow related to the activity of the additives during plating, the possibility is raised that the important role of additives in producing bottom-up fill profiles might also exhibit radial variations.

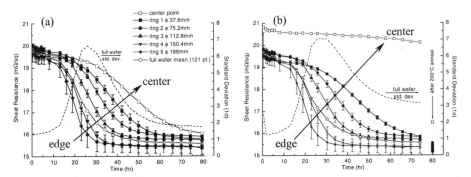

Figure 1. Sheet resistance 1 μm electroplated Cu film (*plus* 0.1 μm PVD Cu seed layer) as a function of time after plating and radial position for duplicate set of 200mm wafers. Data points and bars represent mean sheet resistance and standard deviation values, respectively, for points measured in 5 equally spaced radial rings. The mean sheet resistance and standard deviation full wafer 121 pt. measurement are also plotted.

B. Thickness/Volume Dependence of the Recrystallization of Electroplated Cu Films

In our previous studies on the influence of damascene trench width on the kinetics of the recrystallization process, we observed distinct localization of nucleation at the trench corners of the wider (2, 5 μm) trenches, as shown in Figure 2a.[7] Grain growth then proceeds across the trenches, before extending across the field between the trenches. We now extract quantitative information regarding the difference in rates in these two regions by separately analyzing 5 μm wide segments in the plan-view FIB images corresponding to trench versus field regions. The percentage of Cu recrystallized determined from these image analyses is plotted in Figure 2b, showing a factor of about two difference in the rates for the two regions. Differences in Cu film thickness in the different regions may account for these different rates. While the Cu at the center of these wide trenches is almost equal in thickness to that in the field, the thickness increases from the center to the edge of the trench. An increasing recrystallization rate with thickness is also consistent with the nucleation of grain growth at the trench corners. Another factor that may contribute to the difference in rates that we observe is a difference in concentration of those species (impurities, defects, twins, etc.) responsible for driving the recrystallization as a result of different activities of the accelerator and suppressor species dominating within and outside the trenches, respectively.

The acceleration in recrystallization with film thickness is counteracted, within damascene features by a surface-to-volume constraint. As trench widths narrow to sub-micron dimensions, the Cu within the narrowest trenches does not fully recrystallize, while the surrounding Cu is fully transformed. In manufacturing, however, the EP Cu films will be annealed to 200-400°C after plating and before chemical mechanical polishing to stabilize the microstructure and hardness.

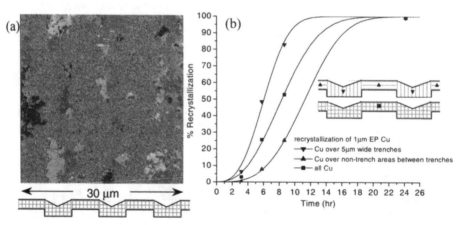

Figure 2. (a) Representative plan-view FIB image and schematic representation of plated trenches and (b) recrystallization rate curves derived from these images for 1μm EP Cu deposited on wafers with equal 5μm trench/space arrays, 1.0μm deep (TaN barrier layer). The overall rate of recrystallization (■) and rates over trenches (▼) and between trenches (▲) are plotted separately. Curves represent fits of the form $X(t) = 1-\exp(t/t_R)^\alpha$, where $X(t)$ is the fraction recrystallized, t_R is the 1/e characteristic recrystallization time, and α is the Avrami exponent.

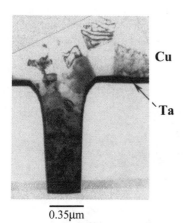

Cu

Ta

0.35µm

Figure 3. Cross-section TEM of EP Cu fill in 0.35µm trench over 500Å PVD Ta barrier/1,000Å PVD Cu. Note that the PVD Cu is indistinguishable from the EP Cu layer after recrystallization.

C. Room Temperature Recrystallization of Sputtered Cu Films

The process sequence for dual damascene Cu metallization involves deposition of an adhesion and diffusion barrier layer, such as Ta, and a seed layer of Cu by sputter deposition. The PVD Cu seed layer acts as the cathode for plating the remaining thickness of Cu to fill the damascene features. In our early studies of the recrystallization process, we noted that after the EP Cu film had fully recrystallized, the initially small-grained PVD Cu layer had also recrystallized and was indistinguish-able from the overlying EP film, as shown in the TEM in Figure 3.[14] Clearly, the driving force for the recrystallization of the 1 µm EP layer is sufficient to extend into the 0.1 µm PVD layer. This observation suggested another approach to examining the energetics of the recrystallization process.

We report here preliminary results on the recrystallization of a thicker, 1 µm PVD Cu film that would remain unchanged in the absence of an overlying EP Cu. A cross-section FIB image recorded 25 hours after plating shows the nucleation and growth of large grains within the EP layer that have grown faster laterally through the EP layer than vertically through the PVD layer. (Figure 4) Recrystallization of PVD Cu at room temperature has not been observed except for films deposited under extreme bias to produce high stress.[15] The driving force for grain growth in these films is not the dominant factor in the EP Cu films. While the EP and PVD Cu layers cannot easily be distinguished in the cross-sections, the horizontal interface between the large and small grains halfway through the Cu film clearly indicates a barrier to grain growth at that interface. Eventually, after the surface appears fully recrystallized, grain growth continues into the PVD Cu layer until both layers appear continuous and large-grained, as seen in Figure 5.

We see no indication in the FIB cross-sections of the original interface between the two Cu layers. Note that although the PVD Cu film is exposed to air before plating and forms a surface oxide, the sulfuric acid plating bath quickly etches the oxide upon contact with the wafer so that we don't see evidence of an oxide in the SIMS spectrum. What we do see is a small jump in secondary ion yield for the additive-derived impurities of C, N, O, S, and Cl halfway through the 2 µm recrystallized Cu film. (Figure 6) The same features are observed, at the corresponding depths, in the SIMS spectra of different combinations of PVD and EP Cu film thicknesses, leading us to conclude that this change in yield is due to a matrix effect that persists despite the recrystallization. Further analyses by high resolution TEM are in progress.

Another surprising observation from the SIMS spectra is that the impurity concentrations do not decrease rapidly at the EP/PVD interface but appear to diffuse into the PVD film with two distinct profiles. The C, N, and O signals decrease slowly and with the same slope in both Cu layers, suggesting a high mobility of these molecular "dopants." This is not inconsistent with the extent of recrystallization that is observed but does indicate higher than expected diffusivities.

The Cl and S signals, on the other hand, do decrease rapidly going in to the PVD layer. One caution regarding SIMS analysis of Cu films is that these films sputter extremely unevenly and a set of consistent and reliable standards are still being developed. Nevertheless, the qualitative results do suggest interesting directions for further investigations.

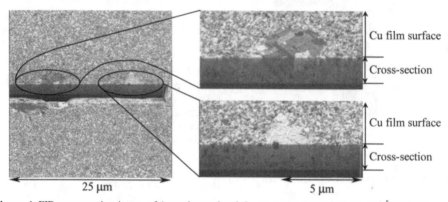

Figure 4. FIB cross-section image of 1μm electroplated Cu over 1μm PVD Cu with 500Å PVD Ta diffusion barrier underlayer after ~25 hr. at room temperature after samples were removed from dry ice storage used to prevent recrystallization. (Images were recorded with samples at 45° tilt to the imaging plane.)

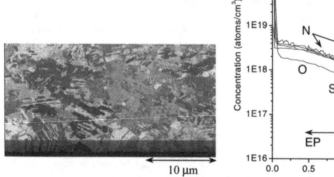

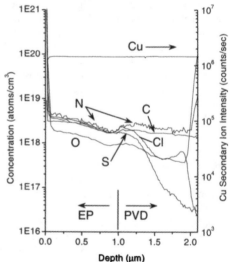

Figure 5. Same sample as in Figure 4 after several days, showing complete recrystallization of the 1μm PVD Cu layer driven by the recrystallization of the overlying 1μm electroplated Cu layer.

Figure 6. SIMS depth profile of 1 μm PVD Cu// 1 μm EP Cu film stack after recrystallization at room temperature. The step at 1 μm indicates a matrix effect in the ion yield at the interface between the two layers.

IV. CONCLUSIONS

The influence of plating bath additives on the room temperature recrystallization of EP Cu can be examined from numerous perspectives to give insights into the mechanisms. We have found a radial variation in recrystallization decreasing from the edge of the wafer in that is not related to thickness, but likely to processing conditions that affect the surface interactions of the additives. A higher recrystallization rate with thickness, or volume, is seen for Cu deposited in damascene trenches. Finally, and perhaps most remarkably, we have demonstrated that there is sufficient driving force in the EP Cu film to recrystallize an equivalent 1 μm thick PVD Cu film. SIMS spectra suggest that differences between the two layers may persist although FIB cross-sections show large, continuous grains extending from through the thickness of the combined layers.

REFERENCES

1. P.C. Andricacos, C. Uzoh, J.O. Dukovic, J. Horkans, and H. Deligianni, IBM J. Res. Devel. **42**, 567 (1998).
2. J.M.E. Harper, C. Cabral, Jr., P.C. Andricacos, L. Gignac, I.C. Noyan, K.P. Rodbell, and C.K. Hu, J. Applied Phys. **86**, 2516 (1999).
3. C. Ryu, K.-W. Kwon, A.L.S. Loke, V.M. Dubin, R.A. Rahim, G.W. Ray, and S.S. Wong, Digest IEEE 1998 Symp. VLSI Tech., p. 156.
4. C.-K. Hu, R. Rosenberg, and K.Y. Lee, Appl. Phys. Lett. **74**, 2945 (1999).
5. I.V. Tomov, D.S. Stoychev, and I.B. Vitanova, J. Appl. Electrochem. **15**, 887 (1985).
6. T. Ritzdorf, L. Graham, S. Jin, C. Mu, and D. Fraser, in *Intl. Interconnect Technol. Conf. Abstracts*, pp. 166-168 (San Francisco, CA, June 1998).
7. C. Lingk, M.E. Gross, J. Appl. Phys. **84**, 5547 (1998).
8. Q.-T. Jiang and K. Smekalin, in *Adv. Metalliz. Conf. 1998*, edited by G.S. Sandhu, H. Koerner, M. Murakami, Y. Yasuda, and N. Kobayashi (Mater. Res. Soc., Pittsburgh, PA, 1999), pp. 209-215.
9. C. Lingk, M.E. Gross, and W.L. Brown, J. Appl. Phys. **87**, 2232 (2000).
10. C. Cabral, Jr., P.C. Andricacos, L. Gignac, I.C. Noyan, K.P. Rodbell, T.M. Shaw, R. Rosenberg, J.M.E. Harper, P.W. DeHaven, P.S. Locke, S. Malhotra, C. Uzoh, and S.J. Klepeis, in *Adv. Metalliz. Conf. 1998*, edited by G.S. Sandhu, H. Koerner, M. Murakami, Y. Yasuda, and N. Kobayashi (Mater. Res. Soc., Pittsburgh, PA, 1999), pp. 81-87.
11. E.M. Hofer and H.E. Hintermann, J. Electrochem. Soc. **112**, 167 (1965).
12. A. Gangulee, J. Appl. Phys. **43**, 867 (1972).
13. F. J. Humphreys and M. Hatherly, *Recrystallization and Related Annealing Phenomena* (Pergamon, New York, 1996).
14. M. E. Gross, C. Lingk, T. Siegrist, E. Coleman, W.L. Brown, K. Ueno, Y. Tsuchiya, N. Itoh, T. Ritzdorf, J. Turner, K. Gibbons, E. Klawuhn, M. Biberger, W.Y.C. Lai, J.F. Miner, G. Wu, and F. Zhang, Proc. Mater. Res. Soc. **514**, 293 (1998).
15. J.W. Patten, E.D. McClanahan, and J.W. Johnston, J. Appl. Phys. **42**, 4371 (1971).

Microstructure of Electroplated Cu Films

H. Okabayashi*, K. Ueno**, N. Ito**, S. Saitoh**, and E. Nomura***
*Research and Development Group, NEC Corp., Miyukigaoka, Tsukuba 305-8501, Japan,
okabayas@lbr.cl.nec.co.jp
**ULSI Device Development Labs., NEC Corp., Sagamihara 229-1198, Japan
***Fundamental Research Labs., NEC Corp., Tsukuba 305-8501, Japan

ABSTRACT

The microstructures and microtextures of electroplated Cu films for interconnect lines in ultralarge scale integrated circuits and their evolution during room-temperature and 300 ℃ annealing were measured by the electron backscatter diffraction technique. The existence of a high twin density in the films was revealed and most microstructural and texture features, for example irregularly shaped grain boundaries and a high <511> texture, were found to be due to the high twin density. Possible favorable effects of twinning on electromigration are briefly discussed.

INTRODUCTION

In wide Cu interconnect lines such as power lines in LSIs, mass transport caused by electromigration is controlled by both grain boundary and surface diffusion [1]. It is important therefore to characterize and control the microstructures of Cu films. Copper films, particularly electroplated films which have been extensively investigated for LSI-interconnect application, have unique microstructural properties compared with sputtered Al films: One is that their microstructures evolve even during storage at room temperature (RT). This so-called RT annealing of Cu films and lines has attracted much attention [2-5]. Another noticeable microstructural difference is the existence of twins in the Cu films, which was revealed by transmission electron microscopy (TEM) [6,7] and electron backscatter diffraction (EBSD) [8-10]. These studies implied that twinning played an important role in grain growth. However, the details of the microstructures of Cu films, including twinning, and their evolution during annealing have not yet been clarified well.

This paper reports the microstructures and microtextures of Cu films and their evolution during room-temperature and 300 ℃ annealing measured by EBSD.

EXPERIMENT

Copper films (0.8 μ m thick) were electroplated on wafers with a Cu seed-layer (100 nm thick) / Ta barrier-layer (30 nm thick) / SiO₂ / Si structure. The electroplating was carried out by using a commercial Cu-sulphuric acid based solution with additives. The Cu seed-layer and the Ta barrier-layer were deposited by sputtering. In the standard θ -2 θ XRD measurements carried out with Cu-K α radiation, the Ta (β -Ta) and Cu seed layers showed only 002 and 111 peaks, respectively. Chips several mm in size were cut from the wafer and used for the EBSD measurements [11].

In the RT annealing experiments, the chips were introduced into the EBSD system after the electroplating and kept there in vacuum at RT for approximately 22 days. EBSD measurements were carried out during the storage. After 22 days in the EBSD system, the chips were taken out and stored in air at RT until they were measured again. Before the EBSD measurements, natural oxide grown on the Cu film surface was etched away with a dilute HF acid solution. In the 300 ℃-annealing experiments, the chip stored in a freezer for 270 days was used. EBSD measurements were carried out before and after annealing at about 300 ℃ in an nitrogen flow for about 30 minutes.

The EBSD system consisted of a standard SEM and EBSD-measurement attachments including Orientation Imaging Microscopy™ analysis software purchased from The TexSem Laboratory, Salt Lake City, U.S.A., and a multi-channel-plate-based detector [12]. The

93

probe-electron beam was scanned with spacings of 0.15 - 0.2 μ m over mostly 10 μ m square areas to produce a hexagonal measurement grid. One measurement point covers an area of approximately 0.01 μ m^2 for a 0.15-μ m-step scan.

RESULTS AND DISCUSSIONS

Grain and Orientation Maps

Figure 1a shows the grain and orientation map of the Cu film stored for 1 day at RT. In this map, grains are defined as the areas with the same crystal orientation within the misorientation tolerance angle of 5 degrees and as the areas greater than those corresponding to 2 measurement points, and grain boundary (GB) lines were drawn perpendicularly to the line connecting the two consecutive measurement points with different crystal orientations at the middle of the line. The GB curves in the maps are not smooth but zigzag because of the hexagonal scan of the probe beam: the GBs consist of chains of some sides of consecutive hexagons. Erroneous determination of orientations often results from areas with poor crystallinity and/or small grain size. Such erroneously determined orientations are usually random and thus there is very little probability that two consecutive erroneously determined orientations will coincide. Most of such erroneous determinations can therefore be removed by defining a grain as the area larger than two measurements points. Figure 1b shows the grain map taken after 264-day storage of the same chip as that for Fig. 1a. The mapping, however, was made in a different area.

Grain maps taken before and after 300 ℃ annealing are shown in Fig. 2. Part of their mapped areas overlap, and only the overlapped area is shown in the before-300 ℃-annealing map. These maps are for a different chip than that shown in Figs. 1a and 1b. The grains labelled with the same number in Figs. 2a and 2b indicate the same grains, which were identified from their crystal orientations and their relative geometric positions. The grains in

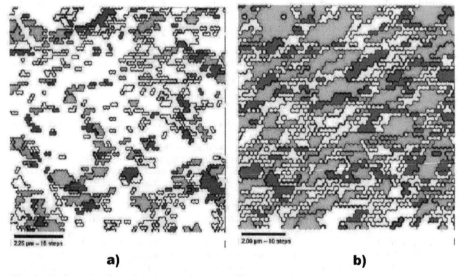

a)   b)

Fig. 1. Grain and orientation maps of the Cu film stored at RT a) for 1 day and b) for 264 days. Grains shaded with light gray and dark gray are <111> and <511> oriented, respectively (within 10 degrees from the substrate normal). In Fig. 1a, grains with other orientations are unshaded and the areas without fringes are fine grained. In Fig. 1b, unshaded areas include both grains with other orientations and the unidentified orientation areas.

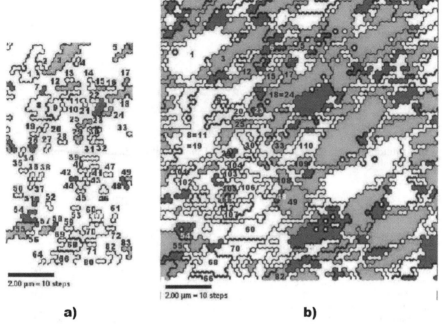

2.00 μm = 10 steps

2.00 μm = 10 steps

a)   **b)**

Fig. 2. Grain and orientation map (a) before and (b) after 300 °C annealing. Grains shaded with light gray and dark gray are <111> and <511> oriented, respectively (within 10 degrees from the substrate normal). The grains labelled with the same number in these maps are the same grains.

the maps tended to be elongated along the diagonal direction from the lower left to the upper right. Part of the elongated grain shape was probably caused by drift of the specimen stage and/or the probe beam, which often occurred during our EBSD measurements and was not corrected for in the software for drawing the maps.

Grain Growth

One of the features of the 1-day-storage map (Fig. 1a) is that there are large grains grown to 1 μ m or larger in size, but the majority of the measured areas, that are not fringed with GB curves, remain fine grained, i.e., the grain size in these areas is the one-measurement-point size or less. The average grain area is 0.09 μ m^2 for the grains larger than 2 measurement points. The map taken after 264-day storage (Fig. 1b) clearly shows that grain growth proceeded over almost the entire surface, and thus the average grain area increased to 0.24 μ m^2. After the 300 °C annealing (Fig. 2) the average grain area further increased to 0.37 μ m^2. Comparison of the maps before and after 300 °C annealing revealed the following results concerning the grain evolution during the 300 °C annealing. Some of near-random-oriented grains (as well as the <111> grains) underwent a marked grain growth (Grains 1, 8, 11, 19, 60, 66, 68, 70). Some closely located grains with the same orientation merged into a large grain (Grains 8, 11 and 19, Grains 18 and 24). Many small grains which existed before the 300 °C annealing were not able to be identified after the 300 °C annealing, which indicates that they were consumed by other grains during their growth process (Grains 4, 7, 10, 13 etc.). Many new grains, including large grains, appeared (Grains 100-112). The 300 °C-annealed-film shows narrow grains and many irregularly shaped GBs. The narrow grains must be lamella twins. The irregularly shaped

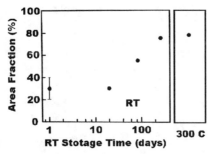

Fig. 3. Evolution of grain growth: Area fraction of grains with 2 measurement-point area for different annealing conditions.

GBs are caused by a high twin density, as described later. These features were seen in all the films regardless of annealing conditions.

Figure 3 shows the evolution of grain growth, i.e., the area fraction of grains larger than two measurement points for different annealing conditions. The measurements were made in different areas. When the RT storage was less than 20 days, approximately 30% of the measured surface area was covered with grains larger than the 2-measurement-points area. However, the fraction increased to about 55% at 82 days and about 75% at 264 days. Erroneous determinations of orientation often occurred at GBs or their very vicinities where EBSD patterns from both grains overlap and/or EBSD-pattern contrasts decrease. If this effect is taken into consideration, the true fractions should be larger than those given in Fig. 3. The lower grain-growth rates at the early stages (shorter than 21 days) may be due to storing the sample in vacuum, which prevents oxidation of the sample surface.

Figure 3 indicates that the annealing rate in the present film is much slower than those reported [2-5]. However, the resistivity of the film measured after several-day storage at RT was decreased to the value only 15% higher than the bulk resistivity. There are several possible reasons for the difference in the annealing rates derived from the grain growth and the resistance decrease. One is the difference in the measured properties. The resistivity might be rather insensitive to the grain size when it is larger than several hundred nm, because the electron mean free path in Cu is approximately 40 nm [13]. However, in the present EBSD measurements, grains larger than approximately 100 nm x 200 nm were investigated. The other possible reason is that only surface area (less than about 0.2 μm from the surface) was measured in the present EBSD, but the resistivity is influenced by the microstructure of the entire film. If grain growth occurs from the bottom to the top surfaces, grains may be large enough to produce the near bulk resistivity deep in the film, but they are necessarily large near the surface.

Crystal Orientation of Grains

In Figs. 1 and 2, the <111> and <511> grains are highlighted by the gray scale, where <uvw> grains are defined as grains in which the <uvw> axis is within 10 degrees from the substrate normal. These maps clearly show that the area fraction of <111> grains is dominant, and that there are only a few <100> and <110> grains. This is consistent with XRD measurements. The electroplated layer seems to have inherited the seed layer texture. In contrast with the XRD measurements, the <511> grains are the next most dominant. Most <511> grains are in the twin relationship with the <111> grains, and the higher area fractions of the <511> grains are due to higher twin density as described in the following section. Since the area fractions of the grains with the orientations other than these four directions are also high (comparable to that of the <511> grains), the film cannot be said to be highly textured, although the θ-2θ XRD measurements indicate a strong 111 texture. The apparent strong 111 texture derived from the XRD measurements is caused by the fact that the grains with larger

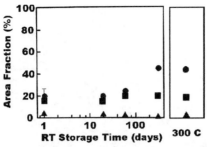

Fig. 4. Evolution of texture: Area fraction of grains with <111> (●), <511> (■), <110> and <100> (▲) for different annealing conditions.

Miller indices produce essentially weak X-ray diffraction. Furthermore, no 511 diffraction peaks can be detected when the Cu-K α radiation is used, because Bragg's relation cannot be satisfied (wavelength/2 times interplaner spacing becomes larger than 1).

Another texture feature is that texture tends to vary locally as seen in the overlapped areas in the before- and after-300-annealing maps (Figs. 2a and 2b), where near random-oriented texture was very apparent.

The evolution of the area fraction of the grains with the above-mentioned four orientations are shown in Fig. 4. Although the relative area fractions vary with storage time, as far as the present experimental conditions are concerned, <111> grains are dominant and the evolution of the crystal orientation is rather independent of annealing time during early stages of the RT annealing. However, the higher <111> to <511> ratios in the 264-day-storage map and the 300 ℃-annealed map suggest that the <111> grains nucleated earlier and/or grew faster. However, more observations are necessary to confirm this suggestion.

Twins

The thick curves drawn between many consecutive grains in Figs. 1 and 2 indicate the GBs between grains whose relative crystal orientations are 60 degrees rotated about one common 111 axis. This 60 degrees <111> condition is the condition for $\Sigma 3$ grain boundaries and also meets the requirement for rotation twins of the fcc lattice (Rotation twins are at the same time mirror twins in a cubic lattice). Thus, in this paper, we call the grains that meet this condition twins. Note that the twin boundaries drawn in thick curves in the maps include both coherent and non-coherent twin boundaries, but they cannot not be distinguished by EBSD. Most grains larger than about 0.5 μm in size are in a twin relationship with other grains, but smaller grains are not necessarily so. Such a high density of twins in the Cu films are in contrast with Al films, where very few twins were observed. This difference is reasonable because the stacking fault energy for Cu, 60 mJ/m^2, is much smaller than that for Al, 200 mJ/m^2 [14]. After 1-day storage, 15 grains among 56 grains in part of Fig. 1a were not twinned. After 21-day storage, however, 11 grains among them became twinned and only 4 grains remained non-twinned. Therefore, non-twinned grains tend to twin when annealing proceeds.

In fcc crystals, one of the {111} planes is the twinning plane. Thus <111> grains twin with <511> grains. As shown in Figs. 1a and 1b, most <111> grains are twinned, mostly with <511> grains. This explains the high area fraction of <511> grains in Fig. 4. Some <111> grains were also in the twin relationship with adjacent <111> grains, as can be seen in Figs. 1b and 2b. This suggests that one of the <111> grains extends under the bottom of the other. If this suggestion is not the case, there is no common 111 axis between them and thus they are not twins. The <511> grains can twin with grains oriented to the directions other than <111>. However, as shown in Fig. 5, some <511> grains are sandwiched and also twinned with <111> grains with the same crystal orientation. This structure seems to be a special type of multiple twins, where grains with orientations A and B form a twin chain as A/B/A/ · · ·. The authors call this structure a special multiple twin (SMT) structure. It also resembles a twin

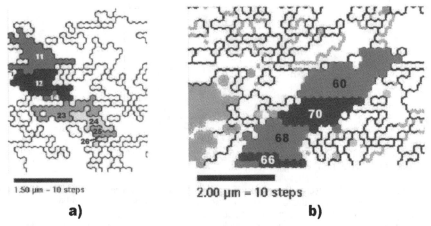

1.50 μm - 10 steps

2.00 μm = 10 steps

a)   b)

Fig. 5. Examples of special multiple twins. (a): after 1-day storage at RT, (b) after 300 °C annealing. Grains 11 and 14: same <111> orientation, Grains 23 and 25: same <111> orientation, Grains 24 and 26: same <511> orientation, Grains 60 and 68: same orientation, Grains 66 and 70: same orientation.

band, where twinned bands with crystal orientation B are created across a grain with orientation A. The SMT structures were observed both in RT- and 300 °C-annealed films, as shown in Fig. 5. The SMT structure consisting of grains 23-26 in Fig. 5a evolved from the twins consisting of grains 23 and 24 in Fig. 1a, i.e., twinned grains 25 and 26 were newly created and joined, or, they grew from a previously undetectably small size (less than one measurement point) during storage between these measurements. Thus, this SMT structure seems not to be created by the twin band formation mechanism. The SMT structures did not necessarily consist of <11̄1> and <511> grains. If the surface energy plays an essential role in forming the SMT structures, the structures may mainly consist of <111> and <511> grains, because {111} planes have the lowest surface energy in a fcc lattice. Multiple twinning in bulk materials do not proceed randomly. It is believed that some selection rules govern the process [15], but it is unclear what selection rules govern the SMT formation.

Twinning reduces the GB energy by dissociating a GB with higher GB energy into coherent twin boundaries and non-coherent twin boundaries with lower energies [16]. The diffusivity of atoms along the coherent twin boundaries are essentially much less than that along the original GB. Thus, the existence of coherent twin boundaries can be neglected in terms of a rapid diffusion path. Therefore, the effective grain size becomes larger. The diffusivity along the non-coherent twin boundaries is expected to be less or at least not larger than that along the original GB because the non-coherent twin boundaries are considered to have lower boundary energy than the original GB. Therefore, twinning is expected to have favorable effects on electromigration of Cu lines.

CONCLUSIONS

We measured the microstructures and microtextures of electroplated Cu films for ULSI lines and their evolution during room-temperature and 300 °C annealing by EBSD. It was revealed that there is a high twin density in the films and that most microstructural and texture features, for example irregularly shaped grain boundaries and a high <511> texture, are due to the high twin density. Twinning may have favorable effects on electromigration of Cu lines, because it is expected to increase the effective grain size and also reduce grain boundary energy and thus the diffusivity of atoms along grain boundaries.

REFERENCES

[1] C.-K. Hu, R. Rosenberg, and K.L. Lee, Appl. Phys. Lett. **74**, p. 2,945 (1999).
[2] K. Ueno, T. Ritzdorf and S. Grace, Proc. of Advanced Metallization Conf. 1998 (MRS Conf. Proc. XIV, Pittsburgh, 1999), p. 69.
[3] C. Cabral, Jr., P.C. Andricacos, L. Gignac, I.C. Noyan, K.P. Rodbell, T.M. Shaw, R. Rosenberg, J.M.E. Harper, P.W. DeHaven, P.S. Locke, S. Malhotra, C. Uzoh and S.J. Klepeis, Proc. of Advanced Metallization Conf. 1998 (MRS Conf. Proc. XIV, Pittsburgh, 1999), p. 81
[4] Q-T. Jiang R. Mikkola, B. Carpenter and M. E. Thomas, Proc. of Advanced Metallization Conf. 1998 (MRS Conf. Proc. XIV, Pittsburgh, 1999), p. 177.
[5] C. Lingk, M.E. Gross, W.L. Brown, W.Y.-C. Lai, J.F. Miner, T. Ritzdorf, J. Turner, K. Gibbons, E. Klawuhn, G. Wu and F. Zhang, Proc. of Advanced Metallization Conf. 1998 (MRS Conf. Proc. XIV, Pittsburgh, 1999), p. 89.
[6] L.A. Giannuzzi, P.R. Howell, H.W. Pickering, and W.R. Bitler, J. Electron. Mat. **22**, p. 639 (1993).
[7] E.M. Zielinski, R.P. Vinci, and J.C. Bravman, J. Appl. Phys. **76**, p. 4,516 (1994).
[8] D.P. Field, Vacuum and Thin Films, p. 36, Jan. (1999).
[9] H. Okabayashi, K. Ueno, N. Itoh, S. Saito, E. Nomura, Extended Abstracts (The 46th Spring Meeting, 1999) of The Japan Society of Applied Physics and Related Societies, No. 2 (The Japan Society of Applied Physics, Tokyo, 1999), p. 889.
[10] D.P. Field and M.M. Nowell, Proc. of the fourth Int. Conf. on Recrystallization and Related Phenomena (The Japan Institute of Metals, Tokyo, 1999), p. 851.
[11] See, for example, Microtexture Determination and its Applications, V. Randle (The Institute of Materials, London, 1992).
[12] D. L. Barr and W.L. Brown, Rev. Sci. Instrum. **66**, p. 3,480 (1995).
[13] A.J. Dekker, Solid State Physics (Prentice-Hall, New Jersey, 1957), p. 284.
[14] P. Haasen, Physical Metallurgy, Second edition (Cambridge University Press, London, 1986), p. 120.
[15] F.J. Humphreys and M. Hatherly, Recrystallization and Related Annealing Phenomena (Pergamon, 1995), p. 218.
[16] R.L. Fullman and J.C. Fisher, J. Appl. Phys. **22**, p. 1,350 (1951).

PATTERN-DEPENDENT SURFACE PROFILE EVOLUTION OF ELECTROCHEMICALLY DEPOSITED COPPER

Tom Ritzdorf, Dakin Fulton, Linlin Chen
Semitool, Inc., 655 West Reserve Dr., Kalispell, MT 59901, tritzdorf@semitool.com

ABSTRACT

Copper bumps formed on top of inlaid copper were often observed during bottom-up plating and the existence of bumps with various heights makes it very difficult for CMP to planarize the surface. This momentum plating effect can be related to the interaction of organic additives and the deposition of copper from the plating solution. Two approaches were investigated in this investigation with the purpose of eliminating the bump formation. It was found that although the bump can be reduced by adding a leveling agent, the presence of the leveling agent degrades significantly the feature-fill capability and results in seam voids in the feature. The best approach was to use plating parameters such as flow rate, wafer spin rate, current density and waveform to achieve bump elimination without affecting organic performance for gap-fill.

INTRODUCTION

The semiconductor industry is transitioning away from the use of physical vapor deposited aluminum and towards the use of electrochemically deposited (ECD) copper for metal interconnects. This transition is leading to an unprecedented concentration on a detailed understanding of the electrochemical deposition process as it relates to filling very fine features. The ability of the copper electroplating bath to completely fill submicron features is largely determined by the organic additives, which typically have roles of suppressing or enhancing the deposition, based on the local additive concentration. The widespread acceptance of ECD processes as the standard deposition method for filling etched features to produce inlaid interconnects is due in large part to the fact that ECD is the only method capable of depositing preferentially inside the features, a characteristic that has been called "superfilling."[1]

The optimization of the organic additives involves understanding the effects of each constituent on the properties of the deposited film, including the fill capability for deep submicron features, the electromigration resistance of the interconnect, and the deposited film topography. When optimizing the additives for feature fill capability, it becomes apparent that topography is also strongly affected. A typical electrodeposited copper film is presented in Figure 1. As seen, bumps on top of the feature were observed and bump heights were strongly dependent on the feature size and feature density with large bumps on top of the small, dense features. Since the next step after electroplating is chemical mechanical polish (CMP) to planarize the wafer surface, these pattern-dependent bumps can lead to uniformity problems for the CMP process.

The effect seen in Figure 1, which has been referred to as the "momentum plating" effect, is very interesting, in that initial concave features have been turned into convex topography. While this behavior may seem counterintuitive, we believe that it is a relatively straightforward result of the additives that have been used to produce the super-conformal deposition profile, which is desired in the filling of deep sub-micron inlaid features.

101

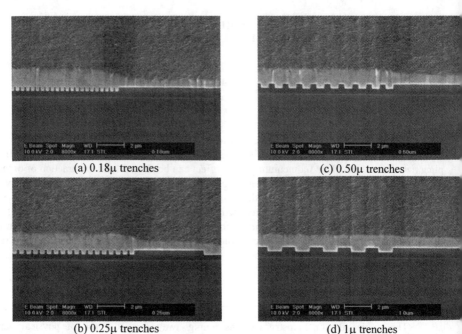

(a) 0.18μ trenches

(c) 0.50μ trenches

(b) 0.25μ trenches

(d) 1μ trenches

Figure 1. The "momentum plating" effect on various feature sizes, where copper is deposited to an inverted profile, immediately above a recessed feature.

EXPERIMENTAL

The electrochemical deposition used in this work was performed in a Semitool ECD reactor, identical to that on an LT210c copper ECD system for copper interconnect manufacturing. All wafers were 200 mm silicon wafers with trench and via features etched in silicon dioxide, and with barrier and copper seed layers deposited over these features to provide electrical conductivity. The copper plating additives used in this study were commercially available and/or experimental additives available from Enthone-OMI, and Shipley Company. Typical acid copper plating solutions, which contains copper sulfate, sulfuric acid and hydrochloric acid, were used for all the investigations.

RESULTS AND DISCUSSION

The ability to fill deep sub-micron, high aspect ratio features used in inlaid copper interconnect technology is largely determined by the presence of organic additives to modify the electrolytic deposition process. Although there is a large number of names that have been used to describe these organic additives, they can be generally classified in two groups. These categories are suppressors and accelerators, and are named after the effect they have on the deposition rate at a given deposition potential. Conventional wisdom holds that the suppressors typically adsorb preferentially to the surface of the wafer in the field area, compared to the insides of the recessed features. This is probably due to an elevated diffusion coefficient for these large molecules inside the microscopic features [2,3]. The accelerators, which are usually small molecules, are then free to adsorb on the interior of the feature, and accelerate the deposition rate locally.

This effect, which results from the organic additives in the copper sulfate plating bath, can be utilized to develop what is referred to as a "bottom-up" deposition profile. This means that the deposition rate within the etched feature, and especially at the bottom, is much greater than the deposition rate on the top surface, or field area. It is exactly this ability of organic additive interaction that provides greater than 100% step coverage, or super-conformal deposition. This super-conformal deposition allows high aspect ratio, or even re-entrant, trenches and vias to be filled without a seam. Figure 2 demonstrates such a "bottom-up" fill sequence.

As can be seen, the trenches were preferentially filled during the initial deposition (Figure 2a) and the preferential deposition in the vicinity of the inlaid feature does not stop once the copper surface has become planarized (Figure 2b and 2c). It is almost as if the deposition process exhibits a "momentum" that carries this increased deposition rate over the nominal surface to produce a convex feature where there was initially a concave one. This effect is unique to electrodeposition in the presence of organic additives. Some clue as to the mechanism behind this effect can be derived from the fact that the bump height generated over a set of features is pattern dependent and exhibits a behavior similar to the pattern-dependent "loading" effect seen in plasma RIE processes.

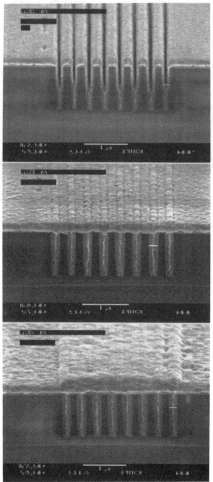

Figure 2: "Bottom-up" fill sequence: after 21 coulombs (top), after 30 coulombs (middle), after 84 coulombs (bottom).

The pattern dependence of the bump height can be seen in Figure 3, which represents the difference in copper thickness between the area immediately above a set of etched trenches and the field copper thickness, as measured using a stylus profilometer. As seen in the figure, the bump height is strongly dependent on the dimensions associated with the underlying features as well as with the recipe used to plate the copper film. With the old recipe, the highest bump height of 2 microns was obtained on the smallest trench (0.5µ) with a smallest pitch size of 1µ. The bump height decreased as the trench size increased. For the same trench of 0.5µ, the bump height was reduced roughly from 2μ to 1μ when the pitch was increased from 1μ to 2 μ. This indicates that there is an interaction between the bump height and the feature and pitch widths. For comparison, the bump heights as a function of trench and pitch widths, which was obtained from an improved plating process with the same organic additives, are also included in Figure 3.

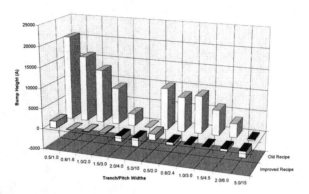

Figure 3. Bump heights as a function of the underlying trench and pitch width

A mechanism is proposed which demonstrates the physical effects associated with the topography of the features that are produced by the momentum effect. If the suppressor adsorbs to the field area, above the etched features, and the accelerator is able to diffuse into the features and promote deposition there, it is not difficult to understand how the raised features may be formed. As seen in Figure 4a, the accelerator is more concentrated inside the feature due to the inability of the suppressor to migrate to this area, and its occupation of the active sites in the field region. The suppressor does continue to diffuse into the feature at some rate, however, where it is incorporated into the film as copper is deposited [2,3]. This causes a localized depletion in the concentration of the suppressor near the feature top, and a relative abundance of suppressor over the majority of the field region of the wafer. Because this creates a situation as seen in figure 4b, where the concentration of suppressor very near the feature opening is depleted at the moment the surface becomes planar, the deposition rate in this area is greater than that in the field region. This dynamic situation caused by the diffusion gradients in the system explains the effects we have observed.

If the mechanism is as discussed in the preceding paragraph, we should expect to see the bumps eliminated by simply pausing the deposition at or near the point of planarizing the features for a time sufficient to allow diffusion to create a uniform surface concentration of additives, then proceeding with the deposition. In fact, when we perform this experiment, we see that there is little or no effect on the bump height above the trenches. Actually, the wafer can be completely removed from the plating solution, rinsed and dried, then returned to the reactor for completion of deposition, with no obvious reduction of the bumps. This indicates that a property of the deposited film itself contributes to the profile evolution. We postulate that it is the incorporation of the additives in the copper film as it is being deposited that causes the effect to be so persistent.

The incorporation of the additives into the film as it deposits is represented much as the adsorbed additives in figures 4a and 4b, with the exception that this mechanism does not rely on the dynamic concentration gradients to produce the effect we are talking about. In this case the additive species may remain at or near the surface and continue to accelerate or suppress the local deposition rate, or an affinity of accelerator in the solution for the accelerator incorporated into the film could produce a similar effect. We believe that the accelerator, in particular, is carried at the surface of the depositing film, where it continues to aid in the deposition of copper in the areas where it has the greatest concentration (i.e. immediately over the features).

(a)   (b)

Figure 4. (a) Additive adsorption during deposition in a high aspect ratio feature, with letters representing molecules of suppressor and accelerator. (b) Additive adsorption just as the feature is filled and the surface becomes planarized.

Another possible mechanism for bottom-up deposition and bump formation may be related to the interaction of chloride and suppressor. For all the plating solutions, the chloride concentration is in the ppm level and a concentration gradient can be easily formed inside small features. It's known that the suppressor needs to interact with chloride to provide suppressing effect [4]. Therefore, the concentration gradient of chloride can lead to enhanced deposition rate in the feature. If the interaction rate between the suppressor and chloride is slower than the generation of fresh copper surface, the copper deposit on top of the feature continues to grow to form the bumps.

We would like to reduce the size of the bumps plated over inlaid interconnect features in order to allow the CMP process to have a reliable process without dishing and erosion in the polished interconnects. Since we know that the deposition of bumps over etched features is due to the effect of adsorbed additives, it is possible to modify this effect through the use of process parameters. In general, there are two methods of modifying the process in order to reduce the size of the bumps. The first method is by a change in the additive package, and the second is through a modification of the deposition parameters (other than bath composition), typically after the fine features have been filled.

When modifying the additive package in order to reduce the size of the bumps over interconnects, there are two approaches that are feasible. The first approach is to modify the suppressor and/or accelerator to reduce the effect, and the second is to add an additional leveling component to the additive package. In practice, each of these methods tends to degrade the ability of the chemistry to fill very high aspect ratio features, which is an unacceptable tradeoff in order to produce improved post-polish interconnect thickness uniformity. Figure 5 compares the effect of leveling agent on bump formation and gap fill. In the absence of leveling agent, good gap fill was achieved for both large and small features as shown in Figs. 5a and 5b. After adding leveling agent, although the bump was reduced (Figure 5a to Figure 5c), tiny seam voids were observed in small trenches (0.18µ 4.9:1AR) in Fig. 5d. This implies that a degradation in gap-fill was caused by the presence of the leveling agent in the plating solution. Furthermore, the presence of an additional organic leveling agent significantly complicates the additive package, making it difficult for on-line analysis and control of individual components.

The second method of modifying the final surface topography is to modify deposition parameters such as current density and waveform, fluid flow, and wafer rotation velocity. Although these parameters may be adjusted at any time during the deposition process, it is usually preferred to utilize parameters that have been optimized for feature fill capability and electromigration resistance until the small features are filled, then switch to a set of parameters that are optimized for the minimization of these surface features.

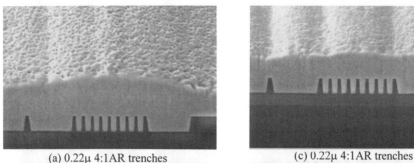

(a) 0.22μ 4:1AR trenches   (c) 0.22μ 4:1AR trenches

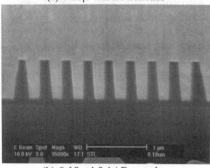

(b) 0.18μ 4.9:1AR trenches   (d) 0.18μ 4.9:1AR trenches

Fig. 5: Effect of leveling agent on bump formation and gap fill: (a) and (b) without leveling agent; (c) and (d) with leveling agent

An example of the capability of modifying the surface topography through the use of process parameters while maintaining a consistent bath composition is seen in Figure 6. The data represented in this figure were produced by using different process parameters (waveform, current density, and fluid flow) with the same copper plating chemistry as in Figure 1. As seen in the figure, the process parameters investigated had a large effect on the surface topography. The utilization of the knowledge presented here has allowed us to minimize the surface topography, while not sacrificing the beneficial aspects of a particular set of plating additives. The only difference between Figure 6 and Figure 1 was a different set of process parameters that produced no deleterious effect on the feature fill capability of the process. The bump heights obtained with improved process parameters as a function of feature size are included in Figure 3. As can be seen, significant improvement in bump heights was obtained. For 0.5μ trench with 1μ pitch, the bump heights were reduced by roughly 10 times with the improved plating parameters.

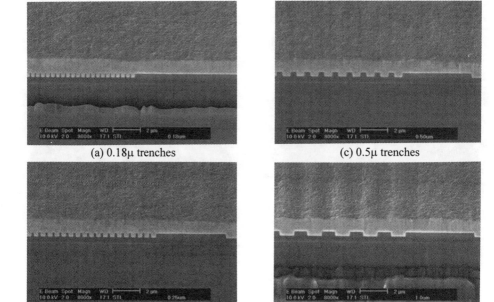

| (a) 0.18μ trenches | (c) 0.5μ trenches |
| (a) 0.25μ trenches | (d) 1μ trenches |

Figure 6. The effect of changing process parameters on the surface topography, as a function of feature size. The electrolyte and additives used were the same as in Fig. 1.

SUMMARY

The momentum plating effect with bump formation is produced by the interaction of the diffusion of organic additives and the deposition of copper from the electrolytic plating bath. Addition of leveling agent in the organic additive package often degrades the gap-fill capability. The topographical features are relatively easily controlled by the modification of process parameters to produce a locally planar surface, which is desirable for process integration, especially when trying to optimize downstream CMP processes.

REFERENCES

1. Andricacos, P. C.; Uzoh, C.; Dukovic, J. O.; Horkans, J.; Deligianni, H.; in *ULSI Fabrication I and Interconnect and Contact Metallization: Materials, Processes, and Reliability*; ECS Proc. 98-6.
2. K. Takahashi and M. E. Gross, in *Advanced Metalization Conference (AMC 1998)*, edited by G. S. Sandu, H. Koerner, M. Murakami, Y. Yasuda and N. Kobayashi (Mater. Res. Soc. Proc. ULSI XIV, Pittsburgh, PA 1998), p. 57-63.
3. M. E. Gross, K. Takahashi, C. Lingk, T. Ritzdorf, K. Gibbons, in *Advanced Metalization Conference (AMC 1998)*, edited by G. S. Sandu, H. Koerner, M. Murakami, Y. Yasuda and N. Kobayashi (Mater. Res. Soc. Proc. ULSI XIV, Pittsburgh, PA 1998), p. 51-56.
4. J.J. Kelly and A. C. West, J. Electrochem. Soc., **145**(10), p. 3472(1998) and **145**(10), p. 3477(1998)

THE INFLUENCE OF ADDITIVES ON THE ROOM-TEMPERATURE RECRYSTALLIZATION OF ELECTRODEPOSITED COPPER

G.R. Stafford, M.D. Vaudin, T.P Moffat, N. Armstrong and D.R. Kelley
Materials Science and Engineering Laboratory, National Institute of Standards and Technology
Gaithersburg, Maryland 20899, USA

ABSTRACT

The recrystallization behavior of copper, electrodeposited from a copper sulfate - sulfuric acid plating bath into which various combinations of NaCl, sodium 3-mercapto-1propanesulfonate (MPSA), and polyethylene glycol (PEG) had been added, was examined by x-ray diffraction and 4-point resistivity. Significant room temperature recrystallization and resistance decrease are observed only when MPSA and PEG are present in the electrolyte together, particularly in the presence of chloride. The rate of recrystallization is linked to the grain refinement brought about by different bath chemistries. The significance of halide adsorption on copper additive plating is discussed.

INTRODUCTION

Electrodeposited copper is rapidly being introduced into chip interconnection technology as a replacement for aluminum. Implementation of the copper damascene process requires the use of inhibitors to ensure complete filling of vias and trenches. These addition agents are also responsible for grain refinement in the deposited material which subsequently leads to recrystallization, resulting in a 20-25 % decrease in resistivity and an order of magnitude increase in grain size [1-5]. In structural materials, the avoidance of such abnormal grain growth is an important aspect of grain size control; however, in the case of copper metallization, recrystallization leads to enhanced performance. Abnormal grain growth not only lowers the electrical resistivity, but the large grains result in grain boundaries which are perpendicular to the flow of electrons, leading to favorable electromigration resistance.

Abnormal grain growth results in a lowering of the free energy due to the reduction in total grain boundary energy, although in thin films additional factors such as the orientation dependence of the surface energy may come into play [6,7]. The main factors that lead to abnormal grain growth are second-phase particles, texture and surface effects, all of which tend to inhibit normal grain growth. TEM studies of copper films plated in the presence of organic species bearing sulfur functional groups reveal the incorporation of inclusions in both the bulk and along grain boundaries [8]. These inclusions have been reported to exert a strong effect on the recrystallization behavior [8-10]. A model, based on grain boundary pinning by particles, has also been developed to describe the microstructural evolution of these copper electrodeposits [2]. Since the additive chemistry is the primary source of particles and is generally responsible for the grain refinement of the as-deposited films, an understanding of additive chemistry is central to understanding and perhaps controlling the copper recrystallization behavior.

Addition agents used in the electrodeposition of copper are typically chloride ions [11-13] and combinations of three classes of organic compounds which have generally been termed as carriers, brighteners and levelers [14-19]. Carriers are typically polyalkylene glycols, brighteners are often molecules with thiol and sulfonic acid groups, while levelers are typically molecules having amine functionality. The role of the brighteners and leveling agents is to improve deposit planarity, while the carrier acts to enhance the effectiveness of the brighteners and levelers. Carriers such as polyethylene glycol are known to inhibit the copper deposition reaction and improve throwing power, particularly in the presence of chloride [20-25]. The synergistic behavior of these addition agents towards the inhibition of copper deposition is also reflected in the recrystallization behavior, although the precise correlation with known chemistry has yet to be reported [26]. This report will describe a variety of experiments, ranging from X-ray diffraction to scanning tunneling microscopy, which are focused on understanding the effect of additives on copper deposition.

Conference Proceedings ULSI XV © 2000 Materials Research Society

EXPERIMENTAL

Copper single crystals were cut from a 2.5 cm diameter boule and aligned using Laue x-ray diffraction. The crystals were then progressively polished to a 0.1 μm diamond finish followed by electropolishing in 85% volume fraction orthophosphoric acid at 1.6 V versus a large platinum wire mesh electrode. The voltammetric experiments were performed in 0.01 mol/L $HClO_4$ into which 0.001 mol/L KCl was added, while scanning tunneling microscopy (STM) experiments were performed in 0.01 mol/L HCl using a Molecular Imaging[*] scanning probe microscope. Tungsten tunneling probes were fabricated by etching in 1 mol/L KOH followed by coating with polyethylene in order to minimize faradaic background currents. The sample chamber and electrolytes were purged with argon before each experiment. A copper wire was used as a quasi reference electrode in the STM experiments.

The influence of additive chemistry on recrystallization behavior was examined on copper films which were electrodeposited onto copper seed layers supported on n-type silicon (100). The silicon substrates were first sputter cleaned, followed by evaporation of a 3.0 nm adhesion promoting layer of chromium followed by a 100 nm layer of copper. The resulting copper seed layer was polycrystalline with a slight (111) preferred orientation. For electroplating, the silicon wafer was supported on a sheet of OFHC (oxygen free high conductivity) copper with a uniform ohmic contact being formed with a liquid gallium/indium eutectic alloy. The wafer and copper current collector were masked with electroplaters tape. A 2.54 cm diameter hole was punched in the tape, exposing 5.1 cm² of the copper seed layer to the electrolyte. The counter electrode was a flat platinum sheet placed parallel to and positioned laterally 8 cm from the working electrode. Electrodeposition was conducted at room temperature and the films were maintained at room temperature during the characterization period.

The base electrolyte was a copper sulfate - sulfuric acid plating bath containing 0.25 mol/L $CuSO_4 \cdot 5H_2O$ and 1.8 mol/L H_2SO_4. Into this electrolyte, various combinations of NaCl (10^{-3} mol/L), sodium 3-mercapto-1propanesulfonate ($NaSO_3(CH_2)_3SH$, 10^{-5} mol/L), and polyethylene glycol ($H(OCH_2CH_2)_nOH$, 0.3 g/dm⁻³, average molar mass ≈ 3,200 g) were added. In this paper, the sodium 3-mercapto-1propanesulfonate and polyethylene glycol will be referred to as MPSA and PEG, respectively. The copper films were electrodeposited at a current density of 15 mAcm⁻² to a total thickness of 1 μm. No stirring or agitation of the electrolyte was used during the 3 minute deposition. The electrodeposits were examined by 4-point resistance probe and x-ray diffraction.

X-ray diffraction scans were carried out on a powder diffractometer equipped with a Ge incident beam monochromator to eliminate Cu $K\alpha_2$ radiation. Multiple scans of several peaks were made in order to follow grain size and texture development during recrystallization. To minimize the time required to measure each peak over an adequate 2θ range with a sufficiently large signal to noise ratio, the effective incident slit width was set at 0.68° and the receiving slit at 0.15°. The relatively large receiving slit lead to significant instrumental broadening which was characterized using a LaB_6 powder specimen with large particle size and very little residual microstrain. The integral breadths of the LaB_6 Bragg peaks were plotted against 2θ and the instrumental breadths at the Cu Bragg peak positions were determined by interpolation. The specimen integral breadths were determined from scans obtained with step width 0.02° and dwell times ranging from 6 to 20 seconds. Integral breadths were determined directly from the raw intensity data. Attempts were made to analyze diffraction line profiles using modified Williamson-Hall and Warren-Averbach [27] approaches using contrast factors to study the density, type and arrangement of dislocations. The results were inconclusive and it became clear that further investigations of this sort will need to employ a high intensity x-ray source with a very narrow instrument profile, such as a synchrotron.

---

[*] Certain trade names are mentioned for experimental information only; in no case does it imply a recommendation or endorsement by NIST.

RESULTS AND DISCUSSION

The voltammetric behavior of three Cu crystals with low index surfaces, in the presence of chloride, is presented in Figure 1. Copper dissolution occurs at potentials positive of -0.1 V while the onset of hydrogen evolution occurs at potentials negative of -0.7 V. The redox waves shown in Figure 1 are associated with chloride adsorption. The irreversible nature, i.e., separation of the oxidation and reduction waves, of the adsorption process on Cu(111) is in strong contrast to the reversible response observed for Cu(100), while a degree of irreversibility is apparent on Cu(110). The formation of a halide overlayer is consistent with LEED-AES experiments which demonstrate specific adsorption of chloride at potentials below the equilibrium potential of the Cu/Cu$^+$ reaction [28,29]. This result is also consistent with the negative potential of zero charge (pzc) reported for copper in KClO$_4$ solutions [30] and work function data [31] as well as the tendency for underpotential reactions in group IB-halide systems [32,33].

Figure 2 shows a high resolution STM image of a Cu(100) surface in 10 mmol/L HCl, held at a potential of -0.169 V vs. Cu/Cu$^+$. The terraces are completely covered by a ($\sqrt{2}$ x $\sqrt{2}$)R45° chlorine adlattice as determined by LEED and in situ STM [29,34,35]. The image reveals a series of terraces bounded by <100> oriented steps, as illustrated in the schematic. The <100> step orientation is in sharp contrast to the close packed <110> orientation associated with clean copper surfaces in UHV systems [36]. The <100> step edge corresponds to the close packed direction of the chlorine adlattice which stabilizes the underlying kink saturated metal steps. The adlayer has been shown to float on the surface during metal deposition thereby acting as a template guiding step flow [32,37,38]. Ordered chloride adlayers have also been observed at saturation coverage on Cu(110) [38] and Cu(111) [38,39] surfaces. Moving the potential towards negative values leads to an order-disorder transition and eventual desorption of the chloride adlayer.

In the case of polyether-sulfide-chloride electrolytes, the ordered chloride adlayers formed on immersed copper surfaces likely facilitate the formation of a well ordered organic layer which inhibits copper deposition. The blocking nature of

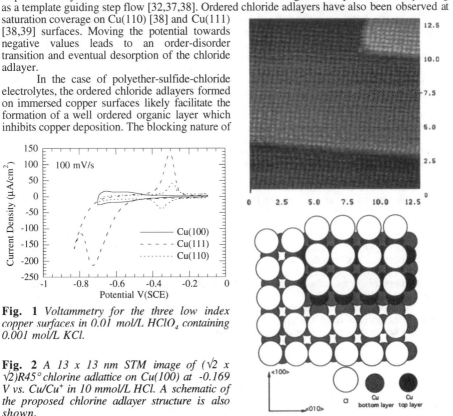

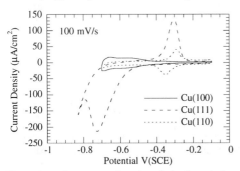

Fig. 1 *Voltammetry for the three low index copper surfaces in 0.01 mol/L HClO$_4$ containing 0.001 mol/L KCl.*

Fig. 2 *A 13 x 13 nm STM image of ($\sqrt{2}$ x $\sqrt{2}$)R45° chlorine adlattice on Cu(100) at -0.169 V vs. Cu/Cu$^+$ in 10 mmol/L HCl. A schematic of the proposed chlorine adlayer structure is also shown.*

this organic overlayer may be subsequently disrupted at more negative potentials where the halide layer becomes mobile due to an order-disorder or some other phase transition [38]. Favorable evidence for such a sharp transition has been seen in polarization curves for copper deposition from sulfate electrolytes containing PEG and chloride [20]. Related voltammetric and STM studies in the presence of bromide revealed the same type of phenomena although the phase transition tends to be displaced to more negative potentials [38], consistent with the relative strength of copper-halide interactions [40]. Preliminary experiments in our laboratory have shown that the copper deposition potential is approximately 0.20 V

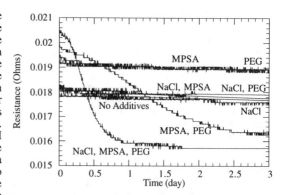

Fig. 3 *In-situ film resistance as a function of time for 1 µm thick copper films electrodeposited from CuSO$_4$-H$_2$SO$_4$ containing (a) no additives, (b) NaCl, (c) NaCl + MPSA, (d) NaCl + PEG, (e) MPSA, (f) PEG, (g) MPSA + PEG, and (h) MPSA + PEG + NaCl.*

more negative when chloride is replaced by bromide in polyether-sulfide-halide electrolytes.

In order to determine the influence of the additive chemistry on the room temperature recrystallization, the sheet resistance of 1 µm thick copper films, electrodeposited from the copper sulfate - sulfuric acid electrolyte containing the various addition agents, was examined. The in-situ measurements for the first 72 hours are shown in Figure 3. Copper films electrodeposited from electrolytes containing (a) no additives, (b) NaCl, (c) NaCl + MPSA, and (d) NaCl + PEG show identical resistance-time behavior for the first 72 hours after deposition. These films have the lowest initial resistance and their resistance decreased by less than 3% over this time period. Films electrodeposited from nominally chloride-free electrolytes containing (e) MPSA and (f) PEG have a 6% higher initial resistance (with respect to films (a) - (d)). The resistance of films (e) and (f) also decreased by less than 3% over 72 hours. Films electrodeposited from electrolytes containing (g) MPSA + PEG had a resistance which initially was 9% higher than films (a) - (d) but decreased 19% in 72 hours. Copper films electrodeposited from electrolytes containing (h) MPSA + PEG + NaCl had an initial resistance which was 12% higher than films (a) - (d) but decreased by 23% within 24 hours.

Electroplated copper films have a much higher electrical resistivity than bulk copper. The excess resistivity may be associated with the defect structure in the electrodeposited films and is the result of scattering from grain boundaries, dislocations, vacancies, impurity atoms in solution as well as a second phase particles [41]. A recent analysis of the possible contributions of these defects to the higher initial resistivity, as well as the subsequent decrease observed during recrystallization, suggests that grain boundaries play the major role [2]. The resistivity increment due to dislocations has been estimated to be 2.1 x 10^{-13} µΩcm^3 [41 and references therein]. With dislocation densities reported to be of the order of 10^{10} to 10^{11} cm^{-2} [42,43], this would result in a dislocation contribution of about 0.04 µΩcm or about 2-3% increase in resistance compared to well annealed bulk copper [2].

Equation (1) has been reported to describe the contribution of grain boundary scattering on film resistance assuming the grain boundaries are randomly distributed and the only grain boundaries which cause diffuse scattering are those transverse to the electron flow [2,44].

$$\rho_g/\rho_o = 1 + (1.40) \, \lambda/G * C_R/(1- C_R) \tag{1}$$

where ρ_g and ρ_o are the resistivities with and without grain boundaries, λ is the intrinsic electron mean free path, taken to be 39 nm for single crystal copper [45], G is the grain size and C_R is the grain boundary reflection coefficient which is assumed to be 0.2 to 0.4, based on values for aluminum [2,44]. From the above expression, one would expect a 12 to 25% decrease in resistivity for an increase in grain size from 0.1 μm to 2 μm which is consistent with experimental observation [2]. Similarly, one would expect to observe different values in the initial film resistance due to variation in grain refinement brought about by different bath chemistries.

X-ray diffraction can be used to assess the grain size of the copper films; however, the observed Bragg peak profiles are the convolution of the instrument profile with the specimen profile. To obtain the integral breadths of the specimen profiles ($ß_{spec}$), the observed intensities were corrected for instrumental broadening in a number of ways. Maximum entropy methods were applied to selected peak profiles [46]; it was found that as the size of the observed broadening ($ß_{obs}$) approached the instrumental broadening ($ß_{inst}$), the maximum entropy method failed to converge on a solution. This is a difficulty encountered by all numerical deconvolution methods under such conditions [47]. Attempts were made to correct the peaks based on different assumptions about the summation of the instrument and specimen profiles: summation in quadrature (Gaussian profiles), simple summation (Lorentzian profiles) and a mixed summation (Lorentzian/ Gaussian), for which:

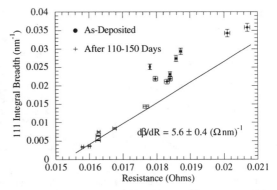

Fig. 4 *Cu 111 integral breadth vs. film resistance for 1μm thick copper films. Data was taken (a) immediately after deposition, and (b) after 110 to 150 days (long term). The solid line represents a linear regression of the long term data.*

$$ß_{spec} = (ß_{obs}^{2} - ß_{inst}^{2}) / ß_{obs} \quad (2)$$

For those peak profiles that were broad enough to be successfully deconvoluted by the maximum entropy method, the integral breadths derived from this deconvolution matched the results from the mixed summation better than the other two summations. Therefore the mixed correction was employed, although the correction method used did not significantly affect the trends that were observed in the data.

The corrected integral breadths of the 111 peak for specimens made with 7 different bath chemistries ((a) and (c) - (h)) were plotted. Figure 4 shows data obtained in the first 100 minutes after film deposition, as well as data obtained 110 to 150 days (long term) following deposition. The initial data do not fall on a smooth curve but the long term data are well fit by a straight line with a slope of 5.6 ± 0.4 (Ωnm)⁻¹. The x-intercept of this straight line fit (0.0154 Ω) represents the film resistance for very large grain material and is quite close to the smallest resistance measured for these copper films. This suggests that the integral breadth and resistance at these later times both have the same functional relationship to the Cu grain size in the films. For dislocation-free materials the integral breadth is inversely proportional to particle size. Equation (1) can be modified by substituting the integral breadths of the x-ray diffraction peaks ($β$) for the grain size ($1/G$); rearranging yields,

$$β = ρ_g(1 - C_R)/(1.40\, ρ_o\, λ\, C_R) - (1 - C_R)/(1.40\, λ\, C_R) \quad (3)$$

A plot of $β$ vs. $ρ_g$ should be linear with a slope of $(1 - C_R)/(1.40\, ρ_o\, λ\, C_R)$. Assuming that $C_R = 0.4$ to 0.2, one would expect a slope of 1.8 to 4.8 (ohm-nm)⁻¹ which is quite close to our long term experimental results of Figure 4. The inference we draw is that once the dislocation content of these films has decreased below the level at which it affects line breadth, a relationship exists between the integral breadths of the x-ray diffraction peaks and film resistance, based on a simple grain boundary scattering model [2]. The 111 and 222 integral line breadths for specimen (g) were analyzed using the Williamson-Hall method [48] which showed that although the microstrain decreased significantly with time, there was still a microstrain contribution to the integral breadth, such that $ß = A / G$, where A was of the order of 2 (in the absence of microstrain, A=1 as in the Scherrer model). If this applied to all the long term specimens in Fig. 4, it would increase the expected slope of $ß$ vs. ρ_g to the range 3.6 to 9.6, in even better agreement with the data.

The initial film resistances shown in Figure 3 and the integral breadth - resistance relationship shown in Figure 4 provide some insight into the effectiveness of the various addition agents with regards to grain refinement. The integral breadths of films that were deposited from electrolytes containing (a) no additives, (b) NaCl, (c) NaCl + MPSA, and (d) NaCl + PEG were very similar. If one assumes that the microstrain contribution to line broadening is similar for each of these deposits, then one can conclude that these films have a very similar initial grain size. In the additive free case, grain size is strongly influenced by the seed layer, and based on previous literature, a few grains begin to dominate during growth, eventually giving rise to a macro rough faceted surface. The addition of halide leads to more marked faceting due to halide induced stabilization of certain crystallographic planes. The addition of either MPSA or PEG appear to have little influence on the film resistance or x-ray integral breadth indicating that chloride minimizes the incorporation of these species, although some alteration of the surface roughness is apparent.

In contrast, when (e) MPSA and (f) PEG additions are examined in the nominally chloride-free electrolyte, an increase in the resistance and integral breadth are apparent, indicating significant grain refinement, and perhaps incorporation of the molecules and/or their constituents in the film. Additional grain refinement is seen when (g) MPSA + PEG are added together indicating synergistic interactions between these molecules and the surface. These effects are further enhanced in the presence of chloride, (h) MPSA + PEG + NaCl. This clearly indicates the importance of the interplay between competitive and co-adsorption effects.

The largest resistance changes and highest recrystallization rates are seen in those films which have the highest initial resistance or the highest degree of grain refinement. The (g) MPSA + PEG film undergoes an 18% decrease in resistance over three days while the copper film electrodeposited from (h) MPSA + PEG + NaCl is similar to the recrystallization behavior reported in the literature; i.e., a 23% decrease in resistance in less than 24 hours. This inverse relationship between initial grain size and recrystallization rate is consistent with reports in the literature [2] and identifies the grain refinement brought about by different bath chemistries as a critical parameter in controlling subsequent changes in film properties.

ACKNOWLEDGMENTS
The authors gratefully acknowledge the contributions of Daniel Josell, Ugo Bertocci, Vladimir Jovic, James Cline, Christian Johnson, John Blendell, and Carol Handwerker.

REFERENCES
1. C. Cabral, Jr., P.C. Andricacos, L. Gignac, I.C. Noyan, K.P. Rodbell, T.M. Shaw, R. Rosenberg, J.M.E. Harper, P.W. DeHaven, P.S. Locke, S. Malhotra, C. Uzoh and S.J. Klepeis, in Advanced Metallization Conference in 1998, G.S. Sandhu, H. Koerner, M. Murakami, Y. Yasuda and N. Kobayashi Editors, p. 81, Materials Research Society, Warrendale, Pennsylvania (1999).
2. J.M.E. Harper, C. Cabral Jr., P.C. Andricacos, L. Gignac, I.C. Noyan, K.P. Rodbell and C.K. Hu, J. Appl. Phys. 86, 2516 (1999).
3. C. Lingk and M.E. Gross, Journal of Applied Physics 84, 5547 (1998).
4. M.E. Gross, K. Takahashi, C. Lingk, T. Ritzdorf and K. Gibbons, in Advanced Metallization Conference in 1998, G.S. Sandhu, H. Koerner, M. Murakami, Y. Yasuda and N. Kobayashi Editors, p. 51, Materials Research Society, Warrendale, Pennsylvania (1999).
5. T. Ritzdorf, L. Graham, S. Jin, C. Mu and D. Fraser, Proc. International Interconnect Technology Conference, IEEE Cat. #98EX102, 106 (1998).
6. F.J. Humphreys and M. Hatherly, Recrystallization and Related Annealing Phenomena, Elsevier Science Inc., Tarrytown, New York (1995).

7. C.V. Thompson, J. Appl. Phys. **58**, 763 (1985).
8. M.S. Abrahams, S.T. Rao, C.J. Buicchi and L. Trayer, J. Electrochem. Soc. **133**, 1786 (1986).
9. D.S. Stoychev, I.V. Tomov, I.B. Vitanova, J. Appl. Electrochemistry **15**, 879 (1985).
10. I.V. Tomov, D.S. Stoychev, I.B. Vitanova, J. Appl. Electrochemistry **15**, 887 (1985).
11. W.H. Gauvin and C.A. Winkler, J. Electrochem. Soc. **99**, 71 (1952).
12. N. Pradhan, P.G. Krishna and S.C. Das, Plating and Surface Finishing, March, 56 (1996).
13. J. Crousier and I. Bimaghra, Electrochimica Acta **34**, 1205 (1989).
14. W. Plieth, Electrochimica Acta **37**, 2115 (1992).
15. D. Stoychev, I. Vitanova, R. Buyukliev, N. Petkova, I. Popova and I. Pojarliev, J. Applied Electrochemistry **22**, 987 (1992).
16. M. Wunsche, W. Dahms, H. Meyer and R. Schumacher, Electrochimica Acta **39**, 1133 (1994).
17. L. Fairman, Metal Finishing, July, 45 (1970).
18. D. Anderson, R. Haak, C. Ogden, D. Tench and J. White, J. Applied Electrochemistry **15**, 631 (1985).
19. L. Mayer and S. Barbieri, Plating and Surface Finishing, March, 46 (1981).
20. M.R.H. Hill and G.T. Rogers, J. Electroanal. Chem. **86**, 179 (1978).
21. J.P. Healy and D. Pletcher, J. Electroanal. Chem. **338**, 155 (1992).
22. D. Stoychev and C. Tsvetanov, J. Applied Electrochemistry **26**, 741 (1996).
23. J.D. Reid and A.P. David, Plating and Surface Finishing, January, 66 (1987).
24. J.J. Kelly and A.C. West, J. Electrochem. Soc. **145**, 3472 (1998).
25. J.J. Kelly and A.C. West, J. Electrochem. Soc. **145**, 3477 (1998).
26. P.C. Andricacos, C. Cabral Jr., J. Horkans, L. Gignac, K.P. Rodbell and C. Parks, Abs.# 273, 195th Meet. Electrochem. Soc., Seattle,Washington, May (1999).
27. T. Ungar and A. Borbely, Appl. Phys. Lett. **69**, 3173, (1996).
28. G.M. Brisard, E. Zenati, H.A. Gasteiger, N.M. Markovic, and P.N. Ross, Langmuir **13**, 2390 (1997).
29. C.B. Ehlers, I. Villegas and J.L. Stickney, J. Electroanal. Chem. **284**, 403 (1990).
30. J. Lecoeur and J.P. Bellier, Electrochim. Acta **30**, 1027 (1985).
31. K. Giessen, F. Hage, J. Himpsel, J.H. Riess, and W. Steinmann, Phys. Rev. Lett. **55**, 300 (1985).
32. O.M. Magnussen, B.M. Ocko, R.R. Adzic, and J.X. Wang, Phys. Rev. B. **51**, 5510 (1995).
33. O.M. Magnussen, B.M. Ocko, J.X. Wang, and R.R. Adzic, J. Phys. Chem. **100**, 5500 (1996).
34. D.W. Suggs and A.J. Bard, J. Phys. Chem. **99**, 8349 (1995).
35. T.P. Moffat, PV 95-8, p. 225-237, The Electrochemical Society, Inc., Pennington, NJ (1995).
36. M. Poensgen, J.F. Wolf, J. Frohn, M. Giesen and H. Ibach, Surf. Sci. **274**, 430 (1992).
37. T.P. Moffat, Mat. Res. Soc. Symp. Proc. Vol. 404, Materials Research Society, Pittsburgh, PA, (1996).
38. T.P. Moffat, Electrochemical Processing in ULSI Fabrication II, Proc. of the 195th Meet. Electrochem. Soc., Seattle,Washington, May (1999).
39. J. Inukai, Y. Osawa and K. Itaya, J. Phys. Chem. **102B**, 10034 (1998).
40. C.Y. Nakakura, V.M. Phanse and E.I. Altman, Surf. Sci. **370**, L149 (1997).
41. A. Gangulee, J. Appl. Phys. **43**, 867 (1972).
42. E.M. Hofer and H.E. Hintermann, J. Electrochem. Soc. **112**, 167 (1965).
43. H.D. Merchant, J. Electronic Materials **22**, 631 (1993).
44. A.F. Mayadas and M. Shatzkes, Phys. Rev. B **1**(4), 1382 (1970).
45. R.G. Chambers, Proc. Royal Soc. London, **A202**, 378 (1950).
46. N. Armstrong and W. Kalceff, J. Appl. Cryst. **32**, 600 (1999).
47. W. Kalceff, University of Technology, Syndey, Australia, (private communication, 1999).
48. G.K. Williamson and W.H. Hall, Acta Metall. **1**, 22-31 (1953).

USE OF ON-LINE CHEMICAL ANALYSIS FOR COPPER ELECTRODEPOSITION

R.J. CONTOLINI, J.D. REID, S.T. MAYER, E.K. BROADBENT, and R.L. JACKSON
Novellus Systems, Inc., Portland Technology Center, Wilsonville, OR 97070

ABSTRACT

The maintenance of adequate chemical specie concentrations is necessary in copper plating baths used for integrated circuit (IC) damascene interconnect formation. Descriptions of both off-line and on-line chemical analysis procedures for measuring these species show that the on-line methods are usually all electrochemical in nature. Statistical information describing results of these procedures on both standardized baths and baths used for electroplating operations show that the on-line methods have a much tighter precision capability than the off-line (benchtop) methods. Additionally, on-line methods offer the convenience of automated sampling, increased frequency of analysis, reduced safety concerns, and potential overall cost savings. Acceptable, reproducible fill of sub-micron features can be achieved by adequate concentration control of the chemical species.

INTRODUCTION

Copper electrodeposition has become an important, enabling fill method for dual damascene interconnect fabrication. In order to maintain the feature fill consistency, plated thickness distribution, and film resistivity attributes of a plating process, it is important to maintain adequate specie concentrations in the plating bath chemistry. Numerous papers have been written about the chemical baths and additives necessary for improving both fill of small features and uniformity over large area substrates for the ultra-large scale integration (ULSI) and related industries [1-5].

Maintenance of bath chemistry encompasses both accurate and precise metrology to measure chemical concentrations, as well as an accurate and precise methodology for chemical dosing. Chemical concentration analysis can be performed both 'off-line' (where a sample aliquot is taken from the plating tool and analyzed using benchtop equipment) and 'on-line' (where analysis equipment is connected directly to the bath reservoir of a plating tool). With a sampling period of say once per day, the advantages provided by on-line analysis are based in convenience, safety, precision, and cost effectivity. With a shorter sampling period by an on-line analysis system, more detailed specie degradation information can be obtained. This information can in turn be used to further refine the performance algorithm of an automated dosing system.

Significantly improved measurement reliability occurs when an on-line, fully automated chemical measurement system (as, for example, the ECI QLC-5000 system [6]) is utilized vs. an off-line system operated by multiple personnel. When different personnel, each having different levels of experience, operate the same off-line metrology equipment, the measurement results can be significantly different, both from a precision (or reproducibility) standpoint as well as an accuracy standpoint. An on-line metrology system performs the required tasks on a day-to-day basis in a more consistent manner due to the same software and hardware capabilities.

It may be desirable to utilize the same on-line analysis system for measurement of multiple plating tools, thus providing a common metrology basis and reduced analytical costs for a large copper plating operation over that associated with embedding such capability on a per tool basis. Safety can also be less of an issue for the on-line analysis system. Operators do not have to fill nor clean various beakers, test tubes, and burettes for the many titrations and other analysis procedures. With fewer personnel needed to work on the above tasks, an overall savings in cost and time are associated with using an on-line metrology system.

The use of an intelligent dosing algorithm which comprehends actual plating tool usage and other key factors can reduce significantly the required sampling frequency for off-line or on-line chemical monitoring. The patented Novellus Smart-Dose™ chemical dosing system allows for accurate dosing of chemical species based on an algorithm which utilizes both predictive (based on tool usage) and corrective (based on chemical analysis) inputs. The Smart-Dose™ system has been integrated into a complete, closed-loop system with the on-line metrology capability to provide a complete Chemical Management System (CMS).

This paper will first review current techniques for analyzing copper plating bath chemistry. These will include visible-light spectroscopy, titration, HPLC (High Performance Liquid Chromatography), and electrochemical methods such as CVS (Cyclic Voltammetric Stripping) and CVS in combination with AC voltammetry. The use of on-line chemical analysis for control of copper electrodeposition in a manufacturing tool is then discussed. Results obtained from equipment that is connected to the bath reservoir of a Novellus Sabre™ plating system will be presented along with a comparison to off-line metrology results. Finally, some examples of feature fill results observed when specie concentrations go out of range will be discussed.

EXPERIMENTAL

This section describes the techniques associated with off-line, benchtop metrology equipment and on-line, electrochemical equipment.

The off-line measurements of copper and chloride were performed by one of the most common techniques, absorption spectroscopy. For copper analysis, a 20:1 dilution of the copper plating solution was done and the sample was poured into a 1 cm long pathlength quartz absorption cell. Using an absorption spectrophotometer set for 814 nm wavelength, the light absorption by the solution was measured by the detector and compared to a standard DI water blank absorption cell. The spectrophotometer's computer then calculates the concentration of the $Cu2+$ species in the solution. Likewise, for Cl- ion in solution, absorption spectroscopy is used. However, in this case silver nitrate, nitric acid, and ethylene glycol are added to form a colloidal suspension of the silver chloride, vs. a precipitate. At 440 nm wavelength scattering occurs such that the greater the concentration of chloride in the initial solution, the greater the extent of scattering.

The off-line titration procedure for analysis of sulfuric acid in the plating solution is rather straightforward. An aliquot of acidic plating solution is placed into a beaker and neutralized with a known amount of sodium hydroxide (the neutralizing base) of a given normality. The endpoint of the titration is reached when a color indicator added to the solution (e.g., methyl orange) changes the solution color from pink to green. The amount of the base to reach the endpoint is then used to calculate the amount of sulfuric acid in the solution. For the on-line method, a pH electrode is utilized to determine the endpoint as the voltage of the pH electrode quickly changes during the transition from acid to base in the solution.

The off-line measurements of the accelerator component were performed by either of two techniques: HPLC (High Performance Liquid Chromatography) or CVS (Cyclic Voltammetry Stripping). The HPLC procedure utilizes the adsorption of species onto the surface of resin beads packed into a long, thin column. The time constant for adsorption and desorption differs for different molecular species in the plating solution, especially the accelerator component, which in many cases is a sulfur compound containing aromatic rings. The time for the accelerator to finally reach the detector differs significantly from the times for other solution species to reach the detector. The detector can be of several types: electrochemical (measurement of current at a given potential), visible/UV adsorption spectrophotometer, electrical or thermal conductivity, or even measurement of the index of refraction. The detector signal amplitude is integrated over time to obtain the specie concentration [7].

Cyclic voltammetry stripping has been used extensively for measuring both the suppressor and accelerator species in different plating baths for many years [8,9]. This method is more useful than the preceding method primarily because it directly measures the electroactivity effects on plating. There are three electrodes utilized, vs. two electrodes most commonly used in an electroplating cell. The voltage is controlled by applying a current between the working and auxiliary electrodes in order to maintain the 'control' voltage between the working and reference electrodes. The working electrode is usually a rotating disk platinum electrode that is plated with copper and stripped of copper in a repeated (or cyclic) manner. The control voltage is varied (or scanned) linearly as a function of time.

A cyclic voltammogram is a plot of the current between the working and auxiliary electrodes as the voltage is scanned (see Figure 1). A negative current indicates copper plating on the working electrode and a positive current indicates copper dissolution (or stripping) of copper from the working electrode. Initially, as indicated between points 1 and 2, plating occurs as the voltage is scanned to negative values. As the voltage is scanned positively, stripping of copper occurs (between points 3 and 4). Finally, as the voltage becomes very positive, oxidation of organics occurs (point 5). At even more positive potentials, point 6, chloride ion becomes oxidized to chlorine. Finally, at point 7, another scanning cycle is started. The working electrode is now clean and smooth. Repeated cycles are done to improve precision and accuracy. The area under the stripping peak is a measure of the amount of Cu deposited for a given scan, for a particular solution with given concentrations of organic species.

Using the aforementioned CVS method, Figure 2 shows an example of how the concentration of one type of accelerator is measured [10]. The peak stripping area determined from CVS plots is given on the left axis; the lower axis shows the concentration of accelerator. First the stripping area is determined from a fresh plating solution without any accelerator. The solution has a very high concentration of suppressor specie in it to eliminate suppressor effects during the different additions of accelerator. Then, a specific volume of plating solution is added. For calibration purposes, two additions (#1 and #2) are completed from a solution of known accelerator concentration. A straight line is drawn through all the points. The concentration of accelerator in the production solution is then determined by using the stripping area of the production solution. Where this area value intersects the line is also where the concentration of accelerator intersects the line. In this example case, an area value of 1.04 mC (milli-Coulombs) corresponds to 4 ml/L of accelerator in the production solution.

The other on-line measurement techniques involve titrations similar to the previously discussed acid titration done off-line. The on-line copper titration technique utilizes an organic molecule called EDTA (ethylene diamine tetra-acetic acid). In step 1, the EDTA grabs (or chelates) the Cu^{2+} ions out of solution and releases 2 hydrogen ions for each Cu^{2+} chelated. Next, as was done for the off-line acid analysis, NaOH is added to neutralize the acid (or hydrogen ions). The amount of NaOH needed to reach the endpoint is directly proportional to the initial concentration of Cu^{2+} in the solution.

The titration for chloride ion is somewhat similar to the copper titration, but in this case $AgNO_3$ solution is added, precipitating out AgCl. As the amount of Cl is removed from the solution, the measured voltage between two electrodes increases until the curve of potential vs. addition volume of $AgNO_3$ flattens out. At this point, the volume of $AgNO_3$ solution added is

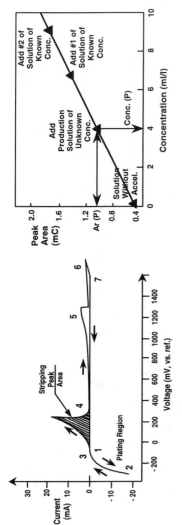

Fig. 1 – Example CVS curve showing sequence of plating and stripping steps on an electrode.

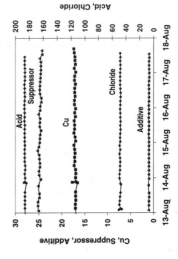

Fig. 2 – Illustrated CVS method for determination of the accelerator concentration in a production plating bath.

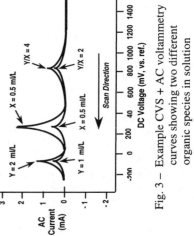

Fig. 3 – Example CVS + AC voltammetry curves showing two different organic species in solution

Fig. 4 – On-line measurements of standard solution obtained using CVS + titration methods. Data was gathered over an approx. five-day period.

used to calculate the amount of chloride ion in the solution.

The CVS + AC voltammetry technique utilizes a dc CVS scan upon which is superimposed an ac voltage which can be from 1 to 100,000 Hz. A phase shift is observed by special electronic instrumentation between the ac applied voltage and the measured ac current signals. The phase shift, which is related to the time it takes for a given surface active specie to cover the electrode, is then used to determine the concentration of a given specie. The concentration of a specie is proportional to the magnitude of the current at a given phase shift.

An example of a CVS + AC voltammetry curve is shown in Figure 3. At a given phase shift θ (for example, 70 degrees) the ac current magnitude is plotted vs. the dc voltage. In this case the dc voltage is scanned from positive to negative values. The peak at -70 mV is empirically correlated with the specie Y. For the lower curve (depicting a scan of a solution of Y = 1 ml/L) the current is ½ of the upper scan curve (where Y = 2 ml/L). For the peak related to specie X (at +300 mV), each solution has X = 0.5 ml/L. However, there is a big change in the respective peak currents, an ~3X magnitude change between the upper and lower curve. Finally, the third peak (at +850 mV) can be used to calculate the concentration of specie X from the ratio Y/X, which changes from Y/X = 4 to Y/X = 2. This example curve shows the intricate empirical data that is used to correlate the scan data with the concentrations.

In order to ensure reproducible plating results, both high precision and high accuracy are necessary for measurement of the chemistry involved in the electroplating process. For example, for a given process parameter and its corresponding control range, it becomes important to establish if a particular metrology method is capable of providing the required measurement precision. One way to assess this is to examine the so-called P/T ratio, which is the relationship between the metrology precision (P) and the process tolerance (T). The P/T ratio is defined as the 6σ of measurement precision divided by the process range (where the range is equal to the upper spec limit minus the lower spec limit). For purposes of this work, a P/T ratio of <0.3 was established as a requirement. Put simply, this means that a 1 sigma standard variation in measurement precision (repeatability) must be on the order of 1/20th that of the total process control range.

RESULTS

On-line CVS + titration measurements are shown in Figure 4 for a standard solution with freshly added chemicals. About 40 measurements were taken for each of the chemical components over a 4½ day period. Concentrations are expressed in arbitrary units. Note that the data indicate very good measurement precision and stability for all species. A very slight degradation is noted for the suppressor and additive species, which can be expected over the course of several days.

The measurement precision and accuracy for the on-line CVS + titration methods represented in Figure 4 is compared in Table 1 to the off-line methods. For the precision data, the best improvements for the on-line method are seen for copper, chloride, and the organic component labeled "Specie B." For the accuracy data, the main improvement for the on-line method is noted for chloride, reduced from 11% to less than 1%. Two components, copper and "Specie B," show slightly less accuracy than the off-line methods.

Figure 5 shows the on-line CVS + AC voltammetry results for a standard solution with freshly added chemicals. About 90 measurements were taken over a 24 hour period. Note that the data also indicates quite good precision for all species. A very slight degradation of the accelerator component is noticed, which is expected.

Figure 6 shows an example automated dosing activity for a given chemical specie along with the on-line CVS analysis of that same specie. The triangular points display the amounts of the additive dosed into a Sabre copper electroplating tool, with their values on the left axis. The diamond points are the measured values of the additive concentration in the bath, with their values displayed on the right axis. This three-day example data set shows that the on-line analysis system does indeed respond quickly and accurately to the 'corrective dosing' events above a background of 'predictive dosing' events.

Table 1: Measurement precision and accuracy data for the on-line CVS + titration methods as compared to off-line methods.

	Acid	Copper	Chloride	Specie A	Specie B
Precision					
Off-Line	0.9%	4.5%	8.9%	2.8%	7.7%
On-Line	1.1%	1.1%	1.4%	2.2%	0.9%
Accuracy					
Off-Line	3.0%	0.8%	11%	8.5%	2.3%
On-Line	1.1%	2.3%	0.6%	3.4%	3.2%

The two sets of micrographs shown in Figure 7 reveal that it is very important to maintain the proper concentrations of chemical species to enable good fill of aggressive features. The FIB-SEM cross-sections are of 0.24 μm wide, 5:1 AR vias that have been plated using different concentrations of accelerator (A) and suppressor (S). Note that there are obvious voids seen in the two micrographs on the left side of the figure. These two sets of vias were plated with inadequate concentrations of accelerator and suppressor. However, the vias on the right side of the figure were plated with adequate specie concentrations. On-line CVS analysis was used to correlate these results with chemical concentrations. Although less challenging than vias, even moderate to high aspect ratio trenches cannot be filled void-free when inadequate specie concentrations are used.

SUMMARY AND CONCLUSIONS

Electrochemical procedures exist for use in automated, on-line analysis of the inorganic and organic components of IC copper plating baths. The benefits of on-line chemical analysis include a tighter measurement precision, the convenience of automated sampling, an increased frequency of analysis, and reduced safety concerns. Additionally, the same on-line analysis system may be used for measurement of multiple plating tools, thus providing a common metrology basis and reduced analytical costs for a large copper plating operation over that associated with embedding such capability on a per tool basis.

ACKNOWLEDGMENTS

We wish to thank T. Walsh, J. Henri, T. Andruschenko, J. Alexy, M. Pereyra, R. Lomeli, and J. Pistilli (of Novellus); G. Chalyt, M. Pavlov, P. Bratin, M. Schneider, and M. Rabinovitch (of ECI Technology); and H. Wikiel, K. Wikiel, and D. Weisberg (of Technic) for their contributions to this work.

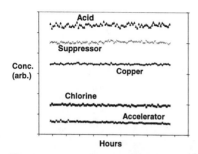

Fig. 5 – On-line measurements of standard solution obtained using CVS + AC voltammetry method. Data was gathered over a 24-hour period.

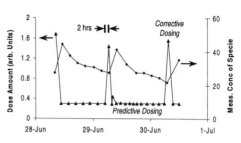

Fig. 6 – An example automated Sabre dosing activity for a given organic specie along with the on-line CVS analysis of that same specie.

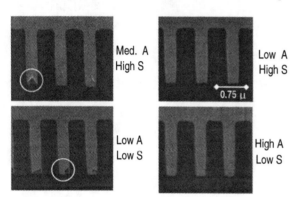

Fig. 7 – Comparative SEM-FIB micrographs showing the importance of adequate specie concentration for damascene feature fill. Via structures possess a 0.24 µm wide, 5:1 aspect ratio. ('A' denotes accelerator, 'S' denotes suppressor.

REFERENCES

1. J. Kelly and A. West, J. Electrochem. Soc. **145**, 3472 & 3477 (1998).

2. J. Dukovic and C. W. Tobias, J. Electrochem. Soc. **137**, p. 3748 (1990).

3. C. Madore, M. Matlosz, and D. Landolt, J. Electrochem. Soc. **143**, 3927 (1996).

4. C. Madore and D. Landolt, J. Electrochem. Soc. **143**, 3986 (1996).

5. A. West, C.Cheng, and B. Baker, J. Electrochem. Soc. **145**, 3070 (1998).

6. The QLC-5000 is the licensed trademark for the on-line chemical monitoring system of ECI, Inc., E. Rutherford, NJ.

7. B. Karger, L. Snyder, C. Horvath, "An Introduction to Separation Science," John Wiley and Sons (1973).

8. P. Bratin, in Proceedings of AES Analytical Methods Symposium, Chicago, Ill., March 1985.

9. D. Tench, C. Ogden, J. Electrochem. Soc., **125**, 194 (1978).

10. W. Frietag and M. Manning, in Proceedings of the 70th AES Technical Conference, Sur/Fin, Indianapolis, IN, June 1983.

CHARACTERIZATION OF ELECTROPLATED AND MOCVD COPPER FOR TRENCH FILL IN DAMASCENE ARCHITECTURE

K. WEISS*[1], S. RIEDEL*, S.E. SCHULZ*, H. HELNEDER**, M. SCHWERD**, T. GESSNER*

*Chemnitz University of Technology, Center of Microtechnologies, D-09107 Chemnitz, Germany, **Infineon Technologies, D-81730 München, Germany

[1]corresponding author: phone: +49 371 531 3673   fax: +49 371 531 3131
e-mail: kristin.weiss@e-technik.tu-chemnitz.de

ABSTRACT

Single damascene trench fill was performed by copper electrical chemical deposition (ECD) on Cu MOCVD and PVD seed layers, respectively. The two types of seed layers were compared with respect to step coverage, electrical resistivity, texture, adhesion behavior and film stress. Different properties induce different properties of ECD copper including grain size, electrical resistivity and adhesion behavior. Chemical mechanical polishing and electrical measurements were feasible on PVD/ECD copper. Alternatively, the MOCVD process was used for trench fill on MOCVD TiN. A comparison of the three fill methods revealed a good fill capability for every case. Electrical data show a good line resistance control and excellent leakage current between unpassivated Cu lines and a tight distribution of the measured values for ECD on PVD seed layers and MOCVD copper fill.

INTRODUCTION

The damascene architecture is mainly used for copper interconnects, which means copper deposition into oxide trenches and subsequent chemical mechanical polishing. Several fill techniques including PVD, CVD and ECD (electroplating) were under investigation, of which ECD is mostly favored due to an exceptional pattern filling capability and the excellent electrical properties compared to conventional Al(Cu) metallizations [1]. The CVD as fill process could become more important at smaller dimensions in next IC generations [2]. PVD Cu is only used to produce seed layers for other fill techniques within the damascene architecture because of its limited fill performance. This paper compares copper fill provided by electroplating on different seed layer types and copper fill by MOCVD, respectively.

EXPERIMENTAL

150 mm substrates were used with different barrier materials being 50 nm PVD Ta as well as 20 nm CVD TiN (nominal thicknesses). The barrier was deposited into oxide trenches with the aspect ratios 1.5 and 3. The Ta deposition was enhanced by IMP and the TiN deposition was performed by MOCVD using 8 cycles of alternating TDMAT pyrolysis (5 s) and N_2/H_2-plasma-treatment (30 s) both at 350°C wafer temperature.
Copper seed layers, which are needed for eletroplating, were deposited by MOCVD and long throw sputtering (LTS). The MOCVD process on an Applied Materials Precision 5000 cluster tool bases on the precursor mixture of Cu(hfac)TMVS with the additives H(hfac) · 2 H_2O (0.4 wt %) and TMVS (5 wt %). LTS Copper was deposited in an Applied Materials Endura PVD tool at 50°C in-situ with the Ta barrier to achieve a smooth and conformal seed layer.

125

Copper fill by electroplating was done with OMI Cubath SC chemistry and forward pulsed conditions using a modified Semitool Equinox Radial equipment. The wafers were annealed afterwards at 150°C for 30 min in forming gas (5 % H_2 in N_2) to avoid self annealing effects and to provide stable layer conditions for CMP processing. Alternatively, the copper fill was performed by MOCVD in-situ with the MOCVD of TiN.

Finally, copper and barrier CMP was performed on a IPEC Avanti 472 Single Head Polisher using a two step polishing process and a touch up at the final platen. The first polishing step is the Cu removal with a high selectivity to Ta. The second polishing step removes the Ta with a higher selectivity to Cu and less selectivity to the oxide to achieve a good degree of planarisation.

The copper is characterized by SEM, FIB, TEM as well as by AFM and AES at selected samples. X cut tape tests with defined adhesion strength (10 N/ 25 mm) were performed to assess the adhesion.

RESULTS

Different Cu seed layers for ECD fill

MOCVD seed layer. The copper MOCVD seed layer was fabricated according to a 3-step method. At first an Ar plasma (100 W, 120 s) was applied to ensure high nuclei density and very thin coherent seed layers, respectively [3]. This pre-treatment is followed by the copper MOCVD. The adhesion of the MOCVD copper is very poor as-deposited. A complete seed layer delamination occurred due to the x-cut tape test. Investigations concerning the adhesion behavior were published within former studies [3, 4]. The approach for adhesion improvement was a subsequent thermal treatment being the third step. It was successfully applied for 50 nm Cu seed layers. This effect was noticed on IMP PVD Ta as well as on PVD TiN as barrier material. Torres [5] and Motte [6], who used in-situ MOCVD TiN as base layer, also observed an improved adhesion of MOCVD copper due to post-annealing. However, annealing at temperatures being to high over a long time can also cause dewetting of copper. As a result holes are formed beside Cu grains of increased diameter. Nevertheless the annealing temperature and duration should ensure a sufficient adhesion. Regarding these findings an optimized method must be found. In [4] was noticed that electroplated copper on a 50 nm MOCVD seed layer, which excellently adhered after MOCVD post-annealing, is completely removable by tape test. CMP processing was not possible on this metal stack. Therefore different seed layer thicknesses (75 nm and 100 nm) were used in this study. They were annealed at 310°C for 5 min. After annealing these CVD copper films adhered excellently to the barrier layer. No copper was removable by the x-cut tape test.

The cross section of a MOCVD Cu seed layer on Ta in a trench (see figure 1) shows a good step coverage of 75 %. An electrical resistivity of 3 - 5 $\mu\Omega$cm was calculated from film thickness and sheet resistance of the seed layers, which may be high due to the surface roughness (5.5 % of film thickness) and the low copper thickness. XRD measurements revealed that the MOCVD copper is not textured. A random orientation of the Cu grains was noticed. Peak displacements due to annealing indicate stress change. According to wafer bow measurements the tensile stress of the 75 nm seed layer decreased from 360 MPa to 99 MPa after annealing. A lower decrease was observed for the 100 nm seed layer, for which 342 MPa before and 211 MPa after annealing was measured. This behavior is assumed to contribute to the adhesion improvement during annealing.

PVD seed layer: The most important differences of the produced PVD Cu seed layer are the excellent adhesion to the barrier and the lower step coverage of about 30 %. That is why a 100 nm thick seed layer is needed at least. It was deposited into high aspect ratio trenches (AR = 4) depicted in figure 2. Further differences in seed layer properties were observed: The specific electrical resistivity was determined to be about 3 µΩcm, which is remarkably lower than for the MOCVD copper seed layer caused by the different film thickness and surface roughness. The surface roughness assessed by AFM was lower compared to MOCVD Cu. The mean roughness value was determined with 2.6 % of the film thickness indicating a very smooth surface. Furthermore, the PVD Cu film shows a strong (111) texture as-deposited. No peaks beside the (111) peak were detected. The adhesion behavior of the PVD copper seed layer is excellent. It passes the tape test at the as-deposited state.

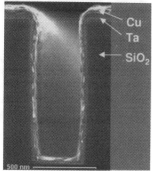

Figure 1: MOCVD seed layer
annealed; aspect ratio: 3

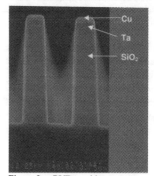

Figure 2: PVD seed layer
as-deposited; aspect ratio: 3

Copper ECD fill on different seed layers:

The fill capability of the electroplating process is excellent regardless the seed layer type. No voids could be found within single damascene trenches filled by ECD copper on PVD seeds as well as on CVD seeds. However, different seed layer properties induce different electroplating fill process attributes (figure 3 and 4). A slightly lower electrical resistivity of 2.0 µΩcm was measured for ECD Cu on a PVD seed layer compared to 2.2 µΩcm for the ECD Cu on MOCVD seed layer. The conformal fill on MOCVD seeds results in grain sizes from 50 to 300 nm. Essential bigger grains with a size of 200 to 600 nm grew on the PVD seed layer. The ECD copper grains on MOCVD seeds have a comparable size if bottom up fill is used.

The adhesion behavior of the Cu films after electroplating is different for the two types of seed layers. ECD copper on the PVD seed layer shows excellent adhesion, whereas the adhesion behavior of CVD seed layers depends on the film thickness and processing. A complete tape test failure was found for ECD on a 50 nm MOCVD seed layer as-deposited as well as after subsequent annealing [4]. For ECD copper on a 75 and 100 nm seed layer a well pass of tape test was not noticed at the as-deposited state and after the pre-CMP anneal of ECD. However, partial delamination of copper occurred due to subsequent CMP processing, which additionally caused scratches at the copper surface. Due to this matter no electrical data could be obtained of this metal stack. A further optimization of the MOCVD Cu deposition with respect to the adhesion is required.

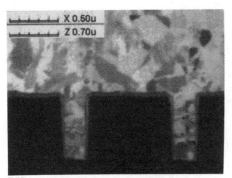

Figure 3: Copper ECD fill on MOCVD Cu seed layer (tilted FIB image)

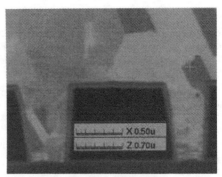

Figure 4: Copper ECD fill on PVD Cu seed layer (tilted FIB image)

Results of electrical measurements for the metal stack Ta barrier/ PVD Cu seed/ ECD Cu on non-passivated comb & serpentine test structures with 0.35 µm line width and 0.35 µm spacings are shown in figures 5 and 6. The goal value of 3.4 kΩ for the line resistance, which was calculated from the designed copper cross section and a copper resistivity of 2.0 µΩcm was well achieved. That indicates no degradation of the Cu resistivity in the trenches during processing. The leakage current between the non-passivated Cu meander structures is very low being comparable to a standard passivated AlCu metallization.

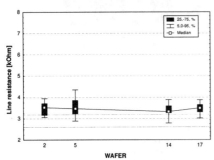

Figure 5: Line resistance of ECD Cu on a PVD seed layer (comb & serpentine structure with 0.35 µm line width)

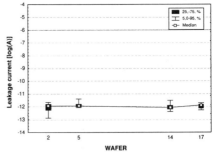

Figure 6: Leakage current of ECD Cu on a PVD seed layer (comb & serpentine structure with 0.35 µm line width and 0.35 µm spacing)

Fill by Cu MOCVD in-situ on MOCVD TiN

This method was applied alternatively to the electroplating fill process. The copper MOCVD as fill process was performed without plasma pre-treatment because it was applied in-situ with the MOCVD of TiN. An annealing step (400°C, 10 min) subsequently follows the Cu MOCVD to ensure a sufficient adhesion behavior, which is not given as-deposited. Complete fill of small trenches with a width of 0.35 µm was observed indicating the good fill capability of this method (see figure 7). From film thickness and sheet resistance a resistivity of 2.1 µΩcm was determined after annealing being comparable to that of electroplated copper.

Microstructure of copper fill before and after annealing was assessed by TEM (see figures 8 and 9). Due to the thermal treatment the grain size of copper has approximately doubled to 100 to 140 nm. The grains grew through the interface at the middle of the trench.

This metal stack passed the chemical mechanical polishing process without any damage due to the excellent adhesion of this copper. Electrical measurements were performed on non-passivated samples (see figures 10 and 11). The line resistance is comparable to the ECD/ PVD Cu. The slight pass of the goal value can be attributed to overpolishing. The leakage current as well as the variation of the measured values is excellent.

Figure 7: Copper MOCVD fill (tilt angle 45°; FIB)

Figure 8: TEM of MOCVD Cu fill on MOCVD TiN, as-deposited

Figure 9: TEM of MOCVD Cu fill on MOCVD TiN, annealed at 400°C, 10 min

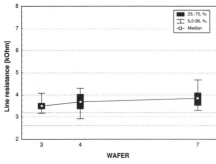
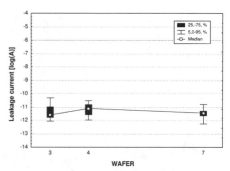

Figure 10: Line resistance of MOCVD Cu fill (comb & serpentine structure with 0.35 μm line width)

Figure 11: Leakage current of MOCVD Cu fill (comb & serpentine structure with 0.35 μm line width and 0.35 μm spacing)

CONCLUSIONS

The most important advantage of the MOCVD seed layer is the high step coverage but the PVD seed layer has a lower electrical resistivity. Both types of seed layers passed the adhesion tape test, but the MOCVD Cu had to be annealed to achieve good adhesion.

The ECD copper properties are different for the two types of seed layers. Due to larger grains the PVD/ECD copper has a slightly lower electrical resistivity of 2.0 μΩcm. The poor adhesion of MOCVD/ECD copper could be improved by using seed layer thicknesses up to 100 nm. Unfortunately, the adhesion of these films is insufficient with respect to CMP processing. The PVD/ECD copper films passed the tape test after electroplating and CMP processing without any damage. The copper MOCVD as fill process showed an excellent fill capability and adhesion behavior. An excellent step coverage was found in 0.35 μm trenches. The electrical resistivity is comparable to PVD/ECD copper.

In conclusion, the fill capability was found to be good for every fill method. ECD on PVD seed layers and MOCVD copper fill processes are well controlled. Electrical data show a good line resistance control and excellent leakage current between non-passivated Cu lines and a tight distribution of the measured values.

ACKNOWLEDGEMENTS

The authors are indebted to S. Collard (TU Chemnitz, Department of Physics) for XRD measurements and to the Semiconductor Failure Analyses Group for physical analyses. This work was financially supported by European Commission EP n° 29858. The authors are responsible for content.

REFERENCES

1. D. Edelstein, J. Heidenreich, R. Goldblatt, W. Cote, C. Uzoh, N. Lustig, P. Roper, T. McDevitt, W. Motsiff, A. Simon, J. Dukovic, R. Wachnik, H. Rathore, R. Schulz, L.Su, S. Luce, and J. Slattery, Technical Digest IEEE Int. Electron Device Meeting, (1997) 773-776.
2. Jackson R.L., Broadbent E., Cacouris T., Harrus A., Biberger M., Patton E., Walsh T., Solid State Technology, 41/ 3 (1998) 49-59.

3.   K. Weiss, S. Riedel, S.E. Schulz, T. Gessner, Mater. Res. Soc. Conf. Proc. ULSI XIV, Warrendale, PA 1999, 171-175.
4.   K. Weiss, S. Riedel, S.E. Schulz, M. Schwerd, H. Helneder, H. Wendt, T. Gessner, Talk presented at the „European Workshop for Advanced Metallization", March 8-10, 1999, Oostende, Belgium; to be published in Microelectronic Engineering.
5.   Torres J., Morand Y., Demolliens, Palleau J., Motte P., Pantel R., Juhel M., Mater. Res. Soc. Conf. Proc. ULSI XIV, Warrendale, PA 1999, 683-689
6.   Motte, P; Proust, M.; Torres J.; Gobil Y.; Morand, J.; Palleau, J.; Pantel, R.; Juhel, M., Talk presented at the „European Workshop for Advanced Metallization", March 8-10, 1999, Oostende, Belgium; to be published in Microelectronic Engineering.

Copper Electroplating: Processing and Integration

C.S. Hsiung, K. Hsieh, W.Y. Hsieh, and W. Lur

Advanced Technology Development Dept., UNITED MICROELECTRONICS CORP.
No. 3, Li-Hsin Rd. 2, Science-Based Industrial Park, Hsinchu, Taiwan, R.O.C.
Fax: 886-3-5644379, e-mail: eric_hsiung@umc.com.tw

ABSTRACT

The effects of electroplating parameters and post plating annealing on film properties and gap filling capabilities of electroplated Cu were investigated in this paper. Electroplating induction delay, electroplating current density, and electroplating temperature are of great significance in obtaining complete gap filling. Voids were minimized by using lower induction delay, higher current density, and lower electroplating temperature. For Cu film properties, current density, electroplating temperature, and annealing temperature are critical. Higher electroplating current density resulted in lower resistivity and less variation in terms of electroplated film thickness. In addition, higher electroplating and annealing temperatures leaded to more (111) oriented texture.

INTRODUCTION

Copper interconnect technology has become the main stream of back end of line (BEOL) for manufacturing of high speed performance VLSI circuits, especially for 0.18 um and beyond [1]. Damascene process was adopted to overcome the difficulty of conventional dry etching techniques on copper [2]. The challenge of this integration scheme requires Cu to be deposited in small, high aspect ratio trench and via structures without voids. Compared with chemical vapor deposition (CVD) or physical vapor deposition (PVD), electroplating is especially appealing because of low cost, high throughput, and excellent gap filling capability [3]. Copper is electrochemically deposited (ECD) by immersing a conductive Cu seed in an acid copper bath.

The performance and reliability of Cu wiring performance are significantly dependent on gap filling and film properties. Cu wiring provides faster speed due to lower resistivity as compared to currently used Al [4]. Cu microstructure is an important factor affecting the reliability of copper interconnects. For Cu films, (111), (200), and random texture components have been reported [5]. Of which, (111) orientation is preferred for better electromigration performance [6,7].

In this paper, the influence of electroplating processing and integration on the Cu gap filling characteristics and Cu film properties were studied. The role of electroplating current induction delay, electroplating current density, electroplating temperature, and annealing temperature were taken into account.

EXPERIMENT

Blanket and patterned 200 mm wafers were used for this experiment. A barrier layer and a highly (111)-textured copper seed layer were deposited by PVD and then filled by ECD Cu film. The electroplating solution consists of sulfuric acid, copper sulfate, and hydrochloric acid, along with recommended levels of proprietary organic additives. High-temperature post annealing was carried out in a furnace with nitrogen ambient. Blanket wafers were used to characterize Cu film properties, whereas patterned wafers were for gap-filling performance examination.

Sheet resistance (Rs) was measured using a four point probe. The Cu film texture was examined by X-ray diffraction (XRD). The gap filling capability was determined by scanning electron microscope along with focus ion beam.

RESULTS

a) Effects of electroplating current induction delay

Electroplating current induction delay is the queue time for current initiation after the wafer immersed into plating bath, which can dramatically influence the ECD gap-filling

133

characteristics. Fig. 1 shows two cross-sectional micrographs after ECD. Large bottom voids were observed for plating with a 1.5-second induction delay. On the other hand, hot entry with no delaying shows complete filling. The former case was thought to be caused by the discontinuity of Cu seed layer. Thin Cu is to be dissolved into sulfuric acid as immersing in electroplating bath. In order to clarify this phenomenon, a test for Cu seed etching in acid electrolyte was conducted and illustrated in Fig. 2. Both as-deposited and room-temperature aged seed layers were examined. Results showed that Cu films were fast etched at the beginning only. It was thought that the faster etching of PVD Cu seed at the beginning is due to the surface oxidation of Cu films. The etching rate decreases as the oxidized copper is consumed. However, both as-deposited and aged Cu films behave similarly in terms of acid bath etching. Control of queue time to electroplating seems incompetent to prevent this initial quick etching in acid bath. Electroplating current induction delay should thus be minimized to avoid voids at the via bottom and sidewall, where the thickness of Cu seed is marginal,.

b) Effects of electroplating current density

The deposition behavior and film properties were found to be closely related to electroplating current density. Fig. 3 shows the effects of the current density on ECD Cu gap fill behavior. Higher current density benefits bottom-up filling. Pinch-off may occur if Cu film is extensively electroplated at low current density due to the inherent overhang from PVD Cu seed deposition. It was reported that suppressor consumption rate is higher at higher current density [8]. In small area like trench/via, suppressor is difficult to penetrate due to large molecular weight. At higher current density, suppressor would thus be depleted faster than supplied from the bulk solution. Accelerator thus promotes bottom-up filling in the trench and via, where is no longer inhibited by the suppressor. However, results indicate that electroplating chemistry does not control the early deposition behavior until a critical thickness is reached. Bottom-up filling is initiated when an almost uniform step coverage of Cu seed is obtained

Electroplating current density also influence ECD Cu films properties. Lower sheet resistivity of ECD Cu films were formed at higher current density plating, as shown in Fig. 4. It was attributed to less impurities incorated, more closely packed, and more uniform grain structures at higher current density [9]. Similar to the previous report, the film resistivity of ECD Cu is increased as film thickness decreased [3]. In addition, higher-current-density plating results in smaller resistivity variation in terms of film thickness.

Texture of ECD Cu films was found to be affected by film thickness and plating current density. A normalized intensity of Cu x-ray diffraction, $I_{(111)}/I_{(200)}$, is shown in Fig. 5. It was found that lower electroplating current density leads to higher $I_{(111)}$ texture, especially for thinner film. These results indicate that the texture of ECD Cu films follows the underlying seed layer well at the beginning and lower plating current density allows more textured alignment on Cu seed layer, i.e. higher (111) orientation. However, after a critical thickness of about 13000 Å, current density does not play a major role in determining electroplated Cu film texture.

c) Effects of plating temperature

It was reported that the deposition temperature strongly influences the PVD Cu gap filling capability and crystalline texture [7]. The effects of plating temperature on ECD Cu film texture and gap filling capability were studied and shown in Fig. 6. $I_{(111)}$ increases as plating temperature increases. Higher temperature is to enhance Cu migration and allows ECD Cu to align to the underlayer during electroplating. However, voids are formed at high temperature electroplating, probably due to non-optimized electrolyte conditions. Electroplating at lower temperatures is essential for repeatable gap filling.

d) Effects of annealing temperature

It was reported that annealing temperature can greatly influence ECD Cu film hardness, stress, surface morphology, and grain growth [10]. In this study, the effects of annealing temperature on gap filling, Cu film resistance, and texture were also investigated. The annealing temperature shows no effect on gap filling, as shown in Fig. 7, for temperatures lower than 400°C. Results also shows that the Rs reduction is indifferent from room temperature self-annealing to 400°C furnace annealing, In general, an approximate 20% to 25% reduction was observed and similar to the other reported [11]. On the other hand, the Cu film texture is significantly affected by annealing temperature. $I_{(111)}$ is enhanced as the annealing temperature is increased and saturated beyond 300 °C. High temperature furnace annealing at $\geq$300°C results in a Cu film with 50% more (111) texture than room temperature self-annealing.

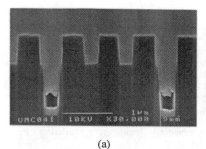

(a)

(b)

Fig. 1. Illustration of ECD Cu gap filling by means of current induction delay. (a): 1.5 second delay; (b) no delay time

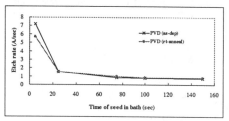

Fig. 2. Etch rate test for PVD Cu seed in acid copper plating bath.

Fig. 3. Effects of the current density on ECD Cu gap fill behavior.

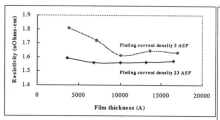

Fig. 4. ECD Cu resistivity at different plating current densities as a function of film thickness.

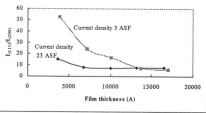

Fig. 5. Cu film texture at different plating current densities as a function of film thickness.

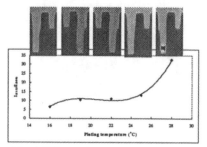

Fig. 6. Effects of the electroplating temperature on Cu gap filling and film texture.

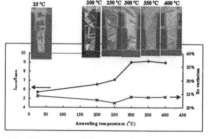

Fig. 7. Effects of the annealing temperature on sheet resistance variation, film texture, and gap filling of Cu films.

CONCLUSIONS

The gap filling capability and film properties of ECD Cu are highly dependent on electroplating processing and integration. Higher electroplating current density results in better bottom-up gap filling, lower sheet resistivity, and less thickness variation. PVD Cu surface is susceptible to acid copper plating bath, resulting in high initial etching rate. ECD process with lower current induction delay is to minimize the void formation as the seed thickness is marginal. High electroplating temperature increases (111) texture but suffers from bad gap filling and easy to form voids at the contact/via bottom. Annealing temperature showed no effect on gap filling and the reduction of sheet resistance. However, more $I_{(111)}$ texture was observed by increasing the annealing temperature, offering better electromigration resistance.

ACKNOWLEDGEMENTS

Authors would like to thank Jen-Hung Wang at Department of Materials Science and Engineering in NTHU for his help in XRD measurements.

REFERENCES

1. K. Rose, R. Mangaser, IEEE/SEMI Adv. Semicon. Manufact. Conf., 347 (1998)
2. P. Singer, Semicon. Intern. 79 (1997).
3. V.M. Dubin, C.H. Ting, R. Cheung, VMIC Conf., 69 (1997)
4. A.R. Sethuraman, J.-F. Wang, L.M. Cook, Semicon. Intern., 177 (1996).
5. D.P. Tracy, D.B. Knorr, J. Electron. Mat. 22(6), 611 (1993).
6. C. Ryu, et al., IEEE Transactions on Electron Develop. 46 (6), 1113 (1999).
7. K. Abe, Y. Harada, H. Onoda, 36th IEEE Ann. Internat. Reliab. Phys. Sym., 342 (1998).
8. P.C. Andricacos, et al., IBM J. Res. Develop. 42, 567 (1998).
9. C.H. Seah, S. Mridha, L.H. Chan, IITC, 157 (1998).
10. H.C. Chen, M.S. Yang, J.Y. Wu, and W. Lur, Proc. of 2nd IITC, 65 (1999).
11. T. Ritzdorf, L. Graham, IITC, 166 (1998).

Crystal Texture of Electroplated Damascene Cu Interconnects

A.H. Fischer*, A. von Glasow*, A. Huot** and R.A. Schwarzer**

*) Infineon Technologies, Reliability Methodology , D-81739 Munich, Germany

**) Physikalisches Institut der TU Clausthal, AG Textur, D-38678 Clausthal-Z., Germany

ABSTRACT

Crystal texture on a meso-scale is supposed to affect local material properties, in particular electromigration resistance and stress migration behaviour. Common X-ray θ-2θ scans or rocking curves, however, are inadequate to get an estimate of crystal texture, whereas Automated Crystal Orientation Measurement (ACOM) with the SEM has become a powerfull tool for the characterization of *microstructure and local texture* of thin metallization layers. The spatial arrangement of the grains is clearly visualized by constructing crystal orientation maps of grains or grain boundaries. From the complete set of orientation data, the quantitative Orientation Density Function (ODF), pole figures or the grain boundary character of neighbouring grains can be determinend.

ACOM measurements were successfully performed on *intact as well as damaged sections* of 4 and 8 μm wide damascene copper lines *after voids* had been formed in an *electromigration test*. A strong correlation between electromigration voiding and grain orientations or misorientations was not found so far, but a tendency of preferred void formation could be observed at neighbouring grains, oriented ⟨100⟩ respectively ⟨111⟩ to the specimen normal direction.

INTRODUCTION

There is clear evidence that reliability of Aluminum based interconnects depends on local crystal texture as well as on mechanical stress. In particular, the resistance of Al(Cu) to electromigration damage and growth of hillocks is affected by individual grains which deviate from the common ⟨111⟩ fiber texture [1]. These grain orientations cannot yet be avoided with certainty. To improve performance of ultra large-scale integrated devices, aluminum is presently about to be replaced by copper for metallization interconnects. Copper is a promising but not easy-to-use alternative since it differs from aluminum in important crystal-related mechanical properties. These are a pronounced anisotropy of Young's modulus in addition to a low stacking fault energy and easy dislocation glide that give rise to inhomogeneous deformation, increased twinning, and a non-uniform distribution of residual shear stress. Therefore meso-scale crystal texture is supposed to have a stronger effect on local material properties than in Al(Cu) metallizations.

The growth mechanism, the resulting microstructure and the texture of Cu thin films and interconnects are strongly dependent on the liner material, seed layer, deposition parameters, metal line design and post deposition treatment. The goal of reliability methodology is to investigate the influence of these process parameters on microstructure, and then to find the correlation between microstructure and reliability behaviour, such as electromigration and stressmigration (Figure 1).

Conference Proceedings ULSI XV © 2000 Materials Research Society

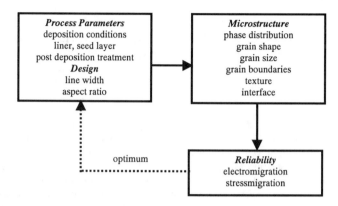

| **Process Parameters**
deposition conditions
liner, seed layer
post deposition treatment
Design
line width
aspect ratio | **Microstructure**
phase distribution
grain shape
grain size
grain boundaries
texture
interface |

optimum

| **Reliability**
electromigration
stressmigration |

Figure 1
Strategy of reliability methodology for process optimization.

LIMITATIONS OF TEXTURE ANALYSIS BY X-RAY DIFFRACTION

Conventional texture analysis by X-ray diffraction with cubic crystal symmetry requires the acquisition of two independent pole figures (or more if they are incomplete) to calculate the full ODF in three-dimensional orientation space. For lower crystal symmetry even more pole figures are needed [2]. With thin metallization layers on single-crystal substrates additional difficulties have to be faced. Since the specimen is tilted through large angles during pole-figure acquisition, the depth of information varies and may exceed the layer thickness for steep incidence of the primary beam. The intensity diffracted from the layer decreases and is then merged with peaks from the substrate. Although they may be used as an internal standard, e.g. for residual stress measurement, they are usually undesired as sources of (unnoticed) peak overlap. Furthermore, metallization layers on single crystals may develop an extremely sharp texture. If step width of pole figure measurement is too coarse, significant peaks may be overlooked or smeared out whereby a much weaker texture is perceived. The high symmetry of the single crystal substrate or a fiber texture must not be taken for granted for the metallization layer since meso-scale texture may be affected by the circuit geometry during the deposition process, annealing and operation. Needless to say that with structured metallization lines a triclinic specimen symmetry has to be considered *a priori* in the measurement, plotting and interpretation of pole-figures as well as in ODF calculation.

A ϑ-2ϑ scan for comparing peak ratios, or a rocking curve about a dominant peak are inadequate on principle, unless the presence of a simple fiber texture has been verified in advance. In ϑ-2ϑ scans only one single direction in the specimen-fixed coordinate system – which corresponds to one single dot on the pole figure spheres – is probed whilst all the other poles are ignored. Often a symmetric X-ray diffraction geometry is used whereby the angles of the incoming and the diffracted beams to the surface are kept equal ($\vartheta_1 = \vartheta_2$). In this case the probed direction corresponds to the surface normal of the specimen. It is, however, not known *a priori* that this direction contains some pole density or is significant. As a consequence the ratio of peak intensities selected from a ϑ-2ϑ scan may be a misleading indicator of texture. A rocking curve about a peak may fail as well, even with a sharp fiber texture, if the fiber axis is, for instance, slightly tilted out of rather than centered on the sample direction.

AUTOMATED CRYSTAL ORIENTATION MEASUREMENT WITH THE SEM

Automated Crystal Orientation Measurement (ACOM) with the SEM by interpreting backscatter Kikuchi patterns [3] has become a powerfull tool for the characterization of the microstructure and texture of interconnect lines. However, this technique is limited by charging effects of non-conducting areas of the specimen, deterioration of pattern quality due to plastic deformation or by thin surface layers (e.g. contaminations, oxide and ARC layers) and the necessity of a steep specimen tilt to the beam. The images are hence excessively foreshortened and affected by surface roughness. With digital beam scan a high speed (typically more than 20.000 orientations per hour) and a high spatial resolution of 0.1 µm or better are obtained on large specimen areas provided that the beam spot is focused dynamically and the pattern center is calibrated automatically during measurement.

Since the crystallographic orientations as well as the positions of each measured point are known from ACOM, the microstructure function $g(x)$ (2) can be constructed as a general description of orientation stereology in six-dimensional space $(x_1, x_2, x_3, \varphi_1, \Phi, \varphi_2)$. Here, $x_{1,2,3}$ are the three spatial coordinates, and $\varphi_1, \Phi, \varphi_2$ the Euler angles describing the grain orientation g by rotating the specimen coordinate system into the crystal system. From this function, several less complex functions are derived, e.g. the Orientation Distribution Function (ODF), texture fields $f(x, g)$, the misorientation distribution function $f(\Delta g)$, pole figures for any reflection (hkl), inverse pole figures and grain boundary character Σ. The spatial arrangement of the grains is visualized by constructing Crystal Orientation Maps (COM) of grain orientations or grain boundaries. In case of ULSI devices the statistical functions of ODF, $f(x,g)$, $f(\Delta g)$ and in particular the well-known pole figures are sensitive indicators of the processes of deposition, heat treatment etc. Circuit reliability, however, is limited by local failures in particular since interconnect line dimensions are in the range or below the average grain size. Hence the orientations of individual grains close to the failure site, their quantitative misorientations and grain boundaries as well as the visualization of microstructure by crystal orientation maps are of interest in failure analysis and reliability methodology.

EXPERIMENTAL

Copper metallization layers about 1 µm in thickness have been electroplated into 500 nm deep trenches, coated with 50 nm Ta liner and a PVD copper seed layer. After 10 min anneal at 120°C the continuous copper layer was removed by chemical-mechanical polishing (CMP). The interconnects had a final effective thickness of about 300 nm. The lines of the S-shaped test patterns were 7.9 mm long and 4 or 8 µm wide, respectively. They have been stressed on package level at current densities of 50 mA/µm^2 and oven temperatures between 250 and 300 °C, until 20% resistance increase. The test structures have been investigated *post mortem* by ACOM in the SEM. In order to obtain good diffraction patterns from the copper line, the SiN cap was carefully removed by plasma etching.

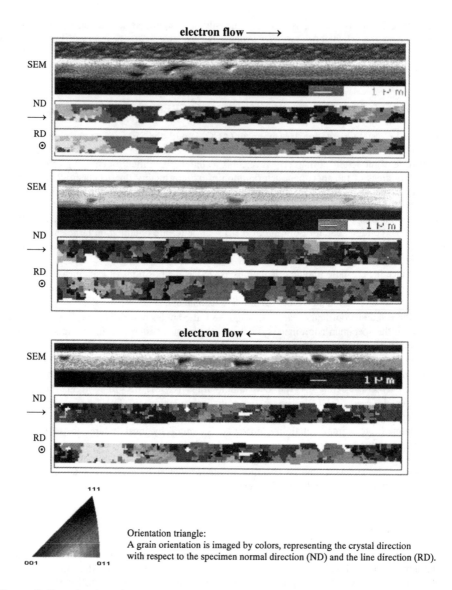

Figure 2 Crystal orientation maps of electromigration damaged Copper lines. The grain structure is clearly revealed with ACOM technique. A tendency of prefered void formation is to be seen at abutting grains, oriented with ⟨100⟩ and ⟨111⟩ parallel to normal direction (ND).

ACOM STUDY OF COPPER DAMASCENE INTERCONNECTS

Figure 2 gives some examples of crystal orientation maps with typical failure sites. The grain structure is clearly revealed. The individual orientations of the grains can be recognized easily by their specific colors (or grayscale). ND represents the specimen normal direction (hkl), and RD the reference direction [uvw] along the interconnect lines. For comparison, conventional SEM micrographs are also given. Prefered $\langle 111 \rangle$ and $\langle 100 \rangle$ directions parallel to ND can be found. But also grains with orientations "in between" are present. This more complex texture of copper differs markedly from that of aluminum metal lines, where a strong $\langle 111 \rangle$ fiber texture is dominant and only a very small number of deviating grain orientations are observed [4]. A clear correlation of void formation with meso-structure (shape, size, arrangement of grains, arrangement of grain boundaries), grain orientations or misorientations is not evident. However, there is a tendency of the voids to be located at sites with two abutting grains, one oriented with $\langle 111 \rangle$ and the other with $\langle 100 \rangle$ in normal direction. On the other hand there are also several sites in the lines with a similar microstructure, but without voiding. It is worth mentioning that the number of twins in these batches was lower than expected.

CONCLUSION

ACOM with it's unique features is a suitable method to visualize and characterize quantitatively the microtexture of electromigration damaged Cu damascene lines. A strong correlation between electromigration voiding and grain orientations or misorientations was not found as yet. However, a tendency of preferred void formation was observed at grains oriented with $\langle 100 \rangle$ and $\langle 111 \rangle$ parallel to ND, respectively.

REFERENCES

[1] Schwarzer, R.A., MRS Symp. Proc. **472**, 281-292 (1997)
[2] Bunge, H.J. (1982) *Texture Analysis in Materials Science.* Butterworths Publ., London, 2nd Ed. Cuvillier Verlag, Göttingen, 1993.
[3] Schwarzer, R.A., *Micron*, **28**, 249-265 (1997)
[4] Lepper, M., von Glasow, A., Piscevic, D., Schwarzer, R. A., *Materials Science Forum,* **273-275**, 573-577 (1997)

EFFECT OF PLATING CURRENT DENSITY on ECD CU GAP FILL AND FILM PROPERTIES

M.H.Tsai, W. J. Tsai, S. L. Shue, C. H. Yu, and M. S. Liang
Taiwan Semiconductor Manufacturing Company Ltd.No. 9, Creation Rd. I, Science-Based Industrial Park, Hsin-Chu, Taiwan, R. O. C.

ABSTRACT

The effects of plating current density on ECD Cu film properties and gap fill were investigated. The film deposited at lower current density was found to be conformal while higher current density films has a higher growth rate in the trench. Films deposited at higher current density were more resistive before annealing, presumably due to higher crystal defects which have higher self-annealing rate. After annealing, the films deposited at lower current density were more resistive owing to higher impurity levels. Stress and XRD results also indicated that low current plating has detrimental effect on film properties. Film deposited at higher current density is preferred because of its higher purity and higher planarity.

INTRODUCTION

ECD Cu has been an important process step in the copper metallization because of cost and technology considerations[1]. Previous works focus on the gap fill study of ECD[2,3]. The more noteworthy requirements include the deposition having a stable low electrical resistivity, the ability to complete fill sub-quarter micro features and consistent physical characteristics to perform similar through CMP process. The work study on the effect of current density on gap fill and their film properties with respect to resistivity, stress, textures, and impurities.

EXPERIMENT

After TaN barrier and Cu seed layer were sputtered, wafers were plated by using an acid copper sulfate bath with organic additives. ECD Cu films were deposited under different current densities in the range of 0.33 to 2.8 A/dm2 (ASD). Cross-sectional SEM was used for gap fill observation. Film impurities were analysis by SIMS. Four-point probe and stress meter were used to measure the sheet resistance and film stress, respectively.

RESULTS and CONCLUSIONS

The trenches filled with ECD Cu deposited at 2.8 A/dm2 were shown in Fig. 1a and Fig. 1d. The highly planarized top surface indicated that higher current has a higher deposition rate in the trench. The leveling capability decreased linearly from 2.8 ASD to 0.67 ASD. The trench topography was still remaining for the Cu deposited at lower current density (0.67 ASD) as shown in Fig. 1c and Fig.1f. Previous report indicated that the suppressor consumption rate was higher at a higher current density.[2] A lower suppressor concentration in the trench would be expected at high current density plating. Since lower suppressor density has a higher plating rate, the bottom up fill would be better for high current plating.

Impurities in the ECD Cu were analyzed by SIMS. Table 1 summarized the Cl, O, C, and Na content in the ECD Cu. The Cl and C were found to be increased with decreasing the plating current density owing to Cl and organic incorporate during low polarization deposition[3]. Fig. 2 shows the dependence of film stress on the plating current. The as-

Conference Proceedings ULSI XV © 2000 Materials Research Society

deposited film deposited under lower current density has a compressive stress and changed to tensile stress as the current density increased. The stress of film after a low temperature annealing increases to 900MPa owing to stress relaxation. The 0.33 and 0.67 ASD films after annealing have higher stress presumably due to higher impurity concentration and smaller grain size. The lower current plated film was more resistive after annealing owing to incorporate more impurity and higher residue stress.

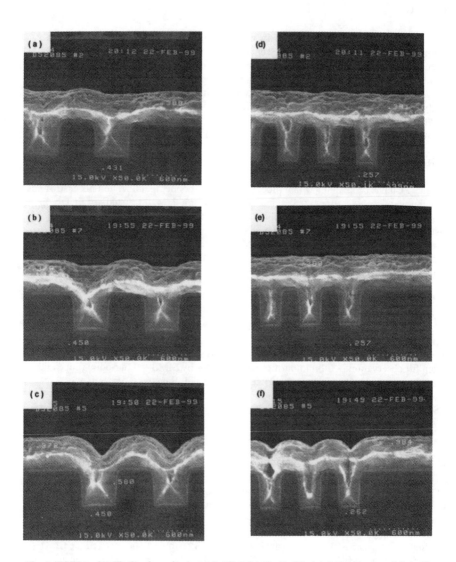

Fig. 1 XSEM of ECD Cu deposited at (a), (d) 2.8A/dm2, (b), (e) 1.5 A/dm2 and (c), (f) 0.67A/dm2.

Table. 1, Impurities of Cl, O, C, and Na in ECD Cu (SIMS)

	Cl	O	C	Na
0.33A/dm2	200000	80	8000	< 10
1A/dm2	50000	30	2000	< 10
1.5A/dm2	6000	< 10	300	120
2.8A/dm2	4000	< 10	400	90

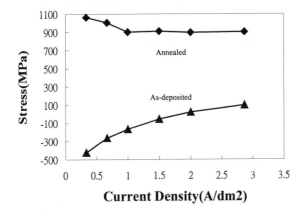

Fig. 2 Effect of plating current on as-deposited and annealed film stress.

Fig.3 shows the sheet resistance as a function of time after deposition.   The Cu film deposited at higher current density, although having less impurity, is more resistive and undergoes a quicker self-annealing.   It was considered that the crystal defects were reduced after annealing.   The XRD of ECD Cu deposited at 1A/dm2 was shown in Fig. 4.   The ECD Cu also follow the seed layer wit highly (111) texture.   Fig. 5 shows the FWHM (full width at half-maximum) of ECD Cu after annealing.   The lowest current density was found to be detrimental to the texture or grain size as compared their FWHM.

CONCLUSIONS

In summary, the effects of plating current density on ECD Cu film properties and gap fill were studied. Films deposited at higher current density were more resistive before annealing, presumably due to higher dislocation which have higher self-annealing rate. After annealing, the films deposited at lower current density was more resistive owing to higher impurity levels.   Stress and XRD results also indicated that low current plating has detrimental effect on film properties.   Film deposited at higher current density is preferred because of its less impurity and higher planarity.

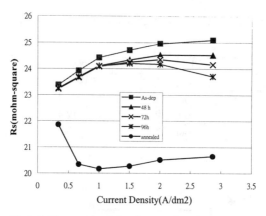

Fig.3 Self-annealing behavior as a function of plating current

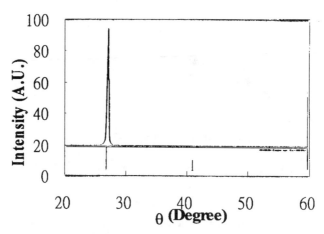

Fig. 4 Typical XRD of ECD Cu deposited at 1 A/dm2

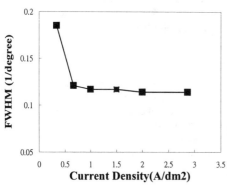

Fig. 5   Effect of plating current on XRD FWHM

REFERENCES

1. D. Edelstein, et al. Tech. Dig. Of international Electron Device Meeting, 1997, p.773.
2. L. Mirkova, St. Rashkov, and Chr. Nanev, Surface Technology, 15 (1982) p 181.
3. T. Ritzdorf, L. Graham, S. Jin, C. Mu, and D. Fraster, Proc. of the first IITS conference, June, 1998, pp, pp. 166-168.
4. C. Ryu, A. L. S. Loke, V. M. Dubin, R. A. Kavari, G. W. Ray, and S. S. Wong, 1998 Symp. on VLSI Technology, June 11, 1998, pp. 156-157.

ROLES OF ADDITIVES DURING FILLING PROCESS OF DAMASCENE STRUCTURES WITH ELECTROCHEMICAL DEPOSITED COPPER

E.RICHARD[*],   I.VERVOORT[*],   S.H.BRONGERSMA[*],   H.BENDER[*],   G.BEYER[*],
R.PALMANS[*], S.LAGRANGE[*,***] and K.MAEX[*,**]
[*]IMEC, Kapeldreef 75, B-3001 Leuven Belgium
[**]Also at E.E. Dept, K.U, Leuven Belgium
[***]On assignment at IMEC for SEMITOOL, Inc. 655 West Reserve Drive, Kalispell, MT 59901

ABSTRACT

In order to be able to get a void-free deposit inside damascene structures, organic contents have to be balanced and a high concentration of organic additive is requested. We have observed that electroplating inside narrow trenches as preferentially bottom-up, which is the most preferred fill mechanism. Unfortunately, when using a high concentration of brightener, an "overfilling" effect has been seen over the narrower trenches as a consequence of the "superfilling" mechanism. Accordingly, the increasing step-height after electroplating could be an issue for Cu-CMP processing. A process solution will be proposed to maintain a void free deposit and limit the increasing step-height. The goal of this paper is then to find out the overfill mechanism related to organic additives.

INTRODUCTION

A candidate for the copper metallization process is electrochemical deposition. Electro-Chemical copper Deposition (ECD) is interesting because it is relatively simple and very cost effective [1] at least compared to sputtering or chemical vapor deposition methods. The amount of copper deposited is dependent only on the current delivered [2]. In reality, the process is much more complicated, especially when we consider various geometries of features. The key aspects of the electroplating process are the different types of organic components in the electroplating bath, their role as inhibitor or catalyst agents and where they act to modify the way in which copper deposits in damascene structures

EXPERIMENTAL DETAILS

Experiments were performed on 6 and 8 inch wafers using a commercially available copper sulphate based plating bath and organic additives developed for IC filling. Wafers received a TaN barrier layer and a Cu seed layer that were plasma sputtered on dielectric patterned substrates. Two kinds of organic additives were used: one called "Additive" and an other one called "Suppressor". In this paper, Suppressor concentration was kept constant. The analysis of those additives in the plating bath was based on the Cyclic Voltammetric Stripping technique. The copper ECD layers were deposited using a Semitool Equinox® fountain plater. A high-resolution profilometer HRP220 from KLA was used to measure surface topography of the Cu ECD layer deposited overlying patterned dies. Structures with various widths of lines were chosen for this evaluation. Focussed ion beam (FIB) imaging and SEM pictures were performed using FEI200 and Philips XL30 respectively.

RESULTS AND DISCUSSIONS

When optimizing filling capability, it becomes apparent that the topography is strongly affected. Voids are observed in 0.35 μm via and 0.4 μm wide trenches at low concentration of Additive whereas it is not seen with high concentration (Fig.1). But while using a high concentration of Additive, an "overfill" effect has been seen over narrow isolated and dense pattern areas (Fig.2).

149

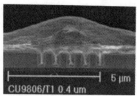

Fig.2: Overfilling over dense 0.4 μm wide trenches

Fig.1: Dual damascene structure in low k for two different concentrations of Additive (Left: low and right: high)

Therefore a question is addressed: where and how does the "overfill" phenomena originate? Answering those questions will provide a key to understanding the mechanism.

A comparison between narrow trenches (e.g below 1.5 μm wide lines) and the widest isolated trenches (10 μm wide) show that apparently there are 2 mechanisms for filling trenches depending on the dimensions of features. A "conformal" (Fig.3 up) and "bottom-up" (Fig.3 down) fill mechanism was observed for 10 μm wide trenches and narrower lines, respectively.

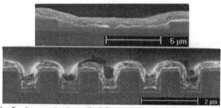 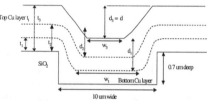

Fig.3: 1 um (up) and 0.2 um (down) top Cu layer deposited showing "conformal" and "bottom-up" fill mechanisms in 10 μm and 0.7 μm wide lines respectively. Both were deposited with a high concentration of Additive.

Fig.4: Schematic of a sequence of a "conformal" fill with Cu ECD deposited inside 10 μm wide trench (d_i and w_i are determined from HRP measurement)

Indeed, the Cu layer is thicker at the bottom of a narrow trench than over a field area. It is interesting to note the lack of build up of Cu at the upper corners where the electrical field is at a maximum. For the widest lines, the bottom and top thicknesses are similar and copper surface is d-translated, "d" corresponding roughly to the depth of the trench (see Fig.4). From the table I, it must be noted that when the thickness t_i (over field area) of copper decreases, the width w_i increases whereas d_i stays constant. Then, the width w_i is lower than the value expected with a conformal fill (Table I). This can be explained only if the deposition rate is conformal and the rate is higher at the bottom corners.

Thickness t_i of Cu layer deposited (μm)	Width w_i (μm)	Depth d_i (μm)
0.35	7	0.74
1.0	4.2	0.74

Table I: Thickness t_i of Cu layer deposited in 10 μm wide trenches and d_i and w_i parameters (3 ml/l).

A different ratio of organic inside trenches could explain the above observations. The same behavior is observed when copper is deposited with a low Additive concentration.

Depending on the pattern-width, the patterns are filled either from the bottom corners (narrow trenches) or the bottom (wide lines). Only in the first case, the "overfill" effect is observed. Additionally, the height of the "overfill" over isolated and dense trenches is strongly dependent on the Additive concentration and the pattern density. The dependency of the height of the "overfill" on the Additive concentration is shown in Fig.5 and Fig.6.

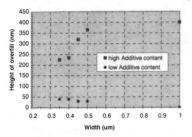
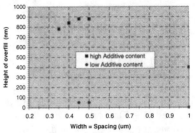

Fig.5: Height of the "overfill" at the top of 1 μm Cu layer overlying isolated lines

Fig.6: Height of the "overfill" at the top of 1 μm Cu layer overlying dense lines

No overfilling is observed for the lowest concentration (0.3 ml/l) of Additive. The "overfill" effect is always less pronounced over the isolated lines. Remarkably, a maximum exists for isolated lines between 0.5 and 0.7 μm wide (Fig.5) and for 0.5/0.5 μm dense lines (Fig.6). In order to understand the "overfill" height dependency with the Additive concentration, SEM cross section pictures were taken at intermediates stages in the plating process with 0.3 (Fig.7 left) and 3 ml/l (Fig.7 right) of Additive.

A "conformal" plating results when using a low concentration of Additive: the 0.3 μm wide trench is quickly closed at the top because of the overhang of the Cu seed layer (Fig.7 a-c). As it can be seen in Fig.7a, the copper layer seems to be thinner at the bottom of the trench due to the low step coverage of the seed layer but the thickness of the Cu ECD is homogeneous.

When a "superfilling" mode is observed with 3 ml/l of Additive; a higher deposition rate in the bottom corners (Fig.3d) than on sidewalls and at the top of trenches is noticed. "Superfilling" mode plating is clearly seen within 1 μm wide lines (Fig.8). When comparing Fig.7a and 7.d, note that the copper layer on the sidewalls is thicker with 3 ml/l. It can also be noticed that the "overfill" effect takes place already after 50 s of deposition (Fig.7f).

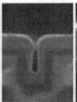
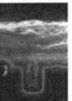

Fig.7: SEM cross-section images of ECD Cu showing progression of filling within 0.5 um wide isolated lines (23s, 38s, 104s and 23s, 38s, 50s deposition time respectively for 0.3 ml/l and 3 ml/l of Additive).

The plating non-uniformity in submicron trenches is usually related to effects of mass transport such as convection, diffusion of cupric ion and additives and potential field effects (solution resistance). Convection within pattern features is considered as negligible towards diffusion of chemical species [3]. Field effects driven by the profile geometry could arise causing strong variations of the flux of species along the profile. However at these dimensions, the effect of solution resistance leading to potential drop in the trench and a non-uniform plating rate is very small [1,4].

Therefore, the most important transport effect is diffusion through the plating solution. Differences in the local rate of metal deposition occur if it is controlled by the diffusion rate either of the copper ions or of organic agents.

Fig.8: FIB cross-section images of ECD Cu showing progression of filling (3 ml/l) within 1 μm wide isolated (23 and 50s deposition time leading to a top Cu layer of 0.25 and 0.35 μm) lines.

In literature [5,6], most mechanisms used to explain the "superfilling" effect assume a greater surface coverage of suppressing molecules in the surrounding field compared to coverage within the trench. So, all the mathematical models have been based on the assumptions that Suppressors are consumed on the electrode surface and are depleted within trenches. But comparing the 2 sets of SEM pictures (Fig.7) obtained with the same Suppressor concentrations and various Additive concentrations, it can not be concluded definitively that only the depletion of Suppressors inside the trench results in a "superfilling". Note that a deposition rate at the lower corners of the trench can proceed more quickly (as seen in Fig.8) only if the concentration of Additive is high enough, meaning that a critical one exists in order to "superfill" the trench. Therefore only a uniform distribution of Additive agents along the topography, combined with a depletion of Suppressors can result in decreasing the concentration polarization according to the Nernst equation [2] and increasing the deposition rate locally. From these considerations, it is more likely that Suppressors are also depleted far from the opening with 0.3 ml/l of Additive. Because of the depletion of both cupric and Suppressor species within the trench, the concentration polarization would be similar at the top and the bottom of the trench. Then, the deposition rate is the same at these locations leading to pinch-off and void formation (Fig.7 left).

The "overfill" effect may be a consequence of the "superfill" plating mode as it was seen only when the trenches are "superfilled". But it is not a straightforward result that, depending on the concentration of Additive, the initially concave recesses are turned into convex topography. Except if the grain size over trenches is completely different offering more nucleation sites for adsorption of organic contents. Some FIB cross section of trenches, partially or completely filled, were taken in order to figure out the "overfill' mechanism.

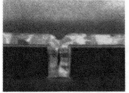 

Fig.9: FIB cross section showing filling of 0.3 μm wide isolated lines with 0.3 (left) and 3 ml/l (right) of Additive (38s of Cu deposition). Pictures taken 1 month after deposition

Comparing these pictures (Fig.9), it must be concluded that in the plane the grains are relatively small, independently of the concentration of Additive (Suppressor is kept constant in both cases). This indicates that the top layer is not self annealed in either case as all the conditions are identical (layer is very thin, high concentration of Suppressors). This is coherent with our previous paper [8] showing that the recrystallisation slows down as the thickness of the layer decreases and as the Additive concentration increases. Therefore, for the lower concentration, e.g 0.3 ml/l if pictures had been taken during plating, the surface grain size over and outside of the trench would have been identical. If the deposit time is longer (to get 1 μm layer plated), as was noticed previously (Fig.5), no overfilling phenomena is observed with these conditions (0.3 ml/l, see Fig.10) and small grains are still present over, inside and outside the trench. So, because the surface grain size is similar over and outside the trench, organic contents might be able to adsorb indifferently everywhere. For the highest Additive contents (3 ml/l), bigger surface grains were observed (see Fig.8 right) only inside the trench meaning that the layer is probably completely self annealed and the ratio Additive/Suppressor is higher inside the trench as compared to the plane.

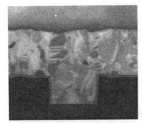

Fig.10: FIB cross section of 1 μm wide lines filled with Cu (0.3 ml/l)

This is consistent with our model (depletion of suppressor and "superfilling" process) and with our study on self anneal [7].

Indeed, the recrystallisation effect inside the trench is faster than in the plane only if the grains inside the trench during plating are smaller than in the surrounding field layer. So it must be concluded that during plating the smallest surface grains may exist inside and over the trench. The smallest surface grains could ease adsorption of additive leading locally to a higher concentration of additive and a higher deposition rate.

A sequence which consists in inserting an anneal treatment after filling the trench to get the same surface grain size and help desorption, is proposed to maintain a void-free deposit and limit the increasing step-height undesirable for CMP process.

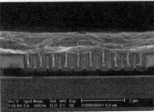

Fig.11: No overfill over dense 0.25 μm lines with a 3 step-approach.

Fig.11 displays the copper layer overlying dense trench structure. The 3-step approach was used as route. It clearly indicates that no "overfill" is observed anymore.

This is expected and consistent with our previous explanation if the surface grain size is identical everywhere after annealing. Some extra cross-section FIB images need to be taken after the filling step to valid our theory.

CONCLUSIONS

From the data presented it becomes clear that a high concentration of brightener is needed to get a void free deposit in damascene structure. But while using high organic content in the plating bath bottom corners fill occurs leading to formation of overfilling. A combination of depletion of Suppressor inside the trench and high Additive content may lead to formation of smaller grains over the trench offering more nucleation sites for adsorption of polymers additives. A process solution which consists in inserting an anneal treatment after filling of the trenches leads to a pattern-independent surface profile.

ACKNOWLEDGMENTS

J.Steenbergen is acknowledged for his SEM support.

REFERENCES

[1] J. Heidenreich et al, Proceeding IITC, June 1-3, 1998, p.151

[2] "Modern electroplating" book, Third edition, Wiley Interscience

[3] K.M. Takahashi, M.E.Gross "Analysis of transport phenomena in electroplated copper filling of submicron vias and trenches" AMC 1998, p.57

[4] A.C.West, Chin-Chang-Cheng and B.C.Baker, "Pulse Reverse Copper Electrodeposition in High Aspect Ratio Trenches and Vias" J.Electrochem.Soc. Vol. 145, No9, Sept. 1998

[5] M.E.Gross, K.M. Takahashi, C.Lingk, T. Ritzdorf, K.Gibbons, "the role of Additives in Electroplating of void-free Cu sub-micron Damascene Features" AMC 1998, p.51

[6] P.C.Andricacos, C.Uzoh, J.o.Dukovic, J.Horkans and H.deliglianni, IBM J.Res.Devel 42, 567 1998

[7] S.H.Brongersma, E.Richard, I.Vervoort, H.Bender, S.Lagrange, G.Beyer and K.Maex, "Two-step room temperature grain growth in electroplated Copper" J.Applied of Physics (to be published)

A NEW POST-ECD ANNEALING TECHNIQUE FOR Cu INTERCONNECTS USING A HIGH-PRESSURE GAS AMBIENT

Kohei Suzuki*, Takuya Masui*, Takao Fujikawa*, Yoji Taguchi** and Tomoyasu Kondo**

*Kobe Steel, Ltd., 1-5-5 Takatsuka-dai, Nishi-ku, Kobe City, Hyogo, 651-2271, JAPAN

Phone : +81-78-992-5614, Fax : +81-78-992-5650

E-mail : kh-suzuki@rd.kcrl.kobelco.co.jp

**ULVAC JAPAN, Ltd., Fuji-Susono factory, 1220-14

Suyama, Susono City, Shizuoka, 410-1231, JAPAN

Phone : +81-559-98-2105, Fax : +81-559-98-1756

ABSTRACT

A novel annealing technique for electrochemical deposition (ECD) copper using a high-pressure gas ambient has been developed. The technique aims to improve the integrity and reliability of copper interconnects by applying hydrostatic force during the post-deposition annealing. A series of experiments using various temperatures and pressures revealed that an electroplating copper film can be extruded into the via under conditions of 350°C and 120MPa. Improvement of adhesion was investigated by an agglomeration test and TEM observation. The high pressure annealing technique can be expected to improve a electromigration reliability, although the related phenomena need further investigation.

INTRODUCTION

Due to the continuous requirement for increasing logic speed, a reduction in RC delay has become a major concern of current multilevel interconnection technology. The use of copper, instead of aluminum, has been intensively studied and implemented as one of the solutions to the problems posed by the above requirement. At the same time, an electroplating technique has been introduced to LSI manufacturing as a copper deposition technique [1]. This electroplating technique demonstrates a superior filling capability over the conventional PVD technique, however, it is reported that the deposition characteristic of electroplating copper is strongly dependent on the seed layer [2]. This means that if the seed layer has insufficient coverage, the copper may grow yielding voids along the interface of the barrier layer. It is also reported that the as-deposited copper is rather unstable and grain growth occurs during several tens of hours after the deposition [3-4]. Thus, a post-deposition annealing is commonly used to stabilize the copper film. The annealing condition is a key issue because vacancies or microvoids that potentially exist in the copper film can easily agglomerate to form larger voids during the annealing. Since the above mentioned voids are supposed to cause a reliability problem, their formation has to be suppressed. To address these problems, we propose a new post-deposition annealing technique using a high pressure gas ambient to suppress the formation of voids.

EXPERIMENTS

The principle of high pressure annealing is an application of the Hot Isostatic Press (HIP) technique used in various industries [5]. The equipment used in this study is a batch-type vertical

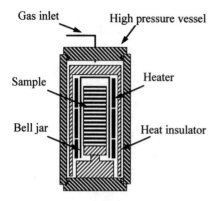
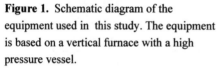

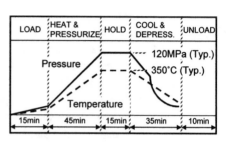

Figure 1. Schematic diagram of the equipment used in this study. The equipment is based on a vertical furnace with a high pressure vessel.

Figure 2. Typical Temperature - Pressure schedule chart. The temperature and the pressure were raised simultaneously in this study.

furnace combined with a high-pressure vessel as shown in figure 1. The maximum temperature and pressure available are 500°C and 200MPa, respectively. Argon is used for the ambient gas. Fig. 2 shows a typical temperature and pressure sc.hedule chart. After the samples are loaded into the chamber, the temperature and the pressure are raised simultaneously to the processing point, and then they are kept constant for 15 minutes (typically). After that, the temperature is lowered to a prescribed value, followed by decompression to atmospheric level. During the high pressure annealing, hydrostatic force is applied to the wafers by the gas pressure. The effect of the high pressure annealing on the wafer is to bring about a plastic deformation of the copper film. This eliminates pore-type defects and promotes adhesion between the copper film and the underlying layer.

To find the temperature and pressure range adequate for copper film, an extrusion test was performed. For this experiment, via patterns with a diameter of 0.3μm and a depth of 2.2μm were etched in CVD SiO_2. A 50nm TaN barrier and a copper seed layer were then deposited by the PVD method, followed by electroplating of 1.7μm-thick copper film. PVD copper was also prepared to fill the via for comparison. Since the via is made with very high aspect ratio, the bottom of the via is not completely filled with copper, as shown in figure 3 (referred to as *"as-deposited"*). This incomplete filling is utilized to observe the extrusion effect clearly. A series of high pressure annealings was performed at various temperatures and pressures, ranging from room temperature to 500°C and atmospheric pressure to 200MPa, respectively. After the high pressure annealing, cross sectional observation was made, using the FIB method. Fig. 3 summarizes the results of the extrusion test. In the case where only pressure was raised without heating, no extrusion was observed for the pressure up to 400MPa. On the other hand, severe

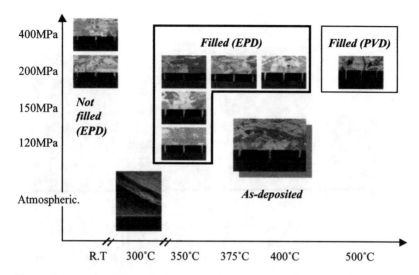

Figure 3. Extrusion test results for various temperatures and pressures. The combination of 350°C and 120MPa was found to be a base condition for electroplating copper. Note that higher temperatures and pressures are required for PVD copper.

peeling was observed in the case where the sample was heated at atmospheric pressure. In the case where the temperature was above 350°C, copper is well extruded into the via by pressure above 120MPa. It must be noted that the minimum temperature of 350°C found in this experiment is below the temperature limit of organic low-k dielectric material [6]. It must also be noted that PVD copper was completely extruded only when a temperature of 500°C and a pressure of 200MPa were applied. In the extrusion test, it was found that significant grain growth usually occurred during the high pressure annealing. Particularly, the copper in the via is almost always single crystal, with some twins.

An agglomeration test was employed to investigate the ability of high pressure annealing to improve adhesion, using the following method. A series of copper films with various thicknesses was deposited by PVD on a 50nm-thick TaN film. Some of the samples were then annealed at a temperature of 350°C and a pressure of 120MPa for 15min. All of the samples were then annealed in a vacuum at a temperature of 400°C for 15min. It is reported that thin copper film tends to agglomerate on the barrier layer [7]. If the copper thickness is lower than a certain value, which is dependent on the adhesion between copper and TaN, the copper film can easily agglomerate to form separate islands during the annealing. Fig. 4 shows the summary of the agglomeration test. The initial film surface is quite smooth and no agglomeration can be observed. For the case without high pressure annealing, severe agglomeration was observed for copper film thicknesses of less than 74nm and the film is no longer continuous, even for the thickness of 111nm. On the other hand, no severe agglomeration was observed for film

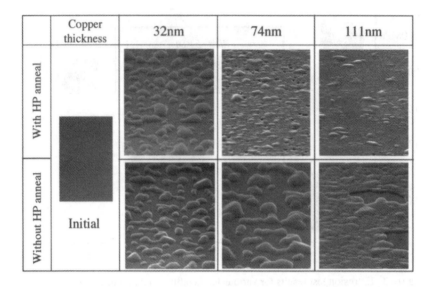

	Copper thickness	32nm	74nm	111nm
With HP anneal				
Without HP anneal	Initial			

Figure 4. The result of agglomeration test showing the improvement of adhesion by high pressure annealing.

thicknesses larger than 32nm even though some protrusion can be observed. This result shows that the adhesion between copper and TaN can be improved by applying high pressure annealing. The improvement of adhesion was also examined in the via structure. Figs. 5a and 5b are TEM

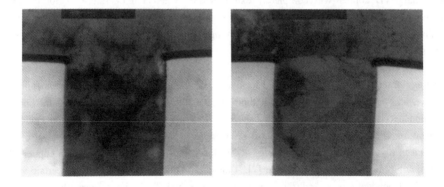

Figure 5. TEM photo showing the elimination of microvoid. Fig. 5a (left) is as-deposited, Fig. 5b (right) is after high pressure annealing. Microvoids along the copper/barrier interface in Fig. 5a were eliminated by high pressure annealing, as shown in in Figure 5b.

pictures showing the microstructure of copper film before and after the high pressure annealing. The specks along the via wall observed in figure 5a are the pore-like lattice defects. Such "invisible voids" are thought to condense during the consequent heat treatment or in the current flow and then grow to be fatal voids. However, by applying the high pressure annealing such atomic-level defects are eliminated, as shown in figure 5b. As a result, improved electromigration reliability is expected in this case.

DISCUSSION

It has been demonstrated that a copper film can be extruded into the via and the void can be eliminated during the high pressure annealing as shown in figure 3. Fig. 6 is another example of void elimination. In this case, the lower half of the via is left as a void and is supposed to be filled with the plating solution. In fact, the thermal decomposition spectroscopy (TDS) detected the emission of H_2O (M=18) from the as-deposition sample as shown in figure 6, while no emission was detected from the sample after high pressure annealing, as shown in Figure 7. It must be noted that copper film was extruded into the via regardless of the existence of remaining solution in the void as shown in figure 6. This fact indicates that the remaining solution has

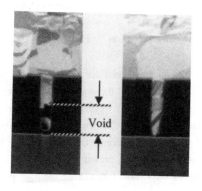

As-deposited   After high pressure
                    annealing

Figure 6. Void elimination example.

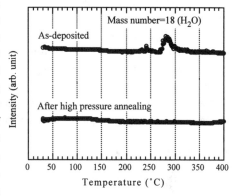

Figure 7. Thermal Decomposition Spectroscopy (TDS) showing the emission of water from void.

diffused out or has been consumed during the high pressure annealing. However, the water in copper, as a general understanding, can hardly diffuse along the grain boundary of copper. This is also supposed to be true in the case where the water is a supercritical fluid under high temperature and pressure. Therefore, the elimination of plating solution cannot be explained by the simple diffusion model.

On the other hand, the reaction of copper with the remaining solution can be assumed to be as follows: As one can see in the potential-pH equilibrium of copper-water system, copper is more "noble" than hydrogen in an electrochemical sense, thus corrosion associated with the generation of hydrogen, as shown in the formula (1), does not occur.:

$$Cu + H_2O = CuO + H_2 \qquad (1)$$

However, one must consider that the plating solution is a mixture of copper sulfate and sulfuric acid. It is known that the corrosion of copper by sulfuric acid can occur with the existence of some oxidant such as oxygen absorbed from the atmosphere or Cl$^-$ ion, which is commonly included as an additive. Since the volume change in the reaction in formula (1) is a reduction by a factor of about 50 percent, assuming that the hydrogen has escaped, the reaction is likely to be promoted under high pressure. However, the above formula also indicates that a significant amount of CuO should exist after the high pressure annealing. Obviously it is inconsistent with the observation shown in Figure 6. This disagreement implies that a copper oxide might be reduced to become a metallic copper, by some reaction. In fact, it is known that the reduction of copper oxide can be usually observed with the existence of a reduction agent such as hydrogen. It must also be pointed out that both the oxidation and reduction of copper are exothermic reactions. This fact implies some connection with the fact that PVD copper, which contains no solution, requires a higher temperature to be filled than electroplating copper, as shown in figure 3. As we have discussed, the reaction of copper and plating solution under high pressure is not well understood so far. However, it is expected that the high pressure annealing technique can improve the reliability of electroplating copper interconnections by eliminating a void or controlling the formation of undesirable copper oxide. On the other hand, the above discussion was made based on thermodynamic data for atmospheric pressure. Thus, further study of the reaction under conditions of high pressure ambient is required for comprehensive understanding.

REFERENCES

[1] D. Edelstein, J. Heidenreich, R. Goldblatt, W. Cote, C. Uzoh, N. Lustig, P. Roper, T. McDevitt, W. Motsiff, A. Simon, J. Dukovic, R. Wachnik, H. Rthore, R. Schulz, L.Su, S. Luce and J. Slattery, IEEE 1997 Intl. Electron Dev. Mtg. Dig., 773 (1997)
[2] S. Lopatin, G. Morales, T. Nogami, A. Preusse, S. Chen, D. Brown, R. Cheung, I. Hashim, S. Ponnekanti, D. Angelo, T. Chiang, B. Sun, P. Ding and B. Chin, Proc. Advanced Metallization Conference in 1998, 35 (1998)
[3] C. Lingk, M.E. Gross, W.L. Brown, W.Y.-C. Lai, J.F. Miner, T.Ritzdorf, J. Turner, K. Gibbons, E. Klawuhn, G. Wu and F. Zhang, Proc. Advanced Metallization Conference in 1998, 89 (1998)
[4] S.H. Brongersma, I. Vervoort, M. Judelwicz, H. Bender, T. Conard, W. Vandervorst, G. Beyer, E. Richard, R. Palmans, S. Lagrange, and K. Maex, Proc. 1999 International Interconnect Technology Conference, 290 (1999)
[5] T. Fujikawa, T. Ishii, Y. Narukawa, T. Masui, K. Suzuki, T. Kondo and Y. Taguchi, Proc. Int'l Conf. HIP (HIP'99), p276-281 (1999)
[6]R.J. Gutmann, W.N. Gill, T.M. Lu, J.F. McDonald, S.P. Muraaka and E.J. Rymaszewsky, Proc. Advanced Metallization Conference in 1996, 393 (199)
[7] T. Nogami, J. Romero, V. Dubin, D. Brown and E. Adem, Proc. 1998 International Interconnect Technology Conference, 298 (1998)

Forcefill applied to ECD copper

G.C.A.M.Janssen

Dimes, Delft University, P.O.Box 5046, 2600 GA Delft, the Netherlands
e-mail: janssen@dimes.tudelft.nl

T.Ohba

Selete, Yokohama 244-0817, Japan

Abstract

As a part of the development of a copper plating process we performed the following experiment: On a 200 mm wafer 0.3 μm (4 : 1 aspect ratio) via holes were etched in SiO_2. A 25 nm TaN barrier and a 200 nm copper seed layer were deposited by PVD. A 750 nm copper film was deposited by electroplating.

After plating all vias in the center of the wafer were filled. Vias at the edge of the wafer show voids.

We demonstrated that the voids could be eliminated by forcefill. Samples from the side of the wafer were subjected to a high temperature, high pressure step (30 min, 365°C, 1500 bar). Since the voids after plating always occur at the side of the via and only at the edge of the wafer, we identify the copper seedlayer deposition step as the most likely cause of the problem.

Introduction

High aspect ratio via fill by ECD copper relies on the succesfull covering of the via sidewall by a TaN barrier and Cu-seedlayer. At high aspect ratios it becomes difficult to cover the sidewall without closing off the via at the top. This problem is most pressing at the edge of the wafer. Since the targets used for barrier deposition and seedlayer deposition are only slightly larger than the wafer, the vias at the edge of the wafer do not receive a radial symmetric flux. Problems associate with PVD barrier and seedlayer therefore are expected to occur at the 'inner' sidewall of the vias at the edge of the wafer.

Experimental

On a 200 mm wafer 0.3 μm (4 : 1 aspect ratio) via holes were etched in SiO_2. We deposited by PVD a 25 nm TaN barrier and a 200 nm Cu seedlayer . Subsequently a 750 nm Cu film was deposited by electroplating. In this film voids were observed in the vias at the edge of the wafer (Fig. 1 left)

Conference Proceedings ULSI XV © 2000 Materials Research Society

As-plated ECD-Cu   After Force-fill Treatment

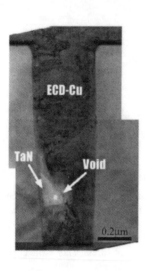 

Fig.1 TEM picture of via at the edge of the wafer after ECD copper and room temperature recrystallization (left) and a via at the edge of the wafer after subsequent high temperature, high pressure treatment (right).

A high pressure, high temperature treatment (365°C, 1500 bar, 30 min.) was sufficient to fill the vias (Fig.1 right). From Fig.1 left one can see that the barrier looks intact. Therefore the problem was diagnozed as an insufficient covering of the sidewall by the copper seedlayer. This incomplete seedlayer is likely to prevent wetting by the copper plating bath. In any case it will prevent deposition of copper from the bath. If wetting does occur, but no growth of copper, then the void will contain plating solution, mostly water. Since the applied pressure surpasses the pressure developped by the superheated water in the void no craters are expected. It is however difficult to immagine water diffusing out of the copper. Probably copper will react with water to form CuO or Cu_2O at high temperature.

Discussion

The fact that forcefill is a method to repair the flaws in the ECD process sheds light on the mechanism of the forcefill process. The forcefill process is either described as plastic deformation driven by shear stress or as a diffusional process driven by the gradient in chemical potential over the depth of the via, caused by the pressure gradient from the top of the film to the advancing metal front in the via[1,2]. Shear stresses deep in a high aspect ratio via are hard to envision. So the fact that forcefill is a method to fill the voids in the ECD copper vias is taken as evidence supporting the diffusional model.

Based on work on forcefill of aluminum vias we will demonstrate that at the length scale of a single via the description in terms of shear stress and plastic deformation reduces to the diffusional model.

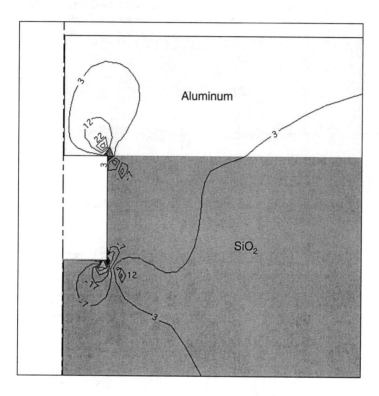

Fig.2 Shear stress in a via structure covered by an aluminum film at the onset of the forcefill process. The shear stresses were calculated by finite elements. The conditions were a compressive stress in the aluminum film due to the difference in thermal expansion of film and substrate. The magnitude of this stress was set by taking into account a thermal expansion equivalent to 50K. The applied hydrostatic pressure was 60 MPa. Via diameter is 0.6 μm, via height is 0.7 μm.

In Fig.2 it can be seen that at the via rim the shear stress becomes comparable to the applied pressure. However in the region were deformation has to occur in order to supply metal to the via the shear stresses are much lower. Since the filling process has to be continuous it will be the lower shear stress that drives the process. From Fig.2 it is concluded that a description of the forcefill process in terms of shear stress should use shear stresses on the order of 3 MPa rather than 60 MPa.

In Fig.3 we present a deformation map for 1 µm polycrystalline aluminum. This map is adapted from the map given by Frost and Ashby[3] for 10 µm polycrystalline aluminum. In the deformation map the deformation rate is given for a specified temperature and shear stress. We have done kinetic forcefill experiments. From those experiments we obtained the deformation rate. These data are plotted as the left bullet in Fig.3 (ref.1). Experiments by Robl. et al[4]. provided the right bullet in Fig.3. From the finite elements caluculations presented in Fig.2 we obtained a value for the shear stress of 3 MPa. This result is plotted in Fig.3 by crosses. The predicted deformation rates from the calculated shear stress are within a factor of three from the observed deformation rates.

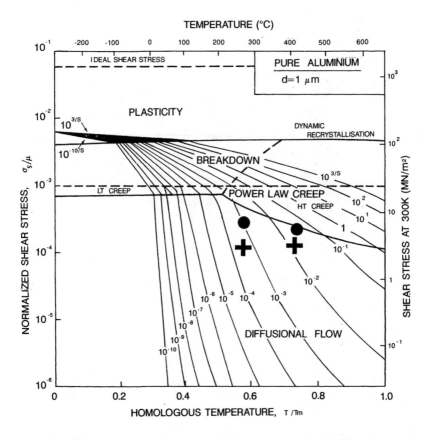

Fig.3: Deformation map for 1 µm polycrystalline aluminum, adapted from Frost and Ashby. From the deformation map it can be seen that the deformation rates predicted from shear stress and temperature, depicted by crosses, are within a factor of three equal to the observed deformation rates, depicted by bullets.

Herring[5] showed that the shear stress leads to differences in chemical potential around a single grain. It are these differences that drive the diffusion through or around that grain. This insight coupled with the fact that the forcefill process is deformation at the level of a single grain reduces the shear stress driven deformation to a diffusional model for forcefill.

Conclusion

Forcefill is able to 'repair' a flaw in the ECD copper via fill process. This practical solution however ought to be replaced by a more elegant solution, increasing the size of the PVD target.
The observation that forcefill is able to fill voids, deep in the via supports the description of forcefill as a diffusion process.

Acknowledgements

We acknowledge the TEM work of ms. Mizushima of Fujitsu and the forcefill work of mr. J.Toth of Delft University.

References

1. J.F. Jongste, X. Li, J.P. Lokker, G.C.A.M. Janssen, S. Radelaar, Th.W. de Loos, in: Advanced Metallization and Interconnect Systems for ULSI Applications in 1996, Eds: Robert Havemann, John Schmitz, Hiroshi Komiyama, Kazuo Tsubouchi, Materials Research Society, Pittsburgh, (1997), p.51-5

2. A.G.Dirks, M.N.Webster, P.Turner, P.Rich and D.C.Butler, J.Appl.Phys, 85, (1999), 571

3. H.J. Frost and M.F.Ashby, *Deformation –Mechanism Maps, Pergamon Press, 1982*

4. W.Robl, J. Förster, M.Frank, H.Cerva, A.Rucki, D.Butler, P.Rich, J.Marktanner in: Advanced Metallization and Interconnect Systems for ULSI Applications in 1997, Eds: R.Cheung, J.Klein, K.Tsubouchi, M.Murakami, N.Kobayashi, Materials Research Society, Warrendale PA, 1998, p.251-255

5. Conyers Herring, J.Appl.Phys, 21, (1950), 437

RAPID THERMAL ANNEALING OF ELECTRO
CHEMICALLY PLATED CU FILMS

G.P. Beyer[1], P. Kitabjian[2], S.H. Brongersma[1], J. Proost[1], H. Bender[1], E. Richard[1], I. Vervoort[1], P. Hey[2], P. Zhang[2], K. Maex[1]

[1]imec, Kapeldreef 75, 3001 Leuven, Belgium
[2]Applied materials, 3050 Bowers Avenue, Santa Clara, California 95054

ABSTRACT

Electro chemically plated copper films, which had been grown to a thickness of 0.7 and 1.3 μm, respectively were annealed in a rapid thermal processing tool. Due to the heat treatment at temperatures ranging from 200 to 400°C the copper grains increased in size. The recrystallisation was accompanied by a sheet resistance decrease. It was found that in the lower temperature range of 200-300°C the degree of recrystallisation depended on the film thickness whereas at 400°C the in-plane grain size was the same for both film thicknesses. The temperature treatment stabilised the sheet resistance of the Cu films.

INTRODUCTION

Electro Chemical Plating or Deposition (ECP or ECD) has emerged as the leading deposition technique to fill high aspect ratio dual damascene structures with Cu. The phenomenon of self-annealing, however, adds to the complexity of processing. During self-annealing, grain growth occurs spontaneously at room temperature. This phenomenon is accompanied by a decrease in resistivity of the Cu [1]. Although both grain growth and resistivity decrease are very desirable phenomena – they improve the properties of the Cu interconnect structure in terms of electromigration and RC delay – they may affect subsequent processing steps. For instance the polishing rate in CMP depends on the grain size and hardness of the blanket Cu layer [2]. As the grain growth can last from hours to weeks, it is very desirable to accelerate it by a thermal treatment after deposition with the intention to control the grain growth and to stabilise the resulting grain structure.

EXPERIMENT

In this study, Cu films were grown by ECP on an IMP Cu seed layer to thicknesses of 0.7 or 1.3 μm. Beneath the seed layer, an IMP TaN barrier had been deposited on an oxide substrate. Between deposition and RTP treatment the samples were stored in dry ice. The RTP treatment was done on 5x5 cm^2 pieces of wafers, which rested on a Si wafer in a N_2 atmosphere. The anneal temperature was controlled by a thermocouple and was varied between 200 and 400°C in 50°C increments. The samples were investigated by top view FIB where the surface was lightly sputtered to remove the surface oxide. The grain boundaries were manually traced on the photo. Imaging software was then used to compute the individual grain diameters. The sheet resistance

167

was measured with a four-point probe. Temperature-stress cycles were measured up to a temperature of 600°C at a ramp rate of 5°/min.

RESULTS AND DISCUSSION

Figure 1 shows the surface of a 0.7 μm thick Cu ECP film after deposition and after a 60 second RTP treatment at 200 and 400°C, respectively. The as-deposited sample contains roughly spheroidal grains with curved grain boundaries. Upon annealing at 200°C the grain boundaries become more faceted and the layer contains grains, which have grown considerably in size in a matrix where grain growth has not yet occurred. Twin formation is observed in the growing grains. This coexistence of large and small grains is indicative of so-called secondary or abnormal grain growth. Note that the appearance of the grains that have not grown in size is different to the grains of the as deposited sample. After the anneal at 400°C the Cu film has undergone complete recrystallisation. Since the intersection of the twin boundaries depends on the orientation of the grains – in (111) grains the boundaries form an angle of 60°, in (100) grains an angle of 90° [3] it is possible to discern from the photos that both Cu(111) and Cu(100) grains are present.

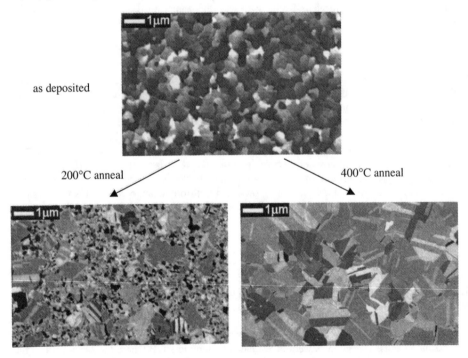

as deposited

200°C anneal   400°C anneal

Fig. 1 FIB image of the surface of an 0.7 μm thick Cu ECP layer after deposition (top) and after RTP anneal at 200°C (left) and 400°C (right).

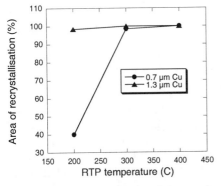

Fig. 2 Area of recrystallisation (in percent) as a function of Cu thickness and RTP temperature.

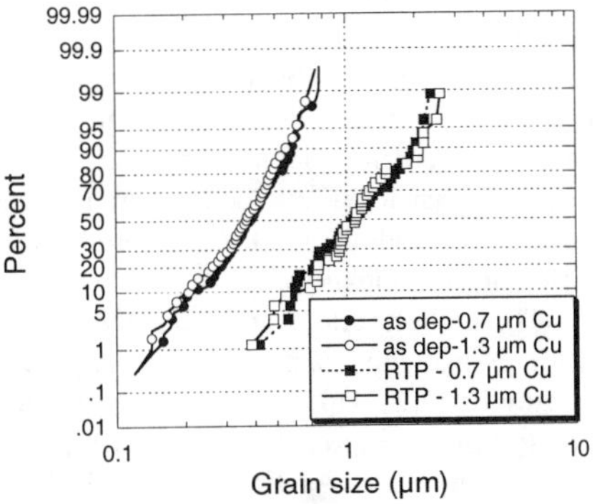

Fig. 3 Grain size distribution of as deposited layers and after RTP treatment at 400°C as a function of Cu thickness.

Due to the difficulties to determine the grain size distribution in a sample such as the one annealed at 200°C a simplified approach was chosen where the percentage of the recrystallised area is measured. Fig. 2 shows that an increase in anneal temperature from 200 to 300°C enhances grain growth in the 0.7 μm thick sample. At 300°C only small islands, containing not yet grown grains remain in a recrystallised matrix. At 400°C the recrystallisation is complete. For the 1.3 μm thick Cu film recrystallisation is already achieved at the lowest temperature. This shows a dependence of secondary grain growth - brought about by a thermal treatment – on the Cu layer thickness in the lower temperature range. Although the mechanism of recrystallisation may differ for thermal and self annealing it should be noted that the film thickness is an important parameter for the recrystallisation during self annealing as well [4, 5].

For a number of conditions the in-plane grain size distribution has been determined. It was found that the anneal at 400°C gives rise to a threefold increase in average grain size. The increase is, however, independent of the film thickness. Moreover the in-plane grain size distribution for either the as-deposited samples or the annealed samples is very similar irrespective of the film thickness (Fig. 3).

Fig. 4a FIB cross-sections of 0.7 μm Cu as deposited films (top row) and after the anneal at 400°C (bottom row).

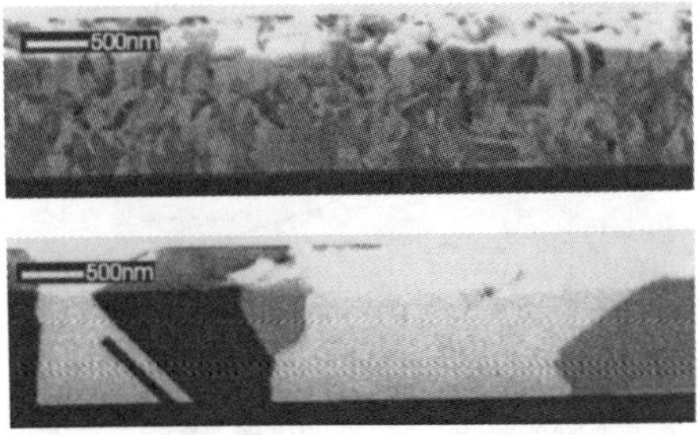

Fig. 4b FIB cross-sections of 1.3 μm Cu as deposited films (top row) and after the anneal at 400°C (bottom row).

More information about the grain structure can be derived from cross-sections of as-deposited films and films annealed at 400°C. In the as deposited samples the Cu films contain grains with a height comparable to the diameter in the surface plane (Fig. 4 top row). Thus the films are grown in the electroplating process in a layered fashion, in which more layers are added, as a larger film thickness is required. After anneal at 400°C the Cu grains extend from the surface to the barrier (Fig. 4 bottom row). Because the average in-plane grain size was found to be the same for both the 0.7 and 1.3 μm thick film the ratio of grain diameter over height differs for different film thickness.

These observations are explained as follows: The make-up of the as deposited Cu films results in grains with a large surface to volume ratio due to the addition of organic compounds to the plating bath [6]. Upon annealing the grain size increases which results in a decrease of the surface to volume ratio. The main driving force for the grain growth is the reduction in grain boundary energy [7]. As Harper et al. [8] have pointed out, in secondary grain growth, the presence of a matrix of small, as-deposited grains governs the growth of the large grains. As a result the boundary velocity of the secondary grain is independent of its size, and the in-plane diameter may exceed the film thickness. The grain growth will draw to a close when the small grains are consumed by the large grains, or in other words, when the low energy grain boundaries of neighbouring large grains come in contact with each other.

The RTP treatment has consequences for the sheet resistance drop associated with the grain growth. A higher RTP temperature gives rise to a larger sheet resistance drop (Fig. 5). For the thin sample the Rs decrease is more pronounced between 200 and 250°C due to the percolation of the secondary grains. A model has recently been proposed in which the reduction of electron scattering at the grain boundaries is linked to the secondary grain growth of electroplated Cu films [8].

Temperature-stress cycles were measured on samples, which had received an RTP treatment at 200 and 400°C, respectively. For the 1.3 μm thick samples the film stress after RTP is independent of RTP temperature. On the other hand the stress level is different for the 0.7 μm thick sample after different RTP temperatures (Fig. 6). For the sample RTP treated at 200°C the

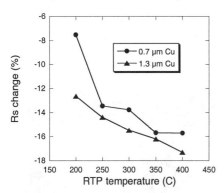

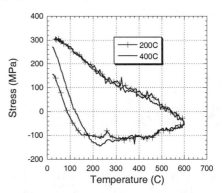

Fig. 5 Sheet resistance decrease (in percent) as a function of Cu thickness and RTP temperature.

Fig. 6 Temperature-stress cycle at 5°C/min of ECP Cu layers after RTP treatment at 200 and 400°C. Cu thickness 0.7 μm.

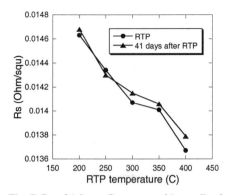

Fig. 7 Rs of 1.3 µm Cu measured immediately after RTP and re-measured 41 days after RTP.

lower stress level is indicative of a lower degree of recrystallisation. As this sample undergoes another thermal treatment grain growth appears to recommence such that the final stress values of the samples RTP treated at different temperatures are the same at the end of the thermal cycle.

The RTP treatment evidently stabilises the Cu films in terms of sheet resistance. The sheet resistance was re-measured on thick samples, which had received an RTP treatment 41 days earlier. This time span is comparable to the time needed to stabilise the sheet resistance of the thick Cu film during self annealing. The sheet resistance after RTP treatment remained stable over the investigated temperature range from 200 to 400ºC as shown in Fig. 7. The deviation between the two sets of data is less than 1% for any given temperature. Note that the recrystallisation was complete for the 1.3 µm Cu film when exposed to this temperature range.

CONCLUSION

A slower recrystallisation of ECP Cu films was observed for the thinner film in the temperature range 200-300ºC. At 400ºC no difference in the in-plane grain diameter was measured for the thin and thick film. A thermal treatment of as-deposited Cu ECP films accelerates the recrystallisation and stabilises the sheet resistance.

REFERENCES
1. T. Ritzdorf, L. Graham, S. Jin, C. Mu, and D. Fraser, Proc. International Interconnect Technology Conference '98, San Francisco, 166 (1998)
2. Q.T. Jiang and K. Smekalin, Proceedings Advanced Metallisation Conference '98, Colorado Springs, 209 (1999)
3. E.M. Zielinski, R.P. Vinci, and J.C. Bravman, J. Appl. Phys. **76** (8) 4516 (1994)
4. Q.T. Jiang, R. Mikkola, and B. Carpenter, Proceedings Advanced Metallisation Conference '98, Colorado Springs, 177 (1999)
5. S.H. Brongersma, I. Vervoort, M. Judelewicz, H. Bender, T. Conard, W. Vandervorst, G. Beyer, E. Richard, R. Palmans, S. Lagrange, and K. Maex, Proc. International Interconnect Technology Conference '99, San Francisco, 290 (1999)
6. M.E. Gross, K. Takahashi, C. Lingk, T. Ritzdorf, and K. Giboons, Proceedings Advanced Metallisation Conference '98, Colorado Springs, 51 (1999)
7. C.V. Thompson, Scripta Metall. Mater 28, 167 (1993)
8. J.M.E. Harper, C. Cabral, P.C. Andricacos, L. Gignac, I.C. Noyan, K.P. Rodbell, and C.K. Hu, J. Appl. Physics. **86**(5) 2516 (1999)

FORMALDEHYDE-FREE ELECTROLESS COPPER DEPOSITION FOR ULTRA-LARGE-SCALE-INTEGRATION (ULSI) METALLIZATION

Y. SHACHAM-DIAMAND*, N. PETROV*, E. SVERDLOV* , R. CHEUNG** and L. CASTRO**,

* Department Of Physical Electronics, Tel-Aviv University, Ramat-Aviv, ISRAEL, 69978, Email: yosish@eng.tau.ac.il, Phone: +972 3 6408064, Fax: +972 3 6423508,
** Applied Materials and Technology Inc., Santa-Clara, California, USA.

ABSTRACT

In this paper we present a study of electroless copper deposition from solutions that do not contain formaldehyde. We discuss the basic chemistry of alternate reducing agents of aldehyde based chemistry. We discuss the effect of the Aldol condensation on potential electroless deposition and the possibility to find other molecules that will induce electroless deposition. We confirmed that glyoxylic acid is an alternative reducing agent that is compatible with deposition for ULSI applications. We present thin film properties of electroless copper deposited from solution with glyoxylic acid. We also present results for deposition inside 0.2 μm features.

INTRODUCTION

On-chip interconnect networks became the critical component for integrated circuit operation and manufacturing. The critical problems of sub - 0.2 μm interconnects evolves from both the extremely small dimensions and the increasingly aggressive aspect ratios of via-contacts lines. An advanced interconnect scheme on a chip will include 5,6, and even more layers each consists with metallic conductors separated with low dielectric constant insulators. Typical metallic conductors include adhesion layer, barrier layer, conducting layer, capping layer and possibly an anti-reflective coating. The main conducting material today is aluminum and copper is replacing it. Copper offers lower specific resistance as well as better electromigration resistance when compared to aluminum.

Thin film copper deposition technologies for ULSI been investigated intensively in the last years and so far electrochemically deposited copper yields the best results. Two electroforming techniques are available today: 1. Electroplating and 2. Electroless plating. Both methods involve the use of ionic solutions with component that might be corrosive, toxic, carcinogenic and, in general, hazardous materials. Electroless deposition solutions are more complex than electroplating solutions, and hence more difficult to handle. However, electroless deposition methods have some unique properties: infinite selectivity and excellent conformality. Electroless deposition systems do not require an electrical contact to the wafer and therefore can solve problems where electrical contact is impossible or increases costs significantly. Therefore, it is worthwhile to improve the existing solutions and to develop stable and environmentally safe solutions with simple and safe waste disposal method.

In this paper we overview the components of electroless deposition solutions and discuss briefly their environmental, safety and health impacts. We focus on replacing formaldehyde and

Conference Proceedings ULSI XV © 2000 Materials Research Society

discuss the chemistry of aldehydes as reducing agents and the possibility to find alternative reducing agents for electroless copper deposition. Finally, we present the application of solutions with glyoxylic acid for ULSI interconnects. We present basic thin film properties, resistivity and deposition rate, and SEM cross section of its application for the metallization of sub-0.25 μm via interconnects.

Electro-forming of Cu interconnects

Electro-forming of copper involves the reduction of copper ions from solution. The reduction current can by supplied externally, i.e. electroplating, or internally by a parallel reaction i.e. electroless plating. There are two processes that occur simultaneously during deposition: 1. Anodic oxidation of reducing agents on catalytic metal surfaces and 2. Cathodic reduction of metal ions. The most common chemistry for electroless deposition is based on the reduction of Copper-ions complexes by formaldehyde in aqueous solution. The deposition is heterogeneous where the process is initiated on special activated surface and continues self-catalyitically. The electroless copper deposition from formaldehyde based solution is well characterized and there are few reviews available regarding basic chemistry [1] and application for VLSI [2]. Such solution yields conformal, high quality layers at relatively low cost and low processing temperature [2] and can be extended to the sub-0.1 μm regime [3,4].

Electroless deposition solution

Electroless plating baths suitable for ULSI contains copper salt, complexing agents, strong base, actrive reducing agent, and various additives including surfactants or wetting agents, stabilizers, step-coverage promoters, ductility promoters and/or additives to retard hydrogen inclusion in the deposit. So far, solutions with formaldehyde yielded the best results for very thin films (below 1 μm) as well as for very thick copper layers. Such solutions were operated at 65 – 70C and yielded high quality films with low defect density and resistivity of about 2 μΩ.cm. Such films have excellent step coverage and are compatible with common ULSI metallization [2]. Note that formaldehyde is also the most widely used reductant for electroless Cu plating in the printed circuit board (PCB) industry. However, formaldehyde is a toxic material that requires special waste disposal procedure and may raise significant working place hazard. In this work our study in we search for alternative aldehyde that can replace formaldehyde in the electroless deposition solution.

Aldehydes serve as reducing agents in electroless deposition. For example, sugars (such as fructose and glucose) are used in electroless deposition of silver and formaldehyde in electroless copper deposition. This activity is due to the high reactivity of the carbonyl group –COH. The carbon-oxygen double bond, C=O, is formed by both σ and π orbitals. Therefore, the C=O bond can be oxidized to form carboxyl COOH group. In the case of formaldehyde the following reaction takes place in high pH solution:

$$HCOH + \left(OH^-\right)_{ads} \leftrightarrow \left[H_2C(OH)O^-\right]_{ads} \tag{1}$$

$$\left[H_2C(OH)O^-\right]_{ads} \leftrightarrow HCOOH + \frac{1}{2}H_2 + e^- \tag{2}$$

The reactivity of the carbonyl group may lead to the wrong assumption all aliphatic aldehydes with formula R-COH, may reduce copper from the electroless copper electrolytes. In fact it is not so since that there are other reactions that may compete with the Cu ion reduction.

Aldehydes undergo a nucleophilic addition reaction with water. In strong basic conditions a dimerization can occur, known as aldol-condensation, for aldehydes with alpha-hydrogen (Hydrogen bonded to the carbon atom adjacent to the carbonyl group). This process seems to lower the reducing potential of the aldehydes and reduce its probability to participate in the electroless deposition process of copper. Aldole condensation process is catalyzed in both alkaline and acidic media. This effect was clearly observed when acetaldehyde was used as a reducing agent in electroless copper deposition process. The deposition rate was very low, a thin (~200 Å) copper layer was deposited after 5 minutes at 70C. We assumed that most of the acetaldehyde has reacted in aldole condensation process to form acetaldol and crotonic aldehyde.

Aldol process takes place by active group CH_2 (aldehydes with α-hydrogen) and can proceed rapidly when longer aliphatic chain is present. Aromatic aldehydes resist the aldole condensation process. However, aromatic aldehydes are toxic and are not good replacement for formaldehydes.

In the alkali media the Cannizzaro reaction takes place between aldehyde molecules. This reaction does not proceed in the presence of active CH_2 group and can be accomplished only with formaldehyde, aromatic, and heterocyclic aldehydes. Cannizzaro process in the example of formaldehyde can be represented as follows:

$$2\,HOOCCOH + 2OH^- \rightarrow C_2O_4^{2-} + HOCH_2COOH + H_2O \tag{3}$$

Cannizzaro process always takes place at elevated temperatures and it limits the maximum deposition temperatures.

The electroless deposition of copper occurs at the solution-deposit interface and it depends on the catalytic properties of the surface. For example, the activity of formaldehyde in electroless copper bath is explained by its special behavior in proximity to the solution-copper interface at high pH [6] Formaldehyde hydrolyzes in water and form methyl-glycolate ions. Those ions may be adsorbed to the surface in a special way the promotes the oxidation reaction and the formation of an available electron.

In the following section we describe the search for alternative reductants. We looked at various chemicals with the carbonyl group (sugars, aldehydes with no α-hydrogen). As a result of the above-mentioned discussion we assume that alternative aldehydes should have simple structure and without alpha-hydrogen. Glyoxilyc acid is an aldehyde where the group R is the carboxylic group. The feasibility of glyoxylic acid for electroless deposition of copper has been proposed first in 1990 by J. Darken at al. [5] and was proven in 1994 by H. Honma [4]. The Glyoxalate ions in the plating bath have low vapor pressure and show good reducing power in the electroless Cu-plating. It was shown that an electroless copper that was deposited from a solution with glyoxylic acid had superior ductility and deposition uniformity when compared over similar deposition from a formaldehyde-containing bath. In this work we follow those previous work and present new data about the application of the glyoxylic-acid bath for ULSI interconnects.

EXPERIMENTAL

The list of chemicals under investigation as reducing agents appears in table 1.

Table 1: List of aldehydes under investigation

Chemicals studied	Reducing capability
1. Formaldehyde (37% sol.)	Good
2. Acetaldehyde (50% sol.)	Medium
3. Propionaldehyde	None
4. Glyoxylic acid (formylformic acid)	Good
5. Glyoxale $C_2H_2O_2$	None
6. Glytaraldehyde (50% sol.)	None
7. Sugars (Sucrose, Maltose, Dextrose, Fructose)	None
8. Glocuse anhydrous	None

Electroless copper baths with the above compounds as the reductants were tested in order to obtain Cu deposits. The most successful effect was reached with glyoxylic acid. During experiments electroless Cu-films were obtained from different electrolytes with formaldehyde and glyoxylic acid as reducing agents on Si and SiO_2 coated with sputtered Co/Ti double layers. We also studied some Si samples activated with $PdCl_2$ /HCl solution.

Table 2 shows solution composition for electroless Cu deposition with glyoxylic acid (formulformic acid) as a reducing agent.

Table 2. Solution composition for electroless Cu deposition with glyoxylic acid

Components	Concentration, gr/l	Concentration, mol/l
$CuSO_4.5H_2O$	7.6	0.0305
EDTA or its salts	14.6	0.0351
Glyoxylic acid solution	Variable	0.054 – 0.11
Surfactant RE-610	0.004	
KOH	Variable	

The bath temperature was 65-75^{0}C.and its pH was in the range of 12-13. (adjusted with the base). The pH was adjusted by adding either KOH or TMAH after mixing the copper sulfate and the complexing agent. In this paper we report on experiment with KOH unless otherwise specified. The deposition was on Cu (20nm)/Ti(10nm) seed layer on oxide or on $PdCl_2$ activated silicon surface.

Finally we deposited electroless copper from glyoxylic acid solution in 0.18 mm via contacts. In this cares the seed layer was a very thin, less than 10 nm, sputtered copper.

RESULTS

In Figure 1 we present the film thickness as a function of time.

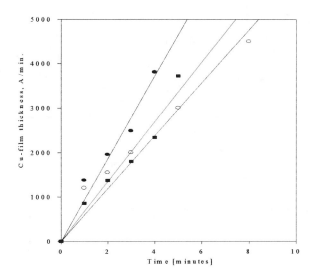

Fig. 1: Electroless copper thickness deposition vs. deposition time.

Table 3: Deposition rates of electroless copper from solutions with glyoxylic acid.

Solution	Substrate	Dep. Rate
15 gr / l glyoxylic acid	Sputterd Co	625 A/min
15 gr / l glyoxylic acid	Pd activated Si	660 A/min
30 gr / l glyoxylic acid	Pd activated Si	900

In table 3 we present the deposition rate for three different conditions.

We measured the resistance of the films as is and after 200C anneal in nitrogen. The results are shown in Fig. 2. The resistivity of thick films is about $2.5 - 3$ $\mu\Omega$.cm as deposited and it drops to about 2 $\mu\Omega$.cm after 200C anneal in nitrogen. The resistivity, ρ, of very thin films, below 100 nm, was inversely proportional to the film thickness, d, as shown by the solid line that is a non-linear fit to the expression $\rho = \rho_b(1+d_0/d)$. ρ_b is the bulk resistivity (~ 1.7 $\mu\Omega$.cm) and d_0 is a characteristic length (about 90 nm in this case).

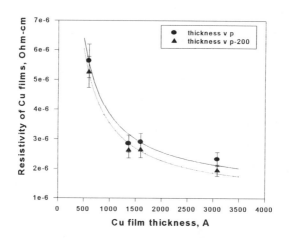

Fig. 2: resistivity of electroless copper as a function of thickness.

Electroless copper was deposited from solution with glyoxylic acid onto via contacts and the results are shown in Fig. 3.

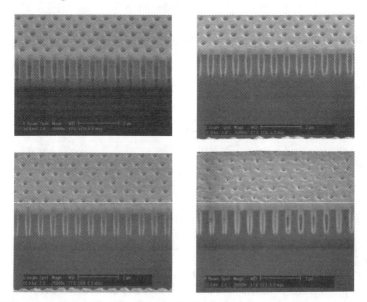

Fig. 3: Electroless Copper film deposited inside via contacts (a as is, (b) after 1 min, (c) after 2 minutes, and (d) after 4 minutes.

SUMMARY AND CONCLUSIONS

Formaldehyde-containing solutions have infinite selectivity and high conformality; therefore, they are even superior to other deposition methods for high aspect-ratio via-contact and trench filling. However, Formaldehyde is both toxic and carcinogenic. Although it is quite simple to neutralize formaldehyde, it is still a hazard. Therefore, we developed new formaldehyde-free solutions that yield copper thin films with properties comparable to that of formaldehyde containing solutions. In this paper we discussed an alternative using glyoxylic acid.

Electroless copper with glyoxylic acid can be applied in a Dual Damascene technology, which utilizes blanket deposition and chemical mechanical polishing. Electroless deposition can be used either for via contact filling. The deposition rate of solutions with glyoxylic acid is in the range of 50-60 nm / min and the resistivity of the deposited film is in the range of 2 $\mu\Omega$.cm.

Further investigating is required to find more reducing agents the yield ULSI quality thin film copper. More investigation is required to develop user-friendly stabilizers, brighteners, step-coverage improving additives (suppressors), wetting agents and others. For example, recent research shows that organic compounds such as 2,2'-dipyridyl can be used as stabilizers and can replace the commonly used cyanides. Some papers have been presented regarding the use of 2,2'-dipyridyl that show that lowers the aging rate of the deposition solution. However, its use for VLSI applications is yet to be determined

ACKNOWLEDGEMENTS

The work was supported by a grant from the Israeli ministry of science and technology. Thanks to Mr. Mark Oksman for the sample preparation and to Mr. Roi Shaviv form Tower Semiconductor Ltd. for the substrate preparation.

LIST OF REFERENCES

[1] Y. Okinaka and T. Osaka, *Advances in Electrochemical Science and Engineering*, Ed. By H. Gerischer and C.W. Tobias, pp. 57-116, vol. 3, VCH publishing, 1994.

[2] Y. Shacham-Diamand, V. Dubin, and M. Angyal, *Thin Solid Films*, Vol. 262, Nos. 1-2, pp. 93-103, June 15th, 1995.

[3] J. Darken, Paper B6/2 presented at Printed Circuit World Convention – V, Glasgow, June 12-15, 1990.

[4] H. Honma and T. Kobayashi, "Electroless copper deposition process using Glyoxylic acid as a reducing agent", J. of ECS, p. 730, Vol. 141, No. 3, March 1994.

[5] Y. Shacham-Diamand and R. Bielski. "Alkali Free Electroless Deposition". US Patent No. 5240497, Aug. 31, 1993.;

[6] S. Gottsfeld, et al., J. of Electrochemical Society, Vol. 133, No. 7, p. 1344, 1986.

ELECTROLESS GROWTH of 10-100 nm COPPER FILMS

S. Lopatin
AMD, One AMD Place, P.O. Box 3453, MS 160, Sunnyvale CA 940088-3453
Tel: (408)-749-5175, Fax: (408)-749-5144, e-mail: sergy.lopatin@amd.com

ABSTRACT

This work focused on deposition of thin electroless Cu layers over thin layers of sputtered Cu seed / Ta barrier (total field thickness ~ 55 nm). Nucleation period of electroless deposition initiation was about 15 seconds after which the autocatalytic Cu film growth was occurring. The average deposition rate was 375 angstrom/min. The electroless/seed Cu thin-film stacks had a relatively low resistivity of 2-2.5 μOhms·cm. A typical effect of decreasing resistivity from 2.5 to 2.0 μOhms·cm with increasing Cu thickness from 19 to 120 nm was observed. The texture gradually formed a stronger cubic (111) orientation with increasing electroless Cu thickness. Grain size distribution was bimodal with small (~ 30-50 nm) and large (~ 100-200 nm) grains indicating compatibility with grain size advantages of electroplated Cu films.

INTRODUCTION

Since the electroplating process relies on the seed layer to carry current from the top of the trench/via to the bottom, discontinuous sputtered Cu seed layers tend to produce voids in the feature and result in low electromigration resistance. The problem indicates the eventual need of highly conformal seed layer deposition. In this work a novel electroless solution was used for conformal copper metallization. Electroless deposition of copper and copper alloys is being considered for 100 nm-scaled damascene metallization because of its high (~ 100%) conformality [1-3], low deposition temperature, high via/trench filling capability [4-6] and low processing cost. The seed layer to initiate electroless Cu deposition is not required to be continuous, and electroless deposition of Cu films on very thin (total field thickness ~ 3-50 nm) seed/barrier layers can be applied with scaling damascene dimensions down to 10-100 nm. The mechanism of electroless Cu deposition can be described by two electrochemical reactions such as anodic oxidation of a reducing agent and cathodic reduction of metal ions [7]. These reactions occur simultaneously on a catalytic surface including Cu film or discontinuous Cu sites and produce highly conformal step coverage in high aspect ratio vias and trenches. This work focused on deposition of thin electroless Cu layers on thin layers of seed/barrier (total field thickness ~ 55 nm) and on characterization of Cu thin-film properties (continuity, roughness, texture, grain size distribution, and resistivity) versus electroless film thickness in the range from 10 to 100 nm.

EXPERIMENTAL

Environmental friendly copper electroless solution without potassium cyanide (KCN is present in some of commercial solutions) and with reduced formaldehyde (HCHO is present in all commercial solutions) concentrations was formulated and

Conference Proceedings ULSI XV © 2000 Materials Research Society

evaluated. Nucleation period of electroless deposition initiation was about 15 seconds after which the autocatalytic Cu film growth was occurring. The average deposition rate was 375 angstrom/min. Thickness dependencies in the low-nm-scale range for electroless Cu were studied in relation to film conductivity and its physical properties using four point probe, X-ray diffraction (XRD), atomic force microscopy (AFM), X-ray fluorescence (XRF), transmission electron microscopy (TEM) plain views, focused ion beam (FIB) and TEM cross sectional views.

RESULTS AND DISCUSSION

Tables 1 and 2 demonstrate measurement results for blanket electroless Cu films of 19.3-122.8 nm thickness deposited on sputtered 30 nm thick Cu seed layer on 25 nm Ta barrier.

Table 1. Average results of roughness, thickness and resistivity measurements for electroless Cu films on 30-nm thick Cu seed layer.

Sample #	Deposition time, min	Surface roughness, nm	Thickness of electroless+ seed films, nm	Thickness of electroless film, nm	Resistivity of electroless +seed stack, μOhms·cm
1	0	1.2	31.2	0	2.54
2	0.5	5.3	50.5	19.3	2.59
3	1	8.2	81.6	50.4	2.41
4	2	11.1	103.8	72.6	2.31
5	3	15	130.0	98.8	2.24
6	4	16.8	154.0	122.8	2.08

Table 2. Texture strength of electroless Cu films in a standard fiber texture plot format.

Sample #	Orientation	$\omega_{0.1}$	$\omega_{0.5}$	$\omega_{0.63}$	$\omega_{0.9}$	% random	Other orientations
1	Cubic (111)	0.24	1.3	1.74	3.36	0	none
2	Cubic (111)	0.20	1.13	1.52	3.01	0	none
3	Cubic (111)	0.19	1.01	1.35	2.74	0	(511)
4	Cubic (111)	0.15	0.82	1.13	2.44	0	(511)
5	Cubic (111)	0.12	0.74	1.04	2.37	0	(511)
6	Cubic (111)	0.11	0.69	0.98	2.28	0	(511)

Thin film resistivity of the electroless/seed stack had a relatively low resistivity of 2-2.5 μOhms·cm, and a typical effect of decreasing resistivity from 2.5 to 2.0 μOhms·cm with increasing Cu thickness from 19.3 to 122.8 nm was observed. Figure 1 shows Cu (111) fiber texture plot and FIB sross sectional view of 122.8 nm thick electroless Cu film. Texture has a strong cubic (111) orientation. The texture strength in Table 2 is in the standard fiber texture plot format. Fiber texture plots provide a convenient way of quantitatively comparing texture strength, and also provide a % random and fraction of other orientations in addition to the primary orientation. The numbers $\omega_{0.1}$, $\omega_{0.5}$, $\omega_{0.63}$, and $\omega_{0.9}$ indicate the number of degrees tilt necessary from the fiber axis (normal to the film) in order to include 10%, 50%, 63% and 90% of the pole intensity. Therefore, the smaller the numbers the stronger the texture. The data in Table 2 show that the texture gradually gets stronger with increasing electroless Cu thickness in the range of 19-120 nm. Increasing (111) intensity from 300 to 700 with increasing Cu thickness from 72.6 nm to 122.8 nm was observed on the representative fiber texture plots. None of the films had a random component, but most had a very weak secondary (511) orientation (twins).

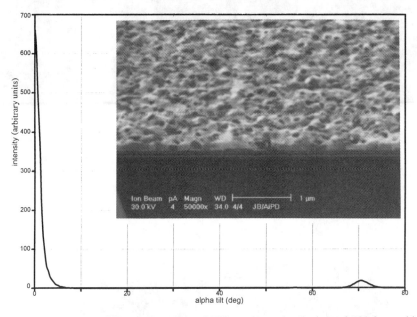

Figure 1. Cu (111) fiber texture plot and FIB cross sectional view of 122.8 nm thick electroless Cu film.

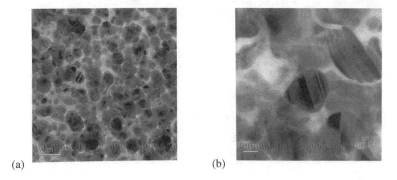

(a)   (b)

Figure 2. TEM plain views of sputtered initial Cu seed layer of 30 nm thickness: (a) uniform grain size distribution, (b) spherical shape of Cu grain with twinning.

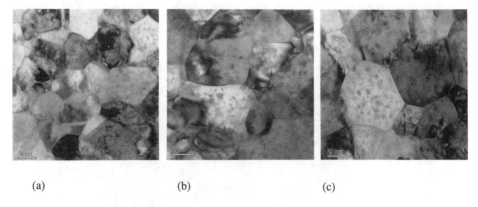

(a)   (b)   (c)

Figure 3. TEM plain views of 50 nm thick (a), 74 nm thick (b) and 99 nm thick (c) electroless Cu films showing evolution of grain size with thickness for (111) textured electroless Cu.

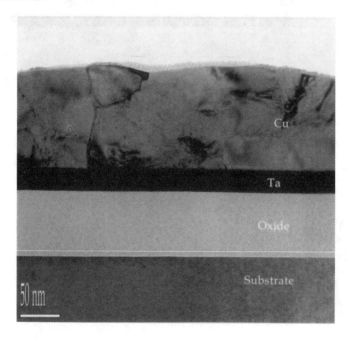

Figure 4. TEM cross sectional image of electroless Cu film on sputtered Cu/Ta seed/barrier.

Correlation between texture strength and thin-film resistivity extracted from above results suggests that increasing in (111) texture intensity with increasing film thickness leads to decrease in resistivity of electroless Cu films.

TEM plain views of sputtered initial Cu seed layer of 30 nm thickness before electroless deposition showed a spherical shape of Cu grains with uniform grain size distribution, some discontinuity between grains, twinning and average grain size of 25 nm (Fig. 2). Comparison of these initial seed films with electroless Cu deposited on the top of the seed layer (Fig. 3) showed that the grain shape changed from spherical for Cu seed to polyhedral shape with curved boundaries for electroless Cu on Cu seed. Grain size distribution for electroless Cu was bimodal with small (~ 30-50 nm) and large (~ 100-200 nm) grains (Fig. 3) having compatibility of electroless Cu grain sizes with electroplated Cu films. Average grain size was increasing with electroless Cu film thickness.

Distinct boundary between electroless and sputtered Cu layers was not observed on TEM cross sections (Fig. 4) indicating a homoepitaxial interface relation between two.

CONCLUSIONS

Low resistivity Cu films of 10-100 nm thickness were grown by electroless Cu deposition on 55 nm thick sputtered Cu/Ta seed/barrier layers. These electroless films exhibited strong (111) texture. The texture gradually formed a stronger cubic (111) orientation with increasing electroless Cu thickness in the range of 19-120 nm.

Electroless Cu films had a bimodal grain size distribution with small (~30-50 nm) and large (~ 100-200 nm) grains. The average grain size increased with Cu film thickness.

TEM cross sectional views indicated a homoepitaxial interface relation between electroless layers and initial sputtered Cu seed films.

ACKNOWLEDGMENTS

The author would like to thank J. Romero, J. Gray, J. Barragan, C. Gonsalves and D. Erb from AMD for assistance with film measurements and helpful discussions.

REFERENCES

1. V.M. Dubin, Y. Shacham-Diamand, B. Zhao, P.K. Vasudev and C.H. Ting, J. Electrochem. Soc. Vol. 144, pp. 898-908, 1997
2. W. S. Min, Y. Lantasov, R. Palmas, K. Maex, and D. N. Lee, in MRS Proc. ULSI XIV, p. 437, 1999
3. S. Lopatin, Y. Shacham-Diamand, V. Dubin, P. K. Vasudev, Y. Kim, T. Smy, in MRS Proc. ULSI XIII, p. 437, 1997
4. Y. Shacham-Diamand, J. Micromech. Microengineering. 1, p.66, 1991
5. Y. Shacham-Diamand, S. Lopatin, Microelectronic Engineering 37/38, p. 77, 1997
6. S. Lopatin, Y. Shacham-Diamand, V. Dubin, P.K. Vasudev, J. Pellerin and B. Zhao in MRS Proc. Electrochem. Synthesis and Modif. of Materials, v.451, p. 463, 1996
7. M. Paunovic, Plating, 55, p. 1161, 1968

NEW COPPER INTERCONNECT TECHNOLOGY
USING HIGH PRESSURE ANNEAL PROCESS

T. FUJIKAWA*, K. SUZUKI**, T. MASUI** , T. OHNISHI**
*Kobe Steel, Ltd.,   2-3-1, Shinhama Arai-cho, Takasago-City, 676-8670, JAPAN, takao-fujikawa@topics.kobelco.co.jp
**Technology Development Dept. Kobe Steel, Ltd., 1-5-5, Takatsukadai, Nishi-ku, Kobe, 651-2271, JAPAN

ABSTRACT

A close study on the difference of the nature of PVD-Cu layer and ECD-Cu layer has led the establishment of new interconnect technology in which the high pressure anneal method[1] is utilized for the complete filling of copper into holes smaller than sub-quarter microns. The whole process comprises, physical vapor deposition(PVD) of copper for seeding and also bridging the openings of the holes, electro-chemical deposition(ECD) of Cu on the PVD-Cu seed and complete filling of Cu into via holes by High Pressure Annealing. In this combination, PVD-Cu prevents the ECD solution invasion into the holes and ECD-Cu helps the PVD-Cu seed to easily deform and fill the holes. As a result, even holes with a diameter smaller than 0.15μ m can be filled with Cu and the high pressure ensures good adhesion of Cu to the barrier layers.

INTRODUCTION

In recent years the copper interconnect technology has been attracting attention, because of the low electrical resistance of copper. The production method, however, has to be greatly modified from the conventional Al interconnect technology, as the copper is not easily etched like aluminum to form the wiring patterns. The dual damascene process using ECD is one of the most promising methods in terms of contact/via hole filling performance and the processing cost. In a couple of years, the designing rule of integrated circuits is deemed to become smaller than 0.15μ m and the ECD method seems not necessarily viable in terms of filling performance and the reliability of the ULSI chips due to the lack of a good seed-layer forming method.   For filling such small holes and trenches with copper, the CVD method is believed to have good potential in terms of conformality. But there are some other problems like the precursor residue, environmental hazardous by-product and its processing cost. If the PVD method or the ECD method with PVD seeding could be used for the copper interconnections for these next generation ULSIs, the total process could become more simple and the processing cost could be kept within a reasonable range.

In order to achieve this, the application of high pressure anneal process after Cu deposition is the most promising approach. The high pressure anneal process uses high pressure argon gas atmosphere in the range of several tens to 200MPa at high temperatures. A similar process, Hot Isostatic Pressing (HIP) has been used to manufacture pore-like defect free products and already has a good reputation in the field of castings, ceramics and powder metallurgy industries, from the view point of fabrication of highly reliable products.

This paper describes the results of the comparison of ECD-Cu film and PVD-Cu film and proposes a new Cu interconnect technology by the combination of PVD, ECD and the high pressure anneal process which could meet the requirements of sub 0.15μ m design rule wiring.

HIGH PRESSURE ANNEAL PROCESS

The high pressure anneal process in which high gas pressure and high temperature are simultaneously applied has good potential for the manufacture of pore-type defect free interconnects. Here, pore-type defects include macro-pores which exist along grain boundaries or insufficient filling of holes, micro-pores which often form in between the micro Cu crystals in the case of ECD or CVD, and on the interfaces of Cu and barrier materials. Most of these defects can be eliminated by this process. More specifically, the high pressure anneal process works in three different ways: 1)Extrusion mode where the Cu layer covering the opening of holes is

187

completely extruded into holes by plastic deformation or creep phenomena is already known for Al interconnects fabrications[2],[3]. 2)High pressure reflow mode in which high pressure gas atoms promote the surface diffusion of Cu to flatten the surface of narrow trenches, 3)Pore-free grain growth mode where high pressure promotes the grain growth of small Cu grains formed by ECD or CVD and suppresses the formation of large voids from nano-scale pores scattered along Cu grains. Fig. 1 shows the concept of Pore-free grain mode in comparison with normal annealing.

In any case the resultant Cu layer has a pore-free structure which is very attractive from the view point of Cu atoms migration because these pores become the source of electron migration voids.

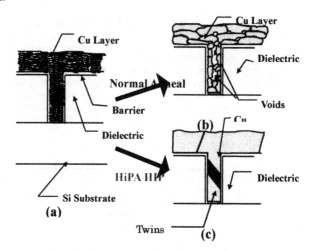

Figure 1. Comparison between Normal anneal and High pressure anneal process in Pore-free grain growth mode.

EXPERIMENT 1

1) Experiment on the high pressure anneal of ECD-Cu and PVD-Cu

Si wafer samples which have contact holes 0.28 μ m in diameter and 2.4 μ m in depth on the BPSG dielectric layer were used. Each sample has a TaN barrier layer with a thickness of 70nm. For ECD samples a seed Cu layer was formed with a thickness of 100nm by the long throw sputtering method prior to the electroplating. For PVD samples the normal PVD technique at 200 °C was employed. In both cases, the aspect ratio of the holes was very large so Cu was not completely filled to the bottom and the total Cu layer thickness was sufficient to bridge over the opening of the holes. After Cu layer formation, samples were high-pressure treated under conditions of 350~500°C, at 120MPa and 200MPa. Table 1(a) and Table 1(b) summarize the results on the filling performance for PVD-Cu and ECD-Cu respectively.

<table>
<tr><td colspan="5">Table 1(a). Results for PVD-Cu</td></tr>
<tr><td>Pres.   / Temp.</td><td>350°C</td><td>400°C</td><td>450°C</td><td>500°C</td></tr>
<tr><td>120MPa</td><td>NG</td><td>NG</td><td>NG</td><td>—</td></tr>
<tr><td>150MPa</td><td>NG</td><td>NG</td><td>NG</td><td>—</td></tr>
<tr><td>200MPa</td><td>NG</td><td>NG</td><td>Partly Good</td><td>Good</td></tr>
</table>

(NG means "Not Good")

<table>
<tr><td colspan="5">Table 1(b). Results for ECD-Cu</td></tr>
<tr><td>Pres.   / Temp</td><td>350°C</td><td>400°C</td><td>450°C</td><td>500°C</td></tr>
<tr><td>120MPa</td><td>Good</td><td>Good</td><td>Good</td><td>—</td></tr>
<tr><td>150MPa</td><td>Good</td><td>Good</td><td>Good</td><td>—</td></tr>
<tr><td>200MPa</td><td>Good</td><td>Good</td><td>Good</td><td>—</td></tr>
</table>

In these tables "Good" means Cu was completely filled down to the bottom, "NG" means Cu was filled only halfway down to the bottom and "Partly good" means some holes were completely filled with Cu down to the bottom. Fig. 2 compares the structure of both samples which were treated at 350 ℃, 200MPa .

Comparing the Tables 1(a) and 1(b), it is clearly seen that PVD-Cu film and ECD-Cu film have different deformabilities. That is, ECD-Cu can be easily filled inside the holes at lower

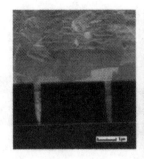

Figure 2. While ECD-Cu can be completely filled into holes (left), PVD-Cu can not be filled at 350℃, 200MPa (right).

temperatures than PVD-Cu. It is very interesting that the same material, Cu, shows completely different behavior when high pressure was applied and it should be noted that in the case of ECD-Cu film the temperature needed for complete filling is lower than 380 ℃ which is supposed to be the maximum allowable temperature when organic low-k dielectric material is used for the insulating layer.

In order to clarify this difference, a series of experiments were carried out on both Cu films . In the case of ECD-Cu insufficient coverage of the PVD-Cu seed layer on the inside surface of the holes was found to have resulted in the inclusion of ECD solutions inside the closed holes. Furthermore it was found that when such a sample, in which a certain amount of water is intentionally enclosed inside the hole (as shown in Figure 4 left.), was annealed at 300℃ under atmospheric pressure, the ECD water solution inside the holes evaporated and as a result the Cu layer was inflated and peeled off.

These series of experiments showed that contamination by ECD solution could occur in the ECD process. This means that ECD-Cu film often contains water, as well as hydrogen which usually gathers to the cathode and may be the ECD-Cu film. A detailed explanation of this phenomenon is discussed in a separate paper[4] by the present authors.

2) Evaluation of PVD-Cu film and ECD-Cu film

In order to understand the cause of the difference in the hole filling performance between PVD-Cu and ECD-Cu, the hardness, microstructure and impurities of both As-Depo and after high-pressure processed samples were closely examined. The result of the micro-hardness measurement is shown in Figure 3. In Figure 3, the values for PVD are those for samples sputtered at room temperature and the values for ECD(A) and ECD(B) are taken from the samples processed by ECD machines manufactured by two different companies. The values for 0 MPa mean before the high pressure anneal treatment. The high pressure anneal treatment was carried out at 350 ℃ with a holding time of 900sec. Although there is little difference between the two ECD materials, the hardness values for PVD are almost two times larger than the values for ECD-Cu. It is also very interesting that PVD-Cu is still considerably hard even after the high pressure annealing. When the hole-filling phenomenon is assumed to proceed in a way similar to plastic deformation, this hardness difference seems the main reason for the poor deformability of PVD-Cu.

In Figure 4, the microstructure of ECD-Cu and PVD-Cu as-deposited conditions is shown. The grain size of PVD-Cu is smaller than that of ECD-Cu. In general the hardness of a metal is known to decrease with the grain size. The above results hold true for this general understanding.

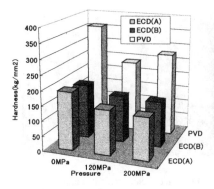
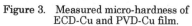

Figure 4. Grain morphology image observed by SIM after FIB milling for As-Depo Cu film. ECD-Cu (left) and PVD-Cu (right). In the left figure, 1/3 from the bottom of the hole is filled with ECD solution.

Figure 3. Measured micro-hardness of ECD-Cu and PVD-Cu film.

It seemed that the outstanding hardness of PVD-Cu could not be explained by only the small grain size, so TEM observations of inside the grains were carried out and it was found that the crystal grain of the PVD-Cu film contains a lot of twins or stacking faults (Figure 5). This unique structure seems to be the main reason for the high hardness of the As-Depo PVD-Cu film.

As for the impurities, the EDX, SIMS and TDS methods were employed to detect the impurities in the as-depo ECD-Cu film and it was found that methane, carbonate and sulfate are detected. In order to detect hydrogen the atmospheric pressure ionization mass-spectroscopy combined with the TDS technique was employed and revealed that 0.078 at% hydrogen was released in the temperature range of 350-500℃. This hydrogen seems to be playing a very important role in the growing and softening the copper grain of ECD-Cu.

NEW CONCEPT FOR SMALLER HOLES

The above experimental results suggest that ECD-Cu film could contain micro-voids at a poor seeding portion and ECD solution could be left in the voids as schematically shown in Figure 6. Of course, the high pressure anneal process can eliminate such voids but there still exists the possibility that some portion of water reacts with copper to form oxides under high temperature and high pressure conditions.

On the other hand, it is very difficult to reflow PVD-Cu film or to pressure-fill the holes, so the chances of insufficient filling could become large, as already explained. In addition to this, the trend of future requirement for holes smaller than 0.15 μm makes it more and more likely that such insufficient filling will happen.

Figure 5. Twins or stacking faults are included inside the PVD-Cu grain (left: Image and right: electron diffraction pattern with mirror pattern).

190

One of the answers to solve these problems may be a different type of combination of PVD, ECD method and the high pressure as shown in Figure 7. In Figure 7, first the PVD-Cu seed layer is formed to close the openings of the holes in order to avoid the inclusion of the ECD solution and second the ECD-Cu layer is formed on the PVD-Cu seed layer. This ECD-Cu layer could change the nature of the PVD-Cu layer below the ECD-Cu layer by supplying hydrogen from ECD-Cu to PVD-Cu and makes the PVD-Cu layer softer than the simple PVD-Cu film. After ECD, the high pressure anneal process is applied to completely fill the contact/via holes at temperatures of 350~400°C. It should be noted that the smaller the diameter of the hole, the more easily such a state of bridging the holes can be created by the PVD process, so if this mechanism could work well, this process could be an alternative to the CVD process which has been so far deemed to be the only process which can meet the future requirement.

The advantages of the present process over the CVD method are 1) no contamination is possible compared with the CVD precursor residue which is one of the concerns with the CVD, 2)the conventional PVD and electroplating equipment can be used with only a little modification of the recipe.

EXPERIMENT 2.

In order to prove that the above concept works well, samples with different combinations of the thickness of PVD-Cu seed layer and ECD-Cu layer were prepared. For holes with an opening diameter of 0.35 μ m, the minimum thickness needed to close the openings of this hole was 0.8 -1.0 μ m, that is, 2.5 -3 times larger than the opening diameter by the room temperature PVD.

Figure 9 shows SEM micrographs for samples with 1.0 μ m PVD-Cu layer and 1.5 μ m ECD-Cu layer. High pressure annealing treatment was performed at 350°C, 400°C, 450°C at 120MPa. The diameter of the hole was as described above and the depth was 2.4 μ m. PVD-Cu was formed by RT sputtering, so it

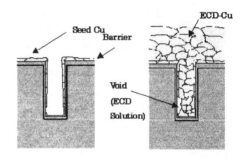

Figure 6. Schematic illustration of the conventional ECD-Cu interconnect When seed layer is incomplete, a void may appear in the course of ECD. HIP can eliminate such voids as well. Some of the residual ECD solution is squeezed out of the structure.

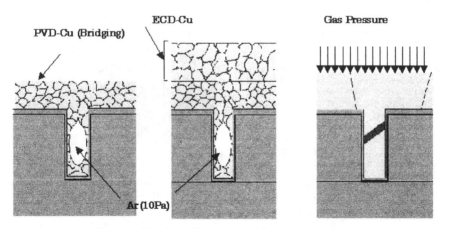

Figure 8. Schematic illustration of the present process

Figure 8. 0.35 μ m diameter hole opening was closed with 1 μ m thickness by RT sputtering and ECD-Cu was formed thereon , total thickness was 2.5 μ m comparative to the hole depth(a). High pressure anneal at 350°C, 120MPa was not sufficient to fill the holes (b), but temperatures higher than 400°C, complete filling was performed . (c)400°C,   (d)450°C.

was rather difficult to close the opening by overhanging of deposited Cu. That is the reason why seed layer thickness of as much as 1.0 μ m was necessary. From Figure 8 (a) after ECD-Cu, no Cu is deposited inside the hole and this means complete closing of the opening was obtained. The high pressure annealing of these samples was carried out at 350, 400, 450°C at 120MPa and 350°C at 200MPa. As shown in Figure 8, complete filling of holes was achieved except for the case of 350°C, 120MPa. In Figure 9 the result for another example is shown. In this case, narrow trenches of   0.13 μ m wide and 0.55 μ m deep were successfully filled with copper. High pressure annealing was performed at   375 °C , 200MPa. These experimental results proved that the above-mentioned concept worked very well and could be a very promising method to realize the wiring technology for sub 0.15 μ m design rules.

Figure 9. Example for an application to narrow trenches, 0.13 μ m wide, 0.80 μ m deep.

DISCUSSIONS

The above experimental results show that high aspect ratio holes can be filled with Cu at temperatures around 350 –400 °C . However the pressure needed at temperatures around 350°C is as high as 200MPa. In order to lower the pressure level needed to achieve complete filling, the mechanism of the hole filling phenomenon under high pressure has to be clarified.

One possibility is the application of the superplastic deformation mechanism which can be seen for materials with very tiny crystal grains. In the above experiment, the sputtering of seed Cu was carried out at room temperature. This was intended to obtain a seed layer with small grains. It was

ECD-Cu

PVD-Cu

Figure 10. Micro-structure of typical sample prepared by RT PVD-Cu and ECD. It is clear that PVD seed layer consists of very small small grains.

expected that if the grain size of seed layer Cu is as sufficiently small as compared with the hole diameter, superplastic deformation can possibly lower the pressure for filling holes. In fact, the grain size of the seed layer was very small as shown in Figure 10, but superplastic deformation seems not to have happened.

As already described in the section of the Evaluation of PVD-Cu and ECD-Cu, the hardness of Cu has a close corelation with the filling performance when high pressure is applied. Moreover the grain size also seems to have some relationship with hardness. In this case, the small grain size seems to have been a negative factor in lowering the pressure level of the process.

Another possible explanation may be the creep mechanism. In this process what makes the understanding of the mechanism very difficult is that this hole filling deformation phenomenon is proceeding in the intermediate temperature range, that is 300 –400℃, which is a little bit lower than the recrystallization temperature of Cu around 450℃. Also, creep phenomenon is not so simple. The fact that even ECD-Cu grains which are larger than the hole diameter can be deformed under high pressures cannot be explained by gliding of the slip planes along (111) nor by the Nabbaro-Herring mechanism in which the vacancy concentration gradient promotes the Cu atoms migration.

At this stage, one thing that the authors can state is that hydrogen atoms included in the ECD-Cu are playing some important role in the deformation of Cu crystals or Cu atoms diffusion. It is very difficult to analyze the hydrogen concentration in the Cu film, but the atmospheric pressure ionization mass spectra analysis combined with heating revealed that 0.078 at% hydrogen was released in the temperature range from 360℃ -500℃ as explained in the previous section. This hydrogen release seems either to be the result or the cause of the Cu atoms migration in this temperature range. In order to understand the mechanism completely further examination of the microstructure inside of Cu grain is necessary.

If the difference between ECD-Cu and PVD-Cu could be perfectly clarified and if the nature of the PVD-Cu could be changed completely to that of the ECD-Cu, a simple combination of PVD and High pressure anneal process could become possible. This is the author's last goal and to reach this goal more intensive study on the micro structure is necessary.

CONCLUSIONS

Close examination of the hole filling performance under high pressure, and some physical properties such as hardness and the grain morphology of the ECD-Cu and PVD-Cu has led to a new hole filling process. This process is not special but a simple combination of seed-Cu layer forming by PVD, ECD to soften the hard PVD Seed-Cu and a high pressure filling method. The potential of filling performance of the present process is far better than the simple PVD method or usual ECD with conformal PVD-Cu seed layer and is expected to be adaptable to 0.1 micron level holes. The advantage over the conventional ECD-Cu process is that forming a seed layer to close the openings, especially of small holes in the present process is much easier than to form a conformal seed film inside the surface of the small cylindrical contour of the holes.

REFERENCES

1. T. Fujikawa, T. Ishii, Y. Narukawa, T. Masui, K. Suzuki, T. Kondo and Y. Taguchi, in Hot Isostatic Pressing edited by LI Chenggong, CHEN Hongxia and MA Fukang (Proc. Int. Conf. HIP, Bejing, China, 1999), p276-281.
2. H. Obinata, Japanese Laid Open Patent No. H2-205678, U.S.Patent No. 5011793 (30 April 1991).
3. G. Dixit et al., Semiconductor International, August (1995), p79.
4. K. Suzuki et al., presented at Advanced Metallization Conference Japan/Asian Session, Tokyo, Japan, 1999(to be published).

Adhesion Study on Copper Films Deposited by MOCVD

S. Riedel*, K. Weiss, S. E. Schulz, T. Gessner

Chemnitz University of Technology, Center of Microtechnologies, D-09107 Chemnitz, Germany

* corresponding author phone: +49 371 531 3673   fax: +49 371 531 3131
e-mail: stephan.riedel@e-technik.tu-chemnitz.de

ABSTRACT

The most common process for Cu-MOCVD is the disproportion reaction of (TMVS)Cu(I)(hfac). But films deposited by this technique adhere poorly to common barrier layers. It is believed that a thin interlayer consisting of impurities (F,C) is the reason for the poor adhesion.

This paper summarizes ideas and experiments concerning the improvement of the adhesion of MOCVD produced copper films by either preventing the formation of the interlayer, its removal after deposition or the application of a glue layer. Different base layers are tested in combination with various pre- or post-deposition treatments. The adhesion of thin Cu-films can be improved by an annealing step after deposition. The combination of a glue layer (Si or Ti) and an annealing step results in well adhering thick copper films.

INTRODUCTION

The current scaling down process in IC technology leads to high ratios of the vertical to the horizontal dimensions of the produced features. It is well known that the CVD technique offers the potential to fill features with high aspect ratios, especially if the Dual Damascene technology is applied. Because copper will be the metal choice in metallization schemes copper CVD is of high interest. Copper MOCVD using the precursor (TMVS)Cu(I)(hfac) is the most widely used process. Unfortunately the films deposited by this technique adhere poorly to common barrier layers. It is believed that a thin interlayer (fig. 1) consisting of impurities (F,C) is the reason for the poor adhesion [1,2].

This study was carried out within a work focused on the development of an optimal deposition of seed layers for electroplating as well as on a complete copper fill process ensuring a good adhesion of copper to the barrier.

Fig. 1: CVD-Cu on PVD-TiN with 3 nm thin interlayer (consisting of F, C) in between

EXPERIMENT

The deposition experiments were carried out using a Precision 5000™ (Applied Materials) cluster tool, equipped with chambers for Cu MOCVD, TiN MOCVD, silane treatment and an etch chamber used for in-situ sputter clean as pre-treatment. The precursor used for all experiments described here was (TMVS)Cu(I)(hfac) (CupraSelect™, Schumacher Corp.) with the additives Hhfac·2H$_2$O and TMVS. The adhesion was evaluated by tape test following ASTM D3359-78.

The wafers (6 inch) were coated with different base layers before the Cu-MOCVD. Ti, TiN, Ta, TaN, W (PVD) and Si (bare substrate) were the tested materials. The Cu-MOCVD with and without pre-treatments was performed ex-situ on all used materials. Additionally, CVD-TiN was covered in-situ with MOCVD-Cu.

The Cu-thickness was 50...100 nm for seed layer applications and 300...450 nm for the complete fill process called thin films or thick films, respectively within the paper.

Different annealing steps after deposition were used in order to improve adhesion. The conditions are as follows:

- vacuum (20 mTorr), 350°C, 5 min for thin films (seed layer application)
- 50 Torr, 900 sccm Ar, 280°C ... 450°C, 7 min for thick films

RESULTS

Influence of the base layer

The impurity layer is formed at the nucleation phase of the Cu-CVD process. Therefore it is likely that it is not found on special materials on which the adhesion should be good. Table 1 shows the adhesion behavior of the MOCVD-Cu films on various base layers. Pre-treatments and in-situ / ex-situ processing are mentioned. The tape test was performed at as deposited films. Thin (50...75 nm) as well as thick (300...450 nm) Cu-films were tested. The adhesion is poor in almost every case. The exception is PVD-Cu. The effect is already known [3] for PVD-Cu seed layers.

Table 1: Results of adhesion test by tape test

Substrate for Cu-CVD	Process flow	Result of tape test
PVD-Ti	ex-situ base layer deposition, in-situ sputter clean prior to Cu-CVD	failed
PVD-TiN		
PVD-Ta		
PVD-TaN		
PVD-W		
CVD-TiN	in-situ deposition, no pre-treatment	
bare Si-wafer	HF-dip, in-situ sputter clean prior to Cu-CVD	
PVD-Cu	ex-situ base layer deposition, without in-situ sputter clean	pass without failure

Thermal treatment

The adhesion could be improved very much by an annealing after deposition for a Cu-film thickness smaller than 100 nm. No improvement was observed for thick films (d ≥ 150 nm). AES depth profiles before and after annealing were used in order to find out whether the impurity layer diffuses out during thermal treatment. Figure 2 shows that the impurities C, F and O remain at the interface.

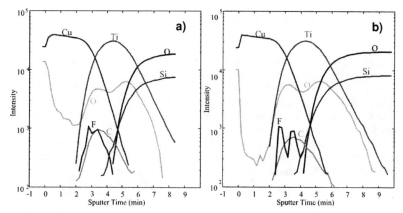

Fig. 2: AES depth profile of CVD-Cu/PVD-TiN/SiO$_2$ stack
a) as deposited, b) annealed 350°C, 5min

The change of mechanical film properties can be a reason for the improved adhesion, too. Therefore the stress behavior of the thin Cu-films was investigated. The stress in the Cu-films on PVD-TiN versus the film thickness is drawn in fig. 3a. The results for as deposited as well as for annealed films are shown. It reveals that thinner films have a lower stress level after the thermal treatment than thicker films although it is the other way around before the treatment. Fig. 3b makes clear that the stress is reduced very much in thin films by annealing, but the stress is remaining at the as deposited value for 150 nm thick films.

Glue layer

If the Cu and the base layer react with each other an interface with a high adhesion strength would be formed. Si and Ti react with Cu at elevated temperatures forming Cu$_3$Si at 200°C and CuTi at 350°C, respectively [4]. Questionable is whether the impurity interlayer would prevent the reaction. The principle was tested by Cu-MOCVD (d$_{Cu}$ = 350 nm) on bare Si-substrate and on 20 nm PVD-Ti. The surfaces were in-situ sputter cleaned. Cu-layers on both substrates failed

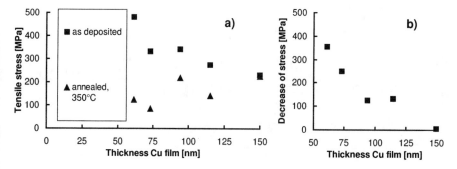

Fig. 3: Stress behavior of thin Cu-films; a) Stress versus film thickness before and after annealing, b) Decrease of stress by annealing step (350°C, 5 min).

the tape test in as deposited state; after thermal treatment (450°C) both passed the test without remarkable failure.

PVD-Ti and bare Si-substrates are no relevant solutions for a CVD based Cu-metallization. Ultrathin Si glue layers or a Si surface coverage can be formed by SiH_4 decomposition at 400°C. The following process flow was tested: CVD-TiN 20 nm, 5 min SiH_4 at 400°C, Cu-CVD 300 nm, thermal treatment. The result concerning adhesion strongly depends on the annealing temperature of the post deposition treatment. The films completely failed as deposited and for 280°C annealing. The adhesion is almost perfect after a 325°C treatment. The films partly or completely failed the tape test after a 450°C treatment.

The formation of Cu_3Si is expected to be the reason for the improved adhesion after a 325°C-treatment. At lower temperatures the reaction could be prevented by the interlayer. The bad result after high temperature treatment (450°C) is not understood yet.

Electric field during deposition

The experiments are based on an article by Lee et al. [5]. After removal of the stabilizing ligand TMVS from the precursor molecule (first part of disproportion reaction) the remaining molecule has a dipole moment (positive side at Cu atom). The molecule would be oriented in an electric field in a way that a negative voltage at the substrate would arrange the Cu atoms towards it. The formation of the impurity interlayer could be prevented. The experimental set-up is drawn in fig. 4.

The deposition of thick Cu-films (350 nm) was investigated. No improvement of adhesion was observed for both polarities of the field with a strength of 167 V/cm. The problem is that the precursor molecules are not perfectly oriented in the field.

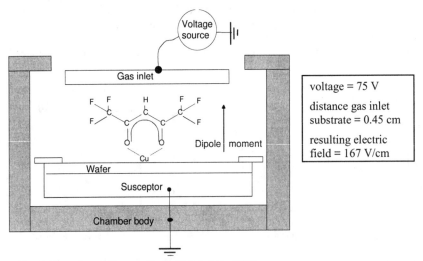

voltage = 75 V

distance gas inlet substrate = 0.45 cm

resulting electric field = 167 V/cm

Fig. 4: Experimental set-up for field aided Cu-CVD

The alignment of a polar molecule (with a dipole moment p) by an electric field (with a strength E) is in competition with its thermal movement. The resulting equilibrium can be estimated by the alignment energy $W_{al} = pE$ and the thermal energy $W_{th} \approx kT$. If ϑ is the angle between the dipole moment and the direction of the electric field, the mean value $\overline{\cos\vartheta}$ of all dipoles gives the average alignment of the molecules. A value of 1 means perfect alignment, 0 means a random orientation of the molecules. One can calculate $\overline{\cos\vartheta}$ by eq. 1 if $kT \gg pE$ [6].

$$\overline{\cos\vartheta} = \frac{pE}{3kT} \quad (1)$$

The product kT is $6.5 \cdot 10^{-21}$ VAs at a deposition temperature of 200°C, pE is $7.8 \cdot 10^{-26}$ VAs ($p = 1.41$ D for Cu^{1+}(hfac) according to [5], $E = 167$ V/cm). Therefore it is possible to use eq. 1. It gives $4 \cdot 10^{-6}$ for $\overline{\cos\vartheta}$. This means the alignment of the precursor molecules is very small. There is no strong alignment possible in this temperature range even for the highest applicable electric field.

CONCLUSION

The adhesion properties of the Cu-films could not be influenced by an electric field because of the weak alignment of the precursor molecules. Thin Cu-layers adhere well to the base layers if they are annealed after their deposition. These layers are applicable as seed layer for electrochemical deposition. The usage of a Si like glue layer in combination with a post deposition annealing results in well adhering thicker films which can be used for the complete fill process of damascene structures. But the mechanism has to be further investigated in detail, especially the influence of the annealing temperature.

A different solution can be a ultrathin titanium film as glue layer plus a post-deposition annealing. Because the impurities remain at the interface or in the film for both solutions the contact resistance could be an issue which has to be solved.

REFERENCES

1. K. Weiss, S. Riedel, S.E. Schulz, M. Schwerd, H. Heleneder, H. Wendt and T. Gessner, talk presented at conference MAM'99 March 8-10, 1999 in Oostende, Belgium, to be published in Microelectronic Engineering

2. N. Yoshida, Y.D. Cong, R. Tao, L.Y. Chen and S. Ramaswami in *Advanced Metallization Conference in 1998 (AMC 1998)*, edited by G.S. Sandhu, H. Koerner, M. Murakami, Y. Yasuda and N. Kobayashi (Mater. Res. Soc. Conf. Proc. ULSI XIV, Warrendale, PA 1999), p. 189

3. T. Gessner, A. Bertz, M. Markert, J. Roeber, J. Baumann, S.E. Schulz and C. Kaufmann in *Advanced Metallization and Interconnect Systems for ULSI Applications in 1996*, edited by R. Havemann, J. Schmitz, H. Komiyama and K. Tsubouchi (Mater. Res. Soc. Conf. Proc. ULSI XII, Pittsburgh, PA, 1997), p. 161

4. J. Li and J.W. Mayer; MRS Bulletin 6/93 (1993), p. 52

5. W.-J. Lee, S.-K. Rha, S.-Y. Lee and C.-O. Park, J. Electrochem. Soc. 144/2 (1997), p. 683

6. Hermann Haken and Hans C. Wolf, *Molekülphysik und Quantentheorie*, Springer Verlag, Berlin Heidelberg New York, second edition, 1994, pp. 26-28

ALCOHOL AND AMINE ADDUCTS OF Cu(hfac)$_2$
FOR CVD COPPER

A. W. MAVERICK,* H. FAN,* M. S. BUFAROOSHA,* Z. T. CYGAN,* A. M. JAMES,* F. R. FRONCZEK,* G. L. GRIFFIN,** J. Y. BOEY**
*Department of Chemistry, Louisiana State University, Baton Rouge, LA 70803
**Department of Chemical Engineering, Louisiana State University, Baton Rouge, LA 70803

ABSTRACT

The CVD performance of the copper(II) precursor Cu(hfac)$_2$ and its adducts with alcohols and amines has been studied. Adducts with simple C$_3$ and C$_4$ alcohols give good films with H$_2$ as carrier gas, and several also work well under N$_2$, because the alcohol can act as reducing agent. Adducts with ether-alcohols, diols, and amines are more stable, but they require H$_2$ for optimum CVD results. Promising results were obtained with the new precursors Cu(hfac)$_2$(propylene glycol) and Cu(hfac)$_2$(allylamine)$_2$. Kinetic studies have been performed in order to compare quantitatively the performance of Cu(hfac)$_2$ with H$_2$O and i-PrOH as co-reactants. Both additives give higher growth rates, with i-PrOH giving a much larger enhancement. However, while H$_2$O provides a smaller growth rate enhancement, it leads to superior film properties.

INTRODUCTION

One of the methods being considered for deposition of copper interconnects for ULSI is chemical vapor deposition (CVD). However, the performance of CVD copper has not yet been as good as that of the combination of PVD Cu seed layer followed by electrochemical Cu fill now being used by several manufacturers. Especially for seed layer deposition, CVD may be more useful as device sizes shrink further and via and trench aspect ratios increase.

Our research with Cu CVD has two goals: (a) to determine whether new precursors show improved performance in deposition rate, film properties, adhesion, or precursor stability compared to existing systems; and (b) to evaluate these precursors by quantitative kinetic studies so as to understand better the deposition mechanisms. We report here recent progress in these areas, as follows. (a) initial CVD studies using alcohol and amine adducts of Cu(hfac)$_2$; and (b) a study of deposition rate enhancements and film properties when water and i-PrOH (2-propanol) are added to a Cu(hfac)$_2$-based Cu CVD system.

EXPERIMENT

Preparation and screening of precursors

Anhydrous Cu(hfac)$_2$ is prepared from commercial Cu(hfac)$_2 \cdot$H$_2$O by dehydration in vacuo over P$_2$O$_5$ or H$_2$SO$_4$ [1]. This material is used directly in kinetic studies. New Cu(hfac)$_2$ adducts are prepared by dissolving Cu(hfac)$_2$ in dichloromethane and adding a slight excess of the appropriate alcohol or amine; the mixtures are allowed to stand for ca. 24 h and the solid adducts isolated by removal of solvent under reduced pressure. CVD screening experiments with the new adducts are carried out in warm-wall glass reactors that have been described previously [2] [3]. Film properties are determined using a stylus profilometer and a four-point probe. Conformality tests employ patterned SiO$_2$-coated Si wafers.

Kinetic studies

Films are deposited using an impinging flow pedestal reactor assembled in-house [4]. The Cu(hfac)$_2$ precursor is introduced into the primary H$_2$ carrier gas stream using a temperature controlled solid sublimator. Either H$_2$O or i-PrOH is introduced into a second carrier gas stream using a liquid evaporator; the precursor and co-reactant streams are mixed just before entering the reactor. The measured weight changes of the evaporators are used to calculate the mole fraction and partial pressure of each component entering the reactor.

Films are deposited onto WN_x-coated Si(100) substrates. The WN_x coating had been deposited by plasma-assisted CVD from WF_6, N_2, and H_2 [5]. Individual samples are prepared by cleaving small rectangular sections parallel to the (010) and (001) axes of the wafer (approximate sample dimensions = 1.5 cm x 1.5 cm). The substrates are cleaned in boiling trichloroethylene, then rinsed with acetone followed by distilled water. After mounting in the reactor, the sample is pre-treated at 400 °C in 80 Torr of pure H_2 for one hour before starting the deposition experiment.

For each set of deposition conditions we determined the growth rate, film microstructure, and film resistivity. Growth rates are determined from the weight change of the substrate as a function of deposition time (see below). The rates are also reported in units of equivalent thickness (i.e., 1 mg cm^{-2} hr^{-1} = 18.6 nm min^{-1}, for fully dense films). The film microstructure is examined using plane view scanning electron microscope (SEM) images. Film resistance is measured using a four-point probe, and is converted to resistivity using the equivalent film thickness.

RESULTS

New Precursors

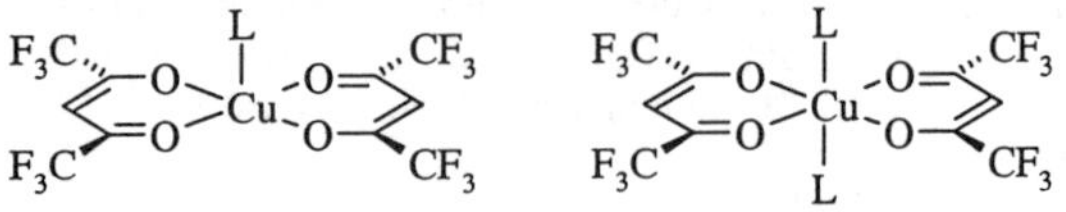

All experiments were performed using adducts of $Cu(hfac)_2$ with alcohol and amine ligands L. They are sketched in Figure 1; their formulas are $Cu(hfac)_2L$ (structure **I**) and $Cu(hfac)_2L_2$ (structure **II**). We

Figure 1. Structures of Cu CVD precursors employed in this study.

$Cu(hfac)_2L$ **I** $Cu(hfac)_2L_2$ **II**

have studied three types of ligands L: (a) monohydric C_1-C_4 alcohols; (b) diols and ether-alcohols; and (c) aliphatic amines. In each case, we envisioned two possible roles for L: accelerating deposition, as has been previously observed for water and alcohols; and acting as the reducing agent (e.g. for carrying out deposition under inert carrier gas).

Adducts with monohydric alcohols. These have the formula $Cu(hfac)_2 \cdot ROH$ and structure (**I**) from Figure 1. We have recently reported initial results with several such precursors [2]; one of the best is $Cu(hfac)_2 \cdot i\text{-PrOH}$. This precursor, like the others, can be used with H_2 as carrier gas; however, it also produces Cu films with an inert carrier gas. Deposition of Cu under N_2 occurs according to equation 1:

$$Cu(hfac)_2 \cdot i\text{-PrOH} \rightarrow Cu(s) + 2\ hfacH(g) + acetone(g) \qquad (1)$$

Cu CVD with $Cu(hfac)_2 \cdot i\text{-PrOH}$ under N_2 occurs approximately four times as fast as with $Cu(hfac)_2 \cdot H_2O/H_2$ under comparable conditions (see data in Table 1). Alternatively, the minimum temperature required for Cu deposition is ca. 40 °C lower for the $Cu(hfac)_2 \cdot i\text{-PrOH}/N_2$ system. (*i*-PrOH vapor must be added to the carrier gas in these experiments in order to inhibit precursor decomposition.)

Figure 2 shows an SEM image of the edge of a cleaved patterned wafer, illustrating the conformality achieved with the $Cu(hfac)_2 \cdot i\text{-PrOH}/N_2$ CVD process.

Adducts with diols and ether-alcohols. We prepared these adducts because they were expected to be more stable toward loss of L than those based on simple alcohols. The adducts prepared were those with ethylene glycol, propylene glycol, 2-methoxyethanol, and 1-methoxy-2-propanol. The best results in this family were obtained with the propylene glycol adduct; see Table 1. The deposition rate under H_2 is nearly as great as for the $Cu(hfac)_2 \cdot i\text{-PrOH}/N_2$ system, and the resistivity of the films is somewhat lower.

Adducts with amines.
Aliphatic amines also react with Cu(hfac)$_2$ to make adducts of the formulas Cu(hfac)$_2$L and Cu(hfac)$_2$L$_2$. Like the diol and ether-alcohol adducts discussed above, all of these amine adducts show significantly improved stability compared to Cu(hfac)$_2$·ROH.

We were interested in how the amine adducts perform as CVD precursors under H$_2$ and N$_2$. All of the amine adducts produce Cu under H$_2$. The best performance is obtained for Cu(hfac)$_2$(allylamine)$_2$ (see Table 1), though this precursor requires a some-

Table 1. CVD results with Cu(hfac)$_2$ adducts[a]

Precursor	Carrier gas	Dep. rate /nm min^{-1}	$\rho/\mu\Omega$ cm
Cu(hfac)$_2$·H$_2$O	H$_2$	5 ± 1	2.8 ± 0.4
Cu(hfac)$_2$·*i*-PrOH	H$_2$[b]	8 ± 2	2.2 ± 0.8
Cu(hfac)$_2$·*i*-PrOH	N$_2$[b]	22 ± 7	2.9 ± 1.0
Cu(hfac)$_2$(propylene glycol)	H$_2$[c]	20 ± 7	2.1 ± 0.5
Cu(hfac)$_2$(allylamine)$_2$	H$_2$[c,d]	20 ± 6	2.6 ± 0.9

[a]200 °C substrate temperature, 1 atm, 800 mL min^{-1}, unless otherwise noted. [b]With *i*-PrOH(g) added. [c]400 mL min^{-1}. [d]Substrate temperature 230 °C.

what higher deposition temperature than the other adducts. Some of the amine adducts also deposit Cu films under N$_2$; an example is the pyrrolidine adduct, which functions as follows:

$$\text{Cu(hfac)}_2(\text{pyrrolidine-NH}) \longrightarrow \text{Cu} + 2\text{ hfacH} + \text{pyrrolidine-N} \tag{2}$$

Thus, it is possible for Cu(hfac)$_2$–amine adducts to function as self-reducing Cu CVD precursors. However, these reactions require higher substrate temperatures than the H$_2$ reactions, and films produced under N$_2$ also show significantly higher resistivities.

Kinetic studies

The kinetic results are summarized in Figure 3. This graph shows the amount of copper deposited as a function of deposition time for three sets of experiments, either with or without an added co-reactant. The other reaction conditions are held constant: H$_2$ partial pressure (P_{H_2} = 80 Torr), Cu(hfac)$_2$ partial pressure ($P_{\text{Cu(hfac)}_2}$ = 2.5±0.3 Torr), and substrate temperature (T_s = 300 °C).

Direct H$_2$ reduction. The lowest symbols in Fig. 3 show the deposition rate obtained when no co-reactant is present (i.e., H$_2$ is the sole reducing agent). The slope through these results yields a deposition rate of 0.36±0.01 mg cm^{-2} hr^{-1} (ca. 7 nm min^{-1} for a fully dense film). This is consistent with our earlier results [6]. The x-axis intercept indicates a nucleation time of 2.5±1.0 min.

Figure 4(a) shows a plane view SEM image of a film deposited for 30 minutes using the baseline conditions. The copper loading of this film is 0.16 mg cm^{-2}, which corresponds to an equivalent thickness of 0.18 μm. The film appears to be discontinuous, with a mean cluster diameter between 0.3–0.5 μm and considerable void volume between clusters. This is reflected in the film resistivity (ca. 20 μΩ-cm), which greatly exceeds the value of bulk copper. The discontinuous morphology, together with the

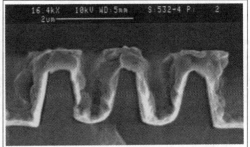

Figure 2. SEM image of Cu film deposited on patterned Si substrate. Conditions: Cu(hfac)$_2$·*i*-PrOH, N$_2$, 10 min, 200 °C; 0.8-μm SiO$_2$ lines.

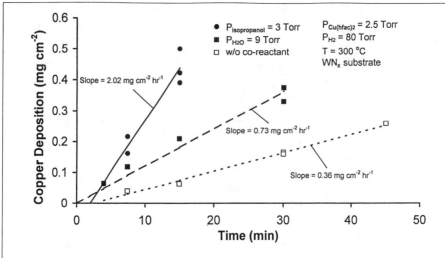

Figure 3. Influence of water and *i*-PrOH on CVD of copper using Cu(hfac)₂ as precursor.

nonzero induction time noted in Figure 3, indicates that film nucleation is a limiting factor under these conditions.

H₂O-assisted reduction. The middle set of results in Fig. 3 shows the effect of adding H_2O as a co-reactant ($P_{H_2O} = 9$ Torr). The slope through the data gives a deposition rate of 0.73 ± 0.03 mg cm^{-2} hr^{-1} (ca. 14 nm min^{-1}). This is a two-fold increase, relative to the baseline conditions without H_2O. The regression analysis also indicates that there is no significant induction time. Thus the H_2O co-reactant also appears to increase the film nucleation rate.

Figure 4(b) shows the SEM image for a film deposited for 15 minutes using $P_{H_2O} = 3$ Torr. The Cu loading is 0.15 mg cm^{-2}, similar to that in Fig. 4(a). However, the morphology is markedly different. The film is nearly continuous, the mean cluster size is at least two-fold smaller, and the void volume between clusters is significantly reduced. The visual appearance of the film is greatly improved, with the color becoming a shiny copper and the reflectivity approaching mirror-like quality. Most significantly, the measured resistivity has achieved the bulk value within experimental uncertainty. These results appear to confirm that H_2O causes a significant increase in the nucleation rate of new Cu clusters on the present substrate.

Alcohol-assisted growth. The uppermost symbols in Fig. 3 show the effect of adding 3 Torr of *i*-PrOH as a co-reactant. The slope yields a deposition rate of 2.02 ± 0.24 mg cm^{-2} hr^{-1} (ca. 38 nm min^{-1}). This is a three-fold increase over the rate obtained with added H_2O, and a six-fold increase over the rate with H_2 alone. However, the *i*-PrOH results also yield an induction time of 2.0 ± 1.4 min, which is similar to that observed in the baseline experiments. Thus *i*-PrOH is found to have a much stronger enhancement effect than H_2O for the steady state growth rate, but it appears not to have a similar enhancement for the nucleation rate.

Figure 4(c) shows the SEM image of a film deposited with 3 Torr of *i*-PrOH for 7.5 min. The Cu loading is 0.22 mg cm^{-2}, which is about 50% thicker than the H_2O-assisted film shown in Fig 2. Despite having a higher Cu loading, the *i*-PrOH-assisted film grown is much less continuous. This is also reflected in the measured resistivity of 22 μΩ-cm.

CONCLUSIONS

Our studies of new precursors suggest that several types of additives lead to improved Cu CVD performance compared to Cu(hfac)₂. In addition to alcohols such as *i*-PrOH, which we

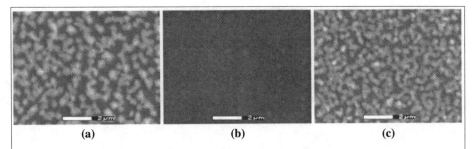

Figure 4. SEM images of CVD Cu films deposited at 300 °C on WN$_x$-coated Si. (a) No co-reactant, 30 min, 0.16 mg cm^{-2} Cu. (b) With 3 Torr H$_2$O, 15 min, 0.15 mg cm^{-2} Cu. (c) With 3 Torr i-PrOH, 7.5 min, 0.22 mg cm^{-2} Cu.

identified previously, some diols and amines are beneficial. The most promising of the new additives are propylene glycol and allylamine.

Kinetic studies have demonstrated the quantitative enhancements available when water and i-PrOH are added to the Cu(hfac)$_2$/H$_2$ system. The rate enhancement is significantly larger with i-PrOH. However, thin films deposited in the presence of i-PrOH tend to be discontinuous, on account of relatively poor nucleation; on the other hand, water improves both nucleation and deposition rates. Therefore, especially in potential applications to seed layer formation, water (or a mixture of water and another additive) may be the most effective in improving film properties.

Preliminary studies of conformality in these systems are also promising. We are now performing quantitative rate studies with the new additives, and determining the effects of the additives on adhesion of the Cu films to underlying substrate layers.

ACKNOWLEDGMENTS

This research was supported by the National Science Foundation through Grant CTS-9612157. Partial support was also provided by the Department of Energy (EPSCoR) and by the Louisiana Board of Regents (graduate fellowship to A. M. J.). We thank Dr. P. M. Smith (Sandia National Laboratories) for a gift of substrate materials.

REFERENCES

1. L. L. Funck and T. R. Ortolano, *Inorg. Chem.* **7**, p. 567 (1968).

2. A. W. Maverick, A. M. James, H. Fan, R. A. Isovitsch, M. P. Stewart, E. Azene, and Z. T. Cygan, *ACS Symp. Ser.* **727**, p. 100 (1999).

3. R. Kumar, F. R. Fronczek, A. W. Maverick, W. G. Lai, and G. L. Griffin, *Chem. Mater.* **4**, p. 577 (1992).

4. N. S. Borgharkar, G. L. Griffin, H. Fan and A. W. Maverick, *J. Electrochem. Soc.*, **146**, p. 1041 (1999).

5. J.-P. Lu, W. Y. Hsu, J. D. Luttmer, L. K. Masgel and H. L. Tsai, *J. Electrochem. Soc.*, **145**, p. L21 (1998).

6. N. S. Borgharkar and G. L. Griffin, *J. Electrochem. Soc.*, **145**, p. 347 (1998).

Effects of Process Variables on Cu(TMVS)(hfac) Sourced Copper CVD Films

D. YANG, J. HONG AND T. S. CALE
Focus Center – New York, Rensselaer: Interconnections for Gigascale Integration
Rensselaer Polytechnic Institute, 110 8th St., Troy, NY 12180

ABSTRACT

This paper presents the effects of substrate temperature, precursor flow, carrier gas flow, total pressure, substrate distance, and water vapor flow in Cu(TMVS)(hfac) sourced Cu CVD on TaN substrates. The measured film properties for the study of the growth stage of deposition are deposition rate, resistivity, surface roughness, and reflectivity. The estimated activation energy for overall film growth in the temperature range of 423 to 498 K is about 0.77 eV at fixed conditions of 20 mg/min precursor flow, 50 sccm carrier gas flow, and 1 Torr total pressure. In order to enhance the nucleation stage of deposition, water vapor is introduced during deposition. The adhesion of Cu nuclei deposited with water vapor is apparently stronger, compared to those deposited without water vapor. The properties of the final, thicker films depend very much on water vapor flow rate and its introduction time. From this study, we conclude that introducing water vapor before or during the initial stage of deposition enhances nuclei density and improves growth rate, resistivity, surface roughness, and adhesion.

INTRODUCTION

Copper based metallization has been introduced into leading edge ICs because of its promising reliability performance, as well as the potential cost savings associated with damascene processing [1-6]. The cost savings and IC performance increases due to Cu introduction will be more fully realized if deep sub-quarter micron, high aspect ratio, features can be filled inexpensively with barrier material and copper. This is one of the most important issues for multilevel metallization (MLM) process flows today [7].

The research reported in this paper focuses on our efforts to develop an engineering level understanding of Cu CVD; *i.e.*, understanding that improves process protocols. Surface morphology and film properties (including adhesion) of deposited films are determined largely by the properties of the initial layers. Therefore, our experiments include studies both of the early stage and the growth stage of Cu deposition. The study of the early stage of CVD processes will help us understand the roles of nucleation and grain growth during continuous film formation. This study focuses on understanding the properties of CVD Cu films, and on finding a method to develop more robust processes in order to deposit conformal Cu films.

EXPERIMENT

In this study, a LPCVD microreactor was used to investigate nucleation, growth, and microstructure evolution during Cu CVD using Cu(TMVS)(hfac) [8] as a Cu precursor. Cu was deposited on a heated TaN substrate (tantalum nitride coated on $SiO_2/Si(100)$) provided by Tokyo Electron Arizona. TaN substrates were chosen for this study because it is a common diffusion barrier for Cu. Cu(TMVS)(hfac) was delivered into the reaction chamber by a direct liquid injection (DLI) system using He as a carrier gas. During the depositions, 2 cm by 2 cm TaN substrates were heated by a resistance heater. A load lock chamber was installed to avoid air contamination inside the reaction chamber. No surface clean or pretreatment was performed on the TaN substrate, but the surfaces were blown with dry nitrogen prior to depositions.

Conference Proceedings ULSI XV © 2000 Materials Research Society

In this study, experiments included studies of the early stage (deposited for 2 and 5 s) and the growth stage (deposited for 10 min) of deposition. For growth during Cu CVD, the selected factors and ranges are summarized in Table 1. The base conditions chosen in this study are 473 K substrate temperature, 20 mg/min precursor flow, 50 sccm carrier gas flow, 1 Torr total pressure, 1 cm showerhead-to-substrate distance, and 333 K process line (from vaporizer to showerhead) temperature. The objectives of this work were to study how each CVD process parameter (substrate temperature, precursor flow rate, carrier gas flow rate, total pressure, and substrate distance from the showerhead) affects selected Cu film deposition phenomena. We examine nucleation and growth rate, and the deposited Cu film properties in terms of morphology, resistivity, and smoothness. In addition, the effects of water vapor as a co-reactant during the initial stage and the growth stage of deposition were studied.

Table 1. Selected study factors and ranges

factor (unit)	range
substrate temperature (K)	348 – 523
precursor flow rate (mg/min)	10 – 50
total pressure (Torr)	0.5 – 4.0
carrier gas flow rate (sccm)	30 – 300
substrate distance (cm)	0.5 – 4.0
water vapor flow rate (sccm)	0 - 30
water vapor introduction time (s)	0 - 600

A surface profilometer (α-step) was used to estimate the average deposited film thickness after making steps in the Cu deposited film at near the center by photolithographic and wet etching methods [9]. Film resistivity was measured with a 4-point probe. Atomic force microscopy (AFM) was used in the contact mode to analyze surface morphologies of the deposited films. The instrument's software [10] was used to analyze the acquired images to determine the nuclei size, density and uniformity resulting from the various process conditions. Surface roughness reported is a root-mean-square (RMS) value and normalized roughness is RMS roughness divided by film thickness. Film reflectivity was determined using a Nanospec/AFT at a wavelength of 480 using a polished Si(100) wafer as the reference. Film texture was determined using x-ray diffraction (XRD).

RESULTS AND DISCUSSION

Fig. 1(a) shows the variation of the deposition rate with substrate temperature. There is no appreciable Cu deposited at temperatures below 348 K for 10 min. Deposition rate increases quickly from 423 to 498 K. Over the studied temperature range, resistivities vary from 2 to 10 μΩ-cm, and the lowest resistivity is about 2.06 μΩ-cm at 473 K, as shown in Fig. 1(b). By increasing the temperature, resistivity first decreases and then it increases. The discontinuities in thinner films deposited at low temperatures can explain the higher resistivities at lower deposition temperatures. An increase in resistivity is noticed at temperatures above 473 K. Normalized surface roughness is shown in Fig. 1(c). The lowest value within the studied range is about 2.7% at 473 K. In general, the adhesion (based on scratch tests) is poor at deposition temperatures above 473 K. Based on the deposition rate information shown in Fig. 1(a), the activation energy for film growth is estimated from an Arrhenius plot to be about 18 kcal/mol (0.77 eV) in the temperature range of 423 to 498 K. It is in rough agreement with 10 – 20

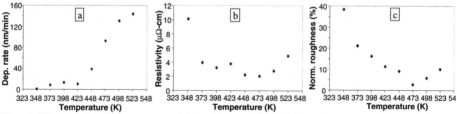

Figure 1. Effects of substrate temperature on: (a) deposition rate, (b) resistivity, and (c) normalized roughness.

kcal/mol on W and TiN substrates reported in the literature [1,2,3], even though the activation energy usually depends on substrate types and deposition conditions.

The effects of precursor flow rate shown in Fig. 2 were studied at 448 K fixed substrate temperature. The mole fraction of Cu(TMVS)(hfac) in the incoming vapor is 1.21×10^{-2} to 6.05×10^{-2} in the studied flow range of 10 to 50 mg/min. The deposition rate increases as the precursor flow increases up to 40 mg/min. This is because the partial pressure of Cu precursor increases as Cu precursor flow increases. However, the deposition rate decreases at flows higher than 40 mg/min; more than 40 mg/min appears to be too much precursor flow for 50 sccm of carrier gas at the studied conditions. The maximum deposition rate within the studied range is about 120 nm/min at a precursor flow of 40 mg/min. Resistivity decreases slightly by increasing precursor flow, and the lowest value is about 1.91 $\mu\Omega$-cm at 30 mg/min. Normalized roughness decreases to 4% as precursor flow increases up to 40 mg/min and then it quickly increases. This result corresponds with the decreased deposition rate above 40 mg/min; the precursor might be partially condensed. Reflectivity decreases as precursor flow increases as the surface roughness increases. RMS surface roughnesses of the films deposited at 10, 20, 30, 40, and 50 sccm precursor flow are 19, 31, 37, 45, and 90 nm respectively. In general, as film thickness increases, reflectivity decreases because of increasing surface roughness.

Total pressure does not affect the deposition rate significantly within the studied range of 0.5 to 4.0 Torr, but resistivities vary significantly. The lowest resistivity and normalized roughness are achieved at 1 Torr total pressure. From the study of substrate distance effects, maximum deposition rate is achieved at 1 cm substrate distance. The deposited film uniformity is poor at distances shorter than 1 cm, and this is possibly because the reactant flow pattern is irregular. Results from the study of carrier gas flow effects indicate that as the carrier gas flow increases, the deposition rate initially increases up to 50 sccm of carrier gas flow, and then the deposition rate decreases. The decrease in the deposition rate above 50 sccm is explained by the decrease in precursor partial pressure. The decrease in the deposition rate below 50 sccm can be due to not enough He flow to carry 20mg/min precursor, resulting in partial condensation or deposition of precursor before the reaction chamber.

Nucleation Stage Study

A set of experiments was performed to study the nucleation stage of CVD Cu. Depositions were conducted for 2 s and 5 s periods with or without water vapor at fixed conditions of 473 K substrate temperature, 20 mg/min precursor flow, 100 sccm carrier gas flow, 0.5 Torr total pressure, and 1.3 cm (0.5 inch) substrate distance. According to the AFM micrographs of Cu nuclei deposited for 2 s without and with water vapor, shown in Fig. 3(a) and (b), the incubation time is significantly reduced by introducing water vapor during the initial stage of deposition.

Fig. 3(c) and (d) show Cu nuclei on TaN substrates after 5 s depositions without and with water vapor. The nuclei sizes and densities vary significantly by introducing water vapor. The 5 s deposition without water vapor (Fig. 3(c)) results in very few sparsely distributed nuclei. This is

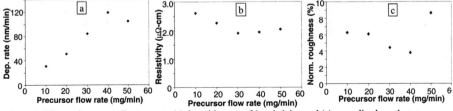

Figure 2. Effects of precursor flow rate on: (a) deposition rate, (b) resistivity, and (c) normalized roughness.

probably due to the low deposition temperature (not enough energy) at the given process conditions. However, the nuclei density increases significantly by introducing 50 sccm of water vapor flow during the initial stage of deposition, which is shown in Fig. 3(d). These results indicate that nucleation rate is enhanced by addition of water vapor, which has been reported by Mermet *et al.* using TiN substrates [5]. The 5 s deposition without water vapor results in a nuclei density of about 30 nuclei/μm^2. The 5 s deposition with water vapor during the deposition results in about 332 nuclei/μm^2. The average nuclei sizes from the micrographs in Fig. 3(c) and (d) are about 63 and 80 nm in diameter, respectively.

During AFM analysis, image scans in the contact mode failed several times for the films deposited without water vapor, which is most likely because the AFM tip moves the nuclei along substrate surface. This result may indicate that the adhesion of Cu nuclei on TaN films is apparently stronger when water vapor is introduced during the initial deposition, compared to those deposited without water vapor.

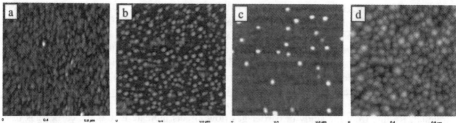

Figure 3. AFM micrographs resulting from CVD Cu depositions at 473 K for: (a) 2 s without water vapor, (b) 2 s with water vapor, (c) 5 s without water vapor, and (d) 5 s with water vapor flow.

Growth Stage Study

Depositions were conducted for 10 min at fixed conditions of 473 K substrate temperature, 20 mg/min precursor flow, 50 sccm carrier gas flow, and 1 Torr total pressure. Water vapor flow rate was about 25 sccm before deposition and/or about 10 sccm during deposition. Fig. 4(a) shows the variation of the deposition rate with water vapor injection conditions. By introducing water vapor, the deposition rate is increased up to about 9 times compared to those without water vapor. This result is in agreement with Jain *et al.* [4]. However, the deposited film qualities deteriorate when water vapor is introduced longer than about 2 min: *i.e.*, resistivity is higher, density is lower. and adhesion is lower (based on scratch tests). Based

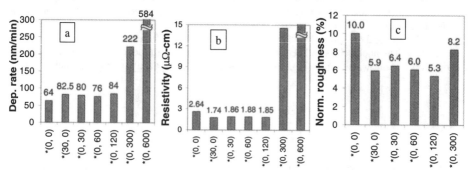

Figure 4. Results from Cu CVD films deposited for 10 min at 473 K: (a) deposition rate, (b) resistivity, and (c) normalized roughness. *Water vapor introduction time (before deposition, during deposition) in seconds.

on XRD data, introducing water vapor longer than 2 min results in Cu_2O formation in the deposited film. Similar resistivity results have been reported by Mermet *et al.* [5]. Introducing water vapor before or during the initial stage (up to 2 min) of deposition results in lower resistivities, smoother films, and higher deposition rates, compared to those without water vapor. The lowered resistivity and improved smoothness is due to improved nucleation stage with water vapor. Introducing 25 sccm of water vapor for 30 s before starting a 10 min deposition provides the best film among the films deposited at the experimental conditions shown in Fig. 4. These conditions result in a deposition rate of 83 nm/min, 1.74 $\mu\Omega$-cm resistivity, 5.9% normalized roughness, and 44% reflectivity. Based on four repeated experiments at a fixed condition, the trends of the results presented in this paper are reproducible within 10% for deposition rate, 6% for resistivity, and 15% for normalized roughness.

CONCLUSIONS

This paper describes an experimental study to establish the effects of various process parameters in Cu(TMVS)(hfac) based Cu CVD on TaN substrates. The results indicate that deposition temperature should be higher than 373 K for appreciable deposition rate. The estimated activation energy for overall film growth is about 0.77 eV in the temperature range of 423 to 498 K. During the nucleation stage, depositions with water vapor result in higher nuclei density, uniform distributions, and apparently stronger adhesion (based on AFM analysis) as compared to those without water vapor. The properties of the final, thicker films are very dependent on water vapor flow rate and introduction time. Introducing water vapor before or during the initial stage of deposition improves growth rate, resistivity, surface roughness, and adhesion (based on scratch tests). The best process protocol for Cu CVD in our reactor system is introducing less than 10 sccm of water vapor before or during the initial stage of deposition at 473 K substrate temperature, 20 mg/min precursor flow, 50 sccm carrier gas flow, and 1 Torr total pressure.

ACKNOWLEDGEMENTS

We gratefully acknowledge Tokyo Electron Arizona for partial support of this project. We would like to thank Dr. M. Tomozawa at RPI for his help with AFM analysis.

REFERENCES

1. A. Jain, K.-M. Chi, T. T. Kodas and M. J. Hampden-Smith, *J. Electrochem. Soc.* **140**, 1434 (1993).
2. E. S. Hwang and J. Lee, *J. Vac. Sci. Technol.* **B16(6)**, 3015 (1998).
3. Y. K. Chae, Y. Shimogaki and H. Komiyama, *J. Electrochem. Soc.*, 145, 4226 (1998)
4. A. Jain, A. V. Gelatos, T. T. Kodas, M. J. Hampden-Smith, R. Marsh and C. J. Mogab, *Thin Solid Films*, **262,** 52 (1995).
5. J.-L. Mermet, M.-J. Mouche, F. Pires, E. Richard, J. Torres, J. Palleau and F. Braud, *Journal De Physique IV*, **5**, C5-517 (1995).
6. A. E. Braun, *Semiconductor International*, **22(9),** August 1999, p. 58.
7. National Technology Roadmap for Semiconductors, SIA (1997).
8. CupraSelect [Cu(TMVS)(hfac)], where hfac = 1,1,1,5,5,5-hexafluoroacetylacetonate and TMVS = trimethylvinylsilane. is a Cu(I) precursor manufactured by Schumacher.
9. S. Kim, J.-M. Park and D.-J. Choi, *Thin Solid Films*, **315**, 229 (1998).
10. PSI ProScan Software Version 1.5 released on June 5, 1998 (Park Scientific Instruments).

New Copper Precursor for Chemical Vapor Deposition of Pure Copper Thin Films

Weiwei Zhuang*, Lawrence J. Charneski, David R. Evans and Sheng Teng Hsu
Sharp Laboratories of America, 5700 NW Pacific Rim Blvd., Camas, WA 98607

ABSTRACT

A new volatile liquid copper precursor, (1-decene)Cu(I)(hfac), has been synthesized. By the introduction of less than 5% extra 1-decene stabilizer ligand, this organometallic compound is stable in a pure liquid phase at room temperature, but is air and moisture sensitive. The structure of this compound has been proposed by the studies on its ^1H and ^{13}C NMR spectra. By further studies on its proton NMR spectrum, the bonding character and the liquid phase stabilization mechanism have been proposed and discussed. The characters of this new copper precursor are its liquid phase stability, inexpensive and fitness for CVD liquid delivery systems for pure copper thin film deposition.

Pure copper thin films have been deposited onto titanium nitride substrates by using this new copper precursor via chemical vapor deposition. The copper thin films exhibit good adhesion to metal and metal nitride substrates, low resistivity, pure copper metal composition and excellent uniformity. In our CVD experiments, water vapor was introduced into the reaction chamber by wet helium gas. The effect of CVD processing conditions on the properties of copper thin films has been studies. The preferred CVD processing conditions have also been explored. The establishment of a very dilute water atmosphere in the reaction chamber not only helps the decomposition of the precursor on the wafer surface, but also decreases the resistivity of the copper thin film. The density of copper thin films varies from 70 to 86% of pure bulk copper metal, and the precursor deposition efficiency is in the range of 9 – 16 wt%. The resistivity was determined by measuring the sheet resistance and the thin film thickness, and plotted with the film thickness. Our primary studies on copper thin films deposited by using this new precursor show that this new synthesized copper precursor is a very promising source for chemical vapor deposition of copper thin films used in copper seed layer and copper interconnect applications.

INTRODUCTION

Copper thin films as an interconnect material in IC devices have been studied over a long time. Currently, the typical copper thin film process is being successfully implemented through two steps: copper seed layer deposition via PVD technology, and then the electroplating. Because of the poor step coverage with PVD technology, the copper thin film deposition will be a problem when the trench size becomes smaller and smaller. On the contrary, chemical vapor deposition process gives excellent step coverage, which enables this technology to be an ideally suitable for copper thin film deposition in narrow and deep trenches.

For CVD copper thin film deposition, many copper precursors have been explored, such as (tmvs)Cu(I)(hfac) (CupraSelect),[1-3] (DMCOD)Cu(I)(hfac) (DMCOD = mixture of dimethyl-1,5-cyclooctadiene and dimethyl-1,6-cyclooctadiene), (3-hexyne)Cu(I)(hfac)[4-7] and (2-butyne)Cu(I)(hfac).[8] Pure copper thin films have been obtained by using all of these precursors, but to meet production requirements, there are still some issues to be solved. The existing problems include poor adhesion of copper thin films to metal nitride substrates, decomposition during liquid delivery, and the cost for precursor synthesis. Thus our interest is directed towards finding new copper precursors, which are stable in liquid phase and can be used in CVD liquid delivery systems for copper thin film deposition.

EXPERIMENT

All precursor synthesis manipulations were carried out in a nitrogen-filled air-free glove box or by using standard Schlenk techniques. Dichloromethane and 1-decene organic stabilizer were purified by refluxing over calcium hydride and distilled before use. Cu_2O and hexafluoroacetylacetone were purchased from Aldrich, and used as received without further purification. Proton and ^{13}C NMR spectra were collected on a QE 300-MHz NMR instrument.

In the synthesis of (1-decene)Cu(I)(hfac), Cu_2O and 1-decene were added into a flask equipped with CH_2Cl_2, and then hfac was slowly dropped in under strong stirring. The color of the solution gradually changed to yellow-green in a short time. The solution was stirred for a while and then filtered through celite. The solvent was removed via dynamic vacuum over several hours, and then the liquid compound was obtained, which was filtered again. The yield of the product is about 85% based on hfac. The proton NMR spectrum of the compound gave the distinct peaks representing all of the protons in the compound.

Copper thin films were deposited on TiN/SiO_2 substrate via CVD. Six-inch wafers were used in all of the experiments. The CVD equipment is based upon a Concept II, made by Novellus. The liquid precursor delivery system and the vaporizer were made by ATMI. Water vapor was introduced into the reaction chamber via the bubbling of helium gas. The CVD processing conditions are listed in Table 1.

Table 1. Copper thin film CVD processing conditions

$F_{precursor}$ (mL/min.)	$T_{vaporizer}$ ($^{\circ}C$)	$T_{showerhead}$ ($^{\circ}C$)	F_{He} (sccm)	F_{WetHe} (sccm)	$T_{dept.}$ ($^{\circ}C$)	$P_{dept.}$ (torr)
0.05 – 0.5	60 - 80	60 – 80	25 – 200	0.5 – 10	160 – 230	0.1 - 2

The resistivity was determined by sheet resistance and film thickness. Precursor efficiency was determined by the amount of copper deposited onto the wafer and the amount of copper in the precursor used for the deposition. The copper thin film deposition rate was obtained by the plot of film thickness via deposition time.

RESULTS AND DISCUSSION

The evaluation of this new copper precursor was carried out for the effect of CVD processing conditions on copper thin film properties. Except for the one parameter which was adjusted during the processing, the other processing parameters are shown in Table 2.

Table 2. Copper thin film CVD processing typical conditions

$F_{precursor}$ (mL/min.)	$T_{vaporizer}$ ($^{\circ}C$)	$T_{showerhead}$ ($^{\circ}C$)	F_{He} (sccm)	F_{WetHe} (sccm)	$T_{dept.}$ ($^{\circ}C$)	$P_{dept.}$ (torr)
0.15	65	70	100	5	190	0.5

The results of the effect of CVD processing conditions on copper thin film properties have been summarized from Figure 1 to Figure 5. Figure 1 shows the effect of deposition temperature. Below 230 $^{\circ}C$, the increase of deposition temperature increases the deposition rate and decreases the resistivity, whereas when the deposition temperature is higher than 230 $^{\circ}C$, the increase of the deposition temperature will result in the decrease of deposition rate and the increase of resistivity. The precursor flow rate has a great effect on the deposition rate, as shown in Figure 2. The thicker the copper thin film, the lower the film resistivity. The copper film resistivity decreases quickly as the film thickness increases to 2,500 Å, and then goes down almost in a linear relationship with the increase of film thickness. In another experiment for the deposition of one micron thick copper thin films for CMP purposes, a resistivity of around 1.9 $\mu\Omega$·cm could be reached. The precursor flow rate has a close linear relationship with the copper thin film deposition rate, as shown in Figure 2(b).

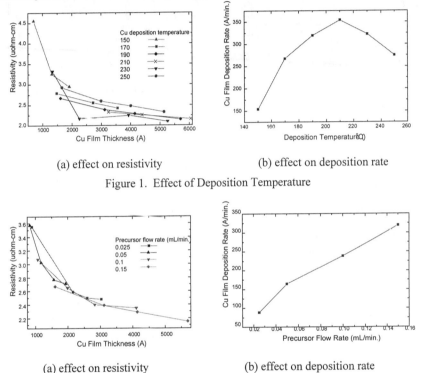

(a) effect on resistivity   (b) effect on deposition rate

Figure 1. Effect of Deposition Temperature

(a) effect on resistivity   (b) effect on deposition rate

Figure 2. Effect of Precursor Flow Rate

    A change in chamber pressure will also effect the copper thin film deposition rate and resistivity. From the Figure 3, lower resistivity can be obtained by increasing chamber pressure, but it decreases the deposition rate at the same time. The increase in chamber pressure will also increase the copper thin film density, which may be the reason for the low resistivity result. The effect of helium carrier gas flow has been summarized in Figure 4, which indicates that the increase in helium carrier gas flow rate induces a decrease of the copper thin film deposition rate, but is not a big influence on the resistivity. Figure 5 gives the water effect on copper film properties. A certain amount of water vapor helps the decomposition of copper precursor on the wafer surface, and improves the copper thin film deposition. The amount of water should be well controlled in the CVD processing. In our case, a low resistivity of copper thin films can be obtained by introducing about 5 sccm wet helium gas. From the Figure 5(b), we can also see that both the copper film density and the precursor efficiency will be increased if some water vapor has been introduced.

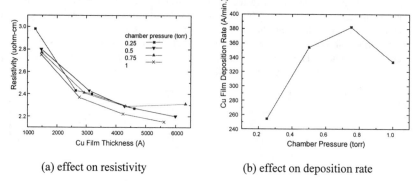

(a) effect on resistivity   (b) effect on deposition rate

Figure 3. Effect of Chamber Pressure

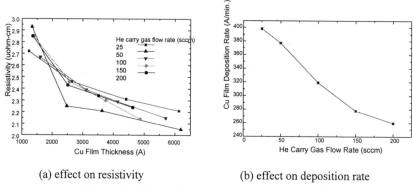

(a) effect on resistivity   (b) effect on deposition rate

Figure 4. Effect of Helium Carrier gas Flow Rate

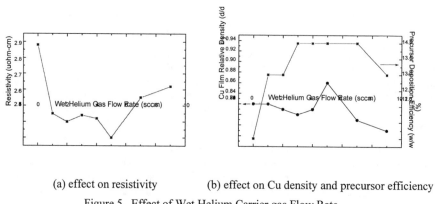

(a) effect on resistivity   (b) effect on Cu density and precursor efficiency

Figure 5. Effect of Wet Helium Carrier gas Flow Rate

The studies on electro-migration properties of copper thin films and the formation of copper seeds by using this new precursor are still on-going, with the results to be published elsewhere.

SUMMARY

A new copper precursor has been synthesized. Pure copper thin films have been obtained by using this new copper precursor via CVD processing. The effect of CVD processing conditions on the properties of copper thin films has been discussed, from which reasonable CVD processing parameters can be summarized.

REFERENCES

1. J. A. T. Norman, B. A. Muratore, P. N. Dyer, D. A. Roberts, and A. K. Hochberg, Journal de Physique IV, **1,** p. C2-271. (1991).

2. J. A. T. Norman, *US Patent, US5098516,* March 24, 1992.

3. J. A. T. Norman, *US Patent, US5187300,* Feb. 16, 1993.

4. H. K. Shin, K. M. Chi, M. J. Hampden-Smith, T. T. Kodas, J. D. Farr, and M. Paffett, Chemistry of Materials, **4,** p. 788. (1992).

5. T. H. Baum, and C. E. Larson, J. Electrochem. Soc., **140(1),** p. 154. (1993).

6. A. Jain, K.-M. Chi, T. T. Kodas, M. J. Hampden-Smith, Farr, J. D. and Paffett, M. F. Chemistry of Materials, **3,** p. 995 (1991).

7. T. H. Baum, and C. E. Larson, *US Patent, US5,096,737,* March 17, 1992.

8. T. H. Baum, and C. E. Larson, Chemistry of Materials, **4,** p. 365 (1992).

MATERIAL ISSUES IN SILVER METALLIZATION

E. Kondoh and T. Asano

Center for Microelectronic Systems, Kyushu Institute of Technology

Kawazu, Iizuka 820-8502, Japan

e-mail: kondoh@cms.kyutech.ac.jp

ABSTRACT

This paper deals with two topics of Ag metallization for ultra-large-scale integrated circuit (ULSI) application. The first topic is the choice of a barrier material. Common barrier metals are studied in view of the reactivity with Ag and of the improvement of Ag (111) texture. The second topic is the choice of a cap material that is used to suppress Ag surface deterioration during plasma processing. Refractory metals and noble metals show good cap performance. It is demonstrated that Cu can be used as a cap material for damascene prototype fabrication, where Cu is chosen because of the compatibility with the current Cu technology, of its low resistivity, and of the feasibility of electroless/selective deposition onto Ag.

INTRODUCTION

The overall performance of ultra-large-scale integrated circuits (ULSIs) is now limited by resistance-capacitance (RC) signal delay in interconnections. Copper has been successfully implemented in functioning ULSIs, replacing conventional Al interconnections. Ag is the only element that has a lower resistivity than Cu at service temperatures. Ag has been widely used as a conductive material in electronics not only because of its low resistivity but also because of its high thermal conductivity. The thermal condition of interconnections is becoming much severer in connection with the introduction of low-density/porous low-dielectric-constant materials. This situation will be extremely important in the coming air-gap/air-bridge interconnection era where thermal conduction within the interconnections would dominated heat dissipation. This paper discusses two crucial issues of Ag metallization after a brief summary of Ag properties.

SILVER PROPERTIES

Table I summarizes properties of Ag, Cu, and Al [1-7] from a metallization point of view. Ag has similar physical and chemical properties as Cu. For instance, Ag is a univalent noble metal, and has a fcc structure, high melting point, and large Young modulus. Ag has the lowest resistivity and highest thermal conductivity among all common industrial metals at IC operation temperatures. The resistivity difference between Ag and Cu increases as temperature increases, because Ag has a lower temperature coefficient of resistivity (TCR) than Cu.

Ag can be deposited by various techniques such as electro/electroless deposition, sputtering, and chemical vapor deposition. Process issues of Ag metallization are similar to those of Cu metallization — 1) Plasma processing: Because of a low vapor pressure of Ag halides, halogen-based conventional dry etching is thought to be not straightforward. Plasma tolerance should also be investigated; 2) Diffusion in SiO_2: Ag atoms can easily diffuse into the SiO_2 network, because the interaction between Ag and oxygen atoms is weak; 3) Adhesion: Ag is not as reactive as Ti or

Conference Proceedings ULSI XV © 2000 Materials Research Society

Al, Ag does not show good adhesion to common insulation materials; 4) Migration: It is known that Ag tends to migrate by an electrochemical mechanism forming 'filaments' or dendrites. Most of these issues are the same as for Cu and have been mostly solved in the current Cu damascene vehicle. ULSI metallization with Ag is thus to be on the extension of the current Cu process technology.

BARRIER METAL SCREENING

The choice of a proper barrier material is discussed as a first issue to be investigated [8,9]. For rough screening, Ag films capped or stacked with commonly used refractory metals are studied. Ag films, 5000 Å thick, were deposited onto barrier (refractory) metals and were then annealed at 400 °C for 30 min in vacuum. The change of the sheet resistance after the annealing is shown in Fig. 1 along with the data for Cu. The Ag films do not show an increase in sheet resistance, whereas a significant increase is seen for the Cu/Ti (Ti liner) specimen. The annealing at 650 °C for 30 min resluted in a resistance increase either for the Ag/Ti or for the Cu/Ti stack, whereas the Ag/Ti stack showed a much smaller increase than the Cu/Ti stack. X-ray diffraction (XRD)

Table I Summary of physical properties of Ag, Cu, and Al.

	Ag	Cu	Al	Ref.
Atomic Weight	107.87	63.55	26.98	1
Melting point (°C)	961.8	1084.6	660.3	1
Density gcm^{-3} (20 °C)	10.5	8.96	2.698	2
Resistivity (μ? •cm) 0 °C	1.47	1.54	2.50	3
150 °C	2.39	2.56	3.14	
Resistivity temperature coefficient (10^{-3} K^{-1})	6.1	6.8	4.29	3
Thermal Conductivity (W/cm K) 25 °C	4.29	4.01	2.37	1
127 °C	4.25	3.93	2.40	
Young modulus (GPa)	82.7	129.8	70.6	4
Thermal Expansion (10^{-6} K^{-1})	18.9	16.5	23.1	1
Self diffusion coefficient at 400 °C (cm^2/s)	2.7×10^{-15}	5.2×10^{-17}	2.0×10^{-11}	5
Diffusivitiy in Si at 400 °C (cm^2/s)	2.1×10^{-15}	2.8×10^{-6}	1.7×10^{-23}	4
Diffusivitiy in SiO$_2$ at 400 °C (cm^2/s)	$\sim 10^{-10}$	$\sim 10^{-11}$		6
Effective Mass (Z*)	-5.5	-5.4	-20.6	7
Temperature at 1mmTorr Vapor Pressure of Chloride (°C)	912 (AgCl)	546 (CuCl)	100 subl. (AlCl$_3$)	2
Electron Affinity (eV)	1.302	1.235	0.411	1
First Ionization Potential (eV)	7.57	7.73	5.99	1
M-O Bond Strength (kJ/mol)	220	269	511	1
Standard potential, E$^\varnothing$ (V)	Ag$^+$ + e$^-$ → Ag +0.80	Cu$^+$ + e$^-$ → Cu +0.52	Al^{3+} + 3e$^-$ → Al -1.66	2

analysis showed strong peaks from Ti compounds such as $CuTi_3$. Cu_3Ti_4, and Cu_4Ti in the annealed Cu/Ti film (650 °C) but a very weak AgTi peak in the Ag/Ti.

Ag films have (111) texture. The 111/200 XRD intensity ratio varies depending on the kind of the underlying barrier metal. The Ag film deposited on TaN shows a superior 111 intensity (Figs. 2 and 3). The Ag films on SiO_2, Ta, and Ti also show good (111) texture. The 111/200 intensity ratio of the Ag on TiN is not very large, however, is still much larger than that of random grains. The full-width-at-half-maximum (FWHM) values of the 111 rocking curves are also shown in Fig. 2. The films formed on TaN and Ti show smaller FWHM values.

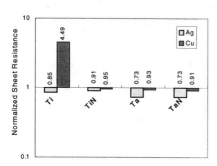

Figure 1   The change of the sheet resistance after annealing (400 °C 30 min) of Ag and Cu films formed on common barrier metals.

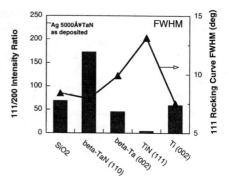

Figure 2   Ag 111 to 100 XRD intensity ratio and 111 rocking curve FWHM for various refractory underlayers.

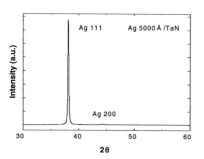

Figure 3 θ-2θ XRD pattern of a 5000Å thick Ag film deposited on TaN.

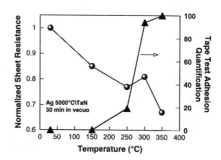

Figure 4   The change of sheet resistance and adhesion strength as function of post-deposition annealing tempearutre.

From these results, we have tentatively employed TaN as a standard liner for experiments. Figure 4 shows the change of the sheet resistance and adhesion strength as function of annealing temperature, where the adhesion strength was evaluated by Scotch™ tape test. The sheet resistance decreases as the annealing temperature increases. The adhesion strength remarkably increases at about 250 °C. The film annealed at 350 °C possessed adhesion strong enough for Al wire bonding. The 111/200 XRD intensity ratio decreases with increasing annealing temperature, whereas the 111 rocking curve FWHM decreases with temperature. Electromigration tests are now being carried out with NIST line structures, and the results will be reported elsewhere.

PLASMA TOLERANCE

The plasma tolerance of Ag is a next clearance to be cared for practical application. Ag can react with group VA and VIA elements especially in plasma. During a damascene process Ag interconnect lines can be directly exposed to a plasma such as F-containing plasma for via definition and O_2-based plasma for photoresist ashing/via cleaning, which can lead to a brutal deterioration of the Ag surface. One way to comfort this issue is to passivate the Ag surface. Capping or encapsulation with a refractory metal is a straightforward approach. Figure 5 shows the change of the sheet resistance of Ag films after a plasma exposure. A large increase in the sheet resistance is seen for uncapped films, whereas no significant change is observed when the surface is covered with cap metals (Ti, TiN, Ta, TaN, Cu, Pt). The reasons for such poor plasma tolerance are thought to be 1) the absence of a protective dense Ag oxide, 2) reaction of Ag-F with moisture; and 3) Ag ion migration.

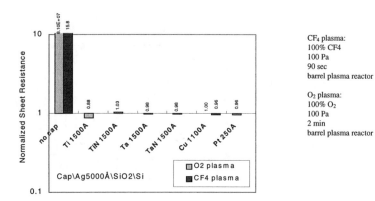

Figure 5   Sheet resistance change of uncapped (far-left) and capped Ag films after O_2 or CF_4 plasma exposure.

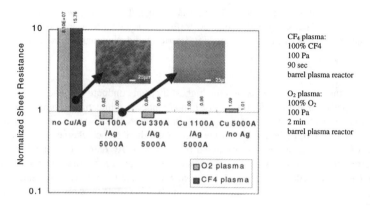

CF4 plasma:
100% CF4
100 Pa
90 sec
barrel plasma reactor

O2 plasma:
100% O2
100 Pa
2 min
barrel plasma reactor

Figure 6   Sheet resistance change of uncapped (far-left) and Cu-capped Ag films after O_2 or CF_4 plasma exposure.

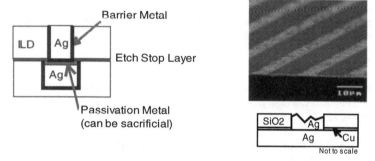

Figure 7 Proposed damascene structure (left) and demonstration of Ag trench formation in a SiO_2 layer pattered with CF_4-based chemistry (right).

We here demonstrate the use of Cu as a cap layer. Cu is chosen because of 1) the compatibility with the current Cu technology, 2) large conductivity, and 3) the possibility of electroless/selective deposition on Ag. Figure 6 shows the change of the sheet resistance of Ag films (5000Å) capped by Cu layers having various thicknesses. The data for bare Ag and Cu films are shown for comparison. It is seen that the capped Ag films show no remarkable increase in sheet resistance and that a 100Å-thick cap is thick enough to suppress the surface deterioration (inset micrographs). A slight increase in sheet resistance is seen for the Cu after the O_2 plasma exposure but no increase is confirmed in the case of the CF_4 plasma exposure.

Figure 7 left shows a conceptual schematic of a proposed damascene structure using a cap layer. We fabricated a prototype of this structure (Fig. 7 right). In this prototype run, a blanket Cu(1500Å)/Ag(5000Å) (Cu top) bilayer was deposited without breaking vacuum, and a SiO_2 film was then sputtered onto it. Trench-type windows are opened in the SiO_2 with lithography, followed by O_2 plasma ashing and isotropic CF_4 plasma etching. The Cu cap was removed by a diluted HCl solution, and finally Ag was deposited in the windows with an electroless plating technique. It is noted that the Ag film surface was not deteriorated during plasma processing.

SUMMARY

Elemental Ag has the lowest resistivity and the highest thermal conductivity. These properties make Ag as an attractive material for future ULSI interconnect architecture where a thermal condition of the interconnections will become of great importance. Because the chemical nature of Ag is similar to that of Cu, the extension of the current Cu damascene scheme seems to be a feasible scenario of Ag metallization. We discussed two issues of Ag damascene metallization: the choice of barrier materials, and the way of surface passivation. From traditional screening tests, we chose TaN as a (tentative) barrier metal in view of its inertness toward Ag and of Ag (111) texturing. The use of Cu cap was studied, and its application to damascene processing was proposed.

ACKNOWLEDGMENTS

Experimental assistance by Y. Nozaki, K. Nakagawa, and T. Yumii is greatly acknowledged. Support by other CMS staff and student members is also acknowledged.

REFERENCES

1. The Handbook of Chemistry and Physics, 76th Ed. (CRC, Boca Raton, FL, 1995).
2. P.W. Atkins, Physical Chemistry, 5th Ed. (Oxford Univ. Press, Oxford, UK, 1994)
3. CRC Handbook of Electrical Resistivities of Binary Alloys, (CRC, Baca Raton, FL, 1983).
4. S. P. Murarka, Metallization Theory and Practice for VLSI and ULSI (Butterworth-Heineman, Boston, UK, 1993)
5. Smithells Metals Reference Book (Butterworth, London, 1983)
6. J.D. McBrayer, R.M. Swanson, and T.W. Sigmon, J. Electrochem. Soc. **133**, 1242 (1986).
7. K.N. Tu, Phys. Rev., **45**, 1409 (1992).
8. Y. Wang and T.L. Alford, Appl. Phys. Lett. **74**, 52 (1999).
9. Y. Zeng, Y.L. Zou, T.L. Alford, F. Deng, S.S. Lau, T. Laursen, B. Ullrich, and J. Manfred. Appl. Phys. **81**, 7773 (1997).

Barriers for Copper

IMPROVEMENT OF DIFFUSION BARRIER PERFORMANCE BY A THIN Al INTERLAYER DEPOSITED BETWEEN BARRIER AND COPPER

Kyoung-Ho Kim, Se-Joon Im, Soo-Hyun Kim, and Ki-Bum Kim
School of Materials Science and Engineering, Seoul National University, San 56-1, Shillim-dong, Kwanak-gu, Seoul, 151-742, Korea, E-mail : kibum@snu.ac.kr

ABSTRACT

It is identified that a thin layer of Al with a thickness of about 10 nm deposited between chemical vapor deposited (CVD) diffusion barrier (TiN and TaN) and Cu significantly improves the barrier property of the layer. For instance, the 20 nm-thick CVD-TiN [deposited by a single source of TDMAT (tetrakis-dimethyl-amido-titanium)] layer which failed after annealing for 1 hour at 500℃ did not fail even after 650℃ annealing with a thin Al interlayer. The improvement of the barrier property is also demonstrated in the case of CVD-TaN [deposited by a single source of PDEAT (pentakis-diethyl-amido-tantalum)] layer.

INTRODUCTION

Copper is now being implemented as a material for interconnection because of its lower resistivity and higher resistance against electromigration than Al and its alloys[1, 2]. However, still there are many problems to be solved for the reliable application of Cu such as the poor adhesion to dielectrics, ease of oxidation[3], and fast migration into Si[4] and SiO_2[5] substrate. In particular, the fast migration of Cu into Si causes the degradation of the device performance by introducing deep level acceptors[6]. Therefore, a suitable diffusion barrier is required to prevent it.

The present process scheme includes sputter deposited Ta or TaN as a diffusion barrier layer[7, 8], sputter deposited Cu as a seed layer for electroplating, and Cu interconnection deposited by an electroplating method. However, the present process scheme based on sputtering process has its own limitation for the conformal deposition. Certainly, it is required to develop a suitable barrier process by chemical vapor deposition (CVD). While there have been many efforts just to make it[9–11], it appears that none of the process satisfies all the required conditions such as low temperature process, low resistivity, good conformality, and, most importantly, good barrier property against Cu migration.

In this study, we will report that a thin layer of Al of about 10 nm thickness deposited between CVD-barrier layer (CVD-TiN and CVD-TaN) and Cu drastically improves the barrier property.

EXPERIMENT

Both CVD-TiN (20 nm thickness) and CVD-TaN (50 nm thickness) films were thermally deposited using tetrakis-dimethyl-amido-titanium (TDMAT) and pentakis-diethyl-amido-tantalum (PDEAT) as a precursor, respectively, on (100) Si substrate. The resistivity of CVD-TiN is about 10^4 μΩ-cm and that of CVD-TaN is about $2×10^4$ μΩ-cm. Then, Al and Cu were sputter deposited onto barrier film without breaking vacuum in a DC magnetron sputtering system. The thickness of Al layer was varied from 0 nm (Cu/barrier/Si), 5 nm, 10 nm, and 20 nm. The thickness of Cu was 300 nm in all the samples. The sputtering conditions were as follow: the base pressure of deposition chamber was lower than $5.0×10^{-6}$ Torr, the deposition pressure was 4 mTorr using Ar as a plasma gas, and sputtering power was 30 W for Al (deposition rate: 20 nm/min) and 100 W for Cu (deposition rate:100 nm/min), respectively.

To test the diffusion barrier properties, all the structures were annealed under the vacuum below $5.0×10^{-6}$ Torr, and at the temperature ranging from 500℃ to 700℃ for 1 hour. Diffusion barrier properties were characterized by sheet resistance measurement with a four-point probe, X-ray diffractometry (XRD), etch-pit observation by scanning electron microscopy (SEM), and cross-sectional transmission electron microscopy (XTEM).

Conference Proceedings ULSI XV © 2000 Materials Research Society

RESULTS

CVD-TiN Case :

Sheet Resistance Measurement and XRD Analysis

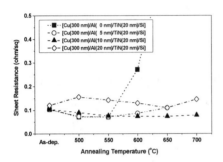

Fig. 1. Sheet resistance of Cu/CVD-TiN/Si and Cu/Al/CVD-TiN/Si structures as a function of annealing temperature.

Fig. 1 shows the sheet resistance of the Cu/Al/CVD-TiN/Si (Al thickness: 0 nm, 5 nm, 10 nm, and 20 nm) structures as a function of annealing temperature. The resistivity of as-deposited Cu film is about 3.0 ~ 3.5 $\mu\Omega$-cm, which is a typical number obtained from our sputtering equipment. The relatively high resistivity of the Cu film is believed to be due to the small grain size and large defect density of the as-deposited Cu film. It should be noted first that the sheet resistance of these structures initially drops upon annealing, except for Cu/Al(20 nm)/CVD-TiN/Si structure. The decrease of sheet resistance is thought to be a decrease in defect density and grain growth of the Cu film during annealing. If one converts this number to the 300 nm thickness of Cu film, one obtains a 2.1 $\mu\Omega$-cm resistivity of Cu. On the contrary, the sheet resistance of Cu/Al(20 nm)/CVD-TiN/Si structure initially increases upon annealing, and stays almost there for the further annealing. Again if one converts this number to the 300 nm thickness of Cu, one obtains a 4.2 $\mu\Omega$-cm resistivity of Cu. The relatively high resistivity of Cu for 20 nm-thick Al interlayer may indicate that some amounts of Al diffused into the Cu layer.

The sheet resistance of the Cu/CVD-TiN/Si structure abruptly rises after annealing at 600 ℃ indicating that a significant barrier failure occurs at this temperature. The X-ray diffraction patterns of the sample annealed at 600 ℃ [Fig. 2(a)] showed the formation of Cu silicide. For the Cu/Al(5 nm)/CVD-TiN/Si structure, the sheet resistance gradually rises at annealing temperatures higher than 600 ℃. This result also indicates that a significant barrier failure occurs, although the degree of failure is much slow as compared to the case of no Al interlayer. Again the X-ray diffraction patterns also demonstrate the formation of Cu silicide after annealing at 600 ℃. On the contrary, for the samples with 10 nm thickness of Al interlayer, there is no increase of sheet resistance even after annealing at 700 ℃. The X-ray diffraction pattern also shows no formation of Cu silicide even after annealing at 700 ℃ [Fig. 2(b)]. The XRD results are similar in the case of 20 nm thickness of Al interlayer.

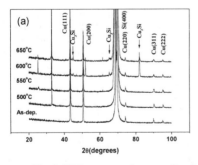
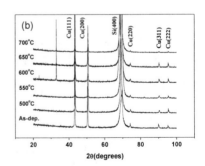

Fig. 2. XRD patterns after annealing for 1 hour at various temperatures;
(a) Cu(300 nm)/TiN(20 nm)/Si and (b) Cu(300 nm)/Al(10 nm)/TiN(20 nm)/Si.

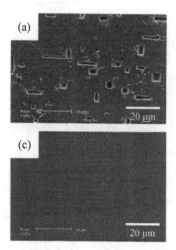
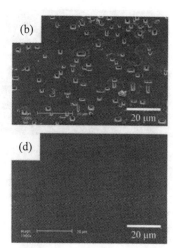

Fig. 3. SEM micrographs of etch-pits after annealing for 1 hour at 650 ℃;
(a) Cu(300 nm)/TiN(20 nm)/Si, (b) Cu(300 nm)/Al(5 nm)/TiN(20 nm)/Si,
(c) Cu(300 nm)/Al(10 nm)/TiN(20 nm)/Si, and (d) Cu(300 nm)/Al(20 nm)/TiN(20 nm)/Si.

Etch-pit Test

To confirm the effect of a thin Al interlayer on the diffusion barrier property of TiN film, the Si surface of the annealed sample was observed after Secco etching. Before Secco etching, the Cu, Al, and TiN layers were selectively removed by using wet-chemical solutions. Fig. 3 is the SEM images of etch-pits after annealing at 650 ℃ for 1 hour. The results clearly show that the barrier failure occurs at this stressing condition for the samples of 0 nm and 5 nm thickness of Al while the samples with 10 nm and 20 nm thickness of Al interlayer survive at this annealing condition.

The density and size of etch-pits observed on Si surface in various structures were summarized in Table 1. As shown in Table 1, etch-pit was not observed up to annealing at 650 ℃ in Cu/Al(10 nm, 20 nm)/CVD-TiN/Si structure. However, in Cu/CVD-TiN/Si structure, etch-pits with a size of a few microns were observed after annealing at 500 ℃ for 1 hour. Again, the data clearly shows that the thermal stability of Cu/Al/CVD-TiN/Si system is better than that of Cu/CVD-TiN/Si system, which is consistent with the results of sheet resistance measurement and XRD analysis.

Table 1. Density and size of etch-pits observed on Si surface after Secco-etching.

Temperature (℃)	[Cu/Al(0)/TiN/Si]	[Cu/Al(5)/TiN/Si]	[Cu/Al(10)/TiN/Si]	[Cu/Al(20)/TiN/Si]
		(nm)		
500	$4.8 \times 10^3/ cm^2$ ($\sim 1.5\ \mu m$)	$5.6 \times 10^3/ cm^2$ ($\sim 2.0\ \mu m$)	none	none
550	$6.3 \times 10^3/ cm^2$ ($\sim 4.5\ \mu m$)	$7.9 \times 10^3/ cm^2$ ($\sim 3.0\ \mu m$)	none	none
600	$9.1 \times 10^5/ cm^2$ ($\sim 9.0\ \mu m$)	$1.3 \times 10^5/ cm^2$ ($\sim 5.0\ \mu m$)	none	none
650	$9.9 \times 10^5/ cm^2$ ($\sim 15\ \mu m$)	$4.0 \times 10^5/ cm^2$ ($\sim 5.5\ \mu m$)	none	none
700	not measured	not measured	not clear	not clear

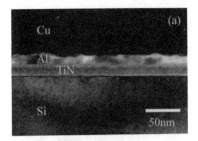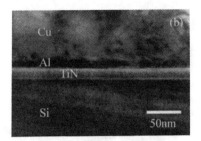

Fig. 4. XTEM micrographs of Cu(300 nm)/Al(20 nm)/TiN(20 nm)/Si structure;
(a) as-deposited and (b) 650 ℃ annealed.

XTEM Analysis

Fig. 4 shows the cross-sectional TEM images in Cu/Al(20 nm)/CVD-TiN/Si system before and after annealing at 650 ℃. No reaction products involving Cu and Si were observed at the interface between Si substrate and the barrier layer. And no significant decrease of Al is observed after annealing as far as we can note although sheet resistance results indicate that some amount of Al might diffused into the Cu layer.

CVD-TaN Case :

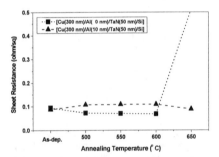

Fig. 5. Sheet resistance of Cu/CVD-TaN/Si and Cu/Al/CVD-TaN/Si structures as a function of annealing temperature.

The similar test was done on CVD-TaN barrier. In this case, we tested only the case where 10 nm thickness of Al is deposited between Cu and CVD-TaN layer. The sheet resistance data [Fig. 5] shows that there also is a significant improvement of the barrier property. Without Al, the sheet resistance increased after annealing at 650 ℃ while, with 10 nm thickness of Al, the sheet resistance stays even after annealing at 650 ℃. The improvement of barrier property is also noted in the results of X-ray diffraction [Fig. 6]. The formation of Cu silicide is clearly observed for the sample annealed at 600 ℃ without Al interlayer while no formation of Cu silicide is observed even after 650 ℃ annealing with Al interlayer.

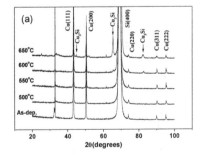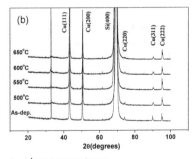

Fig. 6. XRD patterns after annealing for 1 hour at various temperatures;
(a) Cu(300 nm)/TaN(50 nm)/Si and (b) Cu(300 nm)/Al(10 nm)/TaN(50 nm)/Si.

DISCUSSION

There can be three possible mechanisms for the role of a thin Al interlayer. One is the effect of Al stuffing through the grain boundaries of the barrier layer. It has been well known that the diffusion barrier property of TiN film is significantly improved by oxygen "stuffing" for the Al metallization while it is not for the Cu metallization[12]. If the thin layer of Al is diffused into the grain boundaries of barrier layer and forms the Al_2O_3 with the oxygen elements in the barrier layer, then it will acts as a barrier for the grain boundary diffusion of Cu. Another possibility is the formation of thin Al_2O_3 layer with the oxygen which is either on top of barrier layer or in the barrier layer. Finally, it is possible to think that Al itself is acted as a superior barrier for Cu migration. At present, it is not clear which one is the right mechanism for the role of thin Al interlayer. Further study is necessary to resolve it.

CONCLUSION

We have demonstrated the improvement of the diffusion barrier properties by introducing a thin Al interlayer between the barrier layer and Cu film. The Cu/Al/barrier(CVD-TiN and CVD-TaN)/Si structure can be stable up to 650℃ while the Cu/barrier/Si structure without Al interlayer fails after annealing at 500℃ by etch-pit test.

ACKNOWLEDGMENT

This work has been supported by A Collaborate Project for Excellence in Basic System IC Technology through Consortium of Semiconductor Advanced Research (COSAR).

REFERENCES

1. S. P. Murarka, Metallization-Theory and Practice for VLSI and ULSI, (Butterworth-Heinemann, Bosten, 1993), pp. 1-14.
2. H. Takasago, K. Adachi, and M. Takada, J. Electron. Mater. 18, 319 (1989).
3. D. Adams, R. L. Spreitzer, S. W. Russell, N. D. Theodore, T. L. Alfordand, and J. W. Mayer in Advanced Metallization for Devices and Circuits-Science, Technology and Manufacturalility, edited by S. P Murarka, A. Katz, K. N. Tu, and K. Maex (Mater. Res. Soc. Symp. Proc. 337, Pittsburgh, PA, 1994) pp. 231-236.
4. E. R. Weber, Appl. Phys. A, 30, 1 (1983).
5. J. D. McBrayer, R. M. Swanson, and T. W. Sigmon, J. Electrochem. Soc. 133, 1242 (1986).
6. R. S. Muller, T. I. Kamins, Device Electronics for Integrated Circuits, 2nd ed. (John Wiley & Sons, New York, 1986) pp. 1-56.
7. K. H. Min, G. C. Jun, and K. B. Kim, J. Vac. Sci. Tech. B, 14, 3263 (1996).
8. K. Holloway, P. M. Fryer, C. Cabral, Jr., J. M. E. Harper, P. J. Bailey, and K. H. Kelleher, J. Appl. Phys. 71, 5433 (1992).
9. J. O. Olowolafe, C. J. Mogab, R. B. Gregory, and M. Kottke, J. Appl. Phys. 72, 4099 (1992).
10. G. C. Jun, S. L. Cho, and K. B. Kim, Jpn. J. Appl. Phys. 37, L 30 (1998).
11. S. L. Cho, S. H. Min, K. B. Kim, H. K. Shin, and S. D. Kim, J. Electrochem. Soc. 146, (1999) (unpublished).
12. K. C. Park and K. B. Kim, J. Electrochem. Soc. 142, 3109 (1995).

Electrical properties and adhesion of Cu/Zr/TaN layer for Cu metallization

C. J. Uchibori, N. Shimizu and T. Nakamura

Fujitsu Laboratories Ltd.

10-1 Morinosato-Wakamiya, Atsugi 243-0197, JAPAN

uchibori@ccg.flab.fujitsu.co.jp

ABSTRACT

A highly adhesive Cu interconnect structure with low resistivity has been developed by inclusion of a Zr glue layer between the Cu and an underlying TaN barrier metal layer. The mechanism of enhancement of the adhesion strength was found to be chemical and mechanical bonding between the layers by forming a thin diffused region and a roughening of both the Cu/Zr and Zr/TaN interfaces. Since the thickness of the diffused region was only about 5nm and the effect of Zr on the resistivity of Cu was small, the resistivity of Cu/Zr/TaN was not significantly larger than that of Cu/TaN.

INTRODUCTION

Since the electromigration resistance of Cu is higher than that of Al, Cu is an attractive material for use in semiconductor interconnects. However, in order to use Cu metallization for future Large-Scale Integrated (LSI) circuits, further improvement of interconnect properties such as reliability and resistivity is necessary. One approach to improve the reliability is the alloying of Cu with some other metal. The activation energies for electromigration of some Cu alloys have been measured and the addition of Sn[1-3], Pd[4] and Zr[1, 2, 5] are reported to increase the electromigration lifetime. A second approach is control of the grain size and texture of the Cu layer, since the diffusivity at grain boundaries depends on the grain size and the texture[6, 7]. Another method is the improvement of adhesion strength at the Cu and barrier metal interface, since the interface is one of the fastest diffusion paths. High adhesion strength is also important for the Cu metallization process, because damascene structures are mainly fabricated by plating and Chemical Mechanical Polishing (CMP), and the polishing pad causes mechanical stress in the metallization. In this study, therefore, our focus is to improve the adhesion strength.

The fundamental mechanism of adhesion is chemical and/or mechanical bonding at an interface. Interdiffusion between each layer, formation of a rough interface and reaction at the interface will enhance mechanical or chemical bonding at the interface. This is why metals that show excellent barrier properties have poor adhesion values with Cu, because Cu and the barrier metal are generally unreactive. Therefore, high adhesion strength at the Cu/barrier metal interface can be expected by inserting a glue layer at the interface, where the glue layer strongly bonds with both Cu and barrier metal. In order to use a glue layer in the metallization, choice of the glue layer metal is important, because the reaction of the glue layer with Cu can increase the resistivity of Cu and the reaction of the glue layer with the barrier metal may deteriorate the barrier properties.

The requirements for the glue metal are therefore as follows. The glue metal should be reactive with Cu to form a chemically or mechanically bonded interface. In order to avoid an increase in the resistivity of Cu, the effects of the glue metal on the resistivity of Cu should be small and the solubility limit of glue metal in Cu should be low. In this study, Zr is selected as a glue layer. From the Cu-Zr phase diagram, the solubility limit of Zr in Cu is only 0.1at% at 966°C and Zr is reactive with Cu. It is also reported that the increase in the resistivity of Cu is only 1.2% after addition of 0.07at% Zr[8]. Therefore, high adhesion strength with low resistivity is expected by inserting a Zr glue layer at the interface. In addition, as described above, long electromigration

Conference Proceedings ULSI XV © 2000 Materials Research Society

lifetimes are expected because diffusion of Cu atoms at the Cu/Zr interface will be suppressed.

Hence, in this study, the effects of Zr on the electrical properties and the adhesion strength of Cu/Zr/TaN layer were investigated. Subsequently, the mechanism for enhancement of the adhesion strength was investigated by analyzing the interfacial structure and the interdiffusion of Cu, Zr and Ta.

EXPERIMENTAL

The layered structure of Cu/Zr/TaN was deposited in an ultra high vacuum chamber, where the background pressure was 4×10^{-11} Pa. TaN, Zr, and Cu were sequentially deposited by DC magnetron sputtering without breaking the vacuum. TaN was reactively sputtered in a mixed gas of N_2 and Ar. After deposition, samples were annealed from 200°C to 500°C under high vacuum. The adhesion strength was measured by a scanning scratch tester, where a vibrating diamond stylus was pressed against the sample surface[9]. The critical force was defined as the force causing the stylus to peel off the Cu layer. The resistivity was calculated by the sheet resistance measured by a four-point probe method. Interfacial reactions were analyzed by X-ray diffraction (XRD) and observed by Transmission Electron Microscopy (TEM). In order to analyze the interdiffusion at the interface, distribution of Cu, Zr and Ta were measured by energy dispersive X-ray spectroscopy (EDX), where the beam size was 1nm.

RESULT AND DISCUSSION

The resistivity of samples with and without a Zr layer, for different Cu thicknesses, is plotted in Fig. 1 as a function of annealing temperatures. For 200nm thick Cu, after annealing at 400°C, the resistivity of the samples with and without Zr layer show the same value. After

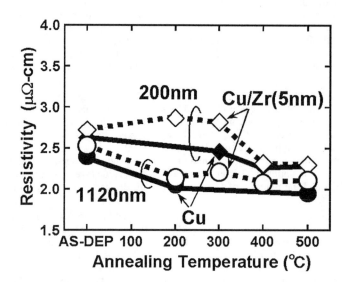

Fig. 1   Resistivity of 200nm and 1120nm Cu, with and without a Zr layer, after annealing at various temperatures.

annealing at 400℃, Cu grain growth occurred and the resistivity decreased. For 1120nm thick Cu, the resistivity is always lower than that of 200nm Cu and the effect of Zr on the resistivity of Cu is also small. The resistivity of 1120nm Cu starts to decrease after annealing at 200℃ which is a lower temperature than for 200nm Cu. The difference between the resistivity for 200nm Cu and 1120nm Cu is due to thin film effects.

To evaluate the effects of Zr on the resistivity, the resistivity after annealing at 400℃ for various Zr thickness was measured. In Fig. 2, the resistivity is plotted as a function of Zr thickness. For comparison, the resistivity for other Cu alloys are plotted. Note that the resistivity values for CuSn and CuPd are for bulk alloys and the value of CuZr is the total value for the layered structure. In this figure, it is clear that the resistivity is not strongly affected by the Zr layer. This is because the effects of Zr on the resistivity of Cu is small and it is believed that the Cu/Zr layered structure remains intact. On the other hand, the resistivity of CuSn and CuPd alloy increase with the composition of alloy metals. This is due to the high solubility limit of Sn (9.1at% at 520℃) and Pd (solid solution) and the large effects of the alloy metal on the resistivity of Cu.

Since the effects of Zr on the electrical properties of Cu/Zr/TaN have already been investigated, the adhesion strength for various Zr thicknesses was evaluated. The critical force, which represents the adhesion strength, is plotted in Fig. 3 as a function of Zr thickness. As shown in this figure, adhesion strength increases with Zr thickness. For these samples, the Zr surface had just been exposed during scratch testing. By scratching with a higher force than the critical force, the Si substrate broke before the surface of TaN appeared. These results indicates that the adhesion strength at the Zr/TaN interface is stronger than that at the Cu/Zr interface.

To investigate the role of Zr in enhancing the adhesion strength, cross sections of the

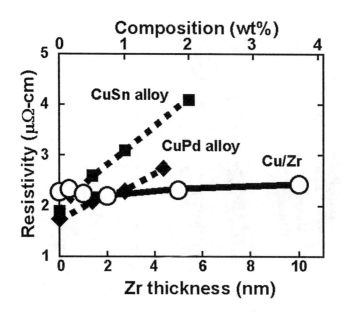

Fig. 2   Resistivity of Cu(200nm)/Zr layer for various Zr thicknesses.

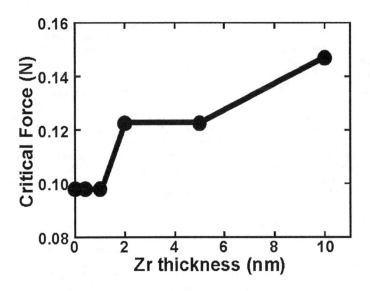

Fig. 3　Critical force in a Cu(200nm)/Zr layer for various Zr thickness.

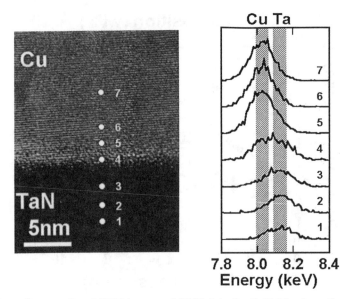

Fig. 4　Cross sectional TEM image and EDX data for Cu/TaN, where the energy peak at 8.01keV indicates Cu and at 8.11keV indicates Ta.

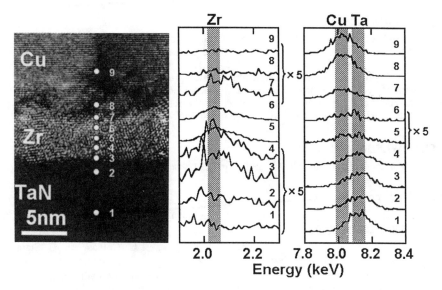

Fig. 5   Cross sectional TEM image and EDX data of Cu/Zr/TaN, where the energy peak at 2.03keV indicates Zr. Some data is amplified to detect weak signals.

samples with and without 5nm Zr were observed by TEM, and the distributions of Cu, Zr and Ta were analyzed by EDX. Fig. 4(a) and 4(b) shows the cross sectional image of Cu/TaN and the corresponding EDX data, respectively. In the cross sectional image, the Cu/TaN interface is clear and no evidence of Cu diffusion into TaN is observed. From the EDX data, the energy peaks at 8.05eV and 8.1eV represent Cu and Ta, respectively. At points 5 to 7, only Cu is detected and Ta is found only at points 1 to 3. Since point 4 is just at the interface, both Cu and Ta peaks are detected.

The cross sections of the sample with a 5nm Zr layer and EDX analysis data are shown in Fig. 5(a) and 5(b), respectively. The energy peak at 2.03eV indicates Zr. From the cross sectional image of this sample, it is seen that the Zr layer has a poly-crystalline structure and the interface between Cu and Zr is rough._Note that the Zr layer has an amorphous structure before annealing and the interface between Cu and Zr is smooth. From EDX analysis, Zr is detected only between the measured points 3 and 7. However, Cu is detected even at point 5, which indicates that it has diffused in the Zr layer. This is probably due to the poly-crystalline nature of the Zr layer, where a high number of grain boundary diffusion paths are present. The roughened interface of Cu/Zr observed by TEM is believed to be due to the formation of a diffused region. Note that a diffraction peak corresponding to Cu-Zr compounds was not detected by the XRD measurement. Although the thickness of the diffused region is only about 5nm, it's presence and the subsequent roughening of the interface is believed to improve the adhesion strength. Moreover, since the diffused region is so shallow, the resistivity of Cu/Zr/TaN layer does not increase. In addition, Ta is detected at point 6, within the Zr layer, but Zr is not detected at point 2, within the TaN layer. This again points to high diffusivity within the poly-crystalline Zr. The top interface of TaN is slightly rough compared with that in the Cu/TaN sample. At this moment, it is not clear why the Zr/TaN interface shows a higher adhesion strength than that of the Cu/Zr interface. However, diffusion of Ta and formation of a roughened Zr/TaN interface also improve the adhesion at the interface.

CONCLUSION

High adhesion strength at Cu/TaN interface is observed by inserting a Zr glue layer between the Cu layer and a TaN barrier layer. At the interface between Cu and Zr, formation of a thin diffused region and roughening of the interface leads to increased chemical and/or mechanical bonding, giving rise to improved adhesion properties. The adhesion strength of the Zr/TaN interface is also improved by the same mechanism. Because of the shallowness of the diffused region and the small influence of Zr on the resistivity of Cu, the resistivity of the Cu/Zr/TaN structure was almost as low as that of Cu/TaN.

ACKNOWLEDGEMENT

Authors are grateful to T. Miyajima for TEM observation and EDX analysis.

REFERENCES

1.   C. –K. Hu, B. Luther, F. B. Kaufman, J. Hummel, C. Uzoh, and D. J. Pearson, Thin Solid Films 262, 84 (1995).

2.   C. –K. Hu, K. Y. Lee, K. L. Lee, C. Cabral, Jr., E. G. Colgan, and C. Stains, J. Electrochem. Soc. 143, 1001 (1996).

3.   C. –K. Hu, K. L. Lee, D. Gupta, and P. Blauner in Advanced Metallization For Future ULSI, edited by K. N. Tu, J. W. Mayer, J. M. Poate and L. J. Chen (Mat. Res. Soc. Proc. 427, Pittsburgh, PA, 1996) pp. 95-106.

4.   C. W. park and R. W. Vook, Thin Solid FIlms 226, 238 (1993).

5.   Y. Igarashi and T. Ito, J. Vac. Sci. Technol. B16, 2745 (1998).

6.   P. Sewmon, Diffusion in Solids, 2nd ed. (The Minerals, Metals & Materials Society, Pennsylvania, 1989), p196.

7.   C. Ryu, A. L. S. Loke, T. Nogami, and S. S. Wong, in Proc. of IEEE International Reliability Physics Symposium, p. 201.

8.   S. Nishiyama, Bull. Jpn. Inst. Metals. 28, 137 (1988).

9.   S. Baba, A. Kikuchi, and A. Kinbara, J. Vac. Sci. Technol. A4, 3015 (1986).

PROCESS DEVELOPMENT AND FILM CHARACTERIZATION OF CVD TANTALUM NITRIDE AS A DIFFUSION BARRIER FOR COPPER METALLIZATION

Se-Joon Im*, Soo-Hyun Kim*, Ki-Chul Park**, Sung-Lae Cho*, and Ki-Bum Kim*
*School of Materials Science and Engineering, Seoul National University, San 56-1, Shillim-dong, Kwanak-gu, Seoul, 151-742, Korea, kibum@snu.ac.kr
**Samsung Electronics Co. Ltd.

ABSTRACT

We have investigated the effect of various ion beam bombardments during the growth of TaNx films by thermal decomposition of pentakis-diethylamido-tantalum precursor. The ion beams used are hydrogen, argon, and nitrogen. The ion beams are applied to the surface of the film with a kinetic energy of about 115 eV and the flux of the ion beam is about 0.1 mA/cm². In case of thermal CVD, the deposition rate is controlled by the surface reaction up to about 350 °C with an activation energy of about 1.07 eV. The activation energy of the surface reaction controlled regime is significantly decreased down to 0.33 eV and 0.26 eV when the hydrogen and argon beam is applied, respectively. In case of nitrogen ion-beam induced chemical vapor deposition, deposition rate levels off in the whole temperature range. By using argon ion beam, the resistivity of the film is reduced from 10000 $\mu\Omega$-cm to 600 $\mu\Omega$-cm and the density of the film is increased from 5.85 g/cm³ to 8.26 g/cm³, as compared to the films deposited by thermal CVD. The use of nitrogen and hydrogen ion beam also considerably lowers the resistivity of films (~ 800 $\mu\Omega$-cm) and increases the density of the films (7.5 g/cm³).

INTRODUCTION

Extensive work during the last decade has proven that Ta and its nitrides, such as Ta_2N and TaN, show excellent barrier properties against Cu metallization.[1-2] In particular, TaN shows the superior barrier properties among them.[1] However, most of the TaN materials tested so far have been deposited by reactive sputtering, which does not yield sufficient conformal coverage to be applied in submicron generation devices. There is no doubt that a chemical vapor deposition (CVD) technique to deposit TaN films should be developed and the barrier properties of those films should be tested. There have been several attempts to develop a CVD-TaN process either by using metallorganic source gases, such as $(NEt_2)_3Ta=Nbu^t$ (tert-butylimido-tris-diethylamido-tantalum, TBTDET),[3] $Ta(NEt_2)_5$ (pentakis-diethylamido-tantalum, PDEAT),[4-5] and $Ta(NMe_2)_5$ (pentakis-dimethylamido-tantalum, PDMAT)[6] or halogen source gas, such as $TaCl_5$[7] and $TaBr_5$[8]. However, none of these films can be successfully integrated to the submicron device scheme either by the high resistivity of the film or by the relatively high deposition temperature. In particular, none of these films shows excellent barrier properties against Cu diffusion, as compared to the one sputter deposited.

In this work, we will report a newly devised CVD system, which we call IBICVD and the film characteristics of the TaNx deposited using this system. During the course of deposition, ion beam (hydrogen, argon, and nitrogen) with a certain kinetic energy is applied to the film surface in order to change the chemistry of the film and/or the physical properties of the film. In comparison, TaNx film is deposited by using thermal decomposition of PDEAT precursor.

EXPERIMENTS

Tantalum nitride film was deposited by chemical vapor deposition with the aid of ion beam (ion-beam induced CVD, IBICVD), which was devised in order to obtain films with different chemistry and with high density. The IBICVD system used in this work has been described in detail, previously.[9] The schematic diagram of the IBICVD system is shown in Fig. 1. This system consists of two parts. One is the deposition chamber and the other is ion beam source. The ion beam source is composed of the plasma tube and two grids. The upper grid is electrically floated to repel most of the electrons and attract ions from plasma. The lower grid is biased to −1

239

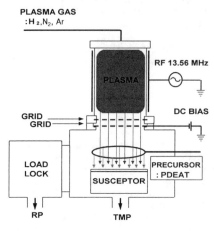

PLASMA GAS
:H₂,N₂, Ar

RF 13.56 MHz

PLASMA

DC BIAS

GRID
GRID

LOAD
LOCK

SUSCEPTOR

PRECURSOR
: PDEAT

RP

TMP

Fig. 1. Schematic diagram of IBICVD system.

kV in order to extract and accelerate the ions. In order to extract nitrogen, argon, or hydrogen ion-beam, N_2, Ar or H_2 was used as a plasma gas, respectively. The energy and flux of the ion beam were measured by using a retarding field ion energy analyzer. Thus, we can identify that the ion energy is 127 ± 13 eV at the –200 V of lower grid bias and 115 ± 25 eV at –800 V.[9] The flux of ion beam is about 57~103μA/cm²[9] as lower grid bias is changed from –200 V to –800 V.

Pentakis-diethylamido-tantalum (PDEAT) precursor was carried by 20 sccm of Ar gas through the bubbler maintained at 80 °C. The mixture of PDEAT and Ar gases was injected into the substrate through the ring-type distributor. The susceptor was resistively heated and was located 7 cm below from the lower grid. TaNx films were deposited using PDEAT as a precursor under the bombardments of nitrogen, argon, or hydrogen ion beam. For comparison, thermally-deposited TaNx films were prepared.

The deposition pressure was fixed to about 10 mTorr and the substrate temperature was varied from 275 °C to 400 °C. All depositions was carried out onto the Si substrates. Before deposition, the Si substrate was cleaned by using the hydrogen ion beams.

The film thickness was measured using a step profilometry. The properties of the deposited films were analyzed by four-point probe for sheet resistance, Auger electron spectroscopy(AES) for composition, Rutherford backscattering spectrometry(RBS) for density, and X-ray diffractometry(XRD) for phase identification. For the diffusion barrier test of TaNx films, 300-nm-thick Cu films are sputter deposited onto 50-nm-thick TaNx films. Then these Cu/TaNx/Si samples were annealed in vacuum ambient for 1 hour at temperature ranging from 500 °C to 650 °C with a 50 °C interval. Barrier failure was estimated by the increase of sheet resistance in Cu(300 nm)/TaNx(50 nm)/Si structure.

RESULTS AND DICUSSION

Deposition Rate

Figure 2 shows the dependence of the deposition rate on the substrate temperature. As has been reported already,[4-5] an Arrhenius plot of the deposition rate of a thermal CVD shows a transition temperature from surface reaction controlled regime to diffusion controlled regime at around 350 °C. The activation energy of the surface reaction controlled regime is about 1.07 eV, which is a typical number for the thermal grown nitride films using dialkylamido metallorganic source gas.[4-5,10-11] When we apply hydrogen ion beam, the overall deposition rate is also divided into two regimes; surface-reaction-controlled regime and mass-transfer-controlled regime. However, the activation energy of the surface-reaction-controlled regime is much smaller (0.33 eV) than that thermally decomposed. Although it is not clear at this stage whether this is due to ions, radicals, or metastables, it shows that the surface reaction mechanism is significantly changed to have lower activation energy. It is also noted that the deposition rate becomes much faster at the lower temperature region. When we apply the argon ions which is the heaviest element in our case, an Arrhenius plot of the deposition rate also shows a transition from surface-reaction-controlled regime with an activation energy of about 0.26 eV to mass-transfer-controlled regime. The lower activation energy means that the surface reaction is enhanced by argon ion beam. With the addition of nitrogen ion beam in the source gas, overall deposition rate was significantly decreased and it rarely changes with the substrate temperature. In other words, the deposition rate follows mass-transfer-controlled regime in the temperature ranges of tested. This behavior is similar to the case of the NH_3 addition to PDEAT.[5] It has

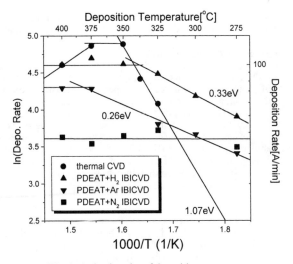

Fig. 2. Arrhenius plot of deposition rates.

been well documented that the transamination reaction[12-13] occurred with the addition of nitrogen source to the dialkylamido metallorganic source gas. It is quite certain that this reaction occurred with the addition of nitrogen ion beams.

Resistivity

The resistivity of TaNx layers as a function of deposition temperature is shown in Fig. 3. In general, the film resistivity decreases with increasing deposition temperature. The decrease of the film resistivity with the increase of deposition temperature is not a surprising result. It is generally observed that the resistivity of the films deposited by CVD process is decreased as the deposition temperature is increased as far as the films forms a continuous layer and is explained due to the formation of large grain size and the densification of the as-deposited film. The effect of ion beam to the resistivity of the film is in the sequence of nitrogen, hydrogen, and argon. The strongest effect of argon ion beam demonstrates that the physical bombardment to make the film

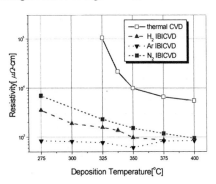

Fig. 3. As-deposited films resistivity.

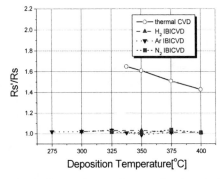

Fig. 4. Sheet resistance changes of as-deposited films after air-exposure of 24 hours.

dense is the most important factor to reduce the film resistivity. The minimum resistivity of TaNx films was about 600 μΩ-cm, which was deposited at 350 °C by using argon ion beam.

Sheet Resistance Increase after Air-Exposure

Figure 4 shows the sheet resistance of TaNx films before and after air-exposure for 24 hours as a function of temperature. This report supports the densification of the film by both increase of deposition temperature and bombardments of ion beam. It is generally believed that the increase of sheet resistance of CVD-TiN film using metallorganic precursor increased by the incorporation of oxygen due to the porous microstructure.[9] Thus, the films deposited at higher temperature show lower resistivity and also reveal better stability in air compared with those deposited at lower temperature. In case of IBICVD, however, all films deposited show little increase of sheet resistance irrespective of deposition temperature, which shows applying ion beam is more effective for film densification than increasing deposition temperature.

Composition

Figure 5 shows AES depth profiles for thermally decomposed and IBICVD deposited TaNx films (150nm) deposited at 325 °C. it is clearly shown that there is about 12 atomic percent of oxygen in the thermally decomposed film. On the contrary, the oxygen content of the IBICVD TaNx film was below detection limit. As explained earlier, this behavior is well consistent with the fact that the thermally grown TaNx films show an aging effect, while IBICVD does not.(See Fig. 4.) We do expect that the hydrogen ion beam affect the chemistry of film deposition in a way of reducing the carbon contents in the film. However, when we compare the results with those of argon ion beam, apparently it is not the case. In case of N_2 IBICVD, the low carbon contents in the film are probably caused by the reaction between these N radicals and PDEAT

Density

The density of the TaNx films deposited at 325 °C is evaluated by RBS and shown in Fig. 6. As is expected, the density of the TaNx films is abruptly increased with the aid of ion beam. The overall densities of the films deposited in this experiment ($5.85 \sim 8.26$ g/cm^3) appear to be much lower than that of bulk TaN (16.3 g/cm^3) or bulk TaC (13.9 g/cm^3). However, it should be pointed out that the stoichiometry of the films is quite different from that of bulk TaN. In other words, the existence of excess light elements results in much lower gram densities relative to that of bulk TaN or TaC.

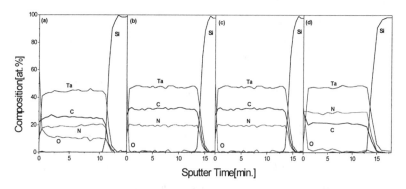

Fig. 5. AES depth profiles of TaNx deposited at 325 °C

(a) thermal CVD, (b) H$_2$ IBICVD, (c) Ar IBICVD, and (d) N$_2$ IBICVD.

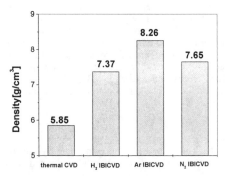

Fig. 6. Density of TaNx films deposited at 325°C.

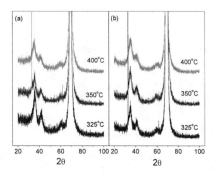

Fig. 7. XRD patterns of TaNx deposited at various temperature (a) thermal CVD and (b) Ar IBICVD.

Phase Identification

In order to identify the phase of deposited films, XRD patterns of as-deposited, 150 nm thick film are taken and shown in Fig. 7. The peaks appear at around 35° and 41° in the XRD patterns can be indexed to either TaN or TaC, because two phases have the same structure (rock-salt) and similar lattice parameter (TaN : 4.33 Å, TaC : 4.45 Å).

Diffusion Barrier Property

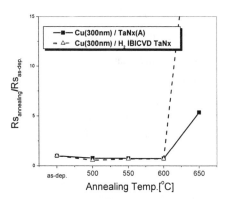

Fig. 8. Sheet resistance change of Cu(300 nm)/TaNx(50 nm)/Si structure as a function of annealing temperature after annealing for 1 hour in a vacuum ambient.

TaNx films(50 nm) deposited were evaluated as a diffusion barrier between Cu and Si. Figure 8 shows the changes of sheet resistance of Cu(300 nm)/TaNx(50 nm)/Si structure after annealing at various temperature. Only two films were tested in this case; thermally decomposed TaNx and hydrogen IBICVD TaNx. After 650 °C annealing, the sheet resistance of all structures drastically increases, indicating the failure of the diffusion barrier. These results show that the improvement of the film density from 5.85 g/cm^3 to 7.37 g/cm^3 is not good enough to show the prominent enhancement of the barrier property. These results are not consistent with our precious reports,[9] which shows the better diffusion barrier performance of TiN film is obtained from the higher density of TiN films processed by argon or nitrogen ion beam bombardment.[9] The Ar IBICVD TaNx film, which has highest density among the films, has not yet evaluated.

CONCLUSION

Tantalum nitride film was deposited by using ion beam induced chemical vapor deposition, which was devised in order to obtain films with different chemistry and with high density. The ions have the energies between 115 eV and 127 eV. Ion current density is about 57 ~ 103 μA/cm^2 as lower grid bias is changed. Pentakis (diethylamido) tantalum was used as a precursor to deposit TaNx film. The use of argon ion beam significantly lowers the resistivity of TaNx film and increases the density of TaNx film ($\sim$ 600 $\mu\Omega$-cm, 8.26 g/cm^3), as compared with the thermally decomposed film ($\sim$ 10000 $\mu\Omega$-cm, 5.85 g/cm^3). The use of nitrogen and hydrogen ion beam also considerably lowers the resistivity of the films ($\sim$ 800 $\mu\Omega$-cm) increases the density of the films (7.37 $\sim$ 7.65 g/cm^3). While the thermally grown TaNx films shows an aging effect, the IBICVD TaNx films do not show aging effect after air-exposure. The change of sheet resistance after annealing shows that thermally-deposited TaNx films and hydrogen IBICVD TaNx film appear to fail at the same temperature, indicating no effect of film densification from 5.85 g/cm^3 to 7.37 g/cm^3.

ACKNOWLEDGMENT

This work has been supported by A Collaborate Project for Excellence in Basic System IC Technology through Consortium of Semiconductor Advanced Research (COSAR) and was partially funded by Applied Materials.

REFERENCES

1. K. H. Min, G. C. Jun, and K. B. Kim, *J. Vac. Sci. Tech.*, **B 14**, p. 3263 (1996).
2. J. O. Olowolafe, C. J. Mogab, R. B. Gregory, and M. Kottke, *J. Appl. Phys.*, **72**, p. 4099 (1992).
3. M. H. Tsai, S. C. Sun, H. T. Chiu, C. E. Tsai, and S. H. Chuang, *Appl. Phys. Lett.*, **67**, p. 67 (1995).
4. G. C. Jun, S. L. Cho, and K. B. Kim, *Jpn, J. Appl. Phys.*, **37**, L 30 (1998).
5. S. L. Cho, S. H. Min, K. B. Kim, H. K. Shin, and S. D. Kim, *J. Electrochem. Soc.*, **146** (10) (1999) (in press).
6. R. M. Fix, R. G. Gordon, and D. M. Hoffman, *Chem. Mater.*, **5**, p. 614 (1993).
7. K. Hieber, *Thin Solid Films*, **24**, p. 157 (1974).
8. X. Chen, G. Peterson, T. Stark, H. L. Frisch, and A. E. Kaloyeros, *in Proceedings of VLSI Multilevel Interconnection Conference*, p. 434 (1997).
9. K. C. Park, S. H. Kim, and K. B. Kim, *Materials Research Society*, San Francisco, CA, p. 410 (1998).
10. M. Eizenberg, K. Littau, S. Ghanayem, M. Liao, R. Mosely, and A. K. Sinha, *J. Vac. Sci. Tech.*, **A 13**, p. 590 (1995).
11. I. J. Raaijmakers and J. Yang, *Appl. Surf. Sci.*, **73**, p. 31 (1993).
12. R. M. Fix, R. G. Gordon, and D. M. Hoffman, *Chem. Mater.*, **2**, p. 235 (1990).
13. R. M. Fix, R. G. Gordon, and D. M. Hoffman, *Chem. Mater.*, **3**, p. 1138 (1991).

INTEGRATION OF VACUUM ARC PLASMA DEPOSITED DIFFUSION BARRIER FILMS WITH LECTROCHEMICAL Cu DEPOSITION PROCESS

O. R. MONTEIRO[*], I. G. BROWN[*], I. C. IVANOV[**], D. PAPAPANAYIOUTOU[**], C. TING[**]
[*]Lawrence Berkeley National Laboratory, University of California, Berkeley, CA 94720, ormonteiro@lbl.gov
[**]CuTek Research, Inc., 2367 Bering Dr., San Jose, CA 95131

ABSTRACT

Electrochemical deposition of copper on thin seed films on different diffusion barriers prepared using filtered cathodic arc (FCA) deposition process has been investigated. To satisfy the trend towards feature size below 0.18 μm and reduction of copper seed layer, we are developing an integrated process using filtered cathodic arc to deposit conformal coatings of a diffusion barrier and a thin copper film in vias with high aspect ratio for electrochemical copper deposition. The filtered cathodic arc process has been shown to successfully produce conformal coatings of Ta diffusion barrier in vias and trenches 130 nm wide and with depth:width aspect ratio up to 8:1. Additional benefit of plasma immersion copper seed deposition is a smooth transition from current Cu electroplating process to thin seed processing. The structure, gap fill, and impurity levels of as deposited and annealed Cu films were investigated using SEM and XRD. The Ta barrier consisted primarily of the β–Ta phase, and the copper seed layer was highly textured. Diffusion barrier properties of the vacuum arc deposited films against Cu diffusion into silicon are presented.

INTRODUCTION

The implementation of copper as the material of choice for interconnects in integrated circuits is well under way. Copper has several advantages over aluminum: the lower resistivity of copper (1.67 μΩ.cm versus 2.65 μΩ.cm for aluminum alloys) allows finer "wires" with lower resistive losses, resulting in shorter on-chip resistance-capacitance (RC) time delays. Also, the higher mass and melting point of copper make it significantly less susceptible to electromigration (transport of atoms of the metal conductor when carrying high current densities) than aluminum, and less likely to fail under stress. Copper, on the other hand, has a high diffusivity in silicon, and promotes the formation of deep-level defects that alter the desired semiconducting properties of silicon. Thus it requires the development of effective diffusion barriers. Tantalum has been demonstrated to successfully prevent the contamination of Cu in Si or SiO_2, and simultaneously improve the adhesion between Si or SiO_2 and Cu.

The damascene process requires that vias and trenches are etched into dielectric layers and filled with a combination of metal nitrides and metals, and finally chemical-mechanical polished to produce a planarized embedded structure. As the width gets smaller and the aspect ratio gets bigger, the greater the challenge to fill these features without leaving internal voids. PVD processes that make use of low energy atoms and ions, such as evaporation and conventional magnetron sputtering, have failed to fill narrow trenches with high aspect ratio. Plugging occurs before filling is complete,

Conference Proceedings ULSI XV © 2000 Materials Research Society

leaving voids inside the metallic conductors. Better filling has been achieved with collimated sputtering [1], or sources with a higher degree of ionization [2,3] such as self-sputtering [4], electron cyclotron resonance (ECR) [5], or vacuum arc plasma sources [6]. These methods however have failed to produce copper filling of trenches with aspect ratios greater than 3:1 in trenches less than 250 nm wide because of inadequate control over the particle velocity distribution and ion energy. Recently a deposition method that makes use of a source of highly ionized plasma with pulse-biasing the substrate was successful in depositing Ta and Cu on narrow trenches with high aspect rations [7]. The proposed method uses a filtered cathodic arc plasma source with tight ion energy control. Particle flux with the combination of high directionality and tight energy control permits tailoring step coverage and deposition mode.

In the present article we present some results of the integration of the technique developed by one of the authors [7} with electrochemical deposition of copper to prepare structures.

EXPERIMENT

The deposition process to prepare the Ta diffusion barrier and the Cu seed layers used in this investigation has been previously described in detail [7]. A metal plasma is produced in a filtered cathodic arc (FCA) plasma source. The plasma is formed at the cathode surface and streams away from the cathode at speeds of $1 - 3 \times 10^4$ m s^{-1} [8]. Cathodic arcs have been used by a number of authors for thin film deposition and other applications [9]. In the process used here, the plasma stream is steered through a magnetic filter to eliminate particulates generated at the cathode [10-12]. The substrate is pulse-biased so as to provide the streaming ions with intermittent additional energy. This additional energy is very important to define the growth mode -degree of conformality - of the depositing film. Deposition was carried out at pressure around 1×10^{-6} Torr, and the surface temperature did not exceed 60°C.

Electroplating of copper was carried out in a CuTek electrodeposition tool (ElectroDep 2000), and morphological and structural information on the films were obtained by using scanning electron microscopy and X-ray diffraction.

RESULTS AND DISCUSSIONS

The net deposition rate during film growth using energetic ions from a metal plasma result from the competition between two processes - ion deposition and ion sputtering. The natural condensation rate of metal ions from the plasma stream on the wafer surface depends on the particle flux magnitude and direction with respect to the substrate, but the fraction of arriving particles that is retained on the surface is modified by sputtering of the deposited layer by the incoming ions. The sputtering rate depends on the mass and energy of the incident particle, the mass of the film atom species, and the angle of incidence. Moreover, different metals have different sticking coefficient on the substrates, and therefore greater mobility at the surface of the film before they come to rest. For all the previous reasons, the deposition process has to be tailored for the specific metal being deposited.

Figure 1 shows an example of a Ta film deposited on trenches patterned on SiO$_2$ when a bias duty cycle of 12.5% is used (duty cycle is defined as the ratio between the duration of the bias voltage-on cycle to the duration of the bias-on plus bias-off pulses)

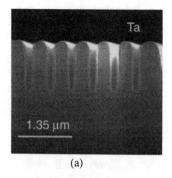
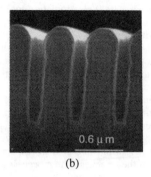

(a)   (b)

Figure 1: Ta film deposited using FCA process with substrate bias of -100 V and duty cycle of 12.5%. (a) and (b) correspond to two different magnifications of 180 nm wide trenches. The wavy appearance of the films is an artifact of the sample preparation.

The Ta films are about 40 nm thick and highly conformal to the trenches. The tapering of the Ta film at the opening of the trench on the top surface of the silicon dioxide is typical of deposition using energetic ions or atoms. The thickness of the Ta film at the bottom of the trenches is about 80% of the thickness at the top of the oxide. Moreover, the thickness on the side-walls is also quite similar to the one at the bottom of the trench. The bias duty cycle in this instance was optimized to minimize the presence of overhangs, that may lead to void formation during the subsequent deposition of copper. Such superior conformality is believed to result from sputtering at the trench opening. The excess material from overhangs that would otherwise form during non-energetic deposition is sputtered into the trenches, where it re-deposits. The only crystalline phase determined by X-ray diffraction was β-Ta. The Ta film is nano-crystalline (grain size around 5 nm) and under slight compressive stress.

In addition to Ta films, FCA was used to deposit copper seed layers for subsequent electroplating. The copper seed layers were deposited on TaN diffusion barriers. Figure 2(a) is a high-resolution (immersion lens JEOL-890 SEM with resolution of 0.7 nm) scanning electron micrograph of a trench after a 15 nm (nominal) thick Cu layer was deposited. To take full advantage of ultra-thin seed (less than a 10 nm average thickness on the high aspect ratio feature side-wall) deposited by the filtered cathodic arc process a two-step advanced electroplating technique was developed at CuTek Research.

A standard copper electroplating process using acidic solution will readily dissolve a thin coper oxide film and part of the metal film. Therefore when the seed layer is too thin, the residual metal layer would not be sufficient to provide acceptable conductivity for the deposition within the feature [13]. In the process developed here, an initial controlled plating produces a mostly conformal and continuous Cu layer about 50 nm thick directly over the ultra-thin seed. This provides the high conductivity seed for the subsequent standard Cu electroplating gap fill process. The deposition rate of 200 nm/min can be modified to satisfy specific process requirements. In case the vacuum arc deposited Cu seed layer is incomplete (spotty) on the side-walls, the deposition of the electroplated seed layer accelerates deposition over edges of areas with missing seed within a side-wall thus creating a conductive Cu-bridge over the high resistivity diffusion barrier film.

Both steps in the electroplating have been performed using a CuTek Research ElectroDep 2000 electroplating tool with inert anode, dry conacts adn wafer positioning "face up". As shown in Figure 2(b), the conformal electroplating seed layer preferential growth direction is normal to the surface of the diffusion barrier anywhere within high aspect-ratio features and on the top surface of the wafer. Minimal variation of the electroplated seed layer thickness and absence of voids allow further reduction of the layer thickness to no more than 20 nm thick, which is suitable for bulk copper gap fill of features 0.15 μm or less.

The diffraction pattern of copper layer deposited by FCA has indicated strong texturing along <111>. Diffraction patterns of FCA copper layers prepared with 3 different bias voltages are shown in Figure 3. The actual seed layer used for the electroplating shown in Figure 2 was carried out using a bias voltage of -100 V. The <111> texturing here is with respect to the top surface of the wafers, and not the side-walls of the trenches. The morphology shown in Figure 2b, however, indicates that even if there is a preferential growth direction other than <111> on the side-walls of the trenches, conformal growth is still achieved.

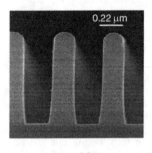 

(a)   (b)

Figure 2: (a) Trenches on SiO_2 with a TaN diffusion barrier and a 20-nm copper seed layer deposited by filtered cathodic arc. (b) Trench structure after electroplated seed deposition.

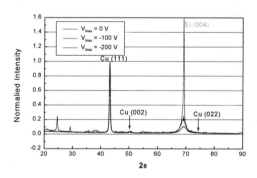

Figure 3: X-ray diffraction pattern of a Cu film deposited by filtered cathodic arc using three different bias voltages.

The diffraction pattern shown in Figure 3 also indicates that the copper films produced under these conditions are under slight compressive stresses. Compressive stresses are common in films prepared using ion-assisted processes due to the ion penning effect.

CONCLUSION

We have shown here an integrated process for the fabrication of copper interconnect lines consisting of a two-step deposition. In the first step, filtered cathodic arc is used to deposit thin diffusion barrier and copper seed layers on pulse-biased substrates. In the second step,   a two-step copper electroplating process provides initial controlled plating of mostly conformal and continuous Cu layer over ultra-thin seed to provide a high conductivity seed for the following standard Cu electroplating gap fill process.

ACKNOWLEDGEMENTS

The authors would like to thank Mr. Ian Ward of Charles Evans & Associates for performing SEM analyses. This work is supported by the U.S. Department of Energy under Contract No. DE-AC03-76SF00098.

REFERENCES

1. S. M. Rossnagel and R. Sward, J. Vac. Sci. Technol. A, **13**, 156 - 158 (1995)
2. S. M. Rossnagel and J. Hopwood, Appl. Phys. Lett. **63**, 3285 - 3287 (1993)
3. S. M. Rossnagel and J. Hopwood, J. Vac. Sci. Technol. B, **12**, 449 - 453 (1994)
4. A. Sano, H. Kotani, H. Sakaue, S. Shingubara, T. Tagahagi, Y. Horike and Z. J. Radzimski, Proc. Advanced Metallization of ULSI Applications, Portland, October 1995, ed: R. C. Ellwanger and Shi-Qing Won, MRJ Pittsburgh, p. 709 - 713
5. C. A. Nichols, S. M. Rossnagel, S. J. Hamaguchi, J. Vac. Sci. Technol. B, **14**, 3270 - 3275 (1996)
6. P. Siemroth, Ch. Wenzel, W. Kliomes, B. Schultrich, and T. Schulke, Thin Solid Films, **308-309**, 455 - 459 (1997)
7. O. R. Monteiro, J. Vac. Sci. Technol. B **17**, 1094 - 1096 (1999)
8. A. Anders, Surface and Coatings Technol. **93**, 158 - 167 (1997)
9. R.L. Boxman, P.J. Martin and D.M. Sanders, eds., "Vacuum Arc Science and Technology", (Noyes, New York, 1995).
10. R. L. Boxman, V. Zhitomirsky, B. Alterkop, E. Gidalevitch, I. Bellis, M. Keider and S. Goldsmith, Surface and Coatings Technology **86-87**, 243 - 246 (1996)
11. D. A. Karpov, Surface and Coatings Technology **96**, 22 - 33 (1997)
12. S. Anders, A. Anders, M. R. Dickinson, R. A. MacGill, and I. G. Brown, IEEE Trans. Plasma Sci. **25**, 670 - 675 (1997).
13. L. Chen and T. Ritzdorf, Proc. 195th Meeting of The Electrochemical Society, Seattle, WA, May 2-6, 1999.

Preclean Barrier and Seed Layers for Dual Damascene Copper Metallisation

JM Gilet, J. Torres*, M. Swaanen**, R. Gonella
STMicroelectronics, F-38926 Crolles Cedex, France jean-marc.gilet@st.com
* France Telecom, CNET/CNS, BP 98, 38243 Meylan cedex France
** Philips Semiconductors working at ST Crolles

ABSTRACT

Reactive Preclean, Ar soft sputter etch, barriers and seed layers have been evaluated to get a successful integration with copper electroplating. Reactive Preclean (RPC) was found to give better via resistance and less copper contamination on the side walls of vias. A DOE was carried out to optimize the influence of the gas flow and RF power. TaN barriers have been evaluated by means of micro Auger analyses and Bias Thermal Stress (BTS) on planar and comb serpentine structures and showed good resistance again Cu diffusion. Dual damascene structures have been fabricated using IMP barrier and Cu seed layer followed by Cu electroplating. Both trench and via chain resistances are reported. Electromigration studies presented in, show a highly significant improvement compared with standard Al metallisation which is a key result for system design.

INTRODUCTION

Copper metallisation is increasingly considered as a necessary solution for ULSI metallisation due to its lower resitivity and its supposed superior resistance to electromigration and stress voiding phenomena compared to currently used aluminium alloys. Several metal deposition solutions are available, but copper electroplating is emerging as the most attractive technique. These electroplating solutions meet the major requirements of high deposition rate, low deposition temperature making it compatible with low-k dielectric materials and low manufacturing costs.

However, the filling capability of the process remains the major challenge. Successful integration of electroplating requires integration of three critical metallisation steps: cleaning [1], barrier and seed layer depositions.

EXPERIMENT

RPC evaluation

The RPC is based on the chemical reaction rather than physical sputtering. It uses a H_2/He gas mixture to reduce the CuO_x generated during previous steps. Therefore, for RPC experiments, wafers with oxidised copper layer were used. Programs have been settled on a ellipsometer in order to measure the Cu oxide thickness. A process has been developed to generate more than 700Å of Cu oxide. These wafers have then been used with different plasma duration and with different parameters setting. The figure 1 shows the sheet resistance as a function of plasma duration compared with the resistance before the oxidation. The 60 seconds duration used to process patterned wafers seems to be more than enough to reduce the copper oxide. After 35 seconds of plasma, the initial sheet resistance measured after metal deposition has been recovered. On the other hand reflectivity never comes back to the initial value, which suggests that the structure of copper changes between the deposition and after RPC treatment. A DOE with the flow of H_2/He gas mixture and RF power as input parameter has been done. The response were reflectivity and sheet resistance. Copper oxide thickness was not a good response because it gave zero or negative thickness values after treatment even for short plasma duration. It can be explained by the fact that the reduction starts from the top of the layer. Therefore a short RPC give a Cu/CuO/Cu sandwich which is not easy to measure. The plasma duration used for the DOE was 20 second because it was the most sensitive to reflectivity and Sheet resistance variation. The R-square obtained was 0.95 for sheet resistance and 0.65 for reflectivity. It indicates that sheet resistance is a more reliable response. The figure 2 is the contour plot obtained with the 2 variables RF plasma power and gas flow.

It shows that the gas flow is the most sensitive parameter and the RF power plays a minor role. So it is clear that the process windows is quit large and therefore we have select the following parameters set: RF power 450W, Bias power 10W, gas flow 100SCCM with a plasma duration of 60 seconds.

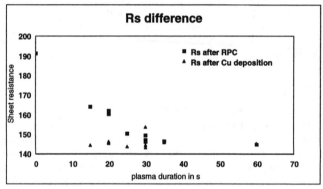

Fig.1. Initial sheet resistance before oxidation and after RPC

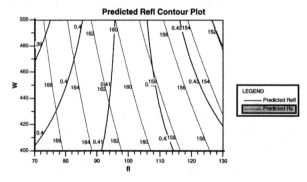

Fig.2. Contour plot of the DOE with gas flow (X axis) and RF power (Y axis) as parameters

Physical characterisation has shown the effect of standard Ar sputter etch. As can be seen in figure 3, it etches the copper underneath the via and resputters this copper on the side walls of the vias and even on the top of the wafer. The effect on unlanded vias are visible on the figure 5 and 6. The over etch on copper and oxide is bigger. The rounding of the via is attenuated with RPC. Micro Auger studies has clearly shown a much lower amount of copper on the top of RPC treated wafers than wafers done with Ar sputtered Preclean. The analyses have been carried between 100µm large lines spaced 200µm and between 0.3µm vias spaced 0.3µm. Two different Ar Preclean recipe were used to check the influence of the 2 RF power. The first recipe was set with RF1 at 300W, RF2 at 300W and the second recipe was set with RF1 at 170W and RF2 at 360W. The plasma duration has been adjusted to have the same amount of oxide etched. In both cases large amount of copper has been detected. Up to 3.5nm of Cu redeposited has been measured between the lines. In the case of RPC all the measurements were under the detection limit. In order to evaluate the effect on the via resistance of RPC compare with Ar sputtering, via resistance chain were fabricated and electrically tested. Figure 4 shows the box plot of 0.28µm via chain showing the significative improvement of more than 10% of the via resistance with RPC and Ar soft sputter etch.

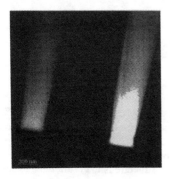
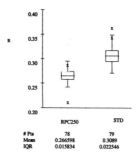

	RPC250	STD
# Pts	78	79
Mean	0.266598	0.3089
IQR	0.015834	0.022546

Fig.3. TEM micro graph of vias etched with Ar Preclean
Fig.4. Box plot of 0.28µm via chain with RPC and Ar soft sputter etch

Barrier reliability and performance

One of the key factor for the Cu metallisation is to be able to find an optimum between a efficient barrier against Cu diffusion and to keep a reasonable line and via resistance. A preliminary evaluation of the step coverage of the IMP TaN barrier has shown that a continuous film could be deposited with a nominal thickness around 250Å. Step coverage between 20% and 25% is currently achieved on 3:1 aspect ratio structures [2]. In order to check if the barrier is sufficient to avoid the diffusion of copper, several evaluations have been carried out. The following stacks were deposited: Si / TaN 75Å / IMP Cu 1500Å / TaN for capping and Si / Ta 75Å / IMP Cu 1500Å / Ta for capping. The capping is useful to avoid the copper oxidation during anneal and also the contamination of the barrier with oxygen. This contamination has been observed at the Si / Ta or TaN interface without the capping layer. The

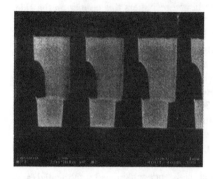 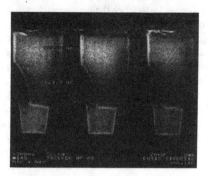

wafers were then annealed at 500°C during 30 mn. No diffusion of copper into silicon was detected by micro-Auger technique for Ta and TaN.

Fig.5. SEM micro graph of structures etched with Ar Preclean

Fig.6. SEM micro graph of structures etch

An other test under Bias Thermal Treatment (BTS) is generally used [3]. In this test, the effectiveness of the barrier is accessed through the oxide reliability. The model used to normalise t_{bd} (time to breakdown) to the stress conditions is:

$$t_{bd} = t_{bd0} \, e^{-\gamma e.E} \, e^{Ea/kT} \qquad (1)$$

where t_{bd0} is a constant, γ_e is the electric field acceleration factor, E the electric field, E_a the activation energy, k the Boltzmann's constant and T the temperature. Two types of structures have been tested: dots and comb serpentine structures.

Dots of stacked barrier copper and capping layers were deposited on an oxide layer through hard masks with different diameters holes. The structure is then submitted to several electric fields and Temperature stresses ranging from 1.5 to 3 Mv/cm and 250°C to 300°C

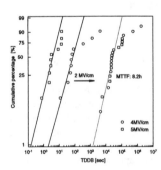

respectively. TaN film Thicknesses were 75Å and 250Å. A test with a barrier without copper and copper without barrier were used as references. On the wafer without barrier, the test could not be performed because the MOS capacitor was leaky from the beginning and even standard C-V curve could not be derived. For the other wafers the figure 7 shows the distribution obtained on TaN 75Å for different test conditions. The wafer without copper has given much longer time fore the same stress. So we can conclude that breakdown for the standard sample is due to copper diffusion in the oxide.

Fig.7. Time to breakdown weibull distribution extrapolated to the supplier conditions for dots

BTS was also applied on serpentine comb structures. These structures are more representative of the final circuit interconnect and therefore should bring more reliable results. The deposition is done on structures and not on full sheet wafers. Wafers with 250Å, 400Å have been tested. The extrapolated life time is far more than our specifications.

<u>Seed layer and via filling</u>

The interaction between seed layer and electroplating has been studied. DC power from 1kW to 3 kW, RF bias from 250W to 450W and pressure from 10mT to 30mT were used as input parameters. First a screening on the IMP copper process has given the following trends:
- Decreasing the bias power and the DC power/RF power improve the bottom coverage.
- The higher the pressure, the better the bottom coverage.
- Concerning the sidewall coverage, no clear trends have been seen due to the uncertainties and small differences seen between the samples.
In order to check if there is some differences 2 different recipes have been used on 2 metal level structures. The box plot on figure 8 shows the results obtained with a recipe using the following parameters compared with the "standard" recipe under bracket: pressure 30mT (20mT), DC power 1kW (2kW) RF bias 450 W (350W). No significative differences have been detected on our structures with 0.28µm via chain.
The process windows for via filling has been evaluated. Different seed layer thickness going from 1000Å to 1300Å have been tried in order to optimise it. The effect of dwell time (the time between the moment the wafer touch the chemical solution and the moment the current start) have been investigated at the same time. The dwell time gives rise to small voids and seams. Therefore it has been suppressed. No voids appeared on the other wafers. So we can conclude that the process windows is quite large between seed layer and electroplating when using an optimised electroplating recipe although, in some cases, the dewetting phenomena starts to appear as can be seen on figure 9.

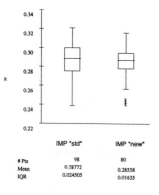
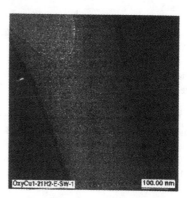

Fig.8. Box plot of the split between different IMP Cu recipes
Fig.9. TEM micro graph showing the dewetting phenomena

Electrical results and Electromigration

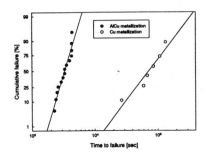

The copper resistivity has been estimated to be about $2\mu\Omega$cm from SEM cross section after metallisation. Electromigration (EMG) test have been carried out on NIST structures at wafer level. More details on integration issues are given elsewhere [4]. Results without annealing are very good and gives activation energy of 1.2eV. Annealing decreases the activation energy to 0.7eV as can be seen in figure 10. Nevertheless, in all cases these results exceed the performances obtained with Al standard metallisation.

Fig.10. Electromigration comparison between Al-Cu metallisation and Cu metallisation

Conclusion

The effect of RPC has been studied and has shown an improvement both in electrical results and in terms of copper contamination in vias and on top of structures. The IMP barrier and seed layer have been studied for their film properties and performances. For the 0.18μm and 0.15μm technologies these processes are clearly workable.

References

1. J. Torres, J. Palleau, F. Tardif, H. Bernard, P. Motte, conference proceedings ULSI XIV 1999 Materials Research Society p. 645

2. M. Moussavi, Y. Gobil, L. Ulmer, L. Perroud, P. Motte, J. Torres, F. Romagna, M. Fayolle, J. Palleau and Plissonnier, IITC June 1998 p. 295

3. M. Vogt M. Kachel, M. Plotner, K. Dreschner. "Dielectric barrieres for Cu metallisation systems" Microelectronic Engineering 37/38 (1997) p. 181 - 187

4. S. Kordic, J. Torres, L. Liauzu, C. Verove, P. Vannier, R. Gonella, P. Gayet, E. Van der Vegt, J.M. Gilet, Sept 7 - 9 1999 VMIC Conference p. 53 - 62

MODIFICATION OF Ta-BASED THIN FILM BARRIERS BY ION IMPLANTATION OF NITROGEN AND OXYGEN

E. WIESER[+], J. SCHREIBER[++], C. WENZEL[*], J. W. BARTHA[*], B. BENDJUS[++], V. MELOV[++], M. PEIKERT[+], W. MATZ[+], B. ADOLPHI[*], D. FISCHER[*]
[+] Forschungszentrum Rossendorf, Germany
[++] Fraunhofer Institut Zerstörungsfreie Prüfverfahren, Aussenstelle Dresden, Germany
[*] Technische Universität Dresden, Germany

ABSTRACT

Ta-based thin film barriers have been treated by ion implantation of nitrogen and oxygen to decrease the density of diffusion enhancing defects and to improve the barrier stability. The implantation changes the composition and the microstructure of the films. Above a threshold dose the original Ta structure is destroyed. Oxygen implantation leads to amorphization. In the high dose regime of N^+-implanted samples a nitride formation is detected. These changes in the microstructure should increase the barrier performance and stability considerably. The thermal stability of the modified layers was tested by annealing experiments. The application to very thin barrier layers requires a shallow implant at low energies. For this a plasma source ion implantation has been tested.

INTRODUCTION

In advanced metallization technology the trend continues to improve the stability of diffusion barriers while decreasing their thickness below 20 nm. The substitution of Al alloys by copper needs a further improvement of the barrier properties. In fact, defects, porosity and impurities influence the barrier stability. The usual deposition technologies (ion metal plasma; long-throw sputtering, metallo-organic chemical vapor deposition) for conductive barriers result in thin films with a high defect density like dislocations, grain boundaries, interfaces and pores. Thus a critical thickness exists of the barrier layer to stop the diffusion of elements effectively. Decreasing the barrier thickness below about 50 nm it begins to fail depending on material, deposition parameters and testing conditions used.

High dose ion implantation is a promising tool to transform the deposited metal film in a nanocrystalline or amorphous microstructure with strongly reduced fast diffusion paths. The main target is to reach thermally stable barriers that can be stressed up to about 700 °C for 1 hour with a thickness in the order of 10 nm.

EXPERIMENT

Thin Ta and TaSi films were deposited onto silicon (100) wafers using long-throw RF magnetron sputtering. Ion implantation of nitrogen or oxygen into Ta layers of 100 nm thickness was carried out using doses from 5×10^{16} to 8×10^{17} ions/cm^2 and ion energies of 65 and 70 keV, respectively. The substrate temperature during implantation was held below 100°C. The thermal stability of the caused struture modifications was studied by annealing in vacuum at temperatures from 650 to 800°C for 1 h. The incorporation of nitrogen and oxygen by plasma source ion implantation using a pulsed voltage of 16 kV (pulse frequency 1500 Hz, pulse duration 5 μs, sample temperature 250 - 300°C) was tested for very thin Ta layers of about 15 nm thickness. The pulse voltage corresponds to an average kinetic energy of 8 keV of the implanted ions taking into account that mainly N_2^+ and O_2^+ molecules are accelerated from the plasma to the substrate. The depth distribution of the implanted species was measured by Auger Electron Spectroscopy

Conference Proceedings ULSI XV © 2000 Materials Research Society

(AES) and Photo Electron Spectroscopy (XPS) in combination with sputter etching by 3 keV Ar$^+$ ions. X-ray diffraction (XRD) of Cu-K$_\alpha$ radiation in gracing incidence was used to detect amorphization or formation of tantalum compounds. The modified layeres were further investigated by X-ray reflectometry, Transmission Electron Microscopy (TEM) and Atomic Force Microscopy (AFM) to study changes of the Ta microstructure.

RESULTS

In Fig. 1 the X-ray diffraction patterns are compared for unimplanted Ta and different implantations. The as-deposited Ta layer is polycrystalline according to Fig. 1a (dominantly β-Ta).

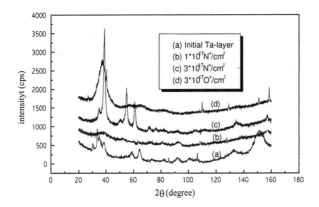

Fig. 1: X-ray diffraction pattern of 100 nm Ta on Si: (a) initial state, implanted with (b)$1x10^{17}$ N$^+$/cm^2, (c) $3x10^{17}$ N$^+$/cm^2, (d) $3x10^{17}$ O$^+$/cm^2 (in the case of $3x10^{17}$ N$^+$/cm^2 reflections, e.g. at about 38^0, 51^0, 61^0, correspond to Ta nitride, the other two sharp reflections are due to the silicon substrate)

By implantation of $1x10^{17}$ N$^+$/cm^2 into 100 nm Ta an essentially amorphous microstructure is formed as shown in Fig. 1b. X-ray reflectometry studies indicate a layered structure of the amorphized tantalum film. Fig. 2 presents the measured reflectometry curves for the tantalum initial layer and for the implantation with $3x10^{17}$ N$^+$/cm^2 and $3x10^{17}$ O$^+$/cm^2. Results of a fit on curve (b) corresponds to a stack of 0,66 nm Ta-O-N on top followed by two Ta-N layers differing in density (2,1 nm with 11,10 g/cm^3 and the main layer of 106,3 nm with 15,21 g/cm^3) on a transition region of 2,07 nm Ta-Si-N. The implantation of doses larger than $3x10^{17}$ N$^+$/cm^2 results in a partial formation of Ta nitrides (see Fig. 1c). The observed strong diffraction peaks based on a smooth background due to amorphous or nanocrystalline parts can be explained by superimposed reflections of the phases Ta$_4$N and Ta$_2$N (not resolved). The crystal structure of both phases is very similar. Ta$_4$N may be considered as a rhomboedric distortion of the hexagonal Ta$_2$N structure. A coexistence of both phases has to be expected because for $3x10^{17}$ N$^+$/cm^2 the average nitrogen concentration within the layer increases from about 20 to 35 at.%. The same fluence of O$^+$ ions gives an oxygen concentration of about 20 at.%. This leads to amorphization of the tantalum lattice as proved by the diffraction pattern of Fig. 1d. In the case of oxygen implantation the amorphous state is observed up to the very high dose of $8x10^{17}$ O$^+$/cm^2 without significant

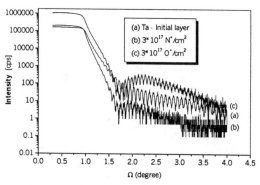

Fig. 2: X-ray reflectometry curves of 100 nm Ta on Si. (a) as-deposited, (b) implanted with 3×10^{17} N$^+$/cm^2, (c) implanted with 3×10^{17} O$^+$/cm^2.

indication of oxide formation in the diffraction patern. However, the existence of nanocrystalline oxides can not be excluded from the X-ray results. By fitting the X-ray reflectometry data (Fig. 2c) a composition of O$^+$-implanted layer was found with a TaO- layer at the surface, an intermediate and a main TaO-layer differing in some parameters. At the interface to the silicon a TaSiO transition region exists.

XPS measurements show indeed that oxidic bindings are formed. For all O$^+$-implantation doses the Ta 4f photoelectron peaks show additional chemically shifted peaks. The deconvolution of the XPS signal resolves two states shifted by 1.5 eV and 4.9 eV respectively. These peaks can be assigned to Ta^{1+} and Ta$_2$O$_5$. (The bulk value for Ta$_2$O$_5$ corresponds to a peak shift of 4.5 eV. The difference can be explained by the amorphous structure.) The peak area of the shifted lines increases with the O content in the layer. Ta$_2$O$_5$ did show up however only at the highest O$^+$ dose of 8×10^{17} O$^+$/cm^2.

AFM investigations confirm the results of the X-ray diffraction. The stronger impact of nitrogen, compared to oxygen implantation on the Ta layer is displayed in Fig. 3. The AFM-image of the initial state (a) shows a very plane and homogeneous surface. Ion implantation of nitrogen (3×10^{17} N$^+$/cm^2) leads to a distinct greater roughness (b). O$^+$-implantation results in formation of valleys in a comparable plane surface like the initial layer (c). The surface roughness RMS for the initial state is 0.22 nm, for the nitrogen implanted surface this value increased to 0.90 nm, while the oxygen implantation created a roughness of 0.29 nm.

Typical results of structural modifications after thermal treatment are shown in Fig. 4. The amorphization induced by oxygen implantation remains stable up to about 650 °C as shown for 3×10^{17} O$^+$/cm^2 (Fig. 4). The single reflection at about 33^0 may indicate the beginning formation of TaO. Annealing of this sample at 750 °C results in significant crystallization (see Fig. 4b).

The most likely interpretation of the observed diffraction peaks is a superposition of contributions from the non stoichiometric phase Ta$_{3.28}$Si$_{0.72}$ and TaO, i.e. the crystallization is correlated with interdiffusion of silicon into the implanted tantalum layer. The N$^+$-implanted sample (3×10^{17} N$^+$/cm^2) which did form nitride already during implantation shows after annealing at 650 °C the superimposed reflections of Ta$_4$N and Ta$_2$N without diffuse intensity distribution from amorphous or nanocrystalline structures. By annealing at 750 °C the diffraction peaks become more narrow and shift to the exact positions of the Ta$_2$N reflections. This observation indicates a transformation of Ta$_4$N to Ta$_2$N.

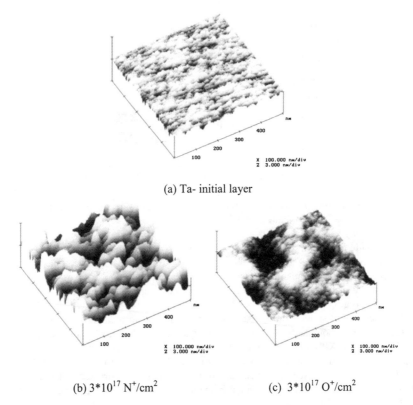

(a) Ta- initial layer

(b) $3*10^{17}$ N^+/cm^2 (c) $3*10^{17}$ O^+/cm^2

Fig.3: Atomic force microscope (AFM) - images of 100 nm Ta on Si, (a) initial state, implanted with (b) $3x10^{17}$ N^+/cm^2 and (c) $3x10^{17}$ O^+/cm^2 (z- range 3 nm).

The 100 nm TaSi layers are already X-ray amorphous after deposition. Here the implantation of N^+ and O^+ does not lead to obvious changes of the microstructure. This is demonstrated in Fig. 5 by comparison of the XRD pattern of an as-deposited layer with the results for implantation of $3x10^{17}/cm^2$ O^+ and N^+, respectively. The two peaks of amorphous TaSi were found in all three curves at the nearly same position. Nitrides and oxides could not be observed.

The X- ray reflectometry curve of the TaSi layer is shown in fig. 6a. We can conclude from these data that the film is clean and uniform over the entire thickness. The implantation of nitrogen or oxygen did not result in a discontinuity over the film cross section.

The intention to implant O^+ or N^+ into the TaSi layers is an expected enhancement of thermal stability of the amorphous structure. This effect is demonstrated by the XRD pattern of post annealed samples presented in Fig. 7. The unimplanted TaSi film recrystallizes already during annealing at 650 °C by formation of silicides as seen in Fig. 7a.. Most diffraction peaks in Fig. 7a correspond to reflections of $TaSi_2$ (marked by stars). The formation of this silicon-rich phase can only be explained by inter diffusion of additional silicon from the substrate. The relatively strong peak at about 37° indicates the coexistence of the phase $Ta_{3.28}Si_{0.72}$ mentioned above. The implantation of both species, N^+ and O^+ respectively, raises the crystallization temperature.

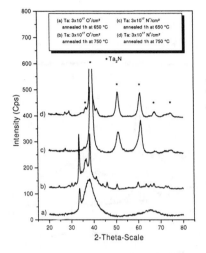

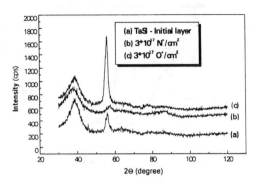

Fig. 4. XRD pattern of 100 nm Ta layers implanted with 3×10^{17} cm^{-2} O$^+$ or N$^+$ ions after annealing at 650 °C and 750 °C, respectively.

Fig. 5: X-ray diffraction pattern of 100 nm TaSi on Si: (a) initial state, implanted with (b) 3×10^{17}N$^+$/cm^2, and (c) 3×10^{17} O$^+$/cm^2 (The strong peaks at about 56 degree are artefacts due to the Si substrate).

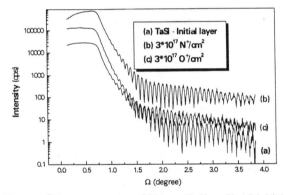

Fig. 6: X- ray reflectometry curves of 100 nm TaSi on Si: (a) initial state, implanted with (b) 3×10^{17} N$^+$/cm^2, and (c) 3×10^{17} O$^+$/cm^2.

After annealing at 650°C the amorphous structure is conserved for both implantations as shown by Figs. 7b and d. Even after annealing at 800 °C/1hr the N$^+$-implanted film is almost amorphous or nanocrystalline, though first indications of crystalline peaks appear in Fig. 7c. This result is in agreement with the high thermal stability of sputtered Ta-Si-N barriers reported in [2]. The layer implanted with 3×10^{17} O$^+$/cm^2 reveals much more pronounced crystallization features. The sharp peaks at about 37° and 42° in Fig. 8e correspond to the phase Ta$_5$Si$_3$. The peak at about 57° corresponds to Ta$_2$O$_5$.

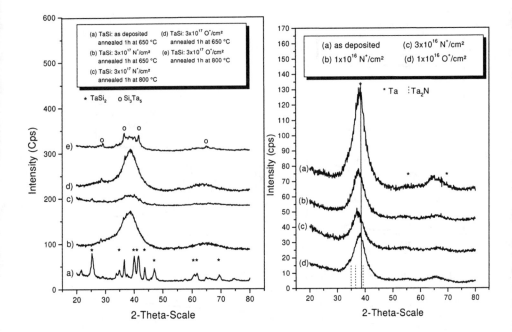

Fig.7. XRD pattern of 100 nm TaSi layers as-deposited and implanted with 3×10^{17} cm^{-2} O$^+$ or N$^+$ ions after annealing at 650 °C and 800 °C, respectively.

Fig. 8. XRD pattern of 15 nm tantalum layers in the as-deposited state (a) and after implantation of 1×10^{16} N$^+$ cm^{-2} (b), 3×10^{16} N$^+$ cm^{-2} (c) and 1×10^{16} O$^+$ cm^{-2} (d)

The stabilization of the amorphous film condition by ion implantation requires an additional process step. Since the deposition of the barrier layer is most likely done by a plasma process, the performance of the implantation step by a plasma based technique seems very applicable.

Plasma source ion implantation (PSII), a technique where the sample is immersed in a plasma from which ions are implanted by applying a pulsed voltage, is an effective method for low energy implantation treatment of very thin films. PSII with 16 kV pulses of N$_2^+$- and O$_2^+$-ions, respectively, and doses in the order of 10^{16} to 5×10^{16} O$^+$/cm^2 into Ta layers of about 15 nm results in oxygen concentrations as discussed above. In the case of nitrogen implantation an oxygen contamination of about 50% of the nitrogen content is detected by AES. It is explained by a low oxygen partial pressure due to contamination of the chamber walls. Considering the positive effects of oxygen implantation discussed above this effect will not be disadvantegeous for the barrier properties of the modified layer.

The PSII treatment does not result in a significant reduction of the tantalum thickness caused by sputtering. X-ray diffraction measurements on layers of about 15 nm show already for the as-deposited sample a diffraction pattern which is characteristic for amorphous or nanocrystalline material. Therefore, XRD is not well suited to investigate possible structural changes due to the implantation. Nevertheless a small shift of the main peak of the N$^+$ implanted layers to lower scattering angles can be observed. This is shown in Fig. 8 where the XRD pattern of a 15 nm as-deposited tantalum layer is compared with that of an implanted specimen. The shift increases with increasing dose and can be explained by nitride formation or by expansion of the tantalum lattice. As indicated by the dotted bars in Fig. 8 two diffraction peaks of Ta_2N are situated at the low angle side of the tantalum reflection.

We have demonstrated that the morphology of a 100 nm Ta film can be changed from polycrystalline to amorphous by ion implantation of O^+ or N^+. The stability of the amorphous state caused by O^+ implantation exceeds 650 °C/1hr. The stability of an amorphous TaSi film treated by N^+ implantation was demonstrated up to 800 °C/1hr. It has been shown that very thin films of 15 nm can be implanted with low energy by plasma source ion implantation. The stabilization of the amorphous state up to such high thermal stress gives hope that very thin and reliable Cu barriers can be realized by this technique.

Conclusions

Implantation of nitrogen and oxygen ions can change the microstucture of Ta thin films dramatically from a polycrystalline to an amorphous-like one. The thermal stability of the amorphous-like microstructure of tantalum thin films after oxygen implantation is stable up to 650 °C/1 hour. The implantation of TaSi with nitrogen increases the thermal stability of the amorphous-like microstructure up to 800 °C/1 hour. The implantation with oxygen leads to a lower thermal stability. Plasma source ion implantation is able to change the microstructure of ultra thin tantalum films (<20 nm thickness). Due to the low ion energy the action of the ions is located in the thin film region.

References

[1]   PDF2 data base, 06-05-69 (N), International Centre of Diffraction Data, Newtown Square, PA 19073-3273, USA.
[2]   Chongmu Lee and Young-Hoon Shin, Materials Chemistry and Physics 57, (1998), p.17.

Effects of Copper Diffusion Barrier on Physical/Electrical Barrier Properties and Copper Preferred Orientation

C.S. Liu, S.L. Shue, C.H. Yu, and M.S. Liang

Research & Development, Taiwan Semiconductor Manufacturing Company, Ltd.
No. 9, Creation Road I, Science Based Industrial Park, Hsin-Chu, Taiwan, R.O.C.
TEL: 886-3-5781688, FAX: 886-3-5773671, E-mail:csliua@tsmc.com.tw

ABSTRACT

Ta-based Cu diffusion barrier properties were widely studied. Cu directly deposited on Ta barrier layer can obtain strongest (111)-textured Cu. However, Ta exhibited the poorest thermal stability compared with other Ta-based barrier scheme. TaN(30 nm)/Ta(5 nm) scheme was found to be better than Ta(5 nm)/TaN(30 nm) in terms of thermal stability because the grains of TaN directly deposited on Ta will grow with columnar structure and keep the same grain boundary diffusion path. Vacuum break between Cu and Ta-based diffusion barrier can improve the Cu thermal stability. However, vacuum break will degrade the growth (111)Cu preferred orientation.

INTRODUCTION

With the demand of low-RC delays and good reliability, copper is considered as an alternate materials for multilevel interconnect. However, Cu is a deep level recombination center in the band gap and can be a device killer if diffused into Si. Hence, a robust diffusion barrier to impede Cu diffuse into devices is required for Cu metallization process. On the other hand, (111)-textured Cu was reported to have 4 times longer electromigration (EM) life time than (001).[1-2] Barrier layer under Cu will influence the growth of subsequent Cu seed layer and electro-chemical plated (ECP) Cu. Tantalum or tantalum nitride is widely used as a copper diffusion barrier material.[3-6] Ta-based barrier materials are PVD compatible, robust and immiscible in Cu. A proper choice on Ta-based Cu diffusion barrier is compared and discussed in this paper.

EXPERIMENT

Diffusion barrier properties of Ta, $Ta_{1-x}N_x$ (x = 0 ~ 0.6), Ta/TaN and TaN/Ta stacks with or without vacuum break between layers prepared by ionized PVD in Cu(200 nm)/ Barrier(15-60 nm)/$CoSi_2$(30 nm)/Si structure were studied. The samples were subsequently annealed in forming gas ambient furnace to the higher temperature to check the barrier performance by electrical performance and analytical tools. RBS was used to define the chemical composition of $Ta_{1-x}N_x$. Four point probe was used to measure sheet resistance (Rs) change of stacked films. SEM, TEM and AES were utilized to analyze the film integrity of Cu(200 nm)/ Barrier(15-60 nm)/$CoSi_2$(30 nm)/Si structure. XRD was used to characterize the phase formation and Cu orientation influenced by barrier layers.

Conference Proceedings ULSI XV © 2000 Materials Research Society

RESULTS AND DISCUSSIONS

Figure 1 showed the Rs varied with different stress temperature among different barrier materials and schemes. Ta barrier has a poorest thermal stability against Cu diffusion. Cu was found to diffuse across Ta and react with Si to form Cu_3Si at 400°C after 30 min N_2/H_2 ambient annealing. TaN as a barrier has better performance than Ta and fails at 450°C. The TaN barrier performance was found to decrease with the x value of chemical composition of $Ta_{1-x}N_x$. For x larger than 0.4, an increase in Rs was noticed.

Vacuum break between Cu deposition and Ta or TaN deposition can improve the barrier properties and the structure remains intact up to 500°C. Thicker TaN also showed improved barrier properties. Furthermore, TaN(30 nm)/Ta(5 nm) scheme was found to be better than Ta(5 nm)/TaN(30 nm) in terms of thermal stability. The grains of TaN directly deposited on Ta will grow along Ta columnar structure and have the same grain boundary diffusion path. On the other hand, Ta deposited on the amorphous-like structure will not grow columnar structure with the same diffusion path of TaN. Cross sectional SEM picture and AES depth profile of TaN and TaN/Ta scheme were shown in Fig.2 and Fig.3, respectively.

A normalized I_{111}/I_{200}Cu XRD intensity was listed in table 1. Vacuum break can improve the Cu thermal stability both on Ta or TaN barrier. However, vacuum break will degrade the (111)Cu preferred orientation of following Cu deposition from relative I_{111}/I_{200}Cu XRD peak intensity as shown in table 1. Cu directly deposited on Ta barrier layer can obtain strongest (111)-textured Cu but Ta barrier exhibited poorest thermal stability against Cu diffusion. A compromise was found in the barrier properties between the growth of (111)Cu texture for improving EM and diffusion barrier against Cu diffusion.

CONCLUSIONS

Ta barrier had poorest thermal stability against Cu diffusion but exhibited the strongest (111)-textured Cu growth. TaN(30 nm)/Ta(5 nm) scheme was found to be better than Ta(5 nm)/TaN(30 nm) in terms of thermal stability. The grains of TaN directly deposited on Ta will grow with Ta columnar structure and keep the same grain boundary diffusion path. On the other hand, Ta deposited on the amorphous-like structure will not grow with columnar structure. A compromise was found in the barrier performance between the growth of (111)-textured Cu to improve EM and diffusion barrier to against Cu diffusion.

ACKNOWLEDGE

J.B. Lai is greatly acknowledged for AES analysis.

REFERENCE

1. K Abe etc, IEEE 36th Annual International Reliability Physics Symposium, Reno Nevada, p. 342 (1998).
2. C. Ryu etc, IEEE 36th Annual International Reliability Physics Symposium, Reno

Nevada, p. 201 (1998).

3. T. E. Gebo etc, Proceedings of the Advanced Metallization for ULSI, p. 103 (1994).
4. K.H. Min, K.C. Chun and K.B. Kim, J. Vac. Sci. Technol. B14, p. 3263 (1996).
5. L. Clevenger, N.A. Bojarczuk. K. Holloway and J.M.E. Harper, J. Appl. Phys. 73, p. 300 (1993).
6. Q. T. Jiang, R. Faust, H. R. Lam and J. Mucha, Proceedings of the International Interconnect Technology Conference, p. 125, (1999).

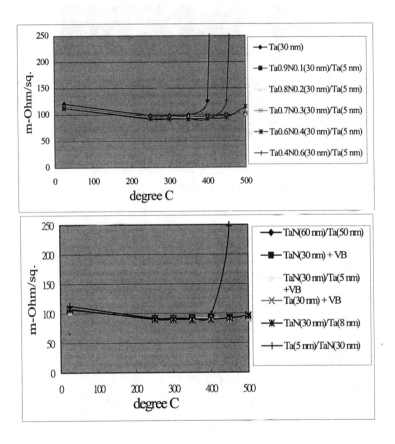

Fig. 1   (a) and (b) showed the Rs change vs different temperature of stress among varied kinds of Ta-based barrier.

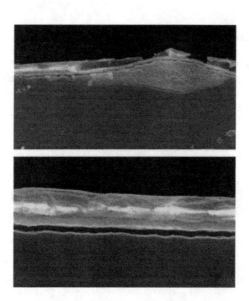

Fig.2 XSEM of (a) TaN(30 nm) and (b) TaN(60 nm)/Ta(5 nm) barriers annealed 500℃
for 30min showing destroyed and intact structures, respectively.

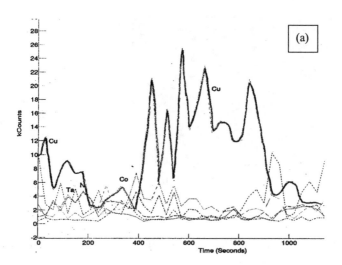

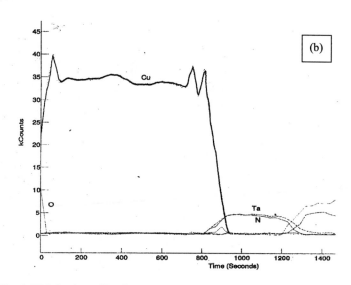

Fig. 3 AES depth profile of (a) TaN(30 nm) and (b) TaN(60 nm)/Ta(5 nm) barriers annealed 500°C for 30min.

Barrier structure	Vacuum break	I_{111}/I_{200} Cu
Ta(30 nm)	No	**2729**
$Ta_{0.8}N_{0.2}$(30 nm) /Ta(5 nm)	No	1695
$Ta_{0.6}N_{0.4}$(30 nm) /Ta(5 nm)	No	1461
Ta(5 nm)/TaN(30 nm)	No	1435
TaN(30 nm)/Ta(5 nm)	No	1375
TaN(60 nm)/Ta(5 nm)	No	1610
TaN(30 nm)/Ta(5 nm)	**Yes**	**19**
Ta(30 nm)	**Yes**	**28**
TaN(25 nm)/Ta(10 nm)	No	1329

Table 1   Normalized XRD I_{111}/I_{200} Cu peak intensity.

IMPROVEMENT OF <111> TEXTURE AND SUPPRESSION OF AGGLOMERATION OF Cu FILM BY Ti/CVD-TiN/Ti UNDERLAYER

Mitsuru Sekiguchi, Haruhiko Sato, Takeshi Harada, and Shinichi Domae
ULSI Process Technology Development Center, Matsushita Electronics Corp.
19 Nishikujo-Kasugacho, Minami-ku, Kyoto 601, Japan
Tel: +81-75-662-7399; Fax: +81-75-681-3189; E-Mail; PAN85044@sel.mec.mei.co.jp

ABSTRACT

Enhancement of <111> texture of and suppression of agglomeration of Cu film is important to improve electromigration endurance of Cu damascene interconnection. These properties are improved by Ti/CVD-TiN/Ti underlayer comparing with the case by CVD-TiN/Ti underlayer. Atom arrangement of the Ti layer on the CVD-TiN/Ti layer is expected to be close to that of (111)-TiN, since CVD-TiN is highly (111)-oriented in the CVD-TiN/Ti structure. This Ti layer is considered to enhance <111> texture of Cu further than that in the Cu/CVD-TiN/Ti structure.

INTRODUCTION

It is reported that good wetting and highly (111)-oriented texture Cu films are effective to enhance electromigration (EM) endurance of Cu damascene interconnection [1, 2]. Sputtered Ta and TaN films are usually used as barrier layers for Cu interconnection since good wetting and highly (111)-oriented texture can be obtained [2]. Though CVD-TiN has an advantage of good step coverage, it is reported that <111> texture and wetting of Cu films on the CVD-TiN films are poorer than those on the Ta and TaN films [2, 3]. To improve EM resistance of Cu/CVD-TiN interconnection, wetting and <111> texture of Cu films should be improved. Recently, we have reported these properties can be improved by depositing a Ti layer under the CVD-TiN film [4].

In this paper, it is found that these properties of Cu films are further improved by Ti/CVD-TiN/Ti underlayer. Atom arrangement of the Ti film on the CVD-TiN/Ti layer is considered to be close to that of (111)-oriented CVD-TiN layer in the CVD-TiN/Ti structure. This Ti layer might enhance <111> texture of the Cu film since a Cu film on (111)-oriented TiN reveals highly (111)-oriented texture [5]. This structure successfully achieves highly (111)-oriented texture of Cu films even in damascene interconnection.

EXPERIMENT

At First, 10nm-thick Ti films were sputter-deposited on P-SiN (Plasma-CVD SiN) films by HDP (high density plasma). Next, 10nm-thick CVD-TiN films were deposited using TDMAT precursor by successive steps of N_2/H_2 plasma treatment. The plasma treatment is effective to form nano-scale crystallites in an amorphous matrix of TiN [6]. For the sample with Ti/CVD-TiN/Ti underlayer, 10nm-thick Ti film was sputter-deposited by HDP again. After exposing the samples to the air, 50nm-thick Cu films were deposited by HDP sputtering. Then, textures of the Cu films were evaluated by XRD (X-ray diffraction) using θ-2θ scanning and rocking curve measurement. For another set of the samples to evaluate "wetting", various underlayers were deposited on P-SiN films in the same way. Then, 15nm-thick Cu films were sputter-deposited and annealed at 400°C for 5 minutes without breaking the vacuum. Surface roughness of the Cu films were measured by AFM in order to check agglomeration.

Single damascene interconnects with 0.38μm wide and 0.30μm deep trenches were formed in 500nm-thick P-SiO$_2$ layer. After various underlayers were deposited, 150nm-thick Cu films were deposited by HDP sputtering as a seed layer. Then Cu electroplating deposition, CMP and passivation with 50nm-thick P-SiN were carried out subsequently. A degree of <111> texture of the Cu films was evaluated by XRD using θ-2θ scanning and rocking curve measurement. Scan direction was perpendicular to the lines.

Via hole resistance of Cu interconnection was measured by Kelvin method. The diameter of via holes was 0.26μm. After via holes were opened by dry etching, Ar sputter etching was applied for all samples before metal deposition. The amount of the Ar sputter etching is equivalent to removal of 25nm-thick SiO$_2$.

RESULTS AND DISCUSSION

Agglomeration and <111> Texture in Blanket films

Figure 1 shows AFM images of Cu surfaces on various underlayers after annealing at 400°C for 5 minutes. To prepare the sample in Fig.1 (a), CVD-TiN surface is exposed to the air before Cu deposition. Note that "//" indicates air exposure of underlayer surface. Thin Cu film on the CVD-TiN easily agglomerated during annealing. Large surface roughness might be observed in the AFM image of the Cu//CVD-TiN structure. Table I shows that mean roughness (Ra) value of the Cu//CVD-TiN structure was 27.0nm. In Cu//CVD-TiN/Ti structure, the film roughness is much reduced to form some pits in the Cu film (Fig.1 (b)). As shown in Table 1, Ra value of this structure is reduced to 1.2nm compared with that of the Cu//CVD-TiN structure. This means wetting of the Cu film is much improved by the CVD-TiN/Ti underlayer [4]. In Fig.1 (c), it should be noted that the sample with Ti/CVD-TiN/Ti underlayer does not reveal any pits and shows continuous Cu surface. Therefore, wetting of the Cu film of this sample is the best in these three samples.   Table I shows that Ra value of

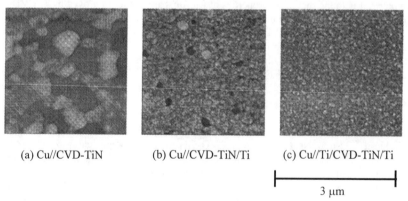

(a) Cu//CVD-TiN   (b) Cu//CVD-TiN/Ti   (c) Cu//Ti/CVD-TiN/Ti

|——————————————————————|
3 μm

Fig.1. AFM images of Cu films on various underlayers after annealing at 400 °C for 5 min. The symbol of "//" indicates air exposure of underlayer.

Table I. Mean roughness of sputtered Cu films on various underlayers after annealing .

Stacked Structure	Ra (nm)
Cu//CVD-TiN	27.0
Cu//CVD-TiN/Ti	1.2
Cu//Ti/CVD-TiN/Ti	0.8

Table II. Cu (111) texture of sputtered Cu films in various underlayers.

Stacked Structure	Intensity of Cu Diffraction : I(111)/I(200)	FWHM of Cu (111) Rocking Curve (degree)
Cu//CVD-TiN	2.7	8.6
Cu//CVD-TiN/Ti	160	4.0
Cu//Ti/CVD-TiN/Ti	190	3.6

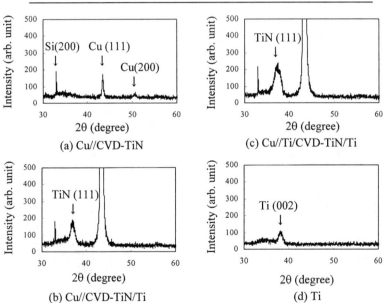

Fig.2. XRD spectra by θ-2θ method of various structures.

the Cu//Ti/CVD-TiN/Ti structure is the lowest. It is found that Ti/CVD-TiN/Ti underlayer improves wetting characteristics of Cu in comparison with those on CVD-TiN/Ti and CVD-TiN.

Table II summarizes results of XRD measurements of sputtered Cu films on various underlayers. The intensity of <111> texture of Cu is the highest in the case of Cu//Ti/CVD-TiN/Ti structure, since FWHM of Cu (111) rocking curve is reduced by 0.4° compared with that of the Cu//CVD-TiN/Ti structure. It is obvious that the Ti/CVD-TiN/Ti underlayer improves <111> texture of the Cu film in comparison with that on the CVD-TiN/Ti underlayer.

To clarify the reason of texture improvement and good wetting of the Cu//Ti/CVD-TiN/Ti structure, XRD analysis was performed. As shown in Fig. 2 (d), single Ti layer has (002)-oriented structure. Figure 2 (a) and (b) shows, this (002)-oriented Ti underlayer enhances crystallization of the amorphous CVD-TiN film to (111)-oriented poly crystals. As the distances between atoms in (002) plane of Ti and atoms of (111) plane of TiN were nearly equal (2.95 and 3.00Å, respectively [7]), coincidence of nearest neighbor atomic distance between Ti-(002) and Ti-(111) plane may assist crystallization. This weak (111)-oriented CVD-TiN film is considered to make (111)-oriented Cu films because of crystal continuity between Cu and TiN [5]. Figure 2 (c) shows there is a higher and broader peak around TiN (111) position in Cu//Ti/CVD-TiN/Ti than that of Cu//CVD-TiN/Ti. The peak is a little shifted to Ti (002) position, but not the same position as the Ti film (Fig. 2(d)). This indicates atom arrangement of the Ti layer on the CVD-TiN/Ti layer tends to be closer to that of (111)-oriented TiN. It is reported that (111)-oriented Cu films grow on (111)-oriented RTN-TiN films [5]. Thus, the Ti layer which shifts close to (111)-TiN is considered to enhance Cu <111> texture. The enhanced crystal relationship between Cu and underlayer is considered to be effective to prevent agglomeration during annealing.

Application to Damascene Interconnection

Table III and Fig.3 show that the Ti/CVD-TiN/Ti underlayer is effective to enhance Cu <111> texture even in damascene interconnection. FWHM of Cu (111) rocking curve is reduced by 0.8° compared with that of the Cu//CVD-TiN/Ti structure. The result is almost the same as that in the case of the blanket films. It has been reported that texture of Cu electroplated films inherits texture of seed layers [2]. Since Cu electroplated films grow from the bottom of the trench preferentially [8], the result for the blanket films is still consistent with the case of damascene structure.

Next, the effect of the Ti/CVD-TiN/Ti underlayer on via contact resistance was evaluated.

Table III. Cu (111) texture in 0.38μm wide damascene interconnection.

Stacked Structure	Intensity of Cu Diffraction : I(111)/I(200)	FWHM of Cu (111) Rocking Curve (degree)
Cu//CVD-TiN	17	6.6
Cu//CVD-TiN/Ti	150	3.7
Cu//Ti/CVD-TiN/Ti	260	2.9

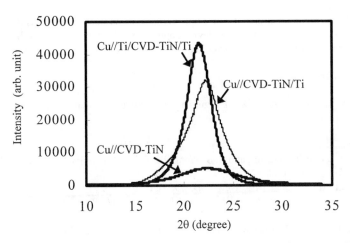

Fig.3. Cu (111) rocking curves of 0.38μm wide damascene interconnection.

Table IV. 0.26 μm via contact resistance.

Stacked Structure	Average (Ω)	Signa (Ω)
Cu//CVD-TiN	2.3	0.46
Cu//CVD-TiN/Ti	1.5	0.19
Cu//Ti/CVD-TiN/Ti	2.0	0.09

Table IV summarizes the results. Contact resistance and its standard deviation (sigma) of the CVD-TiN/Ti underlayer are smaller than those of the CVD-TiN underlayer. In the case of the Ti/CVD-TiN/Ti underlayer, contact resistance is slightly increased compared with that of the CVD-TiN underlayer. This might be due to insertion of Ti layer between Cu and CVD-TiN layers, because Ti has higher resistance (50 μΩcm) compared with Cu (2.0μΩcm). Though contact resistance of this structure is slightly increased, via resistance of 2.0Ω for 0.26μm diameter is acceptable for device application.

CONCLUSIONS

In conclusion, wetting property and <111> texture of Cu on CVD-TiN/Ti are improved by inserting PVD-Ti layer between Cu and CVD-TiN. Atom arrangement of the Ti layer on the CVD-TiN/Ti layer tend to be close to that of (111)-oriented TiN, since Ti-(002) and TiN-(111) plane has almost same nearest neighbor atomic distance. This Ti layer is considered to enhance crystal relationship between Cu and underlayer, because (111)-oriented Cu can grow on (111)-oriented TiN. Therefore, wetting and <111> texture of the Cu film on the Ti/CVD-TiN/Ti structure is improved compared with that on the CVD-TiN/Ti structure. This structure show highly (111)-oriented texture of Cu films even in damascene interconnection. Via resistance of this structure is acceptable for device application. This technology will be useful to realize Cu damascene interconnection with high EM endurance.

ACKNOLEDGEMENTS

The authors would like to thank Dr. M. Ogura and R. Ueda for their encouragement.

REFERENCES

1. C. Ryu, A. L. S. Loke, T. Nogami, and S. S. Wong, IEEE Int. Reliability Phys. Symp. Proc. (IEEE, New York, 1997), p.201-205.
2. S. S. Wong et al., IITC Proc. (IEEE, New York, 1998), p.107-109.
3. T. Nogami, J. Romeo, V. Dubin, D. Brown, and E. Adem, IITC Proc. (IEEE, New York, 1998), p.298-300.
4. M. Sekiguchi, H. Sato, T. Harada, and R. Etoh, IITC Proc. (IEEE, New York, 1999), p.116-118.
5. K. Abe, Y. Harada, and H. Onoda, Appl. Phys. Lett., 71, p.2782 (1997).
6. A. Jain et al., Conference Proceedings ULSI XIII (MRS, 1998), pp.41-47.
7. H. Shibata, N. Ikeda, M. Murota, Y. Asahi, and H. Hashimoto, Symp. on VLSI. Tech. Dig. of Technical Papers (IEEE, New York, 1991), p.33-34.
8. P. C. Andricacos, C. Uzoh, J. O. Dukovic, J. Horkans, and H. Deligianni, IBM J. Res. Develop., 42, p.567 (1998).

Ideal microstructure of thermally stable TiN diffusion barriers for Cu interconnects

M. Moriyama, T. Kawazoe, M. Tanaka, and M. Murakami
Department of Materials Science and Engineering, Kyoto University, Kyoto, Japan
E-mail: moriyama@micro.mtl.kyoto-u.ac.jp

ABSTRACT

One of the major concerns with manufacturing Si-ULSI (Ultra-Large Scale Integrated) devices using Cu interconnects is lack of barrier layers which prevent Cu diffusion with Si or SiO_2 at elevated temperatures. The highest temperature at which the conventional devices are annealed during the fabrication process is 400℃. However, some devices are required to anneal at 700℃ for 30min, and the barriers must prevent Cu interdiffusion into Si or SiO_2 at these temperatures. In addition, in order to minimize loss of Cu wiring conductivity, the barrier thickness should be thinner than 40nm.

The purpose of the present study is to develop a thermally stable 'thin' barrier layer which prevents the Cu interdiffusion at temperatures above 700℃. In the present study, TiN was selected as the barrier layer, because TiN has the low electrical resistivity and the high melting point. The microstructures of TiN films which were prepared by various deposition conditions were analyzed by x-ray diffraction and transmission electron microscopy. Based on these microstructural analysis, a diffusion model of Cu in the TiN barriers was proposed. It. was concluded that the mosaic-spread angle of TiN grains and grain sizes play the key parameters to prepare a barrier layer with high thermal stability.

INTRODUCTION

Copper is an attractive material for interconnects of Si-ULSI devices. In order to apply the Cu interconnects to manufacturing devices, a barrier layer which prevents the Cu diffusion into the Si substrates and SiO_2 insulators at elevated temperatures must be developed. For the conventional Si-devices, the barriers are required to prevent the Cu diffusion at temperatures of around 400℃. However, there are some special devices which are annealed at 700℃ for 30min to activate implanted B^+ after Cu wiring. In addition, the resistivity of the barrier layers must be lower than 300μΩ-cm to maintain low via-contact resistance of devices with 0.25μm wide line. The barrier layer thickness should be less than 40nm to preserve lower effective resistivity of Cu interconnects compared with Al interconnects of these devices.

In order to prepare the barrier layers which satisfy these requirements, TiN was selected as the barrier metal in the present study, because TiN has extensive history in use for barrier layers for Al interconnects of ULSI devices. Also, the sputtered-TiN films have low resistivity and the excellent barrier property against Cu diffusion as demonstrated by Wang et al.[1] However, the barrier property was evaluated for 100nm-thick TiN films, which was thicker by one order of magnitude than those required for ULSI devices.

The purpose of the present investigation is twofold. The first is to understand the diffusion mechanism of Cu in the TiN barriers by analyzing microstructure of TiN films which were deposited at various conditions. The structural analysis was carried out by X-ray diffraction (XRD) and transmission electron microscopy (TEM). These studies will also give us a guideline for development of the barrier layers which are applicable to the future ULSI devices. The second purpose is to challenge to develop a thermally stable 'thin' TiN barrier layer (<40nm) which withstands at

277

temperature above 700℃ to apply this barrier layer to the special devices with 0.25μm wide.

EXPERIMENTAL

The substrates used in the present experiments were (100)-oriented n-type Si wafers. The wafers were dipped in 5% HF solution for 1 min, and rinsed in super clean water prior to loading into the vacuum chamber. The sputter-deposition chamber was evacuated to 3×10^{-8} torr. The 25nm-thick TiN films were prepared at various incident powers by reactively sputtering a Ti target in Ar+N_2 gas with various Ar/N_2 gas ratios, and the substrate temperatures were changed from room temperature to 600℃. Then, the Cu layers with 100nm- thick were deposited on the barrier layers without breaking vacuum. After film depositions were completed, the samples were cooled to room temperature before transferring them to the load-lock chamber.

The Si/TiN(25nm)/Cu(100nm) samples were annealed at temperatures ranging from 700℃ to 900℃ for 30 min in 5% H_2/N_2 mixed gas atmosphere. Microstructural analysis of the samples before and after annealing was carried out by XRD and TEM. TEM observations were performed both in cross-sectional and plan-views using a 400kV electron microscope (JEM-4000EX) equipped with a top-entry goniometer. Samples for plan-view TEM observations were prepared by depositing TiN films (25nm) on Si_3N_4 films (~60nm), which allowed us for in-situ TEM observation.

RESULTS AND DISCUSSIONS

Microstructural analysis

Figure 1(a) shows typical cross-sectional HREM (high-resolution electron microscopy) micrograph of the Si/TiN/Cu sample. The grain boundaries and grain orientations which were determined by contrast of lattice images are indicated by white dots and black arrows, respectively. It is seen that the TiN layer is composed of two layers: columnar grains with low orientation index (i.e. fiber structure) at the top, and small grains with rather random orientation at the bottom. Cross-section of the TiN film structure, which was deduced from TEM observation, is schematically shown in Fig. 1(b).

Because of the high density of grain boundaries in TiN films, the main diffusion paths of Cu in TiN barriers would be the grain boundaries. It is well known that rate of fast grain boundary diffusion depends on the atomic structures of the grain boundaries[2], and diffusion coefficients along the grain boundaries vary from boundary to boundary according to the misorientation between the grains on the either side. There have been a lot of measurements of diffusivity along the grain boundaries as a function of misorientation. Turnbull and Hoffman have experimentally studied the dependence of diffusivity along the grain boundaries on the misorientation angle using Ag bicrystals[3]. The geometrical configuration of these bicrystals is schematically shown in Figure 2. The two grains (grain A and B) of a bicrystal have a common [100] axis within the grain boundary plane and can be brought into coincidence by a rotation of an angle ϕ around the common [100] axis. These grain boundaries are called [100] tilt grain boundaries. In these Ag tilt grain boundaries, the diffusivity increased linearly with misorientation angle ϕ (for $\phi < 16°$). In addition, the diffusivity had anisotropy within the grain boundary plane for parallel ($D_{//}$) and perpendicular directions ($D_⊥$) to [100] axis[4] (Fig. 2). The diffusivity along perpendicular direction was much smaller than along parallel direction.

These experimental results help us to understand the diffusion mechanism of Cu in TiN barrier layers. As shown in Figs. 1, the grain boundaries in the fiber structural layer have low index direction, and are similar to tilt grain boundaries. Thus, it is expected that the Cu diffusivity along

278

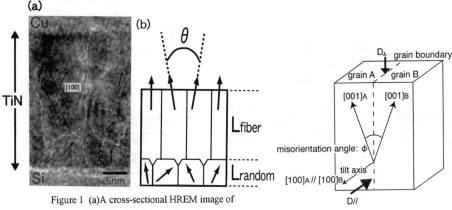

Figure 1 (a)A cross-sectional HREM image of Si/TiN(25nm)/Cu(100nm) sample and (b)corresponding schematic illustration

Figure 2 Schematic illustration of a bicrystal with [100] tilt grain boundary

the grain boundaries in the fiber structural layer is smaller than that in the random-oriented grained layer, and become smaller as decreasing mosaic-spread angle θ (indicated in Fig.1 (b)). The ability to prevent for Cu diffusion is considered to be larger in fiber structural layer which have smaller mosaic-spread angle. This is consistent with the experimental result reported by Murakami and deFontaine[5] that Ag polycrystalline films which have small mosaic-spread angle had very low Cu inter-diffusivity (same as bulk diffusivity). We assume that large L_{fiber}/L_{random} ratio (Fig.1(b)) and small mosaic-spread angle of TiN films is desirable to prepare a thermally stable diffusion barrier. This suggests that control of the film microstructure is very important in improving thermal stability of diffusion barrier layers.

Microstructure of layers with various deposition conditions

Improving adatom mobilities on the substrate surfaces would result in low-index preferred texture and large grain sizes. For the sputter-deposited films, the mobility might increase by increasing the incident power, the Ar gas ratio, and the substrate temperature, and would improve thermal stability of the barrier layers. To investigate the relationship between the deposition condition and film microstructure, the TiN films were prepared with various deposition conditions (different incident power, Ar/N_2 gas ratio, and substrate temperature) and were analyzed by XRD and TEM.

Figures 3(a), and 3(b) are plan-view TEM micrographs of the TiN films deposited on the Si_3N_4 substrates with different incident powers. It can be seen that the mean grain sizes in the TiN films increase with incident power. It is also noted that the films prepared with high power have no voids (or low-density regions) around the grain-boundaries, and fast Cu diffusion paths[6]. The XRD analytical results from the samples deposited at various Ar/N_2 gas ratios also revealed that the grain sizes increase with the Ar/N_2 gas ratio, resulting in large L_{fiber}/L_{random} ratio in TiN films.

Figure 4 shows XRD patterns of 25nm-thick TiN films deposited at different substrate temperatures. The films prepared at high substrate temperatures increase the relative intensity of the (002) reflection. These results indicate that the deposition with high substrate temperature increases (001) preferred orientation, and decreases the mosaic-spread angle.

From these experimental results, it is expected that increasing the incident power, the Ar/N_2 gas

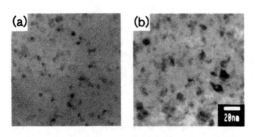
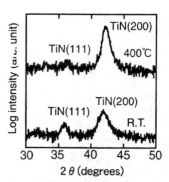

Figure 3 Plan-view TEM micrographs of the TiN films
deposited at differnt incident powers, (a)40W and (b)195W

Figure 4 XRD peak prefiles obtained from
Si/TiN(25nm) samples that were deposited
at different substrate temperatures

ratio and the substrate temperature would improve the thermal stability of the TiN barrier layers.

Thermal stability of barrier layers

In order to investigate relationship between the TiN film microstructure and the thermal stability, Si/TiN(25nm)/Cu(100nm) samples, which were prepared by depositing at various conditions, and annealed at temperatures in the range of 700℃ to 900℃ for 30min were analyzed by XRD. The thermal stability was evaluated by formation of Cu silicide detected by XRD.

Figures 5(a), (b) and (c) show the thermal stability of the TiN barrier layers which were deposited at different incident powers, Ar/N_2 gas ratios, and substrate temperatures. As seen in Figs.5, the thermal stability of the TiN barrier layer increase with these parameters. These results agree well with the above prediction. The highest temperature which the TiN layers prevented Cu diffusion is 850℃ for 30min. The sheet resistance of the 25nm-thick TiN film was measured to be 46μΩ-cm. These results satisfy the requirements for the barrier layer used in the special devices which must be annealed at 700℃ for 30min.

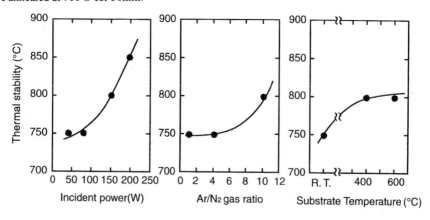

Figure 5 Dependences of thermal stability of TiN barrier layers on different deposition conditions

CONCLUSIONS .

The thermal stability of the 25nm-thick TiN barrier layers and film microstructures was correlated by analyzing the TiN films deposited at various conditions. It was found that the thermal stability was strongly correlated with the film microstructure. It was concluded that the mosaic-spread angle of the films and grain sizes are the key factors to control the thermally stability of the TiN barrier layers. The 25nm thick TiN films with low-index preferred texture and large grain sizes prevented the Cu diffusion after annealing at 850℃ for 30min.

REFERENCES

1 S.-Q. Wang, I. Raaijmakers, B. J. Burrow, S. Suthar, S. Redkar, and K.-B. Kim, J. Appl. Phys. **68**, 5176 (1990)
2 A. P. Sutton and R. W. Balluffi, Interfaces in Crystalline Materials, oxford university press.
3 D. Turnbull and R. E. Hoffman, Acta Metal. **2**, 419 (1954)
4 R. Hoffman, Acta Metal. **4**, 98 (1956)
5 M. Murakami and D. deFontaine, J. Appl. Phys. **47**, 2857 (1976)
6 K.-C. Park and K.-B. Kim, J. Appl. Phys. **80**, 5674 (1996)

Low resistivity TiN films at low deposition temperature
using Flow Modulation Chemical Vapor Deposition (FMCVD)

H. Hamamura* , R. Yamamoto, H. Komiyama* and Y. Shimogaki
Department of Materials Science and Metallurgy, University of Tokyo, JAPAN
*Department of Chemical System Engineering, University of Tokyo, JAPAN
E-mail: hamamura@prosys.t.u-tokyo.ac.jp
TEL&FAX:+81-3-5841-7131

ABSTRACT

A new CVD process, FMCVD (Flow Modulation Chemical Vapor Deposition), for depositing high-quality films in a single CVD process have been proposed [1]. In FMCVD we use deposition followed by a reduction reaction. For TiN FMCVD, we first deposited thin TiN films and then carried out chlorine reduction using NH_3. We repeated this cycle until the thickness that we needed. In this study, we have constructed new FMCVD reactor that has good base pressure around 10^{-7} Torr and performed further study on the FMCVD processes for TiN deposition. With increase of flow modulation cycle number, the residual Cl concentration decreased and crystallization of the films became better, and then the resistivity of the films decreased. Using FMCVD process, we could achieved low resistivity ($400\mu\Omega$cm) TiN films at low deposition temperature (350°C) with good step coverage.

INTRODUCTION

CVD-TiN utilizing $TiCl_4$ and NH_3 are attractive candidates as diffusion barrier layer process in the multilevel metallization for its excellent conformality, especially for small feature size and high aspect ratio contact/via holes. But the films from this process contain high residual chlorine at good step coverage condition at low deposition temperature [1]. To overcome this problem we proposed a new CVD concept (Flow Modulation Chemical Vapor Deposition; FMCVD) and found that FMCVD process was effective for reducing Cl concentration in the films [2].

In this study, we investigated the properties for TiN films deposited by FMCVD with new FMCVD reactor (warm wall reactor) that has good base pressure around 10^{-7} Torr. The effect of processing parameters were also investigated.

EXPERIMENT

We used a warm wall reactor (figure 1) and deposited films on Si substrate. $TiCl_4$ and NH_3 were introduced into the reactor separately to prevent the adduct formation. The reactor wall temperature was set at 200°C. Heater

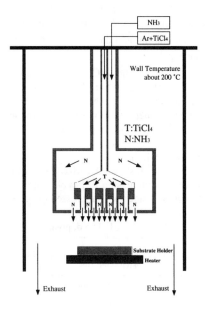

Figure1. The experimental design.

283

temperature was set at 500°C and surface temperature of Si substrate was around 350°C. TiCl$_4$ was introduced by bubbling system with Ar as a carrier gas. Total pressure was 2Torr and initial concentration of TiCl$_4$ and NH$_3$ was 10mTorr and 500mTorr, respectively, which yielded good step coverage. Figure 2 shows the gas flow of the experimental procedure of FMCVD. This gas flow sequence was controlled by the computer which was connected to gas supply system. First, we deposited TiN films

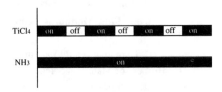

Figure2. The gas flow of the experimental procedure of FMCVD

from TiCl$_4$ and NH$_3$. Then TiCl$_4$ flow was stopped, thus TiN film was annealed under NH$_3$ atmosphere.

The thickness and the sheet resistivity of the films were analyzed by FE-SEM and 4-point probe method. The composition and crystalline structure were investigated by XPS and XRD techniques, respectively.

RESULTS AND DISCUSSIONS

The effect of FM cycle number.

First, we investigated the effect of FM cycle number. Table 1 summarizes the deposition sequences. Total deposition time and annealing time was fixed at 300 sec. and 200 sec., respectively. Normal CVD sample and post annealing sample, which was annealed under NH$_3$ atmosphere after normal CVD process were also made and analyzed.

The deposition rate of FM200 sample was about one-third as large as that of normal CVD sample (figure 3). There are many factors causing the decrease of deposition rate, such as structure, contamination, and the effect of incubation time. We are continuing to study these effects of FMCVD

Figure 4 shows Cl/Ti in the films as a function of the FM cycle number. We can see a decrease of Cl/Ti in the films with increase of FM cycle number. Especially, the values of FM samples were much lower than that of normal CVD and post annealing process. This result suggests that FM sequences with NH$_3$ easily reduced Cl in the films and the thinner TiN films were, the easier TiN films were reduced by NH$_3$.

We also measured the N/Ti ratio of the films, but there was no significant difference in this ratio shown in figure 5.

Figure 6 shows the XRD data of the films, indicating that the crystalline structure became better with increasing FM cycle number. Considering the result of Cl/Ti in the films, a large amount of residual Cl suppressed the crystallization of the films. So, crystalline structure of FM

Table1. The deposition sequences for the effect of FM cycle number.

	Deposition	Annealing
NormalCVD (FM 0)	300s	0s
Post Annealing (FM 0)	300s	200s
FM 20	15s x 20	10s x 20
FM 100	3s x 100	2s x 100
FM 200	1.5s x 200	1s x 200

samples was better than that of normal CVD sample because of low residual Cl concentration.

Figure 7 shows the resistivity of the films as a function of FM cycle number and indicates that the resistivity dramatically decreased with FM cycle number and the value was about 600μΩcm for 200 FM sequences. This result agrees with XPS and XRD analysis in the films as shown before, that is, low Cl concentration and good crystalline structure caused low resistivity TiN films.

We also investigated resistivity change of samples during preservation(figure 8). As shown here, resistivity of normal CVD and post annealing samples was unstable after a few days, we could not measure the resistivity because of extremely high value. On the other hand, resistivity of FMCVD samples was low and stable after one month.

These results show that FMCVD is superior to processes using post deposition annealing.

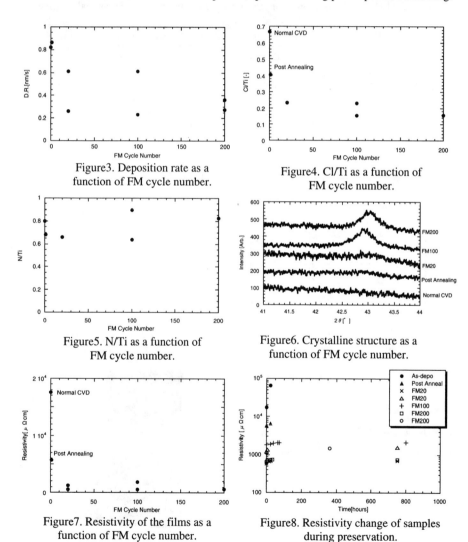

Figure3. Deposition rate as a function of FM cycle number.

Figure4. Cl/Ti as a function of FM cycle number.

Figure5. N/Ti as a function of FM cycle number.

Figure6. Crystalline structure as a function of FM cycle number.

Figure7. Resistivity of the films as a function of FM cycle number.

Figure8. Resistivity change of samples during preservation.

The effect of reduction time in one cycle.

The effect of reduction time in one cycle was also investigated. We changed the reduction time from 1 sec. to 10 sec., and fixed the FM cycle number on 100. Table 2 summarizes the deposition sequences for the effect of reduction time in one cycle.

The Cl/Ti in the films decreased with NH_3 annealing time in one cycle shown in figure 9. This result suggests that FM sequence with long NH_3 reduction time was more effective to reduce the residual chlorine in the films.

Figure 10 shows the crystalline structure of the films as a function of reduction time in one cycle. As shown here, crystalline structure became better with increase of reduction time in one cycle.

Figure 11 shows the resistivity of the samples as a function of NH_3 annealing time in one

Table2.The deposition sequences for the effect of reduction time in one cycle.

	Deposition	Annealing
FM100 (1sAnnealing)	3s x 100	1s x 100
FM100 (2sAnnealing)	3s x 100	2s x 100
FM100 (10sAnnealing)	3s x 100	10s x 100

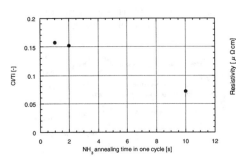

Figure9. Cl/Ti as a function of NH_3 annealing time in one cycle.

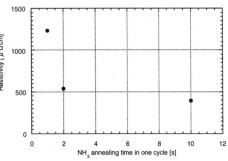

Figure10. The resistivity of the samples as a function of NH_3 annealing time in one cycle.

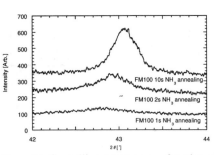

Figure11. Crystalline structure as a function of NH_3 annealing time in one cycle.

Figure12. Step coverage of FM200 sample.

cycle. We can see the decrease of resistivity with increase of NH_3 annealing time in one cycle. The value was about $400\mu\Omega$cm for 10 sec. NH_3 annealing in one cycle. This result agrees with the result of residual Cl concentration in the films and crystalline structure of the films.

Step coverage

Step coverage of the films deposited by FM200 was also investigated. Figure 12 shows the cross section of step coverage with 6:1 AR. As shown here, we could achieve the excellent step coverage TiN film with FMCVD.

CONCLUSION

In summary, we deposited TiN films by FMCVD (Flow Modulation Chemical Vapor Deposition) and achieved low resistivity ($400\mu\Omega$cm) films at low deposition temperature ($350°C$) with good step coverage. FMCVD is attractive for future ULSI processes, because it reduces the overall process time, and reduces the required equipment for a given sequence of process steps, thereby reducing operating costs and increasing throughput.

REFERENCES
1. H. Hamamura, et al., Material Research Society(MRS) Symposium Proceeding, 514, 501-504(1998)
2. H. Hamamura, et al., Advanced Metallization Conference in 1998 (AMC 1998), 345-349(1999)

X-ray photoelectron spectroscopic study of post-treatment on chlorine in CVD-TiN films

Yoko Uchida, Hidekazu Goto, Tsuyoshi Tamaru, and Yoshitaka Nakamura
Hitachi, Ltd., Device Development Center
16-3 Shinmachi 6-chome, Ome, Tokyo 198-8512, Japan

ABSTRACT

We investigated the effect of post-treatment to remove Cl from CVD-grown TiN films through x-ray photoelectron spectroscopy (XPS) analysis. The amount of Cl in a TiN film was decreased by about 30% by increasing the TiN CVD temperature by 100℃ or by post-treatment with NH_3 instead of N_2 after deposition. Three components related to Cl were found in the TiN CVD films through XPS analysis. The amount of the Cl component that existed mainly at the film surface could be significantly decreased by increasing the CVD temperature or by using NH_3 post-treatment. The other components are likely to remain stable up to almost 800℃. We confirmed that the NH_3 post-treatment prevents the degradation of electrical properties due to Cl during device fabrication.

INTRODUCTION

Titanium-nitride (TiN) films are used in ULSI metallization because of their good barrier properties. TiN films formed by chemical vapor deposition (CVD) using $TiCl_4$ and NH_3 as source gasses are especially useful for forming DRAM capacitor electrodes [1] and contact barrier metals [2] because of their good film step coverage and stability during high-temperature processes. However, chlorine (Cl) impurities in TiN films can degrade the electrical characteristics, for example, by corroding the Al interconnects [3], and can increase the contact resistance after annealing. To overcome these problems, we have introduced a post-treatment step in the device-fabrication process to decrease the Cl content in TiN films [4], when N_2 or NH_3 annealing is followed by TiN CVD.

We have used x-ray photoelectron spectroscopy (XPS) to analyze the mechanism by which this post-treatment affects Cl in TiN films grown by CVD. The chemical states of Cl in the CVD films were investigated, and we discuss these in relation to the effects of the post-treatment.

EXPERIMENT

We formed TiN films on Si substrates by low-pressure chemical vapor deposition (LPCVD) using $TiCl_4$ and NH_3 as source gasses. The substrate temperatures were 530, 580

Conference Proceedings ULSI XV © 2000 Materials Research Society

and 630℃. The TiN films were about 70 nm thick. After deposition, the surfaces of the TiN films were exposed to NH_3 gas or N_2 gas in the post-treatment step whose effect we evaluated.

The depth profiles of the Cl content and the amounts of Cl in the TiN films were measured by Auger electron spectroscopy (AES) during etching of the TiN films by Ar^+ ion-beam sputtering at a rate of 3 nm/min. The chemical states of Cl in the TiN films were investigated by x-ray photoelectron spectroscopy (XPS) measurement.

X-ray photoelectron spectra for the various samples were measured on a ULVAC system using a monochromated Al K α-source in an analysis chamber maintained at an ultrahigh vacuum (5 x 10^{-10} Torr). The charging effects were controlled by a low energy electron shower. All binding energy values were corrected by using C1s photoelectron line for adsorbed hydrocarbons [5]. Oxidized surfaces of the samples were etched by Ar^+ ion-beam sputtering before the XPS measurement in this chamber.

RESULTS and DISCUSSION

The behavior of Cl in the TiN films

Figure 1 shows the Cl concentration in the TiN films, relative to the Ti concentration, as obtained by AES measurement. The Cl concentration decreased by about 30% for each 100℃ increase in the deposition temperature for the TiN films when N_2 was the post-treatment gas. NH_3 post-treatment, instead of N_2 post-treatment, also lowered the Cl concentration.

Figure 2 shows the AES depth profiles of samples that received N_2 or NH_3 post-treatment after 530℃ TiN CVD. A pile-up of Cl at the surface of the TiN films was observed when the N_2 post-treatment was used. This also occurred when the deposition temperature of the TiN films was 630℃. The distribution of Cl was fairly uniform over time in the TiN films.

Figure 3 shows an XPS survey spectrum from a CVD-TiN thin film deposited at 530℃ that underwent N_2 post-treatment. We focused on the Cl 2p core emission to study the behavior of Cl in the TiN films.

The emissions of the Cl 2p core level for this sample are shown in Fig. 4. Figure 4(a) shows a spectrum from the surface of the oxidized TiN film which is covered with Ti oxide. Figure 4(b) shows a spectrum of the TiN films after the surface oxide was removed by etching off the top 15 nm of the film by Ar^+ sputtering; this enabled us to obtain information about Cl in non-oxidized TiN films. We obtained similar results from our other samples, which led us to believe that the Cl 2p signal could be deconvoluted into three components, labeled A, B and C in Fig. 4, with respective 2p3/2 binding energy of 198.0 eV, 198.7 eV, and 199.6 eV.

Figure 5 shows the dependence of the Cl concentration, obtained by XPS and relative to the Ti concentration, on the TiN CVD temperature and the post-treatment gas. We calculated the atomic concentration of each element using standard sensitivity factors for our system [6].

More Cl was observed at the surface than in the TiN films when N_2 post-treatment was used. The amounts of Cl at the surface and in the film both decreased when NH_3, rather than N_2, post-treatment was used. Moreover, the amount of Cl at the surface fell to the same level as that in the films with NH_3 post-treatment. These results suggest that there is accumulated Cl at the film surface after N_2 post-treatment, which we observed by AES, but no accumulated Cl remains after NH_3 post-treatment. The total of amount (film and surface) of Cl decreased by about 30% when we increased the TiN CVD temperature by 100°C or used NH_3 post-treatment instead of N_2 post-treatment.

The dependence of the concentration of the three Cl components on the TiN film deposition temperature and the post-treatment gas is shown in Fig. 6. Component A was dominant at the surface and was the main component reduced by an increase in the TiN CVD temperature or NH_3 post-treatment. On the other hand, component C was observed only in the films and remained stable regardless of the post-treatment. Component B at the surface was decreased by NH_3 post-treatment but not by a higher deposition temperature; however, inside the film, component B was decreased as the temperature rose from 530°C to 630°C.

The effect of the reduction treatment

A high deposition temperature or NH_3 post-treatment effectively reduces the amount of Cl at the surface of films, but neither greatly reduces the amount of Cl in the films. This is because these treatments are most effective for component A which mainly exists at the surface. The other two components, B and C, remained stable in our experiments after the post-treatment in the films and at the surface of the films. However, the two components are unstable when subjected to a higher-temperature process because they will resolve near 800°C according to thermal desorption spectroscopy measurements [7].

Figure 4 suggests that chemical structures at the surface differ from these in the TiN films. Probably there are many defects at the surface that are immediately produced when the source gasses are cut off. These defects are likely to be vacancies that are then occupied by Cl; these Cl-occupied defects probably account for component A. Component A of Cl 2p is thought to be the emission corresponding to the Cl-Ti bond of $TiCl_4$[8]. Our results suggest that this Cl component is resolved and desorbs thermally or by chemical reaction with NH_3. One of the reasons is because the Cl-Ti bond is unstable in a solid as Ti chloride usually exists as a liquid at higher temperature than R.T.. These treatments are not able to completely eliminate Cl in the TiN films, but the remaining components of Cl will not desorb in a fabrication process at below 630°C.

Our results indicate that the corrosion of Al interconnects and increased contact resistance can be prevented if the Cl component that is unstable at low temperature is removed. The post-treatment methods we examined can do this. However, a treatment temperature higher than 630°C, probably one near 800°C, will be needed to sufficiently eliminate Cl from the TiN films to improve electrical properties such as the film conductivity.

CONCLUSIONS

We have used XPS to study the effect of post-treatment on Cl in CVD-TiN films made using $TiCl_4$ and NH_3 source gasses. We found three components related to Cl in the CVD-TiN films. Increasing the deposition temperature and NH_3 post-treatment reduced the Cl component that is dominant at the surface. The remaining components in the TiN films were stable and did not desorb during fabrication processes at up to 630℃; these components are unlikely to desorb unless the process temperature rises to almost 800℃. A higher temperature post-treatment is probably necessary to completely eliminate Cl from the TiN films.

ACKNOWLEDGMENTS

The authors would like to thank Yoshifumi Kawamoto, Hideki Tomioka, Nobuyoshi Kobayashi and Yuzuru Ohji for their advice and concurrent encouragement.

REFERENCES

1. I. Asano, M. Kunitomo, S. Yamamoto, R. Furukawa, Y. Sugawara, T. Uemura, J. Kuroda, M. Nakata, T. Tamaru, Y. Nakamura, T. Kawagoe, S. Yamada, K. Kawakita, H. Kawamura, M. Nakamura, M. Morino, T. Kisu, S. Iijima, Y. Ohji, T. Sekiguchi, and Y. Tadaki, IEDM Tech. Dig. 1998, pp.755
2. Y. Nakamura M. Yoshida, H. Goto, T. Fukuda, N. Kobayashi, T. Tamaru, N. Fukuda, H. Aoki, I. Asano, T. Sekiguchi, Y. Tadaki, T. Sekiguchi, Y. Mitsui, and F. Yano, in Adv. Metalliz. Conf. 1998 (Mater. Res. Soc. Proc., Warrendale, Pennsylvania 1999), pp.661
3. M. Hirasawa, Y. Nakamura, T. Tamaru, T. Sekiguchi, and T. Fukuda, Extended Abstract (The 60th Autumn Meeting, 1999), pp.724; The Japan Society of Applied Physics
4. T. Tamaru, H. Sakuma, H. Goto, K. Hosoda, N. Ohashi, M. Kunitomo, I. Asano, and S. Iijima, Proceeding of VMIC 1997 pp.571
5. N. Van Hieu and D.Lichtman appl. Surf. Sci., **201**, pp.186 (1984)
6. Handbook of X-ray Photoelectron Spectroscopy, edited by J. Chastain and R. C. King, Jr., Physical Electronics, Eden Prairie, 1995s
7. T. Jimbo and Y. Ishii, private communication
8. C. Moustly-Desbuguiot, J. Riga, and J. J. Verbist, J. Chem. Phys., **79**, pp.26 (1983)

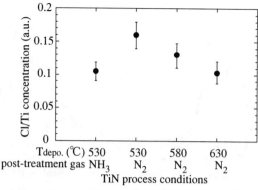

Fig. 1 Dependence of Cl concentration on the deposition temperature
of TiN films and the post-treatment gas

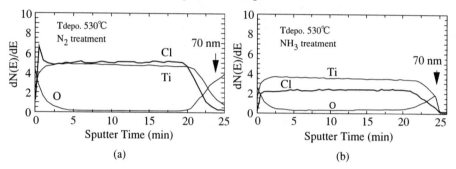

(a)   (b)

Fig. 2 Depth profiles obtained by AES from samples
with TiN films deposited at 530℃.
The post-treatment gas was (a) N_2 and (b) NH_3

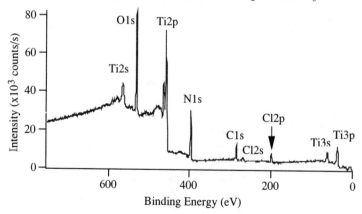

Fig. 3 XPS survey spectrum of a CVD-TiN film
deposited at 530℃ and subjected to N_2 post-trestment

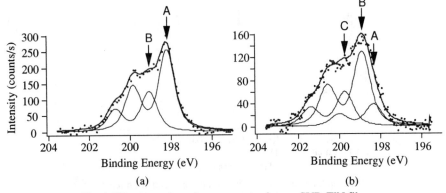

Fig. 4 Cl2p x-ray photoelectron spectra from a CVD-TiN film
deposited at 530°C and subjected to N_2 post-treatment:

(a) before, and (b) after Ar^+ sputtering

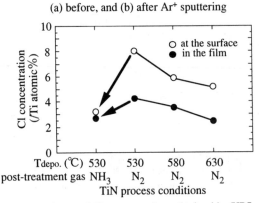

Fig. 5 Dependence of Cl concentration, obtained by XPS,
on the TiN film deposition temperature and the post-treatment gas

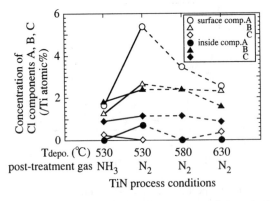

Fig. 6 Dependence of the concentration of Cl components A, B, and C
on the deposition temperature of TiN films and the post-treatment gas

CONFORMAL MOCVD-TiN FILMS DEPOSITED FROM TDEAT AND AMMONIA

S. Q. XIAO, R. TOBE, K. SUZUKI, X. B. XU, A. SEKIGUCHI, H. DOI, O. OKADA, and N. HOSOKAWA
Process Development Laboratory, Anelva Corporation, 5-8-1, Yotsuya, Fuchu-shi, Tokyo, 183-8508, Japan, e-mail: XLT07430@nifty.ne.jp

ABSTRACT

Chemical vapor deposition titanium nitride (CVD-TiN) films were synthesized from Tetrakis (diethylamido) Titanium (TDEAT) and ammonia under a pressure below 10 Pa. Stoichiometries, resistivities, and coverage were studied as a function of temperature and flow rate ratio between NH_3 and TDEAT. Films were characterized by X-ray photoelectron(XPS), Rutherford backscattering Spectroscopy(RBS), Hydrogen Forward Scattering Spectroscopy(HFS), and Transmission Electron Microscopy(TEM). Conformal TiN films in 0.11 μm diameter, aspect ratio(AR) 12 holes were achieved, the mechanism of obtaining conformal barrier film was described. The chemistry of the TDEAT + NH_3 system under pressure below 10 Pa was discussed as well.

INTRODUCTION

As the feature size of ULSI circuits shrinks, the aspect ratio of via and contact holes increases, and the conformality of the diffusion barrier films comes into prominence. Therefore, very thin, conformal film deposition technique has to be established to maximize the cross-section available for Cu in the inlaid structure, and to minimize the dishing of interconnects lines during CMP. Tetrakis (diethylamido) Titanium (TDEAT) and Tetrakis (dimethylamido) Titanium (TDMAT) have been studied as the precursor for barrier films deposition, and fundamental researches have been carried out extensively[1, 2, 3, 4]. However, most of these studies were conducted in the pressure range of 100-3000 Pa or atmospheric pressure. We insist on a low pressure (below 10 Pa) process using TDEAT+NH_3 system to curb the gas phase reaction, and the surface-limited reaction leads to conformal film, less particles, and less frequent cleaning of the chamber. TDEAT rather than TDMAT is used because TDEAT is less reactive than TDMAT due to its steric effect [5]. In this study, We expect to clarify the properties of the films deposited under various conditions, and to explore the optimal process conditions.

EXPERIMENT

TiN films were deposited with an Anelva I-1080 CVD Cluster tool, in which CVD-Cu and plasma treatment modules are integrated. Liquid TDEAT precursor was controlled and introduced into the chamber by a direct liquid delivery and vaporization system. The wafer temperature was altered from 300 °C, 350 °C to 400 °C. The TDEAT partial pressure was kept constant, while NH_3/TDEAT partial pressure ratio (P_{NH3}/P_{TDEATT}) was changed from 6, 12.5, 27, 54, 81, and 162. Deposition pressure was generally 2 Pa (5 Pa when P_{NH3}/P_{TDEAT} was 162). Film compositions and chemical bonding states were analyzed by RBS, HFS, and XPS respectively. XRD, and TEM were also utilized to characterize the films.

Conference Proceedings ULSI XV © 2000 Materials Research Society

RESULTS AND DISCUSSIONS

Deposition Rate

Figure 1 shows that the temperature affected the deposition rate (DR) significantly at low P_{NH3}/P_{TDEAT} regime, the DR increased with increasing temperature. However, under the pressure 5 Pa with P_{NH3}/P_{TDEAT} 162, the DR was independent from temperatures. In this regime, NH_3 might be in excess of TDEAT, and the DR depended only on precursor partial pressure. When the temperature was 300 °C, the DR increased constantly with NH_3 flow rate. However, this increase became much less obvious at 350 °C and 400 °C. This implied that the influence of NH_3 partial

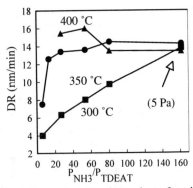

Fig. 1 Deposition rate changed as a function of temperature and P_{NH3}/P_{TDEAT} under 2 Pa.

pressure on DR was stronger at lower temperature than that at higher temperature. One explanation is that more NH_3 decomposed thermally into N_2 and H_2 at higher temperature in vacuum, and there were less reactive species like $-NH_2$ generated. At 400 °C, the DR increased at the beginning and then dropped down in high NH_3 partial pressure regime, gas phase reaction might account for the decrease, when P_{NH3}/P_{TDEAT} was increased to over 81.

Apparent activation energy changed from 0.71 eV, 0.46 eV, 0.24 eV to 0.02 eV, when P_{NH3}/P_{TDEAT} was changed from 6, 27, 81 to 162. This indicated that the reaction changed from surface-limited to precursor transport-limited with increasing NH_3 partial pressure.

Resistivity

The resistivity as a function of temperature and P_{NH3}/P_{TDEAT} is shown in Fig. 2. With increasing temperature from 300 °C to 350 °C at P_{NH3}/P_{TDEAT} 6, the resistivity was observed to fall from 24000 $\mu\Omega$cm to 5700 $\mu\Omega$cm. Large amount of NH_3 also brought about a reduction of resistivity. Under the condition 400 °C with the P_{NH3}/P_{TDEAT} 162(5 Pa), the lowest resistivity of 300 $\mu\Omega$cm was obtained in this study, which was equivalent to the resistivity after plasma densification. The same result was reported by other researchers[1] that excessive NH_3 addition and high temperature led to TiN films of low resistivity .

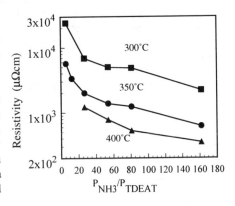

Fig. 2 The resistivity changed as the function of temperature and P_{NH3}/P_{TDEAT}.

Step Coverage

At the temperature of 300 °C and P_{NH3}/P_{TDEAT} 6, we achieved nearly 100% step coverage in holes with 0.2 μm diameter and AR 6.5(Fig. 3). In another case, we deposited the film under the same condition, and it was found that the films remained conformal to 0.11 μm diameter, AR12. However, the conformality deteriorated with further increasing in NH_3 flow rate and temperature. We found that the step coverage in holes with 1.2 μm diameter and AR 1 was 63%, 48%, and 30% at the deposition temperature 350 °C under the P_{NH3}/P_{TDEAT} 27, 81, and 162. The corresponding resistivity was 1992 μΩcm, 1232 μΩcm, and 620 μΩcm respectively. The step coverage became even worse when the temperature was raised to 400 °C. Conformality and resistivity cannot be satisfied simultaneously.

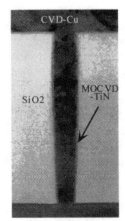

Fig. 3 TEM photo of TiN film deposited at 300 °C, $P_{NH3}/P_{TDEAT} = 6$.

We reported before that compared to the films deposited from TDEAT only, the films deposited with a small amount of NH_3 addition led to a great improvement of conformality[6]. The same result was reported by other researchers[7]. We intend to have a more detailed investigation into this phenomenon.

The non-conformality of the films generally originates from the difference in deposition rates inside and outside the hole. At the deposition temperature of 300 °C, there existed a regime where the deposition rate was independent of the TDEAT partial pressure and only depended on the NH_3 partial pressure (Fig. 4). In this regime, the step coverage was nearly 100% in Ø 0.2 μm, AR = 6.5 hole(Fig. 3).

As suggested by Fix et al.[4], the reaction of TiN formation from TDEAT and NH_3 has two pathways. One is by the pyrolytic decomposition of TDEAT, which is influenced strongly by the temperature. The step coverage follows the distribution of TDEAT, which is generally non-uniform inside and outside the hole due to the large sticking coefficient. This led to less conformal films when the TDEAT was the only reactant even at 300 °C[6]. Another pathway is transamination by NH_3, where the degree of this reaction is in proportional to P_{NH3} partial pressure(P_{NH3}). The step coverage is influenced by the profile of P_{NH3}. When a small amount of NH_3 is added to the chamber, the partial pressure of NH_3 inside and outside the hole

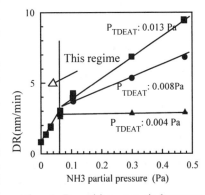

Fig. 4 Deposition rate indepentent of TDEAT partial pressure in low ammonia partial pressure regime (4 Pa, 300°C).

approaches uniform more easily due to the small sticking coefficient, which brought about the conformal deposition. Therefore, low temperature and low NH_3 partial pressure are necessary.

In the high P_{NH3}/P_{TDEAT} regime, the NH_3 is in excess of TDEAT, the step coverage may follow the distribution of TDEAT. It is important to lower the reactivities of both TDEAT and NH_3 in the gas phase. According to the results by Yun et al.[8], IR spectra showed that N_2 could suppress the thermal decomposition of TDMAT in gas phase. In this case, relatively low temperature and large amount of dilute gas may be required.

The trade off between the resistivity and the coverage is difficult in patterns with high aspect ratio. We prefer a good coverage, because thinner barrier film itself can lower the total resistance of the hole. Moreover, the resistivity can be lowered by other method like plasma treatment, which is available in our module. Even without plasma treatment, we lowered the resistivity to 3340 μΩcm with the step coverage of about 60% in holes of 0.34 μm diameter and AR 4.4. If the resistivity increased to 5700 μΩcm, the side coverage still remained to 63% in 0.25 μm diameter, AR 6.5 holes. Continuous TiN film of a thickness of 7 nm at hole bottom was deposited conformally.

Compositions and Chemical Bonding States

The RBS results in Fig. 5 revealed that the film deposited at 400 °C, P_{NH3}/P_{TDEAT} = 81 possessed nearly stoichiometric composition, The film had a resistivity of 300 μΩcm with low C, O impurity levels, and H less than 2.3 at. %. The density was 5.15 g/cm³, near 5.4 g/cm³, the value for bulk TiN. We found that carbon content had an abnormal increase at the temperature of 350 °C, which was in agreement with the result by Musher et al.[7].

We also analyzed the bonding states of carbon at different temperatures(Fig. 6). It was found that the Ti-C bond, organic bonds C-C, C-H had stronger peaks in C1s core level XPS spectra at the temperature 350 °C, which suggested different reaction mechanisms at this temperatures. As proposed by Nakamura et al.[9] and Musher et al.[7], there are three possible different reactions, which have different activation energy. At a low temperature(300 °C), the transamination reaction prevails. Ethylamine ligand -NEt₂ is replaced by -NH₂ and removed from the central metal. This is in agreement with the result in Fig. 1 that NH_3 flow rate affected the DR significantly at 300 °C. While the deposition temperature increases to 350 °C, the part of

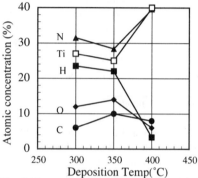

Fig. 5 Compositions of the films deposited at 300°C, 350°C, and 400°C by RBS at P_{NH3}/P_{TDEAT}:81, 2 Pa

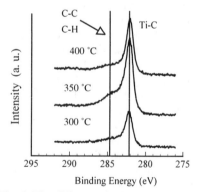

Fig. 6 The difference in C 1s core level XPS spectra depends on wafer temperature at P_{NH3}/P_{TDEAT}: 81, 2 Pa.

298

3-member or 4-member metallcycle formation reactions may become more important, with C bonding to Ti to form more Ti-C. As the temperature is raised to 400 °C, the ß-hydride elimination reaction may increase, which reduces C content by the formation of Ti-H bond. The enhenced migration of adsorbed species on the wafer at elevated temperature may also contribute to the stoichiometry of the clean films.

Barrier Validity

Figure 7 indicates the ratio of the Cu film sheet resistance before and after anneal on MOCVD-TiN(300 °C 10 nm, 350 °C 7.5 nm, 10 nm), sputtered-TiN(10 nm), and sputtered-TaN(10 nm) barrier films. MOCVD-TiN(7.5 nm) showed identical barrier validity to 10 nm sputtered-TaN under the condition of 640 °C x 0.5 h. X-ray diffraction results revealed that the film deposited at 350 °C had an amorphous structure. Dense films without grain boundary may account for the barrier validity.

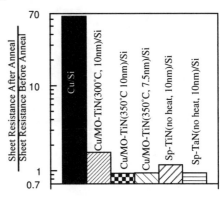

Fig. 7 Comparision of the change in sheet resistance after 640°C*0.5h for various barrier films

CONCLUSIONS

TiN films deposited at 400 °C and P_{NH3}/P_{TDEAT} larger than 27 had the resistivity in the range of 300-1000 µΩcm with compositions close to stoichiometry with poor coverage. When the substrate temperature was 300 °C and the P_{NH3}/P_{TDEAT} below 6, films with nearly 100% coverage was deposited in the holes with 0.25 µm diameter and AR 7, while the resistivity was 24000 µΩcm. At the intermediate temperature 350 °C with P_{NH3}/P_{TDEAT} around 10-30, the resistivity was 2000 µΩcm -5000 µΩcm with acceptable coverage of about 50% in the holes of aspect ratio 4-7. With plasma treatment, the resistivity can be lowered to about 300 µΩcm, and these films are expected to be used in ULSI metallization.

REFERENCES

1. J. N. Musher and R. G. Gordon, J. Mater. Res., 11, 989 (1996).
2. A. Intemann and H. Koerner and F. Koch, J. Electrochem. Soc., 140, 3215 (1993).
3. A. Sekiguchi, S. W. Kim, T. Yoshimura, K. Watanabe, S. Mizuno, S. Hasegawa, O. Okada, N. Takahashi, and N. Hosokawa, in Advanced Metallization and Interconnect System for ULSI Application 1995, edited by C. Russel, Ellwanger, S. Q. Wang, pp.355-361.
4. R. Fix, R. G. Gordon, and D. M. Hoffman, Chem, Mater. 3, 1138 (1991).
5. D. C. Brandley and M. H. Gitlitz, J. Chem. Soc. (A), 980 (1969).
6. S. W. Kim, H. Jimba, A. Sekiguchi, O. Okada, N. Hosokawa, Appl. Surf. Sci. 100/101, 546(1996).
7. J. N. Musher and R. G. Gardon, J. Electrochem. Soc., 143, 736 (1996).
8. J. Y. Yun, M. Y. Park, and S. W. Rhee, J. Electrochem. Soc., 145, 2453(1998).
9. K. Nakamura, A. Tachibana, Advanced Metallization Conference in 1998, Japan/Asia Session proc. edited by M. Murakami, P43.

ELECTRICAL INTEGRITY MONITORING OF ELECTROLESS BARRIERS FOR COPPER INTERCONNECTS

YOSI SHACHAM-DIAMAND, BARAK ISRAEL AND YELENA SVERDLOV,

Department of Physical Electronics, Tel-Aviv University, Ramat-Aviv, ISRAEL, 69978,
Phone: 972 3 640 8064, Fax: 972 3 642 3508, E-mail: yosish@eng.tau.ac.il

ABSTRACT

In this work we present the study of the electrical properties of electrolessly deposited CoWP / copper / CoWP structures using both metal – silicon and metal –oxide-silicon (MOS) capacitors. We present the electrical characteristics as a function of annealing treatments temperature, in the range of 200°C – 600°C. MOS devices were found to be good monitors for the copper-barrier interaction. Capacitance vs. voltage (C-V) and capacitance vs. time (C-t) measurements yield valuable information on the barrier integrity. The electrical measurements were compared to material analysis techniques such as Auger Electron Spectroscopy (AES). The results indicate that electroless CoWP is a good barrier for copper metallization for temperatures up to about 500°C.

INTRODUCTION

Barrier layers are integral part of copper technology for ULSI. As the critical dimensions are getting smaller, the barrier layers should be thinner. Barrier integrity is important factor and it requires special inspection and metrology techniques. Both electrical and material science techniques are used for the evaluation of barrier integrity. Typical barrier studies combined imaging techniques (SEM, TEM) with mass spectroscopy profiling techniques. Techniques such as RBS, AES and SIMS are used for profiling of the elements in the copper-barrier structure. Material science techniques are quite sensitive and can detect copper down to the range of 10^{16} atoms per cm^3. Copper can affect the electrical properties of devices even at lower concentrations. Copper atoms affect both the silicon substrate and the Si-SiO$_2$ interface. When this happens, copper endangers the reliability of the integrated circuit by increasing junctions leakage current, reducing oxide breakdown voltage, and affecting the threshold voltage at the gate or the field of MOS transistors.

In this work, we present results of MOS devices and metal-silicon Schottky junctions that are used as electrical monitors for barrier reliability. The devices are made with an electroless metal gate. We present results for CoWP/Cu/CoWP metallization where CoWP serve as both barrier underneath the copper and capping layer above it. The experiment was limited to thermal anneal without electrical bias. The devices were annealed in vacuum at temperatures up to 600°C. We found the capacitance transient analysis is the most sensitive technique to detect copper penetration into the substrate.

EXPERIMENTAL

The devices were made on <100> oriented, 6" p-type silicon wafers. The wafers were thermally oxidized to form a 14 nm gate oxide. The oxide was than removed for some of the wafers for the formation of Schottky diodes. A 20 nm Co seed layer was sputtered deposited in an ion-beam sputtering tool. Electroless CoWP barrier was deposited on the seed from an aqueous deposition solution at 85°C. Next, electroless copper was deposited [1] and was followed immediately by second CoWP layer, with identical composition as the bottom one,

Conference Proceedings ULSI XV © 2000 Materials Research Society

preventing copper corrosion. The devices were patterned by optical lithography and the metal was etched using by diluted sulfuric acid. Finally, the resist was removed in acetone followed by a cleaning procedure in iso-propyl-alcohol and DI water. Next, the wafer backside was cleaned in buffered HF to remove the oxide. A sketch of the final devices is shown in Fig. 1.

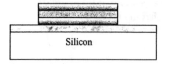

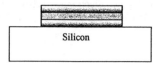

a. Barrier/Copper/Barrier MOS device   b. Barrier/Copper/Barrier Schottky diode

Fig. 1: Schematic cross section of the devices under test.

We fabricated three types of MOS capacitors and Metal-silicon diodes:

(1) Al reference devices – with Al on the Co/Ti seed layer.

(2) Barrier reference – with CoWP on the Co/Ti seed layer.

(3) Full structure with CoWP/Cu/CoWP on the Co/Ti seed layer.

Each sample included the characterization of 72 small devices with area of $3.57 \cdot 10^{-4}$ cm^2, 16 medium devices with area of $3.69 \cdot 10^{-3}$ cm^2 and 8 large devices with area of $14.4 \cdot 10^{-3}$ cm^2. The capacitors were measured at room temperature or 'as deposited'. Next, the samples were split and each piece was annealed in vacuum at temperatures between 300°C to 600°C. Following the annealing, we performed full electrical characterization that includes I-V, C-V and C-t analysis. The devices were characterized in a light-proof box, to prevent light effects. The I-V characteristics were measured using a HP-4140B pico-ammeter. The C-V characteristics were measured using a HP-4280A capacitance meter operating in 1MHz with signal amplitude of 20 mV. Finally, different samples were sent to AES concentration depth profile

RESULTS

Metal-silicon structures

The I-V characteristics of the metal semiconductor devices was found to fit well the following equation:

$$I = I_0 \left[\exp\left(\frac{q(V - IR_S)}{nkT} \right) - 1 \right] \qquad \text{where } I_0 = AA^*T^2 \exp\left(-\frac{\phi_B}{kT} \right) \tag{1}$$

R_s is the series resistance, A is the diode area, A^* is modified Richardson constant and T is the temperature in Kelvin. This model includes both thermionic emission over the barrier and diffusion of the charged particles in the depletion region. $A^* = 32$ A·cm^{-2}·K^{-2} for p-type silicon at 300K [1]. The ideality factor, n, and the effective barrier height, ϕ_s, are shown in table 1.

Table 1: Fitting parameters for diodes annealed for 30 minutes.

Anneal Temp.	As deposited	200°C	400°C	600°C
Experiment #1 : Al(50 nm)/Co(20 nm)/Ti(10 nm)/p-Si				
n	---	2.40±0.42	6.82±0.94	---
ϕ_s (eV)	---	0.44±0.03	0.39±0.01	---
Experiment #2: CoWP(120 nm)/Co(20 nm)/Ti(10 nm)/p-Si				
n	5.72±1.89	2.57±0.32	10.49±2.55	11±3.5
ϕ_s (eV)	0.37±0.01	0.45±0.04	0.35±0.01	0.35±0.09
Experiment #3: CoWP(50 nm)/Cu(180 nm)/CoWP(50 nm)/Co(20 nm)/Ti(10 nm)/p-Si				
n	2.34±0.44	2.2±0.28	8.80±0.85	9.2±2.8
ϕ_s (eV)	0.47±0.01	0.47±0.01	0.37±0.02	0.35±0.15

The fitting parameters for positive bias are shown in table 1. Although we got good numerical fit, the results for devices annealed at temperatures of 400°C and above have excessively high n factor, much beyond what is expected by the physical model. This occurs since the device characteristics became dominated by the serial parasitic resistance that was about 95±15Ω. Next, we present the results of the Auger Electron Spectroscopy (AES).

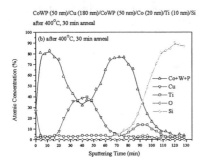
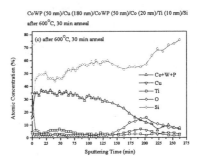

Fig. 2 AES profile of CoWP/Cu/CoWP diodes annealed at 400°C and 600°C for 30 min.

AES spectra of 400°C anneal was essentially the same as the spectra of samples annealed at 200°C. Therefore we assume that the failure of the diodes was not due to the failure of the barriers. At 600°C the cobalt alloy reacts with the silicon and forms cobalt-silicide. This results in a total barrier failure and the copper is completely sucked into the silicon substrate. Annealing the diodes for 400°C degraded the performance of the Schottky contacts. This phenomenon occurred in diodes from the three runs. Apparently, the cobalt diffused through the titanium layer and formed cobalt-silicide contacts on the silicon surface. We found that up to 400°C titanium transports through the cobalt and cobalt-silicide is formed while at the same time the electroless copper was confined between the lower and upper barrier. Therefore, I-V characteristics of Schottky diodes can not be used to study barrier failure.

Metal-oxide -silicon (MOS) structures

The surface state density, which corresponded to the flat band voltage of the as-deposited samples, was about 5×10^{12} cm^{-2} ev^{-1} and it was almost completely annealed after 300°C for 30 minutes in vacuum.

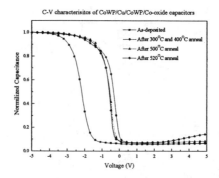

C-V characterisites of CoWP/Cu/CoWP/Co-oxide capacitors

- As-deposited
- After 300°C and 400°C anneal
- After 500°C anneal
- After 520°C anneal

Voltage (V)

Fig.3: C-V plots of Barrier/Copper/Barrier MOS devices annealed at various temperatures

The C-V curves for capacitors annealed up to 500°C were almost identical. It is concluded that there is no significant effect on the fixed interface states or the fast interface state density. After 30 min. annealing at 600°C the C-V curves was completely distorted for some capacitors while remained almost unchanged for others. The C-V characteristics after 600°C anneal indicate on a degradation of the gate oxide. However, note that the flat band voltage did not change much. This indicates that the copper is not ionized at the Si-SiO$_2$ interface or in the oxide. The positive part of the C-V increases which indicates on a decrease of the minority carrier lifetime, possibly by due to enhanced recombination by deep levels associated with copper in the silicon. C-t curves of devices from different runs were recorded. (Fig. 3) and was analyzed using the conventional Zerbst plot.

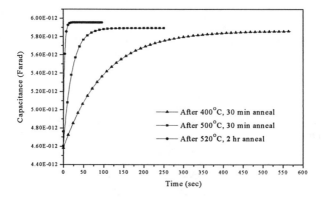

- After 400°C, 30 min anneal
- After 500°C, 30 min anneal
- After 520°C, 2 hr anneal

Time (sec)

Fig. 4 C-t curves of CoWP/Cu/CoWP/Co/SiO$_2$ capacitors annealed at 400°C, 500°C and 520°C. Device area is $3.57 \cdot 10^{-4}$ cm^2.

The minority carrier lifetime that were extracted from Zerbst plots and the recovery time for the annealed capacitors from the three different runs are displayed in table 2. The as-deposited capacitors from different runs displayed very similar generation lifetimes, indicating that there was no degradation of the devices by the electroless deposition of both copper and CoWP. The typical values of generation lifetimes ranged between 40-70 microseconds. Similar results were obtained from samples annealed at 300°C while small degradation was observed at 400°C.

Table 2: Generation lifetime, τ_g (μS), and Recovery time, T_r (sec).

	Al/Co/SiO$_2$/Si		CoWP/Co/SiO$_2$/Si		CoWP/Cu/CoWP/Co/SiO$_2$/Si	
	τ_g (μS)	T_r (sec)	τ_g (μS)	T_r (sec)	τ_g (μS)	T_r (sec)
As-deposited	52 ± 10	400 ± 20	50 ± 6	525 ± 75	48 ± 2	325 ± 75
300°C, 30 min	52 ± 7	350 ± 50	50 ± 5	350-400	58 ± 7	375 ± 25
400°C, 30 min	None	None	None	None	47 ± 4	300± 50
500°C,30 min.	None	None	None	None	22 ± 2	105 ± 15
520°C, 2 hr	None	None	43 ± 7	300-350	12 ± 5	40 ± 20
600°C, 4 hr	None	None	None	None	1.5 ± 0.5	12 ± 2

A significant lifetime degradation was observed for temperatures of 500°C and above for the CoWP/Cu/CoWP/Co capacitors. At 600°C a significant number of the capacitors became very leaky and lifetime measurement was not possible. However, among the capacitors that remained intact, the minority carrier lifetime dropped to about 1 μs.

The degradation in lifetime of the CoWP/Cu/CoWP/Co capacitors, possible due to possible copper penetration at temperatures above 500°C, was supported by the increase of the high-frequency inversion. The high-frequency inversion capacitance of the CoWP capacitors remained constant, at its minimum value, for all annealing temperatures.

AES profiling of the MOS capacitors indicated that the copper profile remained unchanged up to 500°C anneal for 30 minutes. Above 520°C the copper profile became wider and its peak decreased. The are under the profiles remained the same.

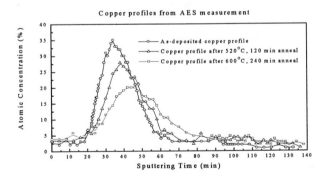

Fig. 4: Copper profiles as measured by AES

DISCUSSION

Copper particles that were introduced to the silicon substrate could have caused the degradation of lifetime by the formation of generation-recombination centers. Therefore, C-t analysis of MOS capacitor is a sensitive monitor to copper penetration. On the other hand metal-silicon devices are very difficult to interpret since in parallel to the copper penetration there is a significant interaction of the barrier materials with the silicon.

Once the copper reaches the silicon, it affects the silicon-oxide interface and the minority carrier lifetime in the bulk. Transient capacitance techniques are sensitive to small quantity of deep-level atoms, such as copper. The minority carrier lifetime τ_g is inversely proportional to the concentration of the recombination centers that are introduced via the interaction of the copper and the silicon lattice. Copper at the surface affects the surface recombination velocity, hence the effective lifetime. A first order model results in a simple relation between the minority carrier lifetime τ_g and the recombination center capture cross section.

$$\tau_g = \frac{1}{\sigma_T \, v_{th} \, N_T}$$ (2)

Where V_{th} is the thermal velocity and N_T is the recombination centers concentration. We found that $1/\tau$ increase with. anneal time for samples annealed in the range of 500 – 520C. This indicates that the number of recombination centers increases with time. Assuming that the recombination centers concentration is proportional to the copper ion concentration we can deduce that the copper penetration is dominated by diffusion through the barrier and the gate oxide.

SUMMARY AND CONCLUSIONS

Measurements indicate that copper penetration into the oxide and silicon substrate can degrade the characteristics of MOS capacitors, and therefore MOS capacitors can be used as sensitive monitors to the diffusion of impurities. The deposition of a CoWP thin film can prevent the damaging effects of copper diffusion at temperatures below 500°C and reduce these effects above this temperature. Transient capacitance measurements of MOS capacitors have been proven to be very sensitive to the early stages in barrier failure. No significant change in oxide charge or interface charge density was observed due to copper penetration to the silicon. Measurements after elevated temperature anneal without electrical bias indicate on the following failure mechanisms:

1. Increased oxide breakdown effects – lower breakdown field, higher leakage current.

2. Decrease of minority carriers lifetime at temperatures above 500°C

ACKNOWLEDGEMENTS

The work was supported by a grant from the Israeli ministry of science and technology. Also thanks to Mark Oksman from Tel-Aviv University and Dr. Roi Shaviv from Tower semiconductors Ltd. for the support in samples preparation.

LIST OF REFERENCES

[1] J. Li et al, "Copper deposition and thermal stability issues in copper based metallization for ULSI technology",. Materials Research Reviews.
[2] M. S. Angyal, "Performance and reliability of ultra-thin diffusion barriers for sub-half micron MOS devices integrated with Cu metallization, Ph.D. thesis, Cornell university, 1995.
[3] A.G. Milnes, "Deep impurities in semiconductors" Wiley, 1973.
[4] J. D. McBrayer et al, "Diffusion of metals in silicon dioxide", JECS. 133 (6), 1986.
[5] G. Raghavan et al, "Diffusion of copper through dielectric films under bias temperature stress", Thin Solid Films 262, 1995.
[6] J. O. Olowolafe et al, "Interaction of Cu with $CoSi_2$ with and without TiN_x barrier layers", Appl. Phys. Lett. 57(13), 1990.

SIMS STUDY OF COPPER QUANTIFICATION AND DIFFUSION IN SILICON, SILICON DIOXIDE, AND SILICON NITRIDE

K. K. HARRIS*, F. A. STEVIE, J. M. MCKINLEY, S. M. MERCHANT, and M. OH
Lucent Technologies, 9333 S. John Young Parkway, Orlando, FL 32819
*Presently at Dept. of Materials Science & Engineering, University of Florida, Gainesville, FL 32611, kharr@mse.ufl.edu

ABSTRACT

The use of copper as a replacement for aluminum in semiconductor devices causes concern for contamination of other layers in the structure. Hence, the ability of layers to resist copper diffusion in the semiconductor process is of great importance. One method that can be used to study inter-diffusion effects is the use of ion implantation of copper into the layers of interest followed by anneals at various temperatures. Secondary ion mass spectrometry (SIMS) has the sensitivity and depth resolution required to measure the copper density distribution in the as-implanted and annealed samples.

Copper was implanted into silicon substrates and into films of silicon dioxide and silicon nitride deposited on silicon. The implanted samples were annealed at 300, 500, 700, 900 and 1000° C. Analysis of copper in all three materials showed that the SIMS parameters must be carefully chosen in order to provide valid results. Previous work has helped determine conditions for analysis of copper in silicon and silicon dioxide, but it is possible for profile distortion to occur under certain analytical conditions. This study shows some movement of copper implanted in silicon with electron bombardment. SIMS analyses of copper in silicon dioxide indicate that the electron beam current and impact voltage can affect the shape of the distribution. Analyses in SiO_2 were further complicated by the need to remove a mass interference with a molecular ion.

SIMS analysis of the annealed samples shows copper has a complex diffusion mechanism in silicon with movement toward the surface at temperatures up to 700° C. There is significant diffusion into the substrate at 900 and 1000° C. The anneals of copper in SiO_2 show little movement until 500° C, but has migrated away from the implanted region of Cu at 700° C and above. Anneals of copper in Si_3N_4 showed limited diffusion until 500° C.

INTRODUCTION

Copper (Cu) has been of increasing interest as an interconnect material for microelectronic devices because Cu has higher conductivity and a larger resistance to electromigration than aluminum. However, Cu atoms diffuse readily into silicon (Si), and can cause degradation of transistor performance. Therefore Cu must be clad with barrier materials on all sides to inhibit its diffusion into regions where it can cause excessive leakage current or shorts. The ability of barrier layers to resist copper diffusion is of significant importance.

Previous studies have helped to determine conditions for analysis of implanted copper in Si and silicon dioxide (SiO_2)[1]; however, under certain analytical conditions profile distortion has been known to occur during analysis. Cu segregates away from the oxide formed by analysis with oxygen primary beam at angles of incidence (near normal) where complete oxidation occurs [2,3,4]. Analyses of Cu at more glancing bombardment angles only partially oxidize the Si and do not distort a Cu implant profile. Previous publications do not cover the range of analysis conditions needed for copper in semiconductor devices.

307

The ion implant and diffusion behavior of Cu in Si, SiO_2, and Si_3N_4 is of interest. Cu is known to be a fast diffuser in Si [5]. Diffusion coefficients (D_o) of $4.7x10^{-3}$ cm²/s at 300 - 700°C at an activation energy of 0.43 eV and $4x10^{-2}$ cm²/s at 800 - 1100°C at 1.0 eV [6] have been estimated. Assuming Cu diffusion from a constant source, the diffusion coefficient of Cu in SiO_2 was estimated at approximately 10^{-5} cm²/s [7].

In the present study, samples will be analyzed using secondary ion mass spectrometry (SIMS). SIMS has the sensitivity and depth resolution to measure the as-implanted and annealed samples. At the present time the literature for SIMS analysis of Cu diffusion is very limited.

The purpose of this study is to determine conditions for SIMS analysis and quantification of Cu in Si, SiO_2, and silicon nitride (Si_3N_4) and to provide additional information on Cu diffusion in these materials.

EXPERIMENT

Cu was implanted into boron doped, p-type, 10 ~ 20 Ω-cm, Czochralski grown (CZ), (100) Si substrates, into 0.31 μm and 0.1 μm plasma enhanced tetraethylorthosilicate (PETEOS) SiO_2, and into 0.5 μm low pressure chemical vapor deposition (LPCVD) Si_3N_4 deposited on Si substrates. Cu ion implantation was done at 115 keV to a dose of $1x10^{14}$ cm^{-2} by R. G. Wilson, Charles Evans & Associates. Pieces from each implanted sample were annealed at 300, 500, 700, 900 and 1000° C in a quartz tube that was placed in a Lindberg furnace for 30 minutes in Argon gas flow ambient.

SIMS analyses were made using a CAMECA IMS-6f magnetic sector SIMS with O_2^+ primary beam. The effective impact energy was 5.5 keV for 4.5 keV sample bias and 10 keV primary voltage. The effective angle of incidence was 42 degrees from normal. Insulator analyses were achieved using a normal incidence electron gun at impact energy of 5 keV and current of 8 μA. The electron gun was used with varied impact energy, current and electron exposure time to study the effect of electron bombardment on profile shape in the SiO_2 and Si_3N_4 layer samples. High mass resolution was used to remove a mass interference with a molecular ion ($^{28}Si^{35}Cl$) for analyses in SiO_2 grown using chlorine. The concentration scale of the annealed samples is based on a relative sensitivity factor (RSF) derived from the dose of the corresponding as-implanted samples. The depth scales are derived from profilometry measurement of the craters.

RESULTS

Cu in Si

Si is normally analyzed without electron bombardment, but the semiconductor structures that need to be measured have many layers, including insulators such as oxides and nitrides. SIMS analysis of Cu in Si devices is typically carried using an O_2^+ primary beam. The use of insulators such as oxides and nitrides requires electron bombardment to avoid sample charging. As shown in Figure 1, no movement of Cu was noted for analysis using only O_2^+ primary beam. However, some movement of Cu was noted in the profile with electron bombardment using O_2^+ primary beam. The movement of Cu is from the peak toward the substrate and collects at a depth that appears to match with the interface between amorphous and damaged silicon caused by the ion implantation. Other workers have noted motion of Cu in Si under electron bombardment [8,9].

Analysis of the annealed samples is shown in Figure 2, where at lower temperatures Cu is seen to migrate toward to the surface. However, the Cu profile was shallower in the 500°C annealed sample than for the 300°C case. This effect is possibly due to surface crystal structure damage during ion implantation, and formation of silicide (Cu_xSi_y) Cu diffused deeper into the silicon as the annealing temperature was increased beyond 700°C.

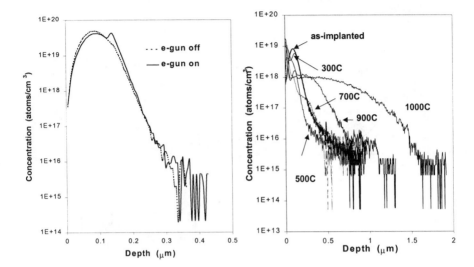

Fig. 1 Effect of addition of electron beam on analysis of Cu implanted into Si

Fig. 2 SIMS profiles of Cu in Si as-implanted and after anneals

Cu in SiO$_2$

The electron impact energy, current and exposure time can affect the Cu profile shape in PETEOS SiO_2 samples. There are two characteristic peaks in the Cu profile for all conditions. At 5 keV electron impact energy and approximately 8 µA, Cu began to migrate toward to the surface immediately after the electron bombardment. The effect of electron bombardment exposure time is shown in Figure 3. The profile was deeper after 5 min, but shallower after 20 min than the profile taken as soon as electrons were present. Both higher current and electron energy pushed the Cu profile deeper. Results using 2 keV electrons on a quadrupole SIMS show that the distortions in the SiO_2 can be removed [10]. It is thought that the lower electron energy is an important parameter.

SIMS profiles of Cu implanted and annealed in 0.31 µm SiO_2 are shown in Figure 4. The copper distribution up to 300°C generally followed a Gaussian profile within the SiO_2 layer, but there were two distinct peaks instead of one. Cu started to migrate toward the surface at 500°C and also diffused through the SiO_2 layer to the SiO_2/Si interface. Cu has moved to the surface and showed drastically reduced concentration at 700°C. There is only a very small amount of Cu detected for both the 900 and 1000°C annealing samples.

In order to duplicate the problem of Cu migration in a semiconductor device, Cu was implanted through a thin oxide (0.1 µm SiO_2) and annealed. The SIMS profiles are shown in Figure 5. The as-implanted Cu profile indicates 0.1 µm SiO_2 is less than the penetration depth for this Cu implantation energy and dose. Therefore, Cu was also implanted into the Si substrate. The profile of the as implanted sample did not follow a Gaussian shape. Some of the Cu in SiO_2 has migrated away from the interface with the Si as a result of the analysis, indicating the profile is not reproduced correctly in the region of the interface. The 300°C anneal showed a similar profile to the as-implanted one. Cu migrated toward to the surface in 500°C annealing, yet Cu is present in both SiO₂ layer and Si substrate. By 700°C, Cu counts were significantly reduced in both SiO₂ and Si, but noticeably high counts were observed at the SiO₂/Si interface. There is only a very small amount of Cu detected near the surface of the sample for both 900 and 1000°C annealed samples.

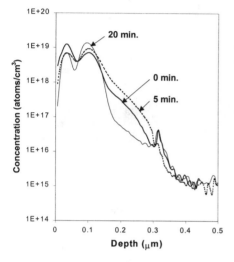

Fig. 3 Cu in SiO₂ for different
Electron bombardment times

Fig. 4 Cu in 0.31 μm SiO₂ as
implanted and after anneals

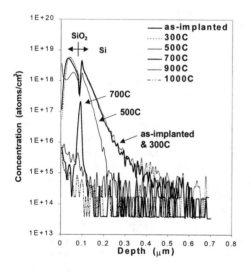

Fig. 5 Cu in 0.1 μm SiO₂ on Si as
implanted and after anneals

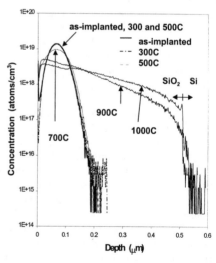

Fig. 6 Cu in Si₃N₄ on Si as
implanted and after anneals

Cu in Si$_3$N$_4$

Analysis of Cu as implanted in 0.5 μm LPCVD Si$_3$N$_4$ showed no indications of Cu movement due to electron bombardment (Figure 6). Analysis of the annealed samples shows no significant diffusion until 700°C, which implies good resistance to Cu diffusion.

CONCLUSIONS

The analysis of Cu implants in Si, SiO$_2$, and Si$_3$N$_4$ showed that the SIMS analysis parameters must be carefully chosen to provide a true representation of Cu distribution in these films. Annealed samples of Cu in Si showed a complex diffusion behavior. Annealing of 0.31 μm PETEOS SiO$_2$ showed there is no significant Cu diffusion until 500°C. Anneals of 0.1 μm SiO$_2$ on Si showed little movement until 500°C; Cu was found only near the surface at temperatures above 700°C. Annealing of 0.5 μm LPCVD Si$_3$N$_4$ showed no significant diffusion until 700°C.

Further study is in progress for understanding Cu diffusion behavior in other barrier materials.

REFERENCES

1. F. A. Stevie, R.G. Wilson, J. M. McKinley, and C. Hitzman, Secondary Ion Mass Spectrometry, SIMS XI, R. Lareau et al., eds., Chichester (1998) 983
2. V. R. Deline, W. Reuter, and R. Kelly, Secondary Ion Mass Spectrometry, SIMS V, A. Benninghoven et al., Springer-Verlag, Berlin, (1986) 299
3. P. R. Boudewijn, H. W. P. Ackerboom, and M. N. C. Kempeners, Spectrochimica Acta **39**B. 1567 (1984)
4. C. J. Vriezema, K. T. F. Janssen, and P. R. Boudewijn, Appl. Phys. Lett. **54** (20), (1989) pp. 1981-1983
5. P. Shewmon, Diffusion in Solid, The Minerals, Metals & Materials Society, Warrendale, PA, (1989) pp. 173-180
6. B. L. Sharma, Diffusion in Semiconductors, Trans. Tech. Pub., Germany, (1970) pp. 87-117
7. Y. Shacham-Diamand, A. Dedhia, D. Hoffstetter and W. G. Oldham, J. Electrochem. Soc., Vol. 140, No. 8, (1993) pp. 2427-2432
8. C. Tian, C. Hitzman, and G. Chao, SIMS XII Workshop Proceedings (1999).
9. C. Tian, G. Chao, C. Hitzman, S. P. Smith, and L. Wang, SIMS XII Proceedings, Sept. 1999 (in press)
10. J. L. Moore, Lucent Technologies, Allentown, personal communication.

Diffusion Properties of Cu in Nanocrystalline Diamond Thin Films

D. Zhou[a,b)], F. A. Stevie[e)], E. Anoshkina[a.b)], H. Francois-Saint-Cyr[a,b,c)] , K. Richardson[a,b,c)]
A. Hussain[a,b)], and L. Chow[a,d)]

[a)] Advanced Materials Processing and Analysis Center
[b)] Department of Mechanical, Materials, and Aerospace Engineering
[c)] School of Optics
[d)] Department of Physics
University of Central Florida, Orlando, FL 32816

[e)] Cirent Semiconductor (Lucent Technologies)
9333 S John Young Parkway, Orlando, FL 32819

ABSTRACT

Diffusion behavior of Cu implanted into nanocrystalline diamond films has been investigated using SIMS. The SIMS depth profiles of Cu in the nanocrystalline diamond films, which have been annealed at the temperatures ranging from 300 to 1000 °C, reveal that no Cu diffusion occurs in nanocrystalline diamond films at the annealing temperature up to 500 °C. SEM analyses show that nanocrystalline diamond films have smooth coating surfaces and that instead of the columnar growth structure nanocrystalline diamond films consist of nanodiamond crystals with a random orientation. Furthermore, the diffusion properties of Li, Na, and K in nanocrystalline diamond films have also been studied by SIMS.

INTRODUCTION

Copper interconnection technology will greatly impact the performance, density, and yield of sub-micron integrated circuits with further reduction of feature sizes [1]. Unfortunately, copper has been considered one of the most dangerous impurities in silicon device fabrication as it can be easily diffused into the active regions of semiconductor devices during heat treatment. Therefore, a barrier for copper diffusion has to be developed in order to realize copper interconnection technology. Many investigations have been carried out using metals and compounds as potential diffusion barrier materials [2-3]. Among these barriers, Ta-based materials have been of increasing interest because Ta does not react with Cu, and diffusion of Cu through a Ta layer normally does not occur up to 700 °C [4]. The diffusion of Cu through microstructural defects such as grain boundaries, protrusions, and voids in the Ta thin film is responsible for the low failure temperature [5-6]. Currently, TiN is one of the most widely used barrier materials in Cu metallization, as well as in aluminum-based metallization [7-9]. Reported failure temperatures of TiN against Cu diffusion are 400-800 °C for furnace annealing [10] and 900 °C for rapid thermal annealing [7]. The difference in failure temperatures is caused by various film microstructures such as grain size, orientation, and composition of the TiN coatings [11]. Therefore, it has been suggested that nanocrystalline diffusion barriers will be more effective for preventing Cu diffusion than the

Conference Proceedings ULSI XV © 2000 Materials Research Society

polycrystalline barriers since the nanocrystalline films do not have large-angle grain boundaries where most of the atomic diffusion generally occurs.

In this paper, we report that nanocrystalline diamond thin films, deposited from plasma enhanced chemical vapor deposition (CVD) with a mixture of CH_4 and Ar as the reactant gas, can be a promising diffusion barrier material. The substrate temperature used for CVD nanocrystalline diamond coating has been reduced to 450 °C. The grain size and the surface roughness of the nanocrystalline diamond coatings, which are independent of the film thickness, range from a few to tens of nanometers and can be controlled by changing CVD processing parameters. SIMS analyses of Cu implanted into the nanocrystalline diamond films, which have been annealed at different temperatures, shows that no Cu diffusion has been observed at the annealing temperature up to 500 °C.

EXPERIMENTAL

Microwave plasma enhanced CVD system has been employed for preparation of nanocrystalline diamond films [12-13]. Mixtures of CH_4 and Ar have been used as the reactant gases for the microwave discharges. N-type single crystalline silicon wafers with an <100> orientation were used as the substrates, and nucleation sites for nanocrystalline diamond growth were provided by a bias enhanced method. The substrate temperatures for the nanocrystalline diamond deposition were kept in the range of 350–800 °C, while total ambient pressure and input power were kept at 100 Torr and 1200 W, respectively. Both plan and cross-sectional views of scanning electron microscopy (SEM) were employed to study the surface morphologies and growth phenomena of the nanocrystalline diamond coatings. Copper was implanted into the nanocrystalline diamond thin films at room temperature with a dose of 1E14 atoms/cm^2. The nanocrystalline diamond samples implanted with Cu were then annealed at temperatures varying from 300 to 1000 °C in Ar environment for 30 minutes. Due to its sensitivity and depth resolution, secondary ion mass spectrometry (SIMS) has been employed to study the behavior of Cu diffusion in the nanocrystalline diamond films after the heat treatments.

RESULTS AND DISCUSSIONS

The nanocrystalline diamond films were first examined by SEM in order to obtain the surface and growth morphologies. Figure 1 (a) shows a top view SEM image of the diamond films prepared from plasma enhanced CVD with a mixture of 1% CH_4 and 99 % Ar as the reactant gas at 450 °C, indicating that the film consists of nanocrystalline diamond structure with a smooth surface coating. The grain size for this particular film ranges from a few to tens of nanometers, which is independent of the film thickness. The grain size and surface roughness of the nanocrystalline diamond coatings depend strongly on the CVD processing parameters, such as the reactant gas, substrate temperature, and input microwave power [14]. A cross-sectional view SEM image of the film, shown in Figure 1 (b), demonstrates a smooth cross-sectional fracture surface, indicating that instead of the columnar structure, which is common for microcrystalline diamond coatings, the nanocrystalline diamond coatings consist of nanocrystalline diamond crystals with a random orientation. This observation suggests that the nanocrystalline diamond does not grow from the initial nuclei at the substrate-film interface, but is the result of very high secondary nucleation rates during the deposition process.

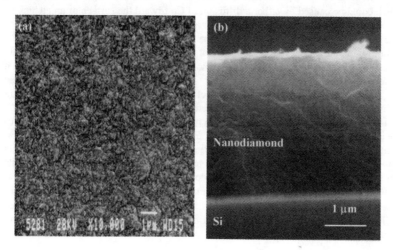

Figure 1 SEM images of a nanocrystalline diamond thin film produced from plasma enhanced CVD with a mixture of CH_4, and Ar as the reactant gas. (a) top view and (b) cross-sectional view.

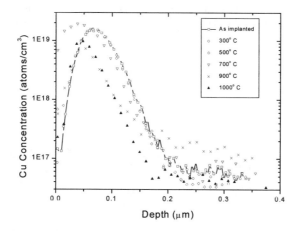

Figure 2 SIMS depth profiles of Cu in nanocrystalline diamond thin films after annealing at different temperatures.

Figure 2 shows SIMS depth profiles of Cu in as-implanted and annealed nanocrystalline diamond samples, indicating that no diffusion activity occurs at least up to 500 °C. A small shift of Cu depth profile has been observed as the sample was annealed at 700 °C, which could be caused by either a diffusion of Cu toward the surface or an etching of the surface of nanocrystalline diamond during the annealing process due to the presence of some oxygen. Finally, some movement of Cu in nanocrystalline diamond occurred when the nanocrystalline diamond samples were annealed at temperatures ranging from 900 to 1000 °C. The results reported here suggest that nanocrystalline diamond can be a potential diffusion barrier material for Cu interconnect technology. Diffusion behavior was additionally studied by the analyses of alkali elements after implanting and annealing nanocrystalline diamond samples. Li, Na, and K were implanted into nanocrystalline diamond films with a dose of $5E13$ atoms/cm^2 for each element at room temperature. The energies used for the implantation of Li, Na, and K were 20, 50, and 80 keV, respectively. Figure 3 (a) shows a depth profile of ^{7}Li, ^{23}Na, and ^{39}K from the as-implanted sample, demonstrating an almost ideal Gaussian distribution for each implanted element and excellent detection limits for these three elements using SIMS. The SIMS depth profile of the same sample annealed at 800 °C for 30 minutes is shown in Figure 3 (b), demonstrating that all three elements are very slow diffusers in nanocrystalline diamond films at 800 °C. This result suggests that nanocrystalline diamond coating can be used as a diffusion barrier not only for copper but also for other contaminants such as Li, Na, and K.

Because diamond with a very tightened tetrahedral structure is the densest material, diffusion coefficiency and solubility of any metallic element in diamond seem extremely low. The only path for diffusions of foreign elements in polycrystalline diamond is the grain boundaries, and the columnar growth structures enhance significantly the diffusions of foreign atoms in microcrystalline diamond. However, the nanocrystalline diamond films with grain sizes ranging from a few to tens of nanometers and random orientation greatly reduced the grain boundary effects on the diffusion of foreign elements, therefore, making nanodiamond films a promising diffusion barrier material.

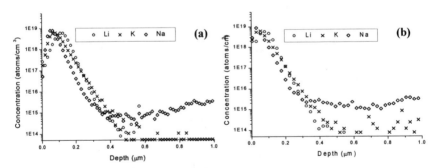

Figure 3 SIMS depth files of Li, Na, and K in (a) as-implanted nanocrystalline diamond film and (b) annealed nanocrystalline diamond film at 800 °C for 30 mins.

CONCLUSIONS

Nanocrystalline diamond thin films have been successfully prepared by microwave plasma enhanced CVD method. The top view SEM image of the film shows that the nanocrystalline diamond films have a smooth surface, which is independent of the film thickness. Cross-sectional SEM analysis reveals that there is no columnar structure in the nanocrystalline diamond coatings. The SIMS depth profile of copper implanted into nanocrystalline diamond films suggests that no copper diffusion occurs when the samples are annealed up to 500 °C. Furthermore, the SIMS analyses demonstrate that Li, Na, and K do not diffuse in nanocrystalline diamond films when the sample is annealed at 800 °C, indicating that nanocrystalline diamond films can also be a good diffusion barrier of contaminants introduced during the device processing.

ACKNOWLEDGEMENTS

The authors would like to acknowledge the technical and financial support of UCF/Cirent Materials Characterization Facility (MCF) and Advanced Materials Processing and Analysis Center (AMPAC) at University of Central Florida. The authors also would like to thank Zia Ur Rahman, Jay Bieber, and Kirk Scammon for their assistance.

REFERENCES

[1]   The National Technology Roadmap for Semiconductors: *Technology Needs*, 1997, pp 99-103.
[2]   D. J. Kim, Y. T. Kim, and J. W. Park, J. Appl. Phys. **82**, p. 4847 (1997).
[3]   T. Iijima, Y. Shimooka, and K. Suguro, Electronics and Communication in Japan part II-Electronics **78** p.67 (1995).
[4]   K. Holloway and P. M. Fryer, Appl. Phys. Lett. **57**, p. 1736 (1990).
[5]   S. Q. Wang, MRS Bulletin **19**, p. 30 (1994).
[6]   J. C. Lee and C. Lee, J. Electrochemical Soc. **146**, p. 3466 (1999).
[7]   I. Sumi, M. Blomberg, and J. Saarilahti, J. Vac. Sci. Technol. **A3**, p. 2233 (1985).
[8]   A. Armigliato and G. Valdre, J. Appl. Phys. **61**, p. 390 (1987).
[9]   S. QW. Wang, I. J. Raaijmakers, B. J. Burrow, S. Suthar, S. Redkar, K. B. Kim, J. Appl. Phys. **68**, p. 5176 (1990).
[10]   J. O. Olowolafe, C. J. Mogab, R. B. Ciregory, M. Kottke, J. Appl. Phys., **72**, p. 4099 (1992).
[11]   M. B. Chamberlain, Thin Solid Films, **91**, p. 155 (1982).
[12]   D. Zhou, T. G. McCauley, L. C. Qin, A. R. Krauss, and D. M. Gruen, J. Appl. Phys. **83**, p. 540 (1998).
[13]   D. Zhou, D. M. Gruen, L. C. Qin, T. G. McCauley, and A. R. Krauss,, J. Appl. Phys. **84**, p. 1981 (1998).
[14]   D. Zhou, F. A. Stevie, L. Chow, J. Mckinley, H. Gnaser, and V. H. Desai, J. Vacuum Sci. Tech. **A17**, p. 1135 (1999).

Studies of Particle Formation Phenomenon observed between Silicon Nitride capping layer and Copper

Kia Seng LOW, Werner Pamler, Markus Schwerd, Heinrich Koerner, and Hans-Joachim Barth, Infineon Technologies AG, Wireless Products, Technology, Munich, GERMANY.
Anthony O'Neill, University of Newcastle Upon Tyne, Department of Electrical and Electronic Engineering, UNITED KINGDOM.

ABSTRACT

Copper (Cu) has to be encapsulated in a damascene structure so as to prevent it from out-diffusion. In this paper, we describe our observation and studies of particle formation on Cu surface when depositing silicon nitride in a SiH_4 based PECVD process as a capping layer on top of Cu. The particle phenomenon is first observed on silicon nitride surface in the bubble-like form. Focused ion beam (FIB) cross section confirms that the bubble phenomenon is due to the hillock-like formation on the Cu surface during the nitride deposition. The results of SEM-EDX on particles and non-particle regions suggested that the particle is due to copper silicide formation. This is confirmed with results from Transmission Electron Microscope (TEM) – EDX measurement of the particle. The measured composition of particle is $Cu_{77\%}Si_{23\%}$. Using various analytical measurements, we illustrate that the particle formation phenomenon is due to an interaction of SiH_4 with Cu and the mass transport of Cu through "weak spots" on the Cu native oxide surface. A model is proposed to explain the phenomenon.

INTRODUCTION

Cu will be used to replace aluminum in the next generation metallisation due to its low resistivity and high electromigration resistance. However, copper diffuses at a faster rate[1] in silicon and silicon dioxide [2], and it is detrimental to the devices if it gets into the active region [3]. Silicon nitride has been chosen to be the capping barrier layer to prevent Cu from out-diffusion in the Cu damascene structure [4]. The common constituents of the silicon nitride deposition are silane, ammonia and nitrogen. Silane remains an indispensable element in the silicon nitride deposition. In this paper, we show that silane has an undesirable effect on Cu surface. A model of the particle formation is described at the end of the paper.

EXPERIMENT

Bubble phenomenon

Using the conventional silane based PECVD silicon nitride deposition on Cu, we observe "black dots" on the nitride surface as shown in Figure 1(a). Under the Scanning Electron Microscope (SEM) observation, the "black dots" appear to be bubble-like form as shown in Figure 1(b). Focused ion beam (FIB) cross section confirms that the bubble phenomenon is due to the hillock-like formation on the Cu surface during the nitride deposition as shown in Figure 2(b). We called this hillock-like formation as particle formation.

Conference Proceedings ULSI XV © 2000 Materials Research Society

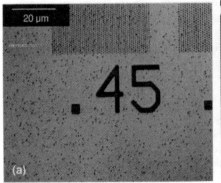
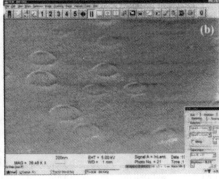

Figure 1: (a) Optical microscope image (b) Bubble phenomenon observed on the silicon nitride surface on Cu.

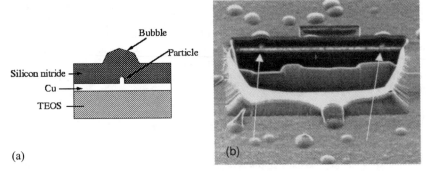

(a)

(b)

Figure 2 : (a) Schematic of the cross sectional Focused Ion Beam (FIB) cut. (b) FIB image of a bubble on the silicon nitride surface. The cross section image shows that the bubble phenomenon is due to particle formation (marked) on the Cu surface.

Oxidation Experiment

In the oxidation experiment, we measure the effect of the exposure time of the copper surface to air on the particle density on the copper surface. The exposure time of the copper surface to air is measured after the copper deposition, and stopped before nitride deposition. The results are shown in Figure 3(a). The results show that the longer the duration the copper surface is exposed to air, the lower is the particle density value. To accelerate the oxidation process, we heated half of a wafer at about 100 °C for a minute, and repeated the nitride deposition experiment. Referring to the schematic in Figure 3(b), we observe no particle found on the heated region, but many particles on the non-heated region after depositing silicon nitride onto the treated wafer. There is a particle transition region where the particle density reduces from the heated region to the non-heated region. This indicates that the oxidised surface is inert to the silane reaction.

320

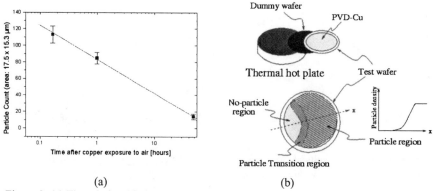

(a)   (b)

Figure 3: (a) The relationship between the particle count and the exposure time of the copper surface to air. (b) Partial wafer oxidation and particle count observed on the wafer.

TEM-EDX analysis of a Particle

Figure 4 shows the Transmission Electron Microscope (TEM) image of a particle. Cu depletion at both sides of the particle, and an interface between the particle and the Cu surface is observed. The Cu depletion or ditches extend through the thickness of the Cu film. The results of the TEM-EDX at the particle and at the non-particle reference position are shown in Figure 5(a) and (b) respectively. The results show that a significant Si signal is detected at the particle, but not at the reference position. The quantification of the TEM-EDX results shows that the composition of the particle is made up of 77% Cu and 23% Si.

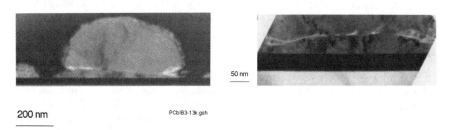

200 nm   PCbIB3-13k.gsh

Figure 4: TEM image of a particle and the interface between the particle and the Cu surface

Effect of Annealing Temperature on Cu surface

Figure 6 shows the effect of annealing temperature on PVD-Cu surface. We observe a hillock-like formation on the Cu surface after temperature anneal at 350 °C. At even high temperature (e.g. 700 °C), we observe giant Cu hillocks formation on the Cu surface. We suspect that this hillock-like formation provides the nucleus for the particle formation. Figure 7 shows that no particle is observed when nitride is deposited below 355 °C.

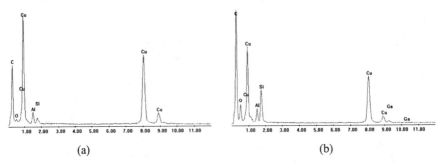

(a)   (b)

Figure 5: TEM-EDX results at (a) a particle and (b) a non-particle region, which indicate that the particle formation is due to the copper silicide formation.

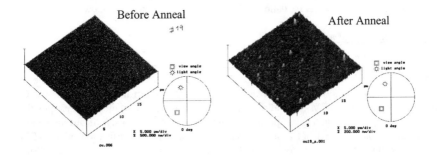

Figure 6: Atomic Force Microscope (AFM) image of a PVD-Cu surface before and after 350 °C anneal of 30 minutes (20 x 20 μm scan area).

Particle Formation Model

In this section, we propose a model to describe the particle formation. The schematic of the model is shown in Figure 8. During the silicon nitride deposition, the wafer is placed on a heater (approx. 400 °C) in the nitride chamber. Miniscule hillocks are formed on the Cu surface, which act as the nucleus for the particle formation. The nucleus refer to the "weak spots" on the Cu surface. The particle grows as the incoming Si diffuses into the Cu film while Cu diffuses through the "weak spot" into the particle. The native oxide on the Cu surface acts as the protective layer. A few nanometer of native oxide is sufficient to prevent the particle formation [4]. When the Cu surface is reduced by removing the native oxide, silane reacts with the Cu and forms a copper silicide layer, which decolorize the copper from red to grey. Oxidation experiments show that the oxidized Cu surface is inert to silane reaction, and there is no particle formation. This is due to the passivation of the "weak spots" during oxidation.

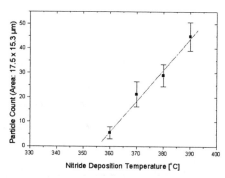

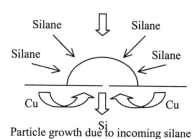

Particle nucleation due to temperature

Silane ⇩ Silane

Silane ↘ ↙ Silane

Cu ⤵ ⇩ ⤴ Cu

Si

Particle growth due to incoming silane

Figure 7: Relationship between particle count and nitride deposition temperature.

Figure 8: Model of particle formation.

CONCLUSIONS

In this paper, we describe the particle formation phenomena that we observe during the PE-CVD silicon nitride deposition on Cu. From the analytical results, we conclude that the particle formation is due to an interaction of SiH_4 with Cu and the mass transport of Cu through "weak spots" on the Cu native oxide surface. The results also show that the particle density is reduced when the Cu surface has been exposed to air for a longer duration. We conclude that there is an effect of the native oxide on the particle formation phenomenon. The results show that when we deposit silicon nitride at less than 350 °C, there is no particle formation. This illustrates the effect of the nitride deposition temperature on the particle density. To prevent the particle formation, it is recommended to deposit silicon nitride at a lower temperature.

ACKNOWLEDGEMENT

The authors would like to thank all the colleagues in the Failure Analysis Group (Perlach, Munich) and the leader, Dr. Weiland, for allowing author using the AFM, FIB, SEM and all other failure analysis equipment. This work was supported by the Federal Department of Education and Research of the Federal Republic of Germany (BMBF projects No. 01M2933 C and 01M2972B). For the content, the authors alone are responsible.

REFERENCES
[1]   R.N.Hall and J.H.Racette, J.Appl.Phys., 35(2), pp.379-397, 1964.
[2]   J.D.McBrayer, R.M.Swanson and T.W.Sigmon, J.Electrochem. Soc., 133(6), pp.1242-1246, 1986.
[3]   K.S.Low, M.Schwerd, H.Koerner, J.H.Barth and A.G.O'Neill," SPIE's 1999 Sym.: Microelectronic Manufacturing, 1999. (To be published)
[4]   W.Pamler, K.S.Low, M.Schwerd, H.Koerner, M.Schwerd and A.G.O'Neill, 196th Meeting of the Electrochemical society, 1999 (To be published).

Low-k Dielectrics

SACRIFICIAL MACROMOLECULAR POROGENS: A ROUTE TO POROUS ORGANOSILICATES FOR ON-CHIP INSULATOR APPLICATIONS

R.D. MILLER, W. VOLKSEN, J.L. HEDRICK, C.J. HAWKER, J.F. REMENAR, P. FURUTA, C.V. NGUYEN, D. YOON, M. TONEY, D.P. RICE, J. HAY*

IBM Almaden Research Center, 650 Harry Road, San Jose, CA 95120-6099
*T.J. Watson Research Center, Yorktown Heights, NY

ABSTRACT

Porous methylsilsesquioxane can be prepared by the templated vitrification of low molecular weight oligomers by macromolecular porogens of controlled structure and architecture followed by pyrolysis of the organic polymers. At intermediate temperatures (e.g., 250°C) optically clear nanoscopic inorganic-organic hybrids are produced which are converted to nanoporous films at higher temperatures. The porous morphology is stable to above 400°C and the process is operationally simple. Dielectric constants of < 2.2 are realized at moderate porosity levels (~20%) suggesting potential applications as on-chip interconnect insulators for advanced semiconductor devices.

INTRODUCTION

Soon advanced logic devices will contain more than 0.5 billion transistors on a single chip connected with over 10,000 m of wiring distributed over 6–8 wiring levels. The increased wiring density causes crosstalk and results in signal delays due to capacitive coupling.[1] The lower intrinsic resistivity of copper metallurgy[2] relative to Al will mitigate the RC delay in the short term, but dimensional scaling requires the introduction of insulating materials with dielectric constants substantially lower than that of the currently employed silicon dioxide (k = 3.9–4.2). This has fueled the search for new low-k materials which are compatible with current on-chip integration demands.[3] New low-k materials will be introduced gradually, reducing the dielectric constant by 20–30% for each new device generation. While there are a plethora of candidates in the dielectric range of 2.6–3.0, no clear winner consistent with all the integration requirements has emerged.[3,4] For ultra-low-k materials (k < 2.2), the list of candidates is short indeed. It is in this dielectric range that porous materials begin to appear,[5-13] presenting a new set of integration difficulties. In spite of this, it is clear that there is no true dielectric device generational extendibility without embracing the concept of porosity. Of the porous materials, porous silica has received the most attention.[6-10] Since porosity is beneficial only in lowering the dielectric constant and adversely effects most other electrical, mechanical and thermal properties, a good rule of thumb seems to be: no more porosity than necessary to achieve the dielectric objectives. It also seems generally agreed that closed-cell porosity is likely to cause fewer integration problems than open-cell interconnected porosity.

Aside from the significant processing difficulties associated with silica aero- and xerogels, the dielectric constant of the densified matrix is too high, requiring the introduction of substantial porosity to achieve the dielectric goals. This is shown in Figure 1 which displays the variation in dielectric constant as a function of the void volume. For comparison, a related calculated plot for methyl silsesquioxane (MSSQ) with a densified dielectric constant of 2.85[14] is also shown. For MSSQ, the void volume necessary to achieve a dielectric constant of 2.0 is reduced from more than 70% for silica to less than 40% for the low-k material. We describe here a simplified process for the production of nanoporous methyl silsesquioxane $(MeSiO_{1.5})_n$ by the blending of a thermally labile, macromolecular pore generator (porogen) with a low molecular weight

Conference Proceedings ULSI XV © 2000 Materials Research Society

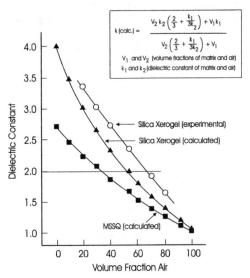

$$k_{(calc.)} = \frac{V_2 k_2 \left(\frac{2}{3} + \frac{k_1}{3k_2}\right) + V_1 k_1}{V_2 \left(\frac{2}{3} + \frac{k_1}{3k_2}\right) + V_1}$$

V_1 and V_2 (volume fractions of matrix and air)

k_1 and k_2 (dielectric constant of matrix and air)

Silica Xerogel (experimental)

Silica Xerogel (calculated)

MSSQ (calculated)

Dielectric Constant

Volume Fraction Air

Figure 1: Dielectric modeling of porous dielectric insulators.

(M_n < 1,000, SEC) MSSQ resin followed by ramping to 430°C to vitrify the matrix and remove the porogen.

RESULTS AND DISCUSSION

MSSQ vitrifies upon heating by a complex sequence of chain extension, crosslinking and silicon-oxygen bond redistribution.[15] An idealized depiction of this process is shown in Figure 2. The actual chemistry occurring during this process is most certainly more complex than depicted in this figure. Since polymer blends almost always undergo macroscopic phase separation[16] resulting in domains which are too large to produce nanoscopic porosity, the MSSQ matrix resin was carefully selected. Low molecular weight resins were chosen to provide a maximum number of polar chain ends (SiOH, SiOEt) for interaction with the blended macromolecular porogen chain ends. The porogen molecules themselves are tailored to provide strong interactions with the low molecular weight MSSQ resin and resist self-aggregation. A generic porogen is depicted in Figure 3. Linear polymer chains are grown by controlled polymerization initiated from a multifunctional core (core-out synthesis). In this regard, we have prepared a number of architecturally defined star-like, dendrimeric and hyperbranched polycaprolactones from hydroxyl functionalized core molecules.[17-19] Branched macromolecular architectures were targeted because of their increased compatibility with other polymers, low tendency toward chain entanglement and the abundance of polar chain ends.

Figure 4 defines the critical mechanical and thermal properties of the MSSQ matrix polymer and porogen. The dynamic mechanical properties of the matrix and blend were probed using an impregnated glass fabric. Upon heating, the resin becomes liquid-like at first, then begins to stiffen significantly as the temperature increases beyond 150°C. The modulus increases steadily to 400°C, although the rate of change decreases significantly above 300°C. After curing, the modulus of the resin is invariant with temperature and the polymer Tg becomes undetectable. The porogen (4-arm PCL star), as shown by thermal gravimetric analysis, begins to decompose around 300°C and is completely converted to lower molecular weight volatile fragments by

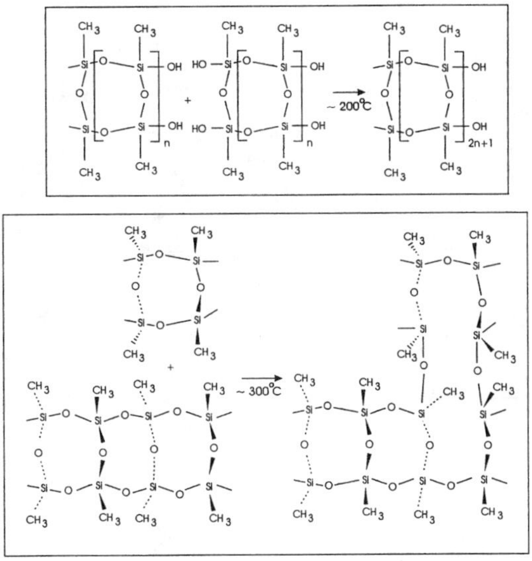

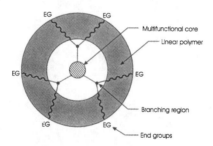

Figure 2: A schematic description of the chain extension and vitrification of low molecular weight MSSQ materials.

EG
EG
Multifunctional core
Linear polymer
EG
EG
Branching region
EG
EG
End groups

Figure 3: A generic description of a 6-arm porogen showing the important sectors of the pore generating macromolecules.

400°C. The complimentary thermal behavior of the porogen and mechanical behavior of the resin is critical to the formation of stable porosity. The strength of the vitrifying resin *must* increase before the porogen decomposes in order to prevent pore collapse. The films remain optically clear during the heating process, consistent with the formation of nanoscopic heterogeneities.

Phase separation of the porogen in the resin after curing to 250°C, a temperature sufficient for vitrification of the MSSQ matrix but below the decomposition temperature of the porogen, was verified by dielectric spectroscopy (porogen $T_g \sim -30°C$). Figure 5 shows a TEM micrograph of the porous film generated from the inorganic-organic nanoscopic hybrid containing 10 wt.% of the porogen. The pores (light areas) appear to be uniformly dispersed and non-interconnecting in the x,y plane. The phase separation process proceeds by a nucleation and growth mechanism and the pores become larger as the porogen loading level increases. Figure 6

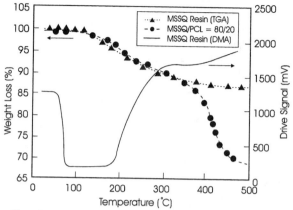

Figure 4: —— DMA analysis of MSSQ resin impregnated into a glass fabric.
····▲···· TGA analysis of MSSQ resin. – ● – TGA analysis of MSSQ
containing 20 wt% of a 4-arm PCL star porogen.

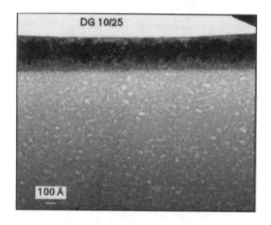

Figure 5: TEM micrograph of porous MSSQ
prepared from a hybrid containing 10% of a 4-arm
star PCL porogen

shows a plot of the dielectric constant of the porous MSSQ as a function of the initial porogen
loading level. As expected, it decreases steadily as the loading level of the porogen increases.
The film densities were determined by x-ray reflectivity.

For a porous material, the porosity is characterized by the volume fraction, pore size and
distribution and interconnectivity. While the first three parameters are easily determined by
TEM analysis, the interconnectivity of the pores and the percolation threshold[20] is more difficult
to define. We have probed the latter by studying both the electrical properties of the hybrid[13] and
the mechanical properties of the porous matrix. Figure 7 shows the AC conductivity of the
hybrid film cured to 250°C. Since the Tg of the porogen is below room temperature, the AC
conductivity varies with temperature but increases markedly at loading levels between 20–25%.
The shape of this curve is somewhat thickness dependent and the onset increase in conductivity

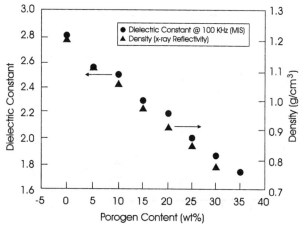

Figure 6: Plots of dielectric constant and density (x-ray reflectivity) as a function of loading level for a 4-arm PCL star porogen in MSSQ.

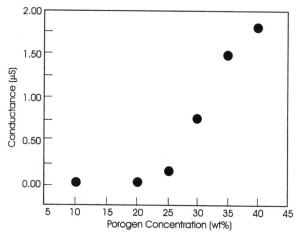

Figure 7: AC conductivity of the nanoscopic hybrids derived from a 4-arm PCL star dispersed in MSSQ. Curing temperature was 250°C.

shifts slightly toward lower loading levels as the film thickness decreases from 1.0 to 0.6 μm. The behavior of the conductivity curve suggests the possible onset of domain interconnectivity in the region where the slope begins to increase markedly.

The modulus and hardness of the porous films as determined by nanoindentation techniques also show a curious behavior as the initial porogen loading level increases. Significant changes in the slopes of both the modulus and hardness plots also occur in the region around 18–20% porosity (Figure 8). For comparison, the modulus and hardness of SiO_2 measured by this technique were 72 GPa and 9.25 GPa respectively. Preliminary experiments using positron annihilation spectroscopy[21] suggest that the positronium lifetimes also increase markedly at porosity levels of 17–20%.[22] The combination of electrical measurements on the hybrid material

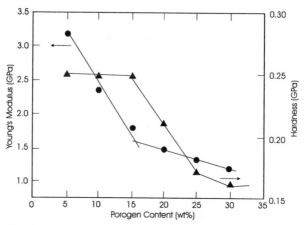

Figure 8: Modulus and hardness of porous MSSQ generated from 4-arm PCL star porogens as a function of loading level, measurements utilized nanoindentation.

and mechanical and spectroscopic measurements on the porous films suggests that important morphology changes occur in the region of 18–20% (porogen loading or porosity level) possibly signaling the onset of interconnection of domains and/or pores.

In summary, an operationally simple route to porous methyl silsesquioxane via templated vitrification using macromolecular pore generators (porogens) of controlled structure and architectures has been described. The porous morphology is stable to 450°C and dielectric constants below 2.2 were achieved at modest porosity levels. These results suggest that sacrificial macromolecular porogens provide a viable route to porous silicates of interest as ultra-low-k insulators for on-chip applications.

ACKNOWLEDGMENTS

C.V.N. acknowledges support from Eloret, Inc. (NASA Contract No. NAS2-1-H031); JPR gratefully acknowledges support from the National Science Foundation Materials Research Science and Engineering Grant DMR-9400354 for the Center for Polymeric Interfaces and Macromolecular Assemblies (CPIMA); the authors also gratefully acknowledge partial support for this work from NIST-ATP Cooperative Agreement No. 70NANB8H4013.

REFERENCES

1. Wilson, S. R.; Tracy, C. J.; Freeman, Jr., J. L. In *Handbook of Multilevel Metallization for Integrated Circuits*; Wilson, S. R., Tracy, C. J., Freeman, Jr., J. L., Eds.; Noyes Publications: Park Ridge, NJ, 1993; Chap. 1.
2. Edelstein, D.; Heidenreich, J.; Goldblatt, R.; Cote, W.; Uzoh, C.; Tustig, N.; Roper, P.; McDevitt, T.; Motsiff, W.; Simon, A.; Durkovic, J.; Wachnik, R.; Rathore, H.; Schultz, R.; Su, L.; Luce, S.; Slattery, J. *Tech. Digest IEEE International Electron Devices Mtg.* 1997, p. 376.
3. Miller, R. D. *Science* **1999**, *286*, 421 and supplementary material.
4. Peters, L. *Semiconductor Int.* **1998** (Sept.), *21*, 63.
5. Hedrick, J. L.; Yoon, D. Y.; Russell, T. P. McGrath, J. E.; Briber, R. M.; Carter, K. R.; Labadie, J. W.; Miller, R. D.; Volksen, W.; Hawker, C. J. *Advances in Polymer Science* **1999**, *141*, 1.

6. Hrubesh, L. W.; Keene, L. E.; Latorre, V. R. *J. Mater. Res.* **1993**, *8(7)*, 1736.
7. Jin, C.; List, S.; Zielinski, E. *Proc. Mater. Res. Soc.* **1998**, *511*, 213.
8. Brunsma, P. J.; Hess, N. J.; Bontha, J. R.; Liu, V.; Baskaran, S. *Proc. Mater. Res. Soc.* **1996**, *443*, 105.
9. Brinker, C. J. *Current Opinion in Solid State and Material Science* **1996**, *1*, 798.
10. Zhao, D.; Feng, J.; Huo, Q.; Melosh, N.; Fredrickson, G. H.; Chmelka, B. F.; Stucky, G. D. *Science* **1998**, *279*, 548.
11. Hedrick, J. L.; Miller, R. D.; Hawker, C. J.; Carter, K. R.; Volksen, W.; Yoon, D. Y. *Adv. Mater.* **1998**, *10(13)*, 1049.
12. Remenar, J. F.; Hawker, C. J.; Hedrick, J. L.; Kim, S. M.; Miller, R. D.; Nguyen, C.; Trollsås, M.; Yoon, D. Y. *Proc. Mater. Res. Soc.* **1998**, *511*, 69.
13. Nguyen, C. V.; Carter, K. R.; Hawker, C. J.; Hedrick, J. L.; Jaffe, R. L.; Miller, R. D.; Remenar, J. F.; Rhee, H.-W.; Rice, P. M.; Toney, H. F.; Trollsås, M.; Yoon, D. Y. *Chem. Mater.* **1999**, (in press).
14. Cross, L. E.; Guruaya, T. R. *Proc. Mater. Res. Soc.* **1986**, *72*, 53.
15. Baney, R. H.; Itoh, M.; Sakakibara, A.; Suzuki, T. *Chem. Rev.* **1995**, *95*, 1409.
16. *Polymer Blends*; Paul, D. R., Seymour, R. P., Eds.; Academic Press: New York, NY, 1978.
17. Trollsås, M.; Hedrick, J. L.; Mecerreyes, D.; Dubois, Ph.; Jérôme, R.; Ihre, H.; Hult, A. *Macromolecules* **1997**, *30*, 8508.
18. Trollsås, M.; Hedrick, J. L.; Mecerreyes, D.; Dubois, Ph.; Jérôme, R.; Ihre, H.; Hult, A. *Macromolecules* **1998**, *31*, 2756.
19. Trollsås, M.; Hedrick, J. L. *J. Am. Chem. Soc.* **1998**, *120*, 4644.
20. Stauffer, D. in *Introduction to Percolation Theory*; Taylor and Francis: Philadelphia, PA, 1985.
21. Simon, G. P. *TRIP* **1997**, *5(12)*, 394.
22. Rodbell, K.; Lynn, K.; Petkov, M.; Weber, M.; Volksen, W.; Miller, R. D. (unpublished results).

OPTIMIZATION OF INTERCONNECTION SYSTEMS INCLUDING AEROGELS BY THERMAL AND ELECTRICAL SIMULATION

R. STREITER *, H. WOLF **, U. WEISS *, X. XIAO *, T. GESSNER *,**
* Chemnitz University of Technology, Center of Microtechnologies,
  D-09107 Chemnitz, Germany
** Fraunhofer Institute Reliability and Microintegration, Dept. Micro Devices and Equipment,
  D-09126 Chemnitz, Germany

ABSTRACT

Nanoporous silica aerogels are promising candidates for ultra low-k dielectrics used in the metallization system of the 0.13 µm IC generation and beyond. The main concern for their integration is the poor thermal conductivity causing considerable temperature increases particularly of global interconnects. In various exemplary cases line temperatures are calculated for different dielectric materials and insertion schemes. To enable the integration of aerogels, heat dissipation can be improved by cooling of global interconnects using heat sinks. Several types are investigated and compared with regard to thermal efficiency, influence on parasitic capacitance, and optimal application by a combined thermal and electrical simulation.

INTRODUCTION

As the speed of a device increases with decreasing feature size, the interconnect delay becomes the major fraction of the total delay. Therefore, the performance requirements of future IC generations can only be met by the introduction of new materials for metal lines and dielectrics together with improvements of chip architecture and interconnect design [1]. Both copper and low-k materials are necessary to restrain the chip size enlargement toward the 0.1 µm IC generation. Values of the effective dielectric constant below 2.0 (required for the 0.13 µm level and beyond) can be achieved by applying aerogels (xerogels) [2] or airgaps [3].

Though nanoporous silica aerogels are a promising candidate both for the interlayer dielectric (IeLD) and the intralayer dielectric (IaLD), the major concerns for their integration are Joule heating of the metal and the poor thermal conductivity of the dielectric. While aerogels are used in industry and space flight as a high temperature insulation material, their poor thermal conductivity leads to a new dimension of thermal problems within ULSI interconnect structures. Thermal transport can be improved by increasing the thermal conductivity of the gaseous component. However, exchanging the gas and optimizing its pressure have only limited effect. Thus, the problem has to be further reduced by an improved interconnect design for more efficient heat transfer or double-sided cooling.

Within single-side cooled chips the minimum temperature of the metallization system is equal to the working temperature of the active chip areas. Any interconnect Joule heating causes heat dissipation to colder adjacent lines and to the substrate. Heat transfer to the substrate strongly depends on the way aerogel is employed. In the case of embedded insertion (aerogels as IaLD, dense (conventional) silica as IeLD), the dielectric constant of the aerogel must be lower than in the case of homogeneous insertion (aerogels both as IaLD and IeLD) to meet the roadmap requirements for the effective permittivity of the metallization system. Thus, embedded aerogels need a higher porosity and have, as a consequence, a lower thermal conductivity compared to the homogeneous insertion scheme.

In the case of homogeneous integration simulations show that the temperature of global interconnects is strongly increased above less occupied lower metal levels. Nevertheless, an optimized thermal design of the interconnect system can enable such an integration scheme.

335

The low thermal conductivity of organic non-aerogel low-k materials does not significantly raise interconnect temperature in an embedded insertion scheme. However, if aerogels are used as IaLD, the heat transfer will be locally reduced, particularly in unoccupied regions of lower metal layers below the global interconnects because of the very poor thermal conductivity of the highly porous embedded aerogels. The most efficient heat transfer from metal level to metal level occurs through via contacts. To improve transport through less occupied metal layers, several additional features are introduced and compared with regard to their thermal efficiency. Such features must be inserted very carefully because of their tendency to increase parasitic capacitance. Therefore, dummy lines are not useful. For optimization, a combined thermal and electrical simulation is performed. The simulation requires careful consideration of material properties and geometrical dimensions of the metallization scheme.

BASIC ASSUMPTIONS

Material Properties

Copper: Within the temperature range from room temperature to several hundred degrees above, the resistivity of bulk copper can be described by the linear approximation

$$\rho(T) = \rho(T_o) + \alpha (T - T_o) \tag{1}$$

using a room temperature resistivity $\rho(298\ K) = 1.67E\text{-}8\ \Omega m$ and a temperature resistivity coefficient $\alpha = 6.8\ E\text{-}11\ \Omega m/K$. For a film thickness $d < 200$ nm surface scattering has to be taken into account resulting in an additional increase of resistivity according to

$$\rho(d) = \rho(\infty)\ (1 + \beta / d) \tag{2}$$

[4,5]. The value of the coefficient β was determined from measurements [6] to be 20 nm. Eq. (2) supplies only a lower limit of resistivity increase, because surface scattering within an interconnect is more intensive than within an extended film of the same thickness. The linear representation

$$\lambda_{Cu}(T) = \lambda_{Cu}(T_o) - \gamma (T - T_o) \tag{3}$$

with $\lambda_{Cu}(273\ K) = 403\ W/(m\ K)$ and $\gamma = 0.08\ W/(m\ K^2)$ fits thermal conductivity data of Cu [7] within the temperature range between 273 K and 573 K.

Silicon dioxide: The thermal conductivity of SiO_2 films depends both on deposition conditions and on film thickness d. Extensive experimental results have verified a thickness dependence of

$$\lambda_{Ox}(d) = \lambda_{Ox}(\infty) - \frac{\delta}{\sqrt{d}} \tag{4}$$

according to the model of Savvides and Goldsmid [8]. Only for thick thermally grown SiO_2 films the known value of $\lambda_{Ox}(1\ \mu m) = 1.2\ W/(m\ K)$ is achieved. For other deposition conditions lower values between 0.45 and 1.0 W/(m K) are applied for simulation. For the dielectric constant of PECVD oxide films a uniform value of 4.1 is used.

Aerogel: Thermal conductivity of aerogels depends on temperature, pressure, and porosity (density). At room temperature and normal pressure values of 0.065 W/(m K) [9] and 0.010 W/(m K) [10] have been measured for porosities of 75 % and 90 %, respectively. Values of the dielectric constant of 1.8 and 1.3 correspond to these porosities [11]. To meet the roadmap requirements for the effective dielectric constant, a value of 1.8 is sufficient down to the 0.10 μm generation in the case of homogeneous insertion. If the aerogel is applied within an embedded insertion scheme, a dielectric constant of not more than 1.3 is required both for the 0.13 µm and the 0.10 µm generation.

Geometrical Dimensions

Table 1 summarizes the structure and the geometrical dimensions of the interconnection schemes used for thermal simulation of future IC generations. The given values of the geometrical parameters are the result of an electrical multi-objective optimization carried out in Ref. [12].

Table 1. Values (in µm) of line width (W), line spacing (S), metal thickness (D), and IeLD thickness (H) in the metal levels M1 to M7 used for the simulation of the 0.13 µm and the 0.10 µm IC generations [12].

Generation	Dim.	M1	M2	M3	M4	M5	M6	M7
0.13	W	0.135	←	←	←	1.23	1.9	-
	S	0.165	←	←	←	1.15	1.94	
	D	0.135	←	0.235	←	2	2	
	H	0.33	←	←	←	2	2	
0.10	W	0.108	←	0.12	←	←	1.15	1.15
	S	0.132	←	0.12	←	←	2.45	3.94
	D	0.108	←	0.209	←	←	2	2
	H	0.264	←	0.24	←	←	2	2

NUMERICAL APPROACH

The temperature distributions within interconnection schemes are calculated by solving the energy balance equation for structures consisting of copper lines and various dielectric materials (dense silica, non-aerogel low-k materials, aerogel). For aerogels both the homogeneous and the embedded insertion scheme are considered. The calculations are carried out for local and for global interconnects, respectively.

The numerical approaches used depend on the occupancy of the metal levels. For arbitrary occupancy, a field solver has to be employed. In the two dimensional and thermally worst case of long lines within a global wiring bus in metal level n above n-1 unoccupied lower metal layers, the analytical solution

$$T_{line} - T_{sub} = \sum_i \Delta T_i = J^2 \frac{A_{line}}{P_n} \left[\frac{H_n}{\lambda_{IeLD}} K_2 + \sum_{i=1}^{n-1} \left(\frac{H_i}{\lambda_{IeLD}} + \frac{D_i}{\lambda_{IaLD}} \right) \right] \rho_{Cu} \qquad (5)$$

can be applied for the case of embedded insertion. J, A_{line} and $P_n=W_n+S_n$ are current density, line cross section, and pitch of the wiring level, respectively. Further geometric items are declared in Table 1. K_2 is the two dimensional spreading resistance factor. Fig. 1 shows its dependence on geometric properties as calculated using a field solver. If applied to homogeneous insertion schemes, H_n must be replaced by the

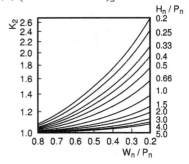

Fig. 1: Dependence of the spreading resistance factor K_2 on geometric dimensions.

distance of lines to substrate and the sum on extreme right of Eq. (5) has to be omitted.

To include the thermal effect of contacts to lower metal levels or substrate and of heat sinks inserted for interconnect cooling, the thermal conductance of the features has been calculated using a field solver. The energy balance equation is then solved for the resulting non-linear network of the respective thermal conductors and sources using an electronic circuit simulator. Such a lumped approach is sufficiently accurate as long as the disturbances of the thermal flow field caused by the features are small compared to their distance.

SIMULATION RESULTS

Temperature of Local Interconnects

Local interconnects are characterized by short distances between direct thermal (metal) contacts to the substrate, low power to length ratios due to small cross sections, and small vertical distances to substrate. Hence, the temperature of local interconnects within lower metal layers is increased for only a few degrees against substrate at current densities up to 1 MA/cm^2 for all dielectric materials considered. The influence of line density on temperature increase is also quite weak. Thus, local interconnects do not thermally limit the performance of the chip.

Maximum Temperature of Long Global Interconnects

For global interconnects the distance between direct thermal contacts to substrate is mostly determined by the distance of repeaters and, thus, comparatively long. The power to length ratio is high because of large cross sections. Longer vertical distances to the substrate and the poor thermal conductivity of the dielectric materials make heat dissipation to the substrate more difficult, especially for global bus wiring above locally unoccupied lower metal layers. Hence, the maximum interconnect temperature can reach intolerably high values. Fig. 2 depicts the dependence of maximum line temperature within global wiring busses above unoccupied metal layers on substrate temperature, current density, and thermal conductance to substrate for the 0.13 and the 0.10 μm generation. The lines are assumed to be sufficiently long, so that the maximum line temperature is not reduced due to cooling by the terminating contacts. The calculations result in a very different degree of applicability of the various dielectric materials and interconnection schemes:

- From the thermal point of view, dense SiO$_2$ and also SiO$_x$F$_y$ (λ = 1.0 W/(m K)) can be employed in the 0.13 μm generation for current densities up to 1 MA/cm^2. The same applies to airgap technologies, because the gaps are formed only in densely occupied and, therefore, well thermally bridged lower metal layers. For the 0.10 μm generation current densities up to 1.5 MA/cm^2 would be possible because of a greater pitch in levels M6 and M7.
- Organic non-aerogels (λ = 0.25 W/(m K)) can be homogeneously inserted up to 0.75 MA/cm^2 and 0.5 MA/cm^2 for the 0.13 μm and the 0.10 μm generation, respectively.
- The application of homogeneously inserted aerogels (λ = 0.065 W/(m K)) leads to tempera-ture increases of about 100 K (0.13 μm) and 50 K (0.10 μm) for the topmost interconnects already at 0.5 MA/cm^2. Thus, the homogeneous insertion scheme is not practicable for aerogels without thermal design improvements.
- For embedded aerogels (λ = 0.010 W/(m K)) the temperature increase of interconnects within the topmost metal level is about 300 K for the 0.13 μm generation at 0.5 MA/cm^2 and, because of greater pitches in levels M6 and M7, 100 K in the 0.10 μm case. This is also too much for typical operation conditions.
- Only a restricted operation of level M6 might become possible in the 0.10 μm generation for aerogels applied in both insertion schemes.

The thermal simulation of global interconnects suggests that a special thermal design for the achievement of an improved heat transfer would be very desirable for the homogeneous as well as for the embedded integration of aerogels.

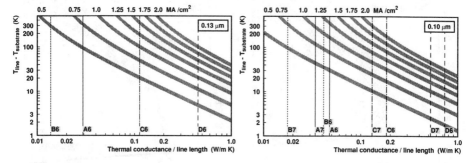

Fig. 2: Dependence of the temperature increase of global interconnects above unoccupied lower metal levels on the thermal conductance per line length between lines and substrate for different current densities. The frames refer to the geometric dimensions of the metallization scheme for the 0.13 μm generation (left) and the 0.10 μm generation (right) given in Table 1. The gray bands cover the range of substrate temperature from 80 °C (lower bound) to 150 °C (upper bound). The vertical lines mark the values of thermal conductance per line length between heated copper lines in the 6th or 7th metal level and the substrate for the different dielectric materials employed, i.e. homogeneously inserted aerogel (A), embedded aerogel (B), homogeneously inserted organic low-k material (C), and dense silica (D).

Temperature Distribution along Global Interconnects

Compared to the worst case discussed above the actual temperature of global interconnects can be locally reduced by high metal densities in lower levels and by line to substrate connections of high thermal conductance. The temperature distribution along global interconnects of level M6 was calculated both above densely packed and above unoccupied lower levels for the 0.13 μm generation at a current density of at 0.5 MA/cm^2. The temperature increase above the densely packed area amounts to only a few degrees. Thus, such regions act very similarly like thermal grounding by direct contacts to driver or load transistors. The maximum line temperature above the unoccupied area depends on the bridged distance between the densely packed functional blocks as shown in Fig. 3. After a distance of about 0.3, 1, and 2 mm from the next block, the limiting temperatures depicted in Fig.2 for very long lines are practically reached for organic low-k material (C), homogeneously inserted aerogel (A), and embedded aerogel (B), respectively. The application of organic low-k materials is not thermally restricted by the line length over regions of less occupied lower metal levels. In the case of aerogels the distance over unoccupied regions is limited for both insertion schemes to at most 0.5 mm. Thus, the connection of adjacent functional blocks can be made by direct crossing of global wiring channels up to a certain width. On the contrary, global lines along the wiring channels have to be equipped with special features (plugs) for improved heat dissipation. The maximum possible distance between points of effective cooling along a global interconnect can be estimated from Fig. 4. Optimal configuration and distribution of cooling features can be achieved by a combined thermal and electrical design.

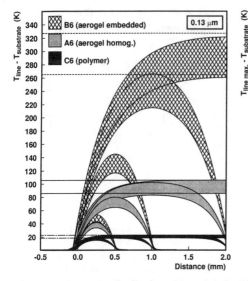

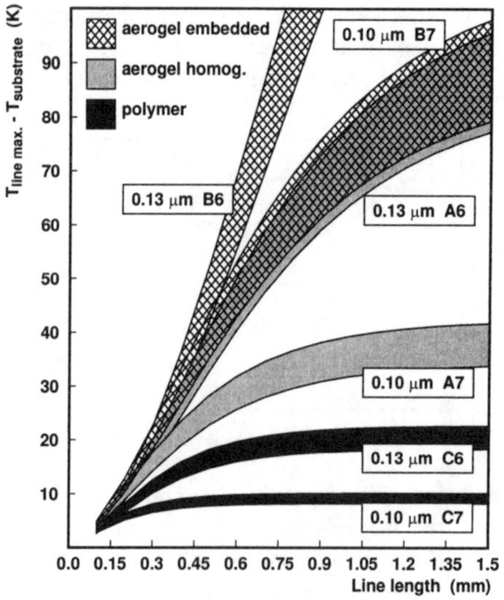

Fig. 3: Temperature distribution along global lines above unoccupied lower metal levels between functional blocks of 0.5, 1, 2, and 4 mm distance for 0.5 MA/cm². Substrate temperatures, dielectric materials, and wiring levels are the same as in Fig. 2.

Fig. 4: Dependence of maximum line temperature of global lines above unoccupied lower metal levels on the distance between ground contacts for 0.5 MA/cm². Substrate temperatures, dielectric materials, and metal levels are the same as in Fig. 2.

Design Optimization

To improve heat transport from global interconnects through less occupied lower metal layers, additional features (metal plugs consisting of W in level M1 and of Cu above) acting as heat sinks can be placed between the repeaters of a global interconnect. The heat sinks depicted in Fig. 5 have been investigated by means of combined thermal and electrical simulation with regard to their thermal efficiency and to their influence on parasitic line to ground capacitance.

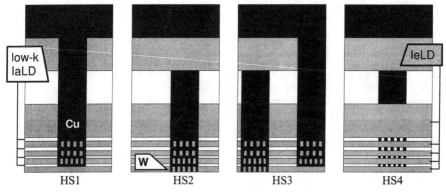

Fig 5: Vertical structure of different types of heat sinks for cooling of global interconnects.

Fig. 6 demonstrates the resulting impact of heat sink insertion on thermal conductance and capacitance between global interconnects and substrate for different material properties of the dielectrics. To calculate the thermal conductance and capacitance of a cooled interconnect, the respective values of the uncooled line and the additional contributions of the heat sinks used have to be added. In the case of the homogeneous insertion scheme, conductance and capacitance are increased by the same factor. On the contrary, for the embedded insertion of an aerogel ($\lambda_{laLD} / \lambda_{leLD} = 0.01 / 1$, $\varepsilon_{laLD} / \varepsilon_{leLD} = 1.3 / 4.1$) the thermal conductance of a 0.1 mm long segment of a global interconnect, cooled by one heat sink of type HS2, is increased 4.6 times, whereas its capacitance is increased only by 1.24. The more effective cooling of the same line segment by heat sink HS1 would increase parasitic capacitance by 1.7, which cannot be accepted. Otherwise, HS1 is useful for cooling of long interconnects of 1 mm length and more. Thus, the way of effective cooling of global interconnects depends on line length and should be tuned by a combined thermal and electrical design.

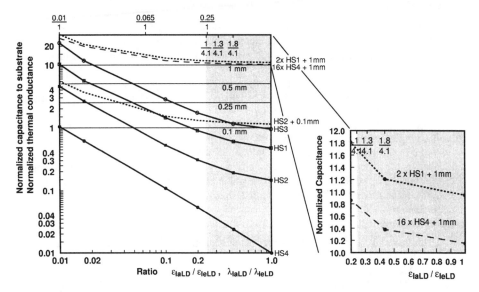

Fig. 6: Dependence of normalized thermal conductance on the ratio of conductivities and of capacitance on the ratio of dielectric constants for different types of heat sinks and cooled lines. The quantities are normalized to the corresponding values of a 100 μm long uncooled interconnect of level M6 (0.13 μm). Horizontal lines refer to uncooled interconnects of different length. The gray area is the possible range of dielectric constants. Special values of conductivities and dielectric constants are marked above and below the top of the frame, respectively.

As shown in Fig. 7, additional cooling of global interconnects is an inevitable requirement to enable the integration of aerogels. The insertion of two heat sinks of type HS1 is nearly equivalent to the insertion of 16 heat sinks of type HS4 with respect to total cooling efficiency and maximum line temperature. However, more distributed cooling appreciably reduces temperature variation along the lines. Moreover, the increase of parasitic capacitance is only 40 % compared to the example of localized cooling with heat sinks HS1 as shown in Fig. 6 (right frame).

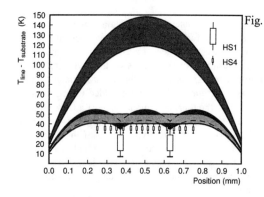

Fig. 7: Temperature distribution along global lines above unoccupied lower metal levels for no additional cooling, cooling by 2 heat sinks of type HS1, and by 16 heat sinks of type HS4. (Calculation for level M6 (0.13 μm), embedded aerogel insertion, current density 0.5 MA/cm^2, and substrate temperatures from 85 °C to 150 °C.)

CONCLUSIONS

- Because of their poor thermal conductivity, the application of nanoporous silica aerogels as ultra low-k dielectrics can considerably increase interconnect temperatures.
- Long global interconnects above less occupied lower metal levels are crucial for the thermal behavior of ULSI circuits.
- The improvement of heat dissipation to the substrate by thermal design is inevitable for the employment of aerogels in future ULSI metallization systems. The same applies to organic low-k dielectrics for higher current densities.
- The insertion of heat sinks is an appropriate means to restrain interconnect temperatures. The various types of heat sinks differ with regard to their thermal efficiency and to their contribution to parasitic capacitance. Optimal choice and placement of the different types depend on the material properties of the dielectrics and on the individual interconnect length.
- The combined thermal and electrical optimization of interconnect cooling becomes more and more important for future integrated circuit design.

REFERENCES

[1] The National Technology Roadmap for Semiconductors, 1997 Edition, Semiconductor Industry Association, San Jose, CA, 1997.
[2] *Proc. SEMATECH Ultra Low K Workshop,* March 16, 1999, Orlando,FL, SEMATECH, Austin,TX, 1999.
[3] B.P Shieh, L.C. Bassmann, D.-K. Kim, K.C. Saraswat, M.D. Deal, J.P. Mc Vittie, R.S. List, S. Nag, L. Ting, *Proc. IEEE IITC 1998*, p. 125.
[4] K. Fuchs, Proc. Cambridge Phil. Soc. **34**, 100 (1938).
[5] E.H. Sondheimer, Advanced Phys. **1**,1 (1952).
[6] F.W. Reynolds, G.R. Stilwell, Phys. Rev. **88**, 418 (1952).
[7] G.W.C. Kaye, T.H. Laby, *Tables of Physical and Chemical Constants*, 16th Ed, Longman, Harlow, GB, 1995.
[8] M. Bourgeois, A.N. Saxena, *Proc. 11th Int. VLSI Multilevel Interconnection Conf. (VMIC)*, June 1994, p. 126.
[9] M. Morgan, J.-H. Zhao, C. Hu, T. Cho, P.S. Ho, T. Ryan, H.-M. Ho, Ref [2], p.237.
[10] P. Scheuerpflug, M. Hauck, J. Fricke, J. Non-Cryst. Solids **145**, 196 (1992).
[11] L.W. Hrubesch, L.E. Keene, V.R. Latorre, J. Mater. Res. 8 (7), 1763 (1993).
[12] M.B. Anand, H. Shibata, M. Kakumu, IEEE Trans. Computer-Aided Design **17** (12), 1252 (1998).

Characterization of Nanoporous Low-k CVD film for ULSI Interconnection

Choon Kun Ryu, Si-Bum Kim, Sam-Dong Kim, and Chung-Tae Kim
Memory Research and Development Division, Hyundai Electronic Industries Co., Ltd.,
San 136-1, Ami-ri, Bubal-eub, Ichon-si, Kyongki-do, 467-701, Korea
Fax) 82-336-630-1313, e-mail) ckryu@sr.hei.co.kr

ABSTRACT

Four SiOCH materials were characterized by various analysis techniques. The materials were prepared by different deposition techniques and hydrocarbon source precursors. FTIR spectra of the SiOCH films showed all SiOCH samples have common peaks corresponding to Si-O, C-H, Si-CH$_3$, and Si-C peaks. The lower temperature-prepared SiOCHs have additional peaks corresponding to Si-H bonds. The chemical composition of the CVD SiOCH films was similar to the one of the Methyl Silsesquioxane (MSQ). To investigate oxygen plasma damage effect, the SiOCH films were exposed under oxygen radicals at a conventional downstream plasma asher and then analyzed by FTIR. C-H and Si-CH$_3$ peaks disappeared almost completely due to oxygen radical attack and the Si-O peak was shifted to higher wave-number position corresponding to PECVD SiOx. The thickness of SiOCH films decreased significantly and the refractive indices increased also.

INTRODUCTION

As integrated circuits become scaled down, interconnect RC delay and crosstalk between metal lines have been the major factors limiting device performance. Reduction of capacitance by the use of low-k dielectric material was considered as the main solution to the device performance degradation problem. Among a variety of low-k materials, low-k CVD SiOCH materials have been evaluated or developed in parallel with low-k polymer spin-on-dielectrics (SOD) by major equipment companies and chip makers since the SiOCH has material properties similar to SiO$_2$ and lower cost of ownership than low-k SOD.[1]

In this paper, the results from the comparative characterization study of various SiOCH films were presented. The oxygen radical damage effect during photoresist stripping was also studied. The analysis results can be used for designing a reliable integration scheme for an ULSI multilevel interconnect.

EXPERIMENTAL

Four SiOCH materials were characterized by various analysis techniques such as FTIR, XPS, etc.. The materials were prepared by different deposition techniques and hydrocarbon source precursors. Samples A and B were deposited by reactive oxidation of methylsilanes around room temperature and annealed at 400 °C. Samples C and D were prepared by 400 °C PECVD of a siloxane-based precursor and tetramethylsilane, respectively.

Conference Proceedings ULSI XV © 2000 Materials Research Society

RESULTS AND DISCUSSION

As shown in Fig. 1, the FTIR spectra of the SiOCH films reveal chemical bond information of the hydrocarbon doped silicon oxide. All SiOCH samples have common peaks corresponding to Si-O, C-H, Si-CH$_3$, and Si-C peaks. The lower temperature-prepared SiOCHs (sample A and B) have additional peaks (at ~2200 cm^{-1}) corresponding to Si-H bonds. As shown in Table 1, the low-k CVD materials except for sample C have similar chemical composition in terms of Si atm. %, O atm. %, and C atm. %. The chemical composition of the CVD SiOCH films was similar to the one of the Methyl Silsesquioxanes (MSQ). [2,3] The MSQ has a peak corresponding to cage Si-O at 1128 cm^{-1}.

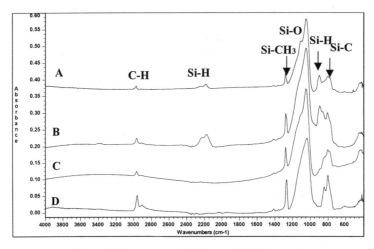

Fig.1. FTIR spectra of different SiOCH samples

Table.1. Chemical composition of different SiOCH samples

Sample	O%	Si%	C%	N%	O/Si	C/Si
A	53	35.9	11.1		1.48	0.31
B	44.8	34.7	14.8	5.7	1.29	0.43
C	50	33	17		1.52	0.52
D	30.9	33.1	30	6	0.93	0.91
MSQ	50	31	19		1.62	0.62

The summary of peak assignments was listed in Table.2. The enlarged FTIR spectra of C-H, Si-CH$_3$, Si-O/ Si-C, and Si-H were shown in Fig. 2(a), (b), (c), and (d), respectively. As carbon content increases, C-H stretching peak was shifted toward lower wavenumber gradually. Si-CH$_3$ symmetric deformation peak was shifted to lower wavenumber when excessive carbon was doped. The Si-O stretching peak of SiOCH film was at lower wavenumber than that (at 1070 cm-1) of undoped silica glass (USG). The peak shift can be explained by the lower electronegativity of carbon than oxygen. Reversely, the Si-O stretching peak of fluorinated silica glass was at higher wavenumber due to high electronegativity of fluorine.[4] The Si-H was shifted to low wavenumber as the silicon content increases.

Table.2. Summary of FTIR peak assignments

Sample	CH(a)	CH(s)	Si-H	Si-CH3	Si-O	Si-H (b)	Si-C
A	2972	2916	2182 (2240)	1275	1042	895.00	801.00
B	2969	2910	2175 (2229)	1273	1044	891.00	804.00
C	2967	2913		1274	1044		804.00
D	2963	2907		1264	1037		804.00

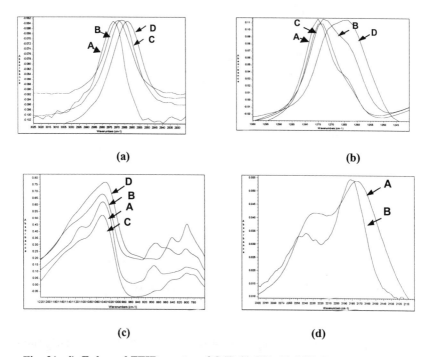

(a)   (b)

(c)   (d)

Figs.2(a-d). Enlarged FTIR spectra of C-H, Si-CH$_3$, Si-O/Si-C, and Si-H peaks, respectively.

To investigate oxygen plasma damage effect, the SiOCH films were exposed under oxygen radicals at a conventional downstream plasma asher for 100 sec and then analyzed by FTIR. As shown in FTIR spectra (Fig.3), C-H and Si-CH₃ peaks disappeared almost completely due to oxygen radical attack and the Si-O peak was shifted to higher wave-number position corresponding to PECVD SiOx. As listed in Table 4, the void % and the thickness of SiOCH films decreased significantly and the refractive indices increased also. These results indicate that the via bowing and poisoning are expected with conventional photoresist stripping scheme. The solution for oxygen sidewall damage to MSQ was the directional ashing with N2/O2.[2] The similar directional ashing can be appropriate for SiOCH film.

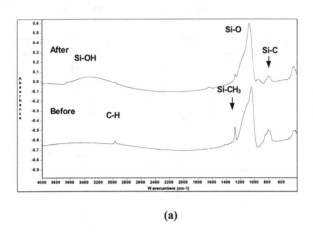

(a)

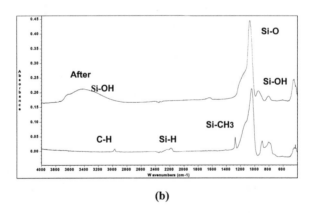

(b)

Fig. 3. FTIR spectra of SiOCH sample A (Fig.3a)and C (Fig.3b) before and after ashing,

Table 3. Properties of SiOCH sample A and C before and after ashing

Sample		SiO2 %	SiC %	Thickness (A)	% Change	R.I.	% Change
A	Before	53.26	10.24	1091.14		1.39	
	After	77.89	4.78	937.10	-14.12	1.44	3.6
C	Before	66.01	7.81	1512.40		1.41	
	After	83.21	3.44	1184.70	-21.67	1.43	1.4

CONCLUSION

In conclusion, even if the various low-k CVD materials were prepared by the plasma or chemical reaction of different hydrocarbon sources, the materials has CHx-doped nanoporous silica glass structure like MSQ. The main integration issue is the oxygen radical attack during ashing and will be resolved by modified etching and directional ashing technique.

REFERENCES

[1] E. Korczynski, Solid State Tech., p43 (May, 1999)

[2] J. C. Sum et al., IITC'99 Tech. Digest, p184

[3] M. Matsuura et al., IEDM'97 Digest, p785

[4] K. K. Singh, C. Ryu, and S. Hong, DUMIC'98 Digest, p261

Evaluation of Nanoglass in a Single Damascene Structure

C. B. Case*, M. Buonanno*, G. Forsythe, H. Maynard*, J. Miner*, W. W. Tai*, J. J. Yang**
* Bell Laboratories, Lucent Technologies, 600 Mountain Ave., Murray Hill, NJ 07974, c.b.case@lucent.com
** formerly of AlliedSignal, 1349 Moffett Park Drive, Sunnyvale, CA 94088

Abstract

Nanoglass, a porous silica dielectric, was evaluated in a Cu single damascene structure. Nanoglass film formation includes a solvent exchange reaction, in which the hydrophilic inner pore surfaces are made hydrophobic by the addition of organic groups. Preservation of hydrophobicity is a key aspect of Nanoglass integration since moisture penetration will raise the effective dielectric constant and may result in metal corrosion. The first part of this work describes experiments designed to identify appropriate Si_3N_4 cap deposition conditions, to examine outgassing (specifically moisture removal), and to look at the effects of a vacuum degas. The second part describes etch, barrier/seed deposition, metal fill and preliminary CMP experiments.

Introduction

Nanoglass [1] has been evaluated as a low dielectric constant inter-level dielectric for sub-0.18 μm CMOS processing in a copper single damascene structure. Nanoglass film formation begins with spin-coating a TEOS pre-cursor. Immediately after coating, the film undergoes gelation, or aging, in the presence of ammonia and water vapor. This is followed by a solvent exchange reaction in which the silanol groups lining the pore surfaces are capped with organic groups, rendering the film hydrophobic;* A primary concern with Nanoglass integration is retention of the carbon moieties during a process which includes multiple thermal cycles and exposure of the film to an etch plasma. Solvent exchange is followed by a hot plate cure (60 seconds each at 175 and 300 °C) and a furnace anneal (60 minutes, 400 °C). The first process step after film formation is capping. The cap provides a physical barrier to penetration of moisture, photoresist or other species into the film. It should be as thin as possible in order to have minimal effect on raising the dielectric constant of the stack and, finally, the deposition should not have an adverse effect on the film. After patterning and etch and prior to metal deposition, wafers are subjected to a vacuum degas to remove adsorbed moisture and volatile surface contaminants. At this step of the process, rapid removal of moisture and surface contamination from the film stack is desirable, in order to maximize throughput and minimize cost of ownership. In doing so, however, care must be taken to maintain film quality by avoiding densification, loss of carbon moieties, or decomposition. In this work, FTIR was used to monitor carbon loss during cap deposition, while outgassing was investigated by residual gas analysis (RGA) and a combination of thermal extraction (TE), gas chromatography (GC) and mass spectrometry (MS). FTIR, thickness and refractive index were used to investigate changes in a Nanoglass film after a vacuum anneal.

Experimental

Si_3N_4 was deposited on Nanoglass in an Applied Materials 5000. Film thickness was either 300 or 1000 Å. FTIR spectra were taken on a Nicolet Magna-IR Spectrometer 550. Water penetration tests were performed by cleaving 1 x 3 cm samples, placing them in petri dishes with water droplets around the edges and looking for naked-eye discernible color changes in the film. Cap layer staining was accomplished by placing water droplets on the cap surface and similarly observing color changes.

---

* Initially, hexamethyldisilazane [HMDZ: $(CH_3)_3SiNHSi(CH_3)_3)$] was employed in solvent exchange; this version of Nanoglass was replaced in late 1998 by a newer version which uses methyltriacetoxysilane [MTAS: $CH_3(CH_3COO)_3Si$]. The nomenclature for the HMDZ version is K2.2-A10 and for the MTAS version: K2.2-A10C. Since this report includes work using both versions, the solvent exchange chemical is indicated by the abbreviations HMDZ or MTAS. The number following the "K" indicates the nominal dielectric constant, either 2.2 or 2.5.

Conference Proceedings ULSI XV © 2000 Materials Research Society

Outgassing was studied using an Inficon Transpector open ion-source residual gas analyzer equipped with an electron multiplier and a modified Hewlett Packard gas chromatograph/mass spectrometer. Films were etched in an Applied Materials eMxP+ etcher using $O_2/CF_4/Ar$. Thickness and refractive index were measured on a Nanometrics 210XP. Ta/Cu barrier/seed layers were deposited in a Novellus M2i sputter tool. Copper was electrochemically deposited on a Semitool Equinox platform with Enthone CuBath M. An IPEC/Planar 472 with a Rodel IC1000/SUBA IV stacked pad was employed for chemical mechanical polishing. For Cu removal the downforce was 4 psi, back pressure was 1.2 psi, platen and carrier speeds were 45 and 65 rpm, respectively. Cabot EC4110 slurry (now discontinued) was used at a flow rate of 50 ml/minute. An On-Trak Synergy was used for post polish cleaning.

Results

Cap Deposition

In previous work, CVD Si_3N_4 was selected instead of TEOS oxide as a cap material, based on a simple test for hydrophobicity. In this work, the effect of CVD Si_3N_4 cap deposition on carbon retention was compared for Si_3N_4 caps deposited at two temperatures and two thicknesses. Carbon retention was quantified by measuring the area under the C-H (2970 cm⁻¹) peaks in the FTIR spectra. The C-H peak area decreased with both deposition time and temperature.

A background FTIR spectrum was run after Nanoglass K2.5-A10C (MTAS) deposition but before cap deposition. The three cap deposition conditions were 300 °C/300 Å, 400 °C/300 Å and 400 °C/1000 Å. For a 300 Å cap, as the temperature increased from 300 to 400 °C, peak area retention decreased from 98 to 89 %. Comparing 400 °C deposition temperatures, as thickness increased from 300 to 1000 Å, peak area retention decreased from 89 to 76 %. (See Table I) A water penetration test was performed on each sample. In this simple visual test, a cleaved sample is exposed to moisture. Observation of a color change indicates diffusion of water into the film. The three films exhibited the same behavior with respect to water penetration, with a 1 mm penetration observed for each. When a water droplet was placed on top of the film, the thin, low temperature cap resulted in the greatest degree of staining while the thick, high temperature cap resulted in no observable color change. The poor performance of the 300 °C cap is not unexpected, due to its' lower density. Based on these results, an effective Si_3N_4 moisture barrier should be deposited at 400 °C, although this will result in a greater degree of carbon loss than a lower temperature cap. A cap thickness of 300 Å will result in less carbon

Table I
C-H Peak Area Retention after Cap Deposition

Temperature (°C)	Thickness (Å)	C-H Peak Area Retention (%)	Staining of Cap Layer
300	300	98	heavy, bad
400	300	89	slight
400	1000	76	none

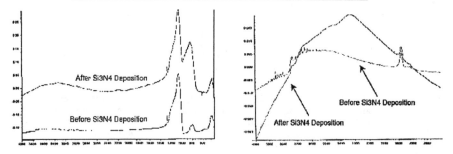

Figure 1: FTIR spectrum of the C-H peak area before and after Si_3N_4 cap deposition.

loss than a thicker cap deposited at the same temperature. Although Si_3N_4 may be compatible with Nanoglass, it will raise the effective dielectric constant of the stack. A CVD C-containing SiO_2 film would be an appropriate choice for evaluation of compatibility with Nanoglass; several films of this type are currently under development. [2]

Outgassing

Outgassing was monitored using the vacuum degas chamber of a Novellus M2i sputter tool equipped with a residual gas analyzer (RGA). This *in situ* observation of outgassing is a convenient, low-cost method for determining the time and temperature requirements for reducing levels of water and other contaminants to acceptable limits. The wafer is in an environment identical to that during degas prior to metal deposition, with one caveat: during residual gas analysis, the backside heater gas is turned off, so the delta between the heater block setpoint and the true wafer temperature is larger than it would be during an actual process. At a 300 °C setpoint, with backside gas, the temperature delta, is in the 30 °C range; without backside gas, the delta may be as high as 50 °C.

A Nanoglass K2.2-A10 (HMDZ) wafer with a 500 Å Si_3N_4 cap and etched vias was introduced to the chamber at a 300 °C setpoint; spectra are shown in Figure 2. After loading, the water peak rises to a partial pressure of 1E-3 Torr (N_2 equivalent); the level of outgassing for species between 50 and 100 amu is 2 orders of magnitude lower. Approximately 5 minutes was required for the water peak return to the mid 1E-4 range, which corresponds to the value observed for a TEOS wafer at the same setpoint. The dominant peak between 50 and 100 amu was at 75 amu, corresponding to dimethylhydroxysylil [$(CH_3)_2OHSi^+$], originating from the HMDZ used in the solvent exchange reaction.

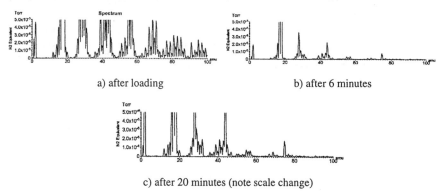

a) after loading   b) after 6 minutes

c) after 20 minutes (note scale change)

Figure 2: RGA spectra of Nanoglass K2.2-A10 at 300 °C setpoint. The y-axis is in units of Torr, normalized to N_2.

An unpatterned Nanoglass K2.2-A10C (MTAS) wafer with a Si_3N_4 cap was then introduced to the chamber (still at a 300 °C setpoint); after loading, the water peak again rose to a partial pressure of 1E-3 Torr (N_2 equivalent); the level of outgassing is two to three times lower than for the etched via wafer and there are some shifts in relative peak intensities. After four minutes, the water level was at 2.7E-4 Torr. Two factors contribute to the lower levels of outgassing in the unpatterned MTAS film: lack of exposed Nanoglass in the sidewalls and the MTAS *vs.* HMDZ solvent exchange chemical. Spectra are shown before loading, after loading, at 4 minutes and at 20 minutes (see Figure 3).

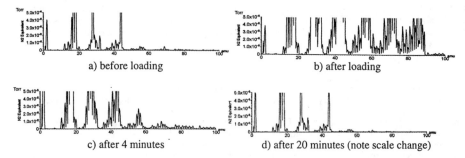

a) before loading

b) after loading

c) after 4 minutes

d) after 20 minutes (note scale change)

Figure 3: RGA spectra of Nanoglass K2.2-A10C at 300†°C setpoint

Outgassing of an MTAS Nanoglass film was also analyzed by a combination of thermal extraction, GC and MS. [3] Two modes: TE-GC-MS and direct mass TE/MS, were employed. In the TE-GC-MS mode, the sample is heated to a set temperature and after reaching temperature, held for a specified time. The system is purged with helium. The thermally extracted compounds are collected on a cryotrap (at -125 °C) and injected onto a GC column for separation. The separated components are then analyzed by MS. In direct mass mode, the sample is also heated to a set temperature and held for a specified time; the difference is the absence of compound separation; the output is abundance of all ions from 10 to 650 amu *vs.* time. Knowledge of the wafer temperature profile allows data conversion to abundance *vs.* temperature.

The analyzed sample was unpatterned and capped with Si_3N_4 (nominally 300 Å) deposited at 400 °C. The sample was heated to 400 °C and held for 30 minutes. Thirteen individual compounds separated on the GC were analyzed by MS. The GC chromatogram is shown in Figure 4. The outgassed species included solvents, hydrocarbons and a siloxane compound.

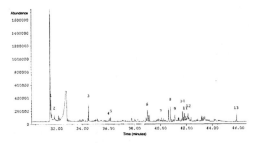

Figure 4: GC chromatogram of unpatterned MTAS Nanoglass with a Si_3N_4 cap; the unnumbered peak near 33 minutes is due to condensation in the chamber

In the direct mass total ion analysis, the wafer was isothermally extracted at 400 °C for 10 minutes; the ramp rate was 25 °C/minute. The direct mass total ion chromatogram and temperature profile are shown in Figure 5. Onset of degassing is ~ 85 °C; there is significant outgassing in the 150 - 175 °C range, which then tapers off but continues up to 400 °C, tapering off during the 400 °C soak. This is consistent with a previous RGA analysis of HMDZ Nanoglass in which an order magnitude increase in partial pressure of outgassed species was observed by a setpoint of 180 °C, as the sample was heated from room temperature. Mass spectral analysis was performed at three points (a - c below) corresponding to temperatures of ~150, 225 and 375 °C. This gives a measure of all ions generated at a specific temperature. When used in combination with the GC analysis, chemical structures can be assigned. There is no indication of elimination of water (18 amu) in the spectrum corresponding to

225 °C. Only in the final spectrum, corresponding to 375 °C, is there a peak at 18 amu (H_2O). This is also consistent with previous RGA data in which the partial pressure of water rose by an order of magnitude between the 300 and 400 °C setpoints. Although not shown, water outgassing reached a maximum in the range of 385 °C. Based on these results, temperatures close to 400 °C are required for rapid elimination of water from Nanoglass. Note, minimizing the interval between process steps will keep both moisture uptake and surface contamination to a minimum.

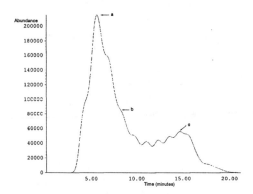

Figure 5: Total Ion Chromatogram for MTAS Nanoglass with a Si_3N_4 cap

Table II lists the mass/charge ratio for the major peaks at each of the three temperatures. One series, beginning with 29 amu and separated by 14 mass units corresponds to saturated aliphatic hydrocarbons. A second series, beginning with 27 amu and also separated by 14 mass units corresponds to aliphatic alkene compounds. The origin of these hydrocarbons is probably a combination of surface contamination and adsorption on the inner pore surfaces. (The surface area for these films is hundreds of square meters per gram.) The peaks at 67, 91 and 149 amu are not included in the saturated aliphatic or the aliphatic alkene series; 67 and 91 are components of five of the thirteen peaks analyzed on the GC chromatogram and 149 is the ion of greatest abundance in peak 8. This analysis shows that of all the outgassed species, the alkane and alkene hydrocarbons are the major contributors. The peak at 15 amu is methyl; the low relative abundance at 150 °C is likely due to incomplete removal of the solvent exchange chemical or environmental contamination of the surface. The RGA spectrum of a TEOS wafer exhibited a peak at 15 amu as well. At 375 °C, methyl bound to silicon at the pore surfaces is likely evolving; this is consistent with the FTIR data of capped wafers: at 300 °C, only a 2 % change in C-H peak area was measured, increasing to 11 % for the same deposition time at 400 °C.

Table II
Mass/charge Ratio for Major Peaks at 3 Temperatures

	15	18	27	29	41	43	55	57	67	69	71	83	85	91	97	99	105	111	113	125	127	141	149
150 °C	**15**		27	29	41	43	55	57		69	71	83	85		97	99		111	113	125	127	141	**149**
225 °C			27	29	41	43	55	57		69	71	83	85	**91**	97	99		111	113	125	127	141	**149**
375 °C	**15**	**18**	27	29	41	43	55	57	**67**	69	71	83	85	**91**		105	111						**149**

Boldface indicates peaks not associated with the alkane or alkene series.

Degas

FTIR, refractive index and film thickness were used to measure the effect of a 300 °C, 300 second degas on unpatterned, uncapped Nanoglass K2.5-A10C. Minimal changes in thickness and refractive index were observed (0.015 and 0.3 %, respectively). No change in organic content was observed by FTIR (as evidenced by the C-H peak at 2970 cm^{-1}). This is consistent with the cap deposition experiment, in which only 2 % carbon loss was measured at 300 °C, which is within the error of measurement. The major change was in the water content of the film. The post degas FTIR spectra exhibits an increase of ~ 50 % in the broad water peak at 3600 cm⁻¹ and a sharpening of the OH peak at 970 cm⁻¹. Since the film was uncapped, it was not surprising to find adsorbed moisture, particularly since the interval between the background FTIR and degas exceeded one month. To reiterate the GC/MS results described above, temperatures in the 400 °C regime are recommended for effective removal of moisture from Nanoglass films.

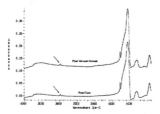

Figure 6: FTIR spectra of Nanoglass K2.5-A10C, before and after a 300 °C, 300 second vacuum degas

Etch and Metal Deposition

Nanoglass trenches are shown in Figure 7. The sidewall profile is vertical and there is no undercut of the Si_3N_4 cap. Eleven wet strippers were evaluated to determine compatibility with Nanoglass K2.5-A10C. FTIR spectra, thickness and refractive index of unpatterned films were measured before and after exposure to 11 solutions; three of these were judged to be compatible with Nanoglass. However, further testing on patterned, etched samples with remaining resist, revealed a shell of resist, hollow down the center, that was not removed. Apparently, exposure to the etch plasma altered the resist properties to such an extent that more aggressive chemistries are required. Alternatively, the combination of a brief oxygen plasma ash with a wet strip may be a viable solution.

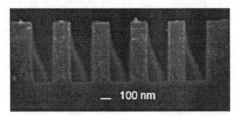

100 nm

Figure 7: 0.18 μm Nanoglass trenches with a 3:1 aspect ratio.

Ta/Cu liner/barrier layers were deposited in a Novellus M2i sputter deposition tool at thicknesses of 500 and 1000 Å, respectively. Ta was deposited at a setpoint of 200 °C and the Cu at sub-ambient. Cu electrodeposition was accomplished on a Semitool Equinox platform with Enthone CuBath M. Following Ta/Cu barrier seed layer deposition, the wafers were visually defect-free. After Cu fill, the only defect observed was slight peeling on some wafers at the Cu/Ta interface, at the flat. This can occur on standard oxide wafers as well, due to handling with tweezers between the plating and rinsing steps.

The wafer shown in Figure 8 was polished, using the standard Cu recipe, and cross-sectioned. Nanoglass remains intact in regions which are high in areal metal density. In an adjacent region of the wafer, in which there is a wide field of dielectric, there are large areas with only a partial thickness of Nanoglass. This may be due to cohesive failure of the Nanoglass or to adhesion failure at the Si_3N_4/Nanoglass interface. Alternatives to conventional CMP, involving wet chemical removal of copper, are ideally suited to porous dielectric structures whose mechanical strength is inherently lower than their full density counterparts. [4] A technique, known as spin-etch planarization, which avoids physical contact of the film with anything other than the process chemicals is in development. In this process, an etch solution is dispensed while the wafer spins. The etchant removes Cu, with high selectivity to Ta and TaN. The barrier, of course, must be removed as well, via a different etchant or by conventional means. Although this work is in its' early stages, it has the advantage of being developed on production-proven hardware.

 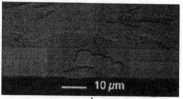

a   b

Figure 8: a) Nanoglass K2.2A10-C after Cu CMP; the Ta liner remains. Trench width is 0.4 μm.
b) An adjacent region in which metal has peeled.

CONCLUSIONS

FTIR was used to evaluate Si_3N_4 cap deposition. A 300 Å cap deposited at 400 °C, was identified as an effective moisture barrier which can be deposited with minimal damage to the organic groups lining the inner surfaces of the Nanoglass pores. Outgassing, and specifically moisture removal, was examined by residual gas analysis in the degas chamber of the sputter deposition tool, as well as by a combination of thermal extraction, gas chromatography and mass spectroscopy. Nanoglass films should be degassed at temperatures near 400 °C to effect complete water removal prior to metal deposition. Narrow (0.18 μm) trenches with a 3:1 aspect ratio and straight sidewalls were etched. Investigation of post etch resist strippers revealed that chemistries which are compatible with Nanoglass and which removed baked resist, did not completely remove resist which had been exposed to the etch plasma. Further work is required to identify a suitable wet resist strip; the solution may lie in a combination of a brief plasma ash plus wet strip. Ta/Cu barrier/seed layers were deposited under conditions standard for an oxide dielectric and filled with electrochemically deposited copper. After CMP of the Cu, areas with a high areal density of metal remained intact. In adjacent areas without metal, there were regions of Nanoglass with thickness loss. CMP alternatives such as spin-etch planarization, involving wet chemical removal of metal films, are well suited for porous films of low mechanical strength.

ACKNOWLEDGMENTS

The authors gratefully acknowledge the work of the Murray Hill wafer fabrication team including Kevin Bolan, Eric Bower, Ray Cirelli, Connie Jankowski, Bill Mansfield, Mark Morris and Al Timko. Thanks to Alka Gupta, both for running the TE/GC/MS experiments and for discussing the results. Thanks to Christopher Case and Jim Drage for careful reading of the manuscript.

REFERENCES

1. Nanoglass is a trademark of AlliedSignal Inc., Sunnyvale, CA. Work on earlier versions of this film can be found in: C. Jin, S. List, W. W. Lee, C. Lee, J. D. Luttmer, R. Havemann, D. Smith, T. Ramos and A. Maskara, Proceedings of *Advanced Metallization and Interconnect Systems for ULSI Applications in 1996*, MRS, p. 463, (1997); R. S. List, C. Jin, S. W. Russell, S. Yamanaka, L. Olsen, L. Le, L. M. Ting and R. H. Havemann, *VLSI Technical Dig*, p. 77 (1977); T. Ramos, K. Roderick, A. Maskara and D. Smith, , Proceedings of *Advanced Metallization and Interconnect Systems for ULSI Applications in 1996*, MRS, p. 455 (1997).
2. Companies developing C-containing low κ films include Trikon Technologies Ltd, Newport, Gwent, U. K., Dow Corning, Midland, MI, Novellus Systems, San Jose, CA and Applied Materials, Santa Clara, CA.
3. Analysis performed at Ramco Technologies, Mountainview, CA, RamcoTGM@aol.com.
4. J. Levert, S. Mukherjee, D. DeBear and M. Fury, presented at the ECS meeting, October, 1999.

SiLK H for Damascene and Gap Fill Application

G.Passemard[2], JC.Maisonobe[1],C.Maddalon[2], A.Achen[3],
M.Assous[1], C.Lacour[2], N.Lardon[2], R.Blanc[2] O.Demolliens[1]

[1] CEA Grenoble LETI GRESSI 17, Rue des Martyrs 38054 Grenoble Cedex 9 - France
[2] ST Microelectronics Central R&D /GRESSI CEA-LETI 17 Rue des Martyrs 38054 Grenoble Cedex 9 - France
[3] Dow Chemical Deutschland Inc. / Industriestrasse 1 / D-77836 Rheinmünster Germany

Abstract

This presentation reports results from work performed on the properties and processing of
SiLK* H Semiconductor Dielectric. SiLK H was evaluated in conventional gap filling and in
Damascene architectures. First gap filling capability was carefully evaluated, then the chemical-
physical properties were compared to the previously studied SiLK I. It was demonstrated that
SiLK H could combine the processing advantages of SiLK G and SiLK I, which are respectively
dedicated to gap fill and Damascene architecture.

Introduction

The reduction of the integrated circuit geometry below sub-quarter micron dimensions, limits
the use of conventional materials like silicon oxide and aluminum in the interconnection stack.
In fact, improvements of device performance in terms of speed and power consumption require
the use of new materials to improve the metal line resistance and to reduce the parasitic
capacitance between lines. Additionally, the ultra large-scale integrated circuits require more
levels of metal interconnect. Due to metal etching difficulties, the choice of integration
architecture will most likely be the Damascene structure with copper metallization. The
dielectric material used in this integration, should ideally offer following properties: Dielectric
permittivity below three and good electrical properties, low moisture absorption, high thermal
and chemical stability, good adhesion to metals, oxides and nitrides, toughness or crack
resistance, easiness to apply, cure and etch on conventional equipment, compatibility to hard
masks, metals and other classical processing chemicals. Another important requirement for the
industrial application of these new dielectrics, is their commercial availability and maturity
level.
Only few materials optimally fulfill most of these requirements. We investigated SiLK*
Semiconductor Dielectric from The Dow Chemical Company as one of the most promising low
K materials. Over the last few years, two product versions, SiLK G and SiLK I were available.
These two formulations of the same polymer backbone had been optimized respectively for
conventional gap fill and the emerging Damascene application. In this work, we have evaluated
a new SiLK formulation, called SiLK H, compatible with a wider equipment base and usable for
both the gap fill and the Damascene architectures.
We have first verified that SiLK H can be deposited on conventional coating tools in normal
clean room temperature and humidity conditions. Then, we have verified if and under which
conditions, the SiLK dielectrics can fill small gaps below 0.4 μm. We also determined the main
chemical-physical properties of cured SiLK H films and compared them to those of SiLK I[1,2]:
Thermal, mechanical and chemical stability, electrical performance. Some details on etching are
presented to finish confirming the transparency of the move from SiLK I to SiLK H.

Conference Proceedings ULSI XV © 2000 Materials Research Society

Experimental

The SiLK I, G and H dielectrics were supplied by Dow Chemical. These materials have the same aromatic hydrocarbon polymer base, but are in solution in different solvents optimized for spinning conditions and film quality. Mesitylene and g-Butyrolactone (GBL) are the carrier solvents for SiLK H, whereas Cyclohexanone and GBL are the solvents for SiLK I. SiLK G which is in solution in NMP, was also evaluated to compare its gap filling performance. However this last material requires specific ambient conditions (25% RH , 27°C) to avoid moisture absorption into the spun film SiLK G film during spinning and between the coating and the hotplate baking steps.
The three materials were deposited on 200mm wafers using a TEL Clean Track Mark 8 coating system. Blank silicon wafers and patterned wafers were used, respectively to evaluate the physical-chemical properties of these materials and the gap filling capability.

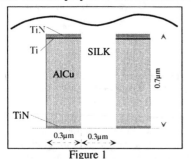

The patterned wafers had the following structure (figure 1): 500 nm of PECVD USG (TEOS precursor) /TiN(40nm) /AlCu(700nm) /Ti(10nm) /TiN(60nm) stack deposited on silicon wafers; then metal etch step to define lines. The most aggressive aspect ratio is 2.3 for a space of 0.3 µm and 0.7 µm of total metal height.

Figure 1

The SiLK H and I deposition was performed with and without the conventional adhesion promoter recommended by the supplier: 3-Aminopropyl triethoxy silane (3-APS) in glycol (AP8000).
After spinning, the SiLK films were baked on hot plates, then cured in a vertical furnace for 10 min at different temperatures. The cure temperatures were chosen between 350°C and 490°C, the atmosphere being nitrogen with less than 2 ppm of residual O_2. For the evaluations, SiLK films with a nominal thickness of 0.6 µm were produced. The thermal stability of the SiLK films was evaluated by Dynamic and Static Thermo Gravimetry Analysis (TGA) under Helium. The static TGA was performed for 1 hour at 450 °C. The chemical evolution of the material was followed by Fourier Transform Infrared analysis. The stress evolution during the thermal cycles was monitored on a Flexus Stress System. The dielectric constant or permittivity was determined by C-V measurement using a SSM Mercury Probe system. The adhesion test was performed using an adhesive tape test. 300nm PECVD hard masks of Si_xN_y and SiO_2 using TEOS and Silane precursors, were deposited directly on SiLK to verify the chemical compatibility, the adhesion and the electrical evolution, before and after a thermal annealing step of 1 hour at 450° C.

Results and Discussion

Gap filling capability :

SILK-I : First, the gap filling capability of SiLK I was evaluated to determine the minimum space and aspect ratio this material can fill. Three different deposition conditions were tested: with and without adhesion promoter, (figure2 and figure 3), then with 50 nm of TEOS oxide liner (figure 4).

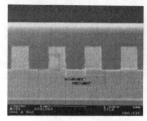  

a) 0.55μm space   b) 0.45μm space   c) 0.36μm space

Figure 2 :SiLK I with Adhesion promoter

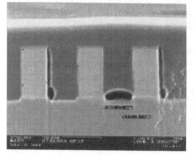
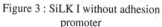
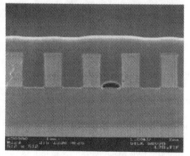

Figure 3 : SiLK I without adhesion promoter   Figure 4 : SiLK I with 50nm of TEOS SiO2 liner

The study of numerous FIB cross sections showed that SiLK I hardly fill spaces below 0.55 μm for the line geometry used in the experiment. Working on the spin parameters helped to improve the SiLK I gap fill capability slightly. Applying a TEOS layer changed the nature of theSiLK – metal interface, also resulting in marginal improvements. The use of adhesion promoter did not reduce the voids left when spinning SiLK I on the structures, but prevented sidewall delamination (figure 3). In conclusion, SiLK I should be dedicated to classic Damascene applications only.

SiLK G : Figure 5 shows the filling capability of SiLK G. Having been designed for this purpose, SiLK G can very well fill narrow spaces (< 0.3μm)

*SiLK H :*It was demonstrated on different spin coaters, that deposition of quality SiLK H films can be performed on conventional coating systems in standard clean room ambient atmospheric conditions. Figures 6 and 7 show the filling capability of SiLK H in conjunction with the AP8000 adhesion promoter. As shown in figure 7 and checked on multiple other FIB cross-sections, SiLK H fills 0.3μm gaps.

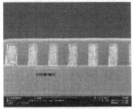 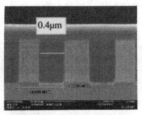 

Figure 5 : SiLK G   Figure 6 : SiLK H   Figure 7 : SiLK H

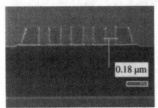

Using a higher concentration SiLK H formulation, a 1.5 μm thick layer was deposited in a single spin over 0.18 μm line/space structures. This shows good gap filling performance useful for classical Al/W BEOL processes and other innovative Damascene architectures. (figure 8)

Figure 8 : gap filling capability of SiLK H

Thermal Stability :

The thermal stability of SiLK H was compared to that of SiLK I. The TGA measurement shows that the two SiLK versions lose less than 1% mass during isothermal annealing for 3 hours at 450°C under helium atmosphere (figure 9)

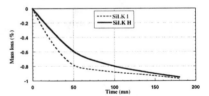

Figure 9 : Static TGA 3hours at 450°C for SiLK H and SiLK I

The thermal stability of SiLK was supported with FTIR measurements, which detect no chemical transformation in SiLK I or SiLK H (figure 10) after one hour anneals at 450 °C. Furthermore, both cured SiLK I and SiLK H films are mechanically stable and present no stress hysteresis after thermal cycling (figure 11). These results demonstrate the good thermal stability of SiLK dielectrics up to 450°C.

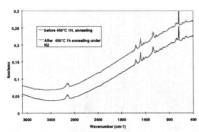

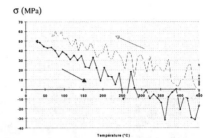

Figure 10 : SiLK H : FTIR evolution
before and after annealing 1H 450°C

Figure 11: SiLK H : Stress hysteresis

Hard mask compatibility :

The values of the various permittivities measured under 300 nm capping layer (after subtracting the effect of the hardmask on permittivity). It was verified that SiO_2 (SiH_4 or TEOS chemistry) and Si_xN_y hard masks depositions do not deteriorate the SiLK I or SiLK H permittivity. The values are close to 2.7.

Furthermore, the adhesion of each hard masks on SiLK H and I films, tested using standard adhesive tape, is excellent (table 1). This test was also performed after annealing (one hour at 450°C) of the encapsulated SiLK layers. The hard mask adhesion to the SiLK films was also tested by performing hardmask CMP with Copper slurries. No delaminations could be observed.

Adhesion	Before 450°C 1h annealing	After 450°C 1h annealing
Silane PECVD SiO₂ deposited on SILK	GOOD No delamination	GOOD No delamination
TEOS PECVD SiO₂ deposited on SILK	GOOD No delamination	GOOD No delamination
Silane PECVD Si₃N₄ deposited on SILK	GOOD No delamination	GOOD No delamination

Table 1

Dielectric Stability :

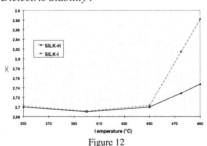

Figure 12

After spinning, SiLK I and SiLK H films were annealed for 10 minutes at temperatures between 360°C and 490°C. Figure 12 shows the permittivity evolution versus the curing temperature. The two SiLK versions show a stable permittivity up to a curing temperature of 450°C. At higher temperatures, the permittivity increases slightly, probably due to thermal decomposition of parts of the organic backbone. The leakage current measured at 0.2MV/cm was 10^{-13} A for each of the two dielectrics..

Etching Behavior :

The comparison of etching process was performed in a LAM 9100 reactor. Etch recipes previously developed for SiLK I were applied to SiLK H. Thin 250 and 300 nm via holes were etched in SiLK I or SiLK H (figure 13). No significant difference was observed between the two materials.

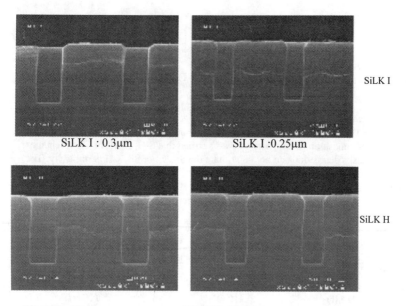

SiLK I : 0.3μm   SiLK I :0.25μm   SiLK I

SiLK H : 0.3 μm   SiLK H : 0.25 μm   SiLK H

Figure 14 : Etching Comparison

Conclusion

SiLK H films demonstrate the same chemical physical properties as SiLK I films: thermal and mechanical stability, dielectric constant, leakage current. Moreover, these two materials have the same process capability in term of etching, hard mask compatibility and adhesion. However, in case of SiLK I, due to poor filling performance of geometries below 0.55 μm, the product is limited to classic Damascene integration. SiLK H Semiconductor Dielectric combines the advantages of SiLK G and SiLK I. SiLK H allows the use of one product for Damascene and gapfill applications without the need of restrictive atmospheric conditions which makes SiLK H accessible to wide installed equipment base.

Acknowledgements

The authors wish to thank the GRESSI and ST/CC people involved in the wafer characterisation .

This work has been carried out within the GRESSI consortium CEA-LETI and France Telecom CNET and by the Centre Commun STMicroelectronics / France Telecom CNET, and has been partially funded within the European Damascene project
• SiLK Semiconductor Dielectric is a trademark of The Dow Chemical Company.

Références :
(1) P.H.Townsend, S.J.Martin, J.Godschalx, D.R.Romer, D.W.Smith, Jr.,D. Castillo, R.DeVries, G.Buske, N.Rondan, S.Froelicher, J.Marshall, E.O.Shaffer, and J-H. Im   Mat.Res.Soc.Symp. Proc. 1997 Vol.476 P.9
(2) J.C.Maisonobe, G.Passemard, C.Lacour, Ch.Lecornec, P.Motte, P.Noël, J.Torres DUMIC Proceeding 1999 p.60

A NOVEL INTEGRATION APPROACH TO ORGANIC
LOW-K DUAL DAMASCENE PROCESSING

K. MIYATA, M. FUKASAWA, T. TATSUMI, M. TAGUCHI,
T. HASEGAWA, K. IKEDA, and S. KADOMURA
LSI Business & Technology Development Group, Core Technology & Network Company,
Sony Corporation, 4-14-1 Asahi-cho, Atsugi-shi, Kanagawa 243-0014 Japan
E-mail: kmiyata@ulsi.sony.co.jp

ABSTRACT

We present a new dual-damascene integration scheme called the dual-hardmask (DHM) approach. This scheme transfers trench/via patterns into low-k organic polymers. We compared it with the via-first-at-via-level process, in which vias are defined at the embedded hardmask before the trench-level coat is applied. The via-first-at-via-level process resulted in insufficient critical-dimension change and left deposition in the vias, which made it difficult for them to be filled. The DHM process, on the other hand, resulted in a fine pattern transfer. Fine dual-level interconnects were obtained by the DHM process and conventional metallization. Furthermore, measurement of the Kelvin via resistance as a function of the metal/via off-axis distance demonstrated that the DHM approach is a good technique for solving the border-less misalignment problem.

INTRODUCTION

Due to the recent demand for high speed and low power consumption IC chips, there is a growing trend for dual-damascene (D.D.) back-end-of-line processing to use low-k dielectric materials. Inorganic, pure organic or mixed organic-inorganic low-k dielectric materials have been proposed. Dielectric materials for D.D. processing are required for high processing capabilities. We chose pure organic polymers because they have good adhesion features and are well etched with an ammonia plasma [1]. We believe that they have great potential for use in back-end-of-line processing.

When we fabricate D.D. interconnects using an inorganic dielectric, we simply use two sets of lithography, etch and resist stripping for vias and trenches. However, organic polymers are easily degraded by ordinary resist stripping techniques, which restrict the degree of integration. A hybrid structure, in which a polymer is used for the IMD and an oxide is used for ILD, allows the polymer to be introduced at an early stage. However, the hybrid structure is less attractive than the full-polymer structure, in which both IMD and ILD are polymers, because it does not adequately reduce interconnect capacitance. Although, several strategies for the full-polymer integration have been invented to overcome this integration problem [2,3], it is still difficult to decide what is the best strategy. In this paper, we demonstrate a new approach for full-polymer integration.

EXPERIMENTAL

We used FLARE™ as organic dielectric polymer. FLARE™ has good adhesion features and a thermal stability of up to 400 °C. The 340-nm-thick FLARE™ films were spin-coated and cured onto wafers for both the via and trench levels. Hardmask deposition was conducted using plasma CVD. Electron-beam lithography was used to obtain photomasks. The hardmask etching was performed in a magnetron etching tool with $C_4F_8/CO/Ar$ gas mixture for oxide and $CHF_3/O_2/Ar$ for nitride. The polymer etching was performed in a high-density plasma etch tool with ammonia gas. When there are photomasks in the polymer etching, they are simultaneously removed at that time. Ammonia is useful for polymer etching because it efficiently provides amino radicals, which are etchants of organic polymers, to etching plasmas. Ammonia etching is an indispensable technology and it has been described elsewhere in detail [1].

RESULTS AND DISCUSSION

Figure 1 shows the via-first-at-via-level approach. In this case, the first polymer film and embedded hardmask of 100-nm-thick oxide were first fabricated on a barrier layer. Via lithography was carried out on the embedded hardmask and the vias were transferred at the embedded hardmask. The first polymer was also etched down to the barrier layer in order to remove the photomask. Subsequently, the spin-on of the second polymer and deposition of the top hardmask were carried out. Next, the trench photomask was fabricated on the hardmask. After the hardmask was etched, the second polymer was selectively etched to the hardmasks. As a result, wiring trenches and via holes were obtained in the full-polymer structure.

This approach resulted in a lot of sputtering at the corner of the embedded hardmask. We found a problem with this approach since large critical dimension (CD) gains caused the formation of crown-like structures in the vias (Fig. 2). To study how the crowns formed, micro Auger electron spectroscopy (μAES) was used. Figure 3 is the μAES spectrum of the crown, showing that Si, O, C, and N are the major elements of the crown. The crown formation is illustrated in Fig. 4. The sputtered hardmask was redeposited on the via sidewall at the time of via definition. The vias were temporarily refilled during the second spin-on. As the vias were etched down again and widened, the crowns appeared.

Figure 5 shows calculated intra-line conductance of a dual-damascene structure as a function of the thickness of the embedded hardmask, along with the calculation preconditions. It is evident that a thinner hardmask is better because it provides a lower capacitance. However, thin embedded hardmasks make the CD change higher and therefore are not suitable for the via-first-at-via-level process.

These disadvantages of the via-first-at-via-level approach forced us to rethink the process flow. Figure 6 shows a novel integration approach that we call the dual-hardmask (DHM) process. It uses another hardmask on the top hardmask. We chose nitride as the upper hardmask material. Trenches were defined beforehand at the upper hardmask. The trench photomask was stripped by an ordinary process since the polymers were covered by the

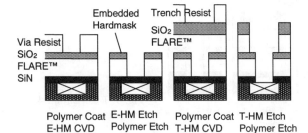

Via Resist
SiO₂
FLARE™
SiN

Embedded
Hardmask

Trench Resist
SiO₂
FLARE™

Polymer Coat   E-HM Etch   Polymer Coat   T-HM Etch
E-HM CVD   Polymer Etch   T-HM CVD   Polymer Etch
Via Litho.   Trench Litho.   Barrier Etch

Fig. 1. Via-first-at-via-level approach.

Fig. 2. Dual-damascene feature after processing by the via-first-at-via-level approach.

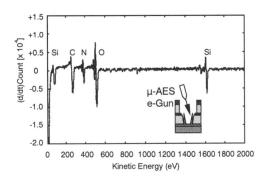

Fig. 3. AES spectrum of the crown in the via.

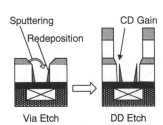

Via Etch   DD Etch

Fig. 4. Crown formation using the via-first-at-via-level approach.

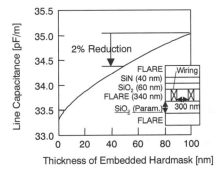

Fig. 5. Intra-line capacitance as a function of the thickness of the embedded hardmask.

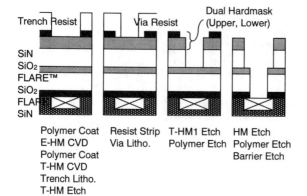

Trench Resist	Via Resist	Dual Hardmask (Upper, Lower)	

SiN
SiO₂
FLARE™
SiO₂
FLAR
SiN

Polymer Coat	Resist Strip	T-HM1 Etch	HM Etch
E-HM CVD	Via Litho.	Polymer Etch	Polymer Etch
Polymer Coat			Barrier Etch
T-HM CVD			
Trench Litho.			
T-HM Etch			

Fig. 6. Dual-hardmask approach.

hardmask. Next, the via photomask was fabricated, and the vias were defined at the hardmask and second polymer. This was followed by hardmask and polymer etching. The hardmask etching was done with an on-nitride selectivity of 5, so that the upper hardmask worked as the etching mask.

Figure 7 shows a cross-section SEM image of the D.D. feature that is formed by the DHM process. The DHM process does not exhibit an embedded hardmask corner until the final etching operation, and therefore results in clean vias that are crownless. The data of the CD change for the two approaches are shown in Fig. 8. The DHM approach has little CD change while the via-first-at-via-level approach has large CD gain. Because the DHM approach etched vias and trenches simultaneously, it decreased the amount of overetching on the embedded hardmask. As a result, thin embedded hardmasks (50-nm thick) could be applied without affecting the CD change. Compared to 100-nm-thick hardmasks that were used in the via-first-at-via-level trial, a 2% reduction in the intra-line capacitance has been easily achieved (Fig. 5). Figure 9 shows copper dual-level interconnects with FLARE™ fabricated by the DHM process. Fine interconnects were obtained after conventional metallization that consisted of TaN sputtering, copper electroplating, and chemical mechanical polishing.

The advantages of the DHM approach can also solve border-less misalignment problems. In the case of misalignment where vias are not within trenches, the via-first-at-via-level approach can not maintain the via shape, which directly affects the subsequent metallization processes and therefore should be avoided. Figure 10 shows the cross-section SEM image of the off-axis vias and trenches. For 2M/2C misalignment, the DHM approach can expand the 2M trenches and resulted in perfectly shaped vias. For 2C/1M misalignment, via etching completely terminated on the top of 1M without any loss of IMD at border-less vias because the polymer etching has fine selectivity to nitride. Figures 11(a) and (b) show Kelvin via resistance as a function of the off-axis distance for 2C/1M and 2M/2C. Here both the via diameter and wiring width are 0.30 μm. The via resistance remained stabilized for more than 0.25 μm for the 2M/2C off-axis distance while it became destabilized at more than 0.1 μm for

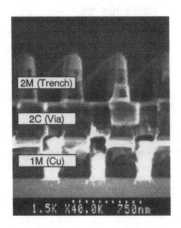

Fig. 7. Dual-damascene feature after processing by the dual-hardmask approach.

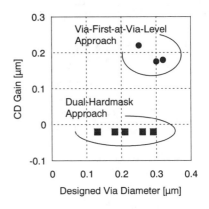

Fig. 8. CD gain for the two approaches.

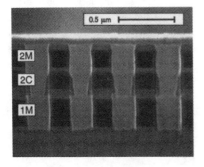

Fig. 9 Dual-level copper interconnects fabricated by the dual-hardmask process.

Fig. 10. Off-axis via holes which were formed by the dual-hardmask process.

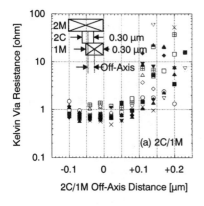 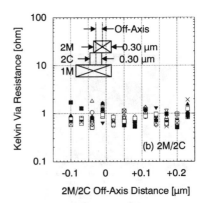

Fig. 11. Kelvin via resistance as a function of the off-axis distance. Figures (a) and (b) show data for 2C/1M and 2M/2C off-axis, respectively.

the 2C/1M off-axis distance. These data for both 2M/2C and 2C/1M showed good off-axis characteristics. This experiment proved that the DHM approach can solve the border-less misalignment problem, especially for 2M/2C misalignment.

CONCLUSION

Two full-polymer dual-damascene (D.D.) integration approaches have been examined by using FLARE™ as low-k dielectric. The via-first-at-via-level approach raised problems of large critical-dimension gain and hardmask redeposition in via holes. On the other hand, the dual-hardmask (DHM) approach resulted in fine D.D. features. The Kelvin via resistance measured as a function of via/metal off-axis distance proved that the DHM approach can solve the border-less misalignment problem, especially for 2M/2C misalignment. The DHM approach is one of the best techniques for fabrication D.D. features in organic low-k polymer structures.

REFERENCES

[1] M. Fukasawa, T. Hasegawa, S. Hirano, and S. Kadomura, 1998 Dry Process Symposium, pp. 175-182.
[2] D. T. Price and R. J. Gutmann, 1998 Advanced Metallization Conference Asia Session, pp. 163-164.
[3] N. Aoi, E. Tamaoka, M. Yamanaka, S. Hirao, T. Ueda, and M. Kubota, 1999 Symposium on VLSI Technology, pp. 41-42.

APPLICATIONS FOR ORGANOSILICON GASES IN PECVD PROCESSES FOR LOW-κ DIELECTRICS

M.J. LOBODA
Dow Corning Corporation, Mail Stop CO41A1, Midland, MI 48686
Email: mark.loboda@dowcorning.com

ABSTRACT

The need for new silicon-based low-k dielectric materials has created interest in developing alternatives to SiH_4 in plasma enhanced chemical vapor deposition (PECVD) processes. Organosilicon gases, such as the family of organosilanes $(CH_3)_xSiH_{4-x}$, can be used in the deposition of low-k oxide and carbide dielectric films. These dielectrics, used separately or together, have properties capable of addressing the ever decreasing capacitance requirement associated with either subtractive aluminum or copper damascene interconnection technologies. But as with any promising new materials technology, there is only a limited experience base associated with the manufacture, usage and integration of organosilicon-based gases and dielectric films in integrated circuit fabrication. This paper will review the properties of the organosilicon gases and their used in PECVD. Gas properties and chemistry, process chemistry, film properties and integration issues will be discussed.

INTRODUCTION

Organosilicon compounds, such as the family of materials $(CH_3)_xSiH_{4-x}$, are emerging as precursor materials in new PECVD processes for low-k dielectric films. This class of materials includes both gas and liquid phase materials, which are supplied to PECVD systems using standard delivery methods. Dielectric films can be deposited from these organosilanes using equipment and process conditions that are comparable to SiH_4-based oxide dielectric deposition to produce a new set of Si-C-based dielectrics. While the traditional dielectric PECVD processes for silicon oxides and nitrides produce films with relative permittivity in the range 4<k<8, the organosilicon precursors can be used to deposit dielectric films with relative permittivity in the range 2.5<k<5. The Si-C-based dielectrics can be used in place of the traditional dielectrics in semiconductor integrated circuit (IC) interconnection schemes (subtractive and damascene) which target signal transmission speed improvements through capacitance reduction.

When compared with traditional dielectrics, associated with Si-C-based dielectrics are a mixed bag of mechanical and integration properties ranging from low stress and reduced hardness/modulus to improved etch selectivity and increased sensitivity in post-etch cleaning. This paper will review the use and issues established to date related to deposition of dielectric films by PECVD of $(CH_3)_xSiH_{4-x}$ compounds. Precursor usage will be described for PECVD applications, and an introduction to plasma chemistry will be discussed. The applications of this class of Si-C-based dielectrics will be covered in the context of advanced IC interconnect technology, and include precursor tradeoffs, film process-property relationships, metal-dielectric interactions and process integration experiment results.

371

PROPERTIES OF $(CH_3)_xSiH_{4-x}$ COMPOUNDS

The compounds $(CH_3)_xSiH_{4-x}$ are synthesized by chlorinating silicon metal to form methylchlorosilanes, and then chemically reducing these materials to replace the chlorine atom with a hydrogen atom. An oversimplified example of this reaction chemistry is as follows:

$$CH_3Cl + Si \text{ (metal)} \rightarrow (CH_3)_xSiCl_{4-x} + (CH_3)_4Si \qquad (1)$$

$$(CH_3)_xSiCl_{4-x} + (CH_3)_4Si + LiAlH_4 \rightarrow (CH_3)_xSiH_{4-x} + \text{reaction by-products} \qquad (2)$$

The yields of the various methylsilanes are affected by the control of the ambient reaction conditions (e.g. pressure, temperature, etc.). At this point the individual methylsilanes are separated and purified by standard distillation techniques.

Table 1 lists the basic physical properties of the methylsilane compounds. All but tetramethylsilane are gas phase at room temperature. The relative flammability and vapor pressure of these materials is proportional to the number of hydrogen atoms bound to the silicon atom. Both $(CH_3)SiH_3$ and $(CH_3)_2SiH_2$ must be handled with extreme care since it is difficult to manufacture these gases without measurable SiH_4 impurity. The bond dissociation data shows that the energy to cleave a H or $-CH_3$ bond is similar for all materials with at least one Si-H bond, and the higher energy is required to dissociate tetramethylsilane.

Table 1. Basic Properties of $(CH_3)_xSiH_{4-x}$ Precursors

	CH_3SiH	$(CH_3)_2SiH_2$	$(CH_3)_3SiH$	$(CH_3)_4Si$
Material Phase at 20 °C	Gas	Gas	Gas	Liquid
Dissociation Energy [1] (kcal/mol)				
Si-H Bond	89.6	89.4	90.3	99.2 (CH_2-H)
Si-CH_3 Bond	88.2	88.1	88.1	89.4
Vapor Pressure at 20 °C (torr)	10249	2964	1218	601
Approximate Boiling Point (°C)	-57	-20	7	27
Hazards	Flammable-Pyrophoric due to SiH_4 impurity	Flammable-Pyrophoric due to SiH_4 impurity	Flammable	Flammable

PLASMA DEPOSITION

The early applications which used organosilanes in PECVD focussed on the deposition of amorphous hydrogenated silicon carbide (a-SiC:H) [2-3] and highly crosslinked polymethylsiloxanes, $(CH_3)_xSiO_y$ [4]. Applications for a-SiC:H films were developed in the areas of enhanced reliability for dual in-line packaging of semiconductor devices [5] and hermetic-like re-passivation for chip-scale packaging [6-8]. It was these packaging applications which highlighted the use of organosilicon-based a-SiC:H as a moisture diffusion barrier compatible with Ti-alloy, aluminum and gold metallization. During this time, deposition of a-SiC:H films by PECVD from methyl- and cyclic- silanes were investigated showing the potential to deposit low-k, high resistivity Si-C alloys from these precursors [9].

All plasma-based processes that use $(CH_3)_xSiH_{4-x}$ as film precursors proceed via reactions of unique Si-C compounds known as *silenes (neutrals) and silylenes (bi-radicals)*. These compounds are generated as a result of the plasma decomposition of the precursor

$$(CH_3)_xSiH_{4-x} + h\nu \rightarrow \begin{array}{c} CH_2{=}SiH_2(\alpha) + CH_2{=}Si(H)CH_3\,(\beta) + CH_2{=}Si(CH_3)_2(\gamma) \\ + \\ CH_2*{-}*SiH_2(\alpha) + CH_2*{-}*Si(H)CH_3\,(\beta) + CH_2*{-}*Si(CH_3)_2(\gamma) \end{array} \qquad (3)$$

These Si-C-based meta-stable compounds are responsible for highly efficient incorporation of carbon into the plasma deposited film relative to process chemistry that employs separate Si and C precursors. The number of methyl groups on the precursor silane increases the energy required for Si-H dissociation and determines the relative gas phase concentrations of the Si-C meta-stable compounds. Because of its molecular structure, only trimethylsilane will dissociate to produce comparable amounts of all three species, α, β, γ. Generation of Si-H containing radicals occurs from all the methylsilanes, monomethylsilane yields more α and larger amounts of SiH_x*, while tetramethylsilane produces γ and $(CH_3)_3SiCH_2$-H. It is both the number of carbons and bonding by which carbon is linked to the silicon in the silane which determines the amount of carbon which is carried to the substrate and available for incorporation into the film. Once the reactive species are adsorbed, they can incorporate into the film network through reaction with surface Si-H or C-H bonds and subsequent elimination of hydrogen gas or methane.

APPLICATIONS

Amorphous Hydrogenated Silicon Carbide (a-SiC:H)

a-SiC:H films are deposited from methylsilanes by direct plasma dissociation in the presence of an inert gas such as helium or argon [9]. The properties of these carbides have been reviewed previously [10-11] and are again summarized in Table 2. In addition to the packaging applications previously described, the a-SiC:H materials are being developed as a diffusion barriers/buried hard mask/etch-stop in copper damascene interconnection technology [12-13].

When compared to plasma deposited silicon nitride films, the a-SiC:H films have lower permittivity, better copper diffusion resistance and slower dry etch rates [12]. Circuit modeling indicates that low permittivity of the a-SiC:H is critical to realization of the speed improvements achievable through the integration of low-k carbon-doped oxide or a-SiCO:H intermetal dielectric materials in the damascene structure [13].

Table 2. Basic Properties of Films Deposited From $(CH_3)_xSiH_{4-x}$ Precursors

Property	a-SiC:H[*]	Carbon-doped SiOx and a-SiCO:H[**]
Process temperature	300-400 °C	0-400 °C
Oxygen Source	-	N_2O or H_2O_2
Oxygen Content	-	18 - 30 atom %
Carbon Content	30 - 40 atom %	12 -40 atom %
Hydrogen Content	25 - 40 atom %	10 - 30 atom %
Equivalent Bulk Density	1.6 - 1.8 g/cc	1.3 - 1.5 g/cc
Relative Dielectric Constant, k	4.0-5.5	2.5-3.0
Leakage Current Density @ 0.5 MV/cm	10^{-10} A/cm^2	$<10^{-10}$ A/cm^2
Breakdown Field @ 10^{-3} A/cm^2	>2.5 MV/cm	2 - 8 MV/cm
Film Stress	<100 MPa compressive	30-100 MPa tensile

[*] trimethylsilane based a-SiC:H   [**] nominal value range for all $(CH_3)_xSiH_{4-x}$ processes

Carbon-doped Silicon Oxide and Silicon Carboxide

Deposition processes using $(CH_3)_xSiH_{4-x}$ precursors for low-k, carbon doped SiO_2 and PECVD low-k silicon carboxide (a-SiCO:H) have recently developed as alternatives to organic-based PECVD low-k materials such as paralyene and a-C:F materials. More compatible with traditional integration processes than organic films, these Si-C-based oxide materials can have relative dielectric constants as low as k~2.5 and are now under evaluation in intermetal dielectric applications. Table 2 lists the range of properties for Si-C-based oxide materials available with various precursors.

The reported $(CH_3)_xSiH_{4-x}$ -based low-k oxide processes take on many forms such as low temperature CVD with anneal [14], common PECVD [15] and cold wafer PECVD with anneal [16]. These processes range 3.1<k<2.5 where lowest reported k values are achieved in films with higher carbon content. At low carbon concentration (c.a. <18 atom %) the film structure is like methyl-doped CVD SiO_2. Effectively termination points in the bond matrix, the methyl groups create volume, without adding porosity. This, in turn decreases the density and permittivity of the oxide. At higher carbon concentrations (c.a. >20 atom %) the film is better described as silicon carboxide, or a-SiCO:H, a material with similar proportions of C-Si-C and O-Si-O bonding. As the acronym suggests, these materials have approximately equal proportions of Si, C, and O. Similar to the lower carbon concentration oxide previously mentioned, the a-SiCO:H material also contains methyl termination points to create volume. The advantages of the increased C in this type of Si-C-based oxide film are lower permittivity, lower moisture absorption and lower film stress, but at the same time C-Si-C bonds are more chemically resilient than O-Si-O, making the film harder to etch compared to the low carbon content films.

INTEGRATION

Since the $(CH_3)_xSiH_{4-x}$ -based low-k films are relatively new, very little integration work has been reported. The thermal stability of the films is comparable to traditional SiH_4-based oxides and nitrides. Because of the carbon bonding in these films, water outgassing is not usually observed at elevated temperatures. Thermal cycling of trimethylsilane-based films has shown minimum stress hysteresis [12,15]. Trimethylsilane-based carbide-SiCO:H-carbide stacks, as would be implemented in dual damascene interconnections, have shown good stability performance in thermal cycling with minimum stress hysteresis, stable <u>net</u> k<2.8 (out of plane), and low leakage currents (J<10^{-10} A/cm^2) as evaluated with copper gates before and after thermal cycling [12]. Figure 1 shows the stress-temperature plots for various anneal cycles for the dielectric stacks.

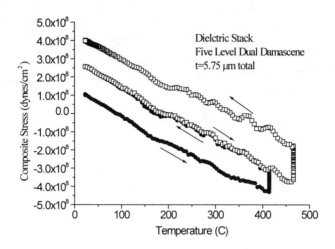

Figure 1. Stress temperature plots for a five level dual damascene dielectric stack of 50 nm a-SiC:H/500 nm a-SiCO:H/50 nm a-SiC:H deposited consecutively without breaking vacuum at 400 ° C. Ramp rate is 5 °C/min and the temperature holds is one hour.

The films can be patterned using fluorine-based etch chemistries. Good etch selectivity has been reported with low-k/carbide stacks using trimethylsilane [13,16,17]. As is the case with many low-k materials, the integration challenges are initially encountered in the area of post etch residue removal. The chemical resistance of the a-SiC:H and a-SiCO:H films insures good compatibility with wet cleaning processes for post etch residue [13,16]. But the aggressive chemical attack of downstream oxygen ashing processes commonly used for SiO$_2$ via and trench clean can oxidize Si-CH$_3$ bonds in the films. This leads to formation of Si-O-Si bonds, densification of the material and bowing in vias and trenches. Nitrogen dilution of these downstream oxygen ash processes has been shown to circumvent the bowing problem in a-SiCO:H film patterning [17].

Generally, all methylsilane-based films show good adhesion to the common metals used in both subtractive aluminum and copper damascene interconnect technologies [12,13,16]. On medium aspect ratio aluminum structures, trimethylsilane-based a-SiCO:H has shown gapfill performance better than that associated with PE-TEOS oxide films [15]. Recently, successful implementation of Si-C based low-k films has been reported in single and dual copper damascene test structures with >25% reduction in capacitance compared to SiO_2 interlayer dielectrics [13,16,18]. This progress demonstrates that the process challenges associated with pattering these materials are not insurmountable, and optimal process integration conditions are likely to develop in the near future.

CONCLUSIONS

New PECVD chemistry to produce low-k dielectrics based on Si-C containing materials has been discussed. Properties of these films allow improved signal transmission speed performance in high density device interconnections. The integration of these films is slightly more difficult than SiH_4-based dielectrics, but the integration methods can employ the same chemistry for patterning and cleaning. The similarities will likely lead to the successful use of Si-C -based low-k materials in state of the art integrated circuit fabrication.

ACKNOWLEDGEMENTS

The author would like to acknowledge extremely helpful discussions with W.J. Chatterton and the efforts of the technical staff of the Thin Film Technology Group of the Dow Corning Corporation, IMEC (Belgium), and Applied Materials.

REFERENCES

[1] R. Walsh, "Bond Dissociation Energy Values in Silicon-Containing Compounds and Some of Their Implications," Acc. Chem. Res., Vol. 14, p. 246-52 (1981)

[2] Y. Catherine and A. Zamouche, "Glow Discharge Deposition of Tetramethylsilane Films," Plasma Chemistry and Plasma Processing, Vol. 5 (4), p.353 (1985)

[3] M.J. Loboda, "Low Temperature PECVD Growth and Characterization of a-SiC:H Films Deposited From Silacyclobutane and Silane/Methane Precursor Gases," in Amorphous and Crystalline Silicon Carbide IV, C.Y. Yang, M.M. Rahman, and G.L. Harris, eds., Springer Proc. in Physics (Springer-Verlag, Berlin Heidelberg), 71, p. 271-280 (1992)

[4] G.W. Hill, "Electron Beam Polymerization of Insulating Films," Microelectronics and Reliability, Vol. 4, p. 109-116 (1965)

[5] G. Chandra, R. C. Camilletti, M.J. Loboda, "Low Temperature Ceramic Films For Microelectronic Applications," Ceramic Transactions (American Ceramic Society), 33, p.209-223 (1993)

[6] R.C. Camilletti and M.J. Loboda, "Thin-film Packaging for MCM Applications,"Proc. 1995 ISHM/IEEE Int'l. Conference on Multichip Modules, April 1995.

[7] M.J. Loboda, R.C. Camilletti, L.A. Goodman, and L. White, "Manufacturing Semiconductor Integrated Circuits With Built-In Hermetic Equivalent Reliability," Proc. of 46th IEEE Electronic Components and Technology Conf., p. 897-901 (1996)

[8] M.J. Loboda, R.C. Camilletti, L.A. Goodman, L. White, H.L. Pinch, J. Shaw, V.K. Patel, C.P. Wu, and G. Adema "ChipScale Packaging with High Reliability for MCM Applications," Int'l Jour. of Microcircuits and Electronic Packaging, 19, 1996

[9] M.J. Loboda, J.A. Seifferly and F.C. Dall, "Plasma Enhanced Chemical Vapor Deposition of a-SiC:H Films From Organosilicon Precursors," J. Vac. Sci. Technol. A, 12 (1), p. 90-96 (1994)

[10] M.J. Loboda, J.A. Seifferly, C.M. Grove, and R.F. Schneider, "Safe Precursor Gas for Broad Replacement of SiH_4 in Plasma Processes Employed in Integrated Circuit Production," Mat. Res. Soc. Symp. Proc. Vol. 447: Environmental Safety and Health Issues in IC Production (1997), p.145-50

[11] M.J. Loboda, J.A. Seifferly, C.M. Grove, and R.F. Schneider, "Using Trimethylsilane to Improve Safety, Throughput and Versatility in PECVD Processes," Electrochem. Soc Proc. Vol. 97-10: Silicon Nitride and Silicon Dioxide Thin Insulating Films (1997), p.443-52

[12] M.J. Loboda, "New Solutions for Intermetal Dielectrics Using Trimethylsilane-Based PECVD Processes," to be published in the J. Microelectronics Engineering, special issue on the Proceedings of the 1999 European Workshop Materials for Advanced Metallization, Oostende, Belgium 1999.

[13] P. Xu, K. Huang, A. Patel. S. Rathi, B. Tang, J. Ferguson, J. Huang, C. Ngai, and M. Loboda, "BLOк™ - A Low-k Dielectric Barrier/Etch Stop Film for Copper Damascene Applications, "Proceedings 1999 IEEE Int'l Interconnect Technology Conf., p. 109 (1999)

[14] S. McClatchie, K. Beekman, A Kiermasz, "Low Dielectric Constant Oxide Films Using CVD Techniques," Proc. Dielectrics for ULSI Multilevel Interconnection Conference (DUMIC), p. 311-18 (1998)

[15] M.J. Loboda, J.A. Seifferly, R.F. Schneider and C.M. Grove, "Deposition of Low-K Dielectric Films Using Trimethylsilane," Electrochem. Soc Proc. Vol. 98-XX: Interconnect and Contact Metallization: Materials, Processes And Reliability, 194th Meeting of the ECS, Boston 1998. *To be published.*

[16] M. Naik, S. Parikh, P. Li, J. Educato, D. Cheung, I. Hashim, P. Hey, S. Jenq, T. Pan, F. Redeker, V. Rana, B. Tang, D. Yost, "Process Integration of Double Level Copper Low-k (k=2.8) Interconnect," Proceedings 1999 IEEE Int'l Interconnect Technology Conf., p. 181 (1999)

[17] S. Vanhaelemeersch, C. Alaerts, M. Baklanov, H. Struyf, "Dry Etching of Low K Materials," Proceedings 1999 IEEE Int'l Interconnect Technology Conf., p. 97 (1999)

[18] T. Gao, W.D. Gray, M. Van Hove, H. Struyf, H. Meynen, S. Vanhaelemeersch, K. Maex, "Integration of the 3MS Low-k CVD Material in a 0.18 um Cu Single Damascene Process," 1999 Advanced Metalization Conference, Orlando, FL, *this conference, to be published.*

Black Diamond™ ---A Low k Dielectric for Cu Damascene Applications

Wai-Fan Yau, Yung-Cheng Lu, Kuowei Liu, Nasreen Chopra, Tze Poon, Ralf Willecke, Ju-Hyung Lee, Paul Matthews, Tzufang Huang, Robert Mandal, Peter Lee, Chi-I Lang, Dian Sugiarto, I-Shing Lou, Jim Ma, Ben Pang, Mehul Naik, Dennis Yost, and David Cheung, Applied Material, Santa Clara, California, USA.

ABSTRACT

The Black Diamond family of low k materials has been developed using conventional parallel plate single-wafer CVD chamber technology. The film is formed by partial oxidation of an organosilane molecule, which resulted in a more open network structure, hence a lower density material than conventional SiO2. The film can achieve bulk dielectric constant of 2.5-3.0, with a post integration IMD stack dielectric constant of <3.0. By maintaining similar adhesion, thermal, and mechanical properties to SiO2, this film provides an evolutionary path from Cu/oxide to Cu/low k device designs. A second-generation of materials with k<2.5 is under development for extension to 0.10μm devices.

INTRODUCTION

New generation integrated circuits will employ a combination of low dielectric constant insulator films and copper metallization for on-chip multilevel inter-connections, primarily because, as feature dimensions are decreased, in-plane capacitance as well as sheet resistivity of interconnects must be reduced.

Multilayer insulation films with lower dielectric constants are required:
- to reduce RC time-constant signal delay (multilevel interconnection propagation delay in VLSI and ULSI circuits is at the point of exceeding gate delay; speed, and in addition, timing and signal integrity, as well as IC design complications)
- to reduce crosstalk (coupling noise)
- to reduce dynamic power consumption (interconnect capacitance is a significant load capacitance in VLSI circuits, and is the dominant load capacitance for CMOS VLSI)

Multilevel interconnection structures comprised of low-κ films integrated with inlaid copper wires in a dual damascene process has been identified as a new generation technology in the National Technology Roadmap for Semiconductors (NTRS).

Silicon oxide-based dielectric films with κ<3 are receiving widespread attention because of a combination of desirable properties, including preservation of film rigidity, superior thermal mechanical properties, and process integration similarity to SiO_2, including extendible plasma etching parameters. These films are deposited by means of chemical vapor deposition (CVD), or by spin casting. Low-κ CVD silicon oxide-based films with dielectric constants below 3 are available for integration, and are expected to find widespread use.

A large number of important properties, including mechanical, electrical, and environmental requirements, must be simultaneously and routinely achieved in a useful and desirable low-κ substitute for standard silicon oxide insulator films. This report describes the

Conference Proceedings ULSI XV © 2000 Materials Research Society

Black Diamond basic film properties and integration results from structures built at Applied Materials.

FILM PROPERTIES

Because the film primarily contains silicon and oxygen, Black Diamond retains many of the thermal and mechanical properties of SiO2 (**Table 1**). The glass transition temperature is well above 450°C as determined by film stress versus temperature experiment(**Fig. 1**). Isothermal weight loss measured by thermal gravimetric is < 1%wt per hour at 450°C (**Fig. 1**). Reduction of the dielectric constant is achieved primarily through maximizing the free volume in the microstructure.

One concern with low k materials is decreased mechanical strength. As shown by hardness data and Young's modulus(**Fig. 1a**), the mechanical strength of the film is similar to oxide and one order of magnitude higher than that of polymeric films. Stress hysteresis is < 20 MPa after 6 cycles to 450°C and no structural or phase transition is observed at these conditions (**Fig. 1**).

Another concern with low k materials is lowered thermal conductivity as compared to SiO2. Joule heating of the high current density circuits depends on the intermetal dielectric to conduct the heat away from the interconnects. A low k material with poor thermal conductivity can lead to excessive heating of the metal lines, which can lead to poor metal reliability. As shown in figure 1a, Black Diamond thermal conductivity near one order of magnitude higher than polymeric materials.

Adhesion to different materials and the ability to withstand CMP are two critical structural issues. Film adhesion was investigated by mechanical stud pull testing and ASTM tape test. Adhesion to TaN, Ta, SiO2, SiN, Ti, and TiN is very good, both when measured as-deposited and after temperature cycling (7 cycles at 400°C). Adhesion was also evaluated after direct CMP of the film and after CMP of copper on the material. No delamination was observed in either experiments. No adhesion layers or under layers were required to achieve these results. A film thickness of greater than 2 microns can be deposited without any cracking or delamination. The electrical properties of the film, such as breakdown voltage and leakage current, were also investigated and show similar values compared to standard PE-TEOS oxide (**Table 1**).

ELECTRICAL PERFORMANCE

The Black Diamond film's dielectric constant was measured by Hg probe and has a bulk k value of 2.5-3.0. The line-to-line capacitance was obtained by in-house electrical test of an integrated Black-Diamond-Cu single damascene structure. Process Flow of this structure is outlined in figure 2. Figure 3 shows a typical SEM cross section of the single damascene structure. Capacitance measurements from inter-digitated comb structures show a 28% reduction in capacitance for the fully integrated film compared to a similar structure using standard SiO2 (**Fig. 2a**). Assuming k = 4.0 for SiO2, this equates to approximately k=2.9 after integration, including etching, resist ashing, solvent clean, and other processing steps.

Black Diamond is an amorphous material. The interdigitated comb capacitance measurements are in agreement with the Hg probe measurement, and confirm the isotropic nature of this film. Leakage current was <5E-9 A/cm2 at 1MV/cm, and the breakdown field was >2.5MV/cm. Thus, the film provides a significant reduction in capacitance, while meeting electrical requirements for dielectric isolation.

FILM INTEGRATION

Integration data are summarized in **Fig. 3** to **7**. Etch, ash, clean, barrier/seed, Cu electroplating, and CMP are all performed on commercially available tools with minimum change to existing baseline recipes. Each step has been evaluated and the results are summarized below.

Figure 3 shows single damascene structure built. Black Diamond etch profile is over 88^0 with minimal etch stop loss. Ashing performance is shown in **Fig. 4** with a comparison between pre and post ash FTIR. No significant Si-H or Si-CH3, or C-H peak loss is observed. CMP performance is shown in Fig 5. There is no significant difference between electrical line width and as drawn line width from 30 microns down to less than 1 micron lines. Also, dishing is minimal as shown in Fig. 3. SEM photo of completed single and dual damascene structures are shown in **Fig. 3** and **6**. Capacitance reduction data is plotted in Figure 2a. A 28% reduction in capacitance is measured when compared with SiO2 baseline. Via resistance for dual damascene structures showed identical performance as compared with the oxide base line. This is true for both the median and the distribution of the via resistance. Manufacturability of the Black Diamond film is demonstrated by Figure 7 with over 10000 wafers run.

EXTENDIBILITY

Under appropriate processing conditions, a lower density film can be deposited without changing the composition. Materials with mass density as low as 1gram/cc can be deposited. The corresponding k for this material is near 2.2. Figure 8 shows a cross sectional SEM of the deposited film. The presence of voids in the film is easily seen. A structural build is underway to understand the post integration performance of this new material.

CONCLUSION

The Black Diamond film demonstrates the thermal, mechanical, and electrical film properties necessary for a viable low k material (k <3.0) for IMD applications. The material has been successfully integrated into single and dual damascene structures using a standard equipment set. Electrical tests indicate a substantial reduction in capacitance as compared to silicon oxide. This technology provides a CVD film capable of meeting advanced device requirements in a production worthy method.

Table 1 - Summary of Black Diamond Film Properties

Description	Blanket Film
Dielectric Constant - Bulk film (Hg Probe)	2.5-3.0 @ 1MHz
Refractive Index	1.42
Uniformity (%, 1 sigma)	<1.5
Stress (dyne/cm2)	4E8 - 8E8 Tensile
Stress Hysteresis (dyne/cm2)	<2.0E8 (RT - 450
Cracking Threshold (μm; blanket film on silicon)>2	
ASTM Scratch Tape Test on SiN, SiON, Ta, TaN	Passed
Stud Pull Adhesion Test (Black Diamond on Silicon)	>12kpsi
Thermal Shrinkage (%), post cure film	<2.0 @ 450
2hrs Isothermal TGA	<1.0% @ 450
n & k @ 633nm	n = 1.42 k = 0
@ 248nm	n = 1.50 k = 0
@ 193nm	n = 1.59 k = 0
Particles @ >0.2μm size	<0.1/cm²
Electrical Properties:	
• Leakage Current (Amps/cm²)	10 @ 1MV/cm
• Breakdown Field (MV/cm)	>2.5

Figure 1 - Thermal Stability of Black Diamond Film

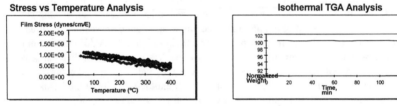

Stress vs Temperature Analysis

Isothermal TGA Analysis

• Stress Hysteresis <2E8
• Estimated Tg >450°C

• No measurable weight loss after 2 hrs at 400°C

Figure 1 a – Thermal Mechanical Properties of Black Diamond

	SiO2	Black Diamond	Spin on Inorganic	High carbon CDO	Organic Polymer
Hardness (Gpa)	7-8	3-4	--	0.4	0.1-0.3
CTE in plane (ppm/C)	~1.0	5-10	20	--	50-70
Biaxial E (Gpa)	70-80	10-20	5-10	--	2-3
Thermal Conductivity (W/mC)	~1.0	0.3-0.4	--	--	<0.1

Figure 2 – Single damascene process flow

1. Grow 150A thermal oxide
2. Deposit 10000A PECVD oxide
3. Deposit 1000A PECVD SiN
4. Deposit 5000A Black Diamond CVD low k film.
5. Apply metal trench mask
6. Etch trench features
7. Remove photo mask
8. Anneal
9. Degas and Preclean
10. Deposit IMP TaN, 250A
11. Deposit IMP Cu seed, 1500A
12. Electroplate Cu
13. Anneal
14. CMP Cu
15. Deposit 1000A SiN
16. Deposit 5000A silicon oxide
17. Apply pad mask
18. Etch pad through oxide
19. Remove photo mask
20. Etch pads through SiN
21. Clean wafer

Figure 2a - Black Diamond Intra-Level Capacitance Comparison
(Single Damascene Structure)

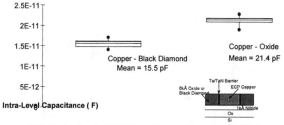

- *28% Reduction in Total Capacitance Compared to Oxide*

Figure 3 - Single Damascene Cu - Black Diamond Structure

Integrated Structure with Black Diamond and Copper Fill

- Black Diamond Low k Film
- Lithography
- IPS™ Oxide Etch
- O_2 Ash and Solvent Clean
- Electra™ TaN Barrier
- DryFill™ Copper
- Mirra™ Copper CMP

Figure 4 - FTIR spectra of pre and post ashing Black Diamond film

Black Diamond Compatibility with Low Pressure, Zero Bias O_2 Ash

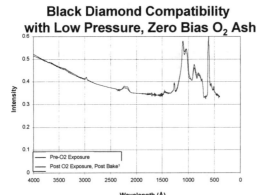

Step	Normalized k
Pre-Oxygen Exposure	1
Post-O_2 Exposure and Bake	0.95

• *No change in film composition or Dielectric Constant after O_2 Plasma exposure*

Figure 5 - As drawn line width vs. post CMP electrical line width

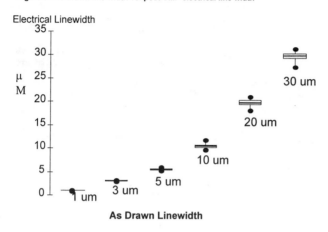

Electrical Linewidth

μ M

As Drawn Linewidth

Figure 6 - Dual Damascene Cu - Black Diamond Structure

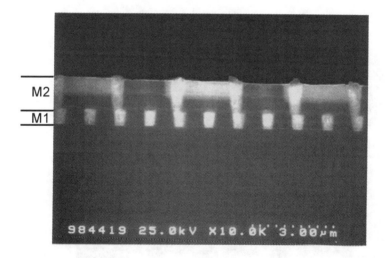

Figure 7 - particle adders performance for 10000 wafers run

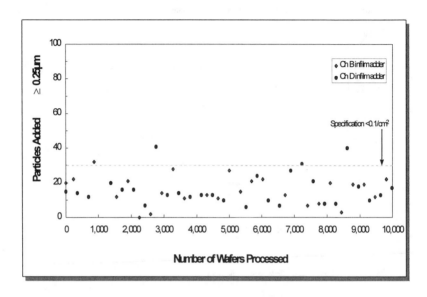

Figure 8 – Cross sectional SEM picture of k < 2.5 Black Diamond film

Effect of thermal conductivity and fluorine of low-k HDP-SiOF dielectric on EM behavior of Al-Cu line

Dong-Chul Kwon, Young-Jin Wee, Yun-Ho Park, Hyeon-Deok Lee, Ho-Kyu Kang, and Sang-In Lee
Semiconductor R&D Center, Samsung Electronics Co., Ltd.
San #24 Nongseo-Ri, Kiheung-Eup, Yongin-City, Kyungki-Do, Korea 449-900
Phone: +82-2-760-6335, FAX: +82-2-760-6299, E-mail: kdckwon@samsung.co.kr

ABSTRACT

The Al corrosion by fluorine and the low thermal conductivity of low-k SiOF dielectric has been a serious concern for intermetallic dielectric (IMD) application. In this study, EM behavior of Al-Cu lines confined in low-k HDP-SiOF dielectric was compared with that in flowable oxide (FOx) which has no fluorine content. From the results, major effect of the fluorine and the thermal conductivity of HDP-SiOF dielectric on EM behavior of Al-Cu lines was identified. Implication for improving reliability performance was also presented.

INTRODUCTION

Low-k HDP-SiOF dielectric with a dielectric constant of 3.5 is one of candidates applicable as an intermetallic dielectric (IMD) to meet gap fill requirement as well as to reduce RC delay of interconnection in 0.25 um regime. However, The Al corrosion by fluorine and the low thermal conductivity of the low-k SiOF dielectric has been a serious issue in Al-Cu interconnection for intermetallic dielectric (IMD) application [1-4]. The low thermal conductivity causes excessive joule heating from Al-Cu lines, and the addition of fluorine results in the attack of Al-Cu lines. Each case causes severe reliability problem for the SiOF application, but extensive study of their impact on EM behavior of Al-Cu lines has not been performed.

In this study, EM behavior of Al-Cu line confined in HDP-SiOF has been compared to that confined in flowable oxide (FOx) without fluorine content. In particular, it is determined whether the Al corrosion of fluorine affects the Cu diffusivity or the mechanical strength of Al-Cu lines.

EXPERIMENT

PE-TEOS oxide, 4000 Å thick, was deposited on <100> Si wafers. The Ti 150 Å and the Al-0.5%Cu alloy, 5700 Å thick, was deposited in-situ at room temperature and 350 °C respectively. The Al capping layer of Ti 150 Å (bottom)/ TiN 400 Å (top) was deposited at room temperature. After line patterning, either 6000 Å thick HDP-SiOF oxide or 6000 Å flowable oxide (Fox) was deposited as intermetallic dielectrics (IMD). Then, PE-TEOS oxide was deposited and planarized by CMP. Via contacts were opened by normal dry-etch process. Ti 100 Å/ TiN 700 Å was deposited as a tungsten barrier layer by collimated sputtering process. Via contacts were filled by blanket CVD tungsten, followed by CMP planarization. Metal-2 structure was built identically to the metal-1 structure and passivated by silicon nitride. All samples were annealed at 450 °C for 30 min for stabilization of microstructure before an accelerated EM testing. The procedure of sample preparation is summarized in Fig. 1.

The accelerated EM tests for Al-Cu lines were performed with various current densities at 180 °C in order to create different amount of joule heating and show the effect of thermal conductivity on EM lifetime. Via-terminated line patterns were used for the EM testing to reduce the test time. Line failures were characterized by focused ion beam (FIB) and the Al attack of fluorine was investigated by secondary electron microscopy (SEM).

Conference Proceedings ULSI XV © 2000 Materials Research Society

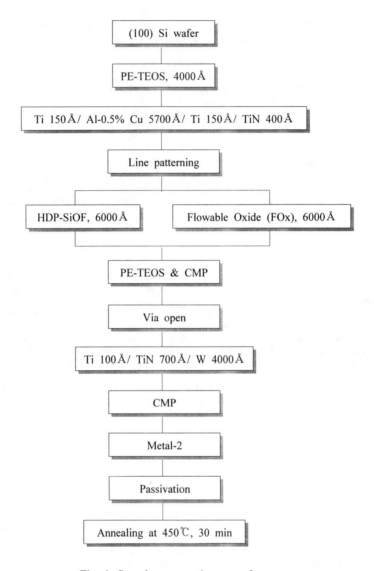

Fig. 1. Sample preparation procedure.

RESULTS AND DISCUSSION

Effect of Thermal Conductivity

In order to show the effect of thermal conductivity of the dielectrics on EM lifetime, various current densities were applied to 0.3 um wide Al-Cu lines. Time to failure (TTF) distribution and corresponding joule heating at various current densities for each dielectric is

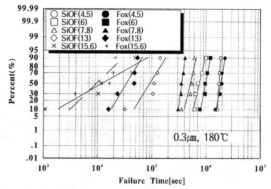

Fig. 2. Time to failure (TTF) of 0.3 um wide lines as a function of current density at 180 ℃. The numbers in parenthesis indicate the current density in MA/cm².

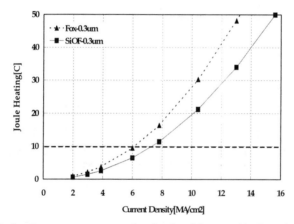

Fig. 3. Joule heating versus current density for 0.3 um wide lines in each dielectric.

shown in Fig. 2 and Fig. 3 respectively. At current densities higher than 7 MA/cm² where corresponding joule heating is higher than ～10 ℃ for both dielectrics (Fig. 3), the SiOF case is shown to provide longer lifetime than the FOx case. However, at lower current densities where corresponding joule heating is lower than ～10 ℃, the lifetime for the SiOF case become shorter than that for the FOx case.

Failures occurred at lower current densities represent a typical EM failure of W-plug structures having a large void under the via at the cathode end as shown in Fig. 4 (a), whereas those occurred at higher current densities show no such voids, as shown in Fig 4 (b), indicative of joule heating effect. This result implies that at relatively high current densities where joule heating effect dominates, thermal conductivity of dielectrics can govern the reliability performance of interconnections. Therefore, application of dielectrics with high thermal conductivity may be beneficial for power lines subjected to high current densities.

389

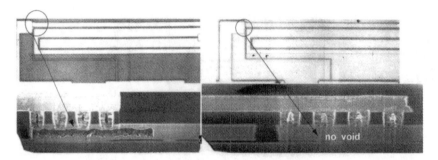

Fig. 4. EM failure occured at (a) lower current densities (< 7MA/cm²) and (b) higher current densities (> 7MA/cm²).

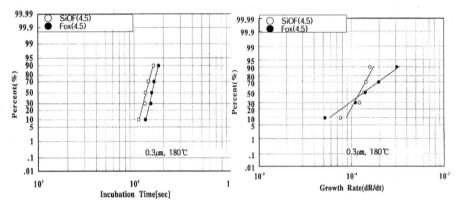

Fig. 5. (a) Incubation time and (b) growth rate of 0.3 um wide lines observed at 180 ℃, with current density of 4.5 MA/cm² in each dielectric.

Effect of Fluorine

Incubation time (t_i) and growth time determined with current density of 4.5 MA/cm² at 180 ℃ for each dielectric case are shown in Fig. 5 (a) and 5 (b) respectively. The incubation time corresponds to the time by which no resistance increase take place during EM testing. From the results, the incubation time determines the overall EM lifetime of 0.3 um wide lines, and the incubation time for the SiOF case is shown to be shorter than that for the FOx case. However, the growth rate, shown in Fig 5 (b), was almost same for both dielectrics. Therefore, it can be concluded that type of dielectric affects predominantly incubation time rather than growth time in EM lifetime behavior.

The incubation time essentially depends on both threshold (critical) current density (J_c) and Cu diffusivity as following [4].

$$t_i = A \ T^2 / \ [D_a \ (J - J_c)^2]$$

where t_i is incubation time, A is constant, T is temperature, D_a is Cu diffusivity, J is current density and J_c is critical current density below which no EM occur due to back- stress induced flux.

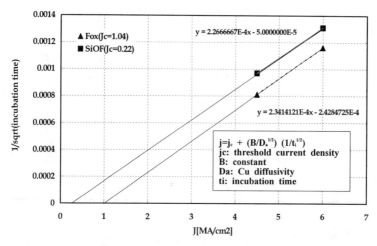

Fig. 6. Jc versus $1/(t_i)^{1/2}$. Relative contribution of the Cu diffusivity and the Jc to the incubation time.

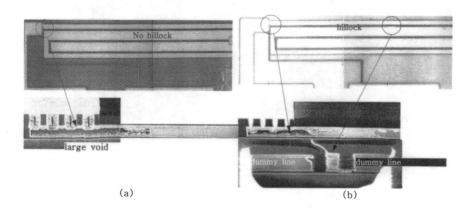

(a)   (b)

Fig. 7. Failure analysis of 0.30 um wide lines in (a) the FOx and (b) the SiOF case. Note that hillock is visible only for the SiOF case.

In order to identify whether the Al attack of fluorine affects the Cu diffusivity or the critical current density, relative contribution of each factor is examined in Fig. 6. It shows J versus $1/(t_i)^{1/2}$, in which the slope and the x-axis intercept corresponds to the Cu diffusivity effect and the J_c effect respectively. The slope is almost same in each dielectric, but the x-intercept is lower for the SiOF case. The same slope indicates no effect of fluorine on Cu diffusivity, and lower x-intercept for the SiOF case indicates lower critical current density. As a result, the shorter lifetime for the SiOF is found to be due to its lower threshold current density (J_c) rather than higher Cu diffusivity.

In addition, the critical current density, J_c, mainly depends on the magnitude of stress that Al can sustain before plastic deformation [6-7]. Thus, the lower J_c above for the SiOF

Table 1. Summary of failure mode, MTTF and compressive yield strength of each dielectric material.

	FOx	SiOF
Failure mode	Void	Hillock
Mean time to failure (MTTF)	$MTTF_{fox} > MTTF_{SiOF}$	
Compressive yield strength	$\Delta \sigma FOx > \Delta \sigma SiOF$	
lower compressive yield strength of SiOF --> lower Jc --> shorter lifetime		

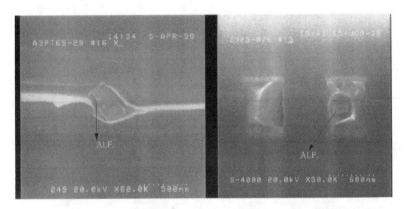

Fig. 8. Formation of aluminum fluoride by the fluorine in the SiOF dielectric.

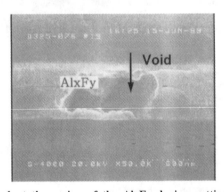

Fig. 9. Void occured at the region of the AlxFy during cutting SEM samples.

case can be attributed to either higher residual tensile stress or lower compressive yield strength. This expectation agrees well with the failure analysis shown in Fig. 7. It shows that

EM-induced hillocks are visible only for the SiOF case, which implies that the SiOF has lower compressive yield strength against hillocks. Consequently, the lower yield strength for the SiOF case is a crucial reason for the decrease in both the J_c and the lifetime of Al-Cu lines. Table 1 summarizes the above results.

The fluorine in the SiOF reduces the compressive yield strength of Al-Cu alloy presumably by the formation of aluminum fluoride as indicated in Fig. 8. Moreover it is quite often observed that the area near the aluminum fluoride is so brittle to be susceptible to voiding in cutting SEM samples, as shown in Fig. 9.

In summary, the lower EM lifetime of Al-Cu lines for the SiOF case results from the shorter incubation time which is attributed to the lower critical current density instead of lower Cu diffusivity. This lower critical current density is originated to the lower compressive yield strength of Al-Cu alloy which is caused by Al attack of fluorine in the SiOF dielectric.

CONCLUSION

Both the thermal conductivity and the fluorine of HDP-SiOF dielectric were found to be dominant factors governing reliability performance of confined Al-Cu lines. Joule heating effect has illustrated that in power lines subject to severe joule heating (> 10 ℃), reliability performance can be limited by the thermal conductivity of dielectric rather than EM behavior of metal lines.

Major influence of the fluorine in the SiOF on Al-Cu lines was found to reduce compressive yield strength of the confined Al-Cu lines against hillocks by the formation of aluminum fluoride. The Al attack of fluorine has no influence on Cu diffusivity. Mechanical properties of Al-Cu lines in application of low-k dielectrics should be considered to enhance EM properties of Al-Cu lines.

REFERENCE

1. A. N. Saxena and M. A. Bourgeois, VMIC Conference, p. 251 (1993)
2. J. H. Choi, B. K. Hwang, H. J. Shin, U. I. Chung, S. I. Lee and M. Y. Lee, VMIC Conference, p. 397 (1995)
3. Minako Murakami and Shinji Matsushita, Semiconductor International July, p. 291 (1996)
4. W.-Y. Shih, M.-C. Chang, R. H. Havemann, and J. Levine, Symposium on VLSI Technology Digest of Technical Papers, p. 83 (1997)
5. Anthony S. Oates, Microelectronic Reliability, Vol 36, No 7/8, p. 925 (1996)
6. I. A. Blech, J. Appl. Phys., Vol 47, p. 1203 (1976)
7. I. A. Blech, Acta Mater., Vol 46, No 11, p. 3717 (1998)

LOW k (≈ 2.0) PLASMA CF POLYMER FILMS MODIFIED BY *IN SITU* DEPOSITED CARBON RICH ADHESION LAYERS

M. UHLIG [1], A. BERTZ , M. RENNAU , S.E. SCHULZ , T. WERNER , T. GESSNER

Chemnitz University of Technology, Centre of Microtechnologies, 09107 Chemnitz, Germany, m.uhlig@chemie.tu-chemnitz.de

ABSTRACT

Fluoropolymer films were deposited by single and dual frequency plasma enhanced chemical vapour deposition (PECVD). Different mixtures of C, F and H containing precursors led to thermally stable (up to 425°C, annealed 0.5 hours in nitrogen atmosphere or vacuum) fluoropolymer films with very low dielectric constant (k = 1.9 ... 2.4). Thermal stability and dielectric constant can easily be varied by deposition process conditions and flow rate ratio C/F. To get well adhering polymer layers, it is necessary to deposit an adhesion layer on barrier as well as on etch stop layers (SiO_2, Si_3N_4, TiN, TiW, WN), respectively. This very thin carbon rich adhesion layer (film thickness d ≈ 10 nm) was deposited *in situ* using the polymer deposition chamber. Compared to the CF polymer films the adhesion layer has a relatively high dielectric constant and a higher leakage current. Advantageously, the electrical properties of the double layer dielectric arrangement are dominated by the CF film. Hence, a very low dielectric constant (k even lower than 2.0) and an outstanding break down field strength (E_{bd} ≥ 5 MV/cm) have been obtained. To study process compatibility with damascene process flow 1 μm thick CF polymer films were successfully polished by CMP (chemical mechanical polishing). Furthermore, other process integration issues of these films were investigated. About 1 μm thick polymer films including the thin C rich film were patterned by high density plasma etching in an oxygen containing atmosphere using a hard mask. The inductively coupled plasma (ICP) etched vias have very smooth, straight sidewalls at an aspect ratio of about 5. Moreover, Cu/TiN filled polymer vias indicate the integration potential of this dielectric material.

INTRODUCTION

The general process of miniaturization in semiconductor device sizes requires the application of new materials in devise fabrication. The conventionally used dielectric material SiO_2 (dielectric constant k = 3.9) will not be able to meet these demands any longer. New dielectric materials with low k value [1], [2] must be investigated with respect to their properties and their integration compatibility. Today's main challenge for the semiconductor industry is the search for new materials that fulfil the demands of a sub 0.18 μm technology. In this process the use of advanced intermetal dielectrics with a dielectric constant k ≤ 2.5 (for sub 0.18 μm) [2] to reduce the capacitance seems to be more important than the decrease of the resistivity provided by the substitution of aluminum by copper [3]. Today three different groups of low k material seem to be applicable: polymers (especially fluorinated C based polymers), inorganic porous material (SiO_2 aerogel / xerogel) and air gap structures [4]. Recently it was pointed out, that among different candidates for low k materials "Amorphous carbon fluoride (a-CF) is one of the more promising CVD materials, with stable 2.3 k" [5].

---

[1] Corresponding author: Fax: +49-371-531-3131,   e-mail: m.uhlig@chemie.tu-chemnitz.de

Conference Proceedings ULSI XV © 2000 Materials Research Society

EXPERIMENT

All investigated CF polymer and carbon rich adhesion layers were deposited on 4" and 6" p-or n-Si wafers by plasma enhanced chemical vapour deposition (PECVD) using single wafer plasma equipment. Deposition on 4" wafers was performed in a single frequency SECON XPL 250 etch tool. RF (13.56 MHz up to 100 W) or LF powered plasma ($\approx$ 300 kHz up to 200 W) could be applied to the heated substrate (T_S = 20 ... 250°C). 6" wafers were coated in a dual frequency (RF 13.56 MHz, 1000 W and LF 100 kHz, 350 W) Tegal 1514e apparatus. To achieve elevated deposition temperatures (20 ... 400°C), the Tegal tool was equipped with a graphite heater, placed within the deposition chamber. Mixtures of fluorinated hydrocarbon gases in combination with other gases were used to produce thin films. Adhesion layers were deposited from CH_4 precursor serving as carbon source. The deposition took place at elevated substrate temperatures ($T_S \approx$ 250°C). Usually a 150 W RF plasma was applied to the parallel reactor plates. The deposition rate for the adhesion layer is very low, about 10 nm/min. XPS studies were performed to determine the structure and the element concentration of the films. To test the thermal stability of the dielectric material, an annealing process was carried out in a vacuum furnace, evacuated and filled with N_2, heated up to 425°C for 0.5 hour and cooled down to room temperature. The film thickness change of certain layers (measured with Tencor alpha-step 200 and Dektak profilometer and Jobin Yvon spectroscopic ellipsometer) was negligible ($\leq$ 1 %).

The thickness of the adhesion layer can be as low as 10 nm. Without interrupting the vacuum conditions the deposition of CF polymer layers on top of the adhesion layer was performed. All electrical measurements were carried out with a capacitance-voltage mercury probe, SSM – 495 CV System (Solid State Measurements Inc.) and in some cases, for comparison, with a MISCAP (metal-insulator-semiconductor capacitor) measuring system using the polymer film stack directly deposited on Si and covered by wet etched metal dots (1.0 mm^2, Al). Usually CV measurements were carried out at a frequency of 0.1 MHz but at higher frequencies (1 MHz) comparable results were obtained.

To qualify the polymer adhesion to silicon, standard and damascene architecture relevant adjacent films (barrier and etch stop layers, copper) the X-cut tape test (according to the American National Standard ASTM) with) was performed using Tesa tape 4130 (adhesion force 10 N).

RESULTS

The thermal stability of fluoropolymer films is a critical issue. It can be improved by increasing substrate temperature , LF plasma power and C content of the process gases. Using the known mixture of C_4F_8 and CH_4 [6], a LF power of 200 W and a deposition temperature T_S = 250°C, layers were deposited offering a good thermal stability for temperatures of more than 425°C. These films are representing a good compromise between low k value (k $\approx$ 2.0) [7] and high thermal stability. The deposition rate for these parameters was about 220 ... 300 nm/min and could be increased by higher plasma density. On the other hand, as shown by tape test (Fig. 1), adhesion of deposited CF films is unacceptable limited. After the insertion of a thin C based adhesion layer in every case tape test were fully passed for underlayers of SiO_2, Si_3N_4, TiN, TiW, WN. The deposition rate is about 10 nm/min and the preferred thickness was 10 nm.

This interlayer is mechanically robust (scratch test by steel) and offers a high thermal stability (> 500°C). Investigations of adhesion layers with Raman spectroscopy pointed out

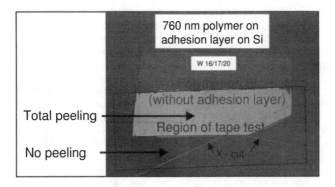

Fig. 1: Adhesion of CF polymer layers, investigated by tape test

that the film has an amorphous carbon structure (including a small amount of sp^3 bonding of carbon) indicating the strong bonding. Nevertheless for application of the double layer film stack the influence of the adhesion layer on the properties of the dielectric stack and its process integration has to be investigated.

Electrical measurements

As expected the dielectric constant of the adhesion film is much higher compared to the CF polymer film (Fig. 2). On the other hand the break down field strength is remarkably

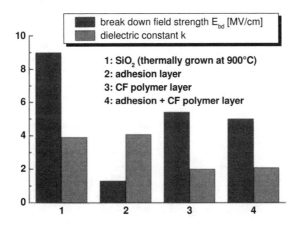

Fig. 2: Break down field strength and dielectric constant of polymer films in comparison to thermally grown SiO_2

lower than for the polymer layer. Fortunately, the electrical properties of the dielectric film stack are strongly dominated by the CF polymer.

The leakage current density of as deposited CF polymer films (film thickness 300 nm, 5 V) is comparable to that of thermally grown SiO_2 and remains below $5 \cdot 10^{-9}$ A/cm^2 for single CF films and CF films in combination with the adhesion layer.

XPS measurements

CF polymer dielectric constant is determined by the fluorine content because the strong C-F bond has a very low tendency to polarize. Thus a decrease of the dielectric constant can be achieved by increasing the fluorine concentration within the CF polymer film [8]. But on the other hand, a high fluorine concentration lowers the thermal stability. High thermal stability can be achieved only by highly cross-linked structures formed by C-C bonds (sp^3 and sp^2 bonded carbon) and CF$_3$ [8].

XPS investigations, performed on the CF polymer films, show the relatively high concentration of CF$_3$ and C-CF$_x$ (CF$_3$: CF$_2$: CF : C-C = 1 : 3 : 4 : 7) (see Fig. 3). These bonds are responsible for the high degree of cross-linking because they are able to form 3-dimensional structures. The element concentration of fluorine in polymer films deposited with a gas flow rate ratio C_4F_8/CH_4 = 6.7 is 38 at % (O: 5 at %, N: 0.5 at %, C: 54 at %). These results were calculated from XPS data (see Fig. 3).

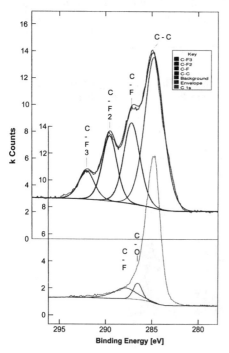

Fig. 3: XPS of CF polymer film and adhesion layer

The high oxygen concentration is detected only at the surface of the polymer layer. XPS measurements after Ar sputtering showed that the amount of oxygen concentration was strongly reduced. That means, oxygen is not significantly incorporated into the polymer. In other words CF polymers have a low tendency for moisture uptake. If water would be incorporated into the polymer, the dielectric constant should increase with time because of the strong polarizability of the water molecule. Sputtered polymer films showed also a lower fluorine concentration on the layer surface compared to as deposited surfaces. In comparison to the CF polymer, the adhesion layer mainly consists of carbon (about 89 at % of investigated elements, see Fig. 3).The relatively high amount of oxygen (9 at %) is detected only at the surface of the adhesion layer. Surface sputtering with Argon (2 kV, 0.6 µA/cm^2, 5 minutes) reduced the oxygen concentration to values lower than the detection limit. Therefore, it is possible to conclude that the detected oxygen is due to adhering water and not incorporated into the adhesion layer itself. The asymmetry of the XPS peak (adhesion layer, not sputtered) centered at a binding energy of 285 eV can be fitted with a C – O peak. This asymmetry disappears after sputtering.

Spectral ellipsometry

In Fig. 4a and 4b the refractive index of a CF polymer and adhesion layer is plotted. Using the known equation:

$$n^2 = k \tag{1}$$

(valid for dielectric materials) it is possible to calculate the dielectric constant. Within the investigated frequency range the refractive index of the examined CF polymer layer reaches n = 1.458 at 630 nm and lowers slightly at higher wavelengths (n → 1.454). k values calculated from this data (k = 2.13 at 630 nm, k = 2.11 at λ → ∞) are in very good

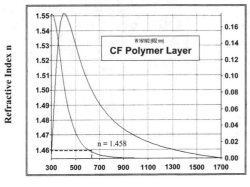

Fig. 4a: Refractive index of CF polymer layer

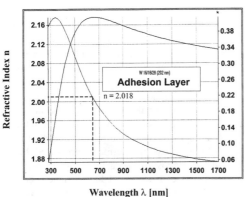

Fig. 4b: Refractive index of adhesion layer

agreement with results from capacitance-voltage measurements performed with the mercury probe (k = 2.12 ... 2.13) on the same layer. On the other hand using the spectroscopic ellipsometry for the determination of the dielectric constant a k value of about 3.5 has been calculated (λ → ∞). However, this could not be confirmed by mercury probe measurements due to the high leakage current density.

Process integration

Experiments showed that CF polymer films, deposited on an adhesion layer, have sufficient hardness and adhere strong enough for CMP processing. The polymers were polished directly (without any cup layer and without copper fill) to test the mechanical stability of the dielectric film stack. CMP was performed with a copper slurry QCTT 1010 + H_2O_2. A polishing rate of ~ 170 nm/min) was achieved [9].

In order to test the process compatibility of CF polymer films the layers were patterned by ICP dry etching (oxygen plasma) using a bilevel masking technique and electron beam exposure. After removing the hardmask a CVD barrier layer (TiN) was deposited and the vias were filled with CVD copper. The results are shown in Figure 5a and 5b.

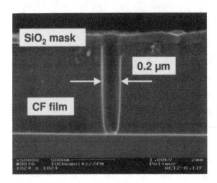 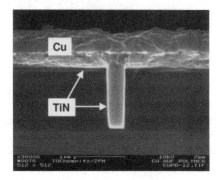

Fig. 5a: ICP etched via (O_2 plasma)   Fig. 5b: Metal filled via (CVD)

CONCLUSIONS

Due to their properties CF polymer layers deposited by PECVD are promising candidates for application as intermetal dielectric material. It is feasible to produce CF polymer films with a dielectric constant $k \approx 2.0$. Films, deposited from C, F and H containing gases are stable up to more than 400°C (annealed in nitrogen atmosphere) meeting the semiconductor industry requirements. Fluorine concentration and degree of crosslinking are responsible for the value of dielectric constant and thermal stability and can strongly be influenced by gas flow rate ratio, plasma power, deposition temperature and annealing process. The break down field strength of the polymer layer is surprisingly high (Ebd ≥ 5 MV/cm) and the leakage current density is comparable to thermally grown SiO_2. Poor adhesion behavior could be overcome by *in situ* deposition of a very thin carbon based adhesion layer. Very good adhesion to Si, several etch stop and barrier layers could be achieved. Advantageously, the adhesion layer has no significant influence on the excellent electrical properties of the CF dielectric material. Good patternability and mechanical CMP stability of the dielectric film stack is also demonstrated.

ACKNOWLEDGMENTS

This work was financially supported by the Federal Ministry of Education and Research of the Federal Republic of Germany (Project No. 01M2972A).

The authors would like to thank D. Dietrich for XPS measurements and the staff of the Center of Microtechnologies for their helpful support.

REFERENCES

1. W.W. Lee, P.S. Ho, MRS Bulletin **22** (10) (1997) p. 19
2. *The National Technology Roadmap for Semiconductors*, Semiconductor Industry Association SIA, 1997, p. 101-105
3. L. Peters, Semiconductor International, **21** (1) (1998) p. 69
4. S.E. Schulz, A. Bertz, R. Streiter, M. Uhlig, U. Weiss, T. Werner, T. Winkler and T. Gessner, 1998 International Conference on „Solid State Devices and Materials", Hiroshima, Japan, Sept. 7-10 1998; Ext. Abstract Vol. by The Japan Society of Applied Physics, p. 264.
5. E. Korczynski, Solid State Technology, **42** (5) (1999) p. 22
6. P. Xu, J. Huang, K. Singh, S. Robles, W.F. Yau, *Plasma enhanced CVD amorphous fluorinated carbon films as low k dielectrics*, DUMIC Conf., Febr. 16-17, 1998, Paolo Alto, Calif.
7. M. Uhlig, A. Bertz, M. Rennau, S.E. Schulz, T. Werner, T. Gessner, MAM`99 Conf., March 8-10, 1999, Oostende, Belgium
8. K. Endo, MRS Bulletin, **22** (10) (1997) pp. 55-57.
9. U. Schubert, private communication

Integration of Cu/FSG Dual Damascene Technology for Sub-0.18 μm ULSI Circuits

C. C. Liu, C. Y. Tsai, Y. M. Huang, J. Y. Wu, and W. Lur

UNITED MICROELECTRONICS CORP.,Advanced Technology Development Dept.
No. 3, Li-Hsin Rd. 2, Science-Based Industrial Park, Hsinchu, Taiwan, R.O.C.
Tel : 886-3-5789158 Ext. 33829, Fax : 886-3-5776889, e-mail : c_c_liu@umc.com.tw

Abstract

Low dielectric constant Fluorine-doped silicate glass (FSG) films were investigated for interlayer dielectrics in copper dual damascene process. Film properties such as deposition rate, refractive index, SiF/SiO ratio, dielectric constant, film stability after CMP process and thermal treatment were studied. Three layers of Cu/FSG dual damascene process with comparable CP yield as traditional Al/FSG process has been demonstrated on 0.18 um ULSI circuits.

Introduction

For higher performance, multi-level interconnection is becoming more and more important in sub 0.18 μm ULSI devices. The RC (resistance x capacitance) delay of interconnect is to limit the operation speed of ULSI devices [1]. It is inevitable to integrate the low-resistivity copper lines with low-dielectric-constant (K) for inter-and intra-metal dielectric materials. FSG film has been of great interest due to its low-K value and easy to be integrated in dual-damascene process. This work investigates the integration concerns on Cu/FSG dual damascene application.

Experiment

FSG films were deposited using High Density Plasma Chemical Vapor Deposition (HDPCVD) reactor with inductively coupled plasma (ICP) source [2]. The Si-F/Si-O ratio of FSG film was measured by Fourier transform infrared spectroscopy (FTIR). The dielectric constant of FSG bulk films were measured by Quantox. The comb-like test structure with 1500 sets of intra-metal dielectric trench was also used to measure the effective dielectric constant of FSG films. The film stability of FSG after Chemical Mechanical Polish (CMP) was examined by Quantox and SIMS. The

403

adhesion between FSG, TaN and SiN was tested by Scotch tape after 450°C thermal stress for 2 hrs.

Results and Discussion

The characteristics of FSG film are summarized in Table 1. The dielectric constant of FSG films measured by both Quantox and comb-like test structure method [3] is compared in Table 2. The result indicates that the dielectric constant is the same for bulk FSG film and intra-metal FSG film. Table 3 lists the refractive index and dielectric constant of FSG film before and after CMP process. The refractive index and dielectric constant were not changed. It is conceived that FSG films do not absorb any moisture during CMP process. The concentrations of Fluorine and Hydrogen were not changed before and after CMP process, as shown in Fig. 1.

In case of adhesion tests [4], Table 4 indicates no peeling observed for any stack films after thermal stress at 450°C for 2 hrs. The inter-metal capacitances for Cu/FSG and Cu/Oxide structure are shown in Fig. 2. The "NB" and "450B" in this figure represent samples without and with thermal stress at 450°C for 30 minutes, respectively. The FSG films showed about 13% reduction in the inter-metal capacitance as compared with conventional oxide films. The effective dielectric constant of FSG was calculated to be about 3.53. No obvious change in the dielectric constant was observed for FSG films after thermal stress. This indicates the FSG film is pretty stable.

For product implementation, XSEM of Cu/FSG dual damascene structures with both landed and unlanded via are shown in Fig. 3. Compared with other low-K materials, FSG is easy to be etched with good etch profile in the case of unlanded via. Cu/FSG can improve chip speed about 8 % compared to Cu/USG process. The cross sectional and top view SEM micrographs of three metal layers Cu/FSG dual damascene structures are shown in Fig. 4.

Conclusion

In conclusion, a manufacturable dual damascene process with copper and FSG has been demonstrated. FSG film can be deposited by current HDP system. The HDP films are feasible for unlanded via etch with sufficient film stability after CMP process and thermal cycle. The process has been implemented into 0.18 um technology and resulted in good device characteristics and comparable yield. The chip speed was improved 8% by using Cu/FSG compared to Cu/Oxide process.

Table 1 Summary of HDP FSG film characteristics

Properties	Deposition Rate (A/min)	Uniformity (% , 1σ)	SiF/SiO (%)	Stress (E9 dynes/cm²)	Refractive Index
HDP FSG	> 5000	< 1.5	3.0	1.5 (Compressive)	1.42

Table 2 Dielectric constant of HDP FSG films

Film	Measurement Method	
	Comb-like test pattern (ref.)	Quantox QV
Tox	-	4.06
FSG	3.53	3.57

Table 3 Film properties comparison of FSG after CMP process

	As Deposition		After CMP	
FSG Film	Refractive Index	Dielectric Constant	Refractive Index	Dielectric Constant
	1.42	3.51	1.42	3.50

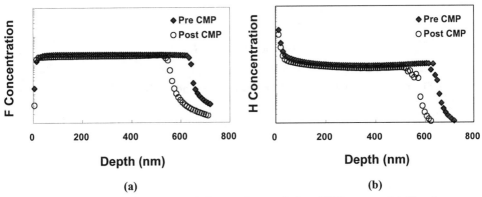

(a) (b)

Figure 1 SIMS analysis of FSG films before and after CMP process (a) Fluorine concentration, (b) Hydrogen concentration.

Table 4 Adhesion test of FSG stack films

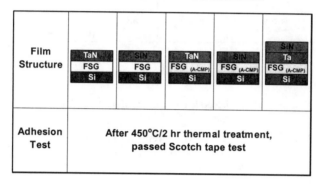

Film Structure	TaN / FSG / Si	SiN / FSG / Si	TaN / FSG (A-CMP) / Si	SiN / FSG (A-CMP) / Si	SiN / Ta / FSG (A-CMP) / Si
Adhesion Test	After 450°C/2 hr thermal treatment, passed Scotch tape test				

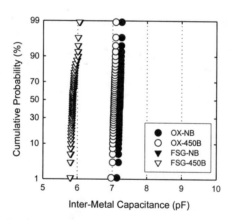

Figure 2. Inter-metal capacitance for Cu/FSG and Cu/Oxide structure.

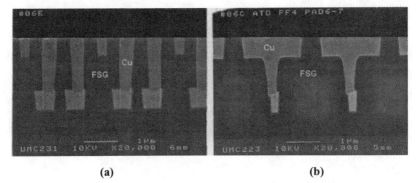

(a)   (b)

Figure 3. XSEM of Cu/FSG dual damascene post Cu-CMP with (a) Landed
and (b) Unlanded via structure.

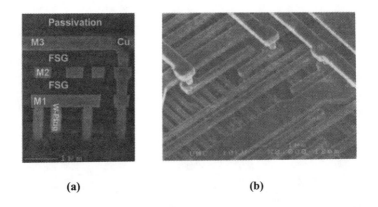

(a)   (b)

Fig. 4. SEM micrographs of three layers Cu/FSG dual damascene structure. (a)
XSEM, (b) Top View after FSG removal.

Reference

1). R. H. Havemann, et. al., "Overview of Process Integration Issues for Low-K Dielectrics", MRS 1998, p3.-14.

2). C. C. Liu, et. al., "An integrated HDP-FSG Process for sub-0.25 micron Device Application", VMIC 1998, p553-555.

3). D.S. Armbrust, et. al., "Integration of HDPCVD FSG for 0.25 um CMOS Technology", DUMIC 1998, P67~74.

4). H. M,Saad, et. al., "Integration of HDP-FSG as ILD Material in Multilevel Interconnect Devices", DUMIC 1999, p210~219.

A modified Capacitance / Voltage Technique to Characterize Copper Drift Diffusion in Organic Low-K Dielectrics

F. LANCKMANS*, L. GEENEN, W. VANDERVORST, KAREN MAEX*
IMEC, Kapeldreef 75, B-3001 Leuven, Belgium
*also at E.E. Dept., K.U.-Leuven, Belgium

ABSTRACT

Low-k dielectric materials have received considerable attention due to their use in microelectronics where capacitance between copper interconnects must be reduced in order to minimize the resistance-capacitance delay. An important issue regarding reliability and barrier requirements is the drift behavior of Cu ions in these dielectrics. The paper investigates the use of an electrical testing technique to address this issue. A conventional method to evaluate Cu penetration in dielectrics electrically is by bias temperature stressing in combination with high frequency capacitance/voltage (C/V) measurements. A Cu/(barrier)/polymer/oxide/Si capacitor structure is used. However charge instabilities during the measurements are seen. These instabilities cause shifts in the flatband voltage -which are not related with Cu drift diffusion-after biasing. To be able, therefore, to investigate the Cu drift diffusion in an organic dielectric electrically, a modification of the conventional test procedure is presented. The modification consists of etching back the polymer and Cu metal after stressing of the capacitor. Next, C/V curves are measured on these samples with the remaining oxide acting as a dielectric. In this way, the presence of Cu ions, which have been penetrated through the polymer, in the thermal oxide is detected. Comparing to the conventional method, electrically stable C/V curves are obtained which show a high reproducibility. An estimation of the drift mobility of Cu in the polymer is possible and a comparison between different low-k dielectrics can be made.

INTRODUCTION

Continuing improvement of microprocessor performance involves a decrease in the device size. This allows a greater device speed, an increase in device packing density, and an increase in the number of functions that can reside on a single chip [1]. However higher packing density requires a much larger increase in the number of interconnects. Propagation delay, crosstalk noise, and power dissipation due to resistance-capacitance coupling become significant due to increased wiring capacitance. Thus, although the speed of the device will increase as the feature size decreases, the interconnect delay becomes the major fraction of the total delay and limits improvement in device performance [2]. To address these problems, new materials for use as metal lines and interlayer dielectrics have been proposed. Cu is considered as the most promising alternative to Al-based alloys for the interconnection material due to its low resistivity and superior resistance to electromigration [3]. Since delays, noise and power consumption all depend critically on the dielectric constant of the separating insulator, much attention has focused on replacing standard silicon oxide with low dielectric constant (k) materials.

For reliable Cu integration with a conventional silicon oxide, appropriate barriers must be used to encapsulate the Cu [4][5]. At process temperatures for back end integration (< 450°C), thermal diffusion of Cu into oxide is negligible [6]. However, in the presence of an electric field, positive Cu ions (Cu^+) can drift rapidly through oxide causing reliability problems [6] [7].

In order to investigate the requirements for barriers in Cu and low-k integration, an elaborate study of the Cu drift diffusion in low-k materials is necessary. Conventional characterization

Conference Proceedings ULSI XV © 2000 Materials Research Society

techniques, including Rutherford backscattering spectroscopy (RBS), secondary ion mass spectroscopy (SIMS), Auger electron spectroscopy (AES) and X-ray photoelectron spectroscopy (XPS) have been used to do this evaluation. However, they lack great sensitivity as well as direct applicability to electrical performance [5]. Electrical characterization provides greater sensitivity for quantifying the drift diffusion. In back end processing, biased temperature stressing (BTS) of metal/insulator/silicon structures using metal contacts to the dielectric provides both a practical and sensitive electrical technique through flatband voltage shifts (ΔV_{FB}). In this technique, a voltage is applied across the capacitor, which is simultaneously being annealed at high temperatures and the resulting effects on the capacitance-voltage (C/V) behavior are subsequently measured at room temperature [8]. As already mentioned, the flatband voltage V_{FB} is an important parameter to investigate the drift diffusion and is defined as

$$V_{FB} = \phi_{ms} - \frac{Q_i + Q_{ss}}{C_{max}}$$ (1)

where C_{max} is the capacitance measured in accumulation and ϕ_{ms} is the difference in workfunction between the metal gate and Si. Q_i and Q_{ss} are the number of charges in the dielectric and at the Si/dielectric interface, respectively. Cu ions who drift in the dielectric cause a change in Q_i (Q_{ss}). This will be reflected in a lower value of the flatband voltage. A sensitivity to the amount of migrated metal atoms as low as 1×10^9 atoms/cm^2 is obtainable from the C/V response [5]. In this paper, the BTS technique is used to investigate the Cu drift diffusion in organic dielectrics. Next to an electrical evaluation, SIMS measurements are performed.

EXPERIMENT

A typical Cu gate capacitor test structure is shown in figure 1. Al metal gate capacitors are also processed and used as a reference. Approximately 20 nm thermal oxide is grown on <100> n-type silicon substrates. A spinning of the low-k polymer on top of the oxide follows this. The thickness of the polymer is 500 nm. Two different low-k polymers are used : a polyarylene ether-based polymer (Allied Signal Flare™, k•2.84) and an aromatic hydrocarbon (Dow Chemical SiLK™, k•2.65). The curing of the low-k polymers is done in a nitrogen environment.

Cu (or Al) is sputtered on top of the dielectric by using a shadowmask. The shadowmask is used to obtain a circular metal gate with an area of approximately 2.2 mm^2. To avoid severe oxidation of the Cu metal during BTS, 30 nm Ti is sputtered on top of the Cu. In order to investigate the effect of a barrier between the Cu metal and the low-k polymer on the drift diffusion behavior of Cu, a TiN layer (10 and 45 nm) is sputtered on the organic dielectric. Before sputtering, a degassing at 350°C is done to release moisture in the polymer. Al backside sputtering is necessary to have a good electrical contact during the measurement. All the test samples receive an initial heat treatment of 10 minutes at 380°C to anneal out the sputter damage to obtain a well-behaved C/V curve [9]. The heat treatment is conducted in a nitrogen environment.

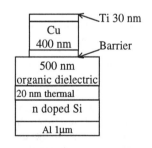

Fig. 1: Experimental capacitor structure to investigate Cu diffusion.

The thermal oxide is necessary to form a good dielectric-to-substrate interface for stable C/V curves [7]. Experiments where the polymer is in direct contact with the silicon, show a very low reproducibility of the C/V curve. In the worst case, it is even impossible to measure a curve.

BTS testing and high frequency (100kHz) C/V measurements are performed on the capacitor test structures using a HP 4156 parameter analyzer and a HP 4275A multi frequency LCR meter. The capacitors are put on a water-cooled thermal chuck in a probe station, which is purged with nitrogen. BTS testing up to 120 hours is done at different temperatures between 200°C and 300°C. The applied voltage is ranged between 20V and 40V.

RESULTS AND DISCUSSION

Electrical Characterization and Related Problems

An electrical evaluation of the Cu drift diffusion experiments on the capacitor structures (fig. 1) is presented and discussed. Figure 2 shows the flatband voltage shift (ΔV_{FB}) as function of BTS time for capacitors with or without a TiN barrier stressed at 0.5MV/cm. The polyarylene ether-based polymer is evaluated here. ΔV_{FB} is defined as

$$\Delta V_{FB} = V_{FB}(before\ BTS) - V_{FB}(after\ BTS) \tag{2}$$

where V_{FB} is the flatband voltage as calculated from the measured C/V curve. Expression 2 indicates that a lower value of the V_{FB} after BTS causes the ΔV_{FB} to be positive. This can be caused by an increase of positive charges (e.g. Cu$^+$) in the dielectric. V_{FB} is taken to be the voltage where the corresponding capacitance C_{FB} on the C/V curve is equal to equation 3 [10].

$$C_{FB} = C_{inv} + 0.95(C_{acc} - C_{inv}) \tag{3}$$

C_{inv} and C_{acc} are defined as the capacitances measured in inversion and in accumulation, respectively. At least three different capacitors are evaluated for every BTS condition (time, temperature and applied electrical field) and the average value is taken.

When applying an electrical field of 0.5MV/cm, it is clear that the ΔV_{FB} increases when the BTS temperature and time increase. The presence of a TiN barrier causes a decrease of the ΔV_{FB} for the same BTS condition. The large deviation on every data point shown in figure 2 is caused by the low reproducibility of the measured C/V curves after stressing. In some cases, it was even impossible to get a stable C/V curve. Such measurements were not taken into account. To determine whether these shifts are caused by Cu$^+$ diffusion, two additional experiments are done.

First, when applying a negative electrical field, a larger negative ΔV_{FB} with increasing time and temperature is seen (fig. 2). These shifts are not related with Cu drift diffusion because Cu ions are not expected to diffuse when applying a negative field. Only a small change of the V_{FB} - due to the elimination of charges - is expected [7]. In a second experiment capacitors with an Al gate are exposed to the same BTS conditions as the Cu gate capacitors. Both the Cu and Al gate capacitors show similar behavior for the same conditions of stressing : a positive ΔV_{FB} when a positive field is applied and a negative ΔV_{FB} for a negative applied field.

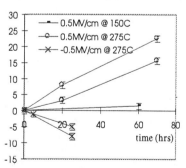

Figure 2 : Flatband voltage shift ΔV_{FB} vs. BTS time for a Cu gate capacitor without (—) and with (---) 10nm TiN barrier.

A spread out of the C/V curves after BTS is seen in both cases, indicating an increase of interface charges and traps [11].These two experiments indicate that possible drift diffusion of Cu in the dielectric is masked by other mechanisms, such as instabilities in the polymer or at an interface. These instabilities cause shifts in the V_{FB} - which are not related with Cu^+ diffusion - after biasing. The presence of dangling bonds in the dielectric or the incorporation of moisture and oxygen in the dielectric are the main cause of instabilities in the polymer [12]. Oxygen and moisture coming from the ambient explain the important influence of the atmosphere during BTS. Ambient control is obtained by purging the probe station with nitrogen, however this can not exclude all traces of moisture and oxygen. Interface related instabilities, as discussed by Wong et al. [4], occur when the low-k polymer is in direct contact with either the gate metal or Si substrate and are caused by bias-induced charges in the dielectric or at the interfaces. In this test structure, the gate metal is in contact with the low-k polymer.

The stressed capacitors were also analysed by SIMS using an oxygen ion beam. To increase the sensitivity, the samples were stripped of the Cu metal using a diluted HNO_3 solution. Due to the insulating behavior of the low-k polymer, around 20 nm Pt is sputtered on top of the polymer to avoid charging of the sample [13]. Prior to SIMS analysis, the system was calibrated by profiling samples with known doses of Cu implanted in the polymer. The detection limit was about 10^{17} atoms/cm^3. The SIMS profile of a capacitor without and with a TiN barrier stressed at 250°C and 0.5 MV/cm is shown after 80 hrs (fig. 3). The profile of a similar capacitor without applying an electrical field is given as a reference. Although the curves of figure 3.a give evidence for Cu (Cu^+) diffusion in the polymer after stressing, the difference in concentration between both profiles is rather small. The measured concentration is close to the detection limit of Cu, indicating that a lower limit is necessary. When the profiles of the capacitors with a 45 nm TiN barrier are investigated (fig. 3.b), no distinction can be made. However, the concentration is much higher as compared to the profiles without a TiN barrier. This effect is due to interferences in the mass spectrum caused by the presence of the TiN barrier. Mass interferences are ionized atomic clusters whose nominal mass-to-charge ratio equals that of the elemental ions of interest [13]. In this case Ti_xO_y molecular ions interfere with Cu ions during the measurement. Therefore it is expected that the profiles of figure 3.b are dominated by the presence of TiN and eventual Cu diffusion is difficult to detect.

In order to investigate the Cu drift diffusion in the polymer, an electrical (C/V) method is preferable because of the greater sensitivity and the applicability to electrical performance. However the instabilities encountered motivate a modification of the experiment to avoid these problems.

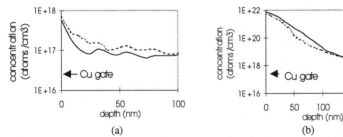

Figure 3 : SIMS profile of Cu in the polyarylene ether-based polymer after 80 hrs of stressing at 275°C and 0.5 MV/cm (—). The profile of a capacitor without applying an electrical field is given as a reference (---); (a) no barrier, (b) 45 nm TiN barrier.

Modified Capacitance – Voltage Technique

In this new modified method, capacitor structures as shown in fig. 4.a are processed. The processing steps are the same as for the structure shown in figure 1 including the annealing. The only difference is that blanket Cu films are sputtered. Al films are processed as well and are used as a reference. Next, the wafers are cleaved in square pieces of approximately 4 cm². These capacitors are stressed at different temperatures (BTS). In a next step, the gate metal(/barrier) and organic dielectric are etched using a sulfuric acid/hydrogen peroxide mixture (fig. 4.b). An Al gate metal is sputtered on the remaining thermal oxide (fig. 4.c) followed by a thermal anneal at 200°C for 10 minutes in a nitrogen ambient. C/V measurements are performed on these Al gate capacitors. In this way, the presence of Cu ions -which have been penetrated through the polymer - in the thermal oxide or at the Si/oxide interface can be detected. Comparing to the conventional method, electrically stable C/V curves are obtained which show a high reproducibility (fig. 5).

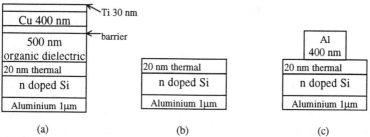

(a)   (b)   (c)

Fig. 4 : Modified test sequence to investigate Cu drift diffusion: (a) initial capacitor structure for BTS test, (b) wet etch of gate metal/barrier and polymer, (c) Al gate metal deposition. The actual C/V curves are measured on test structure c.

Within the window tested, a clear shift in the C/V curves is seen after BTS when no barrier is present (fig.5.a). No shift is seen for the same BTS conditions when a 45 nm thick TiN barrier is present (fig. 5.b). Reference measurements on capacitors with Al instead of Cu show also no change in the C/V curve. These two experiments give a clear indication that Cu ions have been drifted through the polymer in the oxide.

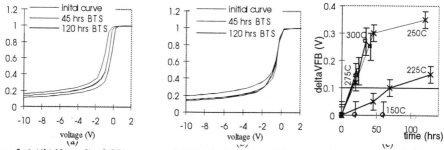

Figure 5: (a)(b) Normalised C/V curves vs. BTS time using the modified test sequence shown in figure 4. BTS conditions are 0.6 MV/cm at 250°C. (a) capacitors without barrier and (b) capacitors with a 45 nm TiN barrier. (c) Flatband voltage shift ΔV_{FB} as a function of BTS time at 0.6MV/cm as measured from the modified test sequence for samples without a barrier. All initial samples contain a Cu gate and a polyarylene ether-based dielectric.

The C/V curves allow calculating the ΔV_{FB} as a function of the stressing time for different conditions of temperature and electrical field. Figure 5.c shows the results for a polyarylene ether-based polymer stressed at 0.6 MV/cm for different temperatures. The general trend is that more stringent BTS conditions (time, temperature and electrical field) cause a larger ΔV_{FB}, indicating an increasing amount of Cu ions in the oxide. For the most severe test conditions (80hrs at 275°C and 0.7 MV/cm) no ΔV_{FB} is seen when a 45 nm thick TiN layer is present between the Cu and the low-k polymer, giving evidence for the barrier properties of TiN.

The modified capacitance/voltage method allows making a comparison between different low-k polymers with respect to the Cu drift diffusion. As already mentioned in the experimental part, two types of polymers are investigated: a polyarylene ether-based and an aromatic hydrocarbon. The ΔV_{FB} as a function of stressing time for four different stress conditions is shown in figures 6.a and 6.b. Using these results, it is possible to estimate the initial drift mobility μ of Cu⁺ in the two dielectrics as function of temperature (fig. 6.c). The initial mobility is calculated here since the electric field in the dielectric changes with time due to accumulation of Cu ions [7]. The actual calculation is performed using

$$t_{drift} = \frac{d^2}{\mu V} \tag{4}$$

where d is the thickness of the low-k dielectric and V is the voltage drop across this dielectric [6]. The drift diffusion time t_{drift} is defined as the time for the Cu ions to penetrate through the polymer. t_{drift} can be approximated as the time to observe a significant change ($\Delta V_{FB}=0.1V$) in V_{FB} of the measured C/V curve in the modified method [6]. This assumes that Cu ions have been drifted as well through the thermal oxide. However, within the window of BTS test conditions the time to diffuse through the thermal oxide is at most one hour [6] (fig. 6.c), which is the same order of magnitude as the error made in the determination of the time where the $\Delta V_{FB}=0.1V$. The last is established by interpolation between the experimental points. Every data point in figure 6.c is an average of experiments done at three different values of the electrical field. At least three capacitors are evaluated for each value of the electrical field. The large scattering on the calculated values of μ is mainly caused by this interpolation. Literature values of the mobility of Cu ions in thermal oxide – using the conventional electrical test procedure - are shown as a reference [6]. Taking the slight difference of test method into account, the initial drift mobility of Cu ions in the tested polymers is lower than in thermal oxide. μ is also slightly smaller for the aromatic hydrocarbon than for the polyarylene ether-based polymer within the tested window.

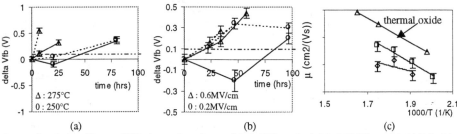

Figure 6 : (a,b) ΔV_{FB} vs. BTS time using the modified C/V method (a) 0.7MV/cm; (b) 300°C; (c) Arrhenius plot of the estimated initial drift mobility μ of Cu⁺ in the aromatic hydrocarbon (•) and the polyarylene ether-based polymer (---) vs. 1000/T. Every data point is an average of at least nine experiments. Literature values of the μ of Cu⁺ in thermal oxide -using the conventional test method- are given as a reference [6].

Loke et al. [14] also determine the initial Cu⁺ drift rate in different low-k polymers. An oxide-sandwiched low-k polymer capacitor structure is used to avoid charge instabilities. The drift rates were extracted by performing gate current versus gate voltage measurements at an electric field of 0.8MV/cm. A higher initial Cu⁺ drift rate is found for a polyarylene ether-based polymer than for an aromatic hydrocarbon. The small differences between the mobility for the two low-k materials are probably related to the interaction between the Cu ions and the chemical structure of the polymer. Loke et al. [14] suggest that the penetration of the Cu ions through the dielectric is retarded by increased crosslinking of the polymer which is the case for the aromatic hydrocarbon showing a highly crosslinked structure.

CONCLUSIONS

BTS and C-V measurements are performed to investigate Cu drift diffusion in low-k polymers because of the important role of Cu diffusion regarding reliability issues in integration. However, instabilities in the polymer or at the interface with the gate metal mask the effect of Cu diffusion in the dielectrics. In order to use these sensitive electrical techniques, a modification of the test method is presented. In this modified C/V technique, the Cu ions, which have been diffused through the polymer in an underlying gate oxide, are measured. The problems encountered using the conventional test procedure are alleviated with this new reliable method. The Cu^+ diffusion in both polymers is significantly lower than in thermal oxide. Cu drifts more readily in polyarylene ether-based polymers than in an aromatic hydrocarbon. A TiN barrier is very effective in slowing down the drift diffusion of Cu.

ACKNOWLEDGMENTS

We thank IMEC P-line for wafer processing. F. Lanckmans acknowledges IWT for financial support. K. Maex is a research director of the National Fund for Scientific Research Flanders.

REFERENCES
[1] W.W. Lee and P.S. Ho, MRS Bulletin, 22, No. 10, p. 19-23 (1997).
[2] D.C. Edelstein, G.A. Sai-Halasz, and Y-J. Mii, IBM J. Res. Dev., 39, No.4, p.383 (1995).
[3] Y.-L. Lee, B.-S. Suh, S.-K. Rha, C.-O. Park, Thin Solid Films, 320, 141-146 (1998).
[4] S.S. Wong, A.L.S. Loke, J. T. Wetzel, P.H. Townsend, R.N. Vrtis, M.P. Zussman, Low-Dielectric Constant Materials IV. Symp. Mater. Soc, Warrendale, PA, USA; xi+386 pp., p. 317-27 (1998).
[5] S. P. Murarka, Materials Science and Engineering, R19, Nos. 3-4, pp. 88-147 (1997).
[6] Y. Shacham-Diamand and A. Dedhia, J. Electrochem. Soc., 140, No. 8, p. 2427-32 (1993).
[7] A.L.S. Loke, C. Ryu, C.P.Yue, J.S.H. Cho and S.S. Wong, IEEE Electron Device Letters, 17, No. 12, p. 549-51 (1996).
[8] W.N. Carr, MOS/LSI Design & Applications, Chapter 1 : MOS device Physics, TI Incorporated, Mc Graw Hill book company, 1972.
[9] T. S. de Felipe, S.M. Murarka, J. Vac. Sci. Technol. B 15 (6), p. 1987-9 (1997).
[10] J.K. Cramer, S.P. Murarka, K.V. Srikrishan, J. Appl. Phys. 73 (5), p. 2458-61, 1993.
[11] M.S. Angyal, Y. Shacham-Diamand, J.S. Reid, Appl. Phys. Lett., 67 (15), p. 2152-4 (1995).
[12] M. Vogt, M. Kachel, M. Plötner, Microelectronic Engineering, 37/38, p. 181-7 (1997).
[13] C.R. Brundle, C.A. Evans, S. Wilson, Encyclopedia of Materials Characterization, Butterworth-Heinemann, Stoneham, 1992, pp. 300-320.
[14] A.L.S. Loke, S.S. Wong, N.A. Talwalkar, J.T. Wetzel, P.H.Townsend,T. Tanabc, R.N. Vrtis, M.P. Zussman, D. Kumar, Low-k/Advanced Interconnect IV. Symp. Mater. Soc., Pittsburgh, Vol. 565, in press.

Oxazole Dielectric (OxD) as a potential low-k dielectric: Properties and preliminary integration results with a Cu damascene architecture

Manfred Engelhardt[1], Hans Helneder, Guenter Schmid[1], Michael Schrenk, Markus Schwerd, Uwe Seidel, Heinrich Körner and Recai Sezi[1]

INFINEON TECHNOLOGIES AG, Technology Development, Wireless Products, 81739 Munich, Germany
1) Infineon Technologies AG, Corporate Research

Abstract

This paper presents a first feasibility study of the low-k material Oxazole Dielectric „OxD" and its chemical, mechanical, and dielectric properties. In particular, the impact of various plasma treatments with inert (Ar) and reactive (CF_4, O_2) species and the resulting consequences for integration schemes are reported. A first assessment of its suitability for a damascene architecture is given by integration of OxD with Cu in a single damascene architecture including the electrical evaluation of this metallization and by patterning dual damascene structures into the polymer.

Introduction

Future devices with ever shrinking dimensions are expected to suffer from enhanced interlevel and intralevel crosstalk between metal lines. Thus, parasitic capacitances and propagation delay times for high signal frequencies will be increasing and device performance will be negatively impacted [1]. One of the current main research focuses is the evaluation and assessment of new materials with a low (< 4) dielectric constant "k" in order to reduce these parasitic effects [2]. Currently several different materials are investigated which have the potential of providing k-values between 3.5 and 2.6 [3]. In addition to a low k value, several other important mechanical, thermal and electrical properties have to be fulfilled and feasible integration schemes need to be worked out.

This paper presents a first study of the low-k material Oxazole Dielectric („OxD"), its chemical, mechanical, thermal and dielectric properties, an investigation of its behaviour during various plasma treatments and a first assessment of its suitability for a damascene architecture.

Experimental data and material properties

The precursor of the spin-on low-k material OxD is a fluorinated poly(o-hydroxy)amide, which is converted to Oxazole Dielectric („OxD") by thermal cure at $T \geq 350$ °C. The chemical formulas and the schematics of this reaction are shown in fig. 1. A whole series of materials can be obtained by modifing the spacer moieties S_{AP} and S_{DA}.

The dielectric constant k of a first OxD material has been determined to be approx. 2.9 in plane and out of plane, after deposition and cure, by means of impedance spectroscopy and contact free inline measurement, respectively. The k-value is almost independent of frequency in the range

Conference Proceedings ULSI XV © 2000 Materials Research Society

of 10 kHz to 10 MHz, as shown in fig. 2 and it does not change with temperature cycling up to 450°C. Meanwhile, there is an improved version of OxD material available (not shown here, but reported in [4]) with a k-value of 2.55 and the same performance properties. Additional material data of the new version is compiled below.

Dielectric properties		Mechanical Properties	
Dielectric constant	2.55	Stress	35 MPa
Dissipation factor	0.005 (MIM 1 MHz)	Therm. exp. coefficient	31 ppm
Dielectric strength	4 MV/cm	Young modulus	2.6 GPa
Leakage	$<10^{-9}A/cm^2$@1 MV/cm	Hardness	0.4 GPa
Volume resistivity	$2.6 \cdot 10^{16}$ Ωcm	Elongation	10%
Surface resisitivity	$10 \cdot 10^{14}$ Ω		
Refractive index	1.60	**Thermal properties**	
Birefringence	0.01	Thermal stability	> 500°C TGA
		Thermodesorption	No outgassing @400°C
Water uptake			
Dipped for 24h@25°C	0.1% (fully reversible)	**Adhesion**	
24h@85°C, 85%RH	0.2% (fully reversible)	Stud -pull	> 60 MPa
		Tape Test	0/100 failure rate
Chemical stability		Thermal cycling	Passed
Excellent against acids, bases, solvents		Substrates	Si, SiO_2, SiN, TaN, TiN

The dielectric layer excellently adheres to PECVD-SiO_2 and –SiN and also provides excellent adhesion for these materials. Thickness non-uniformities are better than 1% (1 sigma) across a 150 mm wafer, and the layers are thermally stable up to 500 °C (3% wt. loss).

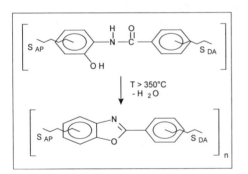 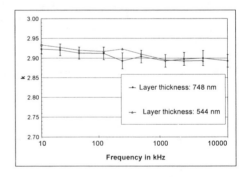

Fig. 1: Chemical formula of the precursor and of the final low-k material and the schematics of the reaction

Fig. 2: Dielectric konstant k of OxD as a function of frequency (10 – 100.000 kHz)

Results and discussion

Impact of plasma treatments

In a Damascene architecture, vias and trenches need to be patterned into the dielectric. During this patterning process, the material might be exposed to plasma environments which typically contain chemically inert (e. g. Ar) and chemically active (e.g. fluorinated hydrocarbons or oxygen) compounds. Therefore, the impact of plasma treatments with such typical etch species like Ar, CF_4 and O_2 on the bulk and surface composition and on the properties of OxD has been investigated.

The *XPS spectra (C1s)* of the untreated *OxD surface* in fig. 3 is in full agreement with its chemical composition. After an enhanced Ar plasma treatment (480 s) the OxD surface shows a decrease of CF_3-type carbon species, which suggests a complete removal of CF_3- moieties from the surface due to the physical bombardment.

A CF_4 plasma contains highly reactive species like radicals i.e. F or CF_3 , ions and non-charged species i.e. CF_2. These species react similar to the SiO_2 dielectric etch chemistry readily with the surface. Besides some etching a non-specific fluorination of the polymer surface occurs, expressed by the emerging C(1s) signals from C-F and $C-F_2$ moieties. In fig. 3 an XPS spectrum of an OxD surface after a 120 s CF_4 plasma treatment is depicted.

In analogy, 120s of oxygen plasma treatment lead predominately to oxidation (etching) and to a change of bonding types, as indicated by the intensity change of the various species. The etching rate of OxD in O_2 plasma is about 4 times higher than in CF_4 plasma.

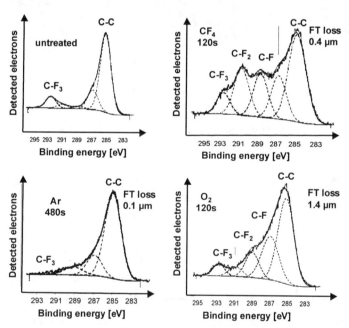

Fig. 3: XPS spectra (C1s) of the OxD surface after certain Ar (480s), CF_4 (120s) and O_2 (120s) plasma treatments in comparison with the untreated surface.

Atomic *Force Microscopy (AFM)* studies exhibit definite modifications of the *surface* after the same plasma treatments, in comparison with the untreated species (Fig 4). After 120 s of reactive CF_4 and O_2 plasma treatment, respectively, a severe increase in OxD *surface roughness* is observed, with the stronger oxidizing/etching O_2-plasma exhibiting the biggest impact.

Significantly longer treatment times are required with Ar plasma in order to roughen the surface. This result is believed to be due to the pure physical bombardement associated with Ar plasma and again indicates that there is also a chemical interaction of the surface with CF_4 and O_2 plasmas, respectively. The AFM investigations also lead to the assumption that plasma treatments might be a suitable means for improving the adhesion of metal layers to the OxD surface due to the increase of surface roughness.

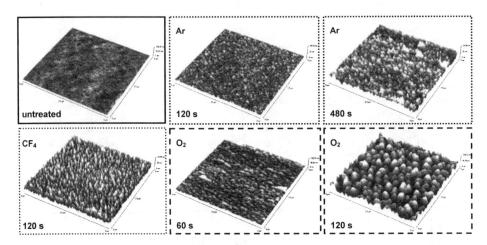

Fig.4: AFM investigations of the OxD surface after certain Ar, CF_4 and O_2 plasma treatments in comparison with untreated samples

These changes in the chemical composition of the surface are also supported by the FTIR spectra of *very thin* films (Fig 5).

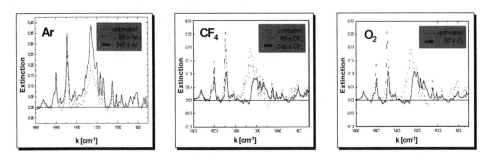

Fig. 5: FTIR spectra of OxD after certain Ar, CF_4 and O_2 plasma treatments in comparison with spectra of untreated samples

After an Ar plasma treatment of up to 240s, no change of the peak position is observed relative to the untreated material. The change of the intensity of certain peaks is presumably due to a modification of the surface structure (in agreement with the XPS investigations) and due to the material loss (thinning) caused by the Ar bombardement. In contrast, significant changes in both peak position and signal intensity are observed after reactive plasma treatments with CF_4 and O_2, respectively.

Possible modifications of the *bulk composition* of OxD after similar plasma treatments have been also investigated by FTIR. Films thicker than 200 nm show no differences in peak positions and intensities relative to the untreated material, indicating that no modification of the bulk polymer occurs.

Fig. 6 shows the impact of plasma treatments on k as a function of treatment time and frequency. Ar and CF_4 plasma treatments result in a decrease of the dielectric constant, which gets more pronounced with longer treatment times. As shown above, these treatments lead to an increased surface roughness and thus to a reduced density especially at the surface. X-ray reflectometry (not shown here) indicated a nominal decrease of the layer density of 6% for Ar plasma treatment and 3.5% for CF_4 treatment, respectively. As shown with FTIR in fig. 5, additional CF_x-species are incorporated into the surface during CF_4 plasma treatment, which explains the less pronounced decrease in material density. As k is related to the density via k ∝ 1/(1-density), the decrease of k due to the density decrease is easily understood. Initially, short O_2 plasma treatments lead to an increased k-value, as shown in Fig. 6. This is due to the incorporation of O-species (partial oxidation of OxD). Longer exposure times lead to a slight decrease of the k-value, which also can be explained by surface roughening and density decrease.

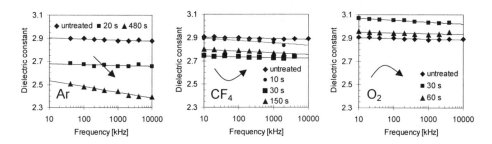

Fig. 6: Impact of Ar, CF_4 and O_2 plasma treatments on the dielectric constant k as a function of treatment time and frequency (10 kHz to 10 MHz).

These investigations with species typically used for etching and resist stripping recommend that a hardmask should be used on top of the low-k material during patterning processes and especially in order to efficiently remove the resist without impacting the surface composition and properties of OxD.

Integration in Single and Dual Damascene architecture
The successful integration of OxD into a single damascene architecture and its electrical characterization is reported below. We have also proven that OxD patterning is feasible for Dual Damascene architecture, as shown in the SEMs in fig. 7.
The SEMs in fig. 7 exhibit a 700 nm thick OxD layer on top of SiO_2, covered by 150 nm of SiO_2 which has been used as a hardmask during via patterning. The via is patterned using i-line litho and the "burried via" approach. The corresponding processing sequence is displayed in fig. 8.

Initially, the hardmask is partially etched using standard resist and oxide etch technology. Next, the resist is stripped using oxygen plasma while the OxD polymer is still protected by the SiO_2 hardmask. In the third step, the hardmask is completely opened; it will allow the selfaligned patterning of the via later. The intermetal dielectric is completed by depositing a second, 500 nm thick OxD layer which will house the metal line afterwards. On top of the OxD, 150 nm SiO_2 has been deposited which again acts as cap layer and hardmask. For the Dual Damascene architecture, a trench is defined into the top hardmask and OxD by applying i-line litho and the same etch processing sequence as shown in fig. 8 (partial open of hardmask, resist removal, complete opening of hardmask). Then, the trench (for the metal line) and the via are simultaneously patterned into the top and bottom OxD layer, respectively, in one etch step using O_2 plasma. This etch chemistry is highly selective towards SiO_2. Thus, the SiO_2 hardmask between the two OxD layers is succesfully used as stop layer for trench patterning, resulting in well defined trenches with highly reproducible thicknesses. Further optimization is in progress in order to reduce hardmask thicknesses.

The SEMs in fig. 7 demonstrate the excellent selectivity of the OxD patterning process towards the SiO_2 hardmasks and simultaneously exhibit smooth via and trench bottoms free of any residues. The sidewalls of the via and the trench are smooth and clean and the dimension control of both features is perfect. Thus, the feasibility of an advanced dual damascene architecture with OxD has been proven.

In fig. 9, copper lines (widths 0.5 μm and 0.7 μm, spacing ≥ 0.3 μm) embedded in OxD after CMP processing are depicted. The SiO_2 hardmask was polished off totally during the CMP sequence and the remaining Cu line thickness is 500 nm, as anticipated. The hardmask loss was no problem neither for subsequent processing nor for leakage current evaluation. Meander line resistances (width 0.55 μm, length 531 mm; figure 10) yielded high (approx. 90%) with exactly the anticipated mean line resistance and an acceptable within-wafer variation. These very promising results are attributed to proper control of the overall process flow, to minimum dishing and erosion during CMP processing as well as to no degradation of copper by the OxD material during processing. Furthermore the leakage current between unpassivated copper comb and serpentine structures (0.5 μm line width and 0.5 μm spacing) shown in figure 11 is comparable to Cu/SiO_2 metallization as well as to standard passivated AlCu metallization. Neither corrosion nor outgassing have been observed due to the fluorine content of the material.

Fig. 7: SEMs of Dual Damascene architecture patterened into OxD

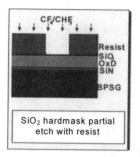 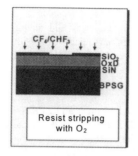 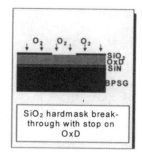

SiO₂ hardmask partial etch with resist	Resist stripping with O₂	SiO₂ hardmask breakthrough with stop on OxD

Fig. 8: Schematic of the OxD patterning sequence

Fig. 9: SEMs of 0.5 μm deep and 0.5 μm as well as 0.7 μm wide lines after CMP (for preparation purposes the surface is coated with tungsten)

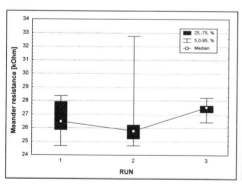 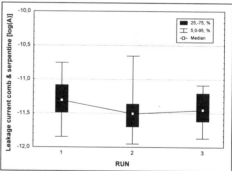

Figure 10: Meander line resistance of 531 mm long meanders (0.55μm line, 0.55μm spacing)

Figure 11: Leakage current of comb and serpentine patterns (0.5 μm width, 0.5 μm spacing)

423

Conclusion:
OxD exhibits very promising mechanical, physical and chemical properties as spin-on low-k material and shows very promising preliminary electrical results when integrated into a Single Damascene architecture with Cu. We have also proven that OxD patterning is feasible for Dual Damascene architecture.

Acknowledgement:
This work has been funded in parts within the European ESPRIT-Project "Damascene" (ESPRIT # 25220) and by the Bundesministerium für Forschung und Technologie (BMBF) within the project number 03N1024. For the content, the authors alone are responsible.

References:

1. W.W. Lee, P. S. Ho, Guest Editors, MRS Bulletin, Vol. 22, No. 10, Oct. 1997, p. 19.
2. P. Singer, „Low k Dielectrics: The search continues"; Semiconductor International, May 1996, p. 88.
3. L. Peters, „Pursuing the Perfect Low-k Dielectric", Semiconductor International, September 1998, p. 64
4. G. Schmid, A.Maltenberger, W. Radlik, R. Sezi, A.Weber, K. Buschik, Proc. 5th int. Dielectrics for ULSI Multilevel Interconnect Conference DUMIC, February 8 – 9, 1999 Santa Clara, 1999 page 35, (IMIC – 444D/99/0035)

Plasma CVD of Low-*k* a-C:F Films Using Substitutional PFC

T. SHIRAFUJI, Y. HAYASHI, S. NISHINO
Department of Electronics and Information Science, Kyoto Institute of Technology
Matsugasaki, Sakyo, Kyoto 606-8585 Japan

ABSTRACT

Low-*k* a-C:F films have been prepared from the low-GWP substitutional fluorocarbon source of C_5F_8 by a PE-CVD method. It has been found that the films prepared under higher flow rate of C_5F_8 show higher thermal stability in addition to higher deposition rate. These results indicate importance of controlling secondary reactions by means of adjusting the mean residence time of gas-phase species in the reaction chamber. *In situ* gas-phase diagnostics using OES and FT-IR were applied for investigating the effects of residence time on gas-phase composition, and revealed that long-residence-time discharge was no longer C_5F_8 plasma but CF_4/C_2F_6 plasma, while inclusion of CF_4/C_2F_6 in short-residence-time discharge was small.

1. INTRODUCTION

Shrinkage of the metal-line spacing in ultra large scale integrated circuits requires low dielectric constant intermetal dielectric (IMD) films for reducing resistivity-capacitance time delay. For the feature dimensions less than 0.13 μm, IMDs have to have the dielectric constant below 2.5[1]. Amorphous fluorinated carbon (a-C:F) films deposited by plasma enhanced chemical vapor deposition (PE-CVD) methods are promising candidates for the IMDs, and the process integration using a-C:F has been successfully demonstrated by Endoh *et al* [2]. However, the most commonly used source gas is C_4F_8, and it possesses quite high global warming potential (GWP) of 8,700 and atmospheric life time of 3,200 years[3]. Therefore, substitutional gases possessing lower GWP and life time are required from the view point of environmental protection. Recently, C_5F_8, which have GWP and life time as low as 100 and 1 year, has been introduced to the reactive etching processes[3]. In this work, C_5F_8 was applied for an IMD film deposition process, and the properties of the films, especially their thermal stability, were investigated. Gas-phase of the plasma was also investigated for explaining the film properties.

2. EXPERIMENTAL

Figure 1 shows schematic representation of the experimental setup. Films were deposited on (100) p^+-type Si substrates using a capacitively coupled RF (13.56 MHz) PE-CVD reactor. The diameter of the RF electrode and the distance between the electrode and grounded substrate holder were 70 and 30 mm, respectively. The diameter of the substrate holder is the same as the RF electrode.

Although C_5F_8 is liquid source at room temperature (25°C), its vapor pressure is as high as approximately 560 mmHg. Therefore, C_5F_8 was supplied to the reaction chamber through a mass

425

flow controller without bubbling by any inert gases. Summary of the deposition conditions is listed in **Table I**.

The thermal stability of the films was investigated by measuring the infrared (IR) absorption intensity at the wavenumber around 1260 cm⁻¹, which corresponds to C-F$_n$ (n=1,2, and 3) bonds in the films, as a function of annealing temperature. The thermal treatment was performed for 1 hour in N_2 ambient at 1 atm. The dielectric constant of the films was determined by measuring the capacitance of a metal-insulator-semiconductor (MIS) structure consisting of Al, film and p^+-Si at the frequency of 1 MHz. Gas phase of C_5F_8 plasma was investigated by optical emission spectroscopy (OES) and *in situ* Fourier transform infrared (FT-IR) spectroscopy[4].

3. RESULTS AND DISCUSSION

3.1 Deposition Rate

Figure 2 shows deposition rate of the films as a function of substrate temperature for the C_5F_8 flow rate of 3.2 and 16 sccm. While the deposition rate decreases down to almost zero at 300°C for the flow rate of 3.2 sccm, it keeps 80 nm/min at 300°C for the flow rate of 16 sccm. Deposition rate for C_4F_8 plasma, which is shown in the same figure for comparison, is fairly low in comparison to that for C_5F_8 plasma. The high deposition rate of C_5F_8 plasma cannot be explained simply by increase in the number of C atoms from 4 to 5, and suggests that the deposition precursors are different from those in C_4F_8 plasma.

In order to increase the deposition rate, hydrogen or hydrocarbon gases are required[5]. However, the use of H-containing gases increases dielectric constant of the films. Therefore, the high deposition rate at elevated temperature is another advantage of C_5F_8 in addition to its original feature of low GWP. Although

Table I. Typical deposition condition

RF:	40 W (φ6)	C_5F_8:	3.2 ... 16 sccm
T_{sub}:	30 ... 400°C		
P:	0.2 ... 0.3 Torr		

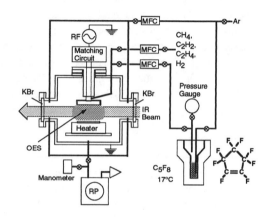

Fig.1 Schematic representation of PE-CVD system.

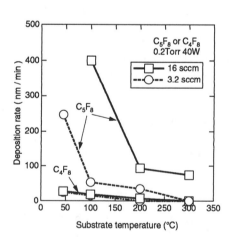

Fig.2 Deposition rate of a-C:F films from C_4F_8 and C_5F_8 under flow rates of 3.2 and 16 sccm as a function of substrate temperature.

the deposition rate at 300°C is not enough for practical use, negative bias on substrate holder has possibility to increase the deposition rate as in the case of C_2F_4[6].

1.2 Film Structure

Figure 3 shows FT-IR spectra of the deposited films under flow rate of 3.2 and 16 sccm. As shown in the figure, there are no marked difference in C-F stretching mode around 1260 cm^{-1} in the spectra for the films deposited under high and low flow rate. However, the films deposited with higher flow rate contains larger peak of C=C bonds[7]. This suggests that deposition precursor contains C=C bonds.

Figure 4 shows XPS spectra of the films. F/C ratio in the figure was calculated from integrated intensity of C(1s) and F(1s) regions considering relative sensitivity for C and F. As seen in the figure, the films deposited under higher flow rate contain higher amount of C-CF bonds and less amount of C-F$_3$ bonds. This means that higher flow rate brings about higher cross-linking of C atoms in the films.

F/C ratio of 1.67 for higher flow rate is close to that of C_5F_8 (1.60), which suggests that structure of deposition precursor is resemble to source molecule.

3.2 Dielectric Constant

Figure 5 shows the dielectric constant of the films prepared from 16 sccm C_5F_8. The films prepared at 100, 200 and 300°C show the dielectric constant less than 2.0 of poly-TFE films[8]. The films prepared at elevated temperature is usually considered to have a rigid structure, which results in higher density and higher dielectric constant. Our re-

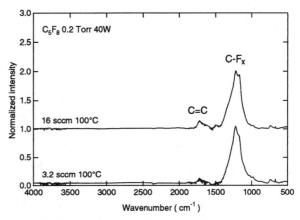

Fig.3 FT-IR spectra of a-C:F films deposited from C_5F_8 under the flow rates of 3.2 and 16 sccm.

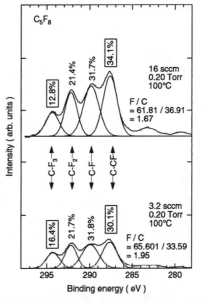

Fig.4 XPS spectra of the films deposited from C_5F_8 under flow rates of 16 and 3.2 sccm. Substrate temperature was 100°C. F/C ratio was calculated by using integrated intensity of F(1s) and C(1s) region.

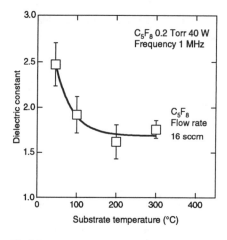
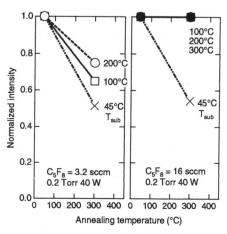

Fig.5 Dielectric constant of a-C:F films from C_5F_8 under flow rates of 16 sccm as a function of substrate temperature. Measurements were performed on the MIS structure of Al/Film/p+-Si at frequency of 1MHz.

Fig.6 Thermal stability of the films deposited under low and high flow rate. Vertical axis indicates IR absorption intensity around 1270 cm^{-1} corresponding to C-F bonds in the films. Annealing was performed in N_2 (1atm) ambient for 1 hour.

sults show opposite characteristics. The films prepared at higher temperature might have porous structure due to desorption of a part of film components, because the deposition rate of the films decreases with increasing substrate temperature and sticking probability of some fluorocarbon radicals (CF_x) are known to decrease by elevating surface temperature[9].

3.3 Thermal Stability

Figure 6(a) and **6(b)** show thermal stability of the films prepared under the C_5F_8 flow rate of 3.2 and 16 sccm. In this figure, decrease of the intensity by annealing means thermal instability of the films.

As seen in **Fig.6(a)**, the films prepared from low flow rate of 3.2 sccm show poor thermal stability although the stability improved by increasing substrate temperature. On the other hand, as seen in **Fig.6(b)**, the films prepared from 16 sccm of C_5F_8 show good stability, except for the film deposited at 45°C. The thickness of the films, which is not shown in this paper, also did not change for these stable films. From these results, it is concluded that the films prepared from higher flow rate show higher thermal stability. Manipulation of flow rate with keeping total pressure corresponds to manipulation of mean-residence-time in the reaction chamber. In the plasma for different residence-time, gas-phase composition might be different. Therefore, we have investigated the gas-phase using OES and *in situ* FT-IR.

3.4 Gas-Phase Diagnostics

Figures 7(a) and **7(b)** show the discharge-time dependence of OES spectra of the plasma under the flow rate of 3.2 and 16 sccm which corresponds to the residence time of 300 and 60 ms. The main peak around 240-300nm arises from CF_2 and CF_2^+[10,11], but we cannot assign the origin of the component around 300-400nm. Regarding the spectra for just after turning discharge on, spectral shape is resemble for both the flow rate. With increasing discharge duration, the spectrum for long-residence-time (3.2sccm) changes, and the peaks corresponding to CO appear[11]. The time for this variation is quite longer than the time-scale of residence-time, which suggests that the change in gas-phase composition is occurring through surface reactions. On the other hand, the spectrum for short-residence-time (16sccm) does not change after turning discharge on. This means that the long-residence-time discharge generates chemical species which extract oxygen on the surface instead of deposit on it.

Figure 8 shows the gas-phase IR absorption spectra as a function of discharge time. Each spectrum were measured under discharge-OFF condition after discharge for the duration indicated in the figure. Assignment of the peaks were performed according to the previous report[12]. As seen in the figure, gas-phase composition changes as follows;

$$C_5F_8 \rightarrow C_2F_4 \rightarrow C_2F_6 + CF_4 \rightarrow CF_4.$$

This means that the discharge with long-residence-time is no longer C_5F_8 plasma, but CF_4 or C_2F_6 plasma in which etching occurs rather than deposition. The extraction of oxygen on the surface for long-residence-time discharge is considered to be due to that the plasma has tendency of etching.

These results suggest that higher thermal stability obtained for higher flow rate is caused by contribution of the deposition precursor which is specific for C_5F_8.

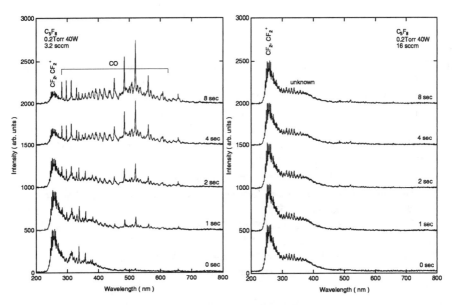

Fig.7 Discharge time dependence of OES spectra for C_5F_8 plasma under low and high source-gas flow rate. Measurements were performed at sheeth edge at which the highest OES intensity was obtained.

4. Conclusion

In conclusion, low-k a-C:F films have been prepared from low-GWP source of C_5F_8 by a PE-CVD method, and it has been found that the films prepared under higher flow rate of C_5F_8 shows higher thermal stability in addition to higher deposition rate. Gas-phase diagnostics revealed that the gas-phase composition for the long-residence-time changes gradually after turning discharge on, and becomes no longer C_5F_8 plasma, but CF_4/C_2F_6 plasma. This result suggests that importance of control of secondary reactions by means of adjusting the mean residence time of gas phase species in the reaction chamber.

Acknowledgments

This work has been partly supported by the Kinki-Area Invention Center.

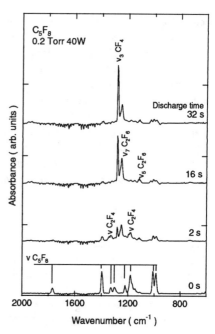

Fig.8 Discharge time dependence of OES spectra for C_5F_8 plasma under low and high source-gas flow rate. Measurements were performed at sheeth edge at which the highest OES intensity was obtained.

References

[1] T. Horiuchi: Proc. 20th Symp. Dry Process (Inst. Elec. Eng. Japan, Tokyo, 1998), p.163.

[2] K. Endo, T. Tatsumi, Y. Matsubara and T. Horiuchi: Jpn. J. Appl. Phys., **37**, 1809 (1998).

[3] Y. Ito, A. Koshiishi, R. Shimizu, M. Hagiwara, K. Inazawa and E. Nishimura: Proc. 20th Symp. Dry Process (Inst. Elec. Eng. Japan, Tokyo, 1998) p.263.

[4] T. Shirafuji, Y. Miyazaki, Y. Hayashi and S. Nishino: Plasmas and Polym., **4**, 57 (1999).

[5] S. Takeishi, H. Kudo, R. Shinohara, M. Hoshino, S. Fukuyama, J. Yamaguchi and M. Yamada: J. Electrochem. Soc., **144**, 1797 (1997).

[6] F. Fracassi, E. Occhiello and J. W. Coburn: J. Appl. Phys., **62**, 3980 (1987).

[7] K. Nakamoto: *Infrared and Raman Spectra of Inorganic and Coordination Compounds 4th ed.* (Wiley, New York, 1986), p.378.

[8] H. Yasuda: *Plasma Polymerization* (Academic Press, Orlando, 1985), p.373.

[9] Y. Hikosaka and H. Sugai: Jpn. J. Appl. Phys., **32**, 3040 (1993).

[10] T. Nakano and S. Samukawa: J. Vac. Sci. Technol. A, **17**, 686 (1999).

[11] R. d'Agostino, F. Cramarossa, S. De Benedictis and G. Ferraro: J. Appl. Phys., **52**, 1259 (1981).

[12] T. Shirafuji, Y. Miyazaki, Y. Nakagami, Y. Hayashi and S. Nishino: Jpn. J. Appl. Phys., **38**, 4520 (1999).

The Use of Fluorinated Oxide (FSG) for Al-Cu and Cu Interconnects

W. Chang, T.I. Bao, C.L. Chang, S.M. Jang, C.H. Yu, and M.S. Liang
Advanced Module Technology Division Research and Development Taiwan
Semiconductor Manufacturing Company
No.9, Creation Rd. I, Science-Based Industrial Park, Hsin-Chu, Taiwan, ROCTEL: 886-3-5781688, FAX: 886-3-5773671

ABSTRACT

FSG film was deposited by inductively coupled high-density plasma CVD process using SiF_4 / SiH_4 / O_2 / Ar gases. F concentration developed in this work is ~ 4 to 5%. Complete filling of 0.20 μm metal line space with aspect ratio over 2.0 was demonstrated with deposition/etching ratio below 2.8. Thermal desorption spectroscopy shows excellent film stability, with F evolving at temperature > 500°C. CMP of FSG exhibits significantly less pattern sensitivity than ozone-TEOS and spin-on dielectrics capped by plasma oxide. FSG process has been integrated with both Al-Cu and Cu conductors. For Al-Cu interconnect, full FSG IMD was implemented without USG as under or cap layer. For Cu damascene, FSG was integrated with SiN as etch stop and Cu diffusion barrier. 10 – 15% reductions in line-to-line capacitance were obtained compared to un-doped oxide. By the improvement of post-CMP clean process, Cu contamination is eliminated and no intra-metal leakage has been detected. For damascene process, the effects of passivation scheme and metal pitch on capacitance reduction and intra-metal leakage have been investigated. Good thermal stability of the FSG/Al-Cu and FSG/Cu has been obtained in this work.

INTRODUCTION

The influence of interconnect RC delay on device performance becomes increasingly important as the interconnect dimension of VLSI device continues to scale down to sub-quarter micron region. To reduce the impact from RC delay, inter-metal dielectric (IMD) using lower dielectric constant has become of great interest.[1-2] To further enhance the interconnect performance, the integration of Cu and low-k IMD is particularly attractive.[3] A wide range of low-k materials produced either by CVD or spin-on process have been developed.[4] In order to incorporate these low-k materials into VLSI technology successfully, detailed study on integration compatibility and reliability are needed. In this work, we investigated the film properties, electrical characteristic, device reliability, and product reliability for FSG low-k using either in IMD or in Cu damascene. The fundamental properties and stability of FSG varied with process tuning will also be presented in this paper.

EXPERIMENT

FSG film was deposited by inductively coupled high-density plasma CVD process using SiF_4 / SiH_4 / O_2 / Ar gases. The process parameters, such as source RF power, bias RF power, SiF_4 flow rate, O_2 flow rate and back-side He flow rate, were optimized to achieve most robust FSG properties. Film thickness, uniformity and refractive index are determined using Nano8000E spectrometer. F concentration for the FSG film is determined by FTIR peak height ratio. Cross-section SEM was used to examine the gap-fill ability in Al-Cu IMD structure.

431

Table 1. Effect of process parameter on fundamental FSG properties

	Bias-RF (W)		O2 (sccm)		SiF4 (sccm)	
Quantity	2750	2250	160	130	63	51
D/E	2.65	3.09	3.08	2.58	2.69	2.97
F%	3.9	4.75	4.03	4.73	4.63	3.81
Gap-fill*	M_clip	Partial	OK	OK	M_clip	partial

1. Base condition: 2500 RF power; SiF_4/O_2 57/145 sccm and D/E 2.8
2. Gap-fill is based on aspect ratio 2.2 to 1

Thermal reliability test was performed on both blanket and patterned FSG wafers at 400°C in N_2/H_2 environment up to 10 cycles with 60 minutes/cycle. The CMP performance for FSG film in subtractive Al-Cu is evaluated using wafers with metal density in the range of 20–60%. Cu CMP process with two-step slurry is used to achieve planarization in Cu/FSG damascene. Metal line-to-line capacitance was determined using a comb-and-serpentine test structure at 100KHz. T/C test is performed from –65°C to 150°C with 10 minutes/cycle up to 500 cycles. PCT test is performed at 121°C, 2 atm, 100% RH for 168 hours.

RESULTS

The effect of key process parameters, such as bias RF power, SiF_4 and O_2 flow rate on FSG basic film properties are summarized in Table 1. F concentration developed in this work is in the range from 3.5 to 5%. Higher F concentration of FSG film can be achieved either by lowering bias RF power and O_2 flow rate or by increasing SiF_4 flow. Complete filling of 0.20 μm metal line space with aspect ratio over 2.0 was demonstrated with deposition/etching (D/E) ratio below 2.8, Figure 1. Metal line clipping phenomenon is found for D/E ratio lower than 2.7. With respect to Bias-RF power and SiF_4 flow rate, O_2 flow has less effect on FSG gap-fill ability even though the same D/E is obtained. Thermal desorption spectroscopy was used to check the film desorption behavior at temperature in the range 25 to 800°C and shows excellent film stability, with F evolving at temperature > 500°C. Stable film properties including thickness/uniformity, refractive index and F% is achieved after 10 hours cycling at 400°C in N_2/H_2, Figure 2 and Figure 3.

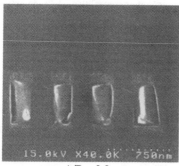

A/R ~ 2.2

A/R ~ 2.8

Figure 1. X-SEM gap-fill for FSG/Al-Cu IMD

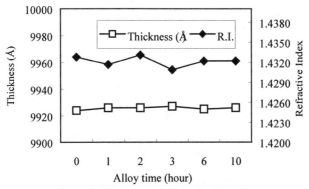

Figure 2. Thickness and R.I. after T cycling

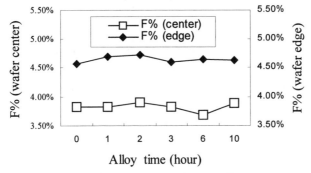

Figure 3. F% of FSG after T cycling

The planarization of different FSG integration schemes is investigated, Figure 4. PE-OX and PE-TEOS are also included for comparison. It is found that usage of thick FSG offered great advantages for chemical mechanical polishing (CMP), including higher removal rate for throughput enhancement and lower pattern sensitivity for manufacturing control. In the metal density range from 20 to 60%, both direct and non-direct CMP of FSG exhibits significantly less pattern sensitivity than PE-TEOS and PE-OX. This CMP characteristic promises very tight control of IMD thickness and intra-metal and inter-metal capacitance for various products and is of great significance to foundry fabs.

The developed FSG process has been integrated with both Al-Cu and Cu conductors. For Al-Cu interconnect, full FSG IMD was implemented without USG as under or cap layer. As a result, 10 to 15% reductions in line-to-line capacitance were obtained compared to un-doped oxide, Figure 5. Lowest L-L capacitance can be obtained by direct CMP on thick FSG wafer. The effect of using different underlayer on L-L capacitance for metal side wall protection purpose is also examined. Increase in L-L capacitance is found when SiON is used.

For Cu damascene, FSG was integrated with SiN as etch stop and Cu diffusion barrier. CMP process optimized with two-step slurry has been developed to minimize dishing, erosion and recess. Good integrity of Cu and IMD surface with eliminated microscratches and metal corrosion was demonstrated with properly designed oxide buff and post-CMP cleaning. By the improvement of post-CMP clean process, Cu contamination is also eliminated and no intra-metal leakage has been detected. The effects of passivation scheme and metal pitch on capacitance reduction and intra-metal leakage are also investigated. Compared with USG, up to 15% reduction in L-L capacitance is obtained with Cu/FSG damascene structure, Figure 6.

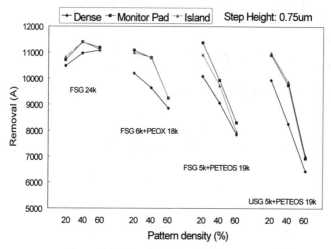

Figure 4. Effect of FSG on CMP pattern sensitivity

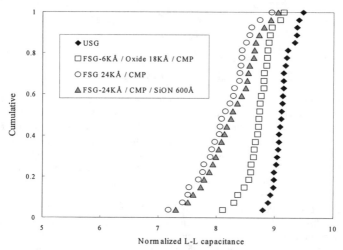

Figure 5. L-L Cp for Al-Cu/FSG with different FSG thickness

Finally, we have investigated the reliability of FSG both in subtractive Al-Cu and Cu damascene. No degradation in metal L-L capacitance is achieved after T/C and PCT tests, Figure 7 and Figure 8. Full FSG IMD has been implemented into a 0.25 μm logic vehicle for product reliability test. Table 3 shows the failure number out of test sample size. Except some failure caused by device defects, FSG film has passed all the tests. This result demonstrates that FSG with direct CMP is reliable for low-k application.

Table 3. Product reliability with FSG IMD

TEST	FSG 24K_/CMP
High temp. oper. Life time (100 hrs)	0/99
Temperature cycle (1000 cycles)	0/77
Temperature shock (500 cycles)	0/77
Pressure Cooking test (168 hrs)	0/55
Temperature humidity bias (1000 hrs)	1/55
High temperature storage (1000 hrs)	1/55

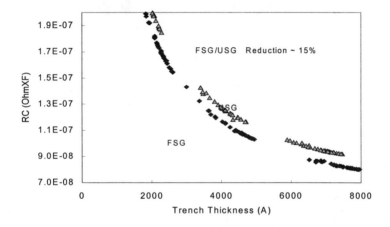

Figure 6. L-L Cp for Cu/FSG with respect to Cu/USG

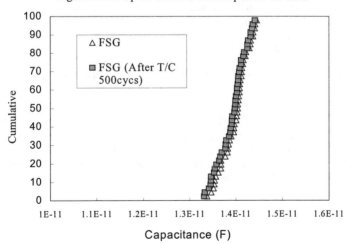

Figure 7. L-L Cp after T/C 500 cycles

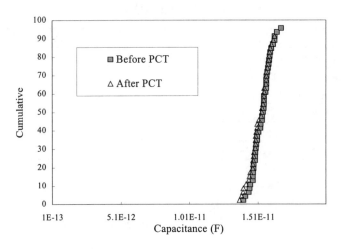

Figure 8. L-L Cp after PCT with passivation

CONCLUSIONS

(1) Good gap fill ability with stable fundamental properties such as thickness/uniformity, refractive index and F% after thermal cycling can be achieved for FSG film through bias RF power, SiF_4 and O_2 flow rate optimization.

(2) Higher removal rate for throughput enhancement and lower pattern sensitivity for manufacturing control can be obtained using either full or embedded FSG IMD.

(3) Good integrity of Cu and IMD surface with eliminated microscratches and metal corrosion can be reached with properly designed oxide buff and post-CMP cleaning.

(4) Excellent thermal stability both in physical and electrical properties indicates good stability of FSG processes developed in this work for both Al-Cu and Cu interconnects. By various product reliability tests, the proposed FSG scheme is proven highly reliable and manufacturable.

REFERENCES

[1]   M.K. Jain et al. MRS. Proc. ULSI, 1997, p. 423.
[2]   A.K. Stamper et al. Proc. 3rd DUMIC 1997, p. 13.
[3]   M. Naik, et al. International Interconnect Technology Conference Proceeding 1999, p.181.
[4]   W.W. LEE et al. Japan IEEE & Appl. Phys. Soc. 20th Dry Proc. Symp. Proc. 1998, P. 167.

BLOκ™ - A LOW-κ DIELECTRIC BARRIER/ETCH STOP FILM FOR COPPER DAMASCENE APPLICATIONS

P. XU*, K. HUANG*, A. PATEL*, S.RATHI*, J. FERGUSON*, B.TANG*,
J. HUANG*, C. NGAI*, M. LOBODA**
*Applied Materials, Inc., 3225 Oakmead Village Drive, Santa Clara, CA 95054
**Dow Corning Corporation, Midland, MI 48686

ABSTRACT

A low-κ dielectric barrier/etch stop film - BLOκ™ - has been developed for use in copper damascene processes. The film is deposited in Applied Materials single-wafer PECVD Chamber using trimethylsilane $((CH_3)_3SiH)$ as a precursor, and has a lower dielectric constant (<5) compared to the SiC film (>7) generated by SiH_4 and CH_4, and plasma silicon nitride (>7). The film characterization, including physical, electrical, copper diffusion barrier properties and etch integration property, demonstrated that this film is a good barrier/etch stop for low-κ copper damascene applications. Because of its low dielectric constant, low effective κ values can be achieved in damascene devices. The integration study shows that the line to line capacitance can be reduced when BLOκ™ is used as a barrier/etch stop layer.

INTRODUCTION

To reduce the RC delay of the interconnection system and improve IC performance, the inter-layer dielectric (ILD) materials with dielectric constants < 3.0 are required to minimize signal delay for the less than 0.25μm technology [1,2]. While many low-κ materials have been studied for ILD copper damascene application, silicon nitride, which has a high dielectric constant (>7), is the primary candidate for the barrier/etch stop layer required in damascene processes. By integrating a low-κ barrier/etch stop film, the effective dielectric constant in the copper damascene structure can be further reduced. As a barrier/etch stop film, it is required that the film should have good stability under thermal cycles, good copper diffusion barrier properties, good insulating properties, good etch integration properties, and compatibility with damascene integration.

The deposition of unique (stable, low stress, low-κ) SiC:H films by PECVD of organosilicon gases, was previously reported [3,4]. Applied Materials has developed a low-κ barrier/etch stop film - BLOκ™ (Barrier **LO**w κ)[1] - based on the PECVD using trimethylsilane $((CH_3)_3SiH)$ [5]. BLOκ™ is amorphous and composed of silicon, carbon and hydrogen (a-SiC:H). Unlike conventional SiC:H film deposited with SiH_4 and CH_4 by PECVD which has high dielectric constant, high leakage current and low breakdown strength, BLOκ™ has a lower k value (< 5) and meets the basic requirements as a barrier/etch stop film in damascene applications [5]. As we will show in this paper, BLOκ™ has also demonstrated the good integration compatibility in Cu damascene structures. Using BLOκ™ integrated with the low k ILD material, the line to line capacitance is much lower compare to using silicon nitride integrated with silicon oxide. The via and trench etch profiles in the dual damascene structure with BLOκ™ as the barrier/etch stop layer show that the high selectivity and good control of critical dimension can be achieved. BLOκ™ is also compatible with ACT and EKC wet strip chemistry.

---

[1] Patent Pending

EXPERIMENTS

BLOκ™ film was deposited using Dow Corning® trimethylsilane ((CH₃)₃SiH) in Applied Materials® DxZ™ or Producer™ PECVD systems. All of the films evaluated were deposited on 200 mm p-type Si<100> substrates. The film thickness and optical constants were measured using a spectroscopic ellipsometer. The leakage current and breakdown strength were measured on capacitors formed using sputtered copper gate electrodes. The dielectric constant was measured by mercury probe and by copper gate electrodes. The silicon to carbon ratio was measured by Rutherford backscattering (RBS), and hydrogen concentration was measured by Nuclear Reactive Analysis (NRA) and by RBS. FTIR spectroscopy was used to investigate molecular bonding in the film. The film density was measured using micro-balance. The copper diffusion barrier properties were studied by applying simultaneous thermal and electrical stress (bias-temperature stress, or BTS), and by secondary ion mass spectroscopy (SIMS) analysis. The scotch tape tests and stud pull tests were performed to examine the adhesion of BLOκ™ film to copper and silicon oxide. The etch performance was studied by etching the via and trench patterned BLOκ™/silicon oxide film stack using Applied Materials Dielectric Etch IPS™ Centura®. The line to line leakage and capacitance were measured using a metal comb structure with BLOκ™ as the barrier/etch stop layer.

RESULTS AND DISCUSSION

BLOκ™ is composed of silicon, carbon and hydrogen (a-SiC:H). From RBS analysis, the ratio of silicon to carbon is about 1:1 (without considering hydrogen concentration). The hydrogen concentration (atomic percentage) has minimal changes after two hours of anneal at 400°C.

Fig. 1 shows the FTIR spectra of BLOκ™ film before and after annealing and of conventional SiC:H film. It indicates that the silicon, carbon and hydrogen are bonded in the forms of C-H, Si-H, Si-CH₃, Si-(CH₂)ₙ- and Si-C. Comparing the FTIR of BLOκ™ and conventional SiC:H, BLOκ™ consists of considerably more Si-CH₃ and Si-(CH₂)ₙ bonds which indicate the polymerization of the film structure. The density of BLOκ™ film is 1.47 (g/cc), thirty percent lower than that of conventional SiC:H film.

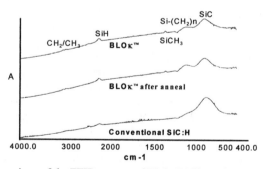

Fig.1: The comparison of the FTIR spectra of BLOκ™ film and conventional SiC film.

The primary film properties of BLOκ™ have been described in detail in reference [5]. The dielectric constant of the film is tunable - the film κ value varies with deposition temperature, RF power and gas flows. The dielectric constant of BLOκ™ is less than 5 and does not increase after two hours of anneal at 400°C. The shrinkage of BLOκ™ films was evaluated by annealing at 400°C for 2 hours in nitrogen. No thickness change was observed, indicating that the film is stable under thermal cycles. The leakage current of BLOκ™ film is 3E-10 (A/cm² @ 0.5 MV/cm), much lower than that of conventional SiC:H (3E-3 A/cm²). The breakdown field is greater than 2 (MV/cm, @/leakage > 1mA/cm²), much higher than that of conventional SiC:H (<0.5 MV/cm), when measured at the same condition. From the bias temperature stress (BTS) test on a 0.38μm BLOκ™ film, the lifetime of BLOκ™ film is greater than 90,000 seconds at 275°C and 2 MV/cm, which exceeds reported values for PE silicon nitride [6]. The BTS results of BLOκ™ film with 600Å thickness also demonstrated no failure after >100,000 seconds at 150°C and 1.0 MV/cm, and at 250°C and 1.0 MV/cm. Following BTS tests, SIMS was performed to check for Cu diffusion. The SIMS depth profiles show no difference among (1) BTS stress capacitor area, (2) capacitor adjacent to BTS area, thermally stressed without bias, and (3) BLOκ™ film not covered by copper. The presence of Cu is only at the gate-BLOκ™ interface, and no diffusion into the film bulk is observed. SIMS analysis was also done for the copper diffusion profile in the stack of silicon oxide, BLOκ™ and copper films after a total of three hours (six cycles of 30 minutes per cycle) annealing. From the SIMS profiles, the copper concentration in BLOκ™ film has dropped three orders of magnitude through 200Å depth. This demonstrates that the interfacial region of copper / BLOκ™ is less than 300Å. The adhesion and stud pull tests were performed after depositing BLOκ™ film on copper films and silicon oxide films. After the deposition and two hours annealing, no peeling was observed using scotch tape test. The stud-pull tests have demonstrated that no failures occurred within the film stack. The only failures that occurred were between the epoxy and film surfaces. The stresses at the failures were greater than 8000 psi.

Fig.2 shows the SEM of via and trench profiles using BLOκ™ as etch stop and silicon oxide as ILD layer in a damascene structure. After 30% over etch of the via, BLOκ™ film has minimal loss at the bottom of the via. After the trench etch, the critical dimension of the via is maintained. The etch study also shows that BLOκ™ can be removed in the subsequent step following the via opening using Dielectric Etch IPS™ chambers.

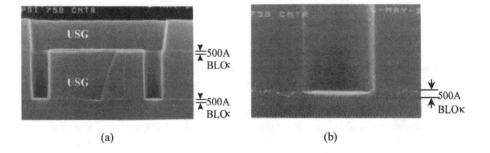

(a)   (b)

Fig.2: (a) Profiles of via and trench with BLOκ™ as the etch stop layer;
(b) Profile of the via (0.35 μm) bottom.

To study the compatibility of BLOκ™ to wet strip chemistry, BLOκ™ was tested in ACT970 (80°C) and EKT265 (65°C) for 30 minutes, and the thickness, refractive index and dielectric constant were measured after the tests. The results show that there is no thickness loss and change of refractive index or dielectric constant after the wet strip.

BLOκ™ enables lower effective κ values in copper damascene structures. Figure 3 is a simulation result that shows the plot of effective κ value versus ILD dielectric constant in a structure with 0.25μm line spacing and 0.40μm metal height. The effective dielectric constant is ≤3.0 when 500Å BLOκ™ is used in conjunction with a Low κ ILD.

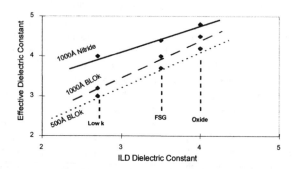

Fig.3: The effective dielectric constant using silicon nitride (κ = 7.2) and BLOκ™ (κ = 5.0) as barrier and etch stop layers in a damascene ILD application (Simulated using Raphael software by Avant!)

To study the effect of BLOκ™ on line to line leakage and capacitance, BLOκ™ was integrated with a low k ILD film (Black Diamond or BD, κ = 2.8 [7]). Fig. 4 shows the side view of metal comb structure used for the test, and the comparison of the RC products between BLOκ™-BD-BLOκ™ and Nitride-Oxide-Nitride structure. The thickness of BLOκ™ in the structure is 1000Å. The line to line leakage current of BLOκ™-BD-BLOκ™ is comparable to that of Nitride-Oxide-Nitride structure. The capacitance of BLOκ™-BD-BLOκ™ has 33% reduction compared to that of Nitride-Oxide-Nitride structure.

CONCLUSION

A low dielectric constant (κ ≤ 5) barrier/etch stop film, BLOκ™, was developed using trimethylsilane in PECVD single-wafer chambers. The film characterizations, electrical properties, and compatibility with damascene schemes (including copper diffusion barrier property, adhesion to ILD films, and etch integration compatibility) show that BLOκ™ is a good barrier/etch stop film for copper damascene applications. When integrated with low κ ILD film (κ = 2.7), an effective dielectric constant of 3 can be achieved based on the simulation. Using BLOκ™ integrated with the low k (=2.8) ILD material, the line to line capacitance reduction is 33% compare to using silicon nitride integrated with silicon oxide.

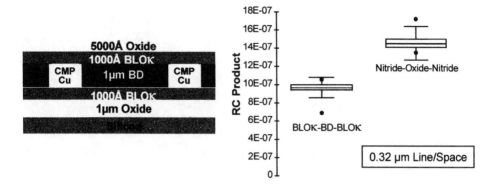

Fig. 4: (a) The side view of metal comb structure for line to line leakage and capacitance test. (b) The comparison of RC products between BLOκ™-BD-BLOκ™ and Nitride-Oxide-Nitride structures.

ACKNOWLEDGMENTS

We would like to acknowledge Mehul Naik, Suketu Parikh and Viren Rana in Process Sequence Integration operations group, and Raymond Huang and Jeremiah T. Pender in Dielectric Etch group at Applied Materials for their support for this study.

REFERENCES

1. P.L. Pai, and C.H. Ting, , Proc. of the VLSI Multilevel Interconnection Conf., p. 258 (1989).

2. Y. Ushiki, H. Kushibe, H. Ono, and A. Nishiyama, Proc. of the VLSI Multilevel Interconnection Conf., p. 413 (1990).

3. M.J. Loboda, J.A. Seifferly, F.C. Dall, J. Vac. Sci Technol. A, 12(1), p.90 (1994).

4. M.J. Loboda, J.A. Seifferly, C.M. Grove and R.F. Schneider, Mat. Res. Soc. Symp. Proc. Vol. 447 (1997), pp.145-50 and Electrochem. Soc Proc. Vol. 97-10 (1997), pp.443-52.

5.   P.Xu, K. Huang, A. Patel, S. Rathi, B. Tang, J. Ferguson, J.Huang, C. Ngai and M. Loboda, Proceedings of the 1999 International Interconnect Technology Conference (IITC99), p.109 (1999).

6. M. Vogt, M. Kachel, M. Plotner, and K. Drescher, Microelectronic Engineering, p.181 (1987).

7. N. Naik, S. Parikh, P. Li, J. Educato, D. Cheung, I. Hashim, P. Hey, S. Jenq, T. Pan, F. Redeker, V. Rana, B. Tang, and D. Yost, Proceedings of the 1999 International Interconnect Technology Conference (IITC99), p.181 (1999).

Low Dielectric Constant Mechanism
of Amorphous FluoroCarbon (a-C:F) Film

E.G. LOH , F.R. HUTAGALUNG , H. KOMIYAMA , Y. SHIMOGAKI*

Department of Chemical System Engineering, School of Engineering, University of Tokyo
*Department of Materials Engineering, University of Tokyo
enggiap@dpe.mm.t.u-tokyo.ac.jp
TEL&FAX: +81-3-5841-7131

I.ABSTRACT

Amorphous fluorocarbon thin films (a-C:F) with permittivity as low as 2.1 were deposited using parallel-plate 13.56MHz RF coupling PECVD reactor(low density plasma). C_2F_4 has been used as source material. In our previous studies, we had known that ions generated in the plasma, contributes the most to the deposition in low density C_2F_4 plasma, instead of radicals. This time, we focused on the mechanism of low dielectric constant, the origin of each polarization, and that of process improvement. It seems that electronic polarization is the predominant feature in the dielectric constant. The only way to achieve low ε with this film is to decrease the orientational polarization derived from -C=O and $-C=CF_2$ groups in the film. And this can be done with high substrate temperature, low pressure and low residence time conditions.

keywords: PECVD, C_2F_4 , a-C:F, low ε, electronic, ionic, orientational, polarization, Kramers-Kronig transformation.

II. INTRODUCTION

As dimensions and wiring intervals of very large-scale integrated devices decrease, wiring resistance and parasitic capacitance increase. To fulfill demands for VLSIs performance improvement, new metal line with a low resistance and alternative interlayer dielectrics(ILD) with low-dielectric-constant(ε) are claimed. Since the RC delay is proportional to ε, a reduction in ε leads to a lower capacitance and RC delay. In this case, amorphous fluorocarbon film (a-C:F) is a possible candidate. Researches focused on a-C:F films show that these films have a low dielectric constant, ranging from 3.0 to 2.0, and a very low moisture uptake. However, the characteristic low dielectric constant of such films are not clear yet. It is important to know how the bonding structure of the films governs the polarization component of dielectric constant. In our previous studies[1], a macroscopic ion deposition mechanism was proposed. However, the variations of bonding structure in the film and the dielectric properties with experimental parameters are still unknown.

III. EXPERIMENTAL

A schematic diagram of the apparatus used in this work, a conventional parallel-plate RF coupling PECVD reactor, is shown in Fig.1. The upper electrode is subjected to 13.56MHz RF power, while the lower electrode was grounded. These electrodes are separated 3.8cm apart. The reaction chamber was evacuated using diffusion pump and mechanical booster pumping system. The chamber was always pre-evacuated to a pressure of 10^{-7} Torr before deposition. The RF-power was fixed at 30W. The substrate temperature was set at 35°C, 150°C, and 300°C, the total pressure at 0.25Torr, 0.50Torr, 0.75Torr and 1.00Torr, whereas the flow rate of C_2F_4 gas was set from 50sccm to 200sccm, during deposition. As Ar+ ion with high energy effect the content of F atoms in the films, no carrier gas was used. The film thickness were determined by a ellipsometer (UVISEL, Jovin-Evon) and FE-SEM (JSM-6340F, JEOL). Concerning the dielectric constant properties and bonding structure of the film,

1)Ellipsometer (UVISEL ellipsometer at a beam incident angle of 74.9°, and from 0.75eV to 4.0eV) for analysis of refractive index (square of this give dielectric constant due to electronic polarization(ε_e)).

2)FT-IR (Nicolet Impact 410 FT-IR Spectrometer), for analysis of structure & chemical bonding of film. From the spectra,dielectric constant due to electronic & ionic polarization ($\varepsilon_e + \varepsilon_i$) can be obtained using Kramers-Kronig transformation.[2]

3)Capacitance-Voltage measurement(HP-4275A multifrequency LCR meter) at 1MHz, for analysis of overall dielectric constant ($\varepsilon_e + \varepsilon_i + \varepsilon_o$).

4)XPS, for analysis of structure & chemical bonding of films, and the ratio between elements of the film. were performed.

443

IV. RESULTS AND DISCUSSION

It is well known that dielectric constant is frequency-dependent as shown in Fig.2[(2)]. The dielectric constant measured at 1MHz is derived from orientational, ionic and electronic polarizations components. We can obtain the total dielectric constant from the C-V measurement at 1MHz. The dielectric constant contributed by the sum of ionic and electronic polarizations can be extracted from FT-IR spectra by using Kramers-Kronig transformation. The dielectric constant due to electronic polarization is calculated from the square of the refractive index at the visible light region. Using this knowledge, we investigated the contributions of orientational, ionic and electronic polarization components to the dielectric constant, and the influence of experimental parameters to each polarization.

Kramers-Kronig Transformation

First, the damping factor, k, which is the imaginary part of a complex refractive index, can be obtained from the absorption coefficient, α, of the IR spectra at each wave number, by using equation

$$k = \alpha\lambda/4\pi \quad ....(1)$$

where λ is the wave length. Then, the Kramers-Kronig transformation in equation (2) is used to calculate the real part of a complex refractive index, n:

$$n_i = n_\infty + \frac{2}{\pi} P \int_0^\infty \frac{\omega k(\omega)}{\omega^2 - \omega_i^2} d\omega \quad ....(2)$$

where P is the principal value of the integral and n_i indicates the real part of a complex refractive index at i-th wave number, respectively. As n_∞ is an unknown value, the refractive index equation can be rewritten as,

$$n_i = n_r + \frac{2}{\pi}\left[P\int_0^\infty \frac{\omega k(\omega)}{\omega^2 - \omega_i^2} d\omega - P\int_0^\infty \frac{\omega k(\omega)}{\omega^2 - \omega_r^2} d\omega \right] ....(3)$$

ω_r indicates wave number at which the refractive index is measured. The refractive index at 632.8 nm is used for n_r in the calculation. However, there is a impossibility to measure the absorption spectra from 0 to infinity. Therefore, equation (3) is modified as below,

$$n_i = n_r + \frac{2}{\pi}\left[P\int_{\omega_a}^{\omega_b} \frac{\omega k(\omega)}{\omega^2 - \omega_i^2} d\omega - P\int_{\omega_a}^{\omega_b} \frac{\omega k(\omega)}{\omega^2 - \omega_r^2} d\omega \right] ....(4)$$

where ω_a and ω_b represent the measured wave number range of the FT-IR spectra. The real IR spectra used in the calculations were in the range of 350~4000 cm^{-1}. The effect of absorption at wave numbers higher than 4000 cm^{-1}, if any, is negligible. The absorption in the range of 0~350 cm^{-1} (where the measurements are impossible) was assumed to be 0, and the integration was carried out from 0 to 4000 cm^{-1}.

From the values of n and k, ε' and ε'', and the corresponding dielectric constant at each wave number can be calculated from the following relationships.

$$\varepsilon' = n^2 - k^2 \quad ....(5)$$
$$\varepsilon'' = 2nk \quad ....(6)$$
$$\varepsilon = \sqrt{\varepsilon'^2 + \varepsilon''^2} \quad ....(7)$$

Figure 3 shows results of the experimental parameters dependency of dielectric constant due to each polarization. As shown in Fig.3a when the temperature increased, ε_e increased about 0.1, ε_i maintained and ε_o decreased about 0.23. The total dielectric constant decreased 0.13. At 35°C, low plasma pressure region(0.25 Torr), when residence time increased, ε_e and ε_i tend to decrease, but ε_o increased(Fig.3b). Whereas at 1.00 Torr, high pressure region, ε_e and ε_i seem independent, and ε_o decreased when residence time increased(Fig.3c). Figure 4 shows the IR spectra of the films. $-CF_x$ bondings are the predominant feature in the film. The large shoulder suggest that the films are highly cross-linked. However, the precise deconvolution and quantitative analysis of $-CF_x$ peaks are somehow impossible, as the vibrations of C-F bonding are easily and highly influenced by adjacent atoms and groups.[3] We had also tried to calculate the vibration modes after optimizing the stucture of the film using Density Functional Theory. And we concluded not to deconvolute the CF_x peak. There are 2 small peaks between 1500cm^{-1} and 1800cm^{-1} which belong to C=O and $C=CF_2$ bonding stretching mode.[3] Area of these peaks decreased with increasing temperature (Fig.5(a)). However, as shown in Fig.5b, these areas increased with residence time when the plasma pressure is 0.25 Torr, and decreased at 1.00 Torr. These trends show good agreement with ε_o. From these facts we can conclude that orientational polarization is originated from C=O and $C=CF_2$ bonding in the film. The impurity O atoms are probably derived from the C_2F_4 polymerization inhibitor in C_2F_4 cylinder or from the atmosphere when the chamber was exposed after deposition.

Figure 6 shows the XPS C1s spectra of a-C:F film obtained using a non-monochromated Mg K_α X-ray source and has not been corrected for sample charging. The peak can be deconvoluted into 6 peaks, which are, in order of decreasing bonding energy, CF_3, $-CF_2-$, -CF-, C=O, -C-CF-, and -C- peaks.[4] Figure 7 shows the variation of each peak area and ε_i with experimental parameters ((a) for temperature dependence, (b) for residence time dependence at 35°C and 1.0 Torr). It seems that ε_i and -CF semi-ionic bonding, have the same trends. From these results, we attribute ε_i to the distortion of -CF semi-ionic bonding in matrix.

Figure 8 shows the variation of F/C ratio and ε_e, the dielectric constant contributed by only eletronic polarization, with experimental parameters. When the temperature increased, ε_e also increased, but F/C ratio decreased (Fig.8(a)). At 35°C, 0.25 Torr(low pressure region), when residence time increased, ε_e decreased, but F/C ratio increased (Fig.8(b)). At 35°C, 1.0 Torr(high pressure region), ε_e and F/C ratio seem independent to residence time (Fig.8(b)). From these results, we therefore concluded that the F/C ratio in the film controls the electronic polarization. Electronic polarization arises from the excursion of electron shell relative to nucleus in the presence of an electric field. Considering the electronegativity difference between F and C, F atoms have a lower polarizability than C atoms. The polarizability of C atoms is about 3 times of the one in F atoms. Consequently, as F/C ratio in the film increased, the displacement of electron shell are restrained, and ε_e will naturally decrease.

V. CONCLUSIONS

The low dielectric constant mechanism of a-C:F films has been explicated. The contribution of each polarization to overall dielectric constant and the experimental parameters dependency of each polarization had been evaluated. The origins of each polarization has also been ascertained. Electronic polarization occupies the most. In these films, as C and F are the main compounds, and C-F bonding is very hard, the ε_e and ε_i are very hardly being decreased. However, the ε_o can be decreased about 0.5 as this polarization results from C=O bondings and $C=CF_2$ bondings in the film. With increasing the deposition temperature, both bondings can be decreased, and a lower dielectric constant can be obtained. Furthermore, decreasing the double bondings means increasing the cross-linking in the films and the thermal stability of the films.

VI. ACKNOWLEDGEMENTS

Finally, we would like to thank Y. Sato from Toshiba Corporation, for technical assistance in Density Functional Calculation.

VII. REFERENCES

(1)E.G.Loh, Y.Shimogaki and H.Komiyama Proc. of The 194th Meeting of ECS(1998)
(2)S.W.Lim, Y.Shimogaki, Y.Nakano, K.Tada, and H.Komiyama, Jpn. J. Appl. Phys., Part 1 **35**, 1468 (1996)
(3)G.Socrates, Infrared Characteristic Group Frequencies, Wiley (1994)
(4)J.F.Moulder, J.Chastain, Handbook of X-Ray Photoelectron Spectroscopy, Perkin-Elmer Corporation, Eden Prairie, MN(1992)

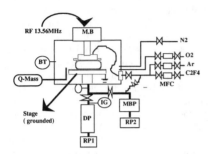

Fig.1 Schematic of the PEVCD reactor

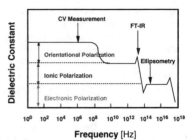

Fig.2 Frequency-dependent of dielectric constant[2]

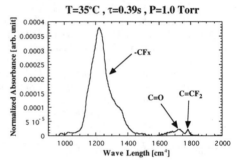

Fig.4 IR spectra of a-C:F film.

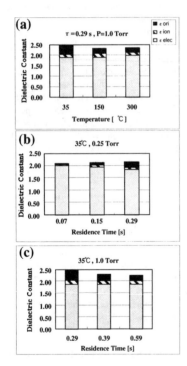

Fig.3 Variation of dielectric constant and each polarization ($\varepsilon_e + \varepsilon_i + \varepsilon_o$) with experimental parameters.
(a) stage temperature effect at 0.29s, 1.00 Torr.
(b) residence time effect at 35°C, 0.25 Torr.
(c) residence time effect at 35°C, 1.00 Torr.

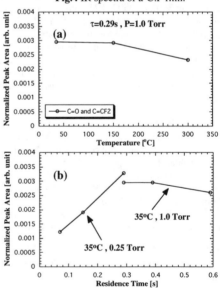

Fig.5 Experimental parameter dependence of normalized peak area of C=CF$_2$, C=O
(a) substrate dependence dependence
(b) residence time dependence

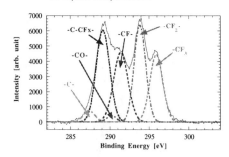

Fig.6 C 1s XPS of amorphous fluorocarbon thin film dielectrics. This peak was decomposed to 6 peaks, which are (in order of decreasing bonding energy) $-CF_3$, $-CF_2-$, $-CF-$, $C=O$, $-C-CF-$, $-C-$ peaks.

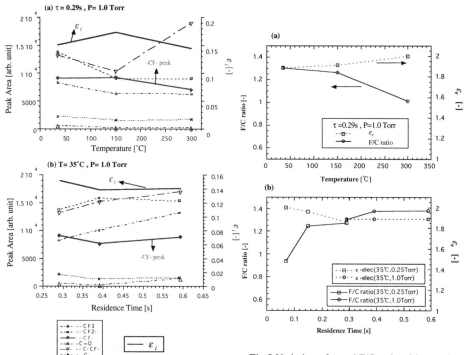

Fig.7 Variation of decomposed C1s XPS peaks with experimental parameters (T, τ).
(a) effect of stage temperature
(b) effect of residence time

Fig.8 Variation of ε_e and F/C ratio with experimental parameters (T, τ).
(a) effect of stage temperature
(b) effect of residence time

Extreme Low Dielectric Interlayers - A Nanoporous Polymers Approach

Thomas J. Markley, Xiaoping Gao, Michael Langsam, Lloyd M. Robeson, Mark L. O'Neill *
Air Products and Chemicals, Inc.
7201 Hamilton Blvd., Allentown, PA 18195-1501
and
Paul R. Sierocki, Shahrnaz Motakef, David A. Roberts
Schumacher
1969 Palomar Oaks Way, Carlsbad, CA 92009

ABSTRACT

For future IC generations reductions in dielectric from that currently used materials (k ~3-4) to much lower levels (k < 2) will be necessary for smaller dimensions to be feasible. The introduction of air (k = 1) via nanoporosity allows for reductions in dielectric while maintaining most of the desirable properties of the dense material. For example, to reduce the K from 3.0 to near 2.0 requires about 35 vol% porosity, while the electrical and mechanical properties should scale with porosity according to theoretical equations. With larger amounts of porosity the dielectric can be reduced further, however this must be balanced with the deleterious effects that the introduction of air will have on mechanical and dielectric strength of the material. The main criteria for a porous dielectric are 1) that pore sizes must be significantly smaller than circuit dimensions so that the dielectric behaves as a continuum, 2) that the nanoporous morphology remain intact to very high temperatures to maintain the reduced K during the high temperature processing stages of chip manufacture (up to 425 °C), and 3) it is preferable that the pores have a closed-cell morphology to prevent channeling of liquids and materials into the matrix during subsequent processing, and to potentially maximize mechanical strength and minimize dielectric breakdown .

Two polymer-based extreme low-k (ELk) products, PolyELk™ and Velox-ELk™, have been developed at Air Products/Schumacher. These materials are produced by novel thin film technology resulting in porous polymer films of ~1 micron thickness with pore sizes < 30 nm. Porosities have been tailored in the range of 0-60 vol% while maintaining excellent mechanical and electrical properties. This technology has been successfully scaled up to 200mm wafers with excellent planarity across the wafer.

INTRODUCTION

With feature dimensions on integrated circuits continuing to decrease there is a need for interlayer and intermetal dielectric materials with extremely low k to reduce crosstalk and allow for gigahertz frequency clock speeds. Dense materials, either organic or inorganic, have not been identified which will meet the materials requirements and have k ≤ 2. As a result, processes to introduce nano-scale or molecular level porosity into existing materials which have acceptable physical properties (other than dielectric constant) has become the primary focus for next generation dielectric materials.

The development of low k materials (k < 3) led to the introduction of poly(arylene ether) (Velox™) as a dense, spin-on organic polymer material offering excellent adhesion, high thermal stability and good mechanical properties, and dimensional stability during chip manufacturing processes [Burgoyne, 1997]. Velox™ can be crosslinked to achieve a glass transition temperature in excess of 425 ºC in air or inert atmospheres; in air the film requires only minutes at 400 ºC, which can be done on an open hotplate. The advantageous process requirements for Velox™ along with the good property profile allow it to compete well for current and future generation including those employing Cu/damascene processing.

With trends towards even lower k for generations, nanoporous versions of Velox™ have been under investigation at Air Products / Schumacher leading to Velox-ELk™ and PolyELk™. While Velox-ELk™ extends directly from the Velox™ parent polymer, the PolyELk™ parent polymer is structurally similar to Velox™ yet does not require a high temperature curing step to lock in the morphology. As well a

Conference Proceedings ULSI XV © 2000 Materials Research Society

templated inorganic material has also been developed by Air Products / Schumacher; MesoELk™ is a nanoporous silica dielectric with well-defined pore structure and property profiles acceptable for future IC generations.

This paper will discuss the efforts involving the organic polymer candidates Velox-ELk™ and PolyELk™. The materials properties will be discussed in detail, however the process developed to make the nanoporous films will only be discussed in general terms as it is patent pending. While Velox-ELk™ has not yet met the requirement of dielectric constant and dimensional stability up to 450 °C, our PolyELk™ material has met this requirement and is thus our currently preferred product. The processing characteristics of Velox-ELk™ and PolyELk™ are quite similar thus the nanoporous characteristics of both materials will be discussed together.

EXPERIMENTAL CHARACTERIZATION

The process for Velox-ELk™ and PolyELk™ involves a typical spin-on of a polymer solution, however it includes a further structuring process step involving a "developer" introduced while spinning. PolyELK films are then subjected to drying (soft-bake) in air to at 200-300 °C followed by curing under N2 to 425 °C for < 5 minutes. The resultant films are nanoporous with pore sizes < 20nm. Dimensional stability of the porous morphology was determined by thermal treatment followed by recharacterization.

The thickness of films were determined by profilometry and by elipsometry. Variations in thickness (or refractive index) across the wafer were assessed as the standard deviation determined by multipoint elipsometric analysis. The dielectric constant and breakdown voltages were determined by mercury probe technique both in air and under dry N2. Adhesion was assessed qualitatively by ASTM standard tape adhesion test and quantitatively by stud pull test employing epoxy. The morphology of the porous films was assessed by SEM and select samples were further tested by Positron Annihilation Lifetime Spectroscopy (PALS) performed by Prof. David Gidley at the University of Michigan and Small Angle Neutron Scattering (SANS)by Dr. Wen-Li Wu of NIST. Modulus and hardness were determined by nanoindentation.

RESULTS

The dielectric constant as calculated from the Clausius-Massotti [Mascia, 1974] or Maxwell [Atkins, 1986] relations using porosities determined by thickness difference and calculated by elipsometry was found to agree quite well with the measured dielectric using mercury probe (Figure 1). As a result simple profilometry measurements were found to be sufficient to provide semi-quantitative information on k. Measurements were also performed in a dry N2 box to assess the quantity of sorbed water in the samples. Data shown in Table 1 were measured in open atmosphere. For PolyELK™ samples with no porosity the measured k ranged from 3.5 to 3.8, however the "dry" k value was found to be 3.1±0.1. Porous films were found to be much less affected by exposure to open atmosphere due to a reduction in water sorption relative to the dense polymer, as observed previously in the literature [Cha, 1995].

The mechanical properties determined for the porous films followed Kerner's equation [Kerner, 1956], Figure 2. Kerner's equation is often employed for modulus prediction of a composite with spherical inclusions, such as voids, in a continuous matrix. The data indicate good agreement with Kerner's equation, implying continuity of the matrix polymer. The polymer used to test mechanical properties as a function of porosity was a surrogate polyimide.

Dimensional integrity of nanoporous films were determined by thermal treatment to temperatures to 500 °C in N2. With PolyELk™ the nanoporous structure is stable up to > 425 °C as films exposed to 425 °C for 30 minutes under N2 showed no measurable change in dielectric constant, refractive index, or thickness.. Film discoloration became apparent at 450 °C, but measurable film deterioration (increase in k, reduction in thickness, and porosity by elipsometry) were only seen after treatment to 500 °C. These results agree with isothermal-thermogravimetric analysis results which indicate very low weight loss at 425 °C (<.10wt%/hr).

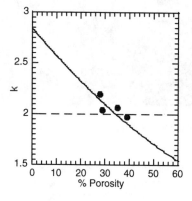

Figure 1. Calculated k versus % porosity for polymer with dense k = 2.85 based on Clausius-Massotti relation (solid line). Points correspond to measured k versus measured porosity by elipsometry.

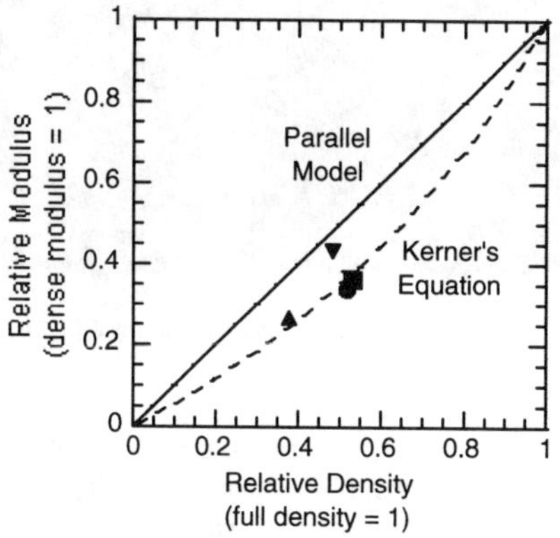

Figure 2. Modulus scaled with density in comparison with theoretical models.

Table 1. Thickness, dielectric constant, and estimated porosity for PolyELK films exposed to various thermal treatments. All measurements were taken in open atmosphere.

Sample	treatment	thickness ± 10 (nm)	k ± .10	est. % porosity
1	325 °C hotplate,	378	3.75	0
2	5 mins.,	580	2.44	30
3	air	650	2.12	40
1	350 °C oven,	377	3.55	0
2	60 mins.,	577	2.40	30
3	N2	650	2.17	40
1	400 °C oven,	370	3.66	0
2	60 mins.,	572	2.44	30
3	N2	657	2.20	40

The nanoporous morphology produced by this technology is shown in the SEM micrographs in Figure 3 for Velox-ELK™ and PolyELK™. Pore sizes of < 30nm are interpreted from these figures, with indications that the morphology is closed cell and that a depth profile exists in the morphology. Dielectric measured for both samples are < 2.0 (dry).

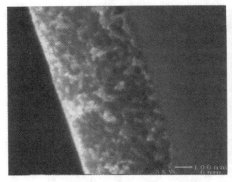

Figure 3. SEM cross-section profiles for Velox-ELk™ (left) and PolyELk™ (right). Both samples have k < 2.0.

Shown in Figure 4 is the result of multipoint elipsometric analysis of a sample of Velox-ELk™ produced during process optimization to minimize pore size. Films produced are optically clear to the naked eye and are homogeneous across 200mm wafer to better than +/- 0.6% thickness (or refractive index) change.

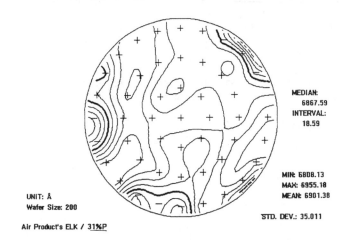

MEDIAN: 6867.59
INTERVAL: 18.59

MIN: 6808.13
MAX: 6955.18
MEAN: 6901.38

STD. DEV.: 35.011

UNIT: Å
Wafer Size: 200

Air Product's ELK / 31%P

Figure 4. Picture showing Velox-ELK™ film uniformity determined by multipoint elipsometric analysis. Samples show less than 0.6% standard deviation across the wafer.

Positron Annihilation Lifetime Spectroscopy (PALS) performed at the University of Michigan with Prof. David Gidley suggests that these films have a closed cell morphology, with a depth profile which indicates a slightly increasing average pore size with depth. Films have a dense surface layer which consists of approximately the top 10% of the thickness of the film. This is followed by a step transition to a porous region with average pore sizes in the range of 10-15nm, which increases to ~20nm at the wafer-film interface. These results agree with preliminary data from Small Angle Neutron Scattering (SANS). The morphology produced in these porous films seems to be unique to this process[Fodor, 1999 #17]. There may be inherent benefits to closed cell morphology and a dense surface layer such as the reduction or elimination of impurities from penetration into the film.

CONCLUSIONS

The overall property profiles for Velox-ELk™ and PolyELk™ are listed in the table below. Velox-ELk™ has not yet met the thermal requirements in regards to dimensional stability when exposed to 450 °C. PolyELk™ has met this criteria and is currently our preferred nanoporous polymer material. The data listed in Table 2 below are measured or known values with the exception of the italicized parameters (mechanical properties and adhesion), which were estimated from the dense polymer properties assuming Kerner's equation is valid for these samples as well. Optimization of the process and scale-up is currently in progress; initial studies on Velox-ELk™ shows promise in achieving the goal of <5nm pore sizes.

Table 2. Key parameters for nanoporous polymer films.

Key Parameter	Velox-ELK™	PolyELK™
k	< 2.0	< 2.0
Thermal Stability	Similar to Velox	< 0.05%hr^{-1} @ 400°C
Modulus	*> 2 GPa*	*> 2 GPa*
Cohesive Strength	*> 40 MPa*	*> 40 MPa*
Adhesion	*> 30 MPa*	*> 30 MPa*
Tg	Target: >425°C x-linked	~490°C
Chemistry	No fluorine	No fluorine
Pore Size	~20 nm (SEM; PA), closed pore	~20 nm (SEM), closed pore
Porosity	> 30% (capacitance, elipsometry)	> 30% (capacitance, elipsometry)

REFERENCES

Atkins, P.W., Physical Chemistry, W.H. Freeman and Company, New York, 1986

Burgoyne, W.F.; Robeson, L.M.; Vrtis, R.N., US Pat. # 5658994.

Carter, K.R., Materials Research Society Symposium Proceedings, *476*, 87-97 (1997)

Cha, H.-J.; Lee, H.; Carter, K.; Hedrick, J.; Labadie, J.; Sanchez, M.; Russell, T.P.; Volksen, W.; Yoon, D.Y., ANTEC '95, 2664 (1995)

Fodor, J.S.; Briber, R.M.; Russell, T.P.; Carter, K.R.; Hedrick, J.L.; Miller, R.D.; Wong, A., Polymer, *40*, 2547-2553 (1999)

Hedrick, J.L.; Carter, K.R.; Labadie, J.W.; Miller, R.D.; Volksen, W.; Hawker, C.J.; Yoon, D.Y.; Russell, T.P.; McGrath, J.E.; Briber, R.M., Advances in Polymer Science, *141*, 2-43 (1999)

Kerner, E.H., Proc. Phys. Soc. B, *69*, 808 (1956)

Kohl, A.T.; Mimna, R.; Shick, R.; Rhodes, L.; Wang, Z.L.; Kohl, P.A., Electrochemical and Solid State Letters, *2*, #2, 77-79 (1999)

Mascia, L., The Role of Additives in Plastics, John Wiley and Sons, New York, 1974

ACKNOWLEDGEMENTS

The authors gratefully acknowledge the work of Prof. David Gidley of the University of Michigan, Dept. of Physics for PALS studies and Dr. Wen-Li Wu of NIST, Gaithersburg, MD for SANS studies.

Process Optimization of Hydrogen Silsesquioxane (HSQ) Via Etch for 0.18um Technology and Beyond

S.C. Yang, M.H. Huang, Y.H. Chiu, B. R. Young, H.J. Tao, C.S. Tsai, Y. C. Chao, M.S. Liang
Advanced Module Technology Division, R&D, Taiwan Semiconductor Manufacturing
Company, Ltd., 9, Creation Rd. I, Science-Based Industrial Park, Hsin-Chu, Taiwan

Abstract

Hydrogen Silsesquioxane, HSQ, is one of the low K materials used in interconnect for 0.18um technology and beyond. The poison via associated with borderless structure is one of the key challenges to implement this material in production. A possible root cause is the solvent trapping in the damaged HSQ via sidewall as well as the slit in mis-alignment structure. To overcome this problem, etching condition, resist stripping, and wet cleaning have been investigated in this study. The result of this work indicates that the HSQ poison via can be significantly improved by adjusting the etching chemistry and choosing the right resist stripping process. More specifically, adding a passivation gas during the via etching can passivate the HSQ sidewall and can also reduce the veil formation. On the other hand, using low pressure O2 plasma to strip the resist can further minimize the HSQ damage. Finally, the right choice of cleaning agent and the optimization of wet cleaning recipe are both critical to avoid damaging the HSQ sidewall.

Introduction

Low k materials are used in the IMD layers to reduce the interconnect RC delay for 0.18um technology and beyond. Some of these materials, such as Hydrogen Silsesquioxane (HSQ), is subject to the damage by etching process, by inadequate resist dry stripping process, or by poor wet stripping process. Typically, the process damage on the low K material and its solvent uptake will cause via poisoning, especially for unlanded via[1-3]. Although it is possible to reduce the solvent uptake by a good degassing process, it is impossible to completely resolve the poison via problem by optimizing degassing process alone. Actually, it is also necessary to optimize the etching chemistry, the resist stripping process, and the polymer removal process to get a robust process for HSQ material.

For the unlanded via (Fig 1), the exposed area of low HSQ is much larger than the landed via thus it has more chance to trap solvent than landed via. Furthermore, there is also a slit in between the metal sidewall and the via sidewall, which can also easily trap the solvent. The trapped solvent in the damaged HSQ sidewall and the slit will introduce poison via through the metal corrosion mechanism. Consequently, it is very important to minimize the damage of low k material in order to avoid solvent trapping in the damaged HSQ.

In this study, the SiH/SiO ratio of FT-IR on HSQ is used as an index for the HSQ damage with very good correlation (Fig 2).

Experiment

The IMD structure is composed of HSQ and PE-TEOS. An IMP Ti/TiN barrier is used

455

prior to W fill and CMP. The via hole size is 0.26um.

Two types of etchers are evaluated in this study: inductive type and RIE type. For the inductive type etcher, the resist is stripped *in-situ* with low pressure recipe. For the RIE type etcher, the resist stripping is performed either by traditional downstream asher, or by a low pressure process with pressure at 15 to 80 mT.

For the wet process study, Amine based and Fluorine-based cleaning agents were evaluated associated with the soaking time.

Results and discussion

Because of its loosely networking structure, HSQ is vulnerable to the process condition, such as dry clean, resist stripping, and polymer wet cleaning. A possible root cause of HSQ via poisoning is the metal line corrosion induced by the trapped solvent inside the damaged HSQ and slit between via and metal (Fig 1). Thus, it is very important to protect the HSQ via sidewall from the solvent attack during the wet cleaning. This kind of micro damage is difficult to be detected by the cross-sectional SEM or TDS. However, FT-IR can reflect the degree of HSQ damage. The peak ratio of SiH (2250cm-1)/SiO(1033cm-1) shows strong correlation to the E-test result and it can be used as a good index for the HSQ micro damage (Fig 2).

(a)   (b)

Figure 1. (a) TEM of HSQ poison via due to metal corrosion and (b) the proposed damage model.

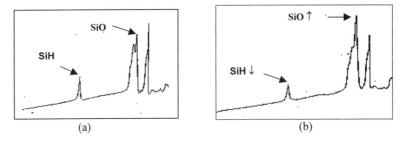

(a)   (b)

Figure 2. FT-IR spectra of HSQ, (a) as cure, SiH/SiO=0.38, (b) after etching, without passivation gas A in the etching recipe, SiH/SiO=0.20.

Based on the physical check, E-test result, and the FT-IR data, several practical ways to prevent the HSQ poison via in the etching stage are summarized as follows:

3.1 Etching chemistry

Adding suitable passivation gas in order to passivate the HSQ via sidewall is very effective to protect the HSQ from solvent attack. The addition of passivation gas during the etching can also reduce the veil formation (Fig3, Table 1,2). However, different kinds of passivation gases show different capability of process improvement. Thus, intensive study should be performed for the chemistry screening.

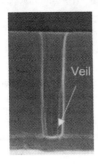

(a)   (b)

Figure 3. (a) Veil formation after via etch, without passivation gas addition, and (b) veil reduction with passivation gas addition during via etch.

Table 1. SiH/SiO ratio vs. etching chemistry

	Pre-etch	With passivation gas	Without passivation gas
SiH/SiO ratio	0.38	0.33	0.20

Table 2.   Normalized via Rc vs. different etching chemistries and PR stripping

Rc (mean/stdev)	Landed	Unlanded (off-set= 0.06um)
Inductive, without passivation gas A	1.6/10.5%	4.1/87.9%
Inductive, with passivation gas A	1.1/6.7%	1.7/0.9%
Inductive, with passivation gas B	1.1/10.2%	1.8/7.7%
Capacitive + downstream asher	1.0/6.1%	1.6/5.8%
		(21%fail)
Capacitive + low pressure O2 ashing	1.0/5.9%	1.5/6.0%

Via Rc is normalized by capacitive + downstream asher landed Rc.

3.2 Etcher

Inductive type and RIE type etchers do not show significant difference for this application if the right chemistry is used (Fig4, Table 2). However, one of the advantages to use inductive type etcher is due to its *in-situ* PR stripping capability which can operate under low pressure and it is effective to reduce the poison via.

(a)   (b)

Figure 4. (a) RIE etcher with downstream resist ashing, and (b) inductive etcher with low pressure resist stripping.

3.3 Resist stripping

Resist stripping by downstream asher can result in the poison via easily. However, this failure is not due to the suspected bowing profile by the lateral ashing since there is no strong evidence to support so (Fig 4). On the other hand, the result of FT-IR study indicates that original HSQ network will be damaged by the downstream asher even though that the via profile is unchanged by cross-sectional SEM check. This micro damage is prone to absorbing the cleaning solvent and result in poison via. Reducing the ashing pressure, such as using low pressure O_2 stripping (<50mT), can prevent this kind of damage and improve the poison via (Table 2, 3).

Table 3. HSQ thickness loss and SiH/SiO ratio vs. PR ashing condition

	Pre-etch	Low pressureO2	Downstream (high pressure)
SiH/SiO	0.38	0.31	0.0046
HSQ thickness. loss	--	100A	300A

3.4 Wet clean

HSQ is susceptible to solvent attack during the wet cleaning stage if its surface has micro-damage induced from previous process steps (Table 2, 3). This problem can be resolved by optimizing the cleaning recipe by choosing a better cleaning agent (Table 4).

Table 4. Normalized via Rc vs. wet cleaning condition. (based on inductive etcher, without passivation gas)

Rc (mean/stdev)	Landed	Unlanded (off-set 0.06um)
Std. agent A process (amine based)	1.4/6.8%	1.6/3.1% (9.1%fail)
Process A with reduced soaking time (amine based)	1.6/6.1%	1.6/7.3%
Agent B process (F-based)	1.0/6.7%	1.7/36%

Via Rc is normalized by agent B process landed Rc.

Conclusions

The key challenge for HSQ is the via poisoning for the unlanded via. A possible root cause is due to the trapped solvent inside the damaged HSQ and the mis-alignment slit. From etching and stripping optimization, poison via can be improved by (1) passivation gas addition in the etching step to reduce the HSQ damage, (2) use low pressure O2 process to strip the resist, (3) reduce the soaking time of cleaning agent or choose suitable cleaning agent.

References

(1) " Low-k materials etch and strip optimization for sub 0.25μm technology", T. Gao et al., 1999 International Interconnect Technology Conference, Proceedings P53, San Francisco, CA. U.S.A.
(2) " Mechanism of AlCu Film Corrosion", K. Siozawa et al., Jpn. J. Appl. Phys. Vol. 36, pp. 2496-2501 (1997).
(3) " Integration of HSQ as IMD in a five metal level, sub quarter micron technology using both W plug and hot aluminum metallisations" K. Barla et al., 1998 VMIC Conference, June 16-18, Proceedings P25, Santa Clara, CA, U.S.A.

O_2 Plasma Treatment of Low-k Organic-Silsesquioxane for Novel Intermetal Dielectric Application

T. Yoshie, S. C. Chen, and J. Kanamori

VLSI R&D Center, Oki Electric Industry Co., Ltd.

550-1 Higashiasakawa, Hachioji, Tokyo, Japan

ABSTRACT

Low-k organic silsesquioxane(k=2.5), methyl-hydrido-silsesquioxane(MHSQ), was successfully integrated with non etch-back Al-W plug IMD(Intermetal dielectric) process. We confirmed that the MHSQ surface oxidation by O_2-RIE plasma treatment was effective to protect the MHSQ against excess oxidation at conventional dry stripping of photo resist. W plugs through MHSQ were successfully formed on silicon substrate.

However, the same process was not suitable for the via holes on TiN/Al layer. Via holes were bowed after dry stripping and MHSQ encroachment occurred after CVD-W process. These problems were caused by titanium contained layer deposited on the via hole sidewall. The titanium contained layer prevented the surface oxidation of MHSQ by O_2-RIE treatment. Furthermore, the titanium contained layer reacted with WF_6 gas and it resulted in the MHSQ encroachment.

In order to solve this problem, we used a newly developed dry stripper for removing the photo resist without via hole bowing. For this, a good yield of 500k via-chains was obtained.

INTRODUCTION

The process integration of low-k material to multilevel interconnection has intensely required due to the device-performance limitation by interconnect propagation delay[1]. Organic silsesquioxane, methyl-hydrido-silsesquioxane($HOSP^{TM}$, Allied Signal Inc.) is an attractive material with a enough low k value(k=2.5) for deep quarter-micron devices. It also has good properties, for example, high thermal stability(450°C), good flowability, good gap-filling property and so on[2].

We tried to use this MHSQ for Al-W plug interconnection. The key point of this process is how to remove the photo resist after via hole etching. The disadvantage of this MHSQ is easy oxidation by O_2 plasma. Three types of O_2 plasma apparatus were evaluated to solve this problem; DF(down-flow) O_2 stripper, via hole etcher for in-situ O_2-RIE treatment combined with following DF-O_2, and O_2-RIE stripper(TCE-4802; new stripper developed by Tokyo Ohka Kogyo Co., Ltd.).

We had obtained good results by using O_2-RIE treatment in via hole etcher in the case of via holes formed on silicon substrate. The condensed SiO_2 was formed on the MHSQ surface by O_2-RIE treatment, which prevented excess oxidation of MHSQ at following DF-O_2 stripping.

461

However, the same process was not suitable for the via holes on TiN/Al layer.

In this study, we describe the reason that this O_2-RIE treatment dose not have any effect in the case of via holes on the TiN/Al layer. To solved this problem, we used newly developed O_2-RIE stripper after via hole etching. We have realized a good yield of 500k via chains using this process.

EXPERIMENTAL

MHSQ(400nm) was formed on a silicon substrate or TiN/Al(100nm/500nm) layer by the conventional spin-on method. Cap-SiO_2[200nm] was deposited on MHSQ. After via hole photo/etch process, the photo resist was stripped using three types of O_2 plasma as shown in Table I. Type I was DF(down-flow) O_2 stripper, which is conventionally used for dry stripping of photo resist. Type II was O_2-RIE treatment in etching chamber. This O_2-RIE treatment was used with following DF-O_2 stripping(Type I). Type III was O_2-RIE stripper which was newly developed by Tokyo Ohka Kogyo Co., Ltd. After barrier TiN deposition, blanket CVD-W film was formed at 430°C

500k via chains of Al-W plug interconnect were fabricated. The sample structure was shown in Fig.1. At this time, O_2-RIE stripper(Type III) was used to remove a photo resist.

Table I Conditions of O_2 plasma to treat MHSQ surface or to remove a photo resist.

	Type I	Type II	Type III
	DF-O_2 stripper	O_2-RIE treatment in etching chamber	O_2-RIE stripper
Gas	O_2	O_2/Ar	O_2
Pressure	Atmospheric	40mTorr	60mTorr
Temperature	250°C	20°C	100°C

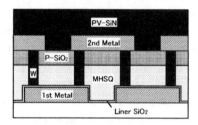

Fig.1 Structure of 500k via chains. Diameter of via holes was 0.36 μ m.

RESULTS & DISCUSSIONS

Figure 2 shows the via hole shapes after via hole etching, DF-O_2 stripping(Type I) and CVD-W deposition. Via holes etched through cap SiO_2/MHSQ were formed on silicon substrate. The via holes had good shapes after etching. However, the via holes were bowed after DF-O_2 stripping. MHSQ on via-hole sidewall was oxidized and shrank at the dry stripping. CVD-W formed on bowing via holes was also observed.

This problem was solved by O_2-RIE plasma(Type II). We confirmed that 50nm-condensed SiO_2 was formed on a blanket MHSQ surface by O_2-RIE treatment. This condensed SiO_2 prevented excess oxidation of MHSQ by DF-O_2 stripping(Type I). This O_2-RIE treatment before DF-O_2 stripping was carried out after via hole etching. W-plugs were successfully formed on a silicon substrate by this process as shown in Fig.3. This result indicates that the condensed SiO_2

1.0µm

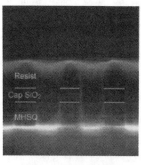 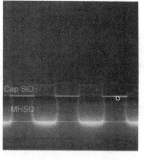 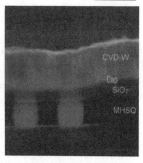

(a) after via-hole etching   (b) after DF-O₂ stripping   (c) after CVD-W deposition

Figure 2   Cross-sectional views of via holes. Films of Cap SiO₂/MHSQ(200/400nm) were deposited on silicon substrate.

layer was also formed on MHSQ surface located on via-hole sidewall.

The same process of O_2-RIE treatment(Type II) before DF-O_2 stripping(Type I) was used for via holes formed on TiN/Al layer. Figure 4 shows the via holes after the dry stripping and CVD-W deposition. Via holes were bowed after the dry stripping and MHSQ was encroached after CVD-W. From these results, there were following issues;

1.   Why were the via holes bowed even though O_2-RIE treatment was carried out?
2.   What did happen during CVD-W process?

To clarify the first issue, the depth profile of titanium deposited on via hole surface was evaluated by SIMS(Secondary Ion Mass Spectroscopy). A large amount of titanium was detected in via hole. We think that this titanium contained layer prevented the formation of the condensed SiO_2 on MHSQ surface by O_2-RIE treatment, which resulted in the excess oxidation of MHSQ during DF-O_2 stripping.

1.5um   1.5um

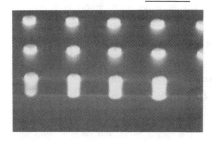   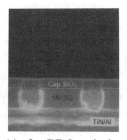 

(a) after DF-O₂ stripping   (b) after CVD-W
with O₂-RIE treatment.

Fig.3 W plugs through cap SiO₂/MHSQ formed on silicon substrate. O₂-RIE treatment(Type II) was carried out before DF-O₂ stripping(Type I).

Fig.4 Cross-sectional views of via holes formed on TiN/Al layer. (a) after dry stripping, (b) after CVD-W deposition followed by dry stripping(a).

463

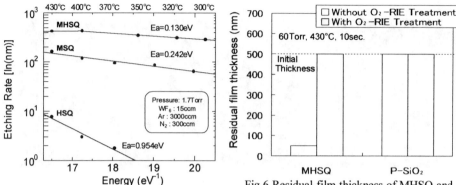

Fig.5 Temperature dependence of the etching rate of various silsesquioxane-type films. Films were exposed in WF_6 gas.

Fig.6 Residual film thickness of MHSQ and Plasma SiO_2 after exposure in WF_6 gas. One of MHSQ and P-SiO_2 was treated by O_2-RIE plasma before WF_6 etching respectively.

To solve the second issue, the reaction between MHSQ and WF_6 gas was studied. Figure 5 shows the etching rate of silsesquioxane films by WF_6 gas. Every silsesquioxane film reacted with WF_6 gas. MHSQ had the highest etching rate and the lowest activation energy of 0.13eV. Figure 6 shows the residual thickness of MHSQ and plasma SiO_2 after exposure in WF_6 gas. One of MHSQ and P-SiO_2, respectively, was treated by O_2-RIE treatment before WF_6 treatment. MHSQ without O_2-RIE treatment was etched off more than 400nm in only 10 seconds. It means a few microns thick of MHSQ is encroached during the CVD-W process. On the other hand, silicon dioxide, such as P-SiO_2 or oxidized MHSQ, did not react with WF_6 gas. Therefore, thin oxidation on MHSQ is good for suppressing MHSQ encroachment.

In the case of via holes on TiN/Al layer, MHSQ encroachment occurred, although via holes were bowed due to the MHSQ oxidation. The encroachment is considered to be caused by the non-oxidized MHSQ behind the titanium contained layer. Non-oxidized MHSQ remains behind titanium contained layer even after DF-O_2 stripping. The encroachment proceeds from the titanium contained layer, and then non-oxidized MHSQ is encroached. On the contrary, in the case of via holes on silicon substrate, MHSQ encroachment dose not occur due to the complete oxidation of MHSQ by DF-O_2 stripping.

From these results, it was found that DF-O_2 stripper was difficult to use for MHSQ interconnection. Therefore, we used newly developed O_2-RIE stripper(Type III) to remove via hole resist. This O_2-RIE stripper removed the photo resist without MHSQ oxidation. Good shape of W plugs was formed on Al interconnects as shown in Fig.7. No bowed via hole after O_2-RIE stripping led to good coverage of barrier TiN before CVD W process. This resulted in the good CVD-W filling.

Figure 8 shows the resistance distribution of 500k via chains using O_2-RIE stripper(Type III). Conventional plasma SiO_2 and MHSQ were used for IMD. Etching time difference is also

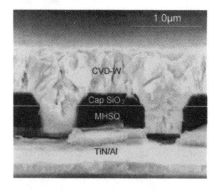

Figure 7 W plugs through cap SiO_2/MHSQ formed on TiN/Al layer. O_2-RIE stripper(Type III) was used for removing via hole photo resist.

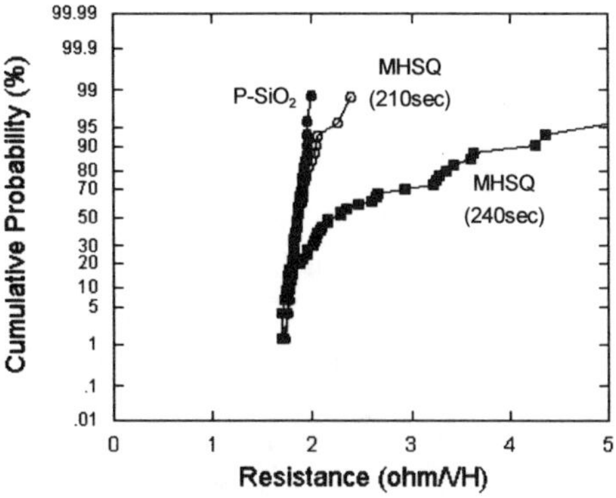

Figure 8 Cumulative probability of 500k via chain resistance. P-SiO_2 or MHSQ was used for IMD. Via hole etching was varied in the case of MHSQ process.

shown in Fig. 8. The resistance of via chains with short etching time was low and almost the same as that of conventional plasma SiO_2. The resistance distribution of the via chains with long etching time became wide. It was confirmed that the titanium increased with increase in etching time by SIMS analysis. We considered the increase in the titanium contained layer on via holes enhanced the possibility of MHSQ encroachment. This result indicates the reduction of the titanium contained layer is important for high reliability of MHSQ IMD process.

CONCLUSIONS

The integration of organic silsesquioxane with low-k for Al interconnection with W plugs was studied. It was confirmed that the condensed SiO_2 thin film on MHSQ surface formed by O_2-RIE treatment is effective to prevent the excess oxidation during down-flow type O_2 plasma treatment was performed. However, the condensed SiO_2 is difficult to be formed when vias were fabricated on TiN/Al underlayer because of the titanium-contained layer exist, which was deposited on via surface during via-hole etching step. This titanium-contained layer also leads to the MHSQ encroachment occurred by WF_6 gas during the CVD-W deposition. The newly developed O_2-RIE stripping process in this study was confirmed to be effective to achieve the low resistance of interconnection structured in via-chains, since this O_2-RIE stripping process remove photo resist without excess oxidation of MHSQ from via-hole sidewall. Furthermore, it was also confirmed that more clean surface, lower contamination of titanium-contained layer of via sidewall leads to obtain higher reliability of this developed process.

REFERENCES

[1] M. T. Bohr, IEDM Tech. Dig., (1995), p.241.
[2] H. Hacker, L. K. Figge, V. Flores, and S. P. Lefferts, Proc. of IITC, (1998), p.286.

VAPOR DEPOSITION POLYMERIZATION OF PARYLENE INTEGRAL FOAM THIN FILMS

James Erjavec, John Sikita, Stephen P. Beaudoin and Gregory B. Raupp
Department of Chemical, Bio, Materials Engineering, Arizona State University, Mail Code 6006, Tempe, AZ 85287-6006

ABSTRACT

A process has been developed to vapor deposit porous Parylene-N polymer films that are sealed or continuous at the substrate-solid and gas-solid interfaces, thus creating a vapor deposited integral polymer foam. In this process the substrate temperature is programmed through three separate stages while all other process conditions including monomer flow into the deposition chamber are held constant. The temperature in the first stage is constant at 261 K; under these conditions a non-porous Parylene-N film is produced. In the second stage substrate temperature is rapidly decreased to 80 K, where a porous film is deposited. In the third stage the temperature is returned to 261 K. Here deposition occurs preferentially in the outermost pores, with closing of the pore mouths until they are fully capped and a continuous, albeit rough, film layer is produced.

INTRODUCTION

According to the Technology Roadmap for Semiconductors, advanced generations of microchips will require interlayer dielectrics (ILD) with dielectric constants below 2.0. In recent years a wide variety of spin-on and vapor deposited organic polymers have been considered as low κ materials options [1-4]. However, if polymers are to be employed, this physical property requirement limitation will limit consideration to films that are porous in nature, or which have integral air gaps. For example, consider the family of parylenes, a polymer type that posseses many properties that are well suited for use as an ILD, including high thermal stability and low water absorption. Parylene-N has a dielectric constant of 2.65. To produce a Parylene-N film with an effective dielectric constant below 2.0, a porosity on the order of 40% would be required.

In a recent publication we have demonstrated that highly porous (porosity ≈ 80%) Parylene-N films can be produced through Vapor Deposition Polymerization (VDP) by holding the substrate at or near liquid nitrogen temperature during deposition [5]. The average density of these porous Parylene-N films is 0.195 g/cm^3, compared to the nominal 1.11 g/cm^3 density expected for non-porous Parylene-N [5,6]. The estimated dielectric constant of these films may be as low as 1.3.

The integration of a porous film into a multilevel interconnection process flow may be problematic. Several critical issues would need to be addressed to make the process manufacturable. At the metal-polymer interface, adhesion would likely be quite poor due to the limited contact area between the solid metal interface and the porous polymer, particularly in light of the fact that inadequate organic polymer – metal adhesion is in general a notoriously difficult problem to overcome. Once deposited, a porous layer presents downstream problems in both wet processes (liquids becoming trapped in pores), and in dry processes (resolution and anisotropy in etching, pore-filling in deposition). Both these problems could be circumvented by depositing films that are non-porous at the solid-solid interfaces, with an intervening porous layer between.

In this note we report on a new process developed in our Thin Film Polymer Processing Laboratory to deposit "integral foam" Parylene-N films. In the conventional bulk plastics industry, integral foams are foamed, porous polymers that have an outer skin whose density is

467

virtually equal to that of the unfoamed polymer [7]. Close inspection of the structural detail usually indicates the presence of an intermediate density transition zone between the outer skin and the fully foamed middle zone [7]. These foams are distinctly different from laminates, in that their parts are not separately manufactured and then glued together, but instead they are monolithic pieces formed in a single manufacturing process. Commercial processes include injection molding, extrusion, or rotational molding [7]. Because our films are likewise produced in a single process, we have created a new vapor deposited monolithic thin film analog to bulk integral foams. These unique integral foam thin films are created by performing deposition in a conventional Parylene-N VDP tool with controlled substrate temperature programming, as described in detail below.

EXPERIMENT

In the parylene vapor deposition polymerization process invented by Gorham [8], di-p-xylylene dimer is sublimed at about 450 K in a reduced pressure sublimator. The dimer vapor flows through a pyrolysis furnace held at 873 K or above, where it cracks quantitatively into two monomers. The monomer then flows into a reactor chamber and condenses and polymerizes on surfaces below a threshold temperature that is dependent on the chemical nature of the monomer. For Parylene-N VDP, this threshold temperature is 303 K.

Deposition experiments were performed in the experimental system described in detail elsewhere [5]. Nominal process conditions for all runs reported here were as follows: sublimator temperature equal to 403 K, pyrolysis furnace temperature equal to 958 K, chamber pressure equal to 0.08 Torr. Substrate temperature was programmed to provide sequential non-porous film deposition, porous film deposition, and non-porous film deposition in a single run. Several example temperature programs are given quantitatively in the RESULTS section.

Rectangular substrates were cut from undoped polished silicon (100) wafers obtained from Silica Source and were employed without pretreatment (e.g., no adhesion promoter was employed). Following deposition, the substrates were freeze fractured with a diamond scribe for cross-sectional scanning electron microscope (SEM) analysis. Cross-sectional SEM micrographs were taken with a JEOL JSM-840 instrument operating at an accelerating voltage of 15 keV. In order to minimize electron beam charging during SEM imaging, samples were coated with gold in a Denton sputter coater for 180 s at 20 mA and 2.5 kV in Ar.

RESULTS

Figure 1 shows the substrate temperature – time history for three representative deposition runs. These temperature programs are similar in that three distinct temperature plateaus or steps are employed, with the programs differing only by the length of the soak times at each step and the transition rates between steps. All programs presented here were initiated at 261 K. In this first stage of the process, a non-porous parylene film is deposited. Following deposition of this initial adhesion layer, liquid nitrogen was circulated through the substrate holder to cool the substrate to near liquid nitrogen temperature, 80 K. During this cool down phase the pressure in the reactor drops as monomer condensation rate on the substrate and holder increases, and then gradually returns to 0.08 Torr. In the second deposition phase the porous layer of the integral foam is deposited. The linear and mass deposition rates for Parylene-N at these substantially reduced temperatures is much faster than that at higher temperatures [5], so the time interval for this phase of the process is relatively short compared to the other phases. The substrate temperature is then returned back to 261 K, with the rate of transition controlled by substrate holder heaters. In this final phase, deposition occurs preferentially on the surfaces of the outer pores, ultimately closing off these pores and providing a continuous, pore-free capping layer.

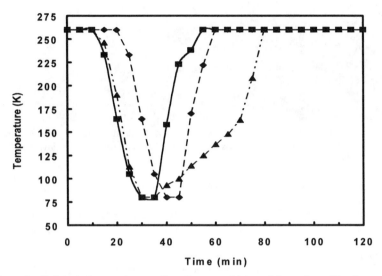

Figure 1. Substrate temperature *vs.* time for several integral foam deposition temperature programs: T-Program 1 (diamonds), T-Program 2 (squares), T-Program 3 (triangles).

Figures 2 and 3 show SEM micrograph cross-section, top plane, and bottom plane views of integral foam films deposited using temperature program for T-Program 1. This schedule has the longest soak time in the initial deposition stage, and therefore yields the thickest initial nonporous film layer. The bottom plane views in the figures show that the film at the Si-polymer interface is reasonably smooth. The porous layers are characterized by large, irregular shaped pores with characteristic dimensions as large as 10 μm. This microstructure is consistent with that previously reported in low temperature isothermal VDP [5]. The top capping layer is very rough, and at least for this temperature program, it appears that not all pores are completely capped (closed off).

Figure 2. *Left-to-right:* cross-section, top, and bottom view of Parylene-N integral foam film using T-Program 1. The film is free standing; in the cross-section view the film top is on the left.

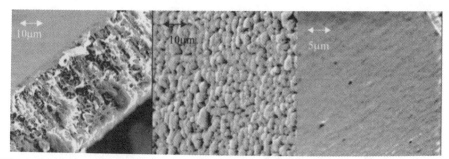

Figure 3. *Left-to-right:* cross-section, top, and bottom view of free standing Parylene-N integral foam film using T-Program 1. In the cross-section view the film top is on the lower right.

To increase the thickness and enhance the smoothness of the capping layer, and to ensure complete capping of the intervening porous layer, the duration of the third deposition stage was increased, as indicated in Figure 1, T-Program 2. In addition, the duration of the initial stage was decreased, since a relatively thin initial non-porous layer will satisfy the enhanced adhesion requirement.

Figures 4 and 5 show two films that were deposited in this modified temperature program. It is evident that the initial seed layer is substantially thinner than the films in Figures 2 and 3. The top layer is in turn thicker, and it appears that the pores of the middle layer have been completely sealed.

During a few integral film deposition runs, we observed rather violent foaming of the growing film during the transition from the second deposition stage to the third deposition stage. This violent action is apparently associated with the induced temperature rise, which results in a rapid increase in the film polymerization rate. To control this foaming effect, we performed some runs with a more gradual increase in film temperature. Figure 1 illustrates this temperature program as T-Program 3. This program successfully minimizes the foaming effect while maintaining complete sealing of the porous layer with the capping layer.

The films that we have deposited are much thicker than those that would commonly be used in the microelectronics industry for interlevel dielectrics. Future experiments and process development will focus issues related to this application, including creation of thinner films with smaller, controlled and reproducible pore size in the porous region, as well as conformality in layers deposited between metal lines.

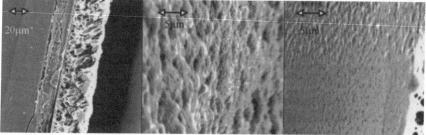

Figure 4. *Left-to-right:* cross-section, top, and bottom view of integral foam film from deposition using T-Program 2. The cross-section shows the film deposited on a silicon wafer.

Figure 5. *Left-to-right:* cross-section, top, and bottom view of free standing Parylene-N integral foam film using T-Program 2. The cross-section view shows the film top on the right.

CONCLUSIONS

A new technique has been developed to seal, or cap, vapor deposited Parylene-N porous films to create integral foams. In this process substrate temperature programming enables sequential deposition of a solid, continuous non-porous parylene interface layer, followed by deposition of a porous layer and then a solid capping layer.

AKNOWLEDGEMENTS

This work was supported by a grant from the National Science Foundation and U.S. EPA, grant no. CTS-9613377.

REFERENCES

1. C. L. Lang, G. R. Yang, J.A. Moore, T-M Lu, Mat. Res. Soc. Symp. Proc., **381**, 45 (1995).

2. J. V. Crivello, Mat. Res. Soc. Symp. Proc., **381**, 51 (1995).

3. W. F. Beach & T. M. Austin, 2nd International SAMPE Electronics Conf., p. 25 (1988).

4. G.A. Dixit, K.J. Taylor, A. Singh, C.K. Lee, G.B. Shinn, A. Konecni, W.Y. Hsu, K. Brennan, M. Chang, 1996 Symposium on VLSI Technology Digest of Technical Papers, p. 86, 1996.

5. J. Erjavec, J. Sikita, G. Raupp, S. Beaudoin, J. Materials Letters, **39**, 339 (1999).

6. W. F. Beach, C. Lee, D. Basset, T. Austin, R. Olson, Encyclopedia of Polymer Science, **17**, 990 (1988).

7. Shutov, F.A., *Integral/Structural Polymer Foams*, Springer-Verlag, 1986.

8. W. F. Gorham, Journal of Polymer Science: Part A-1, **4**, 3027 (1966).

OPTICAL METROLOGY FOR MONITORING THE CURE OF SILK* LOW-K DIELECTRIC THIN FILMS

F. YANG*, W. A. McGAHAN*, C. E. MOHLER**, L. M. BOOMS**
*Nanometrics Inc., 310 DeGuigne Drive, Sunnyvale, CA 94086, fyang@nanometrics.com
**The Dow Chemical Company, 1712 Building, Midland, MI 48674

Key Words: thin films, optical metrology, low-k, SiLK, cure, reflectometry, ellipsometry

ABSTRACT

Thin film optical metrology provides fast and precise real-time measurement on thin film thickness and optical constants. Its non-contact nature makes it ideal for in-line monitoring of product wafers in semiconductor manufacturing processes. In this presentation, it is shown that optical metrology can be applied in monitoring the thermal curing process of the SiLK dielectric thin films. It is found that at the wavelength of 314 nm, indices of refraction, $n(\lambda)$, of the SiLK dielectric thin films change systematically with the curing parameters (cure time and cure temperature). Based on the relationship between the SiLK resin optical constants and its curing condition, a single-parameter empirical interpolation model for SiLK resin optical constants is developed. With this interpolation model, the cure of the SiLK resins can be readily monitored using an automated thin film optical metrology tool, which provides prompt feedback on the condition of the thermal processing equipment.

INTRODUCTION

As device features of ultra-large-scale-integrated (ULSI) circuits continue to shrink, the capacitance of the inter-level dielectric (ILD) material becomes an increasingly limiting factor on overall performance of ULSI chips, and an industry-wide effort is underway to search for a low-k ILD material. One of the leading candidates of low-k ILD materials is Dow Chemical's SiLK dielectric. SiLK dielectric thin films can be readily deposited using conventional spin-coaters. Ideal mechanical, thermal, and electrical properties are achieved after the thin films are cured at 400-450 C in thermal-processing equipment (furnaces, ovens, or hot-plates). Cured SiLK resin has a dielectric constant at 2.65, and can withstand temperature as high as 490 C. Its high thermal stability permits integration with current multi-level interconnect processes. Because the polymerization of SiLK resins is thermally activated, the control of the cure temperature and cure time is critical to the quality of the cured SiLK dielectric thin films[1,2]. Thus, it is essential to have immediate feedback from metrology tests on the process tool, in order to detect and correct tool drift promptly.

Spectroscopic reflectometry and ellipsometry have been widely used as the in-line optical metrology for monitoring thin film thicknesses and optical constants in semiconductor IC production. A modern thin film metrology tool measures the reflectance and/or ellipsometric spectrum of thin film(s), and extracts thickness(es) and/or optical constant(s). One of the biggest advantages of optical metrology arises from its non-destructive nature, which allows measurements on product wafers and on active device areas. With in-line optical metrology, the performance of the thermal processing equipment can be monitored by real-time measurements on product wafers. Monitor wafers can be eliminated. Rework or loss of product wafers due to the out-of-control equipment can be minimized.

Conference Proceedings ULSI XV © 2000 Materials Research Society

This paper reports use of optical metrology for monitoring SiLK resin curing process. It demonstrates that the optical constants n(λ) and k(λ) of the SiLK dielectric thin films can be determined from reflectance measurements, in addition to the film thickness. It shows that the values of refractive indices, n, in the ultra-violet (UV) spectral region correlate to the curing conditions. By measuring the variation of n, the degree of cure of the SiLK resin can be monitored. This method can be implemented on a regular thin film metrology tool such as the Nanometrics NanoSpec 8000XSE®, providing critical information about the cure quickly and precisely.

SAMPLES

For this study, a matrix of SiLK-I dielectric thin film samples were prepared by varying cure temperature and cure time. SiLK-I resins were first deposited on bare silicon wafers by a spin-on process. They next went through a hard bake step (310 C, 90 seconds). After hard bake, they were cured on hot plates. The cure temperature varied from 400 to 470 C, and the cure time from 30 to 360 seconds. The sample thicknesses were about 7400 Angstroms.

INSTRUMENTATION

The optical constants n(λ) and k(λ) of these samples were determined using a J. A. Woollam Co. variable angle spectroscopic ellipsometer (VASE®) [3]. VASE® is a very powerful analysis instrument for characterizing optical properties of thin films[4]. It is particularly useful for accurately determining n(λ) and k(λ) of an organic polymer with complicated dispersion in the UV spectral region[5]. By choosing an appropriate type of the dispersion model for optical constants, the complicated n and k spectra can be parameterized from the VASE® measurement result.

The parameterized dispersion model can be implemented on a high-throughput thin film metrology tool, NanoSpec® 8000XSE. The NanoSpec® 8000XSE can be used to measure and fit spectroscopic reflectance spectra, spectroscopic ellipsometric spectra, or the combination of two. Its sophisticated data fitting algorithm allows the analysis of data from a wide range of materials and layered structures, and the simultaneous determination of thickness and optical constants[6].

RESULTS AND DISCUSSION

Correlation between Cure Parameters and Optical Constants of SiLK Resins

From the VASE® measurements, a spectral window between 280 nm and 340 nm was found where the refractive index (R.I.) n(λ) of SiLK resin changes monotonically with cure temperature and cure time as shown in Fig. 1, where only the UV spectra are included. In the visible wavelength region, the variation of R.I. is not significant. It can be seen in both figures that the sensitivity to the cure parameter is the highest at the wavelength of 314 nm. The total magnitude of the change of R.I. is 0.065 at this wavelength. In Figure 2, R.I. at 314 nm, denoted as n(314 nm), is plotted for all test samples. It is clear that n(314 nm) decreases with either increasing time or temperature. Thus, n(314 nm) can be used as an indicator for the degree of cure of a cured SiLK thin film. The reduction in n(314 nm) decreases as the temperature or time approaches the high end of the curing process window, indicating near completion of the cure.

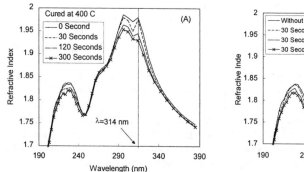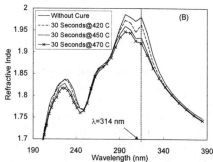

Fig.1   UV refractive indices n(λ) of SiLK resins cured at different conditions: (A) cure-time dependence; (B) cure-temperature dependence. The measurements were performed by VASE®.

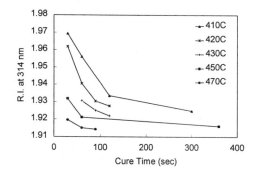

Fig.2   The relation between the refractive indices at 314 nm and the cure conditions (time and temperature), as measured by VASE®.

Single-Parameter Empirical Interpolation Model for Optical Constants of SiLK Resins

For a process-sensitive thin film, process parameter variation can lead to change in the film's n(λ) and k(λ) values, which is the case for cured SiLK resins. To monitor the degree of cure of a SiLK dielectric thin film on an automated metrology tool, it is necessary to have a parameterized dispersion model for SiLK resin optical constants. A parameterized dispersion model represents functional dependence of n(λ) and k(λ) on the wavelength in terms of a small number of adjustable parameters. This model should be relaxed enough to represent n(λ) and k(λ) of the SiLK resins cured within the whole process window, and yet restricted enough to induce minimum uncertainty from the correlation between thickness and optical constants.

An empirical interpolation model is an effective single-parameter dispersion model[7]. Its adjustable parameter normally corresponds to a process-dependent physical quantity (crystallinity of polysilicon, alloy fraction ratio of SiGe, etc.). It consists of a database of optical constant spectra for a given material. Each spectrum corresponds to a specific process condition. When fitting reflectance and/or ellipsometry data of a sample prepared at an unknown condition (within the process window), its optical constants are interpolated between known spectra in the database.

To develop an empirical interpolation model for SiLK resins, optical constant spectra of five samples cured at different conditions were chosen. The adjustable parameter was denoted as "cure index". For the uncured sample, the value of the "cure index" was assigned to 0, which has the highest n(314 nm). For the cured sample having the lowest n(314 nm), its value was assigned to 1. The other three cured samples have their n(314 nm) values equally distanced between the highest and lowest n(314 nm) values (see Table I). Their "cure indices" were determined by proportioning their n(314 nm) values according to those of 0 and 1, i.e.,

$$\text{Cure Index} = [n(314 \text{ nm}) - 1.9791]/[1.9161 - 1.9791] \qquad (1)$$

Table I
The information for the five samples whose optical constant spectra make up the SiLK resin interpolation dispersion model.

Sample #	Cure Condition	R. I. at 314 nm	Cure Index
1	Without Cure	1.9791	0.00
2	30 Seconds @410 C	1.9694	0.15
3	60 Seconds @400 C	1.9513	0.44
4	60 Seconds @430 C	1.9307	0.77
5	360 Seconds @450 C	1.9161	1.00

Automated Production Metrology for Monitoring SiLK Resin Curing Process

The empirical interpolation model for SiLK resins was tested on an automated thin film metrology tool. A measurement recipe was set up to measure UV-visible reflectance only using the reflectometer mode of a NanoSpec® 8000XSE, and calculate both film thickness and optical constants. Figure 3 displays an example of a reflectance scan and the fit result using the

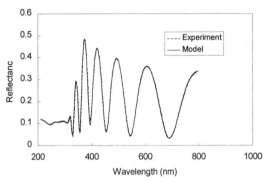

Fig. 3 Fit result of reflectance data measured on a NanoSpec® 8000XSE. The SiLK resin was treated at 400 C for 300 seconds.

empirical interpolation model. It shows an excellent match between the experiment and the model. However, to become a production-worth metrology, following requirements must be met: 1) high precision; 2) high stability; 3) high accuracy; 4) high sensitivity; 5) high throughput.

The measurement precision and dynamic stability were tested on a series of samples with different curing conditions and/or thicknesses. As one of the worst cases, test results of a thin SiLK sample is presented here. This sample was loaded ten times. For each load, five sites (one at center, four around the edge) were measured. The statistical results of the average values of each load are listed in Table II. The standard deviation of n(314 nm) for 10 loads at each site is less than 0.001, and the variation of the average value of n(314 nm) of 5 sites is also less than

Table II
Stability test result: the wafer was loaded ten times; each time 5 sites on the wafer were measured.

	Mean	Standard Deviation
Thickness (A)	1064.8	1.14
R. I. at 314 nm	1.929	< 0.001

0.001 from load to load. Such uncertainty is less than 2% of the total range (about 0.063) of the variation of n(314 nm) that may be induced by curing. Therefore, this measurement technique provides sufficient precision and stability to monitor the cure process of SiLK resins.

The measurement accuracy was tested by comparing n(314 nm) values measured by the NanoSpec® 8000XSE to the benchmark values from the VASE® measurement (see Fig. 4). In the figure, the data points can be fit by a straight line with a slope of 0.9788 and an intercept of 0.0441, indicating excellent correlation between the results of the two types of measurements. Such a close match assures the accuracy of refractive index measurement using the reflectometry mode of the NanoSpec® 8000 XSE.

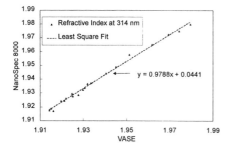

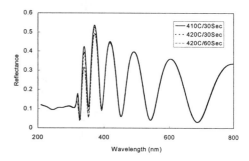

Fig. 4 Correlation between R.I. at 314 nm measured by the VASE® and by the NanoSpec® 8000XSE.

Fig. 5 Reflectance sensitivity predicted by the empirical interpolation model. Spectra are generated at a fixed thickness of 7400 Angstroms.

As for the sensitivity, Fig. 5 plots three model-predicted reflectance spectra, corresponding to three cure conditions: 30 seconds at 410 °C, 30 seconds at 420 °C, and 60 seconds at 420 °C. All 3 spectra were generated based on the same thickness value. The difference in the spectra is purely caused by the difference in the degree of cure. It is clear from the graph that significant difference among these spectra exists in the UV spectral region (between 310 nm to 400 nm), which can be readily measured by a reflectometer. The SiLK dielectric thin films become opaque in the wavelength region below 310 nm. In the visible region, all three spectra are nearly identical. Thus, it is necessary to include UV spectra in order to monitor the cure-induced change in SiLK resins.

The throughput of the measurement depends on the speed of the reflectometer data acquisition and analysis, and the focusing speed of the sample stage (robotics). Currently a 5-site measurement on an 8" wafer takes ~ 1 minute with focusing on each site.

CONCLUSIONS

Organic spin-on polymers need to go through a proper curing process to become an adequate low-k ILD material. In-line characterization of the cure of the spin-on low-k dielectrics is of great importance because it improves overall equipment efficiency, eliminates monitor wafers, and reduces wafer scrap and wasted work. In this paper, we have shown that broad-band reflectometry can be used to monitor the cure process of SiLK resins. Both film thickness and degree of the cure can be simultaneously determined. Because of its fine measurement spot size and pattern recognition capability, this technique can be used to monitor the cure of SiLK dielectric thin films on product wafers.

REFERENCES

[1] P. H. Townsend, S. J. Martin, J. Godschalx, D. R. Romer, D. W. Smith, Jr., D. Castillo, R. DeVries, G. Buske, N. Rondan, S. Froelicher, J. Marshall, E. O. Shaffer, and J. –H. Im, Mater. Res. Soc. Symp. Proc., vol. 476, 9 (1997).
[2] S. Allada, Proc. IITC, 161 (1999)
[3] F. Yang, W. A. McGahan, C. E. Mohler, and L. M. Booms, to be presented at The 46th International Symposium of American Vacuum Society, October 25-28, 1999, Seattle.
[4] J. A. Woollam and P. G. Snyder, Mater. Sci. Eng., vol. B5, 279 (1990).
[5] F. Yang, M. Tabet, and W. A. McGahan, Proc. SPIE, vol. 3332, 403 (1998).
[6] W. A. McGahan, B. R. Spady, J. A. Iacoponi, and J. D. Williams, Proc. ASMC 1996, 359.
[7] W. A. McGahan, B. R. Spady, B. D. Johs, and O. Laparra, Proc. SPIE, vol. 2725, 450 (1996).

SiLK* is a trademark of The Dow Chemical Company.

AN INTEGRATED LOW κ HDP-FSG FOR 0.15 μm COPPER INTERCONNECTS

Hichem M'saad, Manoj Vellaikal, Wen Ma, Kent Rossman
Dielectric Deposition Products Division
Applied Materials, Santa Clara, CA 95054

ABSTRACT

A low κ dielectric has been developed for use in copper damascene processes. The film is deposited using silane, silicon tetrafluoride, and oxygen as precursors in a single-wafer HDP-CVD chamber. The film has a dielectric constant of 3.3. The film has been successfully integrated with silicon nitride and BLOκ™ in single and dual damascene structures.

INTRODUCTION

Silicon dioxide (SiO_2) has traditionally been the dielectric material of choice in multilevel interconnect materials. However, as device dimensions shrink, propagation delays associated with RC time constant become a high fraction of the total delay. Therefore, there is a strong emphasis on reducing both the resistance and capacitance of materials used in multilevel interconnects. In order to decrease the capacitance, different low dielectric constant materials have been investigated. The most promising low dielectric constant material currently used in production with Al interconnects is high density plasma fluorinated silicate glass (HDP-FSG) [1,2]. This material, having a κ value of 3.5, is attractive because it retains the structure of SiO_2 with the addition of small amounts of fluorine. Contrary to common belief, the limit in κ value to 3.5 is not a stability limitation, but in actuality, a gap-fill limitation issue. It is not possible to increase F content, thereby lowering κ below 3.5, without compromising gap-fill of high aspect ratio Al metal lines, typical in 0.18μm technology. However, due to the nature of the HDP-FSG gap-fill mechanism, F concentration between metal lines is up to 40% higher than above metal lines [3]. The FSG between Al lines is very stable and has been integrated in 0.18μm devices. Hence, it is possible to achieve stable κ<3.5 HDP-FSG films for damascene applications where gap-fill is no longer a requirement. With the advent of 0.15-0.13μm technology, the semiconductor industry will shift to copper damascene technology for back-end applications [4]. An HDP-based FSG with dielectric constant values of 3.3 has been developed and integrated with silicon nitride and BLOκ™ [5] in single and dual damascene structures.

EXPERIMENTAL

FSG films have been deposited in a 200mm high density plasma CVD (HDP-CVD) reactor using Ar, O_2, SiH_4 and SiF_4 as the constituent gases. The fluorine concentration in these films were determined using Fourier Transform Infrared Spectroscopy (FT-IR). The intensity of the Si-F peak (occurring at 937 cm^{-1}) was divided by the Si-O stretching peak intensity (occurring at 1095 cm^{-1}). From a FT-IR spectrum, it is also possible to detect the presence of SiF_2 bonding. Thermal Desorption Spectroscopy (TDS) was used to determine the temperature

479

stability of these FSG films. The dielectric constant values were determined using MOSCAP and mercury probe. The Pressure Cooker Test (PCT) was used to determine the extent of fluorine reduction in the film. In this test, the film was exposed to steam at a pressure of 2 atmospheres at 120°C for 10 hours and the extent of fluorine reduction was evaluated by comparing F concentrations before and after PCT using FTIR. Adhesion tests were performed in a tube furnace. The Si/TEOS/500Å SiN/HDP-FSG/500Å SiN structures were annealed for twelve cycles at 410°C for 30 minutes each cycle in a nitrogen ambient. Both visual and confocal microscopic observations were carried out in order to determine the integrity of the stack after anneal.

RESULTS AND DISCUSSION

TDS can be used as a preliminary indicator of film stability. It is used to check the onset temperature for degassing of different species. Figure 1 shows the TDS profile for an FSG film with a κ value of 3.3. The species that are monitored include H_2, H_2O, F, HF, Ar, SiF_2, SiF_3 and SiF_4. No desorption took place at temperatures below 500°C in this film. The species that desorb after 500°C include H_2, F, HF and Ar. This spectrum indicated that the film was stable up to a temperature of 500°C. One of the advantages of high density plasmas over other technologies is the ability to produce denser films with low hydrogen contents.

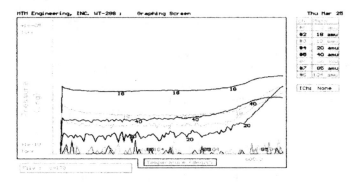

Figure 1: Thermal desorption spectrum of a κ=3.3 HDP-FSG process showing no desorption up to 500°C.

We have investigated different process conditions for obtaining stable FSG films. Gas Mixing Ratio (GMR) was defined as the ratio of SiF_4 gas flow to the sum of SiF_4 and SiH_4 gas flows $\{GMR = SiF_4/(SiF_4 + SiH_4)\}$. The sum of SiH_4 and SiF_4 is important as this determines the amount of silicon that is available for FSG formation. Four different GMR ratios were investigated (0.4, 0.5, 0.6 and 0.7). Intensities of SiF_2, SiF and SiO stretching peaks were measured from FTIR spectra of these films and the ratio of SiF_2 to SiF + SiO has been plotted as a function of GMR (Figure 2). Presence of SiF_2 makes moisture absorption into the structure easier thereby reducing the stability of the film. This is the reason why SiF_2 formation is not desirable. In this figure, two different regions can be observed. Region 2 showed evidence of SiF_2 formation which was not present in Region 1. As the GMR ratio increased, the relative amount of SiF_2 became higher. This was due to the fact that with increasing GMR ratio, the relative concentration of fluorine with respect to silicon increases (with other factors being fixed) thereby rendering the possibility of SiF$_2$ formation higher.

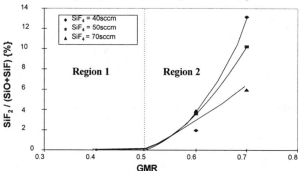

Figure 2: SiF$_2$ content in HDP-FSG vs. gas mixing ratio showing a stable film in Region 1.

One of the major stability issues with low dielectric constant films relate to moisture uptake when left under ambient conditions for an extended period of time. PCT is an aggressive test that can be used to simulate the effect of moisture uptake on the properties of FSG films. The films discussed above were subjected to PCT testing for 10 hours and the fluorine concentration was monitored before and after the test. Figure 3 indicates the change in fluorine concentration after PCT as a function of GMR.

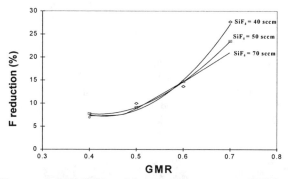

Figure 3: Decrease in Si-F FTIR peak height after exposure to PCT for 10 hours

The reduction in %F after PCT with increasing GMR is apparent from Figure 3. This correlated with increasing SiF$_2$ content in the film. A more open structure is much more susceptible to moisture uptake which hydrolyzes the Si-F bonds, thereby decreasing the Si-F peak strength. The dramatic increase in %F reduction when the GMR increased from 0.6 to 0.7 also correlated with the increase in SiF$_2$ content.

While TDS and PCT testing indicate to some extent the relative stability of FSG, annealing studies with multiple layers have to be carried out in order to establish the integration performance of FSG with different films. In a single or dual damascene structure, an FSG film has to adhere to HDP or PE SiN and BLOκ™ films, be able to withstand the etch, resist ash, and solvent clean processes, should not be affected by the slurry used for copper CMP and must be stable with Ta/TaN barrier layers. In this study HDP-FSG films were integrated with PE and HDP CVD silicon nitride barrier/etch stop layers and the silicon carbide based BLOκ™.

The structure used for anneal was Si/TEOS USG/500Å SiN/HDP FSG/500Å SiN. The underlying TEOS USG was used in order to eliminate the effect of defects on Si surface on the annealing results. The anneal was performed at 410°C for 30 minutes each cycle in an N_2 ambient for 12 cycles. Two different types of FSG films with the same F concentration and similar TDS performance were used. Figures 4(a),(b) show the confocal microscope results after anneal on the two wafers. Figure 4(a) showed bubbling on the surface which was indicative of delamination occurring in the structure. Figure 4(b) showed no bubbling for the optimized FSG process. From a comparison of Figures 4(a) and (b) it was deduced that a good TDS profile is a not sufficient to achieve good adhesion properties with SiN.

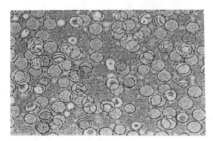

Figure 4(a): Confocal view of an HDP-FSG -SiN structure after two anneal cycles. Circular bubbles are present.

Figure 4(b): Same as Fig. 4(a) but after 12 anneal cycles. Here, the FSG process does not cause delamination.

With an optimized FSG process, adhesion testing was performed with PECVD BLOκ™. FSG passed adhesion with this material for both films in Fig. 4. BLOκ™ adhesion is not dependent on the nature of the HDP-FSG process as silicon nitride. The etch properties of HDP-FSG/SiN structures have also been determined. The etch process was carried out using a CF_4/Ar chemistry. As seen from the SEM cross sectional image of the etch profile in Figure 5, the profiles are slightly tapered with notching which is typical of partial etch.

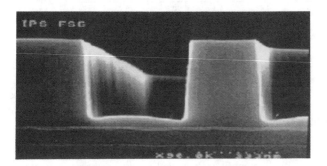

Figure 5: SEM cross section of a single damascene HDP-FSG/SiN etch.

The HDP-FSG film has been successfully integrated in a Cu dual damascene structure as shown in Figure 6. PE SiN of 600 Å in thickness was used as a barrier and etch stop. TaN was the copper barrier. The SEM micrograph shows no nitride blowout and excellent adhesion between FSG and TaN. There is minimal via etching in M1 barrier and Cu. The via chain and leakage performance of the structure in Fig. 6 was similar to Cu/PE TEOS USG.

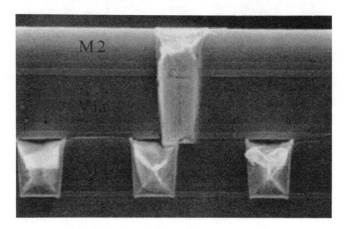

Figure 6: Two-level Cu/HDP-FSG interconnect.

CONCLUSIONS

FSG films with dielectric constant values less than 3.3 have been prepared using high density plasma CVD which was successfully integrated with HDP and PE SiN and BLOκ™ films. In order to obtain SiF_2-free FSG films under the conditions tested, the GMR ratio has to be less than 0.5. The increase in SiF_2 content led to larger amounts of fluorine loss from FSG during PCT testing. Some FSG films with no degassing at temperatures below 500°C as determined by TDS failed annealing test with SiN, thereby implying that a good TDS performance is a necessary but not sufficient condition for ensuring good adhesion. The HDP-FSG film was successfully integrated in a Cu dual damascene structure.

REFERENCES

1. Hichem M'saad, Manoj Vellaikal, Lin Zhang and Derek Witty, DUMIC, Cat. # 99IMIC - 444D, p. 210 (1999).
2. Hichem M'saad, Manoj Vellaikal, Lin Zhang and Derek Witty, Advanced Interconnects and Contacts, edited by D. C. Edelstein, T. Kikkawa, M. Ozturk, K-N.Tu and E. Weitzman. MRS Proceedings, 564, San Francisco, (1999),
3. Y. L. Wang, W. Chang, S. Chen, S. Wang, R. Liao, T-A. Yeh and A. Chen, DUMIC, Cat. # 99IMIC - 444D, p. 91 (1999).
4. Peter Singer, Semiconductor International, June (1998), p. 91.
5. Ping Xu, John Ferguson, Judy Huang, Kegang Huang, Chris Ngai, Anjana Patel, Sudha Rathi, Betty Tang and Mark Loboda, IITC Conference, p. 109 (1999).

Aluminum, Tungsten, and
DRAM Metallization

HIGH TEMPERATURE STABILITY OF CONDUCTING Ir-Ta-O FILM IN OXYGEN AMBIENT

Fengyan Zhang, Sheng Teng Hsu, Jer-shen Maa, Shigeo Ohnishi[*], and Norito Fujiwara[*]
Sharp Laboratories of America, Inc., 5700 NW Pacific Rim Blvd. Camas, WA 98607, USA
[*]Sharp Corporation, Tenri-city, Nara632-8567, Japan

ABSTRACT

The Ir-Ta-O/Ta structure with Ir-Ta-O as the electrode and Ta as the diffusion barrier layer on silicon substrate has been fabricated. The Ir-Ta-O film was deposited by reactive sputtering using separate Ir and Ta targets in oxygen ambient. Annealing results performed from 500-1000°C in oxygen ambient showed that the Ir-Ta-O film exhibited extraordinary high temperature stability. This film showed good conductivity and integrity even after 5min annealing at 1000°C. It is believed that the Ir-Ta-O film can be used as electrode in FeRAM and DRAMs devices, in which the ferroelectric material sometimes is deposited and annealed at high temperature in oxygen ambient in order to achieve good ferroelectric properties.

INTRODUCTION

Thin films of Ir, Pt, Ru, IrO_2 and RuO_2 have been extensively studied for the application as electrode materials in FeRAM and DRAM devices [1-10]. In addition to the advantages of chemical stability and low resistivity, noble metal oxides can also improve the fatigue property of ferroelectric material such as PZT by preventing space charge formation at the electrode and ferroelectric material interface. However, some problems, such as hillock formation and film peeling caused by stresses, poor adhesion, and oxidation of the barrier layer, have limited the application of these electrodes in very high temperature in oxygen ambient, which is the required deposition and annealing condition for some ferroelectric materials such as $SrBi_2Ta_2O_9$.

It has been reported previously that grain boundary precipitation can inhibit hillock formation, improve the structural stability, and enhance the barrier property of thin films[11]. Yoon et al [12, 13] found RuO_2 stuffed Ta barrier was more resistant to oxidation and oxygen diffusion than N stuffed polycrystalline nitride. There was no increase in sheet resistance even after 800°C annealing for 30 min in air. They also found RuO_2 stuffed Pt can prevent the oxygen diffusion up to 650°C for 30 min. A Ru-Ti alloy with high Ru composition was found to have better thermal stability and better barrier properties against interdiffusion of Si and oxygen than Ru metal alone[14].

Comparing with Pt and Ru, Ir has been reported to have better barrier property against oxygen diffusion. Ta and TaN are more resistant to oxidation than Ti and TiN are. The refractory nature of both Ta and Ir, and the high formation temperature of their compounds have assumed their increased thermal stability of the Ta-Ir system [15]. It is expected that the oxygen diffusion resistance and structural stability of the Ir-Ta system can be further improved by oxygen incorporation. The purpose of this paper is to study the effect of Ta and oxygen addition into the Ir film. Discussions on the thermal stability,

487

microstructure and phase changes, and electrical properties of Ir-Ta-O conducting oxide film will be presented.

EXPERIMENT

The silicon (100) p-type wafer was dipped in dilute HF solution, DI water rinsed and spin dried before loading into a sputtering chamber. The base pressure of the sputtering chamber was around 3×10^{-7} Torr. The sputtering chamber was equipped with two separate sputtering targets. The purity of the 4-inch diameter Ta and Ir targets are 99.95% and 99.8% respectively. Ta film was deposited by DC sputtering in pure Ar. Ir-Ta-O film was deposited by reactive sputtering in Ar-O_2 mixture with flow ratio of Ar: O_2 at 1:1. The chamber pressure was maintained at 10 mTorr for Ir-Ta-O film deposition. The DC power on both Ir and Ta targets were about 300 W. The deposition sequence was as follows: Ta layer was deposited on the Si substrate first, then the Ir-Ta-O film was deposited on the Ta layer. The deposition was performed at room temperature. After deposition of the films, the Ir-Ta-O/Ta/Si structure was annealed in a furnace in oxygen atmospheric ambient. The annealing temperature was 500-1000°C. The oxygen flow rate was 2500 sccm and annealing time was 5min at each annealing step. Long time annealing was also performed at 800°C in oxygen atmospheric ambient. The annealing time was 30, 60, and 90 min respectively. The microstructure and phase changes of the Ir-Ta-O film were characterized by X-ray diffraction. The morphology changes of the structure were examined by scanning electron microscopy and transmission electron microscopy. Sheet resistance changes of the films were measured by four-point probe. The composition and the interdiffusion between different layers were analyzed by auger electron spectroscopy depth profile.

RESULTS AND DISCUSSION

Figure 1 is the TEM picture of the as deposited Ir-Ta-O film . In contrast to the typical large grain and columnar structure observed in e-beam evaporated Ir and Pt films, this Ir-Ta-O film exhibited a finer polycrystalline granular structure.

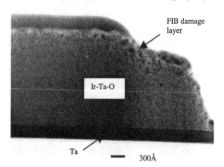
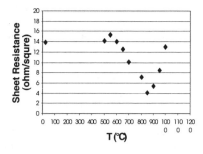

Figure 1. TEM cross-section of Ir-Ta-O/Ta/Si Structure

Figure 2. Sheet resistance changes of Ir-T -O/Ta/Si during annealing from 500-1000°C

This film was then put in the furnace for annealing in oxygen atmospheric ambient. From the experiment results that the Ir-Ta-O /Ta/Si structure can sustain at least 1000°C oxygen annealing for 5min without losing its conductivity and structural stability. No large hillocks and peeling off were observed. This is very different from the annealing results of Ir and IrO_2 films, in which large hillock formation and severe peeling started to occur only after a moderate 700°C 5min annealing in oxygen ambient.

The sheet resistance changes of the Ir-Ta-O structure before and after annealing in oxygen is shown in Figure 2. It can be seen that the sheet resistance increased slightly after 500 –550°C annealing and then started to decrease with increasing temperature in the temperature range of 550-850°C. This was believed to be due to the grain growth of the films. The sheet resistance reached its lowest value after 850°C annealing. Above 850°C annealing, the sheet resistance started to increase again. The sheet resistance of the film after 1000°C annealing was still slightly lower than the original sheet resistance of the as-deposited film.

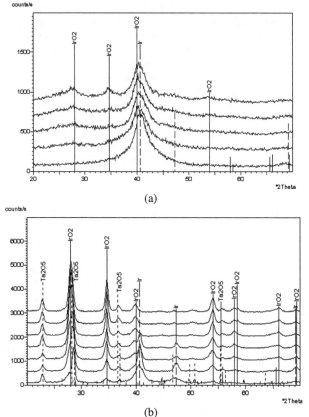

(a)

(b)

Figure 3. XRD spectra of Ir-Ta-O /Ta/Si during annealing from (a) 500-650°C (b) 700-1000°C (—— IrO_2 , ---- Ir , ········· Ta_2O_5)

The X-ray diffraction spectra of the Ir-Ta-O/Ta/Si structure are shown in Fig. 3(a) and Fig. 3(b). Figure 3(a) shows that a very fine polycrystalline phase of Ir was detected in the as-deposited film. After oxygen annealing at temperatures below 650°C, intensities of both Ir and IrO_2 peaks were found to increase, indicating the onset of crystallization and grain growth of Ir and IrO_2. Above 700°C annealing, as shown in Fig. 3(b), sharper IrO_2 peaks were observed, corresponding to the further oxidation of Ir and grain growth of the IrO_2 crystals. The intensities of IrO_2 peaks continued to increase and Ir peaks continued to decrease until about 900°C. After 900°C annealing, the Ir peaks became very weak and further increases of the IrO_2 peaks were no longer observed. Apparently, most of the Ir was oxidized to IrO_2. The phase and microstructure changes observed in X-ray diffraction spectra were found consistent with the sheet resistance changes of the films.

In the XRD spectra, crystalline phases of tantalum and tantalum oxide peaks were not observed in the as deposited film and the film annealed below 650°C. Above 650°C, Ta_2O_5 peaks started to appear. The intensity of the Ta_2O_5 peaks continued to increase up to about 750°C annealing and then stabilized. This indicated that the Ir-Ta-O film after high temperature annealing in O_2 ambient converted to a mixture including IrO_2 and Ta_2O_5. Although Ta_2O_5 is a high permitivity material, the conductivity of the film was not too much degraded even after 1000°C oxygen annealing for 5min.

Figure 4 (a) and Figure 4(b) are the AES depth profiles of the Ir-Ta-O/Ta/Si film before and after 1000°C annealing. From Fig 4(b), it is seen that the Ta barrier layer was oxidized after oxygen annealing. This could be a disadvantage for some type of FeRAM and DRAM devices, in which low contact resistance is required between the bottom electrode and polysilicon substrate. More oxidation resistant barrier layer is under research. However, the interface between Ir-Ta-O and Ta and between Ta and Si remained sharp after annealing. This indicated no significant interdiffusion occurred between these layers. Ta silicide peaks were not detected. This was probably due to the presence of oxygen at the interface of Ta and Si, which have suppressed the formation of tantalum silicide.

It is also found that the thickness of both Ir-Ta-O and Ta films increased about one-third after 1000°C annealing due to the oxidation of the Ir and Ta films. This also contributed to the sheet resistance decreases between 550°C and higher temperature annealing. Similar increases of the thickness were observed for Ru enriched Ru-Ti alloy during oxygen ambient annealing[14]. But their results showed that the oxidation only occurred at the surface and that prevented further oxidation of Ru-Ti layer. In contrast, the Ir in the Ir-Ta-O film was almost totally oxidized. This was probably due to high Ta ratio in the Ir-Ta-O film. By increasing the relative composition ratio of the Ir, or decreasing the oxygen partial pressure during sputtering, partial Ir was left un-oxidized after high temperature annealing, and the barrier property of the Ir-Ta-O film was found to be improved. The disadvantages of increasing the Ir composition ratio and decreasing the oxygen partial pressure are the increased hillock formation and structural failure possibility.

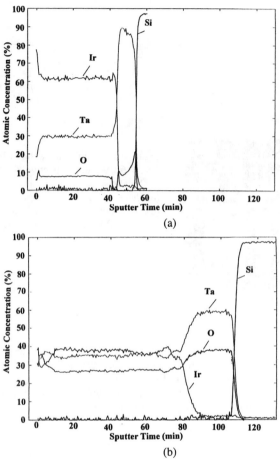

(a)

(b)

Figure 4. AES spectra of Ir-Ta-O/Ta/Si structure (a) before and (b) after oxygen
annealing at 1000°C for 5min

Both the thickness of the film, the conductivity and phases exhibited rather stable
once most of the Ir was oxidized to IrO_2 and Ta was oxidized to Ta_2O_5. Fig 5 are the
XRD spectrum, sheet resistance changes and SEM cross-section of the Ir-Ta-O film after
800C annealing for 30, 60, and 90 min annealing in oxygen ambient. No significant
phase changes are observed and the sheet resistance remained almost the same. It is
further approved that the Ir-Ta-O film is very stable after the initial oxidation stage.

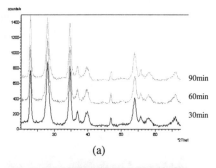

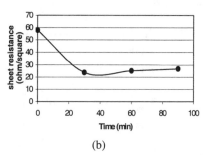

(a)

(b)

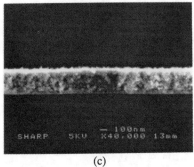

(c)

Figure 5 Long time annealing of Ir-Ta-O for 30, 60, 90 min respectively.
(a) XRD spectra
(b) sheet resistance changes
(c) SEM cross-section of Ir-Ta-O/Ta/Si after 800°C annealing in oxygen for 90 min

CONCLUSION

The Ir-Ta-O film showed thermal stability up to 1000°C oxygen annealing. High conductivity was maintained and no structural failure such as large hillock formation and film peeling were observed. The Ir in the Ir-Ta-O film was oxidized during annealing. The microstructure of the Ir-Ta-O film after high temperature oxygen ambient annealing is a polycrystalline mixture including IrO_2 and Ta_2O_5 phases.

REFERENCES:

1) P.D. Hren, S. H. Rou, H. N. Al-Shareef, K.D. Gifford, O. Auciello, and A.I. Kinggon, *Integarated Ferroelectrics*, **2**, 311 (1992)
2) J.O. Olowolafe, R. E. Jones, A.C. Campbell, P.D. Maniar, R.I. Hegde, and C.J. Mogab, *Mat Res Soc. Symp. Proc.*, **243**, 355 (1992)
3) T. Nakamura, Y. Nakao, A. Kamisawa and H. Takasu, *Appl. Phys. Lett.* **65**, 1522 (1994)
4) T. Nakamura, Y. Nakao, A. Kamisawa and H. Takasu, *Jpn. J. Appl. Phys. Lett.* **33**, 5207 (1994)
5) Fengyan Zhang, Tingkai Li, Tue Nguyen, Sheng Teng Hsu, Ferroelectric Thin Films VII, *Mat Res Soc. Symp. Proc* (1998) (to be published)

6) Seung-Hyun Kim, J.G. Hong, J.C. Gunter, H. Y. Lee, S.K Streiffer, and Angus I. Kingon, *Mat Res Soc. Symp. Proc.* **493**, 131(1997)

7) G. –R. Bai, A. Wang, I-Fei Tsu, C.M. Foster, and O. Auciello, *Integarated Ferroelectrics,* **21**, 291 (1998)

8) A. Grill, R.Laibowitz, D. Beach, D. Neumayer and P.R. Duncombe, *Integarated Ferroelectrics,* **14**, 211 (1997)

9) Katsuhiro Aoki, Yukio Fukuda, Ken Numata, and Akitoshi Nishimura, *Jpn. J. Appl. Phys.* Vol. **34** , 5250 (1995)

10) J.J. Lee, C. L. Thio, and S.B. Desu, *J. Appl. Phys.* **78**(8) 5073 (1995)

11) S. Wolf, *Silicon Processing,* Lattice Press (Sunset Beach, California, 1990), Vol. **2**, p271

12) Dong-Soo Yoon, Hong Koo Baik, Sung-Man Lee, Sang-In Lee, Hyun and Hwack Joo Lee, *J. Vac. Sci. Technol.* B **16**(3), May/Jun , 1137(1998)

13) Dong-Soo Yoon, Hong Koo Baik, Sung-Man Lee, Sang-In Lee, Hyun and Hwack Joo Lee, *Appl. Phys. Lett.* V **73**, 324 (1998)

14) Ray-Hua Horng, Dong-Sing Wuu, Luh-Huei Wu, Ming-Kwei Lee, Shih-Hsiung Chan, Ching-Chich LEU, Tiao-Yuan Huang and Simon Min SZE, *Jpn. J. Appl. Phys.*, **37**, L1274 (1998)

15) D.K. Wickenden, M. J. Sisson, A. G. Todd, and M.J. Kelly, *Solid State Electronics*, **27**, 515 (1984)

A Study on the Metallorganic Chemical Vapor Deposition (MOCVD) of Ru and RuO₂ Using Ruthenocene Precursor and Oxygen Gas

Hyun-Mi Kim, Sung-Eon Park*, and Ki-Bum Kim

School of Materials Science and Engineering, Seoul National University
San 56-1, Shillimdong, Kwanakgu, Seoul, 151-742 Korea
Tel) 82-2-880-7095 Fax) 82-2-886-4156
kibum@snu.ac.kr
*Hyundai Electronics Industries Co. Ltd.

ABSTRACT

Thin films of Ru and RuO₂ were deposited using ruthenocene and oxygen as a precursor and reaction gas, respectively. Two phases, Ru and RuO₂, could be deposited using the same source and reaction gas. The phase of the as-deposited film was found to be critically dependent on the deposition process parameters such as precursor partial pressure, oxygen partial pressure, and substrate temperature. An increase in oxygen partial pressure led to RuO₂ formation. It was also identified that there exists a substrate temperature range in which a Ru film is deposited at a given processing condition. When the temperature is either higher or lower than this temperature range, RuO₂ film deposition results. From careful analysis of the film deposition rate in conjunction with the evolution of the film phase, we conclude that film deposition occurs by two independent processes. One is the source gas decomposition process resulting in Ru film deposition and the other is the oxidation of the Ru film resulting in RuO₂ formation.

INTRODUCTION

Ruthenium (Ru) is a transition metal with hexagonal structure and ruthenium dioxide (RuO₂) is a conducting oxide with rutile structure and low electrical resistivity (46 μΩ-cm). They have good etching properties due to the well-known volatile compounds (RuO₄, RuCl₄, etc.) and good diffusion barrier properties against oxygen.[1-4] Therefore, Ru and RuO₂ have been studied as capacitor electrode of high dielectric materials such as barrium-strontium-titanates(BST) or lead-zirconium-titanates(PZT) in dynamic random access memory(DRAM) and nonvolatile memory devices, respectively.[5,6] In this study, we have investigated the deposition mechanism of metallorganic chemical vapor deposition (MOCVD) Ru and RuO₂ with ruthenocene precursor and have characterized the properties of deposited films.

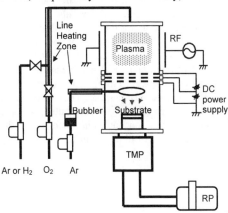

Fig.1 Schematic diagram of RuO₂ CVD equipment.

EXPERIMENTS

Figure 1 shows a schematic diagram of the deposition system used in this study. A detailed explanation is given elsewhere.[7,8] Si(100) wafers with a 100 nm thickness of thermal oxide are used as substrates. Prior to deposition, O₂ plasma treatment is conducted for 5 min at 40 mTorr in order to enhance the nucleation process. The plasma treatment is carried out at RF power of 100 Watts and DC grid bias of 500 V at the

Conference Proceedings ULSI XV © 2000 Materials Research Society

same temperature as with deposition [See ref. 7 for the detailed conditions and the effect of plasma treatment]. Right after the plasma treatment of the substrate, films are deposited by using ruthenocene [Ru(C$_5$H$_5$)$_2$] precursor and oxygen reaction gas by thermal CVD. The deposition pressure is fixed at 40mTorr and the substrate temperature (T$_{sub}$) is changed from 225 °C to 500 °C. In order to change the precursor and oxygen partial pressure, Ar flow rate, oxygen flow rate, and the bubbler temperature are changed. Deposited films are characterized by X-ray diffractometry (XRD), Auger electron spectroscopy (AES), scanning electron microscopy (SEM), α-step profilometer, and 4-point probe.

RESULTS

The Effect of Reaction Gases on Film Deposition

In order to investigate the precursor decomposition behavior, an attempt to deposit a film using only the ruthenocene precursor was made. Under these conditions, no film was deposited within the substrate temperature range of 250 to 500 °C. This result shows that ruthenocene source gas is stable within this temperature range. As a second attempt, we tried to deposit films using hydrogen reaction gas. Again, no film was deposited within the same temperature range. This showed that hydrogen is not effective in decomposing the source gas within this temperature range. Finally, when oxygen gas was used, a specula and well adherent film was deposited on the substrate. This result indicates that oxygen is a far more effective reaction gas in decomposing the ruthenocene source gas resulting in film deposition.

The Effect of Surface Treatment on Film Deposition

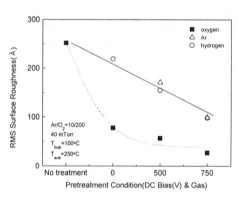

Fig. 2. The change of RMS roughness obtained by AFM results of as-deposited RuO$_2$ thin films with various plasma pretreatment conditions

To investigate the effect of plasma treatment on the deposition behavior, the voltage of DC grid bias was varied from 0 V to 750 V using O$_2$, H$_2$ and Ar as plasma gas. Figure 2 shows the decrease of surface roughness of as-deposited film. The RMS (root mean square) value of the film is dramatically decreased from about 25 nm to 3 nm by applying the oxygen plasma for 5 min prior to the deposition of about 200 nm thickness of film. Other plasma gases such as hydrogen and Ar also decrease the RMS value of the surface roughness but are not as effective as O$_2$ plasma. The decrease of surface roughness is attributed to the enhancement of nucleation density at the initial stages of film growth.

Phase Analysis of the As-Deposited Film

To systemically investigate the effect of oxygen partial pressure on the deposited film phase, the oxygen flow rate was varied from 2 sccm to 200 sccm while other parameters such as Ar flow rate, substrate temperature, and bubbler temperature were fixed at 10 sccm, 400 °C, and

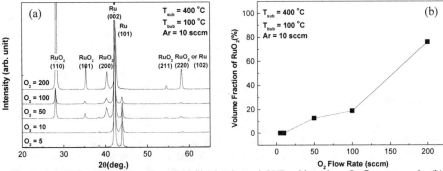

Fig. 3 (a) XRD patterns of as-deposited film by thermal CVD with various O_2 flow rate and (b) the change of volume fraction of RuO_2 in deposited films as a function of oxygen flow rate; T_{sub} = 400 °C, T_{bub} = 100 °C P = 40 mTorr, and flow rate of Ar = 10 sccm.

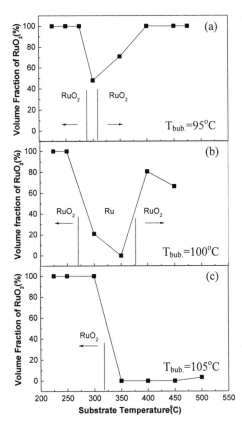

Fig. 4 The change of volume fraction of RuO_2 calculated by XRD quantitative analysis as a function of substrate temperatures; (a) T_{bub} = 95 °C (b) T_{bub} = 100 °C and (c) T_{bub} = 105 °C.

100 °C, respectively. Figure 3(a) shows the changes in the XRD pattern as a function of oxygen flow rate. This figure clearly shows that the only Ru phase is deposited below 10 sccm (O_2 partial pressure < 4 mTorr) of oxygen flow rate, while the intensity of RuO_2 peaks increases as the oxygen flow rate is increased from 50 to 200 sccm. This result shows that both phases (Ru and RuO_2) can be deposited and the phase of the as-deposited film changes from Ru to RuO_2 with the increase of oxygen partial pressure.

The volume fraction of RuO_2 was calculated from the relative ratio of X-ray intensity taking into account the structure factor, multiplicity factor, Lorentz-polarization factor, etc.[9] Figure 3(b) shows the volume fraction of RuO_2 obtained from the results of Fig. 3(a). As expected, the data clearly reveals that the volume fraction of RuO_2 phase gradually increases up to 80 % as the oxygen flow rate is increased to 200 sccm.

Figure 4 shows the effect of bubbler and substrate temperature on the deposited film phase at fixed O_2/Ar flow rate. The bubbler temperatures used were at 95, 100, and 105 °C, and the substrate temperature was varied from 225 to 500 °C at each bubbler temperature. The results suggest that only RuO_2 is deposited at either low substrate temperatures (< 275 °C) or at high substrate temperatures (> 400 °C) when the bubbler temperature is 95 °C. In between these two temperatures, the layer is composed of a mixed phase with varying ratio of each phase. The results appear similar at 100 °C bubbler temperature although the temperature range of Ru-rich phase is broader. When the bubbler temperature was set at 105 °C, two distinct regimes were observed; one is the low substrate temperature regime where RuO$_2$ is formed and the other is high substrate temperature regime where only Ru is formed. This suggests that there is a temperature range in which a Ru-rich phase is formed. If the substrate temperature is either lower or higher than this temperature range, the formation of RuO$_2$ is preferred over Ru. It is notable that the low temperature boundary of the temperature range is almost fixed at about 275 °C independent of the bubbler temperature, while the high temperature boundary is shifted to a higher temperature with increasing bubbler temperature.

These results demonstrate that the evolution of the as-deposited film phase using the ruthenocene source gas and oxygen gas is quite complicated. Both of the phases can be deposited by varying the deposition conditions such as the partial pressure of oxygen gas, Ar flow rate, bubbler temperature, and substrate temperature. Most notably, it is observed that RuO$_2$ can be deposited at low substrate temperatures as well as at high substrate temperatures. These results differ from those of Green et al.[1] and Si et al.[10] who reported that RuO$_2$ was deposited only at above 550 °C using the same source and reaction gas. Our results clearly show that a high temperature is not a necessary condition for the deposition of RuO$_2$ as has been recently reported by Shin et al.[11]

The Analysis of the Deposition Rate

The variation of deposition rate as a function of oxygen flow rate is shown in Fig. 5 (solid squares). This data shows that the deposition rate increases as the oxygen flow rate is increased from 0 to 10 sccm and is saturated as O$_2$ flow rate is further increased from 10 sccm to 100 sccm. However, the figure also shows that the deposition rate drastically increases as the oxygen flow rate is additionally increased to 200 sccm. Thus, it appears that there is no clear correlation between the oxygen flow rate and the deposition rate. However, it should be noted that the phase of the as-deposited film changes from pure Ru to a mixture phase of Ru and RuO$_2$ by increasing the oxygen flow rate (see Fig. 3). Thus, one has to account the changes of volume resulting from the changes of the film phase as the mere presentation of the deposition rate is meaningless. In

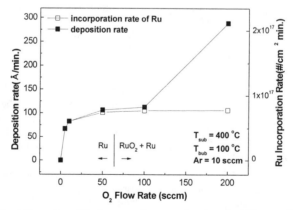

Fig. 5 Apparent deposition rate and Ru incorporation rate of as-deposited film as a function of oxygen flow rate; T_{sub} = 400 °C, T_{bub} = 100 °C P = 40 mTorr, and flow rate of Ar = 10 sccm.

order to understand the effect of the process parameters on film deposition regardless of the phase of the as-deposited film, a new concept called the "Ru incorporation rate" is suggested. The "Ru incorporation rate" is the determination of how many Ru atoms are incorporated in the film per unit time and per unit area ($\#/cm^2$ min.). One can calculate this number from the deposition rate by using the bulk density (ρ_{RuO2} = 6.97 g/cm^3, ρ_{Ru} = 12.30 g/cm^3) and molecular mass (W_{RuO2} = 133.07 g/mole, W_{Ru} = 101.07 g/mole) of each phase and the fraction of them in a phase mixture. In this manner, the deposition rate of the film can be roughly translated into the equivalent "Ru incorporation rate" irrespective of the film phase deposited.

From the results of deposition rate and RuO_2 fraction, the "Ru incorporation rate" was calculated and the result is shown in Fig. 5 as the open square mark. This data clearly shows that the "Ru incorporation rate" is linearly increasing at small O_2 flow rate and becomes saturated as the O_2 flow rate is increased above 50 sccm (i.e., O_2 partial pressure of 83 %). Thus, the oxygen does play a role of promoting source gas decomposition and depositing Ru film up to a certain partial pressure level of oxygen. When the oxygen partial pressure is above that level, the role of oxygen is diminished as the reaction is limited by the source gas supply. This result is similar to that of Ni oxide CVD by Yeh et al.[12] who reported that Ni oxide deposition rate increases linearly and becomes saturated with the increase in the oxygen flow rate.

The apparent deposition rate as a function of substrate temperature (not shown here) showed no systematic variation. However, an Arrhenius plot of the "Ru incorporation rate" as a function of substrate and bubbler temperature clearly shows that the deposition behavior of the Ru element is divided into surface reaction and mass-flow controlled regimes, as is typically the case in CVD (Fig. 6). The surface reaction controlled regime appears below 275 °C, 300 °C, and 350 °C at the bubbler temperature of 95 °C, 100 °C, and 105 °C, respectively. The value of activation energy of the surface reaction controlled regime is about 0.67±0.08 eV at all three bubbler temperatures. In the mass-flow controlled regime, the value of activation energy is less than 0.1 eV, meaning that the "Ru incorporation rate" has almost no temperature dependence. As the bubbler temperature is increased, the "Ru incorporation rate" increases in the mass-flow controlled regime. These results are in agreement with the general CVD process scheme.

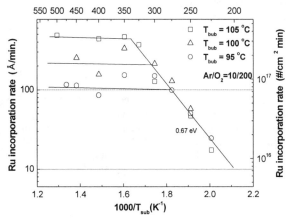

Fig. 6 The Arrhenius plot of Ru incorporation rate calculated by XRD results and deposition rate as a function of substrate temperatures with various bubbler temperatures; P = 40 mTorr, and Ar/ O_2 = 10 sccm/200 sccm .

CONCLUSION

In this study, Ru and RuO_2 thin films were successfully deposited by thermal MOCVD with a ruthenocene precursor and oxygen gas. Gas flow rates, bubbler temperature, and substrate temperature were selected as process parameters, and the effect of these process parameters on the deposition behavior was investigated. The results showed that no film was deposited without oxygen, but Ru, $Ru+RuO_2$, or RuO_2 film deposition occurred in the presence of oxygen. As oxygen flow rate was increased, the phase transition from Ru to RuO_2 occurred with the increasing deposition rate. At a fixed substrate temperature and flow rate of oxygen, a decrease in precursor partial pressure led to RuO_2 formation.

In order to observe the effect of the process parameters regardless of film phase, the "Ru incorporation rate" and Ru oxidation rate were introduced. The results suggest that RuO_2 tends to form if the oxidation rate is higher than the "Ru incorporation rate" as is shown in the Arrhenius plot (Fig. 7).

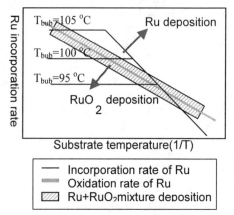

Fig. 7 The schematic diagram of Arrhenius plot of Fig. 6 showing the condition of phase determination.

REFERENCE

1. M. L. Green, M. E. Gross, L. E. Papa, K. J. Schnoes, and D. Brasen, J. Electrochem. Soc. **132**, 2677 (1985).
2. S. Saito and K. Kuramasu, Jpn. J. Appl. Phys. **31**, 135 (1992).
3. L. Krusin-Elbaum, M. Wittmer, and D.S. Yee, Appl. Phys. Lett. **50**, 1879 (1987).
4. K. Yoshikawa, T. Kimura, H. Noshiro, S. Otani, M. Yamada, and Y. Furumura, Jpn. J. Appl. Phys. **33**, L867 (1994).
5. T. Kawahara, M. Yamamuka, A. Yuuki, and K. Ono, Jpn. J. Appl. Phys. **35**, 4880 (1996).
6. R. Ramesh, Thin Film Ferroelectric Materials and Devices, Kluwer Academic Press, 24 (1997).
7. S. E. Park, H. M. Kim, and K. B. Kim, Electrochem. Sol. Lett. **1**, 262 (1998).
8. S. E. Park, H. M. Kim, and K. B. Kim, J. Electochem. Soc. (accepted).
9. B. D. Cullity, Elements of X-ray Diffraction, Addison-Wesley Pub. Company Inc. 407 (1978).
10. J. Si and S. B. Desu, J. Mater. Res. **8**, 2644 (1993).
11. W. -C. Shin and S. -G. Yoon, J. Electrochem. Soc. **144**, 1055 (1997).
12. W. -C. Yeh and M. Matusmura, Jpn. J. Appl. Phys. **36**, 6884 (1997)

OPTIMIZATION OF PRE-METAL DEPOSITION SPUTTER CLEANS FOR 1-GIGABIT DRAMS

R.C. IGGULDEN[I], L.A. CLEVENGER[I], J. GAMBINO[I], R.F. SCHNABEL[S], and S.J. WEBER[S]
DRAM Development Alliance at 1580 Route 52, Hopewell Junction, NY 12533
[I]IBM Microelectronics, [S]Infineon Technologies Inc.

ABSTRACT

As the requirements for high performance, high density circuitry increase and the back end of the line (BEOL) dimensions decrease the influence of pre-metal deposition sputter cleans on yield mounts. The effect of pre-metal deposition sputter cleans on via resistance and continuity was studied on a four-level BEOL 1-Gigabit DRAM (Figure 1).

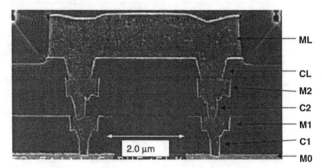

Figure 1: A cross-sectional SEM depicting the 4-level BEOL.

Three via types were investigated: an Al dual damascene interconnect landing on W (C1/M1); an Al dual damascene interconnect landing on Al (C2/M2); and a tapered Al RIE via landing on Al (CL/ML). The dual damascene structures investigated had high aspect ratios (4:1), while the tapered via structures had low aspect ratios (1.1:1 - 1.3:1). Via resistance is dependent on the via type, the via geometry, the via clean, the contact metal, and the sputter clean conditions. In order to avoid a significant yield loss, it is necessary to optimize the sputter clean for any change in via conditions. The two major sputter clean parameters that effect via resistance are the clean time and bias voltage. This paper will investigate the variables that effect via resistance, focusing on the sputter clean conditions.

INTRODUCTION

There have been many recent reports on Al dual damascene technology [1-6] and several of these reports mention the use of in-situ sputter cleans prior to metal deposition [1,4,5]. However, an in depth analysis of the sputter clean is often neglected. This paper will show the importance of customizing the sputter clean upon any change in via conditions.

A sputter clean is used to remove native oxide and post-etch residue without causing electrical device damage to the wafer. Two major parameters control the effectiveness of the sputter clean; the sputter clean time determines the amount of etching and the DC bias, as determined by the plasma power and bias power, controls the angularity of the incoming ion flux.

501

A sputter clean is typically used in conjunction with wet cleans and fusion ashing. Fusion ashing removes excess resist after etch, but can leave behind a metal-rich fluorocarbon film [7-9]. A wet clean is typically executed to remove the fluorocarbons. An in-situ sputter clean follows to remove any remaining contaminants and, more importantly, to remove the metal oxide layer formed upon air exposure [7, 9, 10]. Native oxide removal is especially important for vias landing on Al because most liner materials will not react with Al_2O_3 at typical BEOL processing temperatures [11].

EXPERIMENTAL

Samples consist of test structures used in developing a four metal level, 0.175 μm Al dual damascene technology for 1-Gigabit DRAMs [1,4]. There are three types of vias (Figures 2-5); "C1", a high aspect ratio (4:1) Al via that lands on W; "C2", a high aspect ratio (4:1) Al via that lands on Al; and "CL", a low aspect ratio (1.1:1 - 1.3:1) Al via that lands on Al.

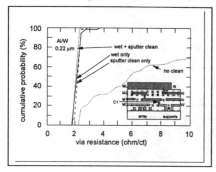

Figure 2: C1 via resistance (Al on W).   Figure 3: C2 via resistance (Al on Al).

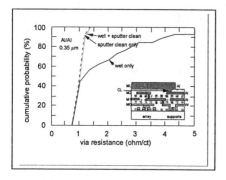

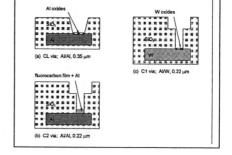

Figure 4: CL via resistance (Al on Al).   Figure 5: Schematics of the 3 vias after fusion ashing.

The diameters at the bottom of the vias are 0.22 μm for C1 and C2, and 0.35 μm for CL. The Al layers have Ti/TiN underlayers, so that the interfaces at the bottom of the vias are W or Al (lower metal) // Ti/TiN/Al (upper metal). The via etches are based on C_4F_8 [2]. The high aspect ratio vias (C1 and C2)

502

use the same etch, whereas the low aspect ratio via (CL) etch has a higher C_4F_8 flow, resulting in more polymer formation and hence a tapered profile [12]. The clean before metal deposition consists of an ozone ash, a wet etch, and an in-situ sputter clean. Some wafers were processed without the wet etch and/or the sputter clean to observe the effect on via resistance.

The resistance of single vias was measured with four-point probe Kelvin structures. The continuity of large numbers of vias (50K or 98K) was measured with via chains. The composition of contamination layers inside the low aspect ratio vias post resist strip and cleans was analyzed by Auger electron spectroscopy. The structure and composition of contamination layers inside vias after metal deposition were analyzed by TEM using electron energy loss spectroscopy (EELS).

RESULTS

The effect of the wet clean and sputter clean on via resistance is shown in Figures 2-4 for the three via types investigated. The cumulative probability plot for a via landing on W (Figure 2) indicates a relative independence to the clean. The W oxides and fluorides have low thermal stability, high volatility, and react readily with many wet clean chemicals [12, 13]. These properties make either a wet clean or a sputter clean sufficient surface preparation for vias landing on W. Additionally, variations in the bias voltage and the time of the sputter clean do not drastically affect the contact yield for vias landing on W. For these type of vias the sputter time must be long enough to etch through the tungsten oxide and the bias voltage cannot be set at a level that could adversely affect the device performance. A bias voltage range of 200-500V had no relative effect on the electrical parameters.

The cumulative probability plot for an Al dual damascene via landing on Al shows that a wet clean and a subsequent sputter clean are needed (Figure 3). When only fusion ashing is used at C2, a large fluorocarbon residue remains (Figure 6).

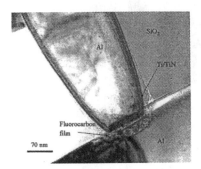

Figure 6: A cross-sectional TEM of a C2 via with only fusion ashing.

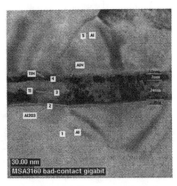

Figure 7: A cross-sectional TEM of a CL via without sputter clean.

These fluorocarbon residues, along with the Al_2O_3 that readily forms upon air exposure of the Al surface (Figure 7) necessitate the use of both a wet clean and a sputter clean. The wet clean is used to eliminate the remaining fluorocarbon residues, while the sputter clean is used to eliminate the Al_2O_3 that intrinsically forms on the exposed Al surface. The sputter etch time has a significant impact on the C2 via resistance (Figure 8).

503

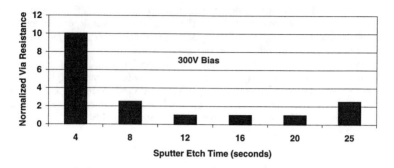

Figure 8: Via resistance vs. sputter etch time for a C2 via.

The observed increase in via resistance for etch times of 8 seconds and below indicates the remaining level of contaminants exceed what the given sputter clean is capable of removing. The increase in via resistance when the time rises to 25 seconds and above indicates that re-sputtering of oxide is becoming significant. There is a large increase in C2 via resistance when the DC bias voltage is in excess of 300 V (Figure 9). Therefore, the optimal C2 sputter clean conditions are a 300V bias with a 16 second etch time.

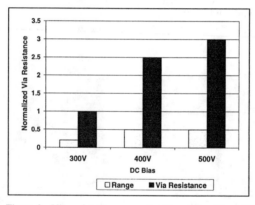

Figure 9: Via resistance vs. sputter bias voltage for a C2 via.

The cumulative probability plot for the tapered via indicates the need for a sputter clean either alone or in conjunction with a wet clean (Figure 4). A continuous Al_2O_3 film remains when no sputter clean is used at the CL level (Figure 7). The larger ground-rules of the tapered via allow for the removal of most of the fluorocarbon film by the fusion ashing. This leaves mainly Al_2O_3 on the surface, which allows the elimination of the wet clean. Similar to C2, the sputter etch time has a significant impact on the via resistance (Figure 10). The increase in via resistance due to re-sputtering of oxide is also similar to the phenomena seen in the C2 via. However, the CL via resistance is improved when using a higher DC bias voltage (Figure 11), contrary to the C2 via result. Therefore, the optimal CL sputter clean conditions are a 400V bias with a 10 second etch time. Figure 12 shows the effect of shrinking dimensions on via resistance for a given set of sputter clean conditions. The larger 0.4 μm (0.2 μm ground rule) vias have equivalent via resistances for any bias voltage ranging from 200-400V, while the corresponding 0.35 μm (0.175 μm ground rule) vias show via resistance degradation below 300V. This illustrates how the sputter clean conditions become increasingly important as via dimensions continue to shrink.

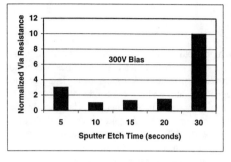

Figure 10: Via resistance vs. sputter etch time for a CL via.

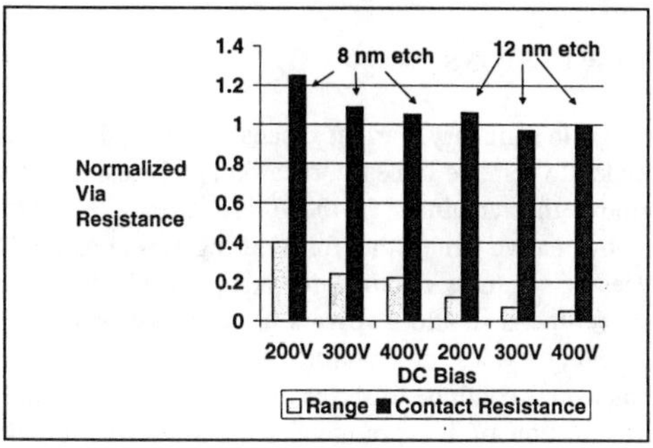

Figure 11: Via resistance vs. sputter clean bias voltage for a CL via.

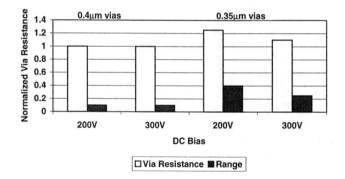

Figure 12: Via resistance vs. bias voltage for 2 different sized CL vias.

The schematic in Figure 5 illustrates the residuals remaining in each of the three via types after the fusion ashing. The tungsten oxides in the C1 via are easily removed with either a wet or sputter clean. The aluminum oxides and fluorocarbons present in the C2 via need both a wet clean and sputter clean to ensure low contact resistance. The aluminum oxides in the CL via are removed with a sputter clean only. Contacts to Al are extremely critical because of the low volatility of Al fluorides and the high thermal and kinetic stability of Al oxides. It is evident that sputter etch time has to be above a minimum to adequately clean contaminants and has to be below a maximum to avoid the re-sputtering phenomena. This signature was seen across all 3 contact types. The common belief that bias voltage should increase with increasing via aspect ratio does not appear to hold true for the CL to C2 aspect ratio increase. However, this discrepancy from commonality can be attributed to a via geometry difference, an etch difference, and a wet clean difference between the CL and C2 vias. However, when examining one via type (e.g. Figure 12 for CL) the direct proportionality between aspect ratio and bias voltage is apparent.

CONCLUSIONS

In summary, various cleans are needed to assure high yield in BEOL contacts. A fusion ashing is needed to remove bulk fluorocarbon etch residues from the vias. A wet clean is sometimes needed to remove the remainder of the fluorocarbon film and to remove oxides. A sputter clean is needed to remove native oxides that form during air exposure. Depending on the via type a combination of these 3 types of cleans is required for adequate yield. A change in the via dimensions, the via taper, or the contact metal should involve a re-optimization of the via clean processes. A sputter clean is needed for removing oxides, and optimization of the process must be done for each new via characteristic. As dimensions continue to shrink at an increased rate and via conditions continue to change accordingly, the optimization of the pre-metal deposition sputter clean conditions become increasingly important in metallization.

ACKNOWLEDGMENTS

The authors would like to thank John Benedict and John Bruley for their TEM work and would also like to acknowledge the support of the staff of the Advanced Semiconductor Processing Center (ASTC). Additionally the support of the DRAM Development Alliance (DDA) is gratefully appreciated. The authors finally would like to thank Elizabeth Morales for her editing advice.

REFERENCES

1. R. Iggulden et al., Proc. VMIC, 19 (1998).
2. M.B. Anand et al., Jpn. J. Appl. Phys., 37, 5526 (1998).
3. K. Sugai et al., Proc. IEDM, 781 (1997).
4. R.F. Schnabel et al., Microelec. Eng., 37/38, 59 (1997).
5. J. Gambino et al., Proc. IITC, 206 (1999).
6. Y. Hayashi et al., Symp. VLSI Tech., 88 (1996).
7. Y. Wang et al., J. Electrochem. Soc., 144, 1522 (1997).
8. K. Honda et al., Proc. 5th Int. Symp. Clean. Tech., Electrochem. Soc. Inc., vol. 97-35, 617 (1997).
9. S. Mayumi et al., Jpn. J. Appl. Phys., 29, L559 (1990).
10. H. Aoki et al., J. Vac. Sci. Tech., A13, 42 (1995).
11. M.B. Chamberlain, J. Vac. Sci. Tech., 15, 240 (1978).
12. T. Ohiwa et al., Jpn. J. Appl. Phys., 31, 405 (1992).
13. T.P. Chow et al., in *Dry Etching for Microelectronics*, Elesevier Sci. Pub., 39 (1984).

DEPENDENCE OF METAL CONTACT RESISTANCE ON CONTACT SIZE DOWN TO 0.14μm AND ON ASPECT RATIO UP TO 12

Hyunchul Sohn, Chang-Young Kim, In-Haeng Lee, Choon-Hwan Kim, Hyung-Soon Park, Kyung-Bok Lee, Woo-Hyun Kim, Soo-Jin Kim, Inn-Cheol Ryu, Hyug-Jin Kwon, Heung-Lak Park Dong-Joon Ahn
Process Development 4, Memory R&D Division, Hyundai Electronics Industries Co.

ABSTRACT

Based on the anticipation of reduction of metal contact size with increasing aspect ratio for higher integration density, the dependence of contact resistance on the contact size and the aspect ratio was investigated for IMP Ti technology. Electrical properties such as contact chain resistance and junction leakage current were investigated to determine the application limits of IMP Ti film.

INTRODUCTION

As the devices design rule shrink to increase integration densities and device speed, concerns on the metal contact are raised since the metal contact dimensions need to be reduced accordingly. The SIA roadmap predicted that the metal contact diameter for 0.13μm technology be reduced below 0.17μm with aspect ratios (AR) above 7.5 in DRAMs with COB architecture [1]. As AR of metal contact increases with decreasing size, the metal contact in memory devices with direct landing on silicon substrate is expected to suffer from a steep increase of contact resistance due to reduction of contact area. Also, the technological requirement on barrier metals becomes severe to achieve appropriate step-coverage at contact bottoms.

It is considered that the step coverage of Ti at contact bottom is the most crucial factor for contact resistance. For deep contacts, Ionized-Metal-Plasma (IMP) Ti and IMP TiN have been used successfully as contact and barrier metal for AR above 5 [2]; however, its adequacy as contact metal appears to be questioned for contacts with AR higher than 8.

Our preliminary experiments [3] and others [4] show that the step-coverage of IMP Ti at contact bottom depends strongly on the ratio of ionized metal ions to neutral atoms and the bias voltage on substrates. The ionization efficiency is affected by the pressure in sputter chamber and the ratio between DC power on the target and the power on RF coil.

In this work, we investigated the relationship between contact resistance and contact diameters from 0.58 μm down to 0.14 μm with the effect of various process parameters. Also we evaluated the limitation of IMP Ti as contact metal for deep contacts with aspect ratio up to 12.

EXPERIMENT

In this work, the samples were prepared as following: After STI isolation process, n+ and p+ source and drain regions were formed using As and BF_2 implantation with implantation energy and dose of 25KeV, 4E15 atoms/cm^2 and 20keV, 3E15 atoms/cm^2 respectively. For the group for contact size effect, thin BPSG film (0.6μm) was deposited as ILD layer, followed by patterning of contacts with various sizes. After contact patterning, the substrate loss due to over-etch was observed to be less than 30nm. After BOE cleaning to remove native oxide at the

507

contact bottom, IMP Ti films of various thickness were deposited and followed by the deposition of MOCVD TiN and RTP annealing to complete contact/barrier metal process. Subsequently, interconnection of W-plug with Al-wiring was formed. For the deep metal contact, 3 μm- and 3.8 μm-thick BPSG films were deposited, then followed by similar subsequent processes.

For the deposition of IMP Ti for deep metal contact, process parameters such as bias power, pressure, DC to RF power ratio and deposition thickness, were varied to investigate effects of those parameters on electrical properties of metal contacts. Contact chain resistance on n+ and p+ regions was measured at room temperature from 5000 chains. Junction leakage current was measured under 4.5V at room temperature from an area pattern of 200μm *200μm. Thickness of TiSi$_2$ layers at the contact bottom was measured by TEM and compared between various deposition conditions.

RESULTS

Figure 1 shows the dependence of metal contact resistance on the contact size on n+ and p+ active region. The chain resistance increased, following inversely proportional relationship to the contact area for contacts between 0.57 μm and ~0.30 μm, but showed deviation from an inverse linear relationship below 0.30 μm. As the contact size was reduced to one fourth from 0.57 μm to 0.15 μm, the chain resistance increased about two times in n+ regions and about three times in p+ regions. In this experiment, the contact height was chosen low in order to minimize the effect of Ti step coverage on metal contact resistance. But it is expected that there was different loading effect during contact etch process, resulting in different silicon loss at contact bottoms, since contacts of various AR were patterned at the same time.

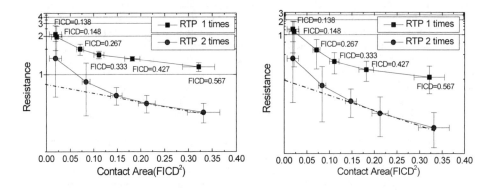

Figure 1. Dependence of contact chain resistance on contact size. Contact size was averaged from 5 measurements of FICD. (a) Chain resistance of metal contact on p+ region (b) Chain resistance of metal contact on n+ region. Normalized values of contact chain resistance are shown in the graphs.

As shown in Figure 1, the uniformity in chain resistance deteriorated significantly as the contact diameter was smaller than 0.3 μm for the same amount of CD variation. Thus, the control of CD variation is considered to be the most crucial factor for a narrow distribution of contact resistance for contact size smaller than 0.3μm.

Figure 2 shows the effect of bias power and thickness of IMP Ti on contact chain resistance of various AR on n+ and p+ regions. For the same deposition condition, the increase of IMP Ti thickness from 50nm to 70nm improved the distribution of contact resistance (not shown). When the bias power was increased from 200W to 400W for the same IMP Ti thickness, the distribution of contact chain resistance was also improved for all AR, indicating that bottom step-coverage of Ti was increased with increasing bias power. The dependence of contact resistance on bias power was more sensitive in the case of AR~10. It was observed that the junction leakage currents on n+ and p+ regions were insensitive on the bias power and Ti thickness for AR~8 with leakage current values similar to one another as shown in Table 1.

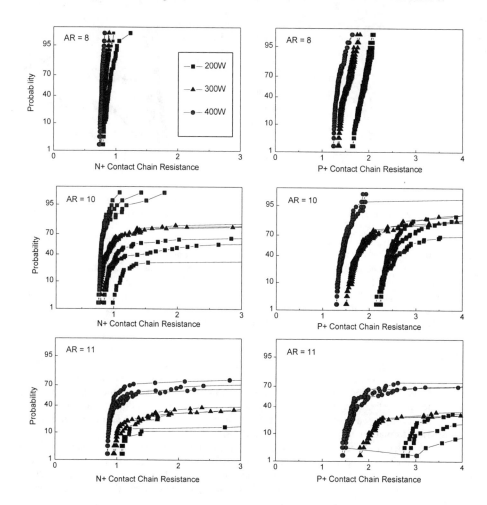

Fig.2 N+ and P+ contact chain resistance at various aspect ratios and IMP Ti deposition conditions. DC/RF = 0.56 and IMP Ti thickness=50nm. Contact chain resistance is shown in normalized scale.

From those results, it is concluded that the bias power up to 400W did not cause any apparent damage on Si substrates in contacts of high AR, contrary to the general belief that high bias power would produce damage on contact bottom. Even though the uniformity of contact resistance was improved for contacts of AR~11 when the bias power was increased to 400W, more than 30% of contact resistance measurements failed to meet the target resistance value.

For the IMP Ti process shown in Figure 2, the ratio of DC power over RF power was set to 0.55 in order to get high ionization of titanium atoms, but by sacrificing the titanium thickness uniformity (1σ ~8%). It was considered that the high failure rate of contact resistance might be due to the non-uniformity of Ti film thickness and the imperfect contact patterning process.

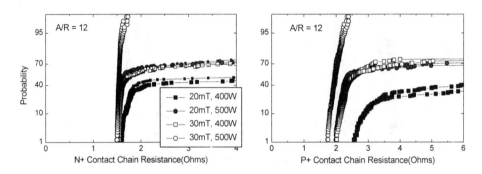

Fig.3 N+ and P+ contact chain resistance at various aspect ratios and IMP Ti deposition conditions. DC/RF = 0.81 and IMP Ti thickness = 65nm. Contact chain resistance is shown in normalized scale.

Figure 3 shows the dependence of chain resistance distribution in contacts with AR~12 on the titanium deposition condition. For this set of data, the contact patterning was carried out at the same condition and only the titanium deposition condition was varied. The ratio of DC to RF was set to 0.81 which produced a Ti thickness uniformity of 4% (1σ) and the pressure and the bias power was varied. For the contacts of AR~12, the resistance uniformity was improved as the deposition pressure and the bias power were increased. When the bias power was set to 500W and the pressure to 30mT, the resistance uniformity was improved drastically. Since the patterned wafers were prepared at the same condition, the difference in resistance uniformity was considered only due to the difference in titanium thickness at the contact bottom. Since the improvement in step-coverage is below 10% when the bias power was increased from 400W to 500W, the contact resistance was considered to be very sensitive to the Ti thickness at the contact bottom. TEM observation shows the non-uniform $TiSi_2$ formation at the contact bottom with a thin unknown interfacial layer between Si substrate and Ti film. Even though the nature of the thin interfacial layer was not identified, it is considered that the interfacial layer caused non-uniform $TiSi_2$ formation, resulting in high contact resistance.

We consider that the existence of the interfacial layer implies the inefficiency of wet chemical cleaning in deep metal contact. The interfacial layer appears to be breakable by a certain thickness of Ti at the contact bottom as shown by Figure 4. Also, it is expected that stable contact properties can be achieved with IMP Ti technology when efficient contact pre-cleaning methods are developed and applied to deep contacts.

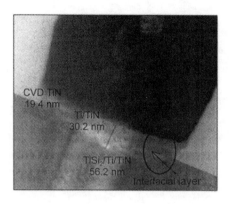

Fig.4 TEM micrograph showing the interfacial layer between the Si substrate and Ti film at the contact bottom and non-uniform TiSi$_2$ formation.

CONCLUSION

As the contact size decreases, a tight CD control is crucial to achieve uniform contact resistance. It was possible to achieve acceptable contact resistance for deep contacts up to AR~12 with IMP Ti technology by process optimization to maximize the ionization of Ti atoms with better bottom step-coverage. However, we observed some indication of inefficiency of conventional wet chemical cleaning in deep contacts and improvement in contact pre-cleaning is essential to achieve stable metal contact properties with IMP Ti technology.

REFERENCES

1. The International Technology Roadmap for Semiconductor, SIA, p101, 1997
2. Hyun-Jin Jang et al., VMIC 98 Proceedings, 109 (1998)
3. C.Y. Kim et al., HEI Internal Report (1998)
4. Shou-Yen Tai et al., VMIC 98 Proceedings, 133 (1998)

Table.1 N+ and p+ junction leakage current (Amp/μm^2)

IMP Ti THK.	500A			700A
Bias Power	200W	300W	400W	300W
N+ LKG(Amp/um2)	1.49E-15	1.38E-15	1.31E-15	1.14E-15
P+ LKG(Amp/um2)	3.21E-16	1.81E-16	1.65E-16	1.74E-16

TUNGSTEN GATE STRUCTURE FORMATION BY REDUCED TEMPERATURE CONVERSION OF TUNGSTEN NITRIDE

C.J. Galewski, A.R. Londergan, C.A. Sans, T.E. Seidel, D. Clarke *, Q. Zhu **
Genus Inc., Sunnyvale, CA 94089
* Steag, RTP Division, San Jose, CA 95134
** Accurel, Materials Analysis Group, Sunnyvale, CA 94086

ABSTRACT

The formation of pure tungsten films by rapid thermal annealing (RTA) of amorphous tungsten nitride (WN_x) is investigated. By adjusting the nitrogen content of the as deposited WN_x film, and changing the annealing ambient from nitrogen to argon, the temperature that fully converts WN_x to tungsten can be reduced from 1000°C to 800°C. Analysis by TEM, XRD, and XRR are used to show that the conversion of the film proceeds from the top surface. For the case of argon ambient a simple activation energy model is derived that predicts the tradeoff between temperature and time to achieve full conversion of WN_x to W.

INTRODUCTION

Tungsten is becoming an important candidate to form a low resistance gate stack around the 0.13 μm generation [1]. In previous work with W/WN_x bi-layer films we found that WN_x converted to tungsten during 30 min 850°C vacuum anneals without silicide formation at the silicon interface [2]. The silicide reaction is prevented by the formation of a W-Si-N interface layer. This interface layer has been documented extensively and is viewed as an integral factor for the success of using tungsten as part of a low-resistance gate stack [3].

More recently it has been proposed that rapid annealing of single layer WN_x films at 1000°C can be used to form high-quality tungsten gate stacks [4]. In this work we extend the single layer process by demonstrating that by choosing appropriate starting films and annealing conditions the temperature required for full conversion to tungsten can be reduced from 1000°C to 800°C. The reduced temperature significantly increases the available process integration options.

EXPERIMENTAL

The conversion process from WN_x to tungsten by RTA is illustrated schematically in Fig. 1. In this study we are depositing the WN_x layer by a PECVD process we have specially developed for the gate application to provide good adhesion and low fluorine content. As deposited the WN_x contains ~30% nitrogen. By XRD the as deposited film is mostly amorphous with a weak W peak suggesting the presence of a small amount of crystalline tungsten.

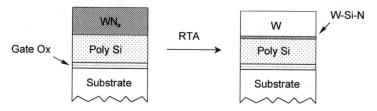

Fig. 1. Schematic representation demonstrating the formation of a W gate stack from WN_x by rapid thermal annealing (RTA).

Conference Proceedings ULSI XV © 2000 Materials Research Society

Substrates used for this study were 200 mm diameter wafers with 1000 Å of thermal oxide and 2000 Å of undoped poly-silicon to simulate the gate stack application without introducing poly resistivity as component of our sheet resistance measurements. High-resistivity bare silicon substrates were used only when needed to provide the smoother interfaces that enhance the ability to use XRR (X-ray reflectivity) analysis. Prior to deposition all substrates were dipped in 100:1 HF for 1 min to remove native oxide. After the clean the wafers were loaded into a Genus *LYNX2™* single wafer cluster tool to deposit about 900 Å of WN_x. The conversion anneals were performed in a Steag 8800 Heatpulse RTA system. A pre-purge time of 60 sec was used to ensure a low background of oxygen in the annealing chamber. The annealing ambient was either nitrogen or argon forming gas. Take out temperature was 450°C.

The electrical and structural characteristics of the films in this study were evaluated by four-point probe (sheet resistance), weight change (thickness), x-ray diffraction (XRD), x-ray reflectivity (XRR), transmission electron microscopy (TEM), and atomic force microscopy (AFM).

RESULTS AND DISCUSSION

Low Temperature Conversion
The first set of experiments we conducted revealed a strong influence of the annealing ambient to enhance the conversion of WN_x to form tungsten. In agreement with published results we confirmed that films convert fully to tungsten at 1000°C when annealed for 60 seconds in a nitrogen ambient [4]. In this case the sheet resistance decreases from the initial 19.2 ohms/sq to 1.28 ohms/sq. However, when the annealing temperature is reduced to 900°C in nitrogen the resulting film is a mixture of 75% crystallized W_2N and 25% crystallized W with a sheet resistance of 10.4 ohms/sq. In contrast, when the same WN_x film is annealed in argon at 900°C for 60 seconds it is completely converted to tungsten reaching a resistivity of 1.33 ohms/sq. Data from this experiment are summarized in Table 1.

	As Deposited	Annealed, 60 sec	
		Nitrogen, 1000°C	Argon, 900°C
Rs (ohms/sq)	19.2	1.28	1.33
XRD			
W Grain Size (Å)	amorphous	510	480
W Crystallized Fraction	N/A	100%	100%
XRR			
Density (g/cm^3)	16.7	19.3	18.9
Interface Thickness (Å)	75	1	30
Film Thickness (Å)	820	775	750
Surface Oxide Thickness (Å)	0	85	75
Resistivity (µohm-cm)	160	9.9	10.0
RMS Surface Roughness (Å)	13	20	14

The above data demonstrate that the tungsten films produced by both annealing conditions exhibit density close to the ideal value of 19.3 g/cm^3 for tungsten. For both type of anneals AFM shows that the films are very smooth with a roughness in the range of 14 to 20 Å. Analysis by

XRR indicates the presence of an interface layer with density that increases from 2.5 g/cm³ to approximately 6 g/cm³ as a result of the annealing. This finding is in agreement with the W-Si-N interface layer reported by TEM analysis [3]. The interface layer appears to be thinner as anneal temperature is increased. XRR data also indicate that the final tungsten thickness is less than the initial WN$_x$ thickness. The tungsten thickness formed by annealing appears to be the balance of decreasing interface layer thickness and some surface oxidation that occurs as a result of the RTA. Overall, the film thickness decrease is on the order of 5%, which was confirmed by TEM. The tungsten resistivity calculated based on XRR thickness is ~10 µohm-cm.

Conversion Process Characterization

An extensive set of experiments was conducted to improve and understand the tungsten conversion process. This study involved systematic variation in ambient, temperature, and time. Film resistivity versus temperature is plotted in Fig. 2 for fixed anneal times of 60 seconds.

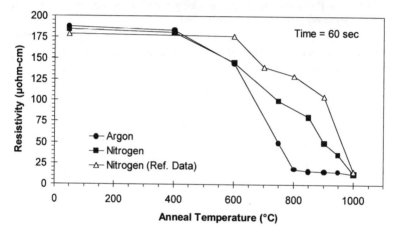

Fig. 2. Sheet resistance versus temperature for annealing time of 60 seconds. The WN$_x$ film is fully converted to tungsten during annealing at 800°C in argon. (Ref. Data from B. H. Lee et al.)

There are several important features that can be observed in Fig. 2. Initially there is little change in resistivity up to 400°C, which agrees with our experience that WN$_x$ films will remain mostly amorphous up to about 500°C. As the anneal temperature is increased above 400°C the resistivity starts to decrease with the transition from a mostly amorphous WN$_x$ film to one also containing a crystallized mixture of the two stable phases W and W$_2$N. Note that in the case of the data published by B. H. Lee, et al. [4] there is no such transition, suggesting perhaps a different W/N ratio, or an as deposited film that already contains some crystallized W$_2$N.

As the anneal temperature in Fig. 2 is increased beyond 600°C there is a dramatic difference in annealing behavior associated with the ambient. In argon, the resistivity decreases smoothly and rapidly to <20 µohm-cm at 800°C, indicating that full conversion to tungsten has been reached. In the case of nitrogen ambient the resistivity decreases more slowly and non-linearly, reaching full conversion only after the 1000°C anneal. A similar non-linear resistivity decrease in the nitrogen ambient is present in both data sets, suggesting a more complex conversion than in argon.

The films in the above conversion study were examined by cross-sectional TEM to further understand the tungsten formation mechanism. Cross-sections from samples annealed at 750°C show that the conversion to tungsten proceeds as expected from the top surface of the film. In the case of nitrogen ambient a distinct boundary can be observed between an upper more crystallized layer and a lower more amorphous layer. In the case of the argon ambient there is not a distinct boundary, the film consists instead of larger grains that extend non-uniformly from the top surface. This result suggests that the nitrogen ambient could be shifting the equilibrium of nitrogen released from the decomposition of the WN_x film. This would be consistent with previously published work were it was found that the amount of nitrogen released from WN_x during vacuum annealing increased significantly above 700°C [5]. After annealing in nitrogen at 900°C the TEM cross sections show fine grains extending throughout the thickness of the film. In the argon ambient the grains extend close to the thickness of the film even after the 60 second 850°C anneal. Annealing at 1000°C for 60 seconds appears to result in a tungsten film with orderly vertical grain boundaries for both the nitrogen and argon ambient. No evidence of tungsten silicide formation at the interface is observed.

The results obtained by TEM match the XRD data shown in Fig. 3. The peaks in the XRD scans indicate the phases present in the films after 60 second anneals at temperatures from 600 to 1000°C in nitrogen and argon ambient.

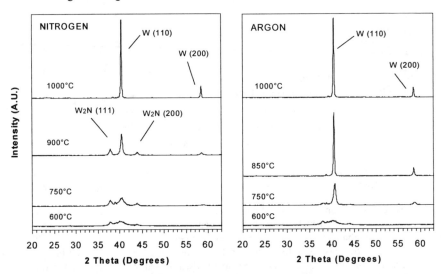

Fig. 3. Glancing angle (1.0°) x-ray diffraction spectra for samples annealed in nitrogen and argon ambient for 60 seconds at the indicated temperatures from 600 to 1000°C.

In the above spectra we can identify W_2N at orientations of (111) and (200) at diffraction angle of 37.7° and 43.8°, respectively. Tungsten is identified for the two orientations of (110) and (200) at angles of 40.3° and 58.3°, respectively. The height and width of a peak correlate to the average grain size. The broad small peaks after the 600°C anneal indicate that there is only a small amount of crystallization and small grain size for either ambient. As anneal temperature is increased there is a large difference between nitrogen and argon ambient. Appreciable amounts of W2N are no longer observed in argon after 850°C annealing, and the tall and narrow W(110) peak indicates increasing grain size. In the case of nitrogen ambient there is W and W_2N

present, and small and broad peaks indicate small grain size. No dependence on ambient is observed in the XRD spectra after 1000°C.

A quantified representation of the conversion process can be obtained from the XRD measurements by estimating the average grain size based on the Scherrer formula. Furthermore, we can also estimate the relative crystallized molar fraction for the W phase, f_a, from the following formula:

$$f_a = I_a / (I_a + I_b C_a / C_b)$$

Where $C_a / C_b = 5.59$. The values for I_a and I_b are obtained by integrating the intensity of the peaks for W(110) and $W_2N(111)$, respectively. We have found that relative amounts of W and W_2N obtained by this method for a fully crystallized film give stable and accurate correlation to process results. In contrast, correlation to process results using the more direct method of RBS to measure the relative amount of W and N has not proven to be as reliable.

Based on the XRD analysis outlined above the conversion mechanism is summarized in the plots shown in Fig. 4.

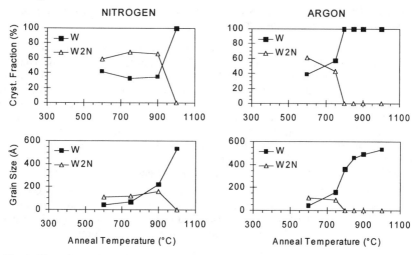

Fig. 4. Plots of crystallized fraction and grain size for W_2N and W obtained after annealing WN_x films for 60 seconds in nitrogen or argon ambient.

As can be seen in Fig. 4, the relative amount of crystallized W and W_2N remains relatively stable to annealing at 900°C in nitrogen. The grain size for both phases increases slowly, remaining at less than 200 Å until W_2N is no longer present. The slowly increasing grain size and presence of higher resistivity W_2N (170 μohm-cm) explains the gradual and non-linear decrease in resistivity observed in Fig. 2. In the case of the argon ambient we find that the amount of W_2N decreases rapidly upon annealing, and that the W grain size increases to 360 Å even after the 800°C anneal.

Activation Energy Model

The above conversion characterization at fixed time of 60 seconds suggests that the formation of W in an ambient without nitrogen may be dominated by thermally activated dissociation of W_2N. Isothermal experiments were performed to assess the effect of time and annealing temperature on the sheet resistance Rs.

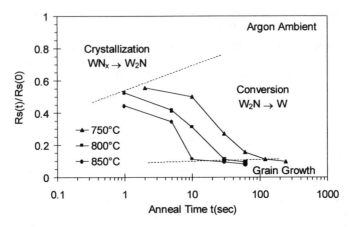

Fig. 5. Fractional decrease in sheet resistance versus anneal time in argon ambient is plotted for annealing temperatures of 750, 800, and 850°C.

In Fig. 5, the fractional decrease in sheet resistance Rs(t)/Rs(0) is plotted versus annealing time. In this plot we see that the decrease in Rs involves 3 distinct phases: crystallization, conversion, and grain growth. During the first phase, labeled crystallization, the initial sheet resistance Rs(0) exhibits a rapid decrease that has taken place even after a nominal annealing time of 1 second. We hypothesize that in this phase the decrease in Rs is dominated by a rapid reorganization of the film from its almost amorphous as-deposited state to a more ordered state. As support for this hypothesis, XRD data suggests that there is formation of crystallized W_2N associated with this phase. The second phase, labeled conversion, is likely to be dominated by the actual dissociation of W_2N and diffusion of nitrogen from the film. The final decrease in Rs once the W_2N is no longer present is associated with grain growth as shown previously in Fig. 4.

The straightforward behavior of resistivity versus temperature as discussed previously indicates that a simple thermally activated model might be sufficient to describe the conversion mechanism. In fact, data for both the conversion and grain growth phase from Fig. 5 does fit very well ($R^2 > 0.998$) to an Arrhenius type expression of the following form:

$$\Delta Rs/\Delta t = A \exp(-Ea/kT)$$

Where ΔRs is the decrease in Rs observed for an annealing time of Δt, Ea is an activation energy, T the absolute temperature, and k the Boltzman constant. The activation energy for the conversion and grain growth phase is 2.46 and 1.73 eV, respectively.

If we define the conversion process from W_2N to W to be complete when sheet resistance is decreased to 10% of the initial Rs we can plot the time and temperature that satisfy this requirement as shown in Fig. 6. A wide range of RTA conditions from 120 seconds at 750°C to 1 second at 1000°C satisfies this requirement. A similar relationship for a further 20% decrease (8% of initial Rs) due to grain growth is also shown in Fig. 6. As seen in the plot, the reduction in Rs due to grain growth is significantly slower than that due to conversion of W_2N to W.

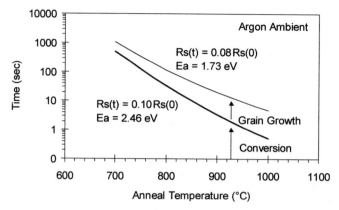

Fig. 6. Activation energy models for sheet resistance decrease as result of RTA of WN_x in argon ambient. Plotted as annealing time versus temperature for conversion to W (10% initial Rs) and grain growth (8% initial Rs).

CONCLUSIONS

Conversion of WN_x to W by RTA is an attractive method to form low-resistance poly-metal gate stacks for scaled devices. High quality tungsten films with a stable interface to poly silicon are obtained even after annealing for 60 seconds at 1000°C. In this paper we demonstrated a significant reduction of required annealing temperature by tailoring the W/N ratio of the as-deposited film, and using argon ambient. The process conditions that result in full conversion to tungsten are summarized in a simple activation energy model that indicates a wide range of conditions from 1000°C for 1 second to 750°C for 120 seconds. This flexibility in choice of annealing temperature allows for many integration options for the conversion process to make future poly-metal gates.

ACKNOWLEDGMENTS

We would like to acknowledge the support of Lawrence Matthysse and Norm Zetterquist for helping prepare the WN_x films for this study. The capable help provided by April Daniel and the Steag Applications Lab staff to anneal wafers is greatly appreciated. Steven Shatas provided a Modular Process Technology RTP 600 bench-top unit for the isothermal annealing of small wafer pieces in an efficient one day effort.

REFERENCES

1. Y. Akasaka et al., IEEE Trans. Electron. Dev., **43**, 11, 1996, 1864

2. C. Galewski et al., "Adv. Metallization and Interconnect Systems for ULSI Applications in 1996", Mater. Res. Soc., 1997, 277

3. K. Nankajima et al., Applied Surface Science, **117/118**, 1997, 312

4. B. H. Lee et al., Tech Dig. IEDM, 1998, 385

5. T. Suzuki et al., "Advanced Metallization and Interconnect Systems for ULSI Applications in 1997", Mater. Res. Soc., 1998, 49

Reliable Pretreatment Technology for Hot Process during W Polymetal Gate Formation

Yasushi Akasaka, Kiyotaka Miyano, Kouji Matsuo*, Kazuaki Nakajima, and Kyoichi Suguro

Microelectronics Engineering Lab., Semiconductor Company, Toshiba Corp., 8 Shinsugita-cho Isogo-ku, Yokohama 235-8522, Japan

*Manufacturing Engineering Center, Semiconductor Company, Toshiba Corp., 8 Shinsugita-cho Isogo-ku, Yokohama 235-8522, Japan

Tel: +81-45-770-3663, Fax: +81-45-770-3577, E-mail: yasushi.akasaka@toshiba.co.jp

Introduction

Development of a new gate structure with reduced gate resistance is strongly required for the realization of higher-performance and higher-density LSIs. W/WSiN/ poly-Si multilayered "polymetal" gate electrode is an attractive candidate for low-resistivity gate.[1]

Regarding the introduction of tungsten in gate electrode, it should be considered that the thermal processes at the temperature higher than 800°C are usually carried out, whereas they are not in the case of back-end processes. It is necessary to overcome several problems caused by thermal processing. For example, as we have already pointed out, whiskers grow on the W surface during thermal processing such as Si_3N_4 LP-CVD. Figure 1(b) shows granular growth of LP-CVD Si_3N_4 film deposited on W. The granular growth occurred when Si_3N_4 film was conformally deposited on whiskers as shown in Figure 1(a). The whisker formation was found to be caused by oxidation of the W surface during loading of the wafer in CVD reactor chamber. Also, it was found that surface morphology of Si_3N_4 film could be improved by decreasing oxide of W surface before deposition.[2]

Figure 2 shows a schematic cross section of a polymetal gate electrode. Short between W and contact metal is expected when whiskers are formed. This is a serious process integration problem for polymetal gate electrode. To prevent this problem, more precise control of whisker formation is required.

In this paper, the impact of slight oxidation of W surface, such as native oxidation, is examined. We propose a pretreatment technology in order to remove tungsten oxide (WO_x). A method of measuring oxide thickness is also proposed.

Conference Proceedings ULSI XV © 2000 Materials Research Society

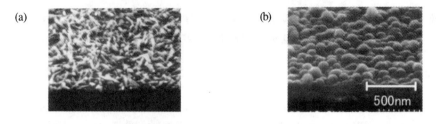

Fig. 1 W surface before (a) and after (b) Si₃N₄ deposition. Whiskers are observed on W surface before deposition. [2]

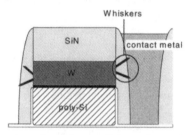

Fig. 2 A schematic cross section of a polymetal gate electrode. Short between W and contact metal could occur when whiskers are formed.

Experiment

W/WN$_X$ (40/5nm) films were deposited on Si wafer with PVD. About 9nm of WO$_X$ was formed on W by O$_2$ ashing. In this case, O$_2$ ashing is used in order to form thick oxide at lower temperature. Hence the difference of with and without H$_2$SO$_4$ treatment was observed in the case of samples oxidized by O$_2$ ashing.

WO$_X$ was removed using diluted H$_2$SO$_4$ (10-20 volume%). Surface condition was analyzed by X-ray Photoelectron Spectroscopy (XPS). Film thickness of native oxide of W after treatment with diluted H$_2$SO$_4$ was also measured using spectroscopic ellipsometry.

Samples were annealed in low-base-pressure furnace with load lock chamber. The exposure-time dependence of surface morphology after annealing at 900°C for 2 min. in N$_2$ was observed by SEM. LP-CVD Si$_3$N$_4$ deposition was also carried out at 780°C to confirm the effect of H$_2$SO$_4$ treatment.

Results and Discussion

The effect of H$_2$SO$_4$ treatment

Figure 3 shows XPS spectra of W 4f before and after treatment with diluted H$_2$SO$_4$. Two peaks at the binding energy of 36 and 38 eV which correspond to the W-O bond decreased and the other two peaks at the binding energy of 31.5 and 33.7 eV which correspond to W-W bonds increased after the treatment. From this result, it is confirmed that the WO_x formed by O_2 ashing can be removed by treatment with diluted H_2SO_4.

Figures 4(a) and (b) show SEM images of W surface after annealing at 900°C for 2min in N_2 with and without H_2SO_4 pre-treatment (oxidized by O_2 ashing). Whiskers (indicated by white arrows) are clearly found in (b), whereas in contrast, no whiskers can be found in (a). From this result, whiskers are formed when WO_x is formed by O_2 ashing. However, the whisker formation can be suppressed by removing WO_x with H_2SO_4 treatment.

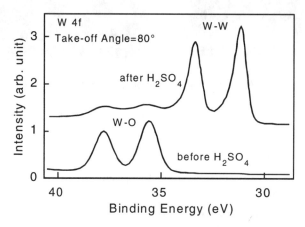

Fig. 3 XPS spectra of W 4f before and after treatment with diluted H_2SO_4

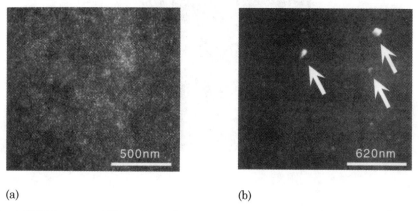

(a)   (b)

Fig. 4 SEM images of W surface after annealing at 900°C for 2 min (a) with and (b) without H_2SO_4 treatment after O_2 ashing.

Figures 5(a) and (b) show SEM images of LP-CVD Si_3N_4 surface deposited on W. As shown in Fig. 5(b), a granular growth occurred without H_2SO_4 treatment after O_2 ashing. Nevertheless, as shown in Fig. 5(a), Si_3N_4 was uniformly deposited on W by using H_2SO_4 treatment. This is thought to be an effect of the suppression of whisker formation by H_2SO_4 treatment.

Fig. 5 SEM images of Si_3N_4 surface deposited on W (a) with and (b) without H_2SO_4 treatment after O_2 ashing.

The effect of native oxide

Figure 6 shows an example of whisker formation after RTA at 900°C for 30 sec. without pretreatment after 1-2 months exposure in the air ambient. It suggests that a native oxidation affects whisker formation.

Fig. 6 An SEM image of W surface after RTA at 900°C for 30 sec. without pretreatment. W film was exposed in the air ambient for 1-2 months.

Figure 7 shows XPS spectra of as-removed W surface and W surface 16 hours after removal. In order to clarify the difference of surface condition, the analysis was carried out with small take-off angle of 15°. The peaks which correspond to W-O bonds obviously had increased after 16 hours.

Figure 8 shows exposure time dependence of film thickness of native WO_x after H_2SO_4 treatment. The film thickness of WO_x rapidly increased to as thick as 0.6nm in 24 hours. It was confirmed that the thickness increased monotonically with time. Therefore it is concluded that this method can be used for the definition of relative thickness of thin WO_x even in the case of the thickness of less than 1nm.

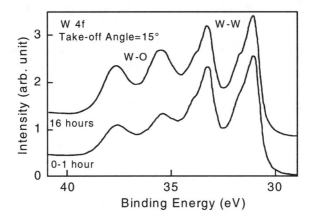

Fig. 7 XPS spectra of W surface at 0-1 hour and 16 hours after removal. The analysis was carried out with small take-off angle (15°).

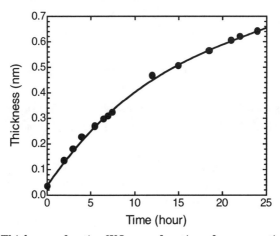

Fig. 8 Thickness of native WO_x as a function of exposure time.

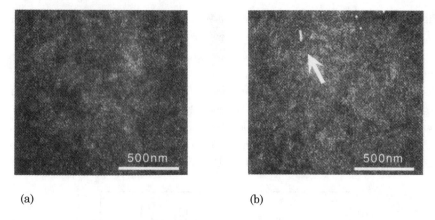

(a)   (b)

Fig. 9 SEM images of W surface after annealing at 900°C for 2 min. The exposure time after treatment with diluted H$_2$SO$_4$ is (a) 2 hours, and (b) 24 hours.

Figure 9 (a) and (b) show SEM images of W surface after annealing after (a) 2 hours exposure and (b) 24 hours exposure (oxidized in by O$_2$ ashing), respectively. Whiskers are clearly found in (b), whereas no whiskers can be found in (a). From this fact, it is supposed that these whiskers are caused by slight native oxidation of W surface whose typical thickness is 0.6nm. However, whisker formation can be suppressed by controlling exposure time after H$_2$SO$_4$ treatment.

Conclusion

The formation of whiskers is a phenomenon which is related to oxidation of the surface of W. Therefore, the relationship between thickness of WO$_X$ and whisker growth was examined. WO$_X$ with a thickness of only 0.6nm was found to affect whisker formation. It was also found that WO$_X$ is effectively removed by diluted H$_2$SO$_4$ and the thickness of WO$_X$ as thin as 0.1nm can be measured by spectroscopic ellipsometry. Whisker formation can be suppressed by controlling exposure time after H$_2$SO$_4$ treatment.

Acknowledgments

The authors thank Mr. H. Harakawa and Ms. T. Kobayashi for helpful discussion. They are grateful to Dr. K. Okumura and Dr. T. Arikado for helpful discussion and encouragement throughout this work.

References

[1] Y. Akasaka, S. Suehiro, K. Nakajima, T. Nakasugi, K. Miyano, K. Kasai, H. Oyamatsu, M. Kinugawa, M. T. Takagi, K. Agawa, F. Matsuoka, M. Kakumu, and K. Suguro, IEEE Trans. Electron Devices, ED-43, 1864 (1996).

[2] Y. Akasaka, K. Miyano, K. Nakajima, M. Takahashi, S. Tanaka, and K. Suguro, Jpn. J. Appl. Phys. 38, 2385 (1999).

Thermally Stable W Bit Line Process for 256M bit DRAM and Beyond

Jeongeui Hong, Youngjun Lee, Won-Hwa Jin, Kyu-Hyun Kim, Tae-Seok Kwon,
Pilsung Kim, Won-Jun Lee, Hong-Seok Kim and Sa-Kyun Rha[+]

Process Development Team, R&D Center, Hyundai Microelectronics Co.,Ltd.
1,Hyangjeong-dong, Cheongju 361-725, Chungbuk Korea, hjea@hmec.co.kr
[+] Dept.of Materials Engineering, Taejon National University of Technology
305-3, Samsung-2dong, Dong-Gu, Taejon 300-717, Korea

ABSTRACT

Thermally stable W Bit line process technology focused on the contact metallization has been reported. Major parameters to have an impact on the process are proved to be the Ti thickness, silicidation temperature, and dopant concentration. TiSix agglomeration and dopant loss have been effectively suppressed by using thinner Ti and higher silicidation temperature, which resulted in good contact characteristics. Among those parameters, silicidation temperature has the strongest impact on the structural and electrical properties. It is believed that TiSix formed at high temperature($\geq 800\,°C$) consists of stable C-54 phase only, not the mixture of C-49 and C-54, which causes very limited out-diffusion of the dopants.

We have successfully integrated W bit line process into a cost-effective and competitive DRAM manufacturing without any issues like lift-off, W abnormal oxidation and chemical attack or cross-contamination.

INTRODUCTION

As the device geometry continuously shrinks, DRAM manufacturers are going to replace conventional W-polycide bit line with W bit line in order to reduce the chip size and use it for a local interconnection. W bit line process could be used in CUB (Capacitor Under Bit line) type DRAMs without any crucial problem except for barrier step coverage and W plug fill issues, because there is no high-temperature thermal process after bit line formation. However, for COB(Capacitor Over Bit line) type DRAMs, high temperature($\geq 700\,°C$) process for the capacitor formation and dielectric anneal can degrade the contact characteristics substantially due to the dopant out-diffusion, TiSix agglomeration and impurities incorporation[1].

In this paper, we demonstrate thermally stable and robust W bit line process with low contact resistance (P+: $<1k\Omega$, N+: $<500\Omega$ @$0.25\mu m$) and junction leakage current ($<10fA/\mu m^2$ @$85\,°C$) after high temperature thermal process ($800\,°C/10min$ + $700\,°C/13hours$) for the advanced COB type DRAMs focused on how to keep the dopant concentration high enough and how to prevent TiSix from agglomerating during post heat cycle.

EXPERIMENT

Ti/TiN thickness (Ti: 7nm~15nm, TiN: 15nm~45nm), silicidation temperature (RTP, 650°C~830°C), and additional implant (with/without) after contact etch were chosen as the variables for this experiment. Processes consist of bit line contact formation (aspect ratio of 3 at the diameter of 0.25μm), and additional P+ ion implantation with BF_2, then wet contact clean using various BOE chemistry followed by Ti/TiN deposition using ionized PVD system, RTP for silicidation of Ti, and CVD W deposition. P+ source/drain implantation condition was BF_2/20keV/2.5x10^{15}. A schematic cross-sectional view of the structure is shown in Fig.1.

Four-point probe, scanning electron microscopy (SEM), transmission electron microscopy (TEM), secondary ion mass spectroscopy(SIMS), and Keithley/HP4062 were used for film analysis and electrical characterization.

RESULTS and DISCUSSION

TiSix agglomeration

It was observed that TiSix formed by the reaction between Ti and Si at low temperature tended to agglomerate during high-temperature ($\geq$ 700°C) process as shown in Fig.3(b) and (d). Since TiSix agglomeration causes to not only reduce the effective contact area, but also enhance the out-diffusion of the dopants, it should be minimized. We have found that the agglomeration has been effectively suppressed by using thinner Ti and higher silicidation temperature, as shown in Fig.2 and 3. After high-temperature annealing and stripping off the entire films, no agglomeration has been observed at the Si surface when 7nm-thick Ti (expected bottom thickness is ~3.5nm) was deposited (Fig.2(a)), and TiSix agglomerated very uniformly when Ti was 5nm thick on the blanket wafers (Fig. 2(b)).

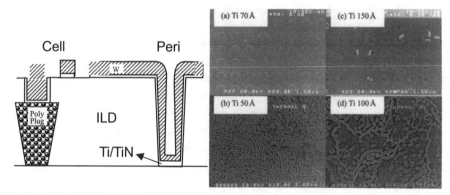

Fig.1 Schematic view of W bit line contact.

Fig.2 SEM Photographs of Si surface after stripping. The thinner Ti (a, b) reveals more uniform silicidation

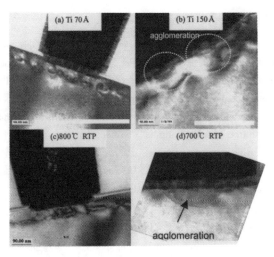

Fig.3. TEM images of the contact bottom showing that TiSix agglomerated when the thicker Ti or the lower silicidation temperature had been used.

Dopant concentration

Metal to silicon contact resistance is largely controlled by the dopant concentration at the interface between metal and silicon. We have focused only on the boron concentration in P+ region because N+ contact resistance has been always stable. Figure 4 shows SIMS depth profiles of boron concentration in the blanket wafers as a function of Ti thickness, silicidation temperature, and additional implantation of BF_2. The peak concentration increases with decreasing Ti thickness and increasing silicidation temperature. It is believed that TiSix can act as a boron sink due to its high solubility into the TiSix[2], which can explain the effect of Ti thickness on boron concentration.

TiSix has two distinct phases which called high resistivity/metastable/heavily faulted C-49 and low resistivity/equilibrium/fault-free C-54 depending on the temperature[3]. At higher silicidation temperature, there is a limited diffusion of the dopants such as boron during post heat cycle, because TiSix is completely converted to C-54. In contrast, at lower silicidation temperature, TiSix may consist of the mixture of both structures, hence the boron can easily diffuse away from the Si accompanied by the phase transformation during high temperature process. When the additional implantation was applied, the boron concentration was over two times higher than that without additional implantation.

Electrical characteristics

Based on the above results, we have investigated the effect of each variable on the contact resistance and the leakage current, as shown in Fig.5 and Fig.6. Low P+ contact resistance was obtained by using thinner Ti and higher silicidation temperature with an additional implantation, which is in a good agreement with the TiSix agglomeration and boron concentration described above. The impact of the additional implantation, however, was smaller than it was expected, even if boron concentration had been almost doubled. This is probably because we did not include activation annealing in process flow.

Junction leakage current measured at $85\,^{\circ}\mathrm{C}$ was lower than $10\mathrm{fF}/\mu m^2$ in any case except for the wafer edge sites due to the edge thin profile of the TiN.

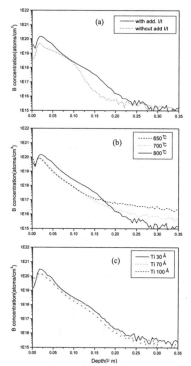

Fig.4 SIMS depth-profiles of Boron concentration as a function of an (a) additional implantation, (b) silicidation temperature, and (c) Ti thickness.

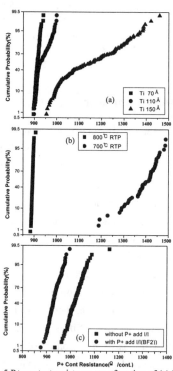

Fig. 5 P+ contact resistance as a function of (a) Ti thickness, (b) silicidation temperature, and (c) additional implantation.

CONCLUSION

Thermally stable and reproducible W bit line process technology has been successfully integrated with DRAM production. Major parameters to have an impact on the W bit line process are proved to be the Ti thickness, silicidation temperature, and dopant concentration. The best electrical properties have been obtained by using thinner Ti and higher silicidation temperature with an additional implantation, which has been possibly explained in terms of the TiSix agglomeration and boron concentration.

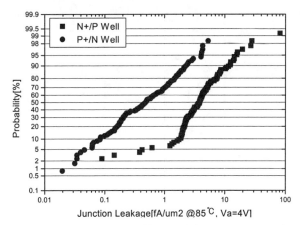

Fig.6. Cumulative probability of junction leakage current measured at 85℃, 4V reverse bias (typical)

REFERENCES

1. Y. Nakamura et al., *Proceeding of Advanced Metallization Conference 1998*, p. 661 (1998).
2. C. M. Osburn et al, J. Electrochem. Soc. **135**, 1490 (1988)
3. R. Beyers and R. Sinclair, J. Appl. Phys. **57**, 5240 (1985)
4. S. R. Wilson and C. J. Tracy, *Handbook of Multilevel Metallization for Integrated Circuits*, (Noyes Publications, 1993) pp.32-67, 82-90.
5. Y. Matsubara, T. Horiuchi, and K. Okumura, Appl. Phys. Lett. **62**, PP.2634-2636 (1993)

Mechanisms of Tungsten Plug Corrosion
In Borderless Aluminum Interconnects

S. Muranaka, I. Kanno, H. Sasai

Mitsubishi Electric Corporation ULSI Development Center

4-1, Mizuhara, Itami, Hyogo 664-8641, Japan

Phone: +81-727-84-7136

Fax: +81-727-80-2675

E-mail: muranaka@lsi.melco.co.jp

ABSTRACT

Tungsten plug corrosion is observed in borderless interconnect structures after wet organic remover treatment for removing resist residue. This phenomenon is strongly dependent on process. The corrosion occurs during ashing process when the aluminum interconnects are exposed by oxygen radicals or ions, and that the usage of neutral chemical, which replaces the alkaline remover, can discharge the interconnects to avoid the corrosion effectively.

INTRODUCTION

Shrinking LSI dimension, the overlap of metal layers for via holes are reduced. And so borderless interconnect structures in which an upper aluminum layer does not fully cover tungsten plugs are common. Tungsten is exposed during aluminum wiring processes, i. e., dry etching, ashing, and wet cleaning. Bothra et. al. reported that tungsten plug corrosion was observed in borderless interconnect structures after wet organic remover treatment for removing resist residue, which adhered on metal surfaces during aluminum dry etching process[1]. The purpose of this study is revealing mechanisms of tungsten plug corrosion.

EXPERIMENT

Figure 1 shows normal process flow from tungsten plug deposition to upper aluminum layer wiring. Tungsten plug are formed by depositing and etching back a tungsten film on a silicon oxide interlayer in which via holes have been fabricated. The upper aluminum layer is patterned by sputtering of AlCu, followed by photo lithography and dry etching, on the tungsten plugs and the interlayer.

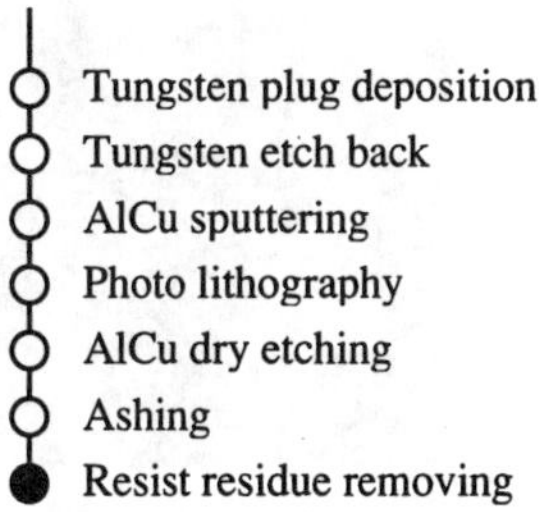

Tungsten plug deposition
Tungsten etch back
AlCu sputtering
Photo lithography
AlCu dry etching
Ashing
Resist residue removing

Figure 1.   Process flow.

Ashing process to remove the bulk photo resist after the dry etching leaves residue, which adhered during and can be removed in chemical remover treatment following the ashing.

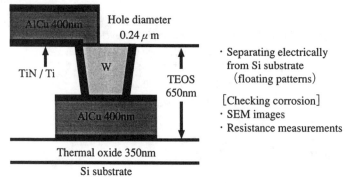

Figure 2.   Test pattern（borderless interconnects）.

We used the Kelvin pattern with a single plug in this experiment. Figure 2 schematically shows vertical structure of the test pattern.

This patterns has a structure in which tungsten plugs with a hole diameter 0.24μm connects the 400nm-thick first aluminum layer with the 400nm-thick second aluminum layer. They are floating patterns and are separated electrically from silicon substrate by a 350nm-thick interlayer of silicon oxide.

Scanning electron microscopy (SEM) observation and resistance measurement for this pattern was performed to examine tungsten plug corrosion.

RESULTS AND DISCUSSIONS

Figure 3 shows SEM images of aluminum interconnects after alkaline(pH～12) and neutral (pH～8.2) resist remover treatment. Tungsten plug corrosion occurred after alkaline remover treatment, while tungsten plugs remained undamaged in the case of the neutral remover treatment. We found that tungsten corrosion depends strongly on pH of resist remover, and neuter organic remover has an ability of preventing corrosion.

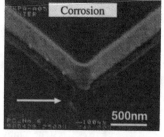

(a) After normal alkaline remover   (b) After neutral remover

Figure 3. SEM images of Kelvin pattern after resist remover treatment.

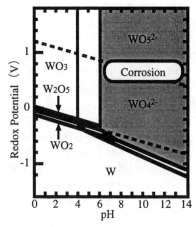

Figure 4.   Pourbaix diagram[2] for W-H2O system at 25℃.

It is assumed that the dependence of resist remover for tungsten plug corrosion has much to do with the nature of tungsten in chemicals.

The Pourbaix diagram[2] can explain the nature of tungsten in chemicals with various pHs. Figure 4 shows Pourbaix diagram for W-H2O system at 25℃. Tungsten dissolves in an alkaline solution with high redox potential.

According to the experimental results shown in fig. 3., redox potential of the tungsten plugs in the alkaline remover with the pH of 12 after ashing should be higher that -0.96V, while it should be lower -0.60V in the neutral remover with the pH of 8.2.

The corrosion phenomenon was dependent on dry processes before remover treatment. Table I. shows effect of processes after AlCu dry etching upon the corrosion. Process A) and B) were indicated in fig. 3.

In the process C) where alkaline remover was treated without ashing after AlCu dry etching, tungsten corrosion did not occurred at the wafer, but resist residue remained.

The wafer of D) was treated ashing and alkaline organic remover treatment after process B), and those tungsten plugs were corroded.

According to the results of the process A) and C), ashing caused tungsten corrosion. As-deposited blanket tungsten was etched off only 3nm during the alkaline treatment.

It is possible that some electrical chemical reaction during alkaline organic remover treatment dissolved acceleratingly tungsten.

Table I.   Dependence on processes after AlCu dry etching.

	Ash	Remove	Ash	Remove		W-plug Corrosion
A	Done	Alkaline	—	—	Yes	Normal Process
B	Done	Neutral	—	—	No	Neutral Remover
C	—	Alkaline	—	—	No	No Ashed
D	Done	Neutral	Done	Alkaline	Yes	Additional Ashing

The fact that the plugs without ashing were not corroded requires the redox potential lower than -0.96V in the alkaline remover. The redox potential of this system in the alkaline remover, therefore, may be affected by the ashing process to vary.

In ashing process, oxygen radicals and ions attack onto a wafer surface. This attack may leave the positive change which provides redox potential. That is to say, the electric charge induced by the incoming oxygen radicals or ions might presumably raise the redox potential to result in the tungsten plug corrosion.

For the verification of the model, the following four counterplans were performed to discharge the wafer. Figure 5 shows experimental flow; E) Touching the Al layer with a metal needle. F) Irradiating UV on the wafer. G) Dipping the wafer in the neutral remover. H) Dipping the wafer in a neutral electrolyte (NH4)2SO4. The wafer treated by the processes E)-H) were dipped in the alkaline remover to examine the tungsten corrosion.

Table II. shows the effects for corrosion of the four counterplans.

In the case of E), where the electric charge are supposed to be removed via the needle, and the redox potential can be reduced, the corrosion did not occur. This result supports our model of charge-up aluminum layer and showed that the corrosion could be prevented by discharging.

Although tungsten plugs of the wafer did not vanish fully in F), resistances of the Kelvin pattern was higher than that of a sample without corrosion. The UV irradiation is lightly effective for corrosion, but is incomplete.

Tungsten plug corrosion does not occur even in G). The result shows that neuter organic remover has also an ability of discharge.

Chemical treatment also has discharge effects, and neutral electrolyte (NH4)2SO4, whose pH is 5~6, was checked at H). Corrosion occurred at the wafer. The chemical can not remove resist residues and polymers on aluminum interconnects. Therefore, electric charge-up parts can not contact the chemical, and exchange of electron did not occur. Interconnects kept charge and high redox potential.

The model that patterns charged up positively by ashing process and tungsten dissolved in alkaline remover treatment can explain these results. Figure 6 schematically shows the possible mechanisms of tungsten plug corrosion.

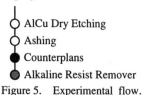

Figure 5.   Experimental flow.

Table II.   Effects for corrosion of the counterplans.

	Counterplans	W-plug Corrosion	
E	Needle Touch for Al	No	The pattern was discharged.
F	UV Irradiation	Slightly	The resistance of pattern raised.
G	Neutral Remover	No	The pattern was discharged.
H	(NH₄)₂SO₄	Yes	Neutral Electrolyte is not effective.

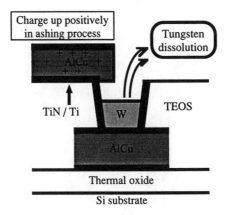

Figure 6.　Mechanisms of tungsten plug corrosion.

CONCLUSION

In conclusion, the mechanisms of tungsten plug corrosion was clarified that aluminum interconnects are electrified positively by oxygen radicals or ions in ashing process and tungsten dissolved acceleratingly in electrical chemical reaction during alkaline organic remover treatment. The neutral remover treatment is effective for corrosion because of having an ability of discharge.

REFERENCE

1.　S. Bothra, H. Sur, and V. Liang: IEEE Annual Reliability Physics Symposium, (1998) 150
2.　C. Fruitman, M. Desai, and D. P. Birnie III: Proceedings of the 12th ᵥLSI Multilevel Interconnection Conference, (1995) 508

BIAS SPUTTERED TUNGSTEN AS A DIFFUSION BARRIER
AND NUCLEATION FILM FOR TUNGSTEN CVD IN OXIDE VIAS

S.B. HERNER, H.-M. ZHANG, B. SUN, Y. TANAKA, K.A. LITTAU, AND G. DIXIT
Applied Materials, Santa Clara, CA 95054

ABSTRACT

The suitability of using bias sputtered tungsten as a diffusion barrier to fluorine to replace titanium nitride for application in vertical metal vias on silicon is evaluated. Tungsten CVD on a film stack of sputtered tungsten/titanium was found to result in less fluorine penetration into the titanium than for a sputtered titanium nitride/titanium film stack. After 380 nm of tungsten CVD at 410°C on 10 nm barrier films, the dose of fluorine in 100 nm titanium films underneath was measured to be 3×10^{13} atoms/cm^2 in the bias sputtered tungsten-protected titanium, vs. 5×10^{14} atoms/cm^2 in the bias sputtered titanium nitride protected titanium. Step coverage in oxide vias with 0.45 μm openings at the surface in 1.2 μm thick oxide was as good or better with similar thickness sputtered tungsten relative to titanium nitride. The resistivity of tungsten CVD films grown on sputtered tungsten/titanium substrates was lower than on sputtered titanium nitride/titanium substrates (8.8 vs. 10.5 μΩ cm). The superior properties sputtered tungsten makes it a good candidate to replace titanium nitride as a barrier film in metal via applications in silicon integrated circuits.

INTRODUCTION

Vertical interconnections within integrated circuits are typically made with chemical vapor deposited (CVD) tungsten in lined oxide vias. The liner is typically composed of titanium, which is either subsequently annealed to form a titanium silicide contact to silicon, or used as an oxide-reducing contact to aluminum. Titanium nitride is deposited between titanium and tungsten to serve as a diffusion barrier.[1] The barrier film reduces diffusion of atomic fluorine and fluorine-containing compounds from the tungsten CVD process into the titanium film. Fluorine indiffusion increases the resistivity of titanium, and can lead to formation of titanium fluoride (TiF$_x$) compounds in extreme cases. The formation of TiF$_x$ results in the "volcano" reaction, in which the diffusion barrier film fractures and a volcano-shaped amalgam of tungsten, titanium nitride, titanium, and titanium fluoride compounds forms above the via.[2] Sputtered titanium nitride barrier films are particularly vulnerable to volcano reactions. The columnar grains provide fast diffusion paths along grain boundaries for fluorine and fluorine compounds.[3] Sputtering methods have poorer step coverage than CVD, and, as such, are less suitable for high aspect ratio vias. The liner and barrier films are much thinner than the radius of the via, and are commonly deposited by sputtering, though CVD methods have recently gained favor.[4] To completely fill vias with high aspect ratios (depths to diameters), tungsten CVD is the most practical and widely employed process in silicon integrated circuits. The relatively small thickness of the Ti and TiN film have allowed sputtering to continue to be used for their deposition long after it had become impossible to completely fill the via with sputtering methods.

The desire to make the diffusion barrier thinner without comprising fluorine indiffusion has led the search for alternatives to TiN. One alternative as a barrier film is sputtered tungsten. By sputtering a thin film of tungsten over titanium, and then filling the via with CVD tungsten, excellent step coverage in high aspect ratio (HAR) vias can be achieved. Replacing titanium nitride with sputtered tungsten has the advantage of lower overall via resistance. The resistivity of a 30 nm film of bias sputtered titanium nitride thin film is measured to be 78 μΩ cm, while 30 nm of bias sputtered tungsten has a resistivity of 28 μΩ cm. With ionization and the application of bias during sputtering of metals, step coverage is improved and fill of the bottom of the via can be achieved.[1] In this paper we compare the fluorine barrier performance, resistivities of tungsten CVD on bias sputtered tungsten and titanium nitride, and step coverage,.

EXPERIMENT

Silicon wafers were prepared for 3 sets of experiments: 1) fluorine penetration (SIMS), 2) texture (x-ray diffraction), and 3) step coverage (oxide vias), andAll wafers had a thermal oxide layer grown on them. Oxide vias were dry etched after lithography on some oxide wafers. A summary of metal deposition conditions for the wafers is in Table 1. Some wafers were loaded into an Applied Materials Endura™ tool and deposited with varying amounts of titanium and titanium nitride by the ionized metal plasma (IMP) process, or biased sputtering. The IMP process uses a high density, inductively coupled plasma between the magnetron source and the substrate to

539

efficiently ionize the sputtered metal atoms. With a bias applied to the wafer substrate, the ionized metal atoms can be deposited onto the substrate with high directionality, yielding improved step coverage.

characterization	condition	liner film	barrier film	CVD W
F penetration	A	100 nm IMP Ti	10 nm IMP TiN	1) + 2)
	B	100 nm IMP Ti	2-40 nm IMP W	2) only
texture	I	-	40 nm IMP TiN	1) + 2)
	II	30 nm IMP Ti	40 nm IMP TiN	2) only
	III	20 nm IMP Ti	40 nm IMP W	2) only
	IV	40 IMP TiN	40 nm IMP W	2) only
	V	30 nm IMP TiN/ 40 nm IMP TiN	40 nm IMP W	2) only
step coverage	1	20 nm IMP Ti	40 nm IMP TiN	1) + 2)
	2	20 nm IMP Ti	40 nm IMP W	2) only

Table 1. Film stack deposition conditions. 1) nucleation film with WF_6 reduction by SiH_4 only 2) via fill film with WF_6 reduction by H_2.

All of these wafers were then loaded into an Applied Materials Centura™ WxZ tool for CVD tungsten depositions. Chemical vapor deposition of tungsten was done at a wafer susceptor temperature of 425°C (wafer surface temperature ~410°C). Tungsten CVD on titanium nitride is accomplished in two steps: (1) reduction of WF_6 by SiH_4 to produce a thin, continuous layer of tungsten ("nucleation film"), followed by (2) reduction of WF_6 by H_2 to produce a thicker film ("via fill film").[6] Wafers with titanium nitride barriers were deposited using both steps, while wafers with tungsten as a barrier were deposited with step (2) only. For CVD tungsten, step (1) was performed by first heating the wafer in argon at 30 Torr, then reducing 30 sccm of WF_6 with 30 sccm of SiH_4 diluted in H_2 and argon to produce a 35 nm film. For step (2), 250 sccm of tungsten hexaflouride (WF_6) was reduced by 4000 sccm of H_2 diluted in argon at 300 Torr, to add ~350 nm of tungsten to the nucleation film. The deposition rate was ~4 nm/sec for the nucleation film and 10 nm/sec for the via fill film. When only H_2 reduction of WF_6 was used on IMP tungsten substrates, the CVD tungsten film was 380 nm. Elimination of step (1) provides another advantage for IMP tungsten: high wafer throughput on subsequent CVD tungsten as it serves as the nucleation film. After CVD tungsten deposition, films were examined by scanning electron microscope (SEM), film surface roughness was determined by atomic force microscopy (AFM) using tapping mode, and fluorine depth profiling was done by secondary ion mass spectrometry (SIMS) with sputtering by Cs^+ beam, and texture by x-ray diffraction (CuK_α radiation). Tungsten CVD film resistivity was extracted by measuring the sheet resistance on 49 points on the wafer of the substrate and film, measuring film thickness by SEM.

RESULTS AND DISCUSSION

A. FLUORINE PENETRATION

The fluorine depth profiles with 10 nm IMP tungsten (condition A) and 10 nm IMP titanium nitride (condition B) barriers are shown in figures 1 and 2, respectively. The IMP tungsten barrier allowed a smaller amount of fluorine to penetrate into the titanium film compared to IMP titanium nitride. The "dip" in the fluorine level in the tungsten film distinguishes the nucleation film from the via fill film. The fluorine dose in the titanium layer with IMP tungsten barrier is estimated to be 3×10^{13} F atoms/cm^2, compared to 5×10^{14} F atoms/cm^2 in the titanium nitride/titanium layers with the IMP titanium nitride barrier. Depth resolution did not allow the TiN film to be accurately distinguished from titanium, though its' presence in figures 2 is denoted by the higher titanium ion yield at the interface. The fluorine dose estimates in the titanium film are conservative: sputter-induced roughening exaggerates the fluorine indiffusion profile in both plots, as evidenced by the broad titanium profile at the titanium/SiO_2 transition. In a separate experiment, the diffusivity of fluorine in tungsten at a wafer susceptor temperature of 425°C was estimated to be $4 \times 10^{-14} cm^2$/sec.[6] While no estimates of the fluorine diffusivity in TiN were found, fluorine diffusivity in titanium at a wafer susceptor temperature of 440°C was estimated at $10^{-12} cm^2$/sec by Ramanath et al.[3] We speculate that the superior barrier performance of bias sputtered tungsten relative to bias sputtered TiN is due to the lower diffusivity of fluorine in tungsten. Much thinner bias sputtered tungsten barriers can be used and still allow less fluorine penetration than thicker TiN counterparts (figure 3).

540

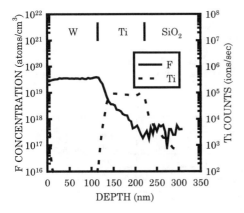

Figure 1. SIMS depth profile of 380 nm CVD tungsten/ 10 nm IMP W/100 nm IMP Ti film stack after CMP removed most of the CVD W film.

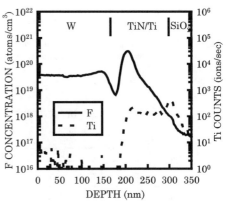

Figure 2. SIMS depth profile of 380 nm CVD tungsten/10 nm IMP TiN/100 nm Ti film stack after CMP removed most of the CVD tungsten.

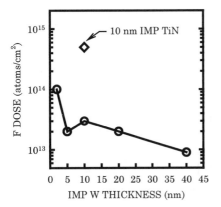

Figure 3. Fluorine dose in 100 nm of IMP Ti underneath various thickness bias sputtered tungsten films with 380 nm tungsten CVD grown on top.

B. GRAIN ORIENTATION/SIZE

Tungsten CVD films on certain IMP tungsten substrates showed a smaller resistivity than those grown on IMP TiN/Ti substrates. Closer investigation showed that tungsten CVD films grown on IMP W/Ti substrates had a lower resistivity, as well as a larger grain size. To determine if this phenomena was strictly a function of the IMP tungsten substrate, tungsten CVD was grown on five different substrates, shown below (figure 4). These provide a variety of substrate crystal orientations for tungsten growth (conditions I-V in Table 1). Properties of 350 nm tungsten CVD films grown on these substrates were measured and summarized in table 2. Only films grown on substrate III showed a low resistivity and large grain size. The mean free path of electrons in single crystal tungsten films at room temperature has been estimated at 40 nm.[7] The larger grain size in films grown on substrate III means fewer scattering centers, hence the lower resistivity.

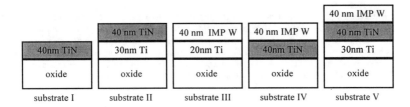

Figure 4. Substrates for CVD tungsten deposition.

The origin of the larger CVD tungsten grain size for films on substrate III is unknown. Atomic force microscopy measurements of the grain size of the sputtered tungsten films on substrates III-V did not show much difference with either each other or with the grain size of the sputtered TiN films on I and II. X-ray diffraction measurements of the tungsten CVD films at on the five substrates are shown in figure 5a-e. The strong (310) orientation of tungsten film grown on substrate III is unique among the films, and may play role in the large grain size. Further investigation of this phenomena is needed to determine the relationship, if any, between grain size and film orientation.

substrate	CVD W dominant orientation	CVD W grain size (μm)	CVD W ρ (μΩ cm)	CVD W reflectivity (% @ 480 nm)
I	(110)	0.06	10.5	100.2
II	(211)	0.07	10.3	106.1
III	(310)	0.17	8.8	87.7
IV	(200)	0.08	10.0	92.3
V	(110)	0.07	10.4	N/A

Table 2. Properties for ~350nm tungsten CVD films on various substrates.

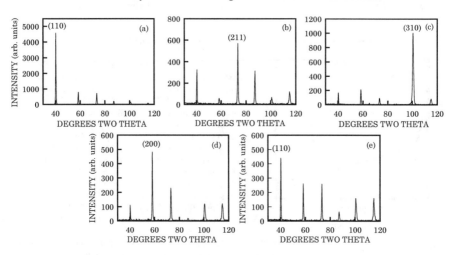

Figure 5. X-ray diffraction spectra of 350 nm tungsten CVD films on a) substrate I b) substrate II c) substrate III d) substrate IV and e) substrate V.

C. STEP COVERAGE

Oxide wafers with 0.45 μm diameter openings at the oxide surface and 3:1 aspect ratio were deposited with 20 nm IMP Ti followed by either 40 nm IMP W or 40 nm IMP TiN. Chemical vapor deposited tungsten completed fill of the vias, using either the one step process for the sputtered tungsten barrier, or the two step process for the sputtered TiN barrier (conditions 1 and 2 in Table 1). Fractured cross sections of the vias are shown above in figure 6. The large grain size of the CVD tungsten film is evident in figure 6a relative to the smaller grains on the TiN in figure 6b. While a small seam is evident in the via with the TiN barrier, the via with the tungsten barrier has complete fill. Bias sputtered tungsten clearly has adequate step coverage to prevent reaction between the WF_6 and underlying Ti.

Figure 6a. Oxide via with tungsten
CVD/40 nm IMP W/20 nm IMP Ti.

Figure 6b Oxide via with tungsten
CVD/40 nm IMP TiN/20 nm IMP Ti.

CONCLUSION

In conclusion, we find tungsten to be a superior barrier to fluorine diffusion from subsequent CVD tungsten than titanium nitride, with both barriers deposited by biased sputtering (IMP). We speculate that the improved barrier performance is due to the relatively low diffusivity of fluorine in IMP tungsten. Tungsten barrier films as thin as 2 nm were found to allow less fluorine penetration into underlying titanium films than thicker titanium nitride barrier films, while at the same time lowering resistance in the overall film stack. Since the IMP tungsten film also serves as the nucleation film, high wafer throughput can be achieved on subsequent CVD tungsten tools. Tungsten CVD films grown on bias sputtered tungsten/titanium film stacks had a larger grain size and lower resistivity than films deposited on TiN substrates. The large grain size of CVD tungsten was found to coincide with the development of a strong (310) orientation of the film, and was not a function of growing on sputtered tungsten substrates. Step coverage of CVD tungsten/IMP tungsten/IMP titanium film stacks on oxide wafers etched with 0.45 μm vias with 3:1 aspect ratio was found to be equivalent to those with CVD tungsten/IMP TiN/IMP titanium film stacks. This allows sputtered tungsten to replace titanium nitride in vias, with improvement in both electrical and diffusion barrier performance.

ACKNOWLEDGEMENTS

We thank Isabelle Roflox and Adli Saleh for their help with the experiments.

REFERENCES

1. S.M. Rossnagel, J. Vac. Sci. Tech. B **16**, 2585 (1998).
2. M. Rutten, D. Greenwell, S. Luce, and R. Dreves, Mat. Res. Soc. Symp. Proc. **239**, 277 (1992).
3. G. Ramanath, J.R.A. Carlson, J.E. Greene, L.H. Allen, V.C. Hornback, and D.J. Allman, Appl. Phys. Lett. **69**, 3179 (1996).
4. J. Hillman, R. Foster, J. Faguet, W. Triggs, R. Arora, M. Ameen, F. Martin, and C. Arena, Sol. St. Tech. **38**, 146 (1995).
5. E.J. McInerney, T.W. Mountsier, B.L. Chin, and E.K. Broadbent, J. Vac. Sci. Tech. B **11**, 734 (1993).
6. S.B. Herner, H-M. Zhang, Y. Tanaka, W. Shi, S.X. Yang, R. Lum, and K.A. Littau, in press.
7. J.E.J. Schmitz, *Chemical Vapor Deposition of Tungsten and Tungsten Silicides* (Noyes, Park Ridge, NJ 1992), p. 107.

Surface Reaction Model of Al Growth using DMAH: Elementary Reaction Simulation and Surface Analysis

M. Sugiyama, H. Ogawa[1], T. Nakajima, T. Tanaka, H. Itoh[2], J. Aoyama[2],
Y. Egashira[3],K. Yamashita, Y. Horiike[1], H. Komiyama and Y. Shimogaki[1]

Department of Chemical System Engineering, School of Engineering, University of Tokyo
7-3-1 Hongo, Bunkyo-ku, Tokyo 113-8656, Japan
TEL&FAX: +81-3-5841-7131
E-mail: sugiyama@chemsys.t.u-tokyo.ac.jp
[1] Department of Materials Science and Metallurgy, School of Engineering, University of Tokyo
[2] Semicondactor Technology Academic Research Center, Japan
[3]Department of Chemical Engineering, School of Engineering, Osaka University

Surface reaction of Al-CVD process from dimethylaluminumhydride (DMAH) was analyzed by surface elementary reaction simulation and surface analysis of the deposited film. The major purpose is to construct the reaction model based on *ab initio* calculations and to check the model validity through the comparison between simulation results and experimental results. The surface elementary reaction model was postulated based on *ab initio* cluster-model calculations, and their rate constants were estimated based on the transition state theory. The simulated deposition rate profile in a tubular reactor agreed well with experimental results showing the model validity. According to the reaction model, the Al surface is covered with $H(CH_3)Al$-adsorbate, and its existence was suggested by XPS and IR-RAS measurements of the deposited film surface without air-exposure.

Introduction

Al-CVD from dimethylaluminumhydride (DMAH) has been intensively researched for its excellent gas-filling property, but its mechanism has not been fully understood. The rate limiting step of Al-CVD from DMAH is reported to be the surface reaction. Experimental approach to elucidate surface reaction mechanism and to measure surface reaction rates is much more difficult than the case of gas-phase reaction. Recent progress in *ab initio* quantum chemical calculations has enabled the calculation of surface reactions with cluster models. The authors have calculated the energy diagrams for the possible reaction pathways of DMAH surface reactions [1]. Using those calculation results and the experimental results on reaction products [2], it is possible to postulate a qualitative reaction model for DMAH surface reaction. Such a qualitative model, however, cannot make a significant contribution to a process development unless its validity is quantitatively confirmed through the comparison with experimental results.

This work concerns in the experimental validation of the DMAH surface reaction model. Al deposition rate profile was observed in the tubular reactor with a temperature gradient because such a profile is sensitive to the activation energies of surface reactions and is suitable for the model validation. In order to predict Al deposition rates based on the reaction model, the rate constant of each elementary reaction was estimated based on the transition state theory [3] and an empirical method [4, 5] to reduce computational efforts. Elementary reaction simulations using CHEMKIN[TM] package [6] estimated Al deposition rate profiles in the tubular reactor, and it enabled the comparison between simulations and experiments. The simulated deposition rate profile agreed well with experimental data, which confirmed the model validity.

As a further examination of the model, the authors analyzed the Al surface after CVD process by XPS and IR-RAS. These experiments used cluster chambers which enabled deposited films to be transferred to XPS or IR-RAS chamber in high vacuum ($<10^{-7}$ Torr). XPS analysis revealed that the Al film after CVD was covered with a carbon monolayer. IR-RAS analysis suggested the existence of C-H and Al-H vibrations at the Al surface. These facts are consistent

Conference Proceedings ULSI XV © 2000 Materials Research Society

with the simulation result that the surface is covered with $H(CH_3)Al$- adsorbate.

Ab initio calculations of DMAH surface reactions

DMAH adsorption on Al (111) surface was examined by *ab-initio* calculations using Al cluster surface model and density functional theory (DFT) [1]. The dissociative adsorption of DMAH monomer yields H-, H_3C- and CH_3Al- adsorbates. The calculated desorption ΔE for the possible reaction products from those adsorbates suggested that H_2, CH_4 and $(CH_3)_3Al$ (TMA) can be reaction products, among which CH_4 seemed the most preferable products. However, the experiment [2] showed TMA and H_2 were the reaction products. Precise calculations of the desorption process revealed the existence of the highest energy barrier for CH_4 desorption (Fig. 1), which is consistent with the experimental observation.

Surface Reaction Model

The above-mentioned calculations led to the following reaction model (Fig. 2). DMAH monomer dissociatively adsorb on Al surface with no energy barrier to yield $H(CH_3)Al$- and H_3C- surface adsorbates ((1) in Fig. 2). $H(CH_3)Al$- decompose to H- and CH_3Al- (2). H-, H_3C- and CH_3Al- migrate on Al surface, and CH_3Al- decompose to H_3C- and deposited Al at surface steps (8). Those surface adsorbates associatively desorb to yield TMA (3 and 4), H_2 (6). CH_4 cannot desorb because of the high energy barrier (5).

Estimation of Reaction Rate Constants

Elementary reaction simulation is necessary to compare the results of this model with the experimental results, especially with deposition rates. The simulation requires the reaction rate constants for all the elementary reactions. The rate constants were calculated using the transition state theory [3]

$$ k = (\kappa T/h)\exp(-\Delta n^{\neq})\exp(\Delta S^{\neq}/R)\exp(-\Delta E^{\neq}/RT) . \tag{1} $$

$\Delta E^{\neq}$ for all the desorption (Fig. 1) and surface migration processes were calculated by *ab-initio* calculations. Calculation of $\Delta S^{\neq}$ requires all the vibrational frequencies including transition states, which, however, takes enormous efforts and time in the case of cluster model calculations. Therefore, vibrational frequencies were approximately estimated using the spectral data of the similar molecules [4, 5], because the contribution of an entropy term to the total reaction rate is much smaller than that of an energy term. Estimation of desorption rates took account of the migration rates of surface adsorbates.

The estimated rate constants are listed in Table 1 as well as the elementary reaction schemes. The reaction numbers correspond to those in Fig. 2. The reaction schemes obey CHEMKIN™ format. The T, B, F and FF indicate the adsorption sites: T: top site, B: bridge site, F: 3-fold site with the second-layer Al, FF: 3-fold site with the third-layer Al. The O indicate the open sites. When there is no open site, DMAH cannot adsorb on the surface.

Elementary Reaction Simulation: Comparison of Al Deposition Rates

In order to check the model validity, Al deposition profiles in a tubular reactor was observed as shown in Fig. 3. Because of the increasing temperature (Fig. 3), the deposition rate increased with the distance from the reactor inlet. The deposition rate drop after its maximum is due to the depletion of DMAH. Therefore, the maximum deposition rate position is sensitive to both temperature and precursor-concentration dependence of the surface reaction rate. For this reason, this deposition rate profile is suitable for the experimental validation of the reaction model.

The simulation of that deposition rate profile used CRESLAF in CHEMKIN™ package which considered flow, mass-transport, heat-transport, gas-phase reaction (monomer-dimer equilibrium of DMAH) and surface reactions listed in Table 1. Sensitivity analysis revealed that the deposition rate profile was most influenced by the forward reaction rate of the reaction 2, and that the activation energy for the best fit was 55 kJ/mol which is within the calculation error from the original value of 60 kJ/mol. In the viewpoint of the deposition rate, this reaction model successfully reproduced the experimental results.

In vacuum Surface Analysis of Deposited Al Films

According to the simulation results, Al surface during the growth is covered with H(CH$_3$)Al-adsorbate. In order to confirm the reaction model, the authors tried to detect that adsorbate by XPS and IR-RAS. Cluster vacuum system equipped with Al-CVD chamber, XPS chamber and IR-RAS chamber was used. The sample after CVD process was transferred to the analysis chambers without air exposure. Base pressure of the transfer chamber was less than 1×10^{-7} Torr.

Figure 4 shows the XPS spectra of H-terminated Si (before deposition), the substrate exposed to DMAH at room temperature and the substrate after CVD from DMAH at 285 °C. C, O and oxidized Al were observed even after DMAH exposure at room temperature. DMAH seems to have strongly adsorbed on the surface, and that DMAH seems to have oxidized during the transfer even at high-vacuum circumstance. On the continuous Al film, C, O, oxidized Al and atomic Al were observed. The Al peak also contained a shoulder peak whose chemical shift was less than that of an oxidized Al. That peak was attributed to Al-C bond in the surface adsorbate.

In order to examine the depth distribution of those atoms, a takeoff angle was changed in XPS measurements. Figure 5 shows the relation between the XPS peak area and the takeoff angle. If the topmost layer of the Al film is covered with a certain atom, its peak area should show 1/sinθ dependence relative to the takeoff angle, θ. The carbon peak area showed this dependence, which indicates the surface was covered with C. The estimated carbon coverage was almost unity. The oxygen peak area showed more gradual θ dependence than C peak area, which suggests O penetrates a little deeper into Al film than C. This thin oxidized layer seems to have formed during the sample transfer.

The XPS analysis suggested the existence of C-containing adsorbate on the deposited Al films. However, H atoms in the adsorbate cannot be detected by XPS. Therefore, the authors tried to detect infrared adsorption peaks of the surface adsorbate using high-sensitivity infrared reflection absorption spectroscopy (IR-RAS). Figure 6 shows the IR-RAS spectra for Al films deposited at different deposition temperatures. Assignment of those adsorption peaks took account of the gas-phase infrared absorption spectrum of DMAH [7]. Absorption peaks around 2900 cm^{-1} are attributed to C-H stretching vibrations of CH$_3$ group. These peaks are different in shape from the adsorption peaks observed when the surface was exposed to DMAH at room temperature, and therefore cannot be attributed to DMAH adsorbed during the cooling process after the deposition. Furthermore, absorption peaks at 1190 cm^{-1} and 1310 cm^{-1} may be those of Al-H vibrations. Therefore, the surface adsorbate can have CH$_3$ group, Al-H bond and Al-C bond (XPS chemical shift). This is consistent with the simulation result that the Al surface is covered with H(CH$_3$)Al- adsorbate.

Conclusion

The surface elementary reaction model for Al-CVD from DMAH was constructed based on *ab initio* calculations, and the rate constants were estimated for a quantitative prediction of this process. The model validity was checked through the comparison of the simulated deposition rate profile with experimental data. Surface analysis of the deposited films suggested the existence of H(CH$_3$)Al- adsorbate which was predicted by the reaction model.

References

1. T. Nakajima, T. Tanaka and K. Yamashita: THEOCHEM, in press.
2. M. Sugiyama, Y. Shimogaki, H. Itoh, J. Aoyama, T. Yoshimi and H. Komiyama: *Proc. Symp. Fundamental Gas-Phase and Surface Chemistry of Vapor-Phase Materials Synthesys* (Electrochemical Society, Pennington, U.S.A., 1998) **98-23** p. 252.
3. K. J. Laidler and H. Eyring: *The theory of rate processes* (McGRAW-HILL, New York, 1964).
4. R. Arora and R. Pollard: J. Electrochem. Soc. **138**, 1523 (1991).
5. Y. F. Wang and R. Pollard: J. Electrchem. Soc. **142**, 1712 (1995).
6. Reacrion Design Co., 6440 Lusk Boulevard, Suite D-209, San Diego, CA 92121,U.S.A.
7. A. S. Grady, S. G. Puntambeker and D. K. Russel: Spectrochimica Acta, **47A**, 47 (1991).

Table 1 Elementary reactions used in the simulation.

		forward reaction rates			reverse reaction rates		
No.	reactions	A	b	E [kJ/mol]	A	b	E [kJ/mol]
1	DMAH_m+O(T)+O(F) = CH₃(T)+HAlCH₃(F)	1.00 *	0.0	0.0	1.04×10^{20}	0.8	177.4
2	HAlCH₃(F)+O(FF) = H(FF)+AlCH₃(F)	4.74×10^{20}	0.5	55.0	2.60×10^{18}	1.2	20.3
3	TMA+O(B)+O(T) = CH₃(T)+CH₃AlCH₃(B)	3.18×10^{12}	1.6	72.9	1.57×10^{17}	1.6	37.7
4	CH₃(T)+AlCH₃(F)+O(B) = CH₃AlCH₃(B)+O(T)+O(F)	3.04×10^{25}	1.1	11.3	5.76×10^{24}	2.3	49.6
5	CH₄+O(T)+O(FF) = CH₃(T)+H(FF)	7.57×10^{13}	2.4	203.7	1.93×10^{18}	0.6	98.3
6	H₂+O(T)+O(FF) = H(T)+H(FF)	6.61×10^{19}	0.6	100.9	3.51×10^{21}	0.0	56.1
7	H(FF)+O(T) = H(T)+O(FF)	5.52×10^{20}	0.2	59.4	5.00×10^{15}	1.8	0.0
8	AlCH₃(F)+O(T) = CH₃(T)+O(F)+Al(D)	5.52×10^{17}	1.0	13.4	5.52×10^{17}	1.0	13.4

Reaction rates: k=A T^b exp(-E/RT)
Al(D): deposited aluminum
* sticking coefficient

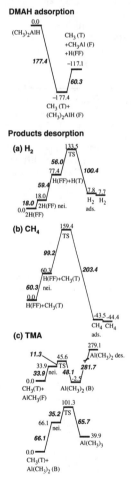

DMAH adsorption

Products desorption

(a) H₂

(b) CH₄

(c) TMA

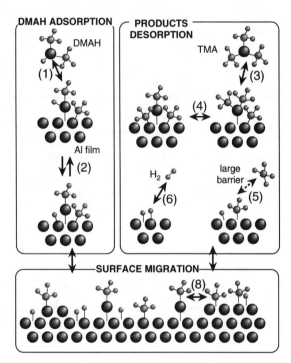

Fig. 2 Schematic of the reaction model. The reaction numbers correspond to those in Table 1.

Fig. 1 (left) Energy diagrams for DMAH adsorption and products desorption.

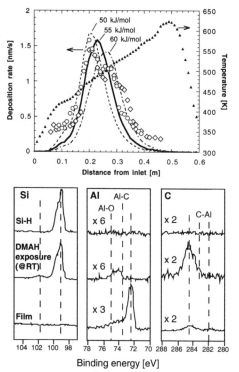

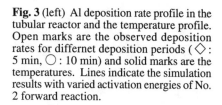

Fig. 3 (left) Al deposition rate profile in the tubular reactor and the temperature profile. Open marks are the observed deposition rates for differnet deposition periods (◇ : 5 min, ○ : 10 min) and solid marks are the temperatures. Lines indicate the simulation results with varied activation energies of No. 2 forward reaction.

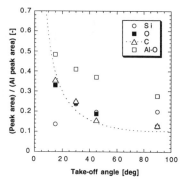

Fig. 5 Take-off angle dependence of XPS peak areas observed for the continuous Al film. The dotted line is the theoretical curve when a monolayer carbon exists on the Al film.

Fig. 4 XPS spectra of the Si and Al surface in the course of Al deposition. Each spectrum was taken after the noted treatment without air exposure.

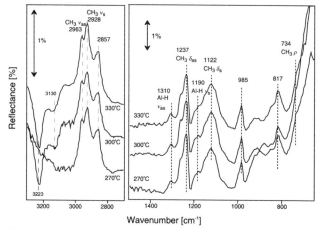

Fig. 6 IR-RAS spectra of the Al surface after the deposition at different temperatures. Peak assignment is based on the analogy to the gas-phase IR spectra of DMAH.

Theoretical study on the reactivity of oxidized aluminum surfaces : Effects of adsorbed metallic atoms (Au, Cu, Ti, V)

T. Tanaka, T. Nakajima and K. Yamashita
Department of Chemical System Engineering, Graduate School of Engineering,
University of Tokyo, Tokyo 113-8656, Japan, yamasita@chemsys.t.u-tokyo.ac.jp

ABSTRACT

The possibility of restoration of reactivity of the oxidized Al surface by adsorption of a metallic atom, such as Au, Cu, Ti, V or Al, was studied theoretically. We studied (i) population distributions, (ii) energy levels of the highest occupied molecular orbital (HOMO), (iii) softness and local softness, and (iv) the adsorption energy and structure of AlH_3, which is a model for dimethylaluminum hydride (DMAH) that is used as the source gas for chemical vapor deposition (CVD). The indicators of reactivity, (i)-(iv), were calculated for three systems: (a) a clean surface, (b) an oxidized Al surface, and (c) an oxidized Al surface adsorbing a metallic atom, by using density functional theory (B3P86/LanL2DZ) with a cluster model for the Al surface. It was found that the adsorption of these metallic atoms can restore the reactivity of the oxidized Al surface. Specifically, the adsorption of Ti can restore all indicators of reactivity.

INTRODUCTION

Aluminum chemical vapor deposition (CVD) is a promising method for finer electrical wiring in ULSI fabrication because of its potential to form aluminum films with low resistivity and good step coverage. However, it is considered that use of the CVD method to form aluminum films without roughness is difficult, due to a decline of reactivity by oxidization of Al surfaces. Many experiments have been employed to realize uniform growth of Al films [1, 2, 3]. These experiments appear to suggest that the reactivity of the oxidized Al surface can be restored by sprinkling metallic atoms or clusters. In this study, we have investigated theoretically the possibility of restoration of reactivity of the oxidized Al surface by adsorption of these metallic atoms. We have considered three systems: a clean Al surface, an oxidized Al surface, and an oxidized Al surface adsorbing a metallic atom. The indicators of reactivity, (i) population distributions, (ii) energy levels of HOMO, (iii) softness and local softness, and (iv) adsorption energies and structures of AlH_3, are analyzed by using density functional theory (DFT). AlH_3 is a model for dimethylaluminum hydride (DMAH), which is used as the source gas in CVD. We have considered Au, Cu, Ti, V and Al as the metallic atoms adsorbed on the oxidized Al surface.

CALCULATION METHODS

We adopted the Al-(12,6)-cluster, where the numbers in parentheses give the number of atoms in each layer, to represent the Al(111) surface. The distance between the nearest Al atoms was taken as the bulk Al lattice value, 2.86 Å. At small exposure of oxygen molecules (from 3 to 10 Langmuir), islands with a (1×1) structure of chemisorbed oxygen atoms are formed on the Al surface and the oxygen atoms are found to occupy threefold coordinated sites at the fcc position, 0.7 Å above the Al surface plane [3, 4]. We used the Al-(12,6)-cluster adsorbing an oxygen atom at the fcc site in order to model the oxidized Al surface. On the other hand, the Al-(12,6)-cluster co-adsorbing an oxygen atom at the fcc site and a metallic atom at the nearest hcp site to the oxygen atom was used to model the oxidized Al surface adsorbing a metallic atom.

We used Becke's three parameter functional B3P86, which is a mixture of Hartree-Fock exchange

551

energy and the exchange-correlation energy based on the density functional theory (DFT). The basis sets adopted here were LanL2DZ, that is, the Dunning/Huzinaga full double zeta (D95) basis for H and O atoms and the Los Alamos effective core potential plus double-zeta basis for Al, Au, Cu, Ti and V atoms. These basis sets were augmented with polarization functions. All calculations were performed using the Gaussian 98 package. Geometry optimizations were executed with the structure of the Al-(12,6)-cluster fixed.

RESULTS

Population Analysis

Figure 1 shows Mulliken populations of the clean Al surface, the oxidized Al surface, and the oxidized Al surface adsorbing a metallic atom. In the case of the clean surface, the three Al atoms that are located at the center of the first layer of the Al-(12,6)-cluster (AL7, AL8, and AL9) are charged negatively. By oxidization of the surface, the negative charge is transferred from these Al atoms to the O atom. The atoms AL7, AL8, and AL9 are consequently charged positively. By adsorbing a metallic atom on the oxidized Al surface, an Au or Cu atom takes a negative charge from the surface, and consequently the populations on the Al atoms remain positive. On the other hand, a Ti, V or Al atom gives a negative charge to the surface, so the populations on AL7 and AL8 change from positive to negative. These results can be explained by the electronegativity of the atoms. It is found that the atoms Ti, V and Al have the potential to restore the population distribution of the oxidized Al surface to that of the clean Al surface.

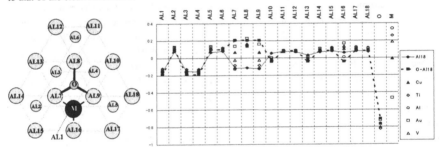

Fig. 1   Population distribution

HOMO Energy Levels

Table 1 shows the energy levels of HOMO of the three systems. The HOMO is stabilized by oxidization, that is, the oxidized Al surface is less reactive than the clean Al surface. By adsorption of a metallic atom other than Al and Au, the HOMO level rises and almost recovers to the original level of the clean Al surface, that is, the reactivity of the oxidized Al surface is recovered by adsorption of a metallic atom. Since the energy levels of the 4s and 3d orbitals of Ti and V, and the 4s orbital of Cu are close to the HOMO level of the oxidized Al surface, the adsorption of Ti, Cu or V results in a significant level shift of the HOMO. On the other hand, the energy levels of the 3p orbital of Al and the 6s orbital of Au are far from the HOMO level of the oxidized Al surface and therefore the adsorption of these atoms gives only a small perturbation to the HOMO. We may conclude that Ti, V and Cu can restore the reactivity of the oxidized Al surface.

Softness and Local Softness

In the DFT formalism, the concepts of softness (global) and local softness are very helpful in discussing the reactivity of surfaces. Table 2 shows the values of softness of the three systems.
The softness (global) is defined as follows:

$$S = \frac{1}{IP - EA} \quad \cdots(1) \qquad IP: ionization\ potential\ (a.u.) \qquad EA: electron\ affinity\ (a.u.)$$

The softness of the Al cluster distinctly decreases by oxidization, and increases by adsorption of a metallic atom. This means the reactivity of the oxidized Al surface is recovered by adsorption of a metallic atom. In particular, adsorption of Ti increases the softness the most among the metallic atoms considered in this study.

The local softness values (s^+, s^-) are plotted in Fig. 2 against each atom of the first layer. The local softness values, s^+ and s^-, behave as follows.

$$s^+(r) = [\rho_{N+1} - \rho_N] * S \qquad in\ the\ case\ of\ nucleophilic\ reaction \quad \cdots(2)$$
$$s^-(r) = [\rho_N - \rho_{N-1}] * S \qquad in\ the\ case\ of\ electrophilic\ reaction \quad \cdots(3) \qquad \rho_N: electron\ density\ of\ N\ electron\ system$$

The local softness of each Al atom of the first layer decreases by oxidization and that of the adsorbed O atom itself is quite small. While the local softness of each Al atom is not restored by the adsorption of a metallic atom, that of the adsorbed Ti atom itself is quite large. From these results, it is found that though the adsorption of a metallic atom does not restore the reactivity of each Al atom of the surface, the high reactivity of the adsorbed metallic atom corresponds to the restoration of reactivity of the oxidized surface.

Table 1   Energy levels of HOMO of a clean surface, an oxidized Al surface and an oxidized Al surface adsorbing a metallic atom

	HOMO level (kcal/mol)
Clean Surface	0.00
Oxidized Surface	-3.62
Au	-2.59
Cu	-0.77
Ti	-0.23
V	+1.25
Al	-3.48

Table 2   Softness of a clean Al surface, an oxidized Al surface and an oxidized Al surface adsorbing a metallic atom

	Softness
Clean Surface	11.22
Oxidized Surface	8.76
Au	9.29
Cu	9.11
Ti	10.00
V	9.25
Al	9.22

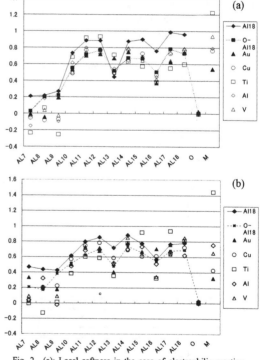

Fig. 2   (a): Local softness in the case of electrophilic reaction, (b): local softness in the case of nucleophilic reaction

Adsorption of AlH₃

We have considered the reaction in which AlH_3 is adsorbed at the fcc site on the Al surface. Figure 3 and Table 3 show the adsorption energy and adsorption structure for the three systems when AlH_3 is adsorbed. When AlH₃ is adsorbed on the clean Al surface, we may note two characteristics. The first is that two Al-H bonds are weakened by the interaction between the Al surface and the H atoms of AlH₃, which is attributable to electron transfer from the surface to AlH₃ (see Table 4). Second, the height of the Al atom of the AlH₃ from the surface is 2.14 Å, which is smaller than the interlayer distance of the cluster, 2.34 Å. This is because negative charge localizes on the Al lattice atom interacting directly with the adsorbed AlH₃ (see Table 4) and a strong interaction is formed between them. The two interaction mechanisms mentioned above contribute to a large adsorption energy, 34.6kcal/mol. When AlH₃ is adsorbed on the oxidized Al surface, only one Al-H bond is weakened

and the height of the Al atom from the surface, 2.41 Å, becomes larger than that for adsorption onto the clean Al surface. The two effects that are important for adsorption onto the clean Al surface, that is, the electron transfer from the surface to the H atom and the localization of negative charge on the Al atom of the surface, are prevented by attraction of negative charge by the O atom. The interaction between the surface and AlH₃ becomes weak, resulting in a smaller adsorption energy, 14.5kcal/mol.

We found that the adsorption energy is restored by adsorption of a

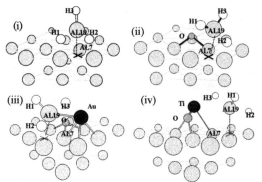

Fig. 3   Adsorption structures and energies of AlH₃:
(i)Clean surface, 34.6kcal/mol;   (ii) Oxidized surface,
14.5kcal/mol; (iii) Au, 22.16kcal/mol, Cu, 32.63kcal/mol,
: (iv) Ti, 34.39kcal/mol, V, 37.38kcal/mol, Al, 35.92kcal/mol

Table 3   Optimized bond length of the systems as shown in Fig. 3   (Å)

	AL19-surface	AL19-H1	AL19-H2	AL19-H3	O-surface	M-AL19	M-H3	M-Surface	O-M
Clean	2.14	1.75	1.72	1.59	—	—	—	—	—
Oxide	2.41	1.60	1.79	1.60	0.86	—	—	—	—
Au	2.45	1.59	1.72	1.76	0.84	2.67	1.93	2.02	3.22
Cu	2.40	1.59	1.72	1.74	0.85	2.54	1.71	1.94	3.10
Ti	2.19	1.59	1.75	1.73	1.47	3.04	1.89	2.07	1.79
V	2.30	1.60	1.81	1.75	1.02	2.97	1.86	2.17	2.02
Al	2.14	1.59	1.72	3.79	1.19	3.10	1.59	2.15	1.90

Table 4   Population analysis of the systems as shown in Fig. 3

	AL19	H1	H2	H3	AL7	O	M
Clean	0.54	-0.19	-0.05	-0.21	-0.63	—	—
Oxide	0.27	-0.06	-0.04	-0.18	-0.15	-0.71	—
Au	0.67	-0.09	-0.23	-0.07	0.07	-0.84	-0.82
Cu	0.64	-0.06	-0.19	-0.20	-0.13	-0.71	-0.83
Ti	0.44	-0.05	-0.21	-0.19	-0.70	-0.73	0.26
V	0.48	-0.05	-0.21	-0.20	-0.48	-0.82	0.19
Al	0.30	-0.03	-0.19	-0.04	-0.87	-0.73	0.47

metallic atom (see Fig. 3). This is due to two reasons. One is common and the other is characteristic of the adsorbed metallic atoms. The former is that the interaction between the H atom and the surface, which is important for adsorption onto the clean surface, is replaced by the interaction between the metallic atom and the H atom. The latter is different for Au and Cu on the one hand and for Ti, V and Al on the other hand. In the case of Au or Cu, negative charge localizes on the Au or Cu atom, and the interaction between these adsorbed metallic atoms and the Al atom of AlH_3 becomes dominant (see Fig. 3(iii), Table 3, 4). On the other hand, the adsorbed Ti, V or Al metallic atom is directly bonded to the O atom resulting in a weak bond between the O atom and the surface. The O atom therefore attracts only a small amount of negative charge from the Al atom and the negative charge remains localized on the Al atom. The interaction between the surface and AlH_3, which is important for adsorption onto the clean surface, has been restored (see Fig. 3(iv), Table 3, 4). Our calculation indicates that while the adsorbed Au or Cu atom is itself highly reactive without restoration of reactivity of the Al surface, the adsorbed Ti, V or Al atom can restore the reactivity of the surface.

CONCLUSIONS

We have investigated the possibility of restoration of reactivity of the oxidized Al surface by adsorption of a metallic atom (Au, Cu, Ti, V or Al), by comparing three systems: (a) the clean Al surface, (b) the oxidized Al surface, and (c) the oxidized Al surface adsorbing a metallic atom. The population analysis, the energy level of HOMO, softness and local softness, and the adsorption energy structure of AlH_3 were used as indicators of reactivity. Our DFT/B3P86 calculations showed that: (i) the adsorption of Ti, V, or Al restores the population distribution of the oxidized Al surface to that of the clean Al surface; (ii) the adsorption of Ti, V or Cu shifts the energy level of HOMO significantly compared with that of the oxidized Al surface; (iii) the adsorption of Ti makes softness much larger than that of the oxidized Al surface; (iv) the adsorption of a metallic atom has no effect on the local softness of the Al surface, but the local softness of the adsorbed Ti is characteristically large; (v) the oxidization prevents the adsorption of AlH_3 by hindering both electron transfer from the Al surface to the H atom of AlH_3 and localization of negative charges on the Al atom of the surface interacting directly with AlH_3; and (vi) the adsorption of a metallic atom promotes the adsorption of AlH_3, especially such metallic atoms as Ti, V or Al, and reduces the influence of the O atom on the Al surface by directly interacting with the O atom. We may conclude that Ti is the most effective atom to be adsorbed on the oxidized Al surface in order to restore the reactivity.

ACKNOWLEDGEMENTS

This work was partially supported by the Semiconductor Technology Academic Research Center (STARC). We are grateful to Profs. H. Komiyama, Y. Shimogaki and Egashira for their valuable comments and discussion. We also thank Drs. J. Aoyama and H. Ito for their useful discussion.

REFERENCES

1. K. Sugai, H. Okabayashi, S. Kishida, T. Shinzawa, Thin Solid Films, **280**, p. 42(1996)
2. T. Shinzawa, K. Sugai, A. Kobayashi, Y. Hayashi, T. Nakajima, S. Kishida, H. Okabayashi, T. Yako, K. Tsunenari, Y. Murao, Symp. VLSI Tech. DIg. Tech. Pap., **6B3**, p. 77 (1994)
3. T. Takahashi, Y. Shimogaki, T. Yoshimi, H. Itoh, J. Aoyama, J. Ueda, H. Komiyama, Proceedings of 13th international VLSI multilevel interconnection conference, p. 257, (1998)
4. M. Kerkar, D. Fisher, D. P. Woodruff, B. Cowie, Surf. Sci., **271**, p. 45 (1992)
5. J. Trost, H. Brune, J. Wintterlin, R. J. Behm, G. Ertl, J. Chem. Phys. **108**, p. 1740 (1998)

Compositional Defect Analysis on a 300 mm wafer by using Auger based defect review tool

R.Oiwa*, T.Ohba**, T.Morohashi*, K.D.Childs***, D.F.Paul*** and S.P.Clough***
* ULVAC-PHI, Inc. 370 Enzo, Chigasaki, Kanagawa, 253-0084 Japan
** Selete 292 Yoshida-cho, Totsuka-ku, Yokohama, 244-0817 Japan
*** Physical Electronics Inc. 6509 Flying Cloud Drive, Eden Prairie, Minnesota 55344

Abstract

Auger electron spectroscopy based defect review system (SMART-Tool) was used for as compositional metrology tool to analyze the particle composition generated during CVD WSi film on the 300 mm wafer. 10 small particles were randomly selected. SEM observation and Auger analysis by SMART-Tool was performed on each particle, optional EDS measurement has been also applied for comparison. Five of ten particles were identified as similar particle, assembled small pieces of needle formation and same composition with reference WSi film. Two small particle in the order of 0.3 – 0.4 μm diameter were found and Auger analysis revealed several additional compositional information related to particle root cause compared with EDS measurement, one was identified as unreacted W rich particle and the another was confirmed as SUS source core covered with SiO2 shell. These informations were not revealed by EDS measurement because of strong overlap of W and Si line and also too small particle size for large analysis volume size. In regard to one large defect, optional FIB process was applied and cross-sectional Auger analysis revealed that the thin carbon layer existed under this defect.

Introduction

Learning the sources of particles and/or the cause of defects during device processing and feedback the information to decrease the particle and/or defect generations are an important part of the yield enhancement in the semiconductor device fabrication. So far the particle characterization has been done by optical particle detection tool which shows the number and the location of the particles on the device wafer or the monitor wafer, size and morphology observation by optical microscope review station. SEM/EDX is commonly used method as detailed morphology observation and composition analysis method of the particles. These information are combined into database system and have been used for particle monitoring. Considering a trend of device processing in the near future, this kind of compositional metrology is going to be important because, 1) new process with new materials such as Cu for metal line may generate unknown particles/defects, 2) critical killer particles/defects size will be getting smaller with shrinking the design rules, 3) fast ramping up the yield curve of the new process is one of the key to survive semiconductor manufacturing competition.

In this situation, we have been developing a field emission Auger based defect review tool that is available to introduce 300/200 mm wafers. The benefits of Auger analysis in term of compositional metrology or root cause determination are, 1)

557

higher sensitivity for light elements compared with EDS, 2) high spatial resolution in the order of 10 nm which provides reliable information from the target particles/defects and 3) surface sensitive. In this paper, we report the results of Auger analysis on several particles generated after CDV WSi process on 300 mm wafer.

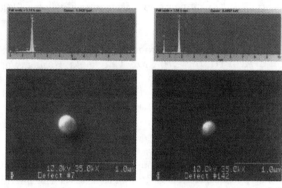

Fig.1 Results of SEM observation and EDS analysis on particle # 7 and # 142.

Experiment

WSi film was deposited on a 300 mm bare wafer by chemical vapor deposition using WF_6 and $SiCl_2H_2$ gas interaction. The thickness of the film was 500 nm. After the deposition, the wafer surface was analyzed by Tencor SP-1 optical particle counter system to detect the number and location of the particles/defects (we call "wafer map"). The wafer map was imported into SMART-Tool. SMART-Tool is a field emission Auger based defect review tool that is available to introduce the 300/200 mm wafer, navigate to the particle position based on wafer map and observe SEM and Auger compositional analysis on the particles. Using this tool, we selected 10 particles for Auger analysis comparing with optional functionality of EDS measurements, one particle for FIB cross-section. The measurement conditions of both Auger and EDS were the same, 10 keV of primary electron beam energy and 10nA of the beam currents.

Results

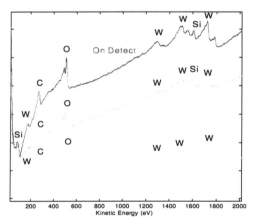

Fig.2 Auger spectra on and off particle # 7. Subtracted spectrum is also shown.

Figure 1 shows the SEM images and EDS analysis results on two similar particles (#142 and #7) founded on the wafer. As shown in the SEM image, the morphology of two particles look very similar, rounded shape, white brightness and around 0.3 μm particle diameter. This means that these two particles may be classified in the same category based on morphology observation. EDS analysis shows slight different, oxygen peak was identified in the spectrum on particle #142. However, EDX analysis has problem to identify correct composition of the particles because of the strong overlap of Si

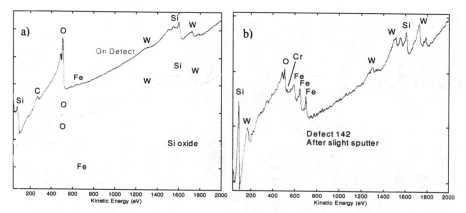

Fig.3 a) Auger spectra on and off particle # 7. Subtracted spectrum is also shown.

b)On particle Auger spectrum after removing surface oxide shell. Fe and Cr peaks are clearly identified.

and W spectrum lines. We compared the spectrum between on particle (#7) and off particle (film) for reference, there was no difference founded. This disables us to discuss detailed composition of the particle #7. In the case of particle #142, same problem is involved in identifying composition of the particle and EDX tells us only the difference of oxygen existence. On the other hand, Auger analysis revealed quiet different information between these two particles. Figure 2 shows the Auger spectra on and off # 7 particle and subtracted spectrum. Auger is a surface sensitive technique so that carbon and oxygen according to surface native oxide are defected in addition to the film material W and Si. Comparing two spectra on and off, it is seen that the peak heights ratio of Si KLL located at 1620 eV and W MNN located 1720 eV is different, on particle spectrum has smaller peak height than that of W peak. To reveal this evidence, once spectrum intensity was normalized by using Si peak height and subtracted spectrum is shown in Fig.2. The subtracted spectrum obviously shows that W is rich in particle #7 compared with film composition. Based on Auger results, we can conclude that the particle #7 was generated as W rich particle during CVD process. Again, in the

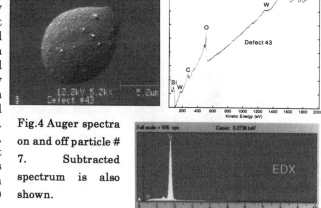

Fig.4 Auger spectra on and off particle # 7. Subtracted spectrum is also shown.

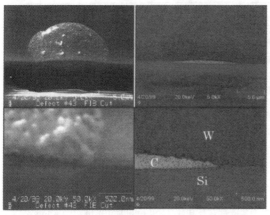

Fig.5 SEM images and overlaid Auger elemental maps of FIB cross-section of the particle #43. Upper images show in low magnification and lower images show high magnification

case of EDX analysis, quantitative compositional analysis was not available because of strong overlap of W and Si line in addition to the possibility of detecting Si substrate information. Particle #142 has been also analyzed by Auger and EDS. Figure 3 shows the results of applying similar data handing shown in Fig.2 (a). In this case, subtracted spectrum shows that the particle surface is mainly covered with Si oxide. This is confirmed by EDS analysis shown in Fig.1, however, Auger analysis extracted quite important information from this particle that is detecting small Iron peak. Fig.2 (b) shows the Auger spectrum after applying slight sputter etching on the particle surface. In this spectrum, Iron and Chromium peaks are clearly observed. So we can characterize the detailed particle root cause from this evidence that the particle #147 has stainless steel based core covered with thin SiO2 shell, high possibility of originating chamber hardware (for example, Gas line). On the other hand, the existence of Fe and Cr were no confirmed by EDS measurement shown in Fig.1. The reasons are, 1) the analysis

Table.1 Summary of particle analysis

ID	Morphology	Auger	EDX
#14, 77, 73, 86, 94		W, Si, O,C Same as film composition	W,Si Strong overlap of W and Si line
# 1		C, O, F, Al, Si	C, O, F, Al, Si
# 91		C, O, Ca, Al, Si	O, Ca, Al, Si Low cross-section of C X-ray
# 7		C, O, W, Si Large amount of W, unreacted W core	W,Si Strong overlap of W and Si line
# 142		C, O, Si, Fe Sputtering revealed SUS particle core	O, Si, W? No signal of Fe, Cr, Ni
# 43		W, Si, O,C Same as film composition FIB cuts revealed C layer under the particle	W,Si Overlap of W and Si line C line was not identified

560

volume of EDS using 10 keV is so large compared with the particle volume so that the total signal attenuated the bulk information of the particle, 2) cross-section of the Fe line was not enough using 10 keV electron beam energy, however, no idea of existing Fe and Cr inside the particle at that moment. Figure 4 shows the SEM image and the results of Auger and EDS analysis on particle #43. As shown in SEM image, this particle has about 10 µm diameter, we expected the combining surface sensitive technique and bulk analysis technique extracts some information from this particle. However, Auger and EDS revealed nothing with the exception of surface native oxide and film composition. Therefore, we decided to attempt the FIB process and applying cross-sectional Auger analysis. Figure 5 shows the SEM image of the particle cross-section and the results of Auger mapping of the cross-section. The Auger mapping clearly shows the excellent evidence that is the existence of Carbon layer under the defect. The question is why the EDS did not detect the carbon signal under the particle? We calculated the wavelength of the carbon characteristic X-ray and resulted the number of 42.9 Angstroms. This is easily absorbed by materials, especially in this case thick layer including heavy metal W terminates the escape of the X-ray through the WSi film. Table 1 summarizes the all results.

Conclusion

This paper reported the powerful use of Auger based defect review tool as compositional metrology for semiconductor manufacturing. Considering the semiconductor device fabrication, most of products include light elements such as photo resists, oxide, nitride, reacted products including Fluorine and Si substrate. Although Auger is known as a surface sensitive technique, it is also suitable technique to detect all elements over Lithium from very small volume. Current and in the future roadmap of the semiconductor design rules, it is getting more and more important to know the origin of the killer particles. For this requirement, we offer to use the Auger based defect review tool and this paper demonstrates the tool is able to extract plus alpha information for yield enhancement.

Acknowledgement

Authors would like to thank Mr.Yamazaki and Mr.Tsutsumi at Selete for sample preparation and optical particle measurement.

Silicides

Thermally Stable SALICIDE Technology Using Epitaxial CoSi$_2$ With Low Junction Leakage and Lower Ideal Factor

Tatsuo Sugiyama*, Ryuji Etoh*, Masato Kanazawa**, Kikuko Tsutsumi** and Shinichi Ogawa*
ULSI Process Tech. Dev. Ctr*, System LSI Div.**Matsushita Electronics Corp.
19, Nishikujyo-Kasugacho, Minami-ku, Kyoto, JAPAN

Abstract

A thermally stable epitaxial Co SALICIDE process has been developed using a TiN /Co /Ti triple-layer. A thin CoSi$_2$ film ($\sim$15nm) formed by this process is thermally stable (Rs$\sim$12Ω/ $\square$@0.2μ m line) even after 900℃, 30min annealing and a low leakage ($\sim$5$\times$10^{-9} A/ cm^2) p-n junction with lower ideal factor has been achieved (Xj$\sim$0.15μ m).

The epitaxial large grain CoSi$_2$ film with smooth morphology at CoSi$_2$/ Si-sub. interface brings about high thermal stability with low junction leakage. In the epitaxial CoSi$_2$ film, Co and Si atoms diffuse predominantly at the CoSi$_2$ film's surface and CoSi$_2$/Si-sub interface because of lower grain boundary density than polycrystalline CoSi$_2$ films, so the interface becomes smoother and thermal agglomeration is suppressed.

Introduction

A cobalt SALICIDE process has an advantage in sub-quarter micron CMOS technology because of no line size dependence of sheet resistance, different from TiSi$_2$. [1] However, CoSi$_2$ films thinner than TiSi$_2$ films have been employed to suppress junction leakage. [1] In application to the embedded DRAM process, the thin CoSi$_2$ film on diffusion layers receive high thermal budget (max 850℃) for memory cell fabrication process. It has been reported that an epitaxial CoSi$_2$ film has high thermal stability [2] , but reported epitaxial CoSi$_2$ films formation methods often lead to high junction leakage by (111) facet formation at CoSi$_2$/(100) Si interfaces and a void formation at an edge of a sidewall spacer and an isolation layer. [3]

In this paper, we propose a new cobalt SALICIDE process with high thermal stability and good junction characteristics by an epitaxial CoSi$_2$ film formation using a TiN/Co/Ti triple layer.

Experimental

Fig. 1 shows a process flow. After MOSFETS formation, Ti, Co and TiN films are deposited sequentially by DC magnetron sputtering method. A CoSi$_2$ film is formed by 2step RTA, and then furnace annealing is performed at temperature range from 650℃ to 900℃ for 30min after CVD-SiO$_2$ deposition. A CoSi2 film is also formed on MOSFETS using a TiN/Co bilayer as a reference.

To examine effects of TiN capping layer on the CoSi$_2$ film formation, samples using a Co/Ti bilayer are prepared.

- 0.2 μ m rule MOSFETS formation
- Ti, Co(8~10nm), TiN sequential sputter deposition
- 1st RTA
- selective wet etching
- 2nd RTA
- CVD-SiO$_2$ deposition
- Furnace annealing   (650~900°C, 30min)

Fig. 1 A process flow of cobalt SALICIDE

Results and Discussion

In the Co/Ti process, it is well known that an underlying Ti layer retards Co diffusion into Si and this controls a reaction between Co and Si. This brings about an epitaxial CoSi$_2$ film formation. [3,4]

In the same manner, the TiN/Co/Ti structure changed into an TiN/epitaxial CoSi$_2$ structure after anneal treatments.

Fig. 2 shows sheet resistance of the epitaxial CoSi$_2$ film on 0.2 μ m width n-diffusion layer versus underlaying Ti films thickness.   Sheet resistance increased as Ti thickness increased. This indicates that Ti films suppressed silicidation, and a slow reaction resulted in the epitaxial CoSi$_2$ formation to minimize energies in the system.

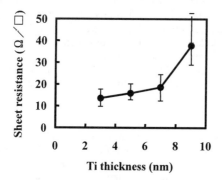

Fig. 2   Sheet resistance of the epitaxial CoSi$_2$ film
on 0.2 μ m width n+ diffusion layer versus
underlying Ti films thickness

Fig. 3 shows a plane TEM image and an electron diffraction (E.D.) pattern of the CoSi$_2$ film on an n+ diffusion layer. The E.D pattern indicates a good quality of epitaxial CoSi$_2$ film is formed and it is estimated that the grain size of the CoSi$_2$ film is larger than a few μ m on n+ and p+ diffusion layers. (grain boundaries are not seen in Fig. 3)

Fig. 4 shows a cross sectional TEM image of a 0.2 μ m NMOSFET with the epitaxial CoSi$_2$ film. The CoSi$_2$/Si-sub. interface is smooth.

0.2 μ m

CoSi$_2$

W plug

poly Si

epitaxial CoSi$_2$

Si-sub.

0.1 μ m

Fig. 3   Plane TEM image of the CoSi$_2$ film from TiN/Co/Ti structure on n+ diffusion layer and electron diffraction pattern

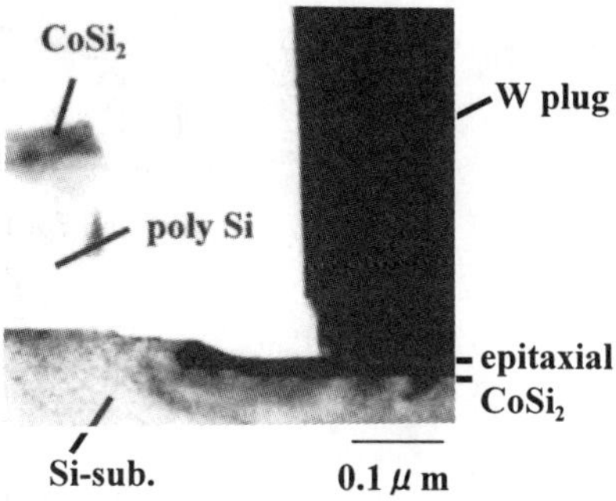

Fig. 4   Cross sectional TEM image of 0.2 μ m NMOSFET

Fig. 5 shows sheet resistance of the epitaxial and polycrystalline CoSi$_2$ film on 0.2 μ m width n+ diffusion layer versus furnace annealing temperatures. Sheet resistance in the epitaxial CoSi$_2$ film did not increased up to 900℃, while sheet resistance of reference (polycrystalline) samples increased abruptly at higher temperatures than 750℃. In reference samples, it is confirmed that a CoSi$_2$ film is polycrystalline by TEM observation. The difference in thermal stability is not caused by difference of those CoSi$_2$ film's thickness because thickness of the epitaxial CoSi$_2$ film (~15nm) is thinner than reference polycrystalline CoSi$_2$ film (~25nm). Gibb's free energy at the epitaxial CoSi$_2$/(100) Si interface is lower than the polycrystalline CoSi$_2$/(100) Si interface. The epitaxial film has large grains (>1 μ m) and fewer grain boundaries than the polycrystalline film. So in the epitaxial CoSi$_2$/Si-sub structure case, Co and Si atoms predominantly diffuse at the epitaxial CoSi$_2$ film's surface and the epitaxial CoSi$_2$/(100) Si interface faster than at the grain boundaries during high temperature anneal. This suppresses thermal agglomeration cause by a transformation at the grain boundaries. Thermal stability of the epitaxial CoSi$_2$ film is higher than those already reported. [4 - 7]

Figure 6(a) and (b) show cross sectional TEM images of the epitaxial CoSi$_2$ films at STI edges for a TiN/Co/Ti triple layer sample and for the Co/Ti bilayer sample after 1st RTA. In Fig. 6(a), no void is seen at the STI edge, while a void is seen below the CoSi$_2$ film at the STI edge and a CoSi film is over-grown onto the STI area in Fig. 6(b), which is similar as is reported in Ref.[3]. We believe that the TiN capping layer completely suppresses excess Co surface diffusion at the area on the STI region and it brings about the uniform formation of the epitaxial CoSi$_2$ film without voids in the TiN/Co/Ti process. This leads to reduction of junction leakage as discussed in below.

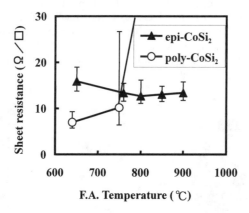

Fig. 5   Sheet resistance of CoSi$_2$ films
on 0.2 μ m width n+ diffusion layer
after furnace annealing for 30min

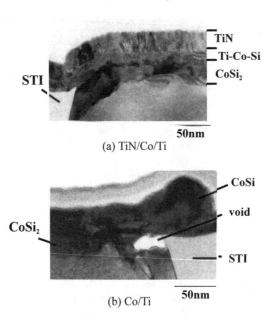

(a) TiN/Co/Ti

(b) Co/Ti

Fig. 6   Cross secitonal TEM images of reactions
at STI edges for (a) TiN/Co/Ti
and (b) Co/Ti structure

Fig. 7 shows Weibull plots of junction leakage for n+/p and p+/n junctions. The junction leakage current for samples with the epitaxial CoSi$_2$ film is lower than samples with the polycrystalline CoSi$_2$ film for n+/p junction, and low junction leakage current has been achieved($\sim 5 \times 10^{-9}$ A/ cm^2) . To examine details of p-n junctions for samples with the epitaxial CoSi$_2$ film and the polycrystalline film, I-V characteristics at forward bias are evaluated. Table. 1 summarized ideal factors obtained for the samples with the epitaxial CoSi$_2$ film, with the polycrystalline CoSi$_2$ film and without SALICIDE process. In the samples with the epitaxial CoSi$_2$ film , ideal factors are similar to the samples without SALICIDE process, and the values are close to 1.00. The results indicate there exists no electrical micro defects as deep level generated by Co diffusion and macro defects such as CoSix spikes.

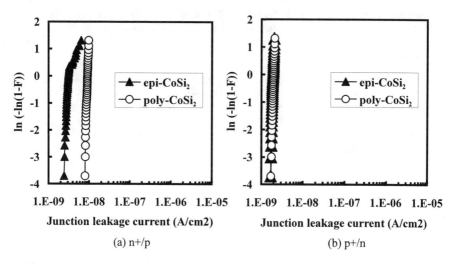

(a) n+/p
(b) p+/n

Fig. 7   Weibull plots of (a) n+/p and (b) p+/n junction leakage area component

Table. 1   Ideal factors of the samples for p-n junctions

	n+/p	p+/n
w/o SALICIDE	1.02	1.00
poly-CoSi$_2$	1.06	1.01
epi-CoSi$_2$	1.02	1.01

Summary

The epitaxial CoSi$_2$ thin film with very smooth CoSi$_2$/Si interface is formed using 2step RTA from the TiN/Co/Ti triple layer.   This SALICIDE process realizes successfully high thermal stability up to 900°C with excellent p-n junction characteristics.   The underlying Ti layer works as a Co diffusion control layer for the epitaxial CoSi$_2$ formation, and the TiN capping layer is for suppression of the void formation.

This process is a promising technology for a deep quarter micron embedded DRAM fabrication process.

Acknowledgements

The authors would like to thank Y. Kato and T. Kouzaki for their fruitful discussions. T. Yamada is also acknowledged for their support in offering the sample preparation.

Reference

[1] K. Goto et al., IEDM Technical digest , p.449, 1995.
[2] S.L. Hsia et al., J. Appl. Phys. , vol. 70, p.7579, 1991.
[3] J.S. Byun et al., J. Electrochem. Soc., Vol. 143. , L56, 1996.
[4] S. Ogawa et al., S.S.D.M. p.195, 1993
[5] T. Ohguro et al., Symp. of VLSI Tech. Dig., p. 101, 1997.
[6] K. Fujii et al., IEDM Technical digest , p.451, 1996.
[7] L.Chon et al., IEDM Technical digest , p.1005, 1998.

Effects of Nitrogen & Ge Ion Implant on Cobalt Silicide Formation

M.Y.Wang, S.M.Jeng, S.L.Shue, C.H.Yu, M.S.Liang
Research and Development, Taiwan Semiconductor Manufacturing Company, Ltd.
No.9, Creation Road I, Science Based Industrial Park, Hsin-Chu, Taiwan, R.O.C.

Abstract

The substrate effect on cobalt silicide transformation behavior has been investigated in this study. The transformation temperature of cobalt silicide remains the same on the different doped substrates. However, the transformation region from Co_2Si to $CoSi$ becomes much narrower for cobalt silicide formation on substrates with nitrogen or Ge implants. This may be due to that facilitates the $CoSi$ to $CoSi_2$ transformation. $CoSi_2$ with nitrogen or Ge pre-amorphization can minimize $CoSi$ spiking to reduce the junction leakage. Bridging between gate and source/ drain can also be improved by one order of magnitude. There is no degradation observed on Rs performance with Ge or nitrogen implant compared with the standard process.

Introduction

Silicides are widely used in silicon integrated circuits as contacts and interconnections. As the design rule shrinks to sub-half micron and beyond, a self-aligned silicide (salicide) technology is demanded for lowering the resistance of poly gates and sources/ drains[1]. A salicide with high stability, low sheet resistance, low stress and less leakage can meet the requirements of device performance. Co salicide is one of the promising candidates that can lower the serial resistance of devices and result in higher switching speeds for the devices. However, the higher junction leakage current is the main concern of cobalt salicide utilization. The nitrogen[2] and Ge implants are then evaluated to improve the phenomenon. The transformation behavior is also investigated to optimize the process condition for Ge/ nitrogen implant and explore the reason for the improvement.

Experimental

This silicide formation study is divided into two parts: transformation temperature evaluation and electrical performance. The transformation curve is characterized for Co silicide on different substrate (N+/P+ implanted (100) p-Si and Poly-Si), nitrogen ion implant, and Ge implant cases. Rs, junction leakage, and bridging of Co silicide are then examined with the optimized process flow from the transformation study.

Blanket wafers were used for Co silicide transformation study. The substrates with poly and (100) p-Si are both implanted with As and BF_2 to a dose of $1E^{17}$ atom/cm^3 to simulate N+ and P+ doped conditions. The nitrogen and Ge implants are processed only on bare Si wafers cleaned with dilute HF solution to remove the native oxide before Co deposition. Co is deposited with a high vacuum PVD system. The silicidation is carried out in N_2 ambient for 30 sec with different temperatures and Rs of the resulting Co silicide film is measured.

Patterned wafers are processed to evaluate the junction leakage, silicide bridging and Rs performance for a standard Co silicide process and a process with extra nitrogen and Ge implants. Co silicidation on device wafers is processed after source/ drain implant. The dilute HF dip is also performed to clean the surface before Co deposition. A two step anneal is used for Co silicide formation in this study[3]. Selective wet etching is used between the two RTA steps to remove unreacted Co.

Results

The transformation curves (Fi   te that the transformation temperature is nearly identical on N+ doped Si (N+   P+ doped Si (P+ OD in breif), N+Poly, and P+Poly substrates. Co$_2$Si can be fo   °C [4]. As the temperature is increased to around 500°C, CoSi phase is gradually formed from Co$_2$Si. As temperatures exceed 575°C, a single CoSi$_2$ phase is formed. There is a two phase plateau between 500°C and 575°C as shown in Figure 1.

The plateau of the transformation curve disappears in both nitrogen and Ge implantation (Figure 2). It means the CoSi phase exists only at the transition temperature and then transform to CoSi$_2$ phase once the temperature is increased. It is suspected that the substrate surface is amorphized by the nitrogen or Ge implantation, which reduces the transformation energy from CoSi to CoSi$_2$.

Both nitrogen and Ge implantation before Co deposition can reduce the junction leakage significantly (Figure 3) because the CoSi phase is more easily transformed to the CoSi$_2$ phase. Therefore, less CoSi formation believed to be the cause of spiking after 1st RTA, can improve the junction leakage. It can also be confirmed by examining the interface roughness in TEM micrographs (Figure 4). The smoother silicide/ Si interface observed for the nitrogen or Ge implant cases, compared with the standard process, can contribute to better junction leakage performance. The silicide bridging from gate to source/ drain can also be significantly improved by Ge or nitrogen implantation (Figure 5). With Si as a moving species during Co silicidation it is easier to induce bridging, compared with Co as a diffuser[5]. Co is the moving species at lower temperatures for Co silicide formation with Ge or nitrogen implant. Bridging current between gate to source /drain is one order of magnitude better (without normalization) than that of CoSi$_2$ without nitrogen or Ge pre-amorphization. Rs of Co silicide on 0.18 um N+ poly line is around 7 ohm/sq as shown in Figure 6. Co silicide Rs on narrow poly lines will not be degraded by nitrogen or Ge implant. The similar C-V curve for NMOS structures showed that gate oxide will also not be impacted by the extra implantations (Figure 7).

Conclusion

There is no difference in Co silicide transformation behavior on different doped substrates. However, sharp transition curves are observed for substrates with Ge or nitrogen implantation. Co silicide with less junction leakage, less bridging and good Rs can be achieved by nitrogen or Ge ion implant before silicidation without impacting other device characteristics.

Reference

[1] Karen Maex, Rob Schreutelkamp, Mat. Res. Soc. Symp. Pro., 260, 133 (1992).
[2] Ho-KT, Nicolet-MA, Thin Solid Films, 127, 313 (1985).
[3] Ken Inoue, Kaoru Mikagi, Hitoshi Abiko, Shinichi Chikaki, Takamaro Kikkawa, IEEE, 45, 2312 (1998).
[4] C.S.Wei, Gopal Raghaven, M. Lawrence, A. Dass, Mike Frost, Teodoro Brat, David B. Fraser, *VMIC Conference*, Santa Clara, June 1989, 241-250.

[5] C. M. Comrie, R.T. Newman, J.Appl. Phys., 79, 153 (1996).

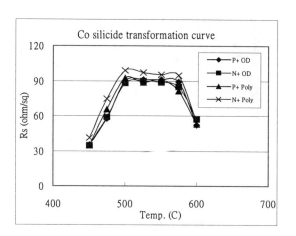

Fig.1 Transformation curves of cobalt silicide on different substrates (N+OD, P+OD, N+Poly, and P+Poly)

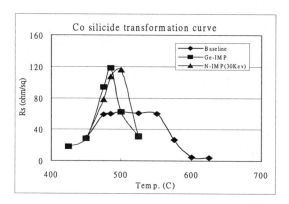

Fig.2 Transformation curves of cobalt silicide on (100) p-Si without and with nitrogen or Ge ion implant.

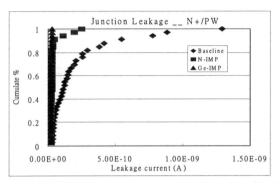

Fig.3 Cumulative plot of N+/PW junction leakage current for with and without nitrogen & Ge implant (Leakage current is not normalized)

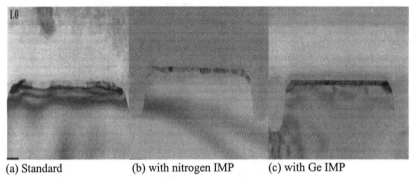

(a) Standard   (b) with nitrogen IMP   (c) with Ge IMP

Fig.4 TEM pictures of Co silicide formation on 1.0μm P+OD

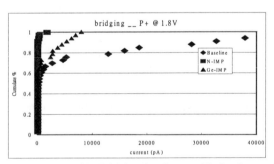

Fig.5 Comparison of salicide Gate to S/D bridging for nitrogen /Ge implant and with no implant case.

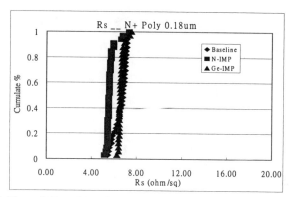

Fig.6 Rs of Co silicide on 0.18μm N+poly line for standard process and with nitrogen, Ge implant.

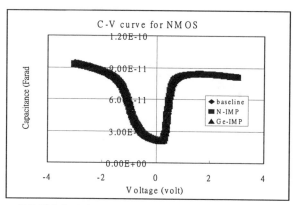

Fig.7 Comparison of NMOS C-V characteristic for standard process and with nitrogen, Ge implant

Thermal Desorption Issues Related to Silicidation and Back-End Metallization

Hua Li, G. Vereecke, M. Schaekers, M. R. Baklanov, E. Sleeckx, K. Maex, and L. Froyen[#]

IMEC, Kapeldreef 75, 3001 Leuven, Belgium
[#] Dept. of Metallurgy and Materials Engineering, K. U. Leuven, 3001 Leuven, Belgium

ABSTRACT

We have used a recent RTP-APIMS system to evaluate thermal desorption behavior of a number of films encountered in Co salicide and back-end interconnection processes. This setup provided an *in situ* measurement at atmospheric pressure, and has high sensitivity with a wide dynamic range. Results on desorption related to Co silicidation show that oxide films can generate significant amounts of H_2O that can have a direct impact on the adjacent silicide formation. Capping the Co with a Ti layer provides an excellent solution. Desorption from a number of oxides is also investigated. The results provide very useful information for the evaluation of these films. It is interesting to note that NH_3 desorption is only associated with oxide deposited using SiH_4 with N_2O.

INTRODUCTION

Extremely low levels of impurities are required in semiconductor manufacturing. For example, impurities adsorbed on the surfaces of wafers can affect the growth and the properties of the subsequent layers.[1] Moreover, desorption can also cause serious reliability problems, such as the so-called "poisoned' via due to desorption of H_2O from dielectrics,[2,3] and the edge thinning effect in Co silicidation.[4] In fact, almost all of the films produce a certain amount of outgassing during fabrication processes, which may degrade device reliability.[2,5] Therefore, studying the thermal desorption behavior of films encountered in the fabrication of integrated circuits (IC) is of great importance.

Atmospheric pressure ionization mass spectrometry (APIMS) is the most sensitive technique for analysis of trace (ppt) gaseous impurities at atmospheric pressure.[6,7] Recently, a RTP-APIMS setup has been demonstrated as capable of quantitatively measuring ambient impurities in a RTP tool *in situ* and with a wide dynamic range (from 0.1 ppb to a few tens of ppm).[8,9] In this work, we use this setup to investigate the thermal desorption of a number of films encountered in Co salicide process, as well as to evaluate several Si oxides used in back-end processes.

EXPERIMENTAL

A single-chamber atmospheric pressure RTP tool (Steag-AST, SHS2800ε) was employed in this study. It is connected to an APIMS (Trace+ from VG Instruments) for gas analysis. Details of the RTP-APIMS system set-up (Fig. 1) and methodology can be found elsewhere.[8,9] Briefly, the chamber ambient gas was sampled downstream of the wafer. In this way the impurities contributed by the chamber and the wafer can all be taken into account.[10] The gas transfer time was about 0.5 s, a value small enough comparing to a gas mean residence time in the chamber of about 20 s in typical operation conditions.

After loading a wafer, the chamber was purged with 4 slm N_2 for typically 0.5 h. This is intended to eliminate the gaseous impurities introduced during wafer loading. Two RTP process

577

recipes were adapted. One is an isothermal run, where a wafer was heated up to 660°C within about 24 s, and held at that temperature for another 120 s before cooling. The second one is a ramping run, where a wafer was ramped to 1000°C in 10 min at a constant rate.

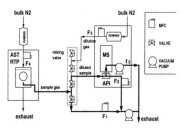

Fig. 1 RTP-APIMS system setup

150 mm p-type (100) Si wafers were used as substrates. For wafers related to Co silicidation, the thicknesses are 150 nm for both TEOS (densified at 850°C) and LPCVD Si_3N_4, 15 nm for Co and 8 nm for Ti. The deposition of a Ti cap on Co was carried out without breaking the vacuum. Details can be found in Ref.11. For the comparison of oxides, four kinds of oxides were prepared. They are SiO_2 deposited by high density plasma (HDP) at 700°C (film A) and 400°C (film B) using SiH_4 and O_2 (1 µm), and at 550°C using SiH_4, PH_3 and O_2 (1 µm, film C), and by PECVD at 400°C using SiH_4 and N_2O (350 nm or 1 µm (film D)).

RESULTS AND DISCUSSION

A. Reduction of the system background

Figure 2 shows the H_2O and O_2 levels in the RTP tool at 660°C, with the RTP door unheated and heated (80°C), respectively. As the wafers do not release any significant amounts of H_2O in these cases, the 3-fold difference in H_2O concentrations with and w/o heating clearly demonstrates the contribution of H_2O from the door, as reported by Kondoh et al.,[8] as well as a solution. In contrast, the O_2 level measured at the same time is low (0.2 – 0.4 ppm) and not influenced by the door temperature.

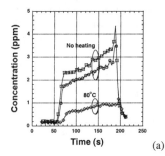

(a)

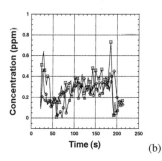

(b)

Fig. 2
Background
(a) H_2O and
(b) O_2 levels
with and
without
heating the
RTP door
during a 120 s
hold at 660°C.

B. Desorption related to Co silicidation

Gaseous impurities can have a detrimental impact on Co silicidation.[12] In this paper, we pay special attention to the desorption behavior of various films encountered in RTP silicidation as desorption has been shown to result in edge thinning effect for Co silicide.[4] Figure 3 (a) shows that indeed the Co film itself is active in adsorbing H_2O and CO_2. Thus a capping layer is needed. Figures 3 (b) and (c) show two cases where the underlayers have very different adsorption ability. The TEOS underlayer desorbs much larger amounts of H_2O (~ 9 ML) than the Si_3N_4 counterpart (~ 1 ML). In contrast, films with a Ti cap do not show a H_2O desorption peak. No O_2 desorption peaks are observed and in all the cases the O_2 level is the same within experimental error. A detailed discussion of the roles of gaseous impurities and capping layers on Co silicidation will be reported elsewhere.[11] From the results shown here one can see that thermal desorption from the films can in some circumstances be very significant.

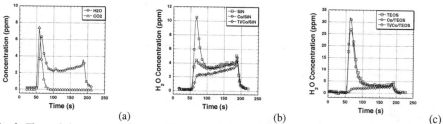

(a)   (b)   (c)

Fig. 3 Thermal desorption of H_2O at 660°C for (a) Co on Si, (b) film stacks having a Si_3N_4 (150 nm) underlayer, and (c) film stacks having a TEOS (150 nm) underlayer. There is no O_2 desorption peak in all these cases.

C. Desorption of various silicon oxides

Si oxides have been used extensively as insulators in the fabrication of multi-level interconnections. Desorption behavior of an oxide is one of the important properties that needs to be evaluated. Otherwise, process and reliability problems may arise as a result of the outgassing. In this section, desorption of a number of oxides is investigated.

1. PECVD oxides deposited using SiH_4 and N_2O

The use of N_2O instead of O_2 can reduce the growth rate of SiO_2, thus allowing the deposition in a more controllable manner. Figures 4 (a) and (b) show the amounts of H_2O and NH_3, respectively, released from this kind of oxide deposited at 400°C. NH_3 is determined from the m/e = 17 intensity after subtracting the contribution of H_2O. As species with m/e=17 other than NH_3 is unlikely in this case, the remaining part is believed to originate from NH_3.

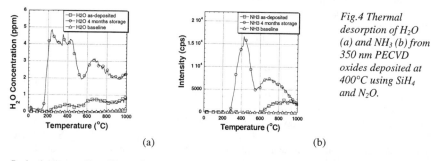

(a)   (b)

Fig.4 Thermal desorption of H_2O (a) and NH_3 (b) from 350 nm PECVD oxides deposited at 400°C using SiH_4 and N_2O.

In both cases, the effect of air storage is clearly visible. For H_2O desorption, two strong desorption regions are observed. The region below approximately 500°C is mainly due to the release of physisorbed H_2O and tightly H-bonded H_2O, while the one above 500°C is from decomposition of H-bonded silanols (Si-OH groups) and isolated silanols.[3] The former may play a greater role as the processes for multilevel interconnection are mainly operated in this temperature range.[13] Desorption of NH_3 also shows two regions. The one at higher temperature is observed in both fresh and stored wafers, but with different magnitudes, while the strong peak in the 300-550°C range is certainly the effect of air storage. It is noted that desorption of NH_3 starts at a higher temperature of about 300°C.

2. Comparison of various HDP and PECVD oxides

Figures 5 (a) and (b) show desorption of these 1-μm-thick oxides after storage in cleanroom air for 1 day and 13 days, respectively. Below 450°C, film D desorbs significant amounts of H_2O. Above 450°C, desorption from film C is very intensive. This is likely to be caused by the breaking of P-OH bonds to form H_2O.[14] HDP SiH_4 + O_2 oxides only desorb H_2O above approximately 600°C, with the film deposited at a higher temperature desorbing less H_2O. For film D, the released amount increases substantially upon prolonged air exposure. In contrast, the increases for other HDP oxides are insignificant within the test duration (13 days). This may indicate the presence of nanoporosity in the case of film D.

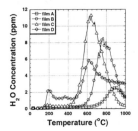

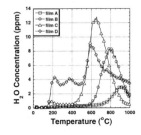

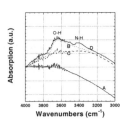

(a)   (b)

Fig. 5 Thermal desorption of H_2O from wafers with different oxide films after (a) 1 day and (b) 13 days storage in cleanroom air.

Fig. 6 FT-IR spectra from the samples in Fig.5

In all these cases, film D is the only one that desorbs NH_3 (see also Fig. 4). A question thus arises as to the source of the NH_3. Further tests (not shown) on this subject were carried out in the following way. A long air exposed 350 nm film of this type was ramped to 800°C. Both H_2O and NH_3 are desorbed from the wafer. This wafer was then stored in cleanroom air for another four months, and a desorption test was carried out by an isothermal run at 660°C for 120 s. Strong NH_3 (and H_2O as well) desorption was observed again. In all these cases the effect of air storage is clearly visible, which might suggest that the desorbed NH_3 is coming from the air. However, the evidence shown in Fig. 4 (b) indicates that desorption of NH_3 starts at rather high temperature (~300°C) and continues beyond 800°C. As NH_3 is a volatile species, adsorption of this species from air should be released at lower temperatures.

It is more logical to think that, due to the different oxidizing gas used (N_2O instead of O_2), the bonding structure of the SiO_2 formed may be somewhat different. Fig. 6 shows part of the FT-IR spectra from the films in Fig. 5. The absorption peak around 3650 cm^{-1} is from Si-OH,[15] while that around 3400 cm^{-1} may be assigned to Si-NH$_x$.[16,17] For film D (SiH_4 + N_2O), the absorption peak around 3400 cm^{-1} is clearly observed. Moreover, the strong Si-O stretch is located at 1065 cm^{-1} for this film while that for other HDP oxides are in the range of 1080-1090 cm^{-1}(not shown). The shift towards lower wavenumbers is an indication of partial replacement of Si-O bonds by Si-N ones.[17] Here one can see that indeed the structure of oxide formed by using SiH_4 + N_2O is somewhat different from the others. In their study of oxidation of oxynitride films in H_2O/O_2 ambient, Kuiper et al.[18] shows the following oxidation reaction scheme:

Si-N-Si + HOH ↔ Si-N-H + Si-OH   (1a);   Si-N-H + HOH ↔ Si-N-H$_2$ + Si-OH   (1b);
Si-N-H$_2$ + HOH ↔ N-H$_3$ + Si-OH   (1c);   Si-OH + Si-OH ↔ Si-O-Si + HOH   (1d).

Though the exact reactions can be very complicated, knowledge from this work indicates that the presence of both Si-NH$_x$ bonds and H_2O are the two preconditions for the formation of NH_3. This can explain the effect of storage (the H_2O adsorbed is the only H_2O source during desorption tests) and the high desorption temperature (rupture of Si-N bonds is an activated process) in our case. Further proof of this explanation can be found from the facts that even though TEOS oxide can adsorb large quantity of H_2O, no NH_3 is released during thermal desorption test due to lack of Si-NH_x bonds; Si_3N_4 does contain Si-NH_x bonds, but no significant amount of NH_3 is released during thermal desorption as this film adsorbs little H_2O.[11] In contrast, release of NH_3 has been reported during oxidation of Si_3N_4 powder in high pressure H_2O since 200°C.[19] Certainly more work is needed to further clarify the exact mechanism.

SUMMARY

(1) The RTP-APIMS setup is capable of doing *in situ* quantitative thermodesorption measurements for a variety of materials encountered in silicide and back-end-of-the-line interconnection processes.

(2) In all these cases, H_2O is the dominant desorption species. The area and temperature range of the desorption peaks can provide useful information for process optimization.

(3) Desorption of NH_3 is only observed in the oxides deposited using SiH_4 and N_2O. It is likely that the interactions of Si-NH_x groups in the film with H_2O cause the desorption of NH_3.

ACKNOWLEDGEMENTS

We thank N. Roelandts for technical assistance and E. Kondoh for part of the initial work. KM is a research director of the Fund for Scientific Research-Flanders (FWO).

REFERENCE

[1] S. R. Kasi, M. Liehr, P. A. Thiry, H. Dallaporta, and M. Offenberg, Appl. Phys. Lett., **59**, 108 (1991).

[2] M. Yoshimaru, T. Yoshie, M. Kageyama, K. Shimokawa, Y. Fukuda, H. Onoda, and M. Ino, IEEE International Reliability Physics Proceedings, (1995) p.359.

[3] J. Proost, E. Kondoh, G. Vereecke, M. Heyns, and K. Maex, J. Vac. Sci. Technol., **B16**, 2091 (1998).

[4] K. Maex, A. Lauwers, P. Besser, E. Kondoh, M. de Potter, and A. Steegen, IEEE Trans. Electron Devices, **46**, 1545 (1999).

[5] K. Shimokawa, T. Usami, S. Tokitou, N. Hirashita, M. Yoshimaru, and M. Ino, Symp. VLSI Technl. Digest of Technl. Papers, 1992, p.96.

[6] Y. Mitsui, T. Irie, and K. Mizokami, Ultra Clean Tech., **1**, 3(1990).

[7] K. Siefering, W. Whitlock, and H. Berger, J. Electrochem. Soc., **140**, 1165 (1993).

[8] E. Kondoh, G. Vereecke, M. M. Heyns, K. Maex, and T. Gutt, J. Vac. Sci. Technol., **A17**, 650 (1999).

[9] G. Vereecke, E. Kondoh, P. Richardson, K. Maex, and M. M. Heyns, submitted to IEEE Trans. Semicon. Manufacturing.

[10] A. Haider, J. J. F. McAndrew, and R. S. Inman, in *Crystalline defects and contamination: Their impact and control in device manufacturing II*, ed. by B. O. Kolbesen, C. Claeys, P. Stallhofer, and F. Tardiff, (ECS Proc. **97-22**, Pennington, NJ, 1997), p.484.

[11] H. Li, G. Vereecke, K. Maex, and L. Froyen, to be submitted to J. Electrochem. Soc.

[12] C. -D. Lien and M. -A. Nicolet, J. Vac. Sci. Technol., **B2**, 738 (1984).

[13] M. R. Baklanov, M. Muroyama, M. Judelewicz, E. Kondoh, H. Li, J. Waeterloos, S. Vanhaelemeersch, and K. Maex, J. Vac. Sci. Technol., **B17 (5)**, (1999) in press.

[14] R. M. Levin, J. Electrochem. Soc., **129**, 1765 (1982).

[15] W. A. Pliskin, J. Vac. Sci. Technol., **14**, 1064 (1977).

[16] W. A. P. Claassen, H. A. J. Th. v. d. Pol, A. H. Goemans, and A. E. T. Kuiper, J. Electrochem. Soc., **133**, 1458 (1986).

[17] R. Koba and R. E. Tressler, J. Electrochem. Soc., **135**, 144 (1988).

[18] A. E. T. Kuiper, M. F. C. Willemsen, J. M. L. Mulder, J. B. Oude Elferink, F. H. P. M. Habraken, and W. F. van der Weg, J. Vac. Sci. Technol., **B7**, 455 (1989).

[19] C. Contet, J.-I. Kase, T. Noma, M. Yoshimura, and S. Somiya, J. Mater. Sci. Lett., **6**, 963 (1987).

Retardation of CoSi2 Formation on Sub 0.18μm Gate Poly due to Oxide Resputtering by *in-situ* RF Sputter Etch Precleaning

Ju-Hyuk Chung, Jang-Eun Lee, *Ja-Hum Ku, Eung-Joon Lee, Jong-Wang Park, Young-Hyun Lee, *Chul-Sung Kim, Sun-Hu Park, U-In Chung, Geung-Won Kang and Moon-Yong Lee
U FAB Process Tech. Team, *Process Development 3 Group,
Process Development Team, Semiconductor R&D, Samsung Electronics
San #24 Nongseo-Ri, Kiheung-Eup, Yongin-City, Kyungki-Do 449-900, Korea
Phone: +82-331-209-4564   Fax: +82-331-209-6120   e-mail:pjking@samsung.co.kr

ABSTRACT
The sheet resistance of Co silicide on sub 0.18μm gates rapidly increases due to poor silicide formation for the gate fabricated on field oxide region. In this study, SEM, TEM, and SIMS analyses were employed to investigate the silicide retardation for the gate on field oxide. It was found that silicide retardation is related to the resputtering of oxide from the field oxide region by *in-situ* rf sputter etch. As a result, with the optimized rf sputter etch and capping processes, oxide resputtering and the sensitivity of the Co salicide process to surface conditions was minimized, giving a larger process window to fabricate deep sub-quarter micron devices.

INTRODUCTION
Co silicide has an advantage over Ti silicide with regard to the relatively linewidth-independent sheet resistance of the gate electrode.[1,2] For the Co salicide process, however, it is very important to maintain a clean Si surface before Co deposition. To minimize the sensitivity of the Co salicide process to the Si surface condition, *in-situ* rf sputter etching is often employed just prior to Co deposition.

In this study, the adverse effect of *in-situ* rf sputter etch on Co silicidation was investigated. From SEM data, it is identified that only 150Å of silicide was formed on poly-Si prepared on the field oxide region, while the thickness of Co silicide on poly-Si fabricated on the active area was measured to be 600Å, indicating full reaction of the deposited Co layer. The sheet resistance of Co silicide on poly-Si formed on a Si substrate was measured to be ≤ 5Ω /□. However, 150Ω /□ of Co silicide sheet resistance was measured for the sample prepared on field oxide. In fact, the poor silicidation on poly-Si fabricated on field oxide causes serious device failures for high speed devices due to increased RC delay time during the operation. TEM analysis of the sample processed by Co/TiN deposition shows a thin layer of oxide at Co/poly-Si interface from the sample prepared on field oxide.

It can conclusively be shown that the oxide is resputtered on poly-Si from the field oxide region during the *in-situ* rf sputter etch, which results in retardation of silicide formation during the annealing. Therefore, by optimizing the *in-situ* rf sputter etch parameters and capping layer, the dependence of Co silicidation on substrate condition (active or field oxide) is eliminated.

EXPERIMENT
Figure 1 summarizes the process flow for Co salicide. After source and drain formation, HF dip and *in-situ* rf sputter etch were performed to remove native oxide and contaminants on the active region. Following Co and TiN capping layer deposition, the samples were RTA annealed at low temperature in a N₂ ambient. After the first RTA, unreacted Co and TiN capping layers were removed

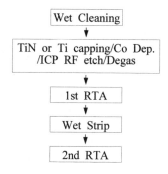

Fig.1 Co silicide module process.

by $H_2SO_4:H_2O_2$ solution etching. To achieve low resistance Co silicide, a second high temperature annealing was conducted. SEM and cross-sectional TEM analyses were performed to explore Co silicide morphology after the Co salicide process. SIMS analysis was also employed to study the behavior of Ti and Co during silicidation. Furthermore, the effect of a new rf etch condition and a capping layer on silicidation was evaluated.

RESULTS
Submaterial dependence on silicidation
Experimental results show that the submaterial (field oxide or Si active) on which the gate poly Si is formed strongly affects the Co silicidation.

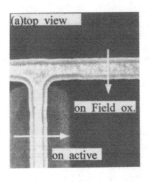 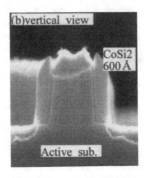 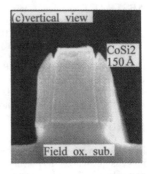

Fig.2 SEM micrograph of Co Silicide on poly-Si patterned on active or field oxide.

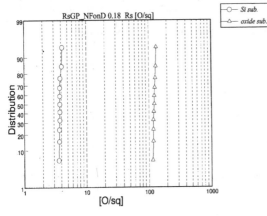

Fig.3 Rs of Co Silicide on 0.18μm Gate poly-Si fabricated on active or field oxide.

Shown in Figure 2 are the Co silicide morphologies on gate poly Si on active or field oxide regions. Co silicide was successfully formed on the gate poly Si formed on the active Si region, while poor silicidation was observed for the gate poly Si on field oxide. The measured sheet resistance in Figure 3 also shows that the sheet resistance of Co silicide for the gate poly Si on field oxide increases drastically compared to that for the gate poly Si on the active region.

TEM analysis for different submaterial
To investigate the silicide retardation, TEM analysis was performed for the gate poly Si prepared on the active region or field oxide region before the first RTA process (wet cleaning + in-situ rf sputter etch + Co deposition + TiN deposition). As shown

in Figure 4, on the oxide layer was identified at the Co/poly Si interface only for the gate poly Si prepared on field oxide. This retards the silicide reaction during annealing.

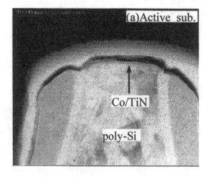 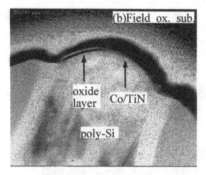

Fig.4 TEM micrograph of TiN/Co/rf etch/Gate poly-Si samples..

Co silicide on isolated active region
Figure 5 showes the SEM picture of Co silicide formed on the active region isolated by field oxide. Co silicide retardation is also observed from the edge of active region, while the center of the active region exhibits successful silicide formation. The poor silicidation on the active region can induce metal-to-contact resistance failure.

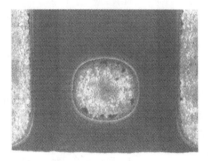

Modeling of silicide retardation
The SEM and TEM analyses suggest that silicide retardation on gate poly Si or Si active regions is closely related to the field oxide. Figure 6 shows a possible mechanism for the

Fig.5 Surface morphology of Co silicide on active Si.

silicide retardation. The oxide is resputtered on poly-Si from the field oxide region during the *in-situ* rf sputter etch, which results in retardation of silicide formation during the annealing.

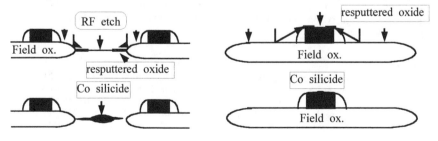

Fig.6 Modeling of RF sputter etch effect on Co silicidation.

Optimization of rf sputter etch condition

During the rf sputter etch, oxide resputtering occurs due to Ar^+ ion bombardment. Plasma power and bias power, which are related to plasma density and self bias, respectively, were optimized to minimize the oxide resputtering during the rf sputter etch. Figure 7 shows the Co silicide morphology when a new rf etch condition was used prior to Co deposition. As can be seen in Figure 7, poor silicidation was observed at the edge of the gate poly Si although the Co silicide morphology was improved by the optimized rf sputter etch condition.

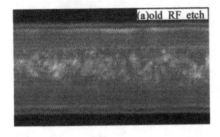 

Fig.7 Co silicidation on Gate poly-Si pattern according to RF etch condition.

New capping process

A Ti capping layer was used for the Co salicide process for successful silicidation on gate poly Si and active S regionsi. In order to evaluate the Ti capping process, the sample was prepared with the optimized rf sputter etch condition. As shown in Figure 8, silicide was successfully formed on gate poly Si although a resputtered oxide layer exists at the Co and poly Si interfaces due to rf sputter etch. In Figure 9, the sheet resistance and junction leakage characteristics of Co silicide formed using the Ti capping process show excellent values and tight distributions.

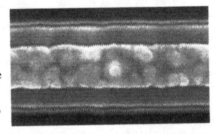

Fig.8 Ti capping effect on Co silicidation on Gate poly Si pattern.
(new RF etch /Ti capping)

To study the behavior of Co and Ti during the silicidation, SIMS analysis was performed after the first low temperature annealing and selective wet etch. SIMS data in Figure 10 show that, compared to a TiN capping process, a large amount of Ti diffuses into the Co/Si interface for the Ti capping process. This suggests that Ti can reduce the thin silicon oxide at Co/poly Si interface, leading to successful silicide formation.[3]

CONCLUSIONS

The SEM and TEM analyses suggest that oxide is resputtered on poly-Si from the field oxide region during the *in-situ* rf sputter etch, which results in retardation of silicide formation during annealing. Therefore, by optimizing the *in-situ* rf sputter etch parameters and capping layer, the dependence of Co silicidation on submaterial (active or field) is eliminated, which gives a larger process window to fabricate deep sub-quarter micron devices.

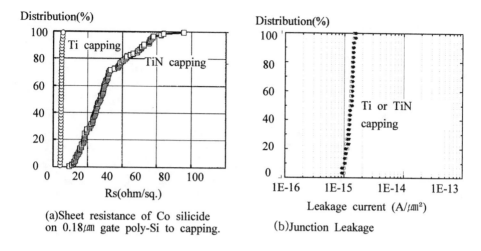

(a)Sheet resistance of Co silicide on 0.18μm gate poly-Si to capping.

(b)Junction Leakage

Fig.9 Electrical characteristics of Co silicide according to capping layer.

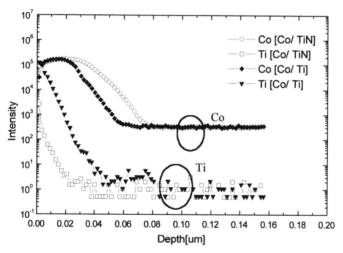

Fig. 10 SIMS depth profile after selective wet strip (Ti or TiN Capping/Co/Si).

REFERENCES

[1] T. Yoshitomi *et al.*, Symp. on VLSI Tech.'96, p34
[2] K. Inoue *et al.*, Tech. Dig. of IEDM.'95, p445
[3] Ja-Hum Ku *et al.*, IITC'99, p256

ICP-Ar/H2 Precleaning and Plasma Damage-Free Ti-PECVD
For Sub-Quarter Micron Contact of Logic with Embedded DRAM

T. Taguwa, K. Urabe, T. Yamamoto, and H. Gomi
ULSI Device Development Laboratory,
NEC Corporation
1120 Shimokuzawa, Sagamihara, Kanagawa 229-1198, Japan
Phone:+81-42-771-0669, Fax:+81-42-771-0938, E-Mail:tagu@lsi.nec.co.jp

ABSTRUCT

Titanium plasma enhanced chemical vapor deposition (Ti-PECVD) was implemented on a $CoSi_2$ layer which is located at the bottom of high aspect ratio contact holes to realize high performance Logic ULSIs for embedded DRAM devices. We achieved extremely low contact resistances by integrating ICP-Ar/H2 precleaning and Ti-PECVD processes. Contact resistances of 11.6 Ω and 12.1 Ω were obtained for φ0.25μm with aspect ratios of 10 on n+ and p+ diffusion layers with $CoSi_2$, respectively. Furthermore, we investigated the plasma damage of Ti-PECVD and successfully eliminated charge-induced damage using a new Ti-PECVD method.

INTRODUCTION

The drive to incorporate more function onto a single chip has created demand for embedded DRAM technologies which offer logic technology and dense memory simultaneously. There are several technical paths for creating an embedded DRAM technology [1,2]. The cobalt salicide process has become a common technology for high performance logic devices [3, 4]. Consequently, logic with embedded DRAM requires high aspect ratio contacts on $CoSi_2$ layer because of the thick interlayer dielectric films in conjunction with the capacitor structure. It is necessary to develop Ti-PECVD technology for high aspect ratio contacts on $CoSi_2$ layers. However, we found that a residual layer was formed on $CoSi_2$ at the bottom of the contact hole after the contact opening process. It is difficult for conventional wet precleaning and Ar sputtering by direct RF (RF etch precleaning) to be used for logic devices, because wet precleaning, such as an HF dip, causes side etching of contact holes. A contact profile change by wet precleaning prevents device geometry from being shrunken. RF etch precleaning sputters oxide from the contact side wall, then forms knocked-on oxide to the bottom of the contact holes.

It was reported in the literature that Ti-PECVD had higher deposition rates on Si than on oxide because $TiSi_x$ grows on the Si surface [5]. We also found the plasma process of regular Ti-PECVD [6, 7] could damage and cause gate oxide degradation. It has become more important to eliminate charge-induced damage for applications to high performance logic devices.

Therefore, we developed an Inductively Coupled Plasma (ICP) Ar/H2 precleaning process to remove the residual layer on $CoSi_2$ layer and a new plasma damage-free Ti-PECVD process.

EXPERIMENT

Figure 1 shows the fabrication procedure of the test device for logic with embedded DRAM. After the formation of gate electrodes and diffusion layers, $CoSi_2$ was formed by high-temperature sputtering followed by in-situ vacuum annealing [3]. Interlayer dielectric (ILD) films were deposited and contact holes with depths between 2.0 - 2.5μm were formed by conventional lithography and fine contact etching. ICP-Ar/H2 was used as the in-situ precleaning before Ti-PECVD. Conventional Ar sputter etch by direct RF plasma was used as a reference for this work. Ti films were deposited by PECVD using $TiCl_4/H_2$ chemistry. Then, TiN films were also deposited by low pressure (LP) CVD using $TiCl_4/NH_3/N_2$ without exposing the substrate to air.

Electrical characteristics were measured using a

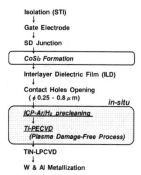

Isolation (STI)
↓
Gate Electrode
↓
SD Junction
↓
CoSi₂ Formation
↓
Interlayer Dielectric Film (ILD)
↓
Contact Holes Opening
(φ 0.25 - 0.8 μ m)
↓ in-situ
ICP-Ar/H₂ precleaning
↓
Ti-PECVD
(Plasma Damage-Free Process)
↓
TiN-LPCVD
↓
W & Al Metallization

Fig.1 Fabrication procedure of the test device for Logic with embedded DRAM.

Conference Proceedings ULSI XV © 2000 Materials Research Society

Al/W/CVD-TiN/CVD-Ti contact structure. Contact resistance was measured using 21000 contact chains. Junction leakage currents were measured using parallel contacts. Transmission electron microscopy (TEM), energy dispersive X-ray (EDX) spectroscopy, and transmission electron diffraction (TED) spectroscopy were used to analyze CVD-TiN/Ti and $CoSi_2$ interfaces.

A new antenna device was prepared to investigate the charge-induced damage for Ti-PECVD. The antenna device with 6nm thick gate oxide was fabricated. Antenna ratio of antenna area to that for gate was in the range from 50 - 10000. Thick ILD films were deposited after gate and antenna electrode formations, and a number of contact holes(ϕ 0.4-3.2µm) was formed on antenna area.

CMOS transistors which has high aspect ratio contact hole on $CoSi_2$ were prepared to evaluate these technologies. Depth of contact hole is 2.5µm and hole diameter was 0.25µm. The thickness of gate oxide is 6nm.

RESULTS AND DISCUSSION

A. Application of ICP-Ar/H2 precleaning to high aspect ratio contact on CoSi2 layer

Figure 2 shows cross-sectional TEM images of the interface between the CVD-TiN/Ti and $CoSi_2$ layers at the bottom of contact holes. A residual layer formed by the contact opening process and air exposure is seen in the case of no precleaning (Fig.2(a)). The thickness of the residual layer is about 5nm. In contrast, the residual layer is removed and a clean surface is obtained using ICP precleaning. We investigated the residual layer using EDX and TED. Figure 3 shows the EDX spectrum and TED patterns of the interface between the CVD-TiN/Ti and $CoSi_2$. It shows an amorphous structure and contains oxygen and carbon, which raise the contact resistance. Therefore, a precleaning technology is necessary to remove residual layers and obtain clean surfaces on $CoSi_2$.

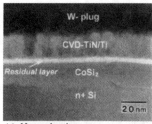

(a) No precleaning

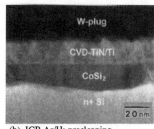

(b) ICP-Ar/H2 precleaning

Fig.2 TEM images of the interface between CVD-TiN/Ti and CoSi2 at the bottom of contact holes.

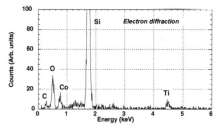

Fig.3 EDX analysis and TED pattern of the interface between CVD-TiN/Ti and CoSi2 in case of no preclean.

We investigated Ti deposition rates after ICP precleaning in comparison with no precleaning. PECVD-Ti deposition rates on n+ and p+ diffusion layers with $CoSi_2$ are shown in fig. 4. In the case of no precleaning, Ti deposition rates are much lower than that for ICP precleaning because impurities such as oxygen and carbon obstruct Ti growth[8]. No obstruction layer exists after ICP precleaning. Consequently, Ti is easy to grow after ICP precleaning because a clean surface is obtained.

Figure 5 shows the ratio of etching rate for $CoSi_2$ to the oxide with different precleaning conditions. RF etch precleaning has a high $CoSi_2$ etching rate and is difficult to control. Consequently it etches not only the residual layer but also the $CoSi_2$ layer. On the other hand, ICP precleanings had much lower etching rates even at higher bias conditions. This result shows that ICP precleaning can effectively remove the impurity layer and prevent etching of the $CoSi_2$ layer.

Figure 6 shows contact resistances of n+ and p+ diffusion layers with $CoSi_2$ versus contact size for different precleaning methods. In the case of no precleaning, contact resistances are extremely high because the residual layer exists as shown in fig.2. RF etch precleaning technology causes large deviations in resistance. Contact resistances after ICP precleaning are lower with smaller deviations. The difference is due to surface conditions between RF and ICP precleaning. RF etch precleaning etches CoSi2 layers as shown in fig.5. It may also resputter ILD materials and deposit them on the bottom of contact holes. On the other hand, the ICP precleaning cleans the CoSi2 surface without etching. Figure 7 shows contact resistances of n-gate and p-gate electrodes with CoSi2 as the function of contact size for different RF bias conditions of ICP precleaning. At low bias conditions, such as 35V, contact resistances are high with large deviations. Excellent contact resistances are also obtained on the gate electrodes at RF bias greater than 60V, because higher bias conditions remove the residual layer effectively with little CoSi2 etching as shown in fig.5. Figure 8 shows the cumulative probability of n+/p and p+/n junction leakage currents after ICP precleaning for different contact sizes. No increase in junction leakage current is observed at any contact size.

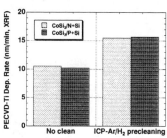

Fig.4 PECVD-Ti depsotition rates on n+ and p+ diffusion layer with CoSi2 .

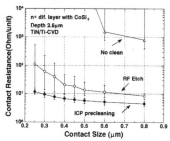

(a) n+ diffusion layer with CoSi2

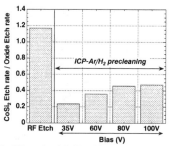

Fig.5 The ratio of CoSi2 etching rate to oxide for different precleaning conditions.

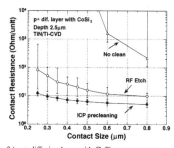

(b) p+ diffusion layer with CoSi2

Fig.6 Contact resistances of (a) n+ and (b) p+ diffusion layer versus contact size for different precleaning methods.

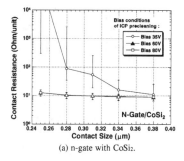

(a) n-gate with CoSi2.

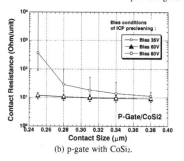

(b) p-gate with CoSi2.

Fig.7 Contact resistance of (a) n-gate and (b) p-gate with CoSi2 versus contact size for different bias conditions of ICP-Ar/H2 precleaning.

591

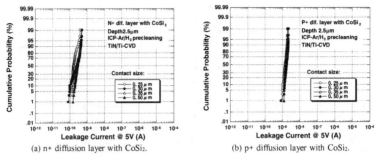

(a) n+ diffusion layer with CoSi2.

(b) p+ diffusion layer with CoSi2.

Fig.9 Junction leakage Current of (a) N+/P and (b) P+/N diode after ICP precleaning.

B. Investigation of the plasma damage during Ti-PECVD

Figure 9 shows a schematic of a new test device to evaluate the plasma damage during Ti-PECVD. The evaluation was carried out using the antenna device with high aspect ratio contact holes. Depth of contact holes and diameter were 2.0μm and 0.4μm, respectively. We investigated the dependence of the leakage current on the ratio of Ti deposition rate on oxide to that for Si, from 0.2 to 0.6. The Ti thickness was fixed at 10nm on the silicon substrate for all deposition conditions. Control samples were prepared without the Ti-PECVD process. Figure 10 shows the cumulative probability of the gate leakage currents for different Ti-PECVD conditions. The control shows no gate oxide degradation despite the use of some plasma processes. It was reported in the literature that Ti-PECVD has a higher deposition rate on Si than on oxide because TiSix forms on Si surface[9]. However, figure 10 indicates that such deposition conditions caused gate oxide degradation by plasma charge-up. Increasing the ratio of the deposition rate on oxide to that on Si can prevent this degradation by diverting electrical charges through Ti layer on the oxide (ILD). The gate oxide degradation decreases with increasing the ratio. In the case of the 0.6 ratio, no plasma damage is observed. Ratios of greater than 0.6 are necessary to eliminate charge-induced damage during Ti-PECVD.

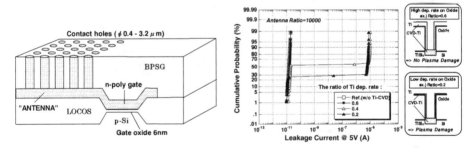

Fig.9 Shematic of new test device to evaluate for plasma damage during Ti-PECVD. (Depth 1.0 - 2.0 μm, Tox=6nm)

Fig.10 Cumulative probability of the leakage current of gate oxide for different Ti-PECVD conditions (The ratio of Ti dep. rate =Dep. on oxide / Dep. on Si).

C. Application to CMOS transistor

We have applied these technologies to CMOS transistors with high aspect ratio contacts on CoSi2. The depth of the contact holes was 2.5μm with hole diameters of 0.25μm. Figure 11 shows Id-Vd characteristics of NMOS and PMOS transistors with 6nm thick gate oxide. Sufficiently good characteristics are obtained.

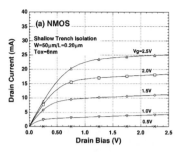 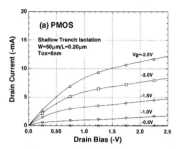

Fig.11 Id-Vd characteristics of (a) NMOS and (b) PMOS transistor with high asepct ratio contacts on CoSi2.

CONCLUSION

In conclusion, we developed ICP-Ar/H2 precleaning and plasma damage-free Ti-PECVD process. As a result, we realized the lowest contact resistances on CoSi2 with high aspect ratio contacts, and no gate oxide degradation from the plasma process during Ti-PECVD.

ACKNOLEDGEMENTS

The authors wish to thank O. Kudo, N. Endo, and S. Saito from NEC corporation, and D.T.C Huo form Lucent Technologies/Bell Labs., for their encouragement through this work. They also wish to thank H. Nanbu, H Kawamoto, K. Inoue, and K. Mikagi for processing necessary to fabricate the experimental samples.

REFERENCES

[1] J. M. Drynan, K. Fukui, M. Hamada, K. Inoue, T. Ishigami, S. Kamiyama, A. Matsumoto, H. Nobusawa, K. Sugai, M. Takenaka, H. Yamaguchi and T. Tanigawa, IEDM Tech. digest., 445 (1998).
[2] K. Kokubun, H. Takato, T. Sakurai, H. Koike, A. Nomachi, H. Ohstuka, H. Harakawa, W. Sato, M. Tanaka, H. Naruse, H. Kamijo, J. Kumagai and H. Ishiuchi, Symposium on VLSI Technology digest., 155 (1999).
[3] K. Inoue, K. Mikagi, H. Abiko and T. Kikkawa, IEDM Tech. digest., 445 (1995).
[4] K. Goto, A. Fushida, J. Watanabe, T. Sukegawa, K.Kawamura, T. Yamazaki and T. Sugii, IEDM Tech. digest., 449 (1995).
[5] K. Ohto, K. Urabe, T. Taguwa, H. Gomi and T. Kikkawa, IEDM Tech. digest., 361 (1997).
[6] J. Hillman, R. Foster, J. Fagout, W. Friggs, R. Arora, M. Ameem, F. Martin and C. Arena, Solid State Technology., 7, 147 (1995).
[7] S. Shibuki, K.Isa, T. Murakami, Y. Morioka and T. Akahori, VLSI Multilevel Interconnection Conf., 405 (1993).
[8] T. Taguwa, K. Urabe, M. Sekine, Y. Yamada and T. Kikkawa, IEDM Tech. digest., 695 (1995).
[9] T. Taguwa, K. Urabe, K. Ohto and H. Gomi, VLSI Multilevel Interconnection Conf., 255 (1997).

CMP/Cleaning/Etching

Copper and Tantalum Dissolution and Planarization in H_2O_2-based Slurries

M. Hariharaputhiran[a], S. Ramarajan[a], Y. Li[b] and S.V. Babu[a,c]
Departments of [a]Chemical and [b]Mechanical Engineering
[c]Center for Advanced Materials Processing, Clarkson University, Potsdam, NY

Abstract

Dissolution and polishing of copper and tantalum in hydrogen peroxide-glycine based slurries using alumina abrasives are discussed. Hydroxyl radicals [*OH] generated by the catalytic decomposition of hydrogen peroxide by the Cu^{2+}-glycine chelate enhance the copper dissolution/polish rates significantly. However, dissolution and polish rates of tantalum are essentially unaffected by the *OH. The effects of pH and ionic strength on tantalum polish rate are found to be more significant.

Introduction

Thanks to technological progress with dual damascene and chemical-mechanical planarization (CMP) technologies, copper, along with Ta/TaN and other barrier/adhesion promotion layers, is now replacing aluminum as the metal of choice for multi-layer interconnects [1, 2] in advanced (sub-0.18 μm) integrated circuit devices. Successful planarization of these films has been the enabling technology behind the inroads made by copper. While products containing copper circuitry are already being shipped, many scientific and technical questions concerning the planarization process for these films remain unanswered. In particular, at a fundamental level the details of the interactions that occur between the abrasive powders and the reagents in the slurry with the films being polished and their relation to defects in the polished film and to pattern erosion and dishing are not understood. While these interactions are important in all planarization processes, whether the film is metallic or dielectric, improved understanding of these interactions can be especially beneficial for integrating copper and Ta/TaN films with next generation low-k films and for production of sub-0.18 μm devices. This paper describes some recent work aimed at elucidating the nature of these interactions during the planarization of Cu and Ta films using silica and alumina abrasive powders in H_2O_2 containing slurries.

While copper can be polished with relatively high removal rates in acidic and near neutral pH conditions using $Fe(NO_3)_3$ and H_2O_2 as the oxidizing agents, respectively, polishing of Ta (and TaN as well) can only be done at a relatively low rate since its oxide is both hard and chemically inert. Since Fe is a deep level contaminant in Si and $Fe(NO_3)_3$ slurries are corrosive, H_2O_2 has become the preferred oxidizer and we will limit our attention here to H_2O_2 containing slurries. Both the dissolution and polishing rates of Cu reach an initial sharp maximum followed by a gradual decrease as the H_2O_2 concentration in the slurry is increased, presumably since an increasingly thick passive and harder oxide layer is formed on the copper surface. Nevertheless, we showed earlier [3] that the addition of glycine and Cu^{++} ions in the form of $Cu(NO_3)_2$ enhances both the dissolution and planarization rates of Cu. The addition of these reagents increases the concentration of OH radicals by catalyzing the dissociation of H_2O_2. Since there is always some excess undissociated H_2O_2, this suggests that the OH radicals facilitate the removal of the passive film formed from the surface of the copper film. Some of the results that validate these observations are presented in the following

In contrast, it was found that Ta polish rates are the highest in de-ionized water and that the addition of oxidizers always lowers the Ta removal rate with both silica and alumina abrasives. This reduction is presumably due to the inability of either H_2O_2 or *OH to enhance the removal,

Conference Proceedings ULSI XV © 2000 Materials Research Society

either by dissolution or by facilitating mechanical abrasion, of the hard oxide film from the Ta surface and/or due to a modification of the electrostatic interactions between the particles and the Ta surface. As in any CMP process, it is important to recognize that pH and ionic strength play a crucial role in determining the nature of the electrostatic interactions between the abrasive powders and the film surface being planarized. As shown later in this paper, changing the pH of the slurry can modify the electrostatic interactions between the particles and the Ta surface from being attractive to repulsive. These results have several implications for a one or two step slurry process development for the CMP of copper and Ta structures.

Experiment

Chemical-Mechanical Polishing

Polishing experiments were performed using a bench-top Struers DAP-V polisher and 3 mm thick copper and tantalum disks (99.99% pure, Aldrich), each with a cross sectional area of 7.5 cm^2. The table speed was set at 90 rpm and the disk holder was held stationary. The applied downward pressure was about 6.3 psi (41.4 KN/m^2). α-Alumina particles (bulk density 3.7 g/cc [4]) obtained from Ferro Electronics and fumed silica particles obtained from Cabot Corporation were used as abrasives. The solids concentration in the slurries was maintained at 3 % by weight and the slurry feed rate was 1 ml/s for all the experiments. The slurry in the supply tank was stirred continuously with a magnetic stirrer to avoid settling of aggregated particles. Suba 500 was used as the polish pad. The pad was hand conditioned prior to each experiment using a 220 grit sand paper and a nylon brush. The polish rate was determined from the difference in the weights of the disk before and after polishing for at least three minutes and the reported values were obtained by averaging over four experiments.

Dissolution

Dissolution experiments were carried out in a glass beaker using 400 ml of the etchant solution and a rectangular Cu coupon (2.3 cm x 2.3 cm x 0.2 cm, 99.99 % pure) as the sample. The Cu coupon was washed with dilute HCl to remove any native oxide from the surface, dried in an air stream and weighed. It was then immersed in the solution for a predetermined length of time. The solution was stirred at 1000 rpm using a mechanical stirrer to minimize mass transfer effects. The coupon was removed, washed repeatedly with DI water, dried in an air stream and reweighed. The weight loss was used to calculate the dissolution rate and the reported rates were obtained by averaging over four experiments, each spanning at least 2 minutes.

In situ Electrochemical Measurements

In situ electrochemical measurements (during polishing) were performed using the same Struers DAP-V polisher and 3 mm thick copper disc with a cross sectional area of 6.16 cm^2. The polishing conditions were maintained the same as those mentioned earlier in section 2.1. A three electrode set up consisting of a working Cu electrode (polishing substrate), a platinum counter electrode and an SCE reference electrode was used. The reference electrode was placed close to the polishing substrate (within 1 inch) and 0.1 M NaClO$_4$ was used as the supporting electrolyte to minimize solution resistance. Under these conditions no IR compensation was necessary. In addition, a slurry build up of at least 2 cms in height was maintained above the pad to eliminate the possibility of obtaining only a sporadic contact between the working electrode and the reference electrode. The details of the corrosion cell and the experimental procedure have been discussed in detail elsewhere [5,6].

Results and Discussions

[OH] Interaction

The polish rates of Cu and Ta in several peroxide containing solutions/slurries are shown in table 1. Irrespective of the presence or absence of the abrasive particles, copper polish rate increased significantly when both hydrogen peroxide and glycine were added (case 4 in table 1). There was a further increase in the copper polish rate with the addition of copper nitrate to the solution/slurry (row 5 in table 1). We had already demonstrated [7] that the decomposition of hydrogen peroxide is catalyzed by Cu^{2+}-glycine chelate yielding high concentrations of *OH, a much stronger oxidizing agent than hydrogen peroxide itself [8]. The high Cu polish rates in the presence of glycine, hydrogen peroxide and copper nitrate is presumably due to the increased concentration of *OH since glycine readily forms a chelate with Cu^{2+} ions in solution [9]. Though no external copper ions were added in the case of hydrogen peroxide and glycine (row 4), the copper abraded from the sample surface can generate copper ions locally which enhance the decomposition of peroxide to yield *OH in the vicinity of the copper surface. These *OH can, in turn, contribute at least to some of the increased Cu removal rates. In contrast, Ta polish rate was zero in all cases in the absence of abrasive particles, perhaps as Ta is much harder than Cu. Furthermore, the maximum tantalum polish rate was obtained in DI water and the polish rate decreased with the addition of chemicals. It should be noted, however, that the pH was not kept constant in all these cases but maintained at the natural values. The variation in pH plays a crucial role in determining the polish rates, as discussed later.

Table 1: Cu and Ta polish rates in various peroxide containing solutions/slurries

Slurry Compositions (all in DI water)	Cu Polish Rate (nm/min)		Ta Polish Rate (nm/min)	
	No Abrasives	With Abrasives*	No Abrasives	With Abrasives*
DI Water	0	290 ± 24	0	67 ± 13
5 wt % H_2O_2	0	210 ± 20	0	31 ± 5
1 wt % Glycine	6 ± 1	327 ± 37	0	37 ± 7
5 wt % H_2O_2 + 1 wt % Glycine	256 ± 15	850 ± 56	0	35 ± 4
5 wt % H_2O_2 + 1 wt % Glycine + 1 wt % $Cu(NO_3)_2$	612 ± 25	1867 ± 168	0	48 ± 3

*α-Alumina particles with a mean size of 340 nm

Figure 1 shows the copper dissolution rate as a function of H_2O_2 concentration in the presence of 1 wt % glycine, both with (1 wt %) and without added copper nitrate. The variation in the open-circuit potential (OCP) measured, in an *ex situ* electrochemical corrosion cell, with respect to SCE at 1000 rpm rotational speed is also plotted as a function of H_2O_2 concentration. The copper dissolution rate has an initial very sharp maximum around 1% H_2O_2 concentration and then decreases with increasing peroxide concentration. If Cu dissolution is controlled only by the *OH concentration at the film/solution interface, then an increase in the H_2O_2 concentration should increase the *OH concentration and the dissolution rate. However, the continuous increase in the OCP towards the anodic direction, i.e., towards a more positive value, with increasing peroxide concentration indicates that the copper surface is transformed into a film that becomes increasingly passive retarding the dissolution process. Furthermore, the *in situ* OCP measured during polishing increased with increasing peroxide concentration (not shown here) confirming the presence of a passive film during polishing. The chemical identity, morphology and the thickness of this passive layer control the magnitude of the copper dissolution/polish rates. Zeidler et al.[10] also made similar observations of decreasing copper dissolution/polish rates with increasing peroxide concentrations (in commercial Rodel 8099 slurry) and attributed the decrease to the formation of a passive film.

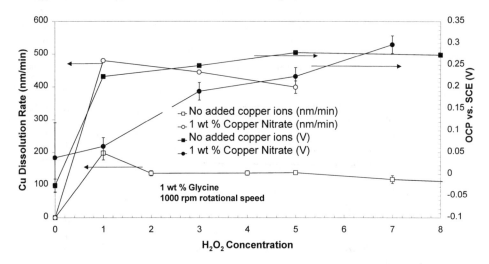

Figure 1: Cu dissolution rate and OCP vs. SCE as a function of peroxide concentration

To further investigate the role of this passive film, *in situ* electrochemical measurements were made during polishing. Figure 2 shows the OCP vs. time curves during Cu CMP using a slurry containing 1 wt % glycine, 5 wt % H_2O_2 and 3 wt % alumina abrasives, both in the presence and absence of 1 wt % $Cu(NO_3)_2$. The experiments were performed for a period of 10 minutes with polishing stopped after 5 minutes. The OCP remains constant during polishing and once the polishing was stopped, shoots up and stabilizes at a value higher than the OCP obtained during polishing. The increase in the OCP once the polishing was stopped is an indication of repassivation of the copper surface. Also the steep rise in the OCP, once the polishing was stopped, indicates that repassivation occurs rather rapidly within a few seconds. Furthermore, with the addition of 1 wt % $Cu(NO_3)_2$ to the slurry, the OCP stabilized at a higher value, when the polishing was stopped, indicating a further increase in the tendency to passivate with the addition of $Cu(NO_3)_2$. Thus copper dissolution and mechanical removal of the passive film are involved during Cu CMP in H_2O_2-glycine based slurries, in addition, perhaps to the mechanical removal of the copper itself by the abrasives. Further, the occurrence of both dissolution and passivation indicates that either the passive film is porous whose properties determine the dissolution rate or the passivation film is an intermediate step in the dissolution process as in the case of Cu dissolution in ammonium hydroxide solution[11]. However, such a hypothesis needs more experimental evidence and is being pursued further.

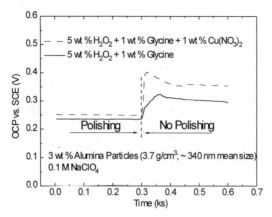

Figure 2: In situ OCP vs. SCE measured as a function of time

Effect of pH on Ta CMP

Figure 3 shows Ta polish rates as a function of pH in alumina and silica slurries. The pH values of the as-dispersed alumina and silica abrasives in DI water were 8.5 and 4, respectively, and were modified by adding either HCl or KOH. With alumina slurries, the tantalum polish rate is very low up to a pH value of 4 and then increases with pH up to a pH of about 8, after which the rate starts decreasing with a further increase in pH. The highest polish rate obtained was about 80 nm/min at pH 8 and the lowest rate was around 10 nm/min at pH values of 2 and 3. The tantalum polish rate reaches a maximum of about 80 nm/min for silica slurries also, but at a different pH of 3.5. The iso-electric points (IEPs), i.e., point of zero charge, of alumina, silica and tantalum pentoxide *particles* were measured and found to be 9.5, 2.2 and 6.5, respectively, and their location is shown in the figure by vertical dashed lines. As indicated in the figure, the surface charge on these particles varies from being positive to negative as the pH is varied around the IEP with the magnitude increasing as we move away from the point of zero charge. These results show that Ta and alumina particle surfaces have opposite charges in the pH range 6.5 - 9.5, while Ta and silica particle surfaces have opposite charges in the pH range 2.2 - 6.5, both coinciding with the range where the maxima in the tantalum polish rate occur.

Of course, what is needed is the IEP of a Ta surface and not that of Ta_2O_5 particles. However, Several investigators have used the zeta-potential values of silica abrasives to qualitatively represent the charges on SiO_2 thin film surfaces [12]. Similarly, we may assume that, since Ta readily forms a protective pentoxide film in aqueous solutions [13], the surface charge on a Ta film surface is qualitatively similar to that of tantalum pentoxide particles. It should be noted, however, that these values should be used to discuss the pH dependence of the polish rate only qualitatively. The precise variation in the rate will, of course, depend on the actual value of the zeta-potential of the oxide surfaces and the width of the electrical double layer, surrounding the abrasive particles and the Ta surface during polishing.

Variation in pH could also result in variation in the mean aggregate size of the abrasive particles and therefore could affect the polish rates. The mean aggregate size of the alumina particles in dispersion increased from about 180 nm to about 260 nm when the pH was increased from 2 to 4 and remained almost constant in the pH range 4 - 8. It was not possible to measure the size of the alumina particles in the pH range 8 - 11 due to the rapid settling of the agglomerated particles as the range was close to the IEP of alumina (9.5). The mean aggregate size of the silica abrasives was around 200 nm at pH values of 2 and 4 and decreased thereafter with an increase in the pH. The observed increase in the mean aggregate sizes of the abrasives at certain pH values could account for some of the increase in the polish rate. However, while the polish rate in the silica slurry increased from 30 to 80 nm/min as the pH is increased from 2 to 3.5, the mean particle size remains almost the same in that pH range. Similarly, the polish rate in the alumina slurry increased when the pH of the slurry was increased from 4 to 6 while the mean particle size remains almost the same. Thus, most of the variation in the tantalum polish rate with pH has to be attributed to the variation in the electrostatic interaction between the abrasive particles and the tantalum film surface.

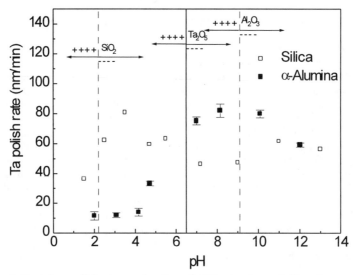

Figure 3: Tantalum polish rate as a function of pH in alumina and silica slurry

The role of electrostatic interactions on removal rates can also be checked by varying the ionic strength in the slurry. Increasing the ionic strength of the slurry, while maintaining the pH constant, decreases the zeta-potential as well as the width of the electrical double layer [14] of both the abrasives and the tantalum surface. This results in a decrease in the electrostatic interaction between the particles and the tantalum surface and also agglomeration of the abrasive particles and hence influences the polish rate. The effect of ionic strength on Ta polishing was studied by varying the concentration of KCl. At pH values where the surface charge on the abrasives and the tantalum film has the same sign (both either positive or negative), an increase in the ionic strength would result in a decrease in the magnitude of the surface charge and hence a reduction in the repulsion between them. This, in turn, would result in an increase in the tantalum polish rate. However, when the charges on the abrasives and the tantalum surface are opposite (one positive and the other negative) an increase in the ionic strength would result in the reduction in the attraction and hence a decrease in the tantalum polish rate. Our results on the effect of ionic strength changes on Ta polish rate, which will be described in detail elsewhere, are consistent with these expectations.

Conclusion

Copper polishing mechanism in hydrogen peroxide-glycine based slurries was studied using electrochemical techniques. Hydroxyl radicals generated by the catalytic decomposition of hydrogen peroxide by Cu^{2+}-glycine chelate increased the copper polish/dissolution rates. However, an increase in the peroxide concentration resulted in a decrease in the copper dissolution rate. Both *ex situ* and *in situ* OCP of copper increased with increasing peroxide concentration indicating an increasing tendency of passivation with increase in the peroxide concentration. The tantalum polish rate was relatively unaffected by the *OH and was highest in DI water. The effect of pH and ionic strength on tantalum polish rate was significant and is consistent with changes in electrostatic interactions between the abrasives and the metal surface.

Acknowledgments

This research has been supported by NY State Energy Research Development Authority through a subcontract from Ferro Electronics, and by grants from NSF and Intel. The authors would also like to acknowledge Rodel Inc., for the supply of silica particles.

References

1. Patrick, W.J., Guthrie, W, L., Standley, C.L., and Schiable, P.M., Journal of Electrochemical Society, 1991, **138**, p. 1778
2. Krishnan, A., Xie., C., Kumar., N., Curru, J., Duane, D., and Murarka, S.P., Proceedings of VLSI Multilevel Interconnection Conference, Santa Clara, CA, 1992, p. 226
3. Hariharaputhiran, M., Zhang, J., Li, Y., and Babu, S.V., MRS spring meeting, Symposium Q, San Francisco, 1998
4. Ramarajan, S., Hariharaputhiran, M., Her, Y.S., and Babu, S.V., proc. Surface Modification Technologies XII, ASM, 1998, p. 415-422
5. Luo, Q., Mackay, R.A., and Babu, S.V., Chemistry of Materials, 1997, 9, 10 p. 2101.
6. Srividya, C.V., Sunkara, M., and Babu, S.V., Journal of Materials Research, 1997, 12, 8, p. 2099, 1997
7. S.V. Babu, Y. Li, M. Hariharaputhiran, S. Ramarajan, J. Zhang, Y.S. Her and J.E. Prendergast, *Proc. 15th VLSI Multilevel Interconnection Conference (VMIC)*, Santa Clara, CA, p. 443, June (1998).
8. Kissa, E., Dohner, J.M., Gilbson, W.R., Strickman, D., J. Am. Oil Chem. Soc., **1991**, 68, 532.
9. Kirk-Othmer, Encyclopedia of Chemical Technology, fourth edition, Vol. 5, p. 764-795, John Wiley & sons, New York, 1996.
10. Zeidler, D., Stavreva, Z., Plotner, M., and Drescher, K., Microelectronic Engineering, 33, p. 259-265, 1997
11. Luo, Q., Mackay, R.A., and Babu, S.V., Chemistry of Materials, 9, 10, p. 2101-2106, 1997
12. U. Mahajan, M. Bielmann and R.K. Singh, Electrochem. Sol. State Let., 2, p. 80, (1999)
13. M. Pourbaix, Atlas of electrochemical equilibria in aqueous solutions, NACE, Cebelcor, Brussels, p. 251, (1974)
14. P.C. Hiemenz and R. Rajagopalan, Principles of colloid and surface chemistry, 3rd edn., Marcel Dekker, Inc., New York, p. 499-533, (1997).

EFFECT OF PLASMA TREATMENT ON THE CMP BEHAVIOR
OF ORGANIC LOW-K MATERIAL

S.J. HONG, H.D. CHUNG, B.U. YOON, S.R. HAH, J.T. MOON and S.I. LEE
Semiconductor R&D Center, Samsung Electronics Co. LTD,
San24 Nongseo-Ri Kiheung-Eup Yongin-City, 449-900, Korea

ABSTRACT

Although a number of studies on low-k materials for high speed VLSI device has been performed, CMP of organic low-k materials has not been successful. In this work, we have investigated the CMP characteristics of organic low-k films and the effect of plasma pre-treatment. The CMP of organic low-k film with silica slurry and ceria slurry gave low removal rates and surface defects. O_2 plasma treatment was found to be effective in improving the polishing rates with removable thickness of 2000 Å and achieving a defect-free surface.

INTRODUCTION

As device feature dimensions decrease, it is necessary to use low-k material for inter metal dielectric (IMD) layers to reduce the capacitance of interconnects. Low-k materials can be classified as organic and inorganic films according to their pre-cursor materials and SOG and CVD films according to their preparation methods. In materials aspects, the dielectric constants of organic low-k materials are much lower than those of inorganic materials such as SiOF and Hydrogen Silsesquioxane(HSQ). Therefore there is increased interest for research on organic low-k materials. In process aspects, spin coating method is being widely used for many low-k materials such as Hydrogen Silsesquioxane(HSQ), Methyl Silsesquioxane(MSQ), and Xerogel, but CVD method which gives good homogeneity and low running cost is also being developed and used for low-k films such as SiOC and SiOF. Since organic low-k films contain carbon, it is difficult to polish them with conventional oxide CMP chemistry and consumables [1]. Therefore improved CMP process with pre-treatment as well as with special slurries is being developed.[2,3] In this work, the CMP defect characteristics of organic low-k film which can be deposited by CVD are studied. The effect of O_2 and N_2O plasma pre-treatment on CMP characteristics and defect is also presented.

EXPERIMENTS

A commercialized organic low-k film was used in this work. The deposited film thickness ranged from 10000 Å to 14000 Å. In order to improve the polishing properties, organic low-k films were treated with O_2 and N_2O plasma. The change in film characteristics was observed by FTIR. The films were polished with a rotary type polisher using commercialized silica-based slurry (SS25 from Cabot Corp.) and experimental ceria-based slurry. The consumable set was perforated IC1000/Suba400 stack pad. The film thickness of PE-TEOS oxide and low-k films was measured by interferometry.

RESULTS AND DISCUSSION

Fig.1 shows the removal rates of PE-TEOS oxide and organic low-k films when polished

Conference Proceedings ULSI XV © 2000 Materials Research Society

with silica-based slurry and ceria-based slurry. The removal rates of low-k film using both slurries are very low when compared to that of PE-TEOS oxide, which means hydrated layer on the surface of organic low-k films has not formed during CMP. Typical surface defects observed on organic low-k film after CMP are shown in Fig.2 The polished layers locally damaged and severely delaminated at wafer edge. These defects suggest a strong mechanical process in film removal.

Fig.3 shows the removal amounts of organic low-k film with and without plasma pre-treatment after CMP of 30 seconds using silica slurry. In the case of 2 minutes O_2 plasma treatment, removal amount increases to 3000 Å.

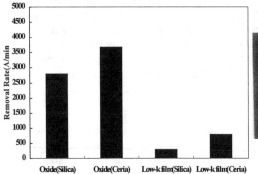

Fig.1 CMP removal rates of PE-TEOS oxide and organic low-k film with silica and ceria slurries.

Fig.2 Surface defects of organic low-k material after CMP with silica-based slurry.

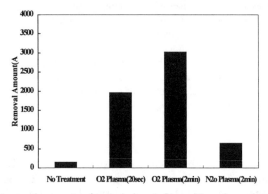

Fig.3 Removal amounts of organic low-k film with various plasma pre-treatment after 30 sec CMP.

However many cracks in the surface is obtained as shown in Fig. 4(b) Whereas in the cases of 20 seconds O_2 and 2 minutes N_2O plasma treatment, defect-free surface can be observed even though the removal amounts are decreased to 2000 Å and 650 Å, respectively.

Fig.5 shows the FTIR spectrum change of organic low-k material with O_2 plasma treatment and CMP. It can be suggested that O_2 plasma treatment breaks the Si-CH$_3$ bonding and transforms it into Si-OH. This change in surface chemistry is believed to enhance the CMP removal rates. As can be seen in Fig.5, the intensity of Si-OH peak is decreased as plasma treatment layer is removed by CMP. After 60 seconds polishing, we cannot find the SiOH peak any more.

Fig.6 shows the removal amount as a function of CMP time for PE-TEOS oxide and plasma pre-treated organic low-k material. The removal amount of low-k film becomes saturated after 30 seconds, which indicates that transformed layer is nearly all removed. Also, thickness of reacted organic low-k films during plasma treatment can be obtained from the dependence of saturated removal amount on the time and gases in plasma treatment in Fig.6

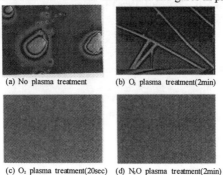

(a) No plasma treatment   (b) O$_2$ plasma treatment(2min)

(c) O$_2$ plasma treatment(20sec)   (d) N$_2$O plasma treatment(2min)

Fig.4 Surface defects of organic low-k film after CMP with and without plasma treatment.

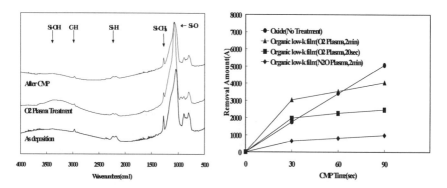

Fig.5 FTIR spectra of organic low-k film after deposition, 2 min O$_2$ plasma treatment, and 30 sec CMP.

Fig.6 Removal amount of PE-TEOS oxide and plasma pre-treated organic low-k films as a function of CMP time.

CONCLUSIONS

CMP characteristics of organic low-k material have been investigated. With commercial silica slurry and ceria slurry low removal rates and severe surface defects such as local lifting-off were obtained due to the mechanical process. Polishing rate of organic low-k film increase as plasma pre-treatment time increase, but cracks in surface can be generated. High CMP removal rates with removal thickness of $2000\,\text{Å}$ and defect-free surface can be obtained with 30 seconds O_2 plasma treatment

REFERENCES

1. C.Chiang. et al., Proc. of   IEEE VMIC,404(1987)
2. H.G.Chiu. et al., Proc. of CMP-MIC,328(1998)
3. E. Hartmannsgruber et al., VMIC Proc.,461(1998)

OXYGEN ENHANCED DRY ETCHING OF LOW-ε ORGANIC SOG (ε =2.9) FOR DUAL-DAMASCENE INTERCONNECTS

Shuntaro Machida, Atsushi Maekawa[1], Takao Kumihashi, Takeshi Furusawa,
Kazutami Tago[2], Takashi Yunogami[3], Takafumi Tokunaga[3] and Kazuo Nojiri[3]

Central Research Laboratory, Hitachi Ltd., Kokubunji, Tokyo 185-8601, Japan
[1]Process Engineering Development Dept., Hitachi ULSI Systems Ltd., Ome, Tokyo 198-8512, Japan
[2]Hitachi Research Laboratory, Hitachi Ltd., Hitachi, Ibaraki 319-12, Japan
[3]Device Development Center, Hitachi Ltd., Ome, Tokyo 198-8512, Japan

ABSTRACT

The chemistry and characteristics of dry etching a methyl-siloxane polymer (organic SOG) using C_4F_8 gas have been investigated. The addition of O_2 to C_4F_8 increased the rate at which the organic SOG was etched six-fold. The O radicals, dissociated from O_2, were shown to have the effect of spontaneously etching CH_3 groups from the organic SOG. The rate of etching of the organic SOG was explained as a superposition of the etching of its organic (CH_3) constituent by O radicals, and the etching of its inorganic constituents by the fluorocarbon. The addition of O_2 also caused the anomalous etching effect called trenching, to appear at the edge of pattern floors. The trenching was attributed to spontaneous CH_3 etching by O radicals at the bottom edge of the pattern because of the thinner fluorocarbon polymer layer at the edge. A dual-damascene structure was realized using the organic SOG by carefully balancing the spontaneous etching and deposition at the edge.

INTRODUCTION

In the recent shrinkage of ultra-large scale integrated devices (ULSIs), the interconnect signal delay becomes a dominant factor in LSI performance [1,2]. To reduce this delay, the use of low-ε interlayer dielectrics (ILDs) [2,3,4] and the Cu wiring damascene process [5-8] have been widely studied. Dry etching of a low-ε dielectric for the damascene process is one of the key processes in realizing high-speed ULSIs.

As a low-ε material, methyl-siloxane polymer is a candidate. It is an organic SOG and has a low dielectric constant of about 2.9 [9]. In this article we report on the dry etching characteristics of the organic SOG in C_4F_8/O_2 plasma, and discuss the role of oxygen in the etching of organic parts of the SOG to realize a low-ε dual-damascene structure.

EXPERIMENT

Dry etching was performed using a dual-frequency narrow gap RIE system, at a pressure of 4 Pa and a total RF power of 3 kW. To C_4F_8/O_2 gases, a constant flow of Ar gas was added. As a methyl-siloxane polymer material, we selected HSG-R7 (Hitachi Chemical Co. Ltd.). The sample was prepared using a typical SOG process [9]. The thickness of the sample was optically measured before and after etching. A photoresist mask was used for pattern etching. Etching profiles were observed with a scanning electron microscope (SEM). Molecular orbital calculations [10] were also performed to clarify the role of oxygen in the etching of HSG-R7.

RESULTS AND DISCUSSIONS

C_4F_8 Flow Rate Dependence of Etch Rate

Figure 1 is a comparison of HSG-R7 and SiO_2 etch rates in a C_4F_8/Ar gas plasma. The etch rate for SiO_2 increased as the C_4F_8 flow rate increased from 0 to 15 sccm. On the other hand, the rate of etching of HSG-R7 decreased in this range. This result indicates that the fluorocarbon polymer (CF polymer), which is deposited on the surface, and can effectively etch for SiO_2 [11,12], cannot etch the methyl groups (CH_3) in HSG-R7 effectively. To etch the CH_3, we examined the addition of O_2.

Effect of O_2 Addition

Figure 2 shows the effect of O_2 addition on the etch rates for HSG-R7 and SiO_2. The etch rate for SiO_2 decreased monotonously as we increased the O_2/C_4F_8 ratio. We attribute this to the decreasing proportion of CF polymer, which acts as an etchant for SiO_2 [11,12].

On the other hand, the etch rate for HSG-R7 peaked at an O_2/C_4F_8 ratio of 1. At this peak HSG-R7 was etched six times than at an O_2/C_4F_8 ratio of 0. This indicates that O_2 had the effect of enhancing the etch rate of HSG-R7. A model for these characteristics is shown in Fig. 3. HSG-R7 could be considered to be a mixture of inorganic SOG constituents and an organic CH_3 constituent. The etch-rate dependence of the inorganic SOG constituents on the O_2/C_4F_8 ratio is probably similar to that of SiO_2. However, the etch rate of CH_3 seems to increase with an increasing O_2/C_4F_8 ratio because the amount of O radicals in the plasma increases. Thus, the etch rate of HSG-R7 shows superposed features of both the inorganic SOG and the organic CH_3 etch rates.

Molecular Orbital Calculation

To clarify the mechanism of CH_3 etching by an O radical, molecular orbital calculations were performed. In these calculations, HSG-R7 was represented by an $(OH)_3Si(CH_3)$ molecule. An O atom in singlet or triplet spin multiplicity was placed near the molecule and a stable configuration was calculated. Calculated reactions and stabilization energies are summarized in Table 1.

In the case of triplet spin multiplicity, the O atom could not react with the $Si-CH_3$ bond. The O atom with singlet spin multiplicity, however, could enter and stabilize as $Si-O-CH_3$. A further singlet O atom could also react with $Si-O-CH_3$ to produce the stable form of $Si-O-O-CH_3$. This peroxide decomposes into $(OH)_4Si$ and H_2CO. These results indicate that O radicals can etch CH_3 spontaneously.

Etching Profile of HSG-R7

Etching profiles of HSG-R7 are shown in Fig. 4. At an O_2/C_4F_8 ratio above 0.45, the unusual effect called 'trenching' appeared. The trenching became deeper as the O_2/C_4F_8 ratio increased. The mechanism which produces this trenching is thought to be as follows: the CF polymer layer may be thinner at the edge of a pattern floor (Fig. 5(a)). Such edges are sheltered from C_xF_y deposition by the sidewall. The CH_3 in the HSG-R7 under the thin CF polymer can thus more easily be etched by O radicals (Fig. 5(b)). After the CH_3 has been etched out, the HSG-R7 becomes porous SOG, which contains only Si and O [13]. This

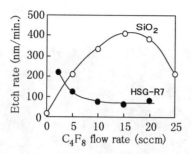

Fig. 1 Dependence of HSG–R7 and SiO$_2$
etch rate as a function of C$_4$F$_8$ flow rate

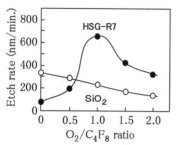

Fig. 2 Effect of O$_2$ addition on the
etch rates of HSG–R7 and SiO$_2$.

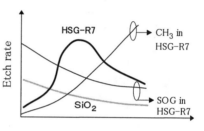

Fig. 3 Diagram to model HSG–R7
etch rates.

Table 1 Calculated reactions and stabilization energies

O spin multiplicity	Reaction			Stabilization energy (eV)
Triplet	(OH)$_3$Si–CH$_3$ + O	$\xrightarrow{\text{repulsion}}$	(OH)$_3$Si–CH$_3$ + O	—
Singlet	(OH)$_3$Si–CH$_3$ + O*	$\xrightarrow{\text{O insertion}}$	(OH)$_3$Si–O*–CH$_3$	7.8
	(OH)$_3$Si–O–CH$_3$ + O*	$\xrightarrow{\text{O insertion}}$	(OH)$_3$Si–O–O*–CH$_3$	4.5
	(OH)$_3$Si–O–O–CH$_3$	$\xrightarrow{\text{decomposition}}$	(OH)$_3$Si–OH + O=CH$_2$	2.0

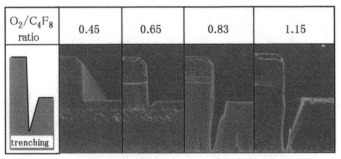

Fig. 4 Etching profiles of HSG–R7 at O_2/C_4F_8 ratios of 0.45, 0.65, 0.83, and 1.15.

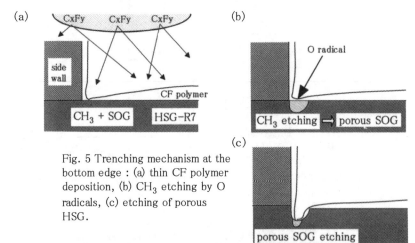

(a)

CxFy CxFy CxFy

side wall

CF polymer

CH_3 + SOG   HSG–R7

(b)

O radical

CH_3 etching ⇨ porous SOG

Fig. 5 Trenching mechanism at the bottom edge : (a) thin CF polymer deposition, (b) CH_3 etching by O radicals, (c) etching of porous HSG.

(c)

porous SOG etching

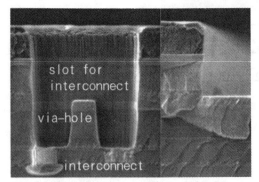

slot for interconnect

via–hole

interconnect

Fig. 6 Dual–damascene profile for HSG–R7 etched at an O_2/C_4F_8 ratio of 0.4.

porous SOG is easily etched by the CF polymer, thus a trench begins to form at the edge of a pattern floor (Fig. 5(c)). Once a trench begins to form, the CF polymer layer within the trench becomes even thinner, therefore, the trench becomes deeper and deeper. Clearly, the O_2/C_4F_8 ratio must be carefully tuned to get both a high etch rate and a good etch profile.

Figure 6 shows a dual-damascene structure made using HSG-R7. The etching was done at an O_2/C_4F_8 flow rate ratio of 0.4. A dual-damascene structure without trenching was satisfactorily realized.

CONCLUSIONS

The dry etching characteristics of methyl-siloxane polymer in a C_4F_8/O_2 environment was investigated. They depended on the spontaneous reactions between oxygen radicals from plasma and methyl groups in HSG-R7. In the etch rate of HSG-R7, features of both methyl-group etching and inorganic SOG etching were superposed. In the case of pattern etching, the unusual effect called trenching appeared at the edge of pattern bottoms as the O_2/C_4F_8 ratio rose above 0.45. The mechanism of trenching was also attributed to methyl-group etching by oxygen radicals; methyl-group etching occurred only at the bottom edge because of the thin CF polymer layer. A dual-damascene structure using HSG-R7 was realized by careful tuning to get both a high etch rate and a good etch profile.

ACKOWLEDGMENTS

The authors would like to thank Tokuo Kure, Yasushi Goto and Hiroyuki Enomoto for useful discussions on the mechanisms of etching and Hiroyuki Maruyama for experimental support.

REFERENCES

1. Mark. T. Bohr, IEEE IEDM Tech. Digest, p.241(1995)

2. L. Peters, Semiconductor International, **10**, 64(1998)

3. S. Crowder et al., Tech. Digest of 1999 Symp. of VLSI Tech., p.105(1999)

4. M. R. Baklanov et al., J. Vac. Sci. Technol., **B17**, 372(1999)

5. S. Venkatesan et al., IEEE IEDM Tech. Digest, p.769(1997)

6. D. Edelstein et al., IEEE IEDM Tech. Digest, p.773(1997)

7. P. Gilbert et al., IEEE IEDM Tech. Digest, p.1013(1998)

8. N. Aoi et al., Tech. Digest of 1999 Symp. of VLSI Tech., p.41(1999)

9. T. Furusawa et al., Proc. of 1994 VMIC, p186(1994)

10. K. Kobayashi et al., Phys. Rev., **A53**, 1903(1996)

11. H. Enomoto *et al.*, Proc. of the 19th Symp. on Dry Process, p.157(1997)

12. M. Schaepkens *et al.*, J. Vac. Sci. Technol., **A17**, 26(1999)

13. T. Furusawa *et al.*, Proc. of the 1996 Int. Conf. on Solid State Devices and Materials, p.145(1996)

CHARACTERISATION OF PLASMA ETCH RELATED RESIDUES FORMED ON TOP OF ECD Cu FILMS

M.R.BAKLANOV*, T.CONARD, F.LANKMANS, S.VANHAELEMEERSCH, D.HOLMES[1], K.MAEX
IMEC, Kapeldreef 75, B-3001 Leuven, Belgium, baklanov@imec.be
[1]EKC Technology, Inc., East Kilbride, UK

ABSTRACT

The plasma etch related residues on top of electrochemically deposited (ECD) Cu film are examined by XPS and ellipsometry. It is established that the native oxide on top of as prepared ECD copper is mainly Cu_2O. The amount of CuO increases during the air storage. Oxygen plasma drastically increases the CuO amount. After a C_2F_6/Ar plasma, a fluorine rich CFx polymer is formed. New Cu compatible EKC chemistries are examined to characterize the cleaning efficiency. The residue formed in oxygen plasma is effectively removed by EKC-525. This solvent is also quite efficient to remove the CFx polymer. However, some carbon fluoride still remains. EKC-505, developed as a Cu compatible resist stripper, does not remove CFx polymer from the copper surface. Combination of EKC-505 and EKC-525 can be used for the resist strip and post dry etch cleaning.

INTRODUCTION

Cu metallisation and low-k interlayer dielectric provide performance improvement in speed and reduce cross-talk noise and power consumption due to the reduction in interconnect capacitance [1]. Most of the presently considered low-k dielectrics are organic polymers. The Cu/organic polymer assembly is quite new from a technological point of view. Therefore all technological procedures should be carefully analysed and optimised.

One of the most important technological steps is via etching. Reactive ion etching (RIE) of the low-k polymer by an oxygen-based plasma has been developed for this purpose [2]. The organic polymer etch process needs an inorganic hard mask (HM: SiO_2, Si_3N_4, SiC etc.) because the low chemical selectivity towards photoresist creating problems related to the resist strip. A diffusion barrier which should be formed on top of the Cu layer to prevent diffusion to the low-k polymer is also an inorganic film [3]. Contrary to the organic polymers, the inorganic HM and diffusion barrier can not be etched by an oxygen plasma. A traditional fluorine based chemistry is used for this purpose.

Because of the chemicals involved in the process, post via-etch residue should include non-volatile Cu oxides and fluorides. CFx polymers should be present at the surface. The composition of this polymer strongly depends on the substrate properties [4]; therefore, removal of this polymer also needs special attention. Moreover, a traditional oxygen plasma cannot be used for the CFx polymer removal because this plasma attacks the organic low-k polymer.

All these reasons stimulate development of new cleaning chemistries that are compatible with both Cu and low-k polymer and that can effectively remove post dry etch residues. Examples of these products are EKC-505™ and EKC-525™ recently developed by EKC Technology for the resist strip and post dry etch residues removal, respectively. EKC-505™ is a blend of organic solvents and amines. The purpose of this product is to remove organic residues and photoresist by dissolution without attacking sensitive substrate materials. EKC-525™ is a three component blend of a quaternary ammonium compound, water and an organic solvent. This

615

product is a highly alkaline solution which demonstrates mild complexing ability. It is specifically formulated to selectively clean copper oxide contamination.

In this work, we studied:

- the chemical composition and thickness of residues formed on top of Cu in O_2/N_2 and C_2F_6/Ar plasmas.

- the removal efficiency and chemical transformation of these residues during the treatment in EKC-505 and EKC-525.

EXPERIMENT

Blanket electroplated copper films are used for the evaluation of the plasma etch residues. The following stack is used: first a PECVD oxide is deposited on a Si substrate, followed by a TaN barrier and PVD Cu seed layer deposition. Next the wafers are electroplated with a 1 micron thick Cu layer.

The wafers were exposed in O_2/N_2 and C_2F_6/Ar plasmas in a LAM TCP 9100 RIE etcher. The plasma exposure time is technically significant in that exposure in a O_2/N_2 plasma was close to standard overetch time after the via opening (during this time Cu can be attacked by an oxygen plasma). The Cu exposure time in the C_2F_6/Ar plasma was close to the etching time of the diffusion barrier. The residue thickness were measured by ellipsometry with λ=6328 A. The chemical composition of the films were studied by a XPS spectrometer Fisons SSX-100 equipped with a monochromatic Al Ka source and a concentric hemispherical electron energy analyser.

Compatibility of EKC-505 and-525 with low-k SiLK and Flare-2 films [2] were also examined. Ellipsometric and FTIR analysis show that no changes of refractive indices and chemical composition in these films occur in the above solutions. Only insignificant film swelling was observed in ellipsometric measurements. Sheet resistance measurement proved that these solutions do not attack the ECD copper film.

RESULTS AND DISCUSSION

The chemical analysis. - Figure 1 presents the Cu2p spectra observed after different technological steps involving the samples stored in air and exposed in a C_2F_6/Ar plasma. The first spectra correspond to a Cu film stored in air for 2 days. The most intense peak corresponds to Cu_2O (932.5 eV). The long air storage increases a peak with binding energy of 935 eV. This peak should correspond to CuO [5]. However, the observed binding energy is slightly higher than for CuO (934 eV), probably, due to hydration of the copper oxide during the storage on air.

The Cu peak intensities decrease after exposure to a C_2F_6/Ar plasma. This phenomenon is related with deposition of a CFx polymer film that does not contain Cu. No significant differences are observed between 8 and 30 sec plasma in regards to the residue formation. The Cu peak intensities increase after EKC-525. This fact suggests that EKC-525 removes the carbon fluoride residues. However the lower intensity of the Cu peaks in comparison with the "as prepared" Cu film suggest that some CFx-residue still remains. The increase of the Cu peaks intensity is not pronounced after EKC-505 treatment. Therefore this solvent is less efficient to remove the post plasma etch residue.

After an O_2/N_2 plasma the Cu surface is mainly covered by CuO (Fig.2). The composition of the film after plasma is not significantly influenced by the plasma etch time (34 or 68 sec). EKC-525 is found to be very effective to remove this oxide but no improvement is observed after a 20 min cleaned compared to a 2 min clean. After EKC-525 the Cu surface contains only Cu_2O

similar to the film stored two days in air. Therefore, after an oxygen plasma the Cu_2O remains only at the Cu/CuO interface. EKC-505 does not effectively change the composition of the post oxygen plasma residues.

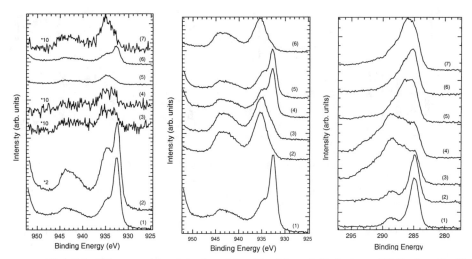

Fig.1. Cu2p XPS spectra of the pristine Cu + 2 days (1) and 1 month (2) air storage; C_2F_6/Ar plasma 8 s (3) and 30 s (4); EKC 525 2 min (5) and 20 min (6); EKC 505, 20 min (7).

Fig.2. Cu2p XPS spectra of the (1) pristine Cu film + 2 days storage; 2: O_2/N_2 plasma, 34 s; 3: O_2/N_2 plasma, 68 s; 4: sample 2 + EKC-525, 2 min.; 5: sample 2 + EKC-525, 20 min.; 6: sample 2 + EKC-505, 20 min.

Fig.3. C1s lines of the photoemission spectra. Pristine Cu + 2 days (1) and 1 month (2) air storage; C2F6/Ar plasma 8 s (3) and 30 s (4); EKC 525 2 min (5) and 20 min (6); EKC 505, 20 min (7).

F1s spectra corresponding to the treatments show that the main fluorine containing compound formed on top of the Cu surface in a carbon fluoride plasma is CFx polymer. Figure 3 shows the C1s spectra after different technological steps after the C_2F_6/Ar plasma etching. The Cu film stored during 2 days shows a peak with binding energy near 285 eV. This peak is typical for wafers stored in the clean room ambient and includes mainly C-H, C-O, C-C bonds [4]. After plasma etching an intense peak near 288 eV is observed. This binding energy is typical for CF_x bonds. The Cu surface is thus covered by teflon-like polymer. The film composition is quite different from the polymers formed on the silicon or metal silicide surface [4]. The CFx polymer on top of Cu is fluorine rich (high ratio of $CF_3/CF_2/CF$) as can be deduced from the fact that the peak intensity increases in the high binding energy region.

One can see that both EKC-525 and 505 change the polymer composition. The intensities measured in the higher binding energy region drastically decrease indicating the removal of C-F bonds. The composition of the remaining films depends however on the treatment. First, the comparison of the two process time with EKC 525 clearly shows that a longer cleaning time increases the removal of the C-F components. This is observed through the decrease of the intensities around 287 eV binding energy between spectra 5 and 6 from Figure 3. Second, it is observed that the remaining film after 20 min EKC 505 is still more C-F rich than the remaining film after 2 min EKC 525.

The residue thickness. Ellipsometric measurements were used to estimate the residue thickness. It was found that during all the experiments mainly the phase angle Δ is changed. The amplitude angle ψ is almost constant. Figure 4 (right axis) shows the change of Δ and of the residue thickness (left axis) during the different technological procedures. For this analysis, it is useful to consider the change of the optical thickness **nd** (product of the film refractive index **n**

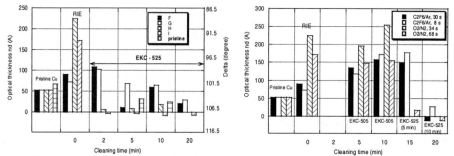

Fig.4. Change of the phase angle Δ and relative residue thickness on top of Cu film during different technological procedures. Post Dry Etch cleaning was done by EKC 525, 65°C.

Fig.5. Change of the relative residue thickness on top of Cu film during different technological procedures. Post Dry Etch cleaning was done by EKC-505 and 525, 65°C.

and thickness **d**) because the values of the refractive indices of the residue film are different for each different plasma etch conditions. As a reference value (zero thickness on the left axis) we used the mean value of the ellipsometric characteristics of the Cu film preliminary exposed in an O-plasma and cleaned by EKC-525. This treatment gives the highest value of Δ. According to the ellipsometric equations [5] the highest value of Δ at d<300 A corresponds to the minimal film thickness. Therefore, our measurements give a relative change of the residue thickness in comparison with a EKC-525 cleaned Cu surface.

The Δ value of the Cu film stored in air during the 2 days was equal to ≈103° which corresponds to a **nd**≈53 A. These data were well reproducible for the 5 wafers used in our experiments. Storage in air during 1 month additionally increases the native oxide thickness (pristine in fig.4). The plasma treatment increases the film thickness. The optical thickness of the residue film after oxygen plasma is almost 3 times higher than after an F-plasma. This fact is related not only to the film thickness but also with lower refractive index of the CFx polymer when comparing with the refractive index of Cu oxides.

The O-plasma related residue thickness decreases drastically after 2 min EKC-525 dipping. No additional significant change was found when the dipping time was increased from 2 to 20 min. It is important that EKC-525 decreases also the native oxide thickness on the Cu surface. Change of the "fluorinated" film thickness are more complicated. After 2 min EKC-525 dipping the optical thickness has increased. This effect is probably related to the film swelling. Long time treatment decreases the residue thickness. However, even after 20 min the residue thickness is still higher than after O-plasma/EKC-525 wafer.

EKC-505 is not so effective for the residue removal (Figure 5). This solvent does not affect the post O-plasma residue thickness. The post F-plasma residue thickness increases in EKC-505. In additional experiments with EKC-505, the treated wafers were dipped in EKC-525. This sequence produced a good effect. The O-plasma related residues were easily removed during the first 5 minutes. F-plasma related residues are more stable. They were removed only after 10 minutes dipping in EKC-525.

CONCLUSIONS

1. Formation, composition, thickness and removal efficiency of the plasma etch related residues on top of the ECD copper film are examined by XPS and ellipsometry. It is established that the native oxide on top of as prepared ECD copper is mainly Cu_2O. The amount of CuO increases during the air storage. The CuO amount drastically increases in an oxygen plasma. After a C_2F_6/Ar plasma a fluorine rich carbon fluoride polymer is formed on the Cu surface.

2. Two new EKC chemistries (EKC-505 and EKC-525) are used to characterise the cleaning efficiency. The residue formed in an oxygen plasma is effectively removed by EKC-525 developed for this precise purpose. This solvent is also quite efficient in removing teflon-like polymer from the Cu surface. However, some carbon fluorides still remain on the Cu surface even after 20 min at 65°C. EKC-505, developed as a Cu compatible resist stripper, does not remove CFx polymer from the copper surface. However, combination of EKC-505 and EKC-525 can be used for the resist strip and post dry etch cleaning of the copper surface. It is also found that these solvents do not attack the SiLK and Flare low-k polymer films.

3. EKC-505 and EKC-525 are potentially effective solvents for the cleaning of the plasma exposed Cu surface. However, some additional optimization experiments with variation of the experimental parameters (temperature, different combination of the strippers) are still needed. Results of these experiments will be reported in our further publication.

ACKNOWLEDGEMENTS

We are thankful H.Struyf for plasma etching. K.Maex is a research director of the National Fund for Scientific research Flanders (FWO).

REFERENCES

1. B.Zhao, D.Feiler, V.Ramanathan, Q.Z.Liu, M.Brongo, J.Wu, H.Zhang, J.C.Kuei, D.Young, J.Brown, C.Vo, W.Xia, C.Chu, J.Zhou, C.Nguyen, L.Tsau, D.Dornisch, L.Camilletti, P.Ding, G.Lai, B.Chin, N.Krisma, M.Johnson, J.Turner, T.Ritzdirf, G.Wu, L.Cook. Electrochemical and and Solid State Letters, 1(6), 276 (1998)

2. M.R.Baklanov, S.Vanhaelemeersch, H.Bender, K.Maex. J.Vac.Sci.Technol. B17(2), (1999).

3. J.Baumann, M.Stavrev, M.Rennau, T.Raschke, S.E.Schulz, C.Wenzel, C.Kaufmann, T.Gessner. Advanced metallisation conference IN 1998, MRS, p.321 (1999).

4. M.R.Baklanov, S.Vanhaelemeersch, W.Storm, Y-B.Kim, W.Vandervorst, K.Maex. J.Vac.Sci.Technol. A15 (6), 3005 (1997)

5. S.A.Chambers, V.A.Loebs, K.K.Chakravorty. J.Vac.Sci.Technol.A8(2), 875 (1990)

6. R.M.A.Azzam, N.M.Bashara. Ellipsometry and Polarized Light. Elsevier Sci.Publ.Co, 1987

Effects of Physical/Reactive Clean, Dry/Wet Clean and Barrier Metal Thickness on Cu Dual Damascene Process

C.S. Liu, S.L. Shue, C.H. Yu, and M.S. Liang
Research & Development, Taiwan Semiconductor Manufacturing Company, Ltd.
No. 9, Creation Road I, Science Based Industrial Park, Hsin-Chu, Taiwan, R.O.C.
TEL: 886-3-5781688, FAX: 886-3-5773671, E-mail:csliua@tsmc.com.tw

ABSTRACT

To achieve low via resistance and higher electromigration (EM), resistance, a proper via clean process has been studied. Wet photo resist (PR) clean process and physical Ar sputtering can effectively remove residual polymer and obtain a better via resistance. H_2/He reactive clean with lower bias power can effectively remove copper oxide layer. However, H_2/He reactive clean doesn't achieve lower via resistance compared with conventional Ar sputtering if etching polymer cannot be properly removed. Ar sputtering clean method is also better than H_2/He reactive clean in terms of EM performance. Thick enough TaN barrier layers were found to be essential for good EM resistance.

INTRODUCTION

With the demand of low-RC delays and good reliability, copper is considered as an alternate materials for multilevel interconnect. Recently, damascene process had been widely implemented in order to eliminate the difficulty of Cu dry etch process.[1]. However, Cu tends to oxidize at room temperature and especially at elevated temperature during post damascene etch PR dry strip. Furthermore, Cu cannot form a self-passivated oxide layer like aluminum. Oxidized Cu will degrade the Cu line performance and became a reliability issue. A well-controlled PR/polymer strip and treatment at exposed Cu surface before pre-barrier metal deposition is hence important to achieve low Cu via resistance and good via EM resistance. The effects on via resistance, EM resistance of via and line, and line-to-line leakage among different clean methods and barrier metals were studied.

EXPERIMENT

A dual damascene structure with 0.28µm/0.28µm line width/space and 0.26µm via size was prepared for line to line leakage and via resistance measurement. To characterize the Cu barrier performance, junction leakage of devices using cobalt silicide as contact was measured.

For H_2 reactive clean copper oxide, a 5% H_2/He mixture gas was utilized in 450W coil induced coupled plasma (ICP) and 10W bias power at about 80mTorr for 1 minute to reduce copper oxide. 2MHz high coil frequency was used to try reduce the copper oxide more effective in this system. To characterize the copper oxygen reducing capability on copper oxide blanket wafer, AES depth profile was used to analyze the chemical composition. After cleaning process, ionized metal plasma (IMP) TaN with

621

different thickness and IMP Cu seed layers were deposited in the same cluster tool followed by an electro-chemical plating(ECP) Cu and a chemical-mechanical polishing (CMP) Cu to obtain a Cu damascene structure.

Cu via and Cu line was stressed by several mega-ampere at elevated temperature to compare their EM resistance performance among different barrier metal and different clean methods.

RESULTS AND DISCUSSIONS

Whole copper layers of damascene process were successfully implemented on 0.18μm logic backend process. Fig. 1 shows the cross-sectional TEM (XTEM) picture of whole copper layers. A proper clean method is critical on dual damascene process to stack the intact whole copper layers.

H_2/He reactive clean with lower bias power can effectively remove copper oxide layer. Fig. 2 showed the AES depth profile of Cu surface with Cu oxide and without Cu oxide layer after H_2/He reactive clean. However, H_2/He reactive clean doesn't achieve lower via resistance compared with conventional Ar sputtering if etching polymer cannot be properly removed. Line to line leakage between 0.28μm line space was measured to be higher in H_2/He reactive clean compared to conventional Ar sputtering as shown in Fig. 3. This conjectured higher leakage comes from stress concentration at the straight trench corner during Cu-CMP process. Furthermore, H_2/He reactive clean also showed the poorer via EM resistance than Ar sputtering clean method as shown in Fig. 4. Polymer remaining was considered as the root cause to early EM failure. Wet or dry clean are alternate processes at post SiN liner removal step in the Cu dual damascene process. Wet clean showed a lower via resistance compared with dry clean process, shown in Fig. 5. A polymer free via hole at bottom after wet clean process contributed to the lower via resistance, as shown in Fig. 6.

Barrier metal thickness, due to the higher resistivity eg. TaN, also play an important role on via/line resistance. The via resistance increases as the barrier layer TaN thickness increasing, via resistance was shown in Fig. 7. However, too thin TaN barrier will cause barrier discontinuity because of poor step coverage and will lead to poor via EM resistance. 100A thick TaN as metal 1 Cu barrier using W-plug as contact hole still showed normal junction leakage current performance compared with thicker one. However, 100A thick TaN was found much poorer than 300A thick TaN deposition in via EM performance, shown in Fig. 8. On the other hand, 0.28μm Cu line showed comparable EM performance between 100A and 300A thick TaN barrier because of less severe aspect ratio on trench line and hence continuous TaN barrier can be achieved.

CONCLUSIONS

The effects of pre-metal deposition clean method and TaN barrier metal thickness on 0.18μm logic copper dual damascene process was studied. H_2/He reactive clean doesn't achieve lower via resistance compared with conventional Ar sputtering if etching polymer cannot be properly removed. This is also reflected to via EM performance. H_2/He reactive clean showed the poorer EM resistance than Ar sputtering clean method. Ar sputtering can round the damascene trench corner, a lower line to line leakage can be obtained due to less stress concentration during Cu-CMP process. Wet cleaning is more effective in removing post SiN liner removal step than dry cleaning only. A lower via resistance can be obtained due to less polymer remained on wet clean process. Thick enough TaN barrier layers are critical to supply a continuous layer in via hole and sustain good via EM performance.

ACKNOWLEDGE

B.R. Young, M.H. Tsai and J.B. Lai are greatly acknowledged for wet clean process, EM testing and AES analysis, respectively.

REFERENCE

1. B. Zhao, etc. Symposium on VLSI Technology 28, (1998)

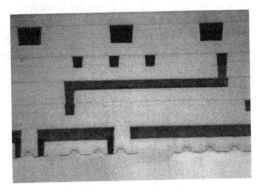

Fig. 1   XTEM picture of 0.18μm logic whole copper layer damascene process.

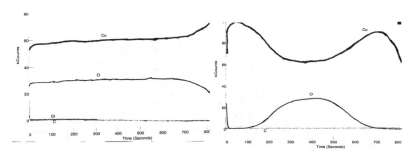

Fig. 2   AES depth profile of (a) Cu oxide and (b) post H₂/He reactive reduced Cu oxide.

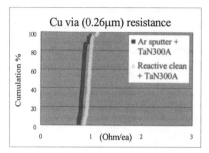
Cu via (0.26μm) resistance

■ Ar sputter + TaN300A

Reactive clean + TaN300A

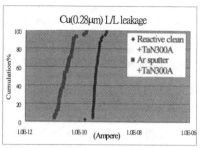
Cu(0.28μm) L/L leakage

♦ Reactive clean +TaN300A

■ Ar sputter +TaN300A

Fig. 3   Cumulative comparison plot of Ar sputtering and H$_2$/He reactive clean (a) via Rc and (b) line to line leakage.

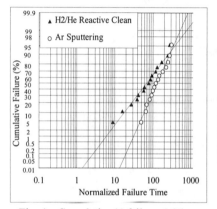
▲ H2/He Reactive Clean

o Ar Sputtering

Cumulative Failure (%)

Normalized Failure Time

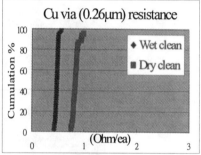
Cu via (0.26μm) resistance

♦ Wet clean

■ Dry clean

Cumulation %

(Ohm/ea)

Fig. 5   Cumulative via Rc comparison plot of dry clean and wet clean.

Fig. 4   Cumulative % failure, 0.26μm Cu via EM plotted on log-normal scale.

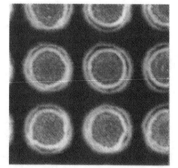
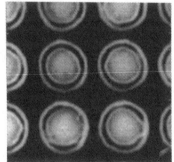

Fig. 6   Top view SEM picture of (a) dry clean and (b) wet clean.

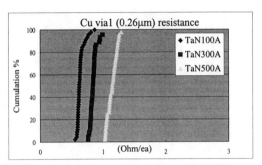

Fig. 7   Cumulative plot of via Rc vs. TaN thickness.

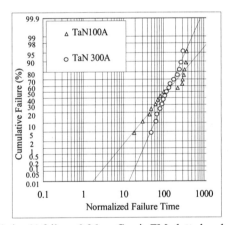

Fig. 8   Cumulative % failure, 0.26μm Cu via EM plotted on log-normal scale.

APPLICATION OF HIGH PRESSURE ARGON SPUTTER ETCH "CONTOURING" FOR THE INTEGRATION OF HIGH RELIABILITY LOW COST VLSI AND ULSI MULTILEVEL METALLIZATION

G.M. GRIVNA   Motorola, Inc., 2200 W. Broadway Rd., Mesa, AZ 85202
Keywords: Planarization, step coverage, gap fill, sputter etch, vias, capacitance, multilevel metal

ABSTRACT

By integrating a new high pressure argon sputter etch process, a low cost alternative to etchback planarization processing for dielectrics has been developed. Results for multilevel metallization show greatly enhanced via step coverage and up to 30% reduction in processing steps. Several other applications for high pressure argon sputter etch will also be discussed.

INTRODUCTION

To achieve adequate step coverage of "cold" (<350°C) sputtered or evaporated aluminum metallization and to reduce the possibility of conductive post etch residues, vertical and re-entrant surface profiles must be minimized prior to metal deposition. Dielectric planarization processes are normally used to achieve this requirement, to reduce the possibility of reflective notching, and to maintain depth of focus. However, when depth of focus is not a constraint, processes which provide local instead of global planararization are usually used due to the reduced cost compared to global planarization techniques [1]. Local planarization techniques typically consist of a spin-on or reflow "planarizing" coat followed by an etchback to the final desired thickness [1].

For VLSI processing with modern g-line and i-line steppers, depth of focus is on the order of 2 to 4 times greater than the dimension of the feature to be resolved [2]. Therefore, for VLSI applications with minimum 3rd metal feature size $\geq 1.0\mu$ and cumulative non-planarity $\leq 2.0\mu$, global planarity should not be a requirement. This is also evidenced by the fact that the predominant planarization processes used on VLSI parts with $\geq 1.0\mu$ minimum feature size consist of local planarization processes such as Boron-Phosphorous-Silicon-Glass (BPSG) reflow or resist etchback (REB). Unfortunately, with most global and local planarization processes, dielectric thickness can still vary by almost as much as the original thickness offset, dependent upon the dimensions of the features planarized [3]. Known exceptions to this would be isotropic etch planarization [4] or metal dummy planarization [5]. Conversely, the ideal situation for forming vias would be to retain the as deposited uniform dielectric thickness over all features while still enabling good metal step coverage. A new technique, "high pressure argon sputter etch" (HPS), provided this capability and a range of other applications.

THEORY

Low pressure argon sputter etch has traditionally been used to remove native surface oxide prior to metal deposition. A low pressure is used to minimize collisions between argon ions and sputtered material, thereby maximizing removal rate from deep vias or contact regions. However, by increasing the pressure at which sputter etching is performed, the attributes of the sputter etching process are changed. For high pressure argon sputter etch (HPS), pressure is increased with a commensurate increase in reflections and re-deposition of the sputtered material. At the higher pressures, the net sputter etch removal rate on flat surfaces drops to essentially zero (Fig. 1,2). However, in crevices and vertical or re-entrant profile areas, sputter etch removal is minimized while sputter re-deposition remains high. This results in a net deposition in these areas which become filled with re-deposited dielectric (Fig. 2). Since the sputtering yield is highest at an approximate angle of incidence of 45° [6] the surface becomes a series of flat areas and 45° oxide facets, resulting in dramatic improvement in subsequent metal step coverage and precise control of dielectric thickness. These results enable a wide range of new low cost processing applications for VLSI and ULSI processing.

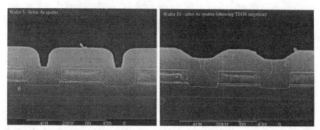

Fig. 1,2 Surface topography before and after HPS etch showing minimal surface oxide loss, re-deposited oxide in narrow gaps, and 45° facet on oxide edge.

RESULTS

Planarization

Dielectric planarization processes are normally used to improve metal step coverage or improve linewidth control of photo patterning. However, when depth of focus is not a constraint, processes which provide local instead of global planararization are usually used due to the reduced cost compared to global planarization techniques.

In standard VLSI CMOS and BiCMOS technologies using reflow or REB planarization techniques, across die contact aspect ratios can range between 0.5:1 to 1.5:1. However, without techniques such as hot metal reflow [7] or W plugs [8], metal step coverage for high aspect ratio contacts has typically been between 0% to 15% (Fig. 3), often relying on the refractory barrier layer of the first level metal to provide a current path. By using HPS to first contour a thin oxide surface followed by a doped oxide capping layer, a controlled contact depth was maintained over all features. Figs. 3,4 show the dramatic improvement in contact step coverage.

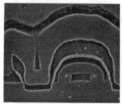

Fig. 3 1.2μ contact with BPSG reflow planarizing (5% step cov.)

Fig. 4 0.8μ contact with HPS etch (70% step cov.).

A common method to reduce the aspect ratio of vias has been to use an isotropic etch to form a "wineglass" shaped via opening [9]. For small vias, the isotropic etch or "flare" must be as deep as possible to minimize any vertical portion in the via which could cause overhang or blocking of the metal deposition. However, the depth of the flare is limited by variations in oxide thickness after the planarization process [3] thereby limiting metal step coverage in the deep vias. By etching the vias first, before planarization, uniform "wineglass" shaped, low aspect ratio vias were created. After resist strip, a HPS etch process was used to transform the wafer surface, including the vias, into a series of 45° angles and flat surfaces [10]. A short in-situ (10-30 sec) reactive ion etch was added to clean up any re-deposited oxide from the bottom of the vias. Fig. 5 shows three levels of metallization with HPS after the via etch process.

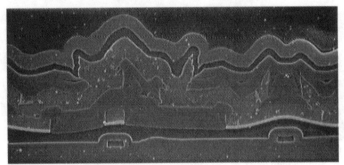

Fig. 5 Three level metallization with >70% metal step coverage using HPS etch after via etch.

Low Capacitance Dielectric Deposition

Conventional chemical vapor deposited (CVD) oxide has a relative dielectric constant of approximately 4.6, and fluorine doped oxides or organic dielectrics have lower relative dielectric constants of approximately 3.3 and 2.6 respectively. Air, by definition, has a dielectric constant of 1.0. Unfortunately, integration of air gaps or voids between interconnect features has proven difficult in the past. Using HPS, a three step process has been demonstrated for the integration of air gaps between interconnect features [11]. The first step consists of a thin deposition of a non-conformal dielectric such as a silane precursor plasma enhanced oxide (PECVD). This step will form a void and begin to pinch off between the narrow features. After the initial non-confromal oxide deposition, a HPS etch fills the wider gaps and stops seams or voids from rising any further above the interconnect surface. A final oxide deposition to the final desired thickness completes the process (Figs 6,7).

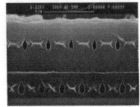

Fig. 4 Four layer metal structure showing air gaps between 1st and 3rd level metallization.

Fig. 5 Top view of 1st metal layer of via chain after dielectric planarization.

ILD Gap Fill

With most PECVD oxide deposition it is possible to form voids between narrow spaced interconnect features. As the gap between features gets wider, the top of the void reaches progressively higher above the metal surface (unless stopped at some point as discussed in the low capacitance section of this paper). Voids which extend above the metal surface have the potential to be opened up during subsequent processing such as chemical mechanical polish (CMP) planarization or "burst open" during subsequent high temperature or low pressure operations. By applying HPS early during the dielectric deposition process, HPS etch can be used as a low cost method to increase the gap fill ability of oxide films well beyond that of typical dep-etch-dep (DED) techniques (Figs. 6,7).

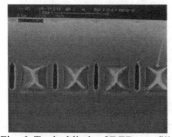

Fig. 6 Typical limit of DED gap fill.
Shown with 2.3:1 aspect ratio.

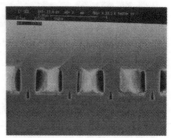

Fig. 7 Gap fill with HPS etch.

Non-Landed Vias

When via sizing or misalignment result in via openings that do not land on the underlying metal interconnect, the via etch can form a small trench next to the interconnect line which can cause problems with both via reliability and via filling. Figures 7,8 show the ability of the HPS etch process to fill the narrow gaps of non-landed vias with re-deposited oxide [10]. With increased sputtering time the entire via can become faceted (Fig.7). A short blanket anisotropic oxide etch removes any re-deposited oxide from the bottom of the via opening.

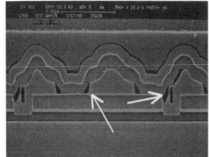

Fig. 7 Non-landed via after HPS etch and metal fill.
Arrows show sites of re-deposited oxide (after
delineation etch for X-section view).

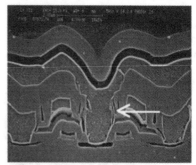

Fig. 8 0.6µ stacked via over contact
showing self-aligning "gap-fill"
feature of HPS etch (arrow shows
site of non-landed edge of via).

CONCLUSIONS

Through the use of high pressure argon sputter etch, a robust, low cost metal interconnect process has been implemented on VLSI MOS devices. By eliminating standard planarization processes and maintaining uniform thickness of dielectric over all features, contact and via processing was simplified while metal step coverage was increased from the typical 0-15% to approximately 70%-100%. HPS etch has shown additional uses for topography reduction, improved gap fill capability, non-landed vias, and low capacitance multilevel metallization.

ACKNOWLEDGEMENTS

The author would like to thank the Motorola analytical labs and acknowledge the support of several manufacturing sites for enabling this development work to be completed.

REFERENCES

1. M. Martinez, Solid State Technology, May 1994, p. 26.
2. S. Okasaki, J. Vac. Sci. Technol. B **9**, 2829 (1991).
3. L.K. White, J. Electrochem. Soc.: Solid State Science, **Vol 130**, No. 7, p. 1543, (1983).
4. G. Grivna, K. Kyler, Semiconductor International, July 1994, p 134.
5. M. Ichikawa, K inoue, K. Izumi, S. Sato, S. Mitarai, M. Kai, K. Watanabe., Proc. VMIC **104,** p.254 (1995).
6. H. Kotani, H. Yakushiji, H. Harada, J. Electrochem. Soc.; Solid State Science and Tech., **130**, 645 (1983).
7. C. Tracy, J. Freeman, R. Duffin, A. Polito, U.S.Patent No. 4 970 176 (13 November 1990).
8. Pete Singer, Semiconductor International, August 1994, p 57.
9. V. Grewel, H.P. Erb, P. Mokrisch, Proc. VMIC **CH-2488-5,** p.298 (1987).
10. G. Grivna, U.S. Patent No. 5 888 901, (30 March 1999).
11. G. Grivna, K. Johnson, and B. Bernhardt, U.S. Patent No. 5 641 712, (24 June 1997).

Optical Emission Diagnostic to Avoid Contact Etch Stop

S.C. McNevin*,K.V. Guinn*, M. Cerullo*, J.Ashley Taylor*, K. Tokashiki**
* Lucent Technologies, Murray Hill, NJ 07974 susanm@lucent.com
** NEC ULSI Laboratory, Sagamihara, Japan

ABSTRACT:

This paper will report on a new optical emission diagnostic for monitoring changes in the etch rate of small contacts.

INTRODUCTION:

The reproducible control of contact etching for small contacts is a major technological challenge. The open oxide area is too small (~1%) to use optical emission endpoint methods reliably to monitor changes in the contact etching. This paper will demonstrate, however, that the time dependence of the optical emission can be used as a simple in-situ manufacturing control. Variations in the time dependence of the optical emission during the etch will be shown to be directly related to variations in the etch rate of small contacts (<0.3 μm).

EXPERIMENT:

A commercial etching tool (Applied Materials HDP 5300) was used for most of the results presented here. [1] Similar optical emission results are also reported for several other commercial etching tools (LAM TCP 9100 and Sumitomo SWP).[2] This optical emission method thus seems to have applicability to a number of commercial etch tools.

The experiments [3-7] were performed with mechanically clamped 150 mm wafers. The baseline conditions for these experiments were: roof temperature (250°C), wall temperature (220°C), wafer temperature (-10°C), C_2F_6 gas flow (25 sccm), pressure (4.5 mTorr), source power (2500 W), bias power (600 W). Following this etch, an oxygen post etch treatment (PET) was done with the wafer still in the chamber. The conditions for this PET step were: O_2 gas flow (40 sccm), source power (2700 W), bias power (150 W), pressure (9 mTorr) with the same wafer and chamber temperatures as the main etch. The duration of this PET step was either 20, 45 or 70 seconds

All of the etching was done in a chamber which was unused for at least 1 hour. The chamber was first warmed up with an Ar/O_2 preheat: source power (2700 W), bias power (0 W), pressure (15 mTorr), Ar/O_2 gas flows (60/40 sccm: 3 minutes, 95/0 sccm: 2 minutes). This preheat was followed by four Si wafers that used the same etch/PET recipe as was used for the patterned wafers which followed. This procedure was used to standardize the starting conditions of the experiments.

The wafers were patterned with Arch 2 DUV photoresist (7400 A) with ~1% open area. These patterned wafers were placed at regular intervals in the cassette to monitor process drift. Blank DUV wafers were used in the intervening wafer positions. The SiO_2 was deposited to a thickness of either 1.02 or 1.55 μm.

633

The optical emission signal was measured by means of a spectrometer. The measurement was through a window along an axis parallel and approximately ~3 mm above the wafer surface.

RESULTS:

Figure 1 shows the optical emission observed for three different commercial contact etchers. All three tools show prominent SiF (440 nm) and C_2 (515 nm) emissions. These are well resolved, and can easily be detected with a simple optical filter arrangement.

By varying the duration of the PET treatment, the chamber can accumulate deposits which decrease the etch rate of the small contacts. This work will show an identical etch recipe but with varying PET duration, which results in a condition where the etch rate in small contacts decreases throughout the processing of the cassette. It will be demonstrated that the time dependence of the SiF and/or C_2 optical emission can be used to detect this variation in the etch rate of small contacts.

Figure 2 illustrates the C_2 emission as a function of etching time (20-90 s) for the three different PET durations. The results for the first wafer processed are compared to those for the last wafer in the sequence. The 20 s PET has the biggest deviation between the first and last wafer. This 20 s PET also had the largest change in the etch rate of small contacts across the cassette ("most etch stop" in figure). While the 40 s PET has better reproducibility over the wafer sequence than the 20 s, it also shows a deviation between the beginning and end wafers. It will be shown that the 40 s PET also has a drift in the etch rate of small contacts, although it is not as bad as the 20 s PET. In contrast to this drift in the optical emission time dependence for the 20 and 40 s PET, there was no change in the optical emission time dependence over 45 wafers for the 75 s PET. It will be shown that this stable optical emission time dependence is correlated with the stable etch rate of small contacts for the 75 s PET clean. Figure 2 illustrates the reproducibility of the C_2 emission, but the results for the SiF emission are similar.

It will now be demonstrated that the emission time dependence reproducibility illustrated in Figure 2 is correlated with the reproducibility of the etching rate in small contacts (<0.3 µm). The results for the 40 and 75 s PET clean will be compared.

The etch depth and emission time dependence results are summarized in Figures 3 and 4. In order to capture the time dependence shown in Figure 2, the ratio of the emission signals at 90/20 s for each wafer are normalized to the first wafer's 90/20 s ratio. If the emission time dependence is constant, there is no change in this ratio. These values are plotted in the bottom half of Figures 3 and 4.

The 40 s PET duration results in a changing optical emission time dependence in Figure 2. This results in the decreasing 90/20 s ratio in Figure 3. The longer 75 s PET duration results in an unchanging optical emission time dependence in Figure 2. This results in the constant 90/20 s ratio observed in Figure 4.

CONCLUSION:

The ability to fabricate reliable interconnect depends on the ability to etch small contacts reproducibly. Either etching too much or too little results in catastrophic failure of the interconnect at the contact. This emission time dependence technique can be used to monitor the contact etch rate in a manufacturing environment using simple optical filters.

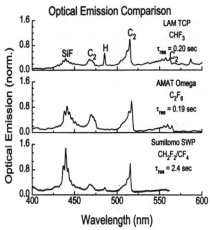

Figure 1: Spectra during contact etch for three commercial etchers. All spectra are dominated by the C_2 and SiF peaks.

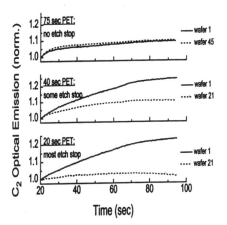

Figure 2: C_2 reproducibility for 20/40/75s O_2 post etch treatment (PET). The longest PET results in steady chamber conditions with constant emission time dependence and etch rate.

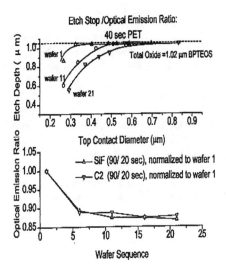

Figure 3: Etch depth and emission for 40 s PET as a function of the wafer sequence. Varying time dependence correlated with varying etch.

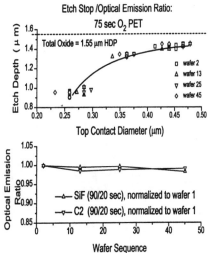

Figure 4: For 75 s PET, time dependence of emission and etch rate are constant.

635

ACKNOWLEDGMENTS:

The authors would like to thank Dale Ibbotson, John T.C. Lee, and Vince Donnelly for helpful discussions. The authors would like to acknowledge the NEC/Lucent Technology Development Agreement which permitted K. Tokashiki to visit Lucent (Murray Hill) and S. McNevin to visit NEC (Sagamihara) and work on this project. They would like to thank the advanced DUV lithography provided by May Chen, Al Timko and John Frackoviak and the SCALPEL projection electron beam lithography provided by Tony Novembre and group.

REFERENCES:

1. J. Marks, K. Collins, C.L. Yang, D. Groechel, P. Keswick, C. Cunningham, and M. Carlson, Proc. SPIE **25**, 1803 (1992).
2. S.C. McNevin, Jpn. J. Appl. Phys., **36**, 2464 (1997).
3. S.C. McNevin and M. Cerullo, J. Vac. Sci. Technol. A **15**(3), 659 (1997).
4. S.C. McNevin and M. Cerrullo, J. Vac. Sci. Technol A **16**(3), 1514 (1998).
5. K. Guinn, K. Tokashiki, S.C. McNevin, and M. Cerullo, J. Vac. Sci. Technol. **14**(3), 1137 (1996).
6. S.C. McNevin, J. Vac. Sci. Technol. A **13**, 797 (1995).
7. S.C. McNevin, K.V. Guinn, and J. Ashley Taylor, J. Vac. Sci. Technol. B **15**(2), 214, 1997.

Characterization of a Novel Method of Cleaning Wafer Back Sides and Effecting a Bevel Etch in a Single Processing Module

C. Dundas, T. Ritzdorf, Gary Curtis, Steve Peace
Semitool Inc., 655 W. Reserve Dr., Kalispell, MT. 59901, cdundas@semitool.com

ABSTRACT

This paper discusses the effects of varying different parameters of a process to achieve copper removal on a wafer back side and bevel, and create a well-defined edge exclusion on the front side of the wafer with a single processing platform. Different etch chemistries were investigated in a novel processing platform, and post clean copper contamination levels, etch intrusion, and etch profile were found to be strongly influenced by etch chemistry, chamber spin speed, and chemical flow rate. After optimization of process parameters, copper back side contamination levels were reduced to less than $1*10^{11}$ atoms/cm^2 and a well-defined, front side edge exclusion was obtained.

INTRODUCTION

As semiconductor manufacturers move to begin volume manufacturing of copper interconnects, integration issues become a major concern as cross contamination of existing process streams can lead to device failures. Copper contamination of wafer back sides has the potential to contaminate processing, metrology, and handling equipment which would in turn contaminate wafers which come into contact with it.

Typically, copper wafers have a seed layer deposited on them by PVD or CVD methods. Both methods can deposit copper onto the wafer bevel, and in many cases this deposit is non adherent and can flake off in subsequent processing steps such as annealing or CMP. Additionally, some manufacturing scenarios deposit layers of material to the edge of wafers. Subsequent layers are deposited with an edge exclusion, leaving the previously deposited layers exposed. Many of these layers allow copper to be deposited on them, but the adhesion is very poor and flaking during post processing is observed. A typical copper example might be an exposed barrier layer such as Ti/TiN being exposed to copper plating solution. Following electrochemical deposition, the barrier layer would have a copper film of low quality which would flake off easily in CMP. Removal of flaking material before CMP processing is desirable as the flakes have the potential to cause scratches in the polished surface, resulting in yield losses.

Covering the exposed barrier layer with a full coverage seed layer would eliminate copper metal from flaking off the barrier and also have the added benefit of increasing usable area on the wafer surface. Even in this case, copper deposited on the bevel during the seed layer and electrochemical deposition would need to be removed, as it too can flake off and/or cause cross contamination of metrology tools. A clear area inboard of the wafer bevel may also be necessary for reliable processing; many clamp rings are very sensitive to surface characteristics.

Processes and chemistries do exist which address these issues. Back side contamination can be removed in immersion and spray type chambers using various chemistries though the front side of the wafer needs to be protected from the cleaning solution. Traditionally, a protective layer of photoresist is used, which must be later removed. Back side contamination removal using spin etching of the wafer is also available. Technology designed to clean bevel areas and edge exclusions is more limited. It may be possible to modify edge bead removal systems for the application.

Rather than add additional processing steps, a single platform which performs both back side cleaning and bevel etching is desirable. In such a platform, suitable chemistries would need to be identified and process parameters would need to be evaluated for their influence on process results. This paper focuses on identification of a back side cleaning chemistry and process characterization of parameters influencing bevel etch using a novel processing chamber.

EXPERIMENTAL DETAILS

Experiments were performed to identify a suitable back side cleaning and bevel etch chemistry, then to characterize process parameter effects on bevel etch/edge exclusion quality in two separate hardware sets manufactured to produce a nominal 1.5mm and 3mm edge exclusion. The general process chamber configuration consists of a lower rotor which holds the wafer and forms the bottom of the processing chamber and an upper rotor which forms the top and sides of the processing chamber. The upper rotor is seated on the lower rotor during processing, and the combination of the two rotors creates a mini environment which provides the fluid flow characteristics suited for back side cleaning and bevel etching.

Unless otherwise noted, all back side cleaning results were obtained using bare silicon wafers with the polished side of the wafer contaminated by placing in an acid copper solution. The wafers were subsequently transferred into a novel Semitool process chamber, rinsed of acid copper solution with copper ion concentration of 17 g/l, back side cleaned, and bevel etched. All variables were held constant except etch chemistry. The evaluated response was post clean copper contamination. Post clean analysis was done using a TXRF with detection limits being roughly 7-9E10 atoms/cm^2. Suitable post clean copper contamination levels were defined as less than 1E11 atoms/cm^2 [1]. An uncontaminated wafer and an un-cleaned, contaminated wafer were included for control purposes.

Bevel etch results were obtained using wafers which had 15,000Å of copper electrochemically deposited using a Semitool LT-210 plating tool equipped with a sealed ring contact. The wafers had a Ta barrier layer of 300Å in thickness and a PVD seedlayer of 1000Å in thickness. The wafers had a post ECD annulus of seedlayer around the perimeter of the wafer roughly 4.5 mm wide. The seedlayer annulus was etched using either hardware manufactured for 1.5mm edge exclusions or 3mm edge exclusions and the quality of the etch as a function of spin speed, flow of chemistry, flow of nitrogen, and flow of water was evaluated. Etch quality evaluation criteria consisted of etch intrusion (edge exclusion) and etch definition (smoothness). Intrusion was defined as the average of the distance from the edge of the wafer to the beginning of the copper film, taken over eight evenly distributed sites, as shown in Figure 1. Smoothness was the amount of roughness of the etched edge as perceived visually, and rated on a scale of 1 to 10, with 10 being perfectly smooth.

RESULTS

Of the various chemistries tested for back side cleaning effectiveness, several appear to be suitable. All chemistries tested resulted in copper contamination levels less than that of the un-cleaned control wafer, which had $1*10^{12}$ atoms/cm^2 copper. Closer examination of the data shows that suitable cleaning levels of less than $1*10^{11}$ atoms/cm^2 copper were obtained generally only as an average of all data points. Figure 2 depicts average post clean contamination levels as columns, with bars which indicate the total measurement range. Post clean contamination results agree with previously published material [2, 3].

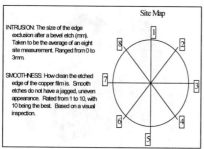

Figure 1: Experimental Responses for Bevel Etch and Edge Exclusion

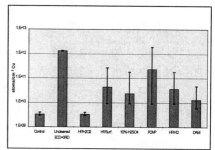

Figure 2: Evaluated Chemistries

HF:Surfactant shows an average contamination level of $6*10^{10}$ atoms/cm^2, but the data range extends from near $1*10^{10}$ to $5*10^{11}$ atoms/cm^2 copper. Both HF:HCl and 10% H_2SO_4 clean back sides to 5-$6*10^{10}$ atoms/cm^2 copper on average; but ranges spread from $1*10^{10}$ to $4*10^{11}$ atoms/cm^2 copper. As can be seen, almost all chemistries, except PCMP (post CMP), yield suitable cleaning results on average. However, only two cases exhibit a data range which is completely below $1*10^{11}$ atoms/cm^2 copper. These two cases, HF:H_2O_2 and DAM (Dilute Acid Mixture), were determined to have the desired cleaning characteristics, effectively removing copper to below $1*10^{11}$ atoms/cm^2 and, in the case of HF:H_2O_2, to detection limits of the analysis equipment. The remainder of this paper focuses on results achieved using DAM. The mixture of HF:H_2O_2 was not chosen as market research has shown it to have some compatibility issues with existing process streams.

Using DAM, characterization of process parameters and their influence on bevel etch and intrusion size and smoothness was performed. Figure 3 contains experimental factors and their respective levels. Two hardware sets were evaluated, designated Set A and Set B. Set A hardware was manufactured to deliver a nominal 1.5mm exclusion zone while Set B was manufactured to deliver a 3mm nominal exclusion zone. In both cases the bevel etch variables which produced the greatest responses included spin speed and chemical flow rate. Nitrogen flow to the bottom of the wafer and DI H_2O flow rates did not exhibit a significant effect on bevel etching and edge exclusion generation and so will not be discussed further.

Hardware Set A was tested first, and showed strong sensitivity to spin speed and chemical flow rate for both intrusion and smoothness; nitrogen flow showed a small influence on intrusion. Figure 4 is a response surface which shows a maximum smoothness achieved near midpoint RPM and greater than midpoint chemical flowrate. Figure 5 shows that increasing chemical flow rate and decreasing spin speed leads to an increase in edge exclusion. This same trend occurs at all nitrogen flows from level -1 to +1, though Figure 6 shows that level -1 N_2 flow rates result in a lower achievable intrusion range. Assuming nitrogen flows to be at the +1 level, Figure 7 is a graphical representation of a process window of Hardware Set A. As the process window indicates, hardware set A should be operated near midpoint RPM and at chemical flow rates greater than midpoint in order to achieve the desired edge exclusion and smoothness of 1.5mm. Hardware Set B underwent similar testing, though DI flow and bottom N_2

variables were eliminated. Intrusion results depicted in Figure 8 show that as chemical flow rate increases, etch intrusion increases; etch intrusion is also increased with decreasing spin speed.

VARIABLES	RANGES	UNITS
DI to top	-1, 0, 1	lpm
DI to bottom	-1, 0, 1	lpm
N2 to top	-1, 0, 1	lpm
N2 to bottom	0, 1	lpm
Chemical flow rate	-1, 0, 1	lpm
Spin speed	-1, 0, 1	RPM

Figure 3: Parameters and Ranges

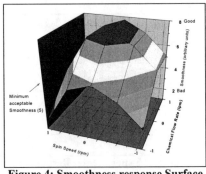

Figure 4: Smoothness response Surface Hardware Set A (1.5mm)

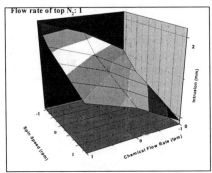

Figure 5: Intrusion Response Surface @ +1 N$_2$; Hardware Set A (1.5mm)

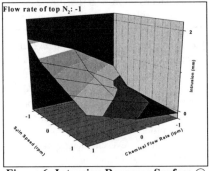

Figure 6: Intrusion Response Surface @ -1 N$_2$; Hardware Set A (1.5mm)

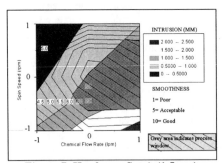

Figure 7: Hardware Set A (1.5mm) Process Window

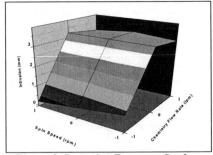

Figure 8: Intrusion Response Surface Hardware Set B (3mm)

In Hardware B, the process shows much greater sensitivity to chemical flow rate than spin speed, though it is not as sensitive as Hardware Set A. As in Hardware Set A, spin speed and chemical flow rate exhibit the strongest effect on smoothness as shown in Figure 9. As chemical flow rate is increased, smoothness increases. Spin speed has a much smaller effect on smoothness in Hardware Set B, but the trend of increasing spin speed to achieve greater smoothness is still present. Figure 10 represents a process window of Hardware Set B. In order to achieve the desired 3mm edge exclusion, Hardware set B should be operated at flow rates greater than the tested midpoint and at spins speeds near the tested midpoint.

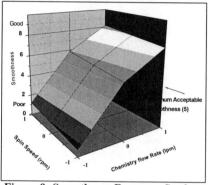

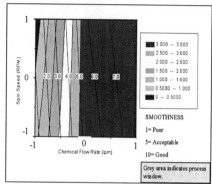

Figure 9: Smoothness Response Surface Hardware Set B (3mm)

Figure 10: Hardware Set B (3mm) Process Window

CONCLUSIONS

Dilute acid chemistry used at optimized flow rates and spin speeds has been shown to clean wafer back sides to less than $1*10^{11}$ atoms/cm^2 while simultaneously cleaning the bevel of the wafer and producing a very smooth edge exclusion of controllable size. Process parameter experimentation has shown spin speed and chemical flow rate to be important factors in bevel etch size and definition. No apparent effect on edge exclusion was found for N_2 to bottom of the wafer or DI flow rate. Using this novel processing chamber it is possible to address both back side copper contamination and bevel etch concerns in a single, controllable platform. This platform facilitates the simultaneous removal of copper from the edge of the wafer and cleaning of the backside of the wafer.

REFERENCES

1. D. A. Ramappa and W. B. Henley, Meeting Abstracts of 193rd Electrochemical Society Meetings **98-1**, Abstract No. 492 (1998).

2. H. Morinaga and T. Ohmi in Proceedings of the 4th International Symposium on Cleaning Technology in Semiconductor Device Manufacturing, edited by R. Novak, and J. Ruzyllo (Electrochemical Soc. Proc. **95-20,** Chicago, IL, 1995) pp. 257-268.

3. V.M. Dubin, E.H. Adem, J. Bernard, D. Schonauer, and J. Bertrand in <u>Advanced</u> <u>Metallization and Interconnect Systems for ULSI Applications in 1997</u>, edited by R. Cheung, J. Klein, K. Tsubouchi, M. Murakami, and N. Kobayashi (Mater. Res. Soc. Proc. , San Diego, CA, 1997) pp. 421-426.

IMPACT OF POST WINDOW ETCH CLEANS PROCESS ON RELIABILITY OF

0.25um VINTAGE WINDOWS

Y. S. Obeng*, J. S. Huang, S. H. Kang, X. Lin, and A. S. Oates
Bell Laboratories, Lucent Technologies,
9333 S. John Young Parkway, Orlando, Fl. 32819
*electronic mail: yobeng@lucent.com, phone: (407)-371-7565

ABSTRACT
In paper, we examine the impact of post-window cleaning chemistry and process flow on the long-term electromigration reliability of 0.25um vintage technology. Specifically, we illustrate how some of the constituents of the cleaning solvent interact with exposed metal at the bottom of the windows to alter the electromigration characteristics of the interconnects.

1.0 Introduction

Currently, most of the back-end-of-line (BEOL) cleaning processes involve the use of liquid based strippers. These strippers are usually organic solvent blends, consisting of a majority solvent base, an amine (or a nucleophille), and additives such as corrosion inhibitors, complexing ligands, and surfactants. The cleaning mechanisms employed by these strippers range from dissolution through redox reactions. In BEOL, the most common cleaning mechanisms involve penetration, solvent swelling, dissolution, and recently reduction. [1]

The cleaning agents with high concentrations of organic solvents are formulated to target organic sidewalls. In high aspect ratio windows this type of sidewall only occurs at the top one-half or third of window. The sidewalls in the lower half of the windows contain mostly inorganic and organometallic residues.[2]

We have examined the impact of post-window cleaning chemistry and process flow on the long-term electromigration reliability of 0.25um vintage technology. Specifically, we will show how some of the constituents of the cleaning solvent interact with the exposed metal at the bottom of the windows to alter the electromigration characteristics of the interconnects.

2.0 Experimental

Two-level metal short-loop testers were used in these studies. Wafers were split for cleaning with various sidewall polymer removers (Solvent-A, Solvent-B, or Solvent-C) after windows were etched. The solvent and / or processes used in the cleaning are described in Table 1.

After processing, the wafers were re-annealed in air at 250C for 500 hours, re-tested, and packaged for electromigration (EM) testing. EM was conducted at 250C, with a current density of 1 MA/cm^2.

Conference Proceedings ULSI XV © 2000 Materials Research Society

TABLE 1: Description of Solvents Used

Solvent / Process	Major Components
A	DMAC-based with 8-hydroxyquninoline corrosion inhibitor
B	NMP based with cathecol corrosion inhibitor
C	NMP based with cathecol corrosion inhibitor
D	A fluoride salt in propylene-glycol
E	Down-stream fluoride plasma

3.0 Results and Discussion

Figure 1 shows the dependence of the contact resistance of a chain of windows, of nominal size, on the post-window etch cleaning chemistry. The windows cleaned with solvents formulated to attack organic residues, e.g., Solvents-A, B, or C, showed about the same contact resistance as uncleaned windows. However, the contact resistance drops when the windows are cleaned with a combination of organic and inorganic residue removers (e.g., Solvent C followed by Solvent D or Solvent C followed by Process E).

Auger spectroscopic analyses of the windows cleaned with Solvent A followed by Process-E and Solvent B followed by Solvent D show that the Solvent A based cleaning process left carbon-rich residues both on the sidewall and on the exposed Al runners at the bottom of the windows. In contrast, the carbon-rich contaminants were found only on the exposed Al runners in windows cleaned with Solvent B. We postulate that the contaminants on the exposed metal are the Al-complexes of the corrosion inhibitors in the solvents used (i.e., 8-hydroxyquinoline in Solvent A and cathecol in Solvent B respectively).[3,4]

These conclusions are consistent with our previous results on 0.5um-vintage metallization schemes.[2] In that study the post window etch clean was split between a DMAC based solvent, containing cathecol, and NMP-based solvent with no corrosion inhibitor. The former solvent left a carbon-rich residue at the interface between the Al line and the Al contact. Also, in that work we observed that the DMAC-based cleans resulted in wider windows due to etchout / erosion.

Figure 2 compares the distribution of electromigration lifetimes as a function of window cleaning chemistry / process. The samples tested were all annealed after processing as described above. The data from unannealed A+E split was included in the figure for reference only. Inspection of Figure 2 reveals the following:

- While uncleaned windows afforded the shortest lifetimes, Solvent A cleaned windows had the longest lifetimes.
- Solvent A cleaned windows produced a wide range of lifetimes.
- The lifetimes of windows cleaned with Solvent A only improved with anneal (i.e., the EM enhancement is thermally activated)
- The lifetimes of annealed samples cleaned with Solvents B and C are comparable to those of 'as processed' samples cleaned with Solvent A.

The resistance increase due to electromigration has been attributed to void growth due to accumulation of vacancies at local vacancy-flux divergence sites. Specifically, for the kind of testers used in this study the increase in resistance is attributable to material depletion around the W-plug. Figure 3 summarizes the results from a finite element (FEA) simulation of the resistance evolution in a 0.25um-vintage plug. Basically, the figure indicates that line resistance increase (ΔR) is not significant when depletion length is less than the W plug width.

Furthermore, sensitivity analysis using interfacial films with a wide range of electrical and thermal conductivities did not show any impact on the rate of resistance change as a function of material depletion as shown in Figure 3. Thus we infer that the contaminants at the W-plug / Al-runner interface did not alter the inherent flux divergence.

From the simulation, the unique enhanced electromigration lifetimes of windows cleaned with Solvent-A is attributed to the retardation of material depletion around the W-plug. This may be due the altered interfacial energy between the W-plug and the Al-runner due to the formation of corrosion inhibitor-Al complexes (e.g., tris-(8-hydroxyquinoline)-aluminum(III)). We propose that the complexation of the exposed Al at the bottom of the window by the corrosion inhibitor retards the material depletion due to flux divergence at the W-plug /Al-runner interface, hence improves the EM performance.

4.0 Conclusions

Care must be exercised when selecting post-etch solvents and processes. This is especially true in the case of post-window etch cleans, as any interfacial residues will have reliability consequences. For example, although Solvent A cleaned windows had the longest lifetimes, they also have a wide range of lifetimes (high lifetime sigma). This will necessarily produce poor FIT rates. Our data also show that interfacial residues may result in altered electromigration characteristics, which may not be apparent during burn-in. The EM changes may be thermally activated and or time dependent.

For Solvent A cleaned windows, we observed a unique enhanced electromigration lifetimes with post-process anneals. We propose that corrosion inhibitor in the solvent binds to the exposed Al at the bottom of the window to form a carbon rich residue. The residue alters the interfacial energy between the W-plug and the Al-runner. This retards material depletion around the W-plug, thus improving the EM performance.

Acknowledgements

We acknowledge the support of the management of the VLSI Technology Lab for material support and the Diagnostic Laboratory of Cirent Semiconductor for analytical support.

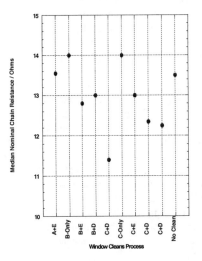

Fig. 1: Contact Resistance Variation With Window Cleans Chemistry and Process

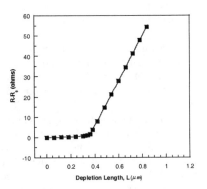

Fig. 3: FEA Simulation of Resistance Change vs. EM Induced Material Depletion

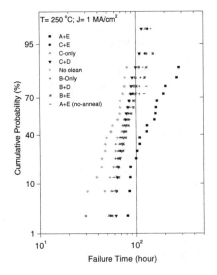

Fig. 2: Electromigration Lifetime vs. Windows Cleans Chemistry and Process

REFERENCES

[1] 'A Proven Sub-Micron Photoresist Stripper Solution for Post Metal and Via Hole Processes', Lee, W. M., EKC Technologies, Inc. technical bulletin, 1993

[2] Obeng, Y.S., Raghavan, R. S., 'Back End Chemical Cleaning in Integrated Circuit Fabrication: A Tutorial' in Mat. Res. Soc. Symp. Proc. Vol. 477, 1997, pp. 145-157

[3] Martell, R.K, Motekaitis, R. J., Smith, R. M., Polyhedron, **1990**, 9, 171

[4] Higginson, K. A., Zhan, X. M., Papdimitrakopolos, Chem. Mater., **1998**, 10, 107-1029

R&D Program for Reduction of PFC Emission at ASET

Hironori Matsunaga, Takuya Fukuda, Nobuo Aoi and Hitoshi Itoh
Association of Super-Advanced Electronics Technologies (ASET)
292, Yoshida-cho, Totsuka-ku, Yokohama Kanagawa 244-0817, Japan

ABSTRACT

The Association of Super-Advanced Electronics Technologies (ASET) has launched a program to develop a new metallization process in order to reduce the consumption of energy and perfluoro compounds (PFCs) in dry etching. It is scheduled to run for 5 years from 1999 to 2003. The program aims to develop a new metallization process employing PFC-free etching to improve the RC delay of ULSI devices. This paper gives an overview of the program and outlines the development approach, which is based on the use of ultralow-k materials ($k < 1.5$) in the metallization process without PFCs.

INTRODUCTION

Nowadays, a large amount of PFCs (CF_4, C_2F_6, etc.) are consumed as the reaction gas in the plasma etching of interlayer dielectrics (ILDs), such as SiO_2. However, PFCs are thought to be a significant factor in the greenhouse effect, in which their global warming potential (GWP) and lifetime are 10,000 times and 100 times greater than that of CO_2, respectively, as shown in Table I. COP3 (The Third Conference of the Parties to the United Nations Framework Convention on Climate Change), which was held in Kyoto in 1997, established targets for the reduction of the emission of greenhouse gases, to which HFC, PFC, and SF_6 were added. Thus, it is very important for the semiconductor industry to develop alternative processes that do not use PFCs in order to preserve the global environment.

The program pursues three lines of research and development aimed mainly at achieving a reduction in PFC emissions in dry etching:

(1) high-efficiency dry-etching technology;
(2) dry etching using alternative PFC gases; and
(3) the use of low-k ILD films and alternative processes without PFCs.

Item (1) is intended to improve the efficiency of dry etching in order to shorten the etching time and thereby reduce PFC consumption. The key point is the selection of reactive species with a high etching yield for ILDs based on the reaction mechanism involved in the etching. The purpose of Item (2) is to clarify the feasibility of using alternative PFC gases with a low GWP in order to reduce deleterious environmental effects. The key point here is the development of alternative etching gases with a low GWP and a high etching yield for ILDs. Even if success is achieved with the first two items, that does not mean that the zero emission of PFCs has been achieved, because any conventional or alternative PFCs that are used will have a GWP. On the other hand, if organic films are used for ILDs, they can be etched with non-PFC gases, such as O2, N2, etc. So, the development of organic ILDs, as specified in Item (3), is very important for the achievement zero PFC emissions.

Table I. GWPs and lifetimes of typical PFCs.

Gas	GWP_{100}	Lifetime(years)
CF_4	6,500	50,000
C_2F_6	9,200	10,000
C_3F_8	7,000	2,600
C_4F_8	8,700	3,200
CO_2	1	200

R&D APPROACH OF ULTRALOW-k ILD MATERIALS

As devices are scaled down, the increase in the parasitic capacitance and resistance of interconnects causes the RC delay to increase. Figure 1 shows the calculated dependence of RC delay time on metal pitch for a scaled metal height and a fixed metal width and interlayer spacing. For conventional Al/TEOS interconnects, the RC delay time rises sharply, as the pitch becomes smaller. But the use of Cu/low-k interconnects greatly suppresses the increase in the delay. These results indicate that the use of a low-k ILDs and Cu metal for high-density interconnects is a very effective way to reduce interconnects delay. Moreover, their use will also reduce the number of metal layers required, as sown in Fig. 2[1], which will help to reduce the energy consumed during LSI fabrication.

The dielectric constant of ILDs should be as low as possible; but currently the most common ILD used in actual LSI processes is SiO_2 ($k = 4.2$). Figure 3 lists several low-k materials reported to be suitable for ILDs. The main focus is on organic silica film, e.g. SOG, whose porous film structure provides a low effective dielectric constant. However, the dry-etching of ILDs, including Si, requires the use of PFCs. So, we will try to find suitable low-k organic polymer materials for use as ILDs. Generally, the thermal and mechanical stability of organic materials is quite poor. This is why very few low-k polymer materials have been reported to be good candidates. The development of ultralow-k materials is a very difficult and challenging subject.

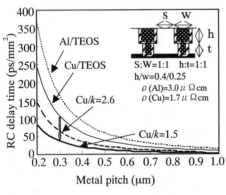

Fig. 1. RC delay time vs. metal pitch for scaled metal height and fixed metal width and interlayer spacing.

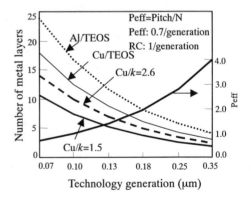

Fig. 2. Number of metal layers in the future[1].

Our target k value is 1.5 because this is desirable for the 0.07-μm generation of technology in the future [2]. Low-k organic ILDs will be developed in two ways. One way is through the structural design of organic molecules. In order to illustrate our concept of material research, Fig. 4 shows an example of a possible molecular model consisting of a benzene ring and carbon, which bond together to form a diamond structure. The molecular frame of this structure has nanometer-level spacing, which is very effective in reducing the density of an ILD. The dielectric constant of ILDs should decrease roughly in proportion to molecular density. Figure 4 shows some calculated values of molecular density and dielectric constant, which decrease as the number of bridged benzene rings increases. In the design of a molecular structure, thermal stability and mechanical strength must be taken into account to ensure suitability as an ILD. So, we will try to devise a simulation method that will enable the estimation of material properties from molecular structure.

The other way is to develop porous organic thin films. Very little research has been done so far on the use of porous polymer films for ILDs [3]. But in the field of chemistry, several methods of forming mezoporous materials have been proposed, for example, by using a molecular template, microphase separation phenomena, and so on. So, it should be possible to develop a new method of forming porous polymer materials. For porous films, the k value of the mother material should be as low as possible. If it is not low, the porosity of the film must be made very high to obtain a low-k ILD, as can be seen in Fig. 5, which shows that the porosity of TEOS must be above 85% to obtain a low enough k. The problem is that high porosity usually means low mechanical strength. So, the mother material should have a low k so that the porosity of the ILD can be kept reasonably low. This is why the molecular design mentioned above is so important.

Ultralow-k materials that are developed must be examined by integrating them into an actual metallization process. The typical issues regarding integration technologies are shown in Fig. 6, and the program will also attempt to resolve them. An overview of the development schedule is shown in Fig. 7. In Phase 1, basic technologies for materials, dry-etching, and other elemental processes will be developed. In Phase 2, these technologies will be evaluated by process integration.

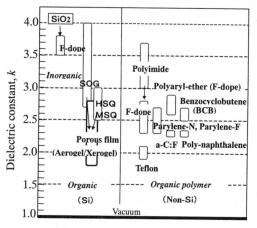

Fig. 3. Dielectric constant of low-k materials.

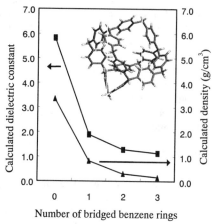

Fig.4. Example of low-k molecular structure.

651

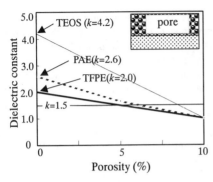

Fig. 5. Dependence of dielectric constant on porosity.

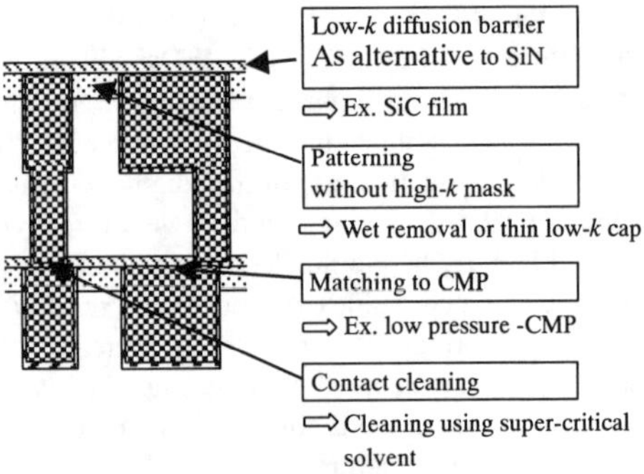

Fig. 6. Issues of integration technologies to be resolved.

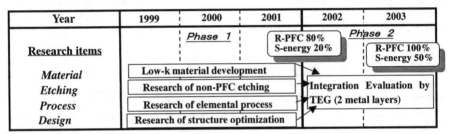

Fig 7. Overview of development schedule.

CONCLUSIONS

A new ASET program for reducing PFC emissions in metallization processes has been launched. The program aims to develop ultralow-k ILDs ($k < 1.5$) that can be etched without PFCs. Furthermore, the feasibility of integrating ultralow-k materials into an actual metallization process will be investigated.

ACKNOWLEDGMENTS

This program is being performed under the management of ASET in Ministry of International Trade and Industry (MITI) and New Energy and Industrial Technology Development Organization (NEDO).

REFERENCES

1. M.T.Bohr, IEDM Tech. Digest,p.241-244(1995).
2. "National Technology Roadmap for Semiconductors: interconnect", 1997 Edition, (Semiconductor Industry Association).
3. S.Ukishima, M.Sato, M.Iijima, Y.Takahashi, S.Sasaki, T.Matsuura and F.Yamamoto, Proceeding of DUMIC Conference, p.267-271(1999).

Process Modeling

EQUIPMENT SIMULATION FOR COPPER METALLIZATION TECHNOLOGY

Christoph Werner and Alfred Kersch, Infineon Technologies, Corporate Research
D-81730 Munich, Germany, Tel +4989 636 45559, FAX +4989 636 47069
e-mail: Christoph.Werner@infineon.com

Abstract:
We discuss a variety of process models and simulation programs that have been used to support the development and optimization of copper metalization technology. The process steps considered include the barrier and the seed layer formation by ionized physical vapor deposition, the copper fill by an electrochemical deposition process and the chemical mechanical polishing process for planarization of the deposited layer. The capability and the limits of currently available models have been evaluated and their extensions to cover the needs of today's copper process development are pointed out.

INTRODUCTION

Copper metalization shows a number of advantages over the more traditional metals tungsten and aluminum, which have triggered a rapid transition in the main stream of high performance CMOS technology. Since the process technologies used for the deposition and the planarization of copper layers are rather different from the aluminum and tungsten technologies, the simulation tools that can support the process development cannot simply be adopted for the new requirements. Moreover, a variety of new physical effects have to be considered in the simulation, such as angular dependent scattering rates of the copper ions at the side walls of contact holes, the shielding of the electrostatic potential in an electroplating bath or the lateral pressure distribution which the CMP pad exerts on a nonplanar surface topography. Depending on the depth of physical understanding the different process steps are simulated by models with a varying position between exact physical accuracy and pure empirical fitting.

SIMULATION MODELS

Barrier and seed layer formation

Recently, ionized physical vapor deposition (IPVD) has become an important technique for depositing films over high aspect ratio features. This process is characterized by a highly directional flux of energetic ions and atom-surface interactions which include scattering, resputtering and adsorption. To determine the ion distribution function as a function of process parameters, a collisional plasma sheath model including a time dependent, selfconsistent field is applied [2]. In some cases, long throw sputtering is used for the seed layer instead of IPVD. In this case the angular distribution of the particles is calculated with a discrete simulation Monte Carlo method taking the gas rarefaction due to energetic collisions into account [8]. For the formulation of the atom-surface interaction model, a rather fundamental approach has been taken where the reaction probabilities of the atoms/ions as a function of incoming angle and energy are first calculated with molecular dynamics calculations [1]. The rates are then incorporated into a feature scale simulator. With these models the bottom coverage and the sidewall coverage in a structure can be determined. The ion energy is controlled by the bias voltage and can be used to modify the bottom/sidewall coverage. Figure 1 shows the barrier deposition into a trench under different energies and Figure 2 the coverage as a function of the bias voltage. The simulation of a Ta-barrier/Cu-seed via structure in comparison with the experiment is shown in Figure 3. Here the Ta was deposited with 70 V bias IPVD and the Cu with long throw sputtering.

Copper fill

The electrochemical deposition process (ECD) used to deposit the copper film upon the seed layer presents a rather unusual technology for process simulation. Nevertheless the fundamental equations for fluid flow and electrochemistry have been well known since a long time [3,4]. In our simulation we have

Conference Proceedings ULSI XV © 2000 Materials Research Society

used the CVD simulator PHOENICS-CVD, where the temperature was replaced by the electrostatic potential. We did not attempt to calibrate all the transport parameters correctly for the electroplating process, but rather adjusted the CVD parameters in a way to achieve the correct dimensionless flow numbers (Peclet number, Reynolds number, Schmitt number, etc.) regardless of their absolute values. This was proven by the qualitative agreement with experimental data on deposition uniformity. Fig. 4a shows simulation results for the fluid flow velocity and the electrostatic potential distribution in an electroplating chamber used for copper deposition. The process uniformity was optimized by the inclusion of specially formed shielding blocks in the electrolyte near the wafer edge. The choice of the position and the form of these blocks could drastically improve the edge uniformity of the deposited layers, as it is shown in Fig.4b. In general we found that the process models could not be calibrated ab initio as for the IPVD process, but rather required a number of empirical model constants which had to be carefully determined by comparison with experimental results to guarantee quantitative predictions from the simulation.

Planarization

The chemical mechanical polishing process (CMP) for planarization seems to be one of the least understood processes from the physical point of view. Models used so far are purely empirical and require a continuous adaptation of both model equations and parameters as soon as process equipment or consumables are changed. Models which describe the pad as an elastic material and calculate the local pressure distribution [5] across a feature or across the whole wafer seem to have some advantage over a pure empirical fitting approach [6].
The data shown in Fig. 6 and 7 were derived from a three dimensional extension of the model described in [5]. It represents the polishing pad by a network of elastic springs in vertical and lateral directions as shown in Fig.5, and calculates the local polishing rate as a linear function of the local pressure applied through the springs onto the wafer surface. Care was taken to correctly describe the surface plane of the pad at high nonplanarities, when the down features are not directly touched by the pad. Moreover, the exact compression of the pad is determined by an iterative procedure from the condition that the total force applied by the pad pressure is equal to the sum of the local forces onto the wafer surface. The lateral and vertical spring elasticities are empirically adjusted to yield a good agreement with experimental results (Fig. 6), while the polishing rates for the metal, liner and oxide materials are determined to yield the correct behavior at planar surfaces. The model has been used to predict planarization data for a variety of features that had not been included in our test chip, as shown in Fig.7. We feel that the copper CMP models today can well predict the planarization behavior of arbitrary layout features after they have been calibrated on a variety of test structures for the same process [7]. Unfortunately, at the present state of the art, any process modification will require an in-depth recalibration of the model.

CONCLUSION
In this contribution we have described simulation models and their recent applications for process optimization in a copper metalization technology. While the IPVD process for barrier and seed layer formation can be accurately modeled from basic atomistic data, the electroplating process requires a number of empirically fitted transport parameters to yield correct results. At the other end, for the planarization with the CMP process even the physical equations solved in the simulation are more or less empirical. We found that modeling can do a good job to predict the planarization behaviour of different features for the same process, but any process modification would require at least a new set of fitting parameters.

References:
[1] A. Kersch and U. Hansen, Phys. Rev. B 1999
[2] R.P.Brinkmann and M.Kratzer, to be published
[3] M. Georgiadou, Journ. El. Chem. Soc. 144, 2732 (1997)
[4] P.C.Andricacos, C.Uzoh, J.O.Dukovic, J.Horgans, H.Deliglianni, IBM J.Res. Develop., 42, 567 (1998)
[5] N. Elbel, B. Neureither, B. Ebersberger, P.Lahnor, Journ. El. Chem. Soc. 1998
[6] P. Warnock, Journ.El.Chem.Soc138, 2398 (1993)
[7] T.H.Smith S.J.Fang, D.S.Boning, G.B.Shinn, J.A.Stefani, CMP-MIC Conference, proceedings p.97, (1999)
[8] A.Kersch, W.Morokoff, C.Werner, J.Appl.Phys. 75 (1994) p.2278

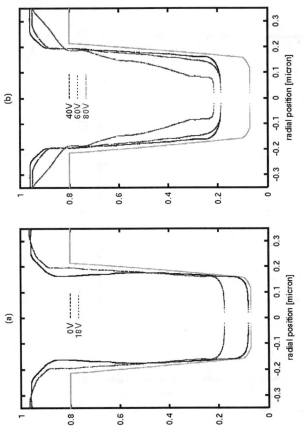

Fig. 1 Simulation of barrier layer formation (a) from PVD and IPVD with 18V self bias (b) from IPVD with 40V, 60V, 80V bias

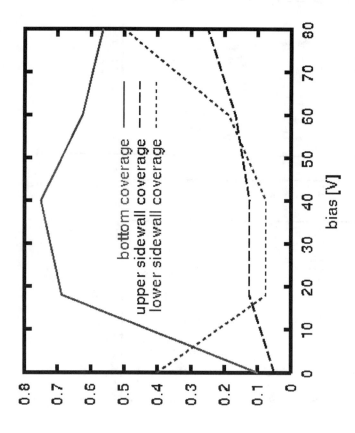

Fig. 2 Simulated bottom coverage, upper and lower sidewall coverage of an IPVD barrier layer. Values at 0V are from PVD

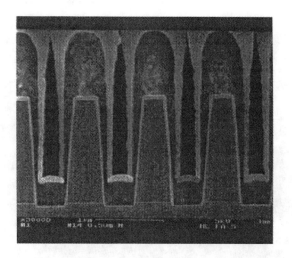

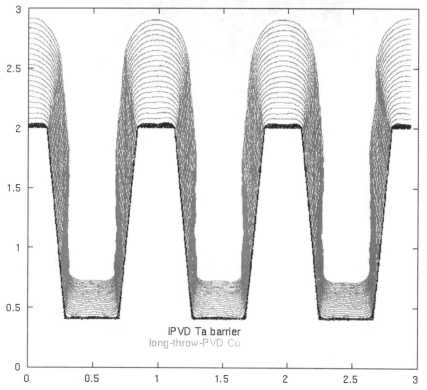

IPVD Ta barrier
long-throw-PVD Cu

Fig. 3 Simulation of Ta barrier/ Cu seed layer formation. Angular distribution of ions and neutrals from Monte Carlo simulation, reaction rates of Ar$^+$, Ar, Ta$^+$, Ta., Cu$^+$, Cu with the surfaces from Molecular Dynamics simulation. The results agree very well with experimental data.

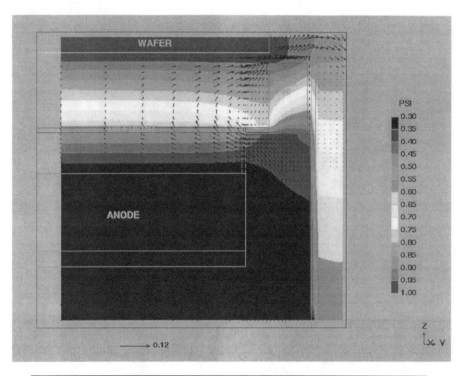

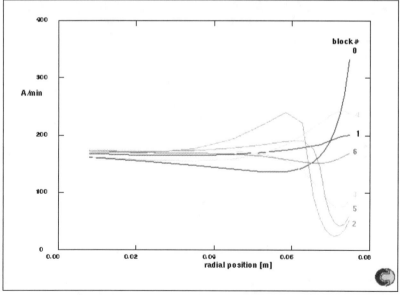

Fig.4 Electroplating chamber for copper deposition. (a) Electrostatic potential and fluid flow velocity; (b) deposition rate uniformity

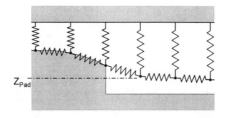

Fig.5 Basics of simulation model for planarization and dishing in a copper CMP process

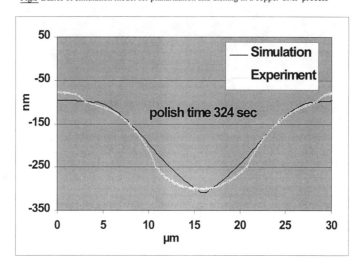

Fig. 6: Experimental and simulated topology results for a test structure after a polish time of 324 sec

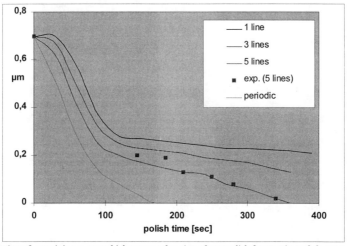

Fig. 7: Prediction of remaining copper thickness as a function of overpolish for a variety of features not covered in the experimentally characterized test structure.

SIMULATION OF IONIZED MAGNETRON SPUTTERING OF COPPER

M.O. BLOOMFIELD, D.F. RICHARDS, T.S. CALE
Rensselaer Polytechnic Institute, Troy, NY, 12180, USA

ABSTRACT

Computer simulations are used to study ionized physical vapor deposition, with ionized magnetron sputtering of copper as the primary system of The effects of sputtering-ion energy and sputtering-ion angular flux distributions on the evolution of sub-micron scale features during IPVD are explored using the EVOLVE simulator. Our goal is to develop semi-quantitative engineering relationships that accurately predict the trends in experimental responses to changes in operating conditions and feature geometry. Sticking-factor based deposition by neutrals is used in which the sticking-factor of incident Cu atoms can depend on arrival angle and energy. Copper ions can also become part of the growing film. Results from molecular dynamic simulations are used to predict energy and angular dependent sputter yields for both copper and argon ions. Sputtered material is ejected from the surface, tracked through the gas phase, and allowed to redeposit. Redeposition is also modeled via a sticking-factor based approach. The redistribution of film material results in non-intuitive profiles and complex relationships between final profiles and process parameters such as sample bias and neutral-to-ion flux ratios.

INTRODUCTION

Ionized physical vapor deposition (IPVD) is a modified physical vapor deposition (PVD) process in which RF induction coils are coupled to the plasma between the source and the substrate. The coils ionize a significant fraction of the in-flight metal atoms, allowing the potential maintained between the target and substrate to accelerate these ions toward the substrate. The energetic impacts of these ions can sputter the film already deposited on the substrate, resulting in redistribution of material over the growing film. We simulate the resulting evolution of the surface of the film on a feature scale via the physically based process simulator, EVOLVE [1, 2].

THEORY

The particular system discussed here is the ionized magnetron sputtering of copper through an argon plasma into 1.1:1 aspect ratio features as described in detail by Cheng and co-workers [3]. Under the operating conditions specified, it is reasonable to assume that collisions in the plasma are sufficient to thermalize the incoming fluxes of both neutral species, Ar and Cu. Accordingly, in the EVOLVE implementation of the system both of these fluxes are represented with a cosine distribution. In contrast, the fluxes of Cu^+ and Ar^+ have tight angular distributions. For a lack of experimental data, these distributions are represented by Gaussians with a standard deviation of 0.02 radians.

Neutral Cu atoms have a sticking factor, S_n. Those that do not stick are re-emitted in a cosine distribution from the surface. For these simulations, S_n is taken to be a constant value of 1.0, independent of angle.

Cu^+ ions deposit with a sticking factor $S_i(\theta, E)$ but also eject atoms from the surface with a yield, $Y_{Cu^+}(\theta, E)$. Ar^+ ions do not deposit, but they do sputter Cu from the surface with a yield, $Y_{Ar^+}(\theta, E)$. (See Figure 1). The angular and energy depedencies of the sticking factor and sputter yields were obtained by recent molecular dynamics simulations [4]. The angular distribution of Cu sputtered from the surface is assumed to be cosine. Sticking factor and sputter yield are considered independently, with the yield being the gross, not the net, number of Cu atoms removed from the surface per impacting ion. Thus a sticking factor of 0 and a yield of 1 indicates that after impact,

663

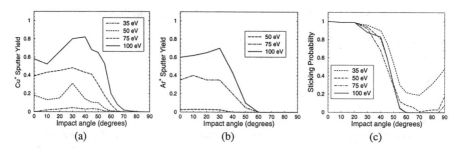

Figure 1: (a) $Y_{Cu^+}(\theta)$, Sputtering yield of Cu^+ as a function of angle and energy. (b) $Y_{Ar^+}(\theta)$, Sputtering yield of Ar^+ as a function of angle and energy. (c) $Y_{Ar^+}(\theta)$, Sputtering yield of Ar^+ as a function of angle and energy.

there is one fewer Cu atom in the material.

Within this model of the system, no attempt is made to account for sputtering or angular-dependent sticking factors of ions after the first impact with the surface. Any ion striking the surface from the source, regardless of its sticking or sputtering is regarded as losing its charge. Thus, any Cu^+ which does not stick upon first impact is re-emitted from the surface at that point as part of a cosine distribution of neutral Cu flux.

There is some evidence in the literature of substantial asymmetric components of the angular distributions of the material ejected from copper films [5]. For lack of better information, our simulations currently assume cosine re-emission of material from the film.

RESULTS

A series of simulations was run to calibrate the model by matching simulation profiles against the experimental data of Cheng and co-workers [3]. Figure 2 shows a series of those results.

Simulations were performed for ion energies of 35.0 eV, 50.0 eV, 75.0 eV, and 100.0 eV with the neutral-to-ion flux ratio (Γ) and the Cu^+-to-Ar^+ ratio (Ω) ranging from 0.2 to 10.0. All simulations were carried out on a 0.44 μm wide infinite trench, to a flat wafer film thickness of 0.5 μm.

In the discussion of responses of the profile to changes in parameters, three profile characteristics closely associated with IPVD processes are considered. These characteristics are formation of bevels at the mouth of the feature, sidewall deposition, and formation of peaked deposits at the bottom of the feature.

Ion Energy Effects

Sixteen series of simulations were performed varying only the ion energy at various combinations of the flux parameters Γ and Ω. Figures 3 and 4 show two of these series and highlight that the dependency on energy is itself strongly dependent on the flux ratios. For high values of both Γ and Ω, that is, systems with a high neutral flux compared to ion flux and an ion-flux dominated by Cu^+, ion energy has a negligible effect on the final profile. However, for the same Cu^+-to-Ar^+ flux ratio, but a moderate value of Γ, there is a strong energy effect on the final profile, as shown in Figure 3.

At values of Γ for which the film profile demonstrates a significant energy dependence, the trend is for an increase in beveling and sidewall coverage with increases in ion energy. The bottom deposit undergoes a shape change and decrease in height with increased ion energy. (See Figure 3.)

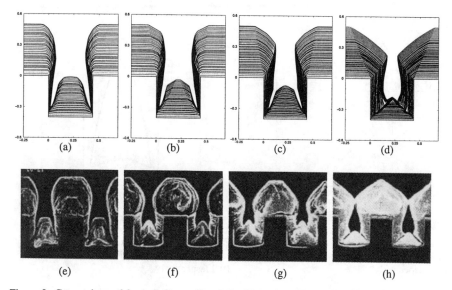

Figure 2: Comparison of final profiles at Γ and Ω of 0.7 and 10.0 respectively, with ion energies of (a) 35 eV, (b) 50 eV, (c) 75 eV, and (d) 100 eV to experimental data SEMs [3] for sample biases of (e) −5.0 V, (f) −20.0 V, (g) −30.0 V, and (h) −50.0 V.

Copper-Argon Ion Flux Ratio Effects

Sixteen series of simulations in which Ω was varied at constant values of ion-energy and Γ were performed. For low energies, there was very little effect on the final profile from changes in Ω, but at 75.0 eV and 100.0 eV, an extreme dependence on Ω was observed. (See Figure 5.) As Ω was varied from 0.2 to 10.0, sidewall coverage decreased significantly, as did beveling. The rounded pedestal bottom deposit at 10.0 first flattened and then became concave as Ω was decreased to 0.2. (See Figure 5).

Neutral-to-Ion Flux Ratio Effects

Sixteen series of simulations in which Γ was varied at constant values of ion-energy and Ω were performed. At a given energy, a significant decrease in sidewall deposition was apparent as Γ was increased from low to moderate values. Over the same range, beveling decreased at a rate sharply dependent on the ion energy. Both of these effects appear to saturate as Γ is increased above 5.0. The shape and height of the bottom deposit is strong function of Γ, ranging from a concave bottom to a mound as Γ increased from 0.2 to 10.0. These changes were qualitatively the same at both 75.0 eV and 100.0 eV, increasing in magnitude with ion energy. (See Figure 6.)

CONCLUSIONS

Quantitative comparison between IPVD simulations and experiment is difficult, as there is not necessarily a one-to-one correspondence between a given simulation parameter and any single reactor operating condition. For example, a correlation between sample bias and ion energy can be established, but it will not necessarily be independent of changes in other reactor setpoints. Moreover, such a correlation is typically reactor specific. Currently, only qualitative and semi-

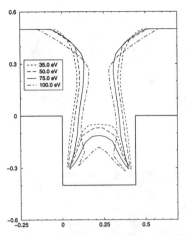

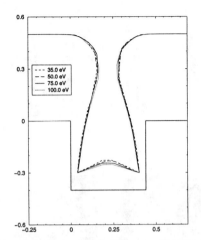

Figure 3: Comparison of final profiles from various ion-energies at Γ and Ω of 1.0 and 10.0 respectively

Figure 4: Comparison of final profiles from various ion-energies at Γ and Ω of 10.0 and 10.0 respectively

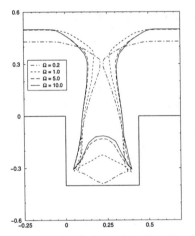

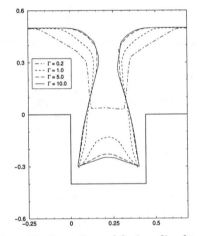

Figure 5: Comparison of final profiles from various values of Ω with a Γ of 1.0 and an ion-energy of 75.0 eV.

Figure 6: Comparison of final profiles from various values of Γ with a Ω of 5.0 and an ion-energy of 75.0 eV.

quantitative relationships between simulation parameters and reactor conditions are available.

These difficulties not withstanding, Figure 2 demonstrates that the trends in sidewall coverage and peaking of bottom deposits with sample bias are captured qualitatively by changes in ion energy. The increase in beveling with sample bias over the given range of bias data seems to be reflected by a slow increase in beveling in the upper half of the range of ion energies. Simulations at energies higher than 100 eV may further demonstate this trend; however, complete sets of yield curves and sticking factors are not yet available for these energies.

One thing is made clear by these simulations: the relationship between surface topography and process parameters is quite complex. The complexity of this relationship is intimately tied to the interplay of the the surface geometry with the angular dependence of sticking and sputtering. Since the ion energy parameterizes the sticking and sputtering curves, the ion energy indirectly affects the developing profile. This interpretation is supported by simulations carried out with very large values of Γ. Under these conditions, the IPVD process is dominated by *angularly independent* neutral deposition so the surface geometry is relatively independent of the ion energy and Cu^+-to-Ar^+ flux ratio.

For more moderate neutral-to-ion flux ratios (Γ), the Cu^+-to-Ar^+ flux ratio (Ω) becomes an important process variable. Comparison of the Cu^+ yield curves to the Cu^+ sticking factor curves (Figure 1) shows that Cu^+ has a positive net deposition rate for all incident angles, except within a small window near $50°$. Thus, Cu^+ tends to deposit on horizontal surfaces and in the bottoms of features, but tends to erode bevels. In contrast, Ar^+ ions have the net effect of removing copper for incident angles less than roughly $60°$. Consequently, Ar^+ flux tends to erode horizontal surfaces but form stable bevels on rounded and diagonal surfaces. It is this difference in sputtering behavior that causes the observed Ω dependence of Cu IPVD.

This Ω dependence may be exploitable. High values of Ω tend to create large bottom deposits as Cu^+ is directed vertically to the bottom of the feature. Figure 6 suggests a potential for complete fillage by making use of the directional capabilities of Cu^+. The final profile for $\Gamma = 0.2$, shows a completely filled feature. With $\Omega = 5.0$, this simulation corresponds to a process dominated by the flux of Cu^+ to the bottom of the feature.

ACKNOWLEDGEMENTS

The authors gratefully acknowledge support for parts of this work from MARCO, DARPA, and NYSSTF.
Additionally, we would like to thank D. G. Coronell for useful discussions.

REFERENCES

[1] EVOLVE is an extensible topography simulation framework. EVOLVE 5.0i was released in June 1999. © 1991–1999, Timothy S. Cale.

[2] T. S. Cale and V. Mahadev, *Modeling of Film Deposition for Microelectronic Applications*, vol. 22 of *Thin Films*, ch. 5, pp. 175–276. San Diego, CA: Academic Press, 1996.

[3] P. J. Cheng, S. M. Rossnagel, and D. N. Ruzic, "Directional deposition of Cu into semiconductor trench stuctures using ionized magnetron sputtering," *J. Vac. Sci. Technol. B*, vol. 13, pp. 203–208, Mar/Apr 1995.

[4] J. D. Kress, D. E. Hanson, A. F. Voter, C. L. Liu, X. Y. Liu, and D. G. Coronell, "Molecular dynamics simulations of Cu and Ar ion sputtering of Cu (111) surfaces," *J. Vac. Sci. Technol. A*, vol. 17, pp. 2819–2825, Sep 1999.

[5] C. Doughty, S. M. Gorbatkin, and L. A. Berry, "Spatial distribution of Cu sputter ejected by very low energy ion bombardment," *J. Appl. Phys*, vol. 82, pp. 1868–1875, Aug 1997.

FULL 3D MICROSTRUCTURAL SIMULATION OF REFRACTORY FILMS DEPOSITED BY PVD AND CVD

T. Smy, R.V. Joshi[1] and S. K. Dew[2]

Dept. of Electronics, Carleton University, Ottawa, ON, Canada K1S 5B6,
ph: 613-520-3967, fax: 613-520-5708: email: tjs@doe.carleton.ca
[1] IBM, T.J. Watson Research Center, Yorktown Heights, NY 10598, USA
[2] Dept. of Electrical Eng. , University of Alberta,

ABSTRACT

This paper reports the use of a novel simulator – *3D-FILMS* – which simulates the creation and growth of 3D microstructure in vapour deposited refractory films. The simulator is used to address a number of issues; including deposition over dual damascene structures, barrier layer deposition and refractory metal CVD processes. The simulator produces a detailed depiction of the film, including grain and columnar structure and is capable of producing density and porosity information.

INTRODUCTION

Modeling of thin film deposition and etching processes has been identified as a important part of the future development of "state-of-the-art" IC fabrication processes. In particular the modeling of microstructural details of the films is seen as crucial. To optimize a deposition a wide number of issues need to be considered such as film uniformity, thickness, stoichiometry and microstructure. All of these issues contribute to the effectiveness of the diffusion barrier or conductive layer. The quality and reliability of backend interconnect technologies are extremely dependant on the grain structure, porosity and texture of the metal thin films used to construct the interconnects. It is also important to note that realistic interconnect structures and film microstructures are inherently three dimensional. These two facts support the need for a fully 3D thin film deposition simulator that predicts both film coverage and microstructure.

Previously, modeling of sputter deposition of refractory barrier layers has primarily been two dimensional (2-D) and consequently limited to deposition either over long features such as trenches, or cylindrically symmetrical features such as circular vias/contacts. Development of 3D models that depict the film as a homogeneous structure and only model the development of the surface have been reported[1], [2]. Atomistic simulations of very small structures in restricted situations have been undertaken [3], however, a model that produces a full depiction of the 3D microstructure of the film including the grain and columnar structure and such information as local density or stichometry has not been previously published.

SIMULATION FRAMEWORK

Microstructural modeling of film deposition in 2D has been extensively reported using the SIMBAD/SIMSPUD framework [4]. *3D-FILMS* takes a similar approach as SIMBAD in representing the film as aggregation of a large number identical building blocks. In SIMBAD the building blocks were discs, however, in *3D-FILMS* the blocks are small cubes. As in SIMBAD the "building blocks" represent the statistically averaged behavior of a large number of film atoms (typically 1000). Particles are serially launched from just above the growing film and follow straight line trajectories until they strike the substrate or the film. The particles aggregate into the film after a short surface diffusion to minimize a surface curvature-dependent chemical potential. The combination of ballistic shadowing and short range surface diffusion successfully accounts for the formation of the columnar thin film microstructure characteristic of refractory metals at typical deposition temperatures. Care is taken during the calculation of the surface potential to ensure that the data structure does not force any artificial anisotropic growth patterns.

In order to minimize memory requirements a data structure consisting of a two tiered array is used.

The first tier in the structure is a "course" 3D array. Each element of the course array is pointer to a 3D "fine" array structure. A "fine" array consists of pointers to individual cube data structures. Each cube data structure contains information on the block material, surface state, material type, surface normal (if appropriate) and deposition history. The course array is allocated at the initiation of the simulation, but fine arrays and block structures are only allocated as needed (fine arrays initially contain NULL pointers). This approach allows for an efficient use of memory particularly in a feature such as via or a porous film. Resolutions of 500x500x500 are possible requiring around 300M of memory. For this situaiton using a single 3D array as a datastructure would require 2.5 10^9 pointers which is impractical on typical work stations. Typically, each fine array is 20-50 elements on a side. To some degree the fine array size can be optimized for a particular film.

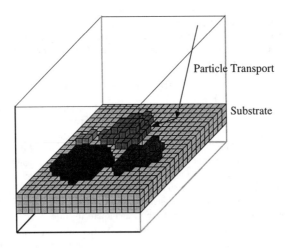

Fig. 1. Basic simulation geometry showing a simple flat substrate, three initial grains and a simple particle transport to the surface and then to a growing grain.

At the start of simulation a substrate is defined along with parameters controlling the deposition. The substrate definition is simply an initial specification of cubes and can be set at run time to a variety of configurations including a flat substrate, trenches, vias, and steps. During simulation particles are launched ballistically from slightly above the growing film and follow straight line trajectories until they strike the substrate or the film. The particles are then incorporated into the film, after searching the nearby exposed surface, minimizing a surface curvature-dependent chemical potential. This process is shown schematically at a very low resolution in Fig. 1. The combination of ballistic shadowing and short range surface diffusion successfully accounts for the formation of the columnar thin film microstructure characteristic of high melting point materials at typical deposition temperatures.

The surface diffusion is modeled by searching the immediate film surface (exposed cube sides) over a region of specified radius. During this search a list of possible nesting sites of minimum surface energy is collected. Care is taken during the calculation of the surface potential to ensure that the data structure does not force any artificial anisotropic growth patterns. Finally, a deposition site is randomly chosen from the list of possible surface sites of minimum energy.

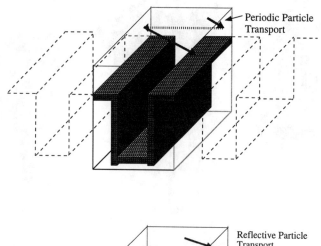

(a)

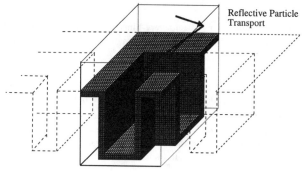

(b)

Fig. 2. Two boundary conditions used in *3D-FILMS*. a) Periodic BC. b) Reflective BC.

The boundaries of the simulation need to be handled carefully during film growth, diffusion and ballistic transport. Two cases are allowed; a boundary can be specified to be periodic or mirrored (shown in Fig. 2. If a boundary is periodic a particle traveling across the simulation from left to right will re-enter on the left side of the simulation region and the film on the left edge will "match" that on the right. For a mirrored boundary particles will "bounce" off the boundary and the film structure will be "virtually" mirrored at the boundary. Periodic boundary conditions must be used for non-cylindrically symmetric incident flux distributions as might be encountered at a wafer edge.

As in SIMBAD a variety of effects including non-unity sticking, resputtering and local annealing phenomena are easily incorporated in the Monte-Carlo framework. The initial distribution of the incident atoms can be specified in a number of ways, including a compressed cosine at an oblique angle (including possible rotation or changes angular orientation with time), a fixed angle can be specified or an angular distribution can be read in from an input file.

Once the simulation is finished it is straight forward to calculate local film densities and surface areas. A rectangular region can be specified, the total amount of matter and exposed surface determined and then

(a)	(b)	(c)

Fig. 3. Three *3D-FILMS* simulations of refractory films deposited over a dual damascene structure. a) Using 1:1 aspect collimation. b) Small bias ionized system. c) TaN deposition depicting stoichiometry as a greyscale.

normalized to the volume (or area) of the specified region. Using this approach it is possible to obtain information about, for instance, the density of a film as a function of height.

To view the film in 3D an openGL viewer was written that displays the film on a monitor. The image can then be captured (or exported) in an image format such as jpeg. During the creation of the viewing information the surface of the film is smoothed to remove the sharp cube edges. The actual film is depicted as a large number of triangles. The viewer allows for the film to be viewed in a number of ways including vertical and horizontal slices.

RESULTS

In Fig. 3a and b two simulations are shown for a refractory barrier deposition over a dual damascene structure consisting of trench and a via. The two simulations differ in the angular distribution of the incoming sputter flux. Fig. 3a uses a angular distribution typical for a 1:1 aspect ratio collimated system, where as the second simulation has a narrow distribution more characteristic of a ionized PVD system with a small bias on the substrate (the bias is assumed to be sufficiently small that no resputtering occurs). It can clearly be seen that the first simulation has poor bottom coverage but the side walls of both the trench and via are well covered. The fine microstructure and overhanging grains at the via and trench top corners are typical of a refractory barrier.

The third simulation in Fig. 3 shows a representation of the stoichiometry of a reactively sputtered TaN film over the same structure. For the purposes of demonstration the Ta angular distribution was assumed to be a directed flux corresponding to a IPVD system and the N to a have a relatively high sticking coefficient ($S_c = 0.5$). The stoichiometry of the film is depicted as a greyscale with a light shade indicating a Ta rich film and a dark shade as N rich film. A Ta rich film is found at the bottom of the via due to poor N transport down the via. On the trench walls underneath the overhanging film on the top corners a N rich film is found that is due a large N/Ta flux ratio in these areas because of shadowing effects.

An aspect of film growth that is of considerable importance for the deposition of barrier layers in the

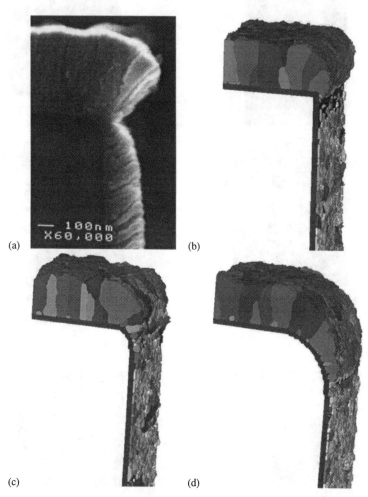

(a)

(b)

(c)

(d)

Fig. 4. Details of refractory film deposition over a via top corner. (a) SEM of a Ti film. Simulation of (a) sharp corner. Rounded corners of (b) 0.005 μm radius. (c) 0.07 μm radius.

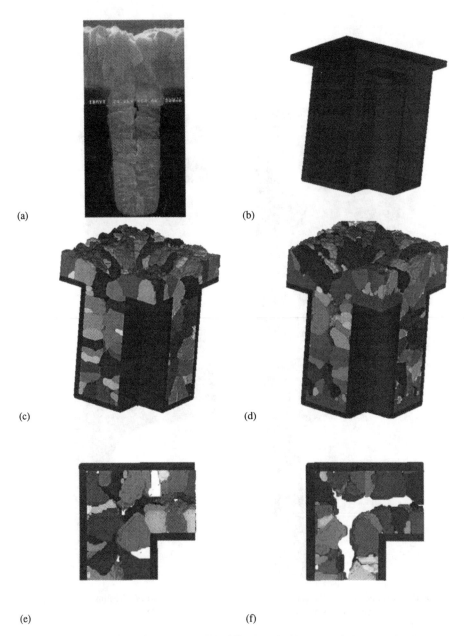

(a)   (b)

(c)   (d)

(e)   (f)

Fig. 5. Simulation of a W CVD process. a) SEM of film deposited into a trench. b) initial simulation topography of a trench corner. c) Simulation over initial topography. d) Simulation over initial topography
plus a barrier deposition. Horizontal cross-sectional views of depositions in c) and d).

674

formation of a "overhang" at the top corners of vias and trenches. A SEM showing the formation of this overhang for a Ti film is shown in Fig. 4. The presence of an overhang can cause poor sidewall and bottom coverage and also affect subsequent filling processes. The creation of this overhang is strongly dependant on the columnar growth of the film at the edge of the feature. A study of overhang creation as a function of the "shape" of the top via corner is shown in Fig. 4. The first simulation clearly shows the formation of an overhang on a "sharp'" corner. This overhang results in a discontinuous film immediately below the overhang and poor bottom coverage due to shadowing. The two following simulations show the effect of "rounding" the via top corner. This smoothing of the corner clearly removes the overhang and results in much better sidewall coverage. A second approach to removing this overhang would to be use controlled amounts of resputtering to "etch" off the overhang during growth. This approach could be implemented in an ionized sputtering system.

The importance of the overhang on filling characteristics is shown in Fig. 5 c-d. This simulation shows the filling of a trench corner using W CVD system. Films deposited by W CVD show a characteristic columnar microstructure (see Fig. 5a). A simulation of the filling using a sticking coefficient of 0.001 over the original structure shows complete filling of the structure (Fig. 5 c). Likewise a simulated deposition over a structure with a overhanging barrier appears to be filled((Fig. 5 d). However, Fig. 5e and f show horizontal slices through the two structures clearly indicating that although the first film does fill the feature the second does not. This lack of filling is due to the 3D nature of the initial structure, the presence of an overhang and the growth of the microstructure over the feature.

CONCLUSIONS

In this paper a novel simulator, *3D-FILMS*, is presented for simulating thin deposition over VLSI topography. *3D-FILMS* produces a full 3D depiction of the film coverage and the internal aspects of the film, including grain and columnar structure, and density and stichometric information. An openGl viewer is used to view the film from and perspective and to view slices and details of the internal film structure.

A number of examples of film growth were exhibited. Barrier layer deposition over a dual damascene structure consisting of a via at the end of a trench was simulated for both collimated sputtering and an ionized system. The simulator was also used to investigate film growth over the top corner of a feature such as trench or via/contact. It was shown that the creation of an overhang during barrier layer deposition could have significant affects on subsequent W CVD deposition into a damascene trench corner. The use of "rounding" the top corner of the initial trench topography was shown to be a possible solution to filling problems.

REFERENCES

[1]   F.H. Baumann and G.H. Gilmer, Proceedings of the International Electron Devices Meeting '95, IEEE, p. 89, 1995.
[2]   Kan, E.C. McVittie J.P., and Dutton R.W.. IEEE Trans. on ED. 1997 Sept. V. 44, No. 9.
[3]   H.Huang, G. Gilmer, And T. Diaz, Journal of Applied Physics, Vol. 84, No. 7, Oct 1, 1998,
[4]   M.J. Brett, S.K. Dew, and T. Smy, *Thin Films: Modeling of Film Deposition for microelectronic Applications.* Editor S. Rossnegal, Chpt.1 , Academic Press, 1996.
[5]   T. Smy, S.K. Dew, and M.J. Brett, MRS Bulletin Special issue on Metalization, Vol. 20, no. 11, p. 1995, Nov. 1995.

MODELING OF FEATURE-SCALE PLANARIZATION IN Cu CMP USING MESA™

T. LAURSEN*, S.R. RUNNELS**, S. BASAK*, M. GRIEF*, K. MURELLA*,
*SpeedFam-IPEC, 305 N. 54th St., Chandler, AZ 54226-1321
**Southwest Research Institute, P.O. Drawer 28510, San Antonio, TX 78228-0510
tlaursen@sfamipec.com

ABSTRACT

MESA™ is a feature-scale planarization model that can simulate chemical mechanical polishing (CMP) of Cu pattern wafers. This model has been validated experimentally by data collected on the Auriga polisher using Sematech 926 Cu wafers, Rodel IC1000/1400 pads, Cabot 4110 and Arch Cu20K slurries. MESA accounts for observed differences in Cu planarization using IC1000 with and without foam backing. It also accounts for the planarization behavior of a two-step slurry process using Cabot 4110 slurry for Cu removal followed by Arch Cu20K slurry for optimal planarization.

INTRODUCTION

In IC manufacturing, the fabrication of Cu interconnect structures involves chemical mechanical polishing (CMP) as one of its most critical process steps. The Cu structures are formed using a dual-Damascene process, whereby the metal is inlaid into trenches and vias of dielectric films separated by a thin barrier film [1]. Several Cu CMP processes are currently being evaluated for their polish rate, non-uniformity, planarity, defectivity etc., which all have to be within tight specifications [2]. The most stringent specifications relate to feature-scale planarization. Dishing and erosion, which describe the recession of the metal and dielectric in the pattern, occur as a result of polishing excess-Cu and barrier layer above the patterned dielectric layer. The excess-Cu removal is carried out until the pattern clears, which is done with minimal dielectric removal (erosion) and minimal depression of the Cu lines (dishing and metal recess).

Dishing and erosion in patterned structures are typically measured across the wafer after polishing using stylus profilometry. For a given process, each structure within a die exhibits its own planarization characteristic, which depends on its precise dimensions. The MESA model developed at Southwest Research Institute has been demonstrated to have predictive capability for feature-scale planarization in oxide CMP [3,4]. The present paper presents an application of MESA to a Cu CMP process and examples are presented where model predictions of surface topography are compared to experimental results.

MESA

MESA is a model for simulating feature-scale planarization of the CMP process. It was originally developed as a three-dimensional model for oxide CMP [3] and was validated in a study using product wafers at Lucent Technologies [4]. MESA has now been extended to handle multi-layer structures, multi-slurry processes and extensive two-dimensional analysis, which enables the model to be applied to most CMP processes, including shallow trench isolation, W and Cu CMP.

Conference Proceedings ULSI XV © 2000 Materials Research Society

MESA's computation of the topographical evolution is based on calculating the local pressure distribution at the pad-wafer interface and the associated material removal. The removal is calculated according to the expected pressure proportionality, i.e., Preston's equation [5]. The topography computation during polish is based on simple input for down force, polish rates, pattern dimensions as well as validated input for the pad's elastic properties.

The polish-rate inputs are obtained from polishing respective Cu, barrier (Ta) and dielectric (TEOS) sheet wafers. The pattern dimensions of the whole structure are entered using either nominal values or as a stylus scan measured prior to polish. In order to ensure a well-defined force distribution on the structure, field is included in order to match the average pattern density of the whole wafer. The validity of matching overall pattern density is still somewhat tentative at this point, and will be explored in more detail in future validation work.

The pad's elastic properties are entered as simple input after an initial validation procedure has established their values. The elastic properties are represented by two spring constants, k_1 and k_2, which are obtained by fitting model predictions to experimental results. The model validation is based on making the model reproduce a set of experimental results and estimate its range of predictive capabilities. k_1 represents the stiffness of the pad stack and is related to its compressibility, while k_2 represents the flexural bending strength of the top pad [3].

EXPERIMENTAL

In this study the surface-topography evolution was measured by recording stylus scans for selected pattern structures for different polish increments across the wafer and associating them with the Cu thickness measured in the adjacent field region. The stylus scan is measured using the Tencor HRP profiler and the adjacent field Cu thickness is determined from a 49-point diameter scan obtained using Tencor RS75 four-point resistivity probe. Model predictions and experimental stylus scans are compared at similar field-Cu thickness levels.

Two validation experiments were carried out on the SpeedFam-IPEC Auriga polisher using two 200-mm Sematech 926 pattern wafers on IC1400 and IC1000 pads. Using the first-step part of the Auriga Cu CMP process of record, the wafers were polished with Cabot 4110 slurry to pattern clearing. This polish step uses 2 psi down pressure, 30-rpm table speed and 29-rpm carrier speed.

The second-step polish, using the Arch Cu50K low-selectivity slurry, was carried out immediately following the validation experiment using Cabot 4110 on IC1400. The other process conditions remained the same.

For all polishes, the 50% pattern-density structures on the center, mid-radius and edge die having line widths of 0.5, 5, 10, and 50 μm were measured after each polish increment. In addition, the 33% pattern density structure with 50-μm linewidth in the center die was measured.

RESULTS AND DISCUSSION

The first validation was based on the Cu fill planarization with Cabot 4110 slurry on IC1400 pad. A good fit between MESA and experiment for all structures on all three dies measured were obtained for $k_1 = 7$ psi/μm and $k_2 = 0.5$ psi/μm. One example of such a comparison is shown in Fig. 1, where 25 50-μm wide trenches are planarized prior to clearing. The final scan shows ~ 1400 Å dishing in the Cu lines after pattern clearing.

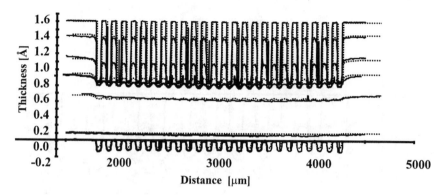

Fig. 1: Cu-fill planarization with high-selectivity slurry on IC1400 pad. MESA prediction (k_1 =7 psi/μm, k_2 =0.5 psi/μm, dotted line) is compared to experimental data (solid line).

Although the overall agreement between MESA and experiment is good, some discrepancies are noted. First, the experimental scans often exhibit a concavity across the 2.5-mm wide pattern. Second, minor deviations between measured step heights and those predicted by MESA are occasionally seen in the transition from field to pattern. In those instances, MESA overestimates the planarization.

While those discrepancies are minor, a more significant discrepancy is observed after clearing, where for the pattern considered, MESA underestimates the dishing by up to 50%. The fidelity of Mesa's Cu planarization prediction along with other detailed validation of Mesa, such as that performed at Lucent Technologies, suggests that Mesa's mechanical model is sound [3]. Yet the fact remains that, in some cases, MESA underestimates dishing and erosion. At this point, explanations for the underestimate are speculative. One cause may in part be due to the pattern structure clearing before the field – this was e.g. observed by visual inspection. Another possible explanation is that that the original Cu surface, having been exposed to air, develops a layer that is initially slightly more resistant to polishing. At the start of the polish, the pad only contacts the up features and so that layer is removed there. But when full contact is reached, the low areas still have the resistant layers. In this way, the initial selectivity to up features would be enhanced. Without knowledge of that effect, Mesa's k values would have to be increased to match that increased planarization rate. Those artificially increased values would then later lead to an underprediction of dishing. These and other possible explanations are being addressed in a future study.

Using the validation from the Cu fill planarization, MESA can predict the planarization behavior of the second step polish with the Arch Cu20K slurry. This slurry has a selectivity of 1:0.8:1.8 to Cu: Ta: TEOS. From the comparison in Fig. 2 it can be seen that MESA predicts the overall planarization behavior. Good planarity was obtained with the low-selectivity slurry, but involved 1500 Å of oxide loss. While this study switched slurries after substantial dishing took place, an optimized process using motor-current-based endpoint detection has resulted in less than 500 Å dishing and erosion with less than 800 Å field oxide loss [6].

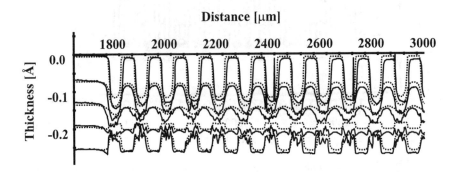

Fig. 2: Pattern planarization with low-selectivity second-step slurry on IC1400 pad. MESA prediction (k_1 =7 psi/μm, k_2 =0.5 psi/μm, dotted line) is compared to experimental data (solid line).

The second validation was based on the Cu fill planarization with the same process conditions as above except for using an IC1000 single pad. A good fit between MESA and experiment for all structures were obtained for k_1 =15 and k_2 =0.5, see Fig. 3. The

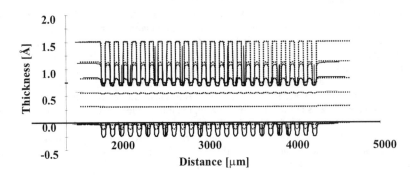

Fig. 3: Cu-fill planarization with high-selectivity slurry on IC1000 pad. MESA prediction (k_1 =15 psi/μm, k_2 =0.5 psi/μm, dotted line) is compared to experimental data (solid line).

effect of increasing pad stiffness, k_1, is to reduce the concavity observed in the stylus scans. Thus, there is strong experimental evidence that is supported by the model to suggest that this concavity is associated with using foam backing. The other effect of removing the foam backing on IC1000 is to increase the pad-stack stiffness k_1 from 7 to 15 and maintain flexural bending strength k_2 at 0.5. This qualitatively behavior is in support of associating k_1 and k_2 with the pad stiffness constants. Furthermore, strength-of-materials analysis also predicts a pad stiffness constant k_1 of 15 psi/μm.

CONCLUSION

In conclusion, MESA has been extended to predict the topographical evolution on a feature scale for most basic CMP applications. The initial validation work has shown that the model can be applied to Cu CMP. When validated by fitting k_1 and k_2 to describe the Cu fill planarization, Mesa describes the planarization trends with the same slurry during pattern clearing for all the pattern structures investigated. One minor deviation was associated with a concavity shape across a pattern, which was attributed to the use of a foam-backing pad. In general, MESA underestimates dishing and erosion with the high-selectivity slurry, but this may in part be due to a pattern-density effect in time to clear or other effects not directly related to MESA's mechanical model. The validated MESA model provided a reasonable prediction of the topographical evolution obtained with the 2nd step low-selectivity slurry.

REFERENCES

[1]   R.L. Jackson, E. Broadbent, T. Cacouris, A. Harrus, M. Biberger,
       E. Patton, T. Walsh, Solid State Technology, March 1998, p. 49 (1998)
[2]   S. Basak, C.O. Fruitman, M. Grief, T. Laursen, K. Murella, MRS Conf. Proc.
       ULSI XIV, p. 231-235 (1999)
[3]   S.R. Runnels, J. Electrochemical Materials **25**, pp. 1574-1580 (1996).
[4]   S.R. Runnels, Inki Kim, F. Miceli, CMPMIC '99 Proceedings, pp. 128-135
       (1999)
[5]   F.W. Preston, J. Soc. Glass Tech., **XI**, p. 214 (1927)
[6]   S. Basak, M. Grief, T. Laursen, K. Murella, CMP Technology for ULSI
       Interconnection, Semicon West 99, p. P1-P14 (1999)

A MODIFIED REMOVAL RATE MODEL FOR BOTH
IN-SITU AND EX-SITU PAD CONDITIONING CMP

ERIC TSENG, MICHAEL MENG, S.C. PENG
United Silicon Incorporated
No.3, Li-Hsin Road 2, Science-Based Industrial Park, Hsinchu, Taiwan, R.O.C.
Tel: 886-3-5789388 Ext:36410, Fax: 886-3-5789758, E-mail: eric_tseng@usic.com.tw

ABSTRACT

A modified removal rate model based on the assumption that removal rate is proportional to the slurry holding and transporting capability of pad is presented to describe the oxide CMP polishing characteristics. This model included the pad degradation coefficients is capable of predicting the removal rate of both in-situ and ex-situ pad conditioning CMP during every moment of the CMP polishing. The excellent agreement between theoretical and experimental results is demonstrated by the experiment of marathon run and different polishing time removal rate test. By using this model to calculate the removal rate of in-situ or ex-situ conditioning CMP, it will make the process more flexible and easy to be controlled.

INTRODUCTION

As device geometry continues to shrink and the levels of metal interconnect continue to increase, global planarization is needed to achieve reliable and robust metal interconnects. CMP is recognized as one of the most important technologies to achieve global planarization. However, in addition to the consumable batch-to-batch variation, the degradation in the oxide removal rate can result in significant CMP process variation. Degradation of removal rate is typically due to insufficient conditioning of the polishing pads, decrease of slurry holding and transporting capability of pads, and the change of groove or hole depth because of pad wear [1~2]. For the commercially available polisher, some used in-situ pad conditioning and others were ex-situ pad conditioning. These two kinds of pad conditioning will obtain different behaviors of removal rate. Unfortunately, the Preston's equation, $R=Kp*(F/A)*V$, did not consider the pad degradation factor [3]. For the absent of an adequate model, it is very hard to optimize and control this process.

In this study, we assumed that the removal rate is proportional to the slurry holding and transporting capability of pad. Discuss both in-situ and ex-situ pad conditioning cases individually, we can obtain a mathematical equation which will fit for any kind of conditioning type. How to calculate the pad degradation coefficients is also presented.

MODELING

According to Preston's equation, the theoretical removal rate is given as:

$$R = Kp \cdot (\frac{F}{A}) \cdot V \tag{1}$$

While, Kp is the Preston's coefficient, F/A is the applied stress, and V is the relative velocity between the wafer and the polishing pad. Considering the pad degradation factor, we can assume that Kp is a time dependent coefficient. The Preston's equation should be rewritten as equation (2).

$$R(t) = Kp(t) \cdot (\frac{F}{A}) \cdot V \tag{2}$$

In the following derivations, we will discuss the $Kp(t)$ in equation (2) for two different cases, one is the in-situ pad conditioning and the other is the ex-situ pad conditioning.

In-situ Pad Conditioning

For the in-situ pad conditioning CMP, the difference of run-to-run removal rate is assumed to cause by the pad conditioning. Pad conditioning removes polishing residues and debris from pad surface and open new pores in the pad to hold and channel slurry between the pad and the wafer surface. But it also changes the thickness of pad and depth of grooves, which will decrease the slurry holding and transporting capability. The removal rate proportional to the pad thickness or groove depth can be expressed as equation (3).

$$R(t) = Kp(t) \cdot (\frac{F}{A}) \cdot V = Kp \cdot [1 - D \cdot C(t)] \cdot (\frac{F}{A}) \cdot V \tag{3}$$

While D degradation coefficient of pad life
             $C(t)$ accumulated conditioning time

Ex-situ Pad Conditioning

For the ex-situ pad conditioning CMP, the removal rate is assumed to correspond to the pores in the pads. The pores can hold and channel slurry between the pad and wafer surface and will decrease by removing oxide from the wafer surface. From Fig.1, we can obtain a removal rate equation as following:

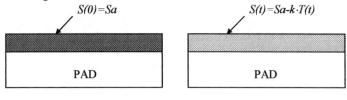

Slurry adsorptive ability of pad   Slurry adsorptive ability of pad
before polishing   during polishing

Fig. 1 The scheme of pad degradation during CMP of ex-situ pad conditioning

$$R(t) = Kp(t) \cdot (\frac{F}{A}) \cdot V = Kp \cdot \frac{S(t)}{S(0)} \cdot (\frac{F}{A}) \cdot V = Kp \cdot \frac{[Sa - k \cdot T(t)]}{Sa} \cdot (\frac{F}{A}) \cdot V \tag{4}$$

While $T(t)$ is the removed oxide thickness
            Sa is the initial slurry holding capability

$$T(t) = \int_0^t R(t) \cdot dt \tag{5}$$

Substitute $T(t)$ into equation (4)

$$R(t) = \frac{Kp}{Sa} \cdot [Sa - k\int_0^t R(t) \cdot dt] \cdot (\frac{F}{A}) \cdot V \tag{6}$$

Differentiate this equation, we can obtain the following equation.

$$dR(t) = -\frac{Kp}{Sa} \cdot k \cdot (\frac{F}{A}) \cdot V \cdot R(t) \tag{7}$$

Solving this differential equation with the boundary condition $R(t)=Kp \cdot (F/A) \cdot V$ at $t=0$ gives

$$R(t) = Kp \cdot (\frac{F}{A}) \cdot V \cdot e^{-Dct} \tag{8}$$

While Dc is the degradation coefficient of pad conditioning

$$Dc = \frac{Kp}{Sa} \cdot k \cdot (\frac{F}{A}) \cdot V \tag{9}$$

Combination of Both Cases

Combine equation (3) and equation (8), we can obtain a modified removal rate model for both in-situ and ex-situ pad conditioning CMP. The removal rate as a function of polishing time can be expressed as the following equation.

$$R(t) = Kp \cdot (\frac{F}{A}) \cdot V \cdot \{[1 - D \cdot C(t)] \cdot e^{-Dc \cdot t}\} \tag{10}$$

for the in-situ conditioning:

$$C(t) = Co + t, Dc = 0$$

for the ex-situ conditioning:

$$C(t) = Co, Dc > 0$$

EXPERIMENTAL

In order to verify this model, two experiments have been done. One is the in-situ conditioning removal rate test of 500 wafers. By this experiment, we can yield the degradation coefficient of pad life. The other is the different polishing time removal rate test for both in-situ and ex-situ pad conditioning. From this experiment, we can obtain the degradation coefficient of pad conditioning and compare the difference of two conditioning types. All wafers for these tests are PETEOS blanket wafers. The film thickness is measured by an optical spectrophotometer.

RESULTS AND DISCUSSIONS

For the marathon run, we used same recipe for every wafer and the removal rate was gradually decay from 2500 A/min to 2200 A/min (shown as Fig.2). Substitute the removal rate and accumulated conditioning time, the degradation coefficient of pad life can be derived.
The results of second test which used different polishing time for several runs and measured the removed thickness is shown as Fig.3 and Fig.4. The theoretical value of removed thickness can be derived from equation (3) and equation (8).
For in-situ pad conditioning CMP

$$THK(in-situ) = \int_0^t Kp \cdot (\frac{F}{A}) \cdot V \cdot [1 - D \cdot C(t)] \cdot dt = Kp \cdot (\frac{F}{A}) \cdot V \cdot D \cdot [\frac{t}{D} - Co \cdot t - \frac{t^2}{2}] \qquad (11)$$

For ex-situ pad conditioning CMP

$$THK(ex-situ) = \int_0^t Kp \cdot (\frac{F}{A}) \cdot V \cdot e^{-Dc \cdot t} \cdot dt = Kp \cdot (\frac{F}{A}) \cdot V \cdot \frac{(e^{-Dc \cdot t} - 1)}{-Dc} \qquad (12)$$

Substitute the removed thickness data into the equation (11) and (12), we can easily obtain the degradation coefficient D and Dc.

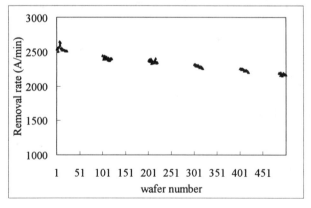

Fig.2 The Removal Rate Data of Marathon Run

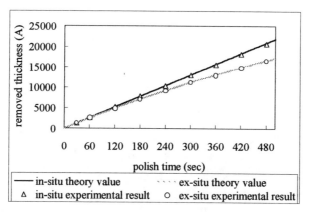

Fig.3 The removed thickness vs polish time for both in-situ and ex-situ pad conditioning

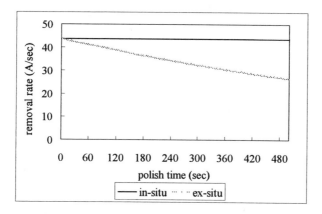

Fig.4 The removal rate vs polish time for both in-situ and ex-situ pad conditioning

CONCLUSIONS

A modified removal rate model for both in-situ and ex-situ pad conditioning CMP has been developed based on the assumption of that removal rate is proportional to the slurry holding and transporting capability. The removal rate of in-situ conditioning CMP is a linear decay function of polishing/conditioning time and the removal rate of ex-situ conditioning CMP will be an exponential function of polishing time. It is confirmed that this model can accurately predict the removal rate of every moment during both in-situ and ex-situ pad conditioning CMP.

REFERENCES

[1] Iqbal Ali et al., Solid State Technology, Vol.40, No.6, June 1997, p.185.
[2] Jia-Zhen Zheng et al., CMP-MIC Conf. Proc., February 1997, p.315.
[3] F. Preston, J. Soc. Glass Tech., 11, 214 (1927)

Reliability

ELECTROMIGRATION RELIABILITY STUDY OF SUBMICRON Cu INTERCONNECTIONS

C-K. Hu,* R. Rosenberg,* W. Klaasen,** A.K. Stamper**
*IBM T. J. Watson Research Center, Yorktown Heights, NY 10598, haohu@us.ibm.com
**IBM Microelectronics Division, Essex Junction, VT 05452

ABSTRACT

Electromigration in 0.15 µm to 10 µm wide PVD Cu unpassivated lines fabricated by lift-off processing stressed in nitrogen ambient, and 0.27 µm wide electroplated Cu damascene lines passivated with SiN_x/SiO_2 has been investigated. Void growth at the cathode end, and hillocks or protrusions at the anode end of the lines, as a result of electromigration, have been found. The dominant diffusion mechanism is a mixture of grain boundary and interface/surface diffusion for the polycrystalline line, while in bamboo-like lines the dominant mechanism is interface/surface transport. The activation energies for electromigration in the bamboo-like lift-off Cu lines tested in nitrogen ambient and Cu damascene lines are found to be 0.8 ± 0.1 eV and 1.1 ± 0.1 eV, respectively.

INTRODUCTION

Copper interconnection in integrated circuitry chips has been commercially manufactured by IBM since 1998. One of the important chip issues is the assurance of electromigration lifetime in on-chip Cu interconnections. Electromigration has been investigated in Cu,[1] as well as Al,[2] thin film wiring for many years. Recently, we reported surface or interface diffusion as the most likely dominant transport mechanism during electromigration of bamboo-like lines.[3-5] In this paper, we extend our investigation to the Cu mass motion when the electrons flow from the Cu via to Cu line in a damascene structure.

EXPERIMENT

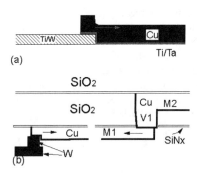

Fig. 1. Schematic diagrams of test structures, (a) unpassivated Cu line on W, (b) three-level Cu damascene line. Arrows on the Cu line show the electron flow.

Two types of test structures, Fig. 1(a) and (b), were used. Fig. 1(a) consists of an unpassivated 200 µm long Cu line overlapping a 10 µm wide Ti(10 nm)/W(200 nm) bar at both ends of the Cu line. The Ti (10 nm)/Ta (15 nm)/Cu (300 nm) test lines were all deposited by an E-gun evaporator at a base pressure of 10^{-7} Torr and fabricated by a lift-off process using e-beam lithography.[6] The Cu lines were 0.15 µm to 10 µm wide and overlap the Ti/W bar by 0.85 µm and 1.0 µm for 0.15 µm to 0.25 µm wide lines and 0.75 µm to 10 µm wide lines, respectively. The samples were annealed in helium at 400 °C for 3 hours before testing to stabilize the microstructure. Fig. 1(b) shows a three-level damascene interconnection with a SiN_x/SiO_2 dielectric which consists of a 400 µm long electroplated Cu test line (M1) connected to the lower level W line (M0) by a W interlevel via and the upper level of the wide Cu line (M2) by a Cu via (V1). The

691

W lines and W vias, and Cu M1 were fabricated by a single damascene process. The upper level of Cu M2 and V1 were fabricated by a dual damascene[1] process. The Cu lifetimes were measured on the 0.27 μm wide M1. The thicknesses of the Cu damascene line at M1/SiO₂ and M1/W overlapping sections are 0.31 μm and 0.21 μm, respectively. The diameters of the top W via and the bottom of the Cu V1 are 0.32 μm and 0.53 μm, respectively. The final passivation on the Cu damascene structure was a 1 μm thick SiO₂ on top of the M2 and the wire bonding pads were covered with 1 μm thick Al(Cu)-system metallization.

The microstructures of the Cu lines were examined by transmission electron micrographs. Bamboo-like, near bamboo, and polycrystalline structures were found in the lift-off Cu line metal linewidths between 0.15 μm to 0.5 μm, 0.75 μm to 1 μm, and 2 μm to 10 μm, respectively.[3] The bamboo-like and near-bamboo grain structures were found in the 0.27 μm wide Cu damascene lines and Cu via, respectively.[7] The bamboo-like structure is mostly single-grain-per-line-width, and the polycrystalline structure has more than two grains-per-linewidth. The near bamboo structure has mixed sections of each.

The samples were tested in a vacuum furnace at temperatures ranging from 255 °C to 405 °C maintaining a chamber pressure of 15 Torr of high purity nitrogen with total impurities CO_2 and H_2O < 1 ppm. Some of the Cu damascene lines were also tested in air. No observed difference in lifetime was found for the Cu damascene lines tested in air or in nitrogen ambient. The current densities were 15 and 22 mA/μm² for lift-off and damascene Cu lines, respectively. However, the value of current density j in the Cu line on the top of the W via section was about 33 mA/μm². Lifetime, τ, was determined as the time for the line resistance R to increase by 10% of the initial value R_o and the failure distribution was assumed to be a log-normal function. The void size in the tested line was measured by a scanning electron microscope (SEM). Mass transport of Cu in the Cu lines as a function of temperature was measured using both void growth rate and resistance measurements.

RESULTS and DISCUSSION

Theoretical Background

The atomic flux produced by an electromigration driving force, F_e, is given by $J_e = n \, v_d$, where n is the density of atoms, v_d is the drift velocity. The drift velocity is related to the product of mobility and force which is usually expressed by the Nernst-Einstein relation, $v_d = (D_{eff}/kT)F_e = (D_{eff}/kT) Z_{eff}^* eE = (D_{eff}/kT) Z_{eff}^* e\rho j$, where e is the absolute value of the electronic charge, Z_{eff}^* is the apparent effective charge number, E is the electric field, ρ is the metallic resistivity, D_{eff} is the effective diffusivity of diffusing atoms through a metal line, T is the test absolute temperature, and k is the Boltzmann constant. The metal line is damaged by electromigration when there is an unbalance of Cu flux at a particular location, x, by the equation

$$- \frac{\partial n}{\partial t} = \frac{\partial J}{\partial x} = (J_{in} - J_{out})/\Delta x \tag{1}$$

where J_{in} and J_{out} represent the Cu flux entering and leaving at x.

For the test structures used, if the W bar or via and liner were complete blocking boundaries for the Cu fluxes, then Eq. (1) at contact interfaces, $x = x_c$, can be simplified:

$$J_{out}(Cu) - J_{in}(W) = J_{out}(Cu) = n \, v_d , \tag{2}$$

where $J_{in}(W)$ and $J_{out}(Cu)$ are the atomic Cu fluxes in and away from W or liner, respectively, and $J_{in}(W) = 0$ because no Cu can diffuse through the blocking W or liner. The edge displacement ΔL (void growth) is due to mass depletion at the cathode end of the line. Long lines and high current density were used in the present investigation, thus the effect of electromigration induced back flow[8] on the drift velocity at the cathode end of the line can be ignored. Then the void growth is $\Delta L = \int^\tau v_d\, dt$. The lifetime τ is obtained by $\tau = \Delta L_f/v_d$ for a constant drift velocity where τ is the amount of time required to grow a critical void size ΔL_f which resulted in $\Delta R/R_o = 10\%$. (The mass accumulation at the anode end of the line causes hillocks or extrusions.)

For the case of a bamboo-like grain structure, the contribution of mass transport by electromigration along grain boundaries, GBs, is negligible and surface diffusion is dominant.[3-5] The contribution from the Cu bulk diffusion is nil due to its extremely low diffusivity.[9] The surfaces at the sidewalls and on top of the line become the fast diffusion paths for the lift-off structure.[4] On the other hand, only the top surface (Cu/SiN$_x$ interface) is a fast diffusion path in the damascene line.[5] Drift velocity can be written as:

$$v_d = \delta_S\,(2/w + 1/h)D_S\,Z_S{}^*e\rho j/(kT) \quad \text{for a lift-off line} \tag{3a}$$

$$v_d = \delta_S\,(1/h)D_S\,Z_S{}^*e\rho j/(kT) \qquad \text{for a damascene line} \tag{3b}$$

where the subscript S refers to the Cu surface (at two sidewalls plus top of the line in Eq. (3a) and Cu/SiN$_x$ interface in Eq (3b)); δ_S denotes the effective width of the surface, w is the line width, h is the line thickness. Eq. (3a) states that the marker velocity (or void growth rate) in the bamboo-like unpassivated lift-off line structure will be increased as the metal linewidth or thickness are decreased, at a fixed sample temperature and current density. It suggests that the lifetime decreases as linewidth decreases for a constant ΔL_f. Eq. (3b), however, indicates that the drift velocity and thus the lifetime for the constant ΔL_f for a damascene line are independent of metal linewidth ($1/h$ only). .

For the case of a polycrystalline lift-off Cu line structure, the effective drift velocity becomes:

$$v_d = [(\delta_{GB}/d)(1 - d/w)D^0{}_{GB}\,e^{(-Q_{GB}/kT)}Z_{GB}{}^* + \delta_S(2/w + 1/h)D^0{}_S\,e^{(-Q_S/kT)}\,Z_S{}^*]\,e\rho j/kT \tag{4c}$$

and for a damascene structure

$$v_d = [(\delta_{GB}/d)(1 - d/w)D^0{}_{GB}\,e^{(-Q_{GB}/kT)}Z_{GB}{}^* + \delta_S(1/h)D^0{}_S\,e^{(-Q_S/kT)}Z_S{}^*]\,e\rho j/kT \tag{4d}$$

where the subscript GB refers to the Cu grain boundary; δ_{GB} denotes the width of grain boundary; d is the grain size; $D^0{}_{GB}$ and $D^0{}_S$ are the pre-exponential factors, and Q_{GB} and Q_S are the activation energies for grain boundary and surface diffusion, respectively. One would expect the drift velocity or lifetime to be less sensitive to linewidth for large w for both damascene and lift-off structures, because $(1 - d/w) \sim 1$ and $(2/w + 1/h) \sim 1/h$.

<u>Lift-off Structure</u>

Fig. 2 shows the mean lifetime for the unpassivated lift-off Cu lines as a function of line width w at 370 °C. Void growth at the cathode end of the line causes the line to fail. Lifetime increases monotonically with linewidth w but goes through a maximum at $w \simeq 1$ µm and then slightly decreases to a constant value. This observed behavior can be explained by the above mentioned model in which the dominant Cu diffusion path is the surface for the bamboo-like and a mixture of grain boundary and surface for polycrystalline line structures. For the bamboo-like structure when $w \leq 1$ µm, grain boundary transport is eliminated and only surface diffusion of the Cu lines is active. Eq. (4a) then predicts that the lifetime is proportional to $1/(2/w + 1/h)$ $\propto w$, (h is held constant). For polycrystalline line structure, Eq. (4c) predicts that the drift velocity or lifetime are less sensitive to linewidth for $w > 1$ µm.

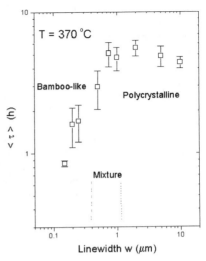

Fig. 2. Plot of mean lifetime as a function of linewidth for unpassivated lift-off Cu lines with $j = 15$ mA/µm² at 370 °C.

Damascene Line

In the case of Cu damascene lines with electron flow from W to Cu M1, the normalized line resistance as a function of time for sample temperatures of 256 °C is shown in Fig. 3. Initially, the line resistance slowly increases, followed by a period of rapid rise. The initial period of a slow resistance change rate was caused by void growth within the W/Cu overlap area. The sharp resistance increase occurred when the void grew and crossed into the line outside of the overlapped area where the current had to pass through the thin, high resistance, liner underlayer to connect the remaining Cu line and the W via. Typical cumulative % failure in a log-normal distribution for the sample temperature at 256 °C is plotted in Figure 4. For comparison, the data for the electron flow from Cu via to Cu M1 line for sample temperature at 350 °C are also plotted in Fig. 4. Reasonably good standard deviation σ in the log-normal distribution of 0.25 was obtained for both cases. As an example of damage, fig. 5 is an SEM micrograph of Cu at the cathode end of a Cu line tested for 133 h at 294 °C. The Cu depletion at the cathode end of the line is clearly shown.

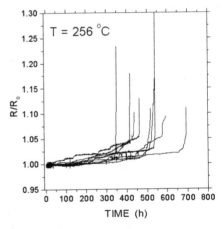

Fig. 3. The typical line resistance change as a function of time.

In the case of electron flow from a dual damascene Cu line/via to Cu line, the lifetime distribution was more complex than that of Cu atoms drifting away from the W via because of a possible instability of the thin liner at V1/M1 to block Cu diffusion. Electromigration induced compressive stress built up at the liner (the anode end of the line) may have forced the Cu atoms

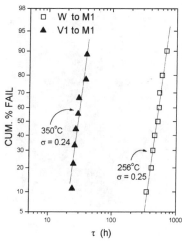

Fig. 4. Typical cumulative % failure data for the Cu damascene line on a log-normal scale.

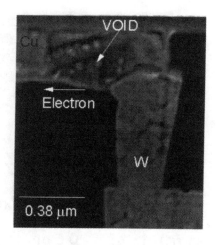

Fig. 5 SEM micrograph of the cathode end of a Cu bamboo-like damascene line.

to punch through to M1. Then Eq. 2 is no longer held ($J_{in} \neq o$) and Eq. 1 should be used. Fig. 6 shows that the normalized line resistance as a function of electromigration stress time can have two distributions with quite distinct lifetimes. For the earlier failure group (dotted lines), it can be postulated that the liner at the bottom of the Cu via can effectively block Cu diffusion from V1 to M1, which resulted in a shorter t_{50} of 220 h. But in some samples (solid lines), the Cu atoms may migrate through the liner from the V1 to M1 resulting in a long lifetime > 2200 h. For simplicity, only the earlier failure lifetimes were used in the estimations of t_{50}; the longer lifetime group was excluded. Figs. 7 (a) and (b) are the SEM micrographs of the cathode ends of M1 which have been tested for 271 h and 2200 h, respectively. Fig. 7(a) shows that the void grew beneath the bottom Cu V1 and beyond the Cu via/line overlapping section. This resulted in a sharp resistance increase and caused the line failure, since the current had to pass through the thin liner to connect the remaining Cu line and the via. The measured Cu drift velocities from Fig. 7(a) coincided with Fig. 4 suggesting that good diffusion barrier liners were present in the samples in the earlier failure group. The location of the voids shown in Fig. 7 (b), at the corner of M2 and near V1 sidewall liner, suggests that Cu atoms leaving from the Cu V1 to Cu M1 are more than Cu V1 receives from the M2 line. It indicates that the damage results from a polycrystalline Cu via adjacent to a bamboo-like M2 line, in which a

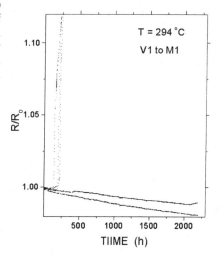

Fig. 6. Normalized line resistance as a function of time. Two distinct lifetime groups (dotted and solid lines) were observed.

695

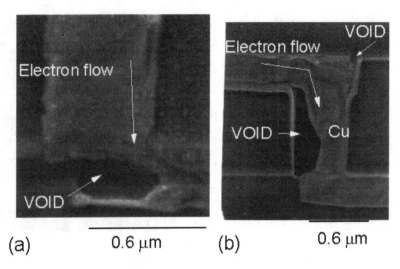

(a)   0.6 μm   (b)   0.6 μm

Fig. 7. SEM micrographs of electromigration damage lines in the case of electron flow from V1 to M1 tested for (a) 271 h, and (b) 2200 h.

discontinuity exists in GB paths. The failure mechanism is similar to electromigration in a single line where voids occur at grain boundary triple point.

Activation energy and Lifetime prediction

The measured t_{50} for both bamboo-like unpassivated lift-off Cu lines and Cu damascene lines passivated with SiN_x/SiO_2 as a function of 1/T is plotted in Fig. 8. The solid lines are the least squares fits. The values of t_{50} estimated from the earlier failure group (Fig. 4) in the case of electron flow from Cu V1 to M1 damascene structure is about 2x of those of electron flow from W to M1, although the measured drift velocity is about the same for both electron flow from W to M1 and V1 to M1. The factor of 2 can be roughly predicted by the volume ratio of the Cu beneath the Cu via (0.4 x 0.31 μm²) to that above the W via of (0.29 x 0.21 μm²). This agrees with the fact that the removal of twice the volume of Cu at the same current density required twice the amount of time. The activation energies for electromigration in the bamboo-like unpassivated lift-off 0.25 μm wide and 0.27 μm wide Cu damascene lines are found to be 0.8 eV and 1.1 ± 0.1 eV, respectively. In the bamboo-like test

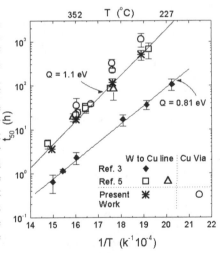

Fig. 8. Plot of t_{50} vs. 1/T. The straight lines are least-squares fits. The data points are obtained with the electron flow from W to Cu lines, but the data with open circles have electrons flow from Cu via to Cu line.

structures, the contribution of mass transport by electromigration along grain boundaries is negligible. The Cu surface or Cu/dielectric interface appear to be the fast diffusion paths for Cu so that the different activation energies obtained for the unpassivated lift-off and passivated damascene Cu lines may be attributed to surface impurities and ambient,[10-13] or SiN_x passivation. From the lifetime data of electron flow from W to M1 in Fig. 8, t_{50} of 1.3 and 100 years are estimated for the lift-off and Cu damascene lines at 125 °C, respectively. Using the σ of 0.3 in the log-normal distribution, one can estimate the dc lifetime at -2σ, $t_{-2\sigma} = t_{50}\exp(-2\sigma)$ of 0.7 years and 55 years for the unpassivated lift-off and Cu damascene lines.

CONCLUSIONS

Electromigration in Cu unpassivated lift-off lines and damascene lines passivated with SiN_x/SiO_2 has been investigated. The mean lifetime in bamboo-like lift-off Cu lines decreases as the linewidth decreases, which is attributed to Cu surface diffusion along the two sidewalls and top surfaces of the lines. For the wide lift-off polycrystalline line, lifetime is roughly independent of linewidth. For the case of Cu damascene, a systematic increase in lifetime was found for the electron flow from Cu via to Cu lines over that of electron flow from W via to Cu line. This is an artifact of the test structure; the volume of the Cu line under the Cu via is larger than that above the W via. A longer lifetime was obtained when the liner at V1/M1 apparently did not block Cu migration for electron flow from V1 to M1. The activation energies for electromigration in the bamboo-like unpassivated lift-off 0.25 μm wide and 0.27 μm wide Cu damascene lines are found to be 0.8 eV and 1.1 ± 0.1 eV, respectively. The different activation energies obtained for the lift-off and damascene Cu lines are most likely due to different nature (environment) of Cu outer surfaces (and their passivation).

ACKNOWLEDGMENTS

We would like to acknowledge the efforts of our colleagues at the IBM Microelectronics Division, Burlington, VT, for fabrication of Cu damascene samples. We are also grateful to S. Boetther L. Kimball, and T. Bauer for providing SEM micrographs.

REFERENCES

1. C.-K. Hu, J.M.E. Harper, Material Chem. and Phys. 51, 5 (1998)
2. A. Scorzoni, B. Neri, C. Caprile, and F. Fantini, Mat. Sci. Report, 7 143 (1991)
3. C.-K. Hu, K.L. Lee, L. Gignac, R. Carruthers, Thin Solid Films, **308-309**, 443 (1997).
4. C.-K. Hu, R. Rosenberg, K.L. Lee, Appl. Phys. Letter, **74**, 2945 (1999).
5. C.-K. Hu, R. Rosenberg, H. Rathore, D.Nguyen, B. Agarwala, Proc. IITC (1999) p.267.
6. K.Y, Lee, C.-K. Hu, T. Shaw, T.S. Kuan, J. Vac. Sci. Technol. B, **13**, 2869 (1995).
7. L. Gignac, private communication.
8. I.A. Blech, J. Appl. Phys., 47, 1203 (1976).
9. S.J. Rothman and N.L. Peterson, Phys. Status Solidi, **35**, 305 (1969).
10. D. B. Butrympwicz, J. R. Manning, and M. E. Read, J. Phys. Chem. Ref. Data, **6**, 1 (1977)
11. H.P. Bonzel, in J.M. Blakly (ed.), Surface Physics of Materials, Vol. II, Academic Press, NY, 1975, Chap. 6.
12. C.W. Park, and R.W. Vook, Thin Solid Films, **226**, 238 (1993).
13. P. G. Shewmon, Diffusion in Solids, 2nd ed. (McGraw-Hill, Warrendal, PA. 1989).

Reliability Improvement of Aluminum Dual-Damascene Interconnects by Low-k Organic SOG Passivation Dielectrics

H. Kaneko, T. Usui, T. Watanabe, S. Ito, M. Kawai and M. Hasunuma
Microelectronics Engineering Laboratory TOSHIBA Corp.
8, Shinsugita-cho, Isogo-ku, Yokohama-city, 235-8522, JAPAN, hisashi.kaneko@toshiba.co.jp

ABSTRACT

The via electromigration(EM) reliability of aluminum(Al) dual-damascene interconnects by using Niobium(Nb) new reflow liner is described. It has been found that the EM lifetime was further improved by introducing low-k organic spin on glass(SOG)-passivated structure than the conventional TEOS-SiO$_2$/SiN-passivated structure. Higher EM activation energy of 1.08 eV was obtained for the SOG-passivated structure than the TEOS-passivated structure of 0.9 eV, even though no significant Al micro-crystal structure difference was found for both structures. It has been turned out that the low-k SOG material has the 1/7 Young's modulus (8 GPa) of TEOS SiO$_2$ (57 GPa) or thermal SiO$_2$ (70 GPa). The small Young's modulus means that SOG is more elastically deformable than TEOS or thermal SiO$_2$. This elastic deformation of the low-k SOG could retard the tensile stress evolution due to the Al atom migration near the cathode via, and elongated the time until the Al interconnect tensile stress exceeds the critical stress value for void nucleation. It has been concluded that the small-RC and reliable multi-level Al interconnect can be realized by the Nb-liner reflow-sputtered process with soft and low-k SOG dielectric materials.

INTRODUCTION

The dual-damascene multi-level interconnect process realizes the planarized metal/insulator surface at each level for keeping depth of focus margin for lithography, and has potential for the process cost down[1]. The key technology for the Al dual-damascene process is the superior hole and trench filling capability with least effective interconnect resistance increase. This filling capability strongly depends on the choice of the liner material for the sputter reflow[2]. By using the Niobium(Nb) liner, the filling capability of aspect ratio of 7.5 with φ 0.17 μm via opening, and 30% decrease in the effective resistance than the conventional Ti liner could be achieved[3]. For high performance ULSIs, on the other hand, the reduction of the parasitic wiring capacitance or RC delay is highly necessary[4]. The low-k SOG dielectrics[5] and/or polyimide are the promising candidate for the low-k inter layer dielectric (ILD) materials. However, the reliability knowledge about the Al damascene and/or dual damascene structure has been limited to the Al micro-crystal structure and texture studies[6-9], or the Al extrusion phenomena due to the poor mechanical properties of low-k materials[10,11].

In this paper, a reliability impact on the via EM phenomena of the Al dual damascene structure by using the low-k SOG material with dielectric constant of 2.5[5] and the conventional TEOS as the passivation has been studied[12]. The Al-0.5wt%Cu dual-damascene via EM test structure was fabricated by reflow-sputtering using Nb liner with the CVD-W lower-level interconnect. The observed EM lifetime difference between SOG- and TEOS-passivated test structure has been discussed in terms of the difference of the Young's modulus of the passivation materials and the tensile stress evolution due to the EM induced Al atom migration.

EXPERIMENT

Figure 1 shows the schematic illustration of the two level via EM test structure. The 0.7 μm thick TEOS SiO$_2$ was deposited onto the 0.3 μm thick lower level CVD-W interconnect fabricated by CMP. The φ 0.25 μm via hole and the 0.3 μm deep and 0.25 μm wide upper groove were simultaneously filled with Al-0.5wt%Cu by the 2-step reflow sputtering using 15 nm thick Nb liner at 450 °C. After the Al dual damascene structure was defined by the Al CMP, the test structures were passivated by the 0.4 μm thick low-k SOG at 450 °C or by the 0.6

Conference Proceedings ULSI XV © 2000 Materials Research Society

μm thick TEOS SiO$_2$ and 0.6 μm thick SiN. The length of the 0.25 μm wide Al damascene interconnect was 100 μm, and the interconnect was directly connected to the anode pad in order to avoid the back-flow effect[13]. The EM accelerated testing was carried out with the current density of 2 MA/cm^2 and at the temperatures of 200 °C-250 °C. The critical length for 2 MA/cm^2 current density is around 30 μm, much less than the 100 μm of the present test structure. The electron flow was from the lower level W interconnect to the upper Al damascene interconnect, and therefore the EM induced void was expected to nucleate around the via. The failure criterion was a 10% resistance shift, and the current supply of each sample was stopped at a 20% shift.

RESULTS AND DISCUSSION

1. EM-INDUCED VOID LOCATIONS

Figure 2 shows cross-sectional TEM photographs of Al dual damascene interconnects with TEOS SiO$_2$/SiN passivation (A and B), and with low-k SOG passivation (C), taken after EM testing. EM-induced voids were always accompanied by the horizontal grain boundary and/or grain boundary triple point. The voids observed in photo B and C were formed upstream of the bamboo grain boundary. This suggests that the dominant Al migrating path is the interface between bamboo Al grain and NbAl reaction layer in the present metallization, and the top edge of the via corner (Al and TEOS and/or SOG interface in photo A) and the grain boundary triple point (in photo B and C) act as the void nucleation sites. No voids were observed at the bottom of the via and/or the interface between W and NbAl reaction layer where the maximum Al atom flux divergence is generated. Many observations of the monolithic Al vias by TEM show that they are single-crystal. This microstructure would be due to the high enough reflow temperature (450 °C) to promote the grain growth in the small via. In addition to this favorable Al microstructure, adhesion of the Al to the NbAl reaction layer, and NbAl reaction layer to the W would be good. Therefore, the critical stress that enables the void nucleation at this interface is higher than the triple point in the Al damascene interconnect, resulting in

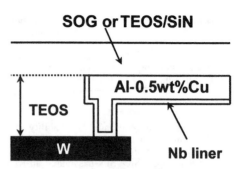

Fig. 1. Schematic illustration for the via EM test structure. The lower level interconnect was CVD-W line, and the 100 μm long upper level Al-0.5wt%Cu interconnect and the via were made simultaneously by reflow sputtering using Nb liner.

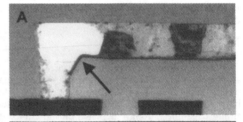

Fig. 2. Cross-sectional TEM photographs after EM testing. Arrows indicate EM-induced voids.
Photo-A,B:TEOS-SiO$_2$/SiN-passivated samples.
Photo-C:SOG-passivated sample.

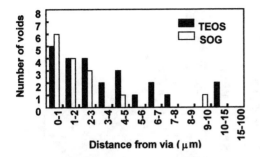

Fig. 3. Distribution of EM-induced-void locations. The EM test condition was 2 MA/cm^2 and 250 °C.

the no void nucleation at the via bottom. It is also noteworthy that there is not apparent grain microstructure difference between the TEOS SiO$_2$/SiN-passivated sample and low-k SOG-passivated sample.

The distribution of EM-induced void locations, measured as distances from the via by the optical microscopy is shown in Fig. 3. 24 TEOS SiO$_2$/SiN-passivated samples and 15 low-k SOG-passivated samples were used for analysis. Each sample had only one void in the test structure. Most of the voids were observed within 10 μm from the via. Voids would nucleate at the top edge via corner and the grain boundary triple point that is nearest to the via. It was also observed that there is no significant difference in the distribution of void locations between TEOS SiO$_2$/SiN-passivated and low-k SOG-passivated samples.

2. EM LIFETIME IMPROVEMENT IN LOW-K SOG PASSIVATED STRUCTURE

Fig. 4 shows the cumulative failure distribution for TEOS SiO$_2$/SiN-passivated and low-k SOG-passivated EM test structures. The observed EM lifetime was longer for the low-k SOG passivated structure at each temperature. Moreover, the difference in MTF between the two test structures becomes larger as the test temperature decreases. And, therefore the MTF difference will by significantly larger at a typical LSI operating temperature of 85 °C to 100 °C. From the cumulative failure distributions, the activation energy of the Al with low-k SOG passivation is 1.08 eV, while that of the TEOS SiO$_2$/SiN-passivation is 0.9 eV

As described in the former section, the Al microstructure was similar for both TEOS SiO$_2$/SiN-passivated and low-k SOG-

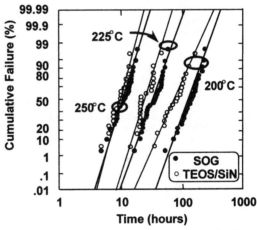

Fig. 4. EM cumulative failure distribution for reflow sputtered Al-0.5wt%Cu/Nb-liner dual damascene interconnects with low-k SOG passivation and TEOS SiO$_2$/SiN-passivation

passivated structures. It is reasonable to expect that the EM-induced Al and also the Cu solute migrating fluxes are the same for both passivated structure, and the Cu-sweep out model[14] that succeeds in explaining the existence of the incubation time by Cu doping to Al interconnect would not be a dominant mechanism in the present metallization. Since the EM-induced void nucleation is also controlled by the tensile stress as the analogy of the stress-migration[15], the incubation time is also defined as the time that is necessary to create the critical tensile stress for void nucleation around the via by EM-induced Al atom migration[16]. Therefore, it is considered that there exists another mechanism that elongates the incubation time for the low-k

SOG passivated structure. Figure 5 shows the thermal stresses in the 0.25 μm wide damascene interconnects with low-k SOG and TEOS-SiO₂/SiN passivation as a function of temperature measured by the X-ray diffraction technique[17]. The initial tensile stress for the low-k SOG passivated sample was much lower than that of the TEOS-SiO₂/SiN passivated sample, even though the passivated temperature of the low-k SOG sample was higher. Moreover, the slopes of the low-k SOG passivated sample was smaller.

In order to clarify the difference in the EM lifetime and also the stress behavior during thermal cycle between the two systems further, the Young's moduli of low-k SOG and TEOS SiO₂ dielectrics were calculated by the micro-Vicker's hardness. Figure 6 shows indentation behavior of low-k SOG, TEOS SiO₂ and thermal SiO₂ measured by the nano-indentation technique. The slope for each material is proportional to the Young's modulus. The Young's moduli for low-k SOG and TEOS SiO₂ were deduced using known Young's modulus of thermal SiO₂ (70 GPa) The observed Young's modulus were 8 GPa for low-k SOG and 57 GPa for TEOS SiO₂, respectively. The Young's modulus of low-k SOG is about one

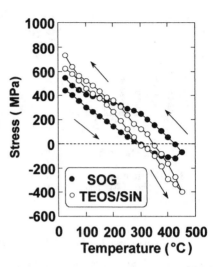

Fig. 5. Stress along the 0.25 μm wide damascene interconnect line with low-k SOG and TEOS-SiO2/SiN passivated structure as a function of temperature measured by the X-ray diffraction technique.

seventh of that of TEOS SiO₂. This indicates that low-k SOG is much elastically deformable than TEOS SiO₂ and/or the low-k SOG is soft. From the results of the lower stress during thermal cycle and the smaller slope, and the smaller Young's modulus for low-k SOG material, it is concluded that the Al damascene interconnect stress has been suppressed by the easy elastic deformation of the low-k SOG passivation.

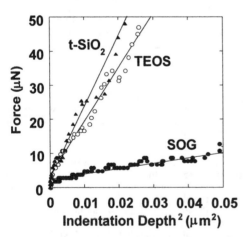

Fig. 6. Indentation behavior of low-k SOG, TEOS SiO₂ and thermal SiO2 dielectrics.

3. MODEL FOR THE EM LIFETIME IMPROVEMENT BY LOW-K SOG ELASTIC DEFORMATION

Figure 7 shows schematic illustration to explain the lifetime difference with low-k SOG and TEOS-SiO₂/SiN passivated test structure. The V is the Al volume that is necessary to produce the critical tensile stress around the via for the TEOS-SiO₂/SiN passivation, and J is the EM-induced Al atom flux. Again, the J is the same for both low-k SOG and TEOS-SiO₂/SiN passivated sample, because the same Al crystal microstructure for both systems. t_{TEOS}, the time to reach the critical stress for void nucleation in the Al interconnect around the via for TEOS-SiO₂/SiN passivated sample is V/J. However, for the low-k SOG passivation, because of the ready elastic deformation, the additional Al volume of V_1 is required to create the same amount of the critical tensile stress. The EM lifetime difference, t_{SOG}- t_{TEOS}, is expressed as V_1/J. Therefore, the EM-MTF for low-k SOG passivation is longer than the TEOS-SiO$_2$/SiN passivation. Moreover, since J has the normal thermal activation component, e. g., $J \propto \exp(-Ea/kT)$, where Ea is real diffusional activation energy for the real migrating path, the lifetime difference becomes large at the lower accelerating temperature, as shown in Fig. 4.

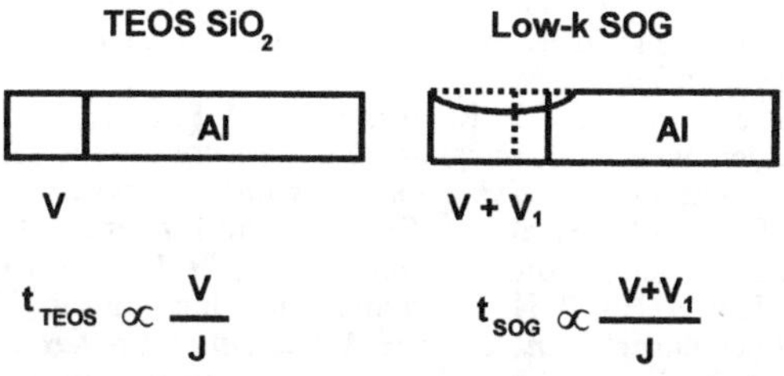

CONCLUSION

The lifetime improvement for the via EM has been found for the low-k SOG passivated Al dual-damascene test structure. It is concluded that this improvement is due to the softness and/or ready elastic deformation of the low-k SOG material with Young's modulus of 8 GPa.

TEOS SiO$_2$ **Low-k SOG**

$$t_{TEOS} \propto \frac{V}{J} \qquad t_{SOG} \propto \frac{V+V_1}{J}$$

Fig. 7. Schematic illustration to explain the EM-MTF difference for TEOS-SiO$_2$/SiN and low-k SOG passivation. V is the Al volume necessary to create the critical tensile stress for void nucleation around the via. V_1 is the additional Al volume required due to the ready elastic deformation of low-k SOG. J is the EM-induced Al atom flux, t_{TEOS} and t_{SOG} are the time to reach the critical stress for TEOS-SiO$_2$/SiN and low-k SOG passivation, respectively.

This elastic deformation can suppress the Al damascene interconnect tensile stress around the via, and elongate the incubation time needed to create the critical stress for the EM-induced void nucleation. The small-RC and reliable multi-level interconnect can be realized by the Nb-liner reflow-sputtered Al-0.5wt%Cu dual-damascene process with soft and low-k SOG dielectric materials.

ACKNOWLEDGMENTS

The authors would like to thank members of ULSI Process Development Department III, Process Engineering Laboratory and Advanced Process Engineering Sec. III, Integrated Circuit Advanced Process Engineering Department for sample preparations and their helpful discussions through this work.

REFERENCES

1. C. W. Kaanta *et al.*, "Dual Damascene : A ULSI Wiring Technology", *in Proceedings of International VLSI Multilevel Interconnection Conference*, 1991, p.144.
2. L. A. Clevenger *et al.*, "A Novel Low Temperature CVD/PVD Al Filling Process for Producing Highly Reliable 0.175µm wiring/0.35µm Pitch Dual Damascene Interconnections in Gigabit Scale DRAMs", *in Proceedings of International Interconnect Technology Conference*, 1998, p. 137.
3. J. Wada *et al.*, "Low Resistance Dual Damascene Process by New Al Reflow using Nb Liner", *in Proceedings of IEEE 1998 VLSI Symp. on Technol.*, 1998, p.48.
4. D. C. Edelstein, G. A. Sai-Halasz and Y. J. Mii, "VLSI On-Chip Interconnection Performance Simulation and Measurements", *IBM J. Res. Develop.*, **39**, p. 383, 1995.
5. R. Nakata *et al.*, "New Low-k Material "LKD^{TM}" for Al Damascene Process Application", *Extended Abstracts of the 1999 International Conference on Solid State Devices and Materials*, Tokyo, 1999, pp. 506-507.
6. J. M. E. Harper and K. P. Rodbell, "Microstructure Control in Semiconductor Metallization", *J. Vac. Sci. Technol*, **B15**, 1997, p. 763.
7. J. L. Hurd et al., "Linewidth and underlayer influence on texture in submicrometer-wide Al and AlCu lines", *Appl. Phys. Lett.*, **72**, 1998, p.326.

8. A. Furuya et al., "Electrical Characterization and Microstructure of 0.32μm-Pitch Aluminum-Damascene Interconnects", *in Advanced Metallization Conference in 1998 Japan/Asia Session*, 1998, p.33.

9. J. E. Sanchez, P. R. Besser and D. P. Field, "Microstructure of Damascene Processed Al-Cu Interconnects for Integrated Circuit Applications", *in AIP Conference Proceedings of the Fourth International Workshop on Stress-Induced Phenomena in Metallization* **418**, 1997, p. 230.

10. S. Foley et al., "A Study of the Influence of Inter-Metal Dielectrics on Electromigration Performance", *Microeletron. Reliab.*, **38**, 1998, p. 107.

11. P. S. Ho, P. H. Wang and J. Kasthurigrangan, "Structure Integrity and Reliability of Low K Interconnects", *in Advanced Metallization Conference in 1998 Japan/Asia Session*, 1998, p.19.

12. T. Usui et al., "Significant Improvement in Electromigration of Reflow-Sputtered Al-0.5wt%Cu/Nb-liner Dual Damascene Interconnects with Low-k Organic SOG Dielectric", *in Proceedings of 37th International Reliability Physics Symposium*, 1999, p.221.

13. I. A. Blech, *J. Appl. Phys.*, **47**, 1976, p. 1203.

14. C. K. Hu, M. B. Small and P. S. Ho, "Electromigration in Al (Cu) two-level structures: effect of Cu and kinetics of damage formation", *J. Appl. Phys.*, **74**, 1993, p. 969.

15. H. Kaneko *et al.*, "A Newly Developed Model for the Stress-induced Slit-like Voiding", *in Proceedings of 28th International Reliability Physics Symposium*, 1990, p.194.

16. M. A. Korhonen *et al.*, "Stress evolution due to electromigration in confined metal lines", *J. Appl. Phys.*, **73**, 1993, p. 3790.

17. A. Tezaki *et al.*, "Measurement of three Dimensional Stress and Modeling of Stress Induced Migration Failure in Aluminum Interconnects", *in Proceedings of 28th International Reliability Physics Symposium*, 1990, p. 221.

RELIABILITY ENHANCEMENT OF COPPER INTERCONNECTS USING WAFER-LEVEL ELECTROMIGRATION TESTING

V.V.S. RANA*, J. EDUCATO*, S. PARIKH*, M. NAIK*, T. PAN*, P. HEY*, D. YOST*, D. PIERCE**
*Applied Materials, Inc., 3320 Scott Blvd, M/S 1148, Santa Clara, CA 95054, viren_rana@amat.com
**Sandia Technologies, Inc., 6003 Osuna Road NE, Albuquerque, NM 87109

ABSTRACT

A wafer-level (WLR) electromigration measurement technique was used to study the failure of electroplated damascene copper lines. The failure characteristics were used to optimize the unit processes and the damascene process flow. A reliable multilevel copper module has been developed that is being implemented on submicron VLSI devices.

INTRODUCTION

With the continuing evolution of integrated circuits to deep submicron technology the use of copper and a low k dielectric becomes necessary to reduce the RC delay and prevent the interconnect delay from affecting the speed of operation of the circuit. However, along with the reduction of RC delay the reliability of the interconnect system becomes critical. This is because with rapidly decreasing design rules the interconnects are subjected to increasingly higher current densities and thermal gradients. While there are many reports on the evaluation of electromigration of Cu interconnects, there are few relevant to the process technology actually going into production. What is required is to determine the reliability of sub-half micron Cu interconnects formed by electroplating in damascene structures with the appropriate dielectric for isolation. Besides electromigration of Cu lines and vias, the reliability evaluation and enhancement of the damascene structure is needed

The reliability measurement and analysis are important tools to improve the performance of the interconnects. Since traditional post-package measurements and feedback are very time consuming and hence not suitable for rapid development of a reliable interconnect system we have used wafer-level electromigration measurements to provide results in a timely fashion to not only understand the elctromigration behavior of Cu but also to enhance the overall reliability of Cu/SiO$_2$ and Cu/low k structures.

EXPERIMENTAL

We report here the results obtained on Cu lines formed using single damascene scheme as outlined in Table I. All processing was carried out in Applied Materials' Equipment and Process Integration Center (EPIC). A 250Å of sputtered, Ionized Metal Plamsa (IMP) TaN was used as the barrier layer and a 2000Å of sputtered, IMP Cu was used as a seed material for electroplating. The Cu lines were 4000-5000Å thick. The passivation consisted of 1000Å of PECVD SiN followed by 5000Å of PECVD SiO$_2$. Pads were then opened in this dielectric layer in order to probe the electromigration test structures.

The structures used for electromigration (EM) testing of lines were ASTM F1259-95 compliant 800 µm long NIST style structures (Figs. 1 and 2). The structures also included horizontal extrusion monitors spaced 0.4 μm from the line under test. The mask set including these test structures was designed by TestChip Technologies [1].

Table I. Copper Damascene Process Flow

Step	Description	Step	Description
1	Dielectric Deposition	6	Copper CMP/Clean
2	Metal Trench Lithography	7	Passivation
3	Trench Etch/Ash/Wet Clean	8	Pad Open Lithography
4	Barrier/Seed/Electroplated Cu-fill	9	Pad Open Etch/Ash/Wet Clean
5	Copper Anneal	10	Electrical Test

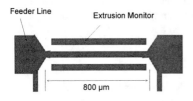

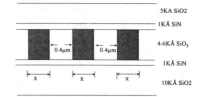

Fig. 1. EM Structure of Copper Lines   Fig. 2. Cross-section of EM Structure

The EM measurements were made using a constant resistance isothermal technique consistent with JEDEC standard JESD 61. In this technique the test structure is heated to the test temperature using Joule heating by the applied current and maintained at that temperature by maintaining a constant resistance using a special test software HP/ST PDQ-WLR [2]. The testing was carried out using an HP 4071 parametric tester with a 1A SMU, that provides capability of testing lines as wide as 4.0 μm. The prober used was 8" Electroglass automatic prober. Once the damage starts other non-electromigration induced failures can occur because of the high current densities and temperatures used in the test. Because of this a very small change in resistance (1% $\Delta R/R$) was used as a failure criterion. The testing was carried out from 200 to 400°C and the current densities to obtain these temperatures were in the range of 10 to 45 MA/cm^2, depending upon the line size, etc. Further details on this testing technique are given in ref. [3].

RESULTS

The electromigration data is analyzed in terms of the well-known Black's equation

$$t_f = \frac{A}{j^n} e^{\frac{Ea}{kT}}$$

(1)

where t_f is the time to failure (sec), A and n are constants, j is the current density (A/cm^2), Ea is the activation energy for diffusion, k is Boltzmann's constant, and T is the temperature (°K).

Since Joule heating was used to achieve the test temperature the current density and temperature are interdependent. A current exponent (n) of 2 has been obtained by other workers

706

when the effects of Joule heating are well accounted for in the failure analysis. Hence, Black's equation becomes

$$t_f \times j^2 = A \ e^{\frac{Ea}{kT}}$$

(2)

The failure can then be described by the failure parameter ($t_f \times j^2$) where the failure times are normalized by current density squared.

Effect of Line Width:

The line-width effect is illustrated using two different line sizes, 4.0 and 0.4 µm. The wide line (4.0 µm) has a polycrystalline structure (Fig. 3) having triple grain boundary junctions. These triple grain boundary junctions act as sources of flux divergence when subjected to test current densities. The electromigration failure (1% ΔR/R) was accompanied by voids at these triple grain junctions (Fig. 4). The 0.4 µm lines on the other hand had a bamboo like structure (Fig. 5) and the failure times were much higher. The failures frequently initiated at defects in the line and not related to intrinsic EM behavior. Defect-free lines did not show any physical signs of degradation for the failure criterion used.

Fig. 3. FIB micrograph of cross-section of 4.0 µm test structure

Fig. 4. 4.0 µm line during test

Fig 5. FIB micrograph of 0.4 µm test structure

With the WLR methodology used here the line under test is selectively heated while the rest of the wafer including the pads remains at room temperature. If WLR testing is carried out for a long enough time the final failure takes place by extrusion of heated line and shorting with the adjacent extrusion monitor line. This is the reason a very low change in resistance (1%) was chosen to define EM failure in order to avoid the stress dependent failures such as extrusions.

The sequence of events leading to extrusion in 4.0 µm lines during WLR testing is shown in Fig. 6 (a). Initially, there was a collection of material in the line that on continued testing resulted in extrusion shorts. Although no accumulation of material was seen in 0.4 µm lines the final failure was typically due to extrusion shorts. As the FIB micrographs of cross-section of test structures in Figs 6 (a) and (b) show the extrusion of copper always occurred across the top

SiO₂/SiN interface. While this was not true electromigration failure it was useful in highlighting this interface as a weak link in the damascene structure. We have optimized the unit processes to improve the strength of this interface.

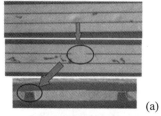
(a)

(b)

Fig. 6. Failure due to Cu extrusion in (a) 4.0 μm lines, and (b) 0.4 μm lines

Application to Process Optimization:

Since EM behavior of wide lines is determined by the grain structure of lines the WLR tests were used to evaluate processes that influence grain structures such as electroplating (ECP) and annealing conditions. Figure 7 shows the evaluation of two different ECP chemistries. Chemistry 1 deposited films that were much finer grained than chemistry 2 and on testing there was rapid failure with voids extending the full width of the line. Chemistry 2 on the other hand resulted in a polycrystalline film with a bigger grain size. In these lines the void growth was much slower and the failure times were longer. On continued testing beyond the failure criterion of (1% $\Delta R/R$) and to the final failure the voids could not extend fully across the line before there was extrusion of Cu across the SiO₂/SiN interface, as shown in Fig. 8. As can be seen not only the failure times are lower for chemistry 1 but also the rate of failure is much higher indicating higher grain boundary diffusivity for this chemistry.

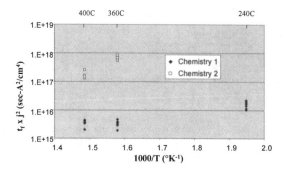

Fig. 7. Failure characteristics of 4.0 μm lines of electroplated Cu deposited using two different bath chemistries.

Optimization of anneal conditions for electroplated copper is important for improved reliability of Cu lines. We have investigated the effect of placement of anneal process in the process flow using failure investigation of 0.4 μm lines that were more susceptible to defects.

Figure 9 shows the failure parameter measured at 360°C for films that were either annealed before CMP or after CMP. The annealing was done at 350°C for 30 min. in argon. It can be seen that for lines not annealed after deposition but after CMP there was rapid failure of lines. On the other hand, annealing the film before CMP results in increased resistance to failure. Copper lines with post-CMP anneal only were found to have hillock growth after annealing. These hillocks were found to lead to extrusions during WLR testing (Fig. 10).

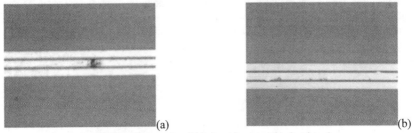

Fig. 8. The final failure modes of (a) chemistry 1 and (b) chemistry 2

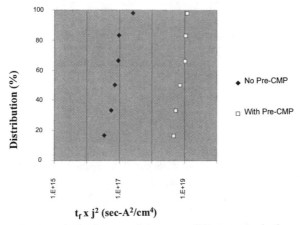

Fig. 9. Effect of placement of copper anneal process at different steps in the process flow.

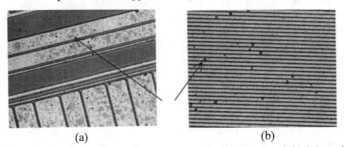

Fig. 10. Hillock growth, shown by arrows, on (a) wide lines, and (b) 0.4 μm lines after post-CMP anneal when there was no pre-CMP anneal.

Oxide vs Black Diamond Dielectric:

The electromigration behavior of Cu lines with Black Diamond (k=2.8) [4] as the dielectric was investigated. As described earlier because of the WLR technique used there was a strong tendency to form extrusions not related to the EM behavior of the material. Because of the low thermal conductivity of Black Diamond films there was an increased tendency to form such extrusions during testing particularly in 0.4 μm lines. The failure parameter for 4.0 μm lines not affected by these extrusions is shown in Fig. 11. It can be seen that the electromigration failure parameter is the same for both SiO_2 and Black Diamond (BD).

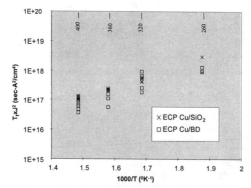

Fig. 11 . Electromigration of 4.0 μm lines with SiO_2 or Black Diamond Dielectric.

Electromigration Behavior:

The electromigration behavior of 0.4 μm lines was studied after the optimization of processes in the single damascene process flow. Some of this optimization has been discussed here. As mentioned earlier, the 0.4 μm lines had a bamboo structure and hence no source of flux divergence. With a current density of 20 to 45 MA/cm^2 the line failure increasingly become dependent on defects that would cause localized concentration of stress. Localized line variations such as line striations caused by improper plasma etch or lithography can lead to rapid catastrophic failures such as extrusions discussed above. To understand true electromigration induced failures that are caused by atom movement due to the 'electron wind' the defect induced failures need to be avoided. The failure time improvement can thus drive the process improvement. The failure parameter (t_f x j^2) for 0.4 μm lines with optimized process and process flow, is plotted in Figure 12. Such measurements on several lots have yielded an activation energy of 0.9 to 1.0 eV for electromigration of these fine lines. These values are close to those that have been recently reported [5,6].

The wafer-level electromigration testing as described here was used in a program to develop multilevel damascene copper module capability. This MLM Cu module capability is now being used to form copper interconnects on wafers where 'front end' processing is completed in a microchip manufacturing fab. Figure 13 shows two level copper interconnects formed on a memory device as a part of a feasibility demonstration study. It is essential that the reliability of such interconnects be understood and maximized.

CONCLUSIONS

A wafer-level (WLR) electromigration measurement technique was used to study the elctromigration behavior of Cu lines. The technique was also helpful in optimizing the various processes and the damascene process flow. A reliable two level metal Cu interconnect module has been implemented on a memory device to demonstrate its capability.

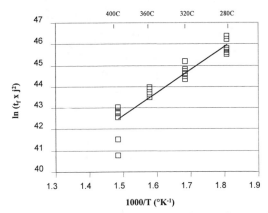

Fig. 12. The electromigration behavior of 0.4 μm lines.

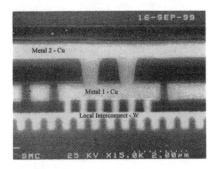

Fig. 13. Two level Cu interconnects in a memory device.

REFERENCES

1. TestChip Technologies, Inc., Dallas, TX.
2. Sandia technologies, Inc., Albuquerque, NM.
3. Donald Pierce, James Educato, Viren Rana, and Dennis Yost, *1998 IEEE International Integrated Reliability Workshop Final Report,* IEEE Catalog No. 98TH8363. Workshop held Oct. 12-15, 1998, Lake Tahoe, CA.
4. M. Naik, et al., 181, IITC (1999).
5. H.S. Rathore, D.B. Nguyen, L. Ashley, G. Biery, J. Hummel, B. Agarwala, T. Kane and P. Faitz, Proc. VLSI Multi. Int. Conf., 89 (1999).
6. C.-K. Hu, R. Rosenberg, H.S. Rahtore, D.B. Nguyen, B. Agarwala, 267, IITC (1999).

3-D ELECTRICAL RESISTANCE MODEL OF VIA AND CONTACT RELIABILITY IN INTEGRATED CIRCUIT METALLIZATION

S. H. Kang, A. S. Oates, and Y. S. Obeng

Bell Laboratories, Lucent Technologies, 9333 S. John Young Parkway, Orlando, Florida 32819

ABSTRACT

We present a three-dimensional thermal-electric finite element model that calculates an electrical resistance change led by electromigration-induced voiding in multilevel interconnects. The model provides a simple methodology for finding the critical void volume (V_c) and the void drift velocity (v_d) for various interconnect configurations at a given failure criterion. The model is useful in understanding the effects of refractory layer (TiN, Ti) arrangements, contact misalignments, and current crowding on the electromigration characteristics of deep submicron contacts and vias. The model is useful also in evaluating the effect of Joule heating by calculating the local current-density distributions through the layers.

INTRODUCTION

One of the primary concerns in the manufacture of advanced microelectronic devices is to ensure that metallic contacts and interconnects do not fail by electromigration through the service life of integrated circuits. Electromigration-induced damage ordinarily appears as voids that can grow to raise the electrical resistance to unacceptable values (e.g. 10% shift in the relative resistance). It is well recognized that electromigration failure of deep submicron contacts and vias is dominated by diffusive displacements (material depletion) at the cathode terminals of the Al(Cu) stripes where vacancies preferentially accumulate under an electric field [1,2]. The total time to failure is the sum of the void nucleation and growth times. The nucleation (or incubation) stage is characterized by a sequence of multiple physical processes: Al_2Cu precipitation, Cu sweeping prior to Al migration, vacancy buildup, and finally void nucleation [1]. But, there is essentially no resistance increase during this stage. Instead, it can rather drop due to the precipitation. The resistance escalation that causes failure is a consequence of void growth [2]. It is well documented that the rate of the resistance increase is small in the beginning of the void growth stage, but with further material depletion it accelerates and then reaches a steady state. Prior work suggests that the void growth stage can be characterized by two parameters [2]. One is the "critical void volume", which is the amount of Al(Cu) that must be depleted to meet the failure criterion. The other is the "drift velocity" of the void front, which describes how fast the nucleated void grows to the point of failure. To understand the failure kinetics by exploring the parameters, we have developed a 3-D thermal-electric finite element model that calculates the resistance variation as the void front drifts away from the contacts or the vias. This model can simulate complex multilevel metallizations that incorporate various refractory layers like TiN and Ti. In addition, the model is useful in studying the Joule-heating issues associated with current crowding in the presence of a void and also with the increasing number of metal layers in integrated circuits. The simulation results are discussed based on the experimental results obtained from the identical test structures.

Conference Proceedings ULSI XV © 2000 Materials Research Society

MODEL

This work presents a three-dimensional (3-D) electrical resistance model developed using ANSYS® [3], a commercial code for finite element analysis (FEA). The elements used are 8-node, thermal-electric bricks that have two degrees of freedom per node, voltage and temperature. The model performs a current conduction analysis to determine the electrical potential and current density distribution under direct-current (DC) loads that simulate typical electromigration experiments. From the Maxwell's equations the constitutive relations for steady current density $\mathbf{J}$ in the absence of nonconservative energy sources are

$$\nabla \cdot \mathbf{J} = 0 \tag{1}$$

$$\nabla \times (\mathbf{J}/\sigma) = 0 \tag{2}$$

where σ is the electrical conductivity. For the geometry that contains more than two conductors, Eqs. (1) and (2) are applied to the interface between different materials with conductivities σ_1 and σ_2, leading to the boundary conditions for the normal (n) and tangential (t) components of $\mathbf{J}$

$$J_{1n} = J_{2n} \tag{3}$$

$$J_{1t}/J_{2t} = \sigma_1/\sigma_2 \tag{4}$$

These relations are implemented into the following finite element formulation

$$\{\mathbf{K}\} \cdot \{\mathbf{V}\} = \{\mathbf{I}\} \tag{5}$$

where $\{\mathbf{K}\}$, $\{\mathbf{V}\}$, and $\{\mathbf{I}\}$ are the coefficient matrix, the nodal electric potentials, and the applied load vector (current), respectively. The simulation is conducted at $J = 10^6$ A/cm^2 and $T = 250°$C. At these conditions, for common electromigration test structures, Joule heating should be negligible. But, it may become significant when current densities are larger or the Joule heat is not effectively dissipated. Hence, the model is programmed also to conduct a thermo-electric "coupled" analysis by adjusting the electrical resistivities at elevated temperatures. The Joule heat is passed to thermal analysis as a heat source, while the computed temperatures are used to revise the electrical conductivities for subsequent electric analysis. This coupled analysis, of course, is a much more time-consuming simulation than the electrical analysis at a fixed T.

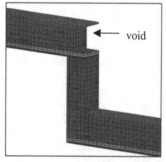

FIG. 1. A finite element model of the W-via structure.

Figure 1 illustrates an example of the 3-D finite element model constructed for resistance calculation as a function of electromigration-induced material depletion. The simulated W-via (0.36x0.36x0.9 µm) structure consists of the lower (M1) and the upper (M2) level Al-0.5 wt.% Cu stripes (thickness: 0.45 µm, length: 240 µm) coated with a 300Å Ti / 600Å TiN underlayer and a 250Å TiN anti-reflection coating (ARC) overlayer. The simulated contact structure has an aligned or a misaligned single-level stripe with the M2 configuration that is connected to a W contact. The electrical data for the materials were measured experimentally from the thin films prepared for the electromigration test structures. The thermal data were taken from various literature sources.

RESULTS AND DISCUSSIONS

Resistance Calculation for the Via Reliability

The simulated initial resistance (R_0) of the W-via structure without voiding is 174.4 Ω at T = 250°C, which is essentially identical to the experimentally measured values (169±6.6 Ω). The model calculates resistance changes as electromigration depletes Al(Cu) in the vicinity of the W via. Figure 2 plots the resistance increase as a function of depletion length (L) when electrons flow from M1 to M2 (depletion proceeds above the via).

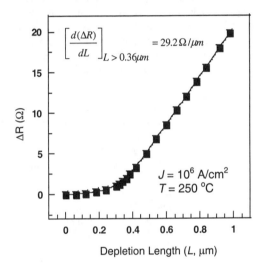

FIG. 2. Resistance variation as a function of Al(Cu) depletion length measured away from the W plug (0.36 x 0.36 x 0.9 μm).

Note that there is only little resistance change while the depletion length is shorter than the W-via size (0.36 μm). Even when there is no Al(Cu) left over the W plug, i.e., L = 0.36 μm and the void volume = 0.06 μm^3, ΔR is 1.87 Ω, which is only 1.1 % increase in the relative resistance ($\Delta R/R_0$). However, the resistance starts to increase sharply as soon as the depletion length exceeds the via size. The resistance change with time becomes linear with the slope $d(\Delta R)/dL$ = 29.2 Ω/μm. When a common electromigration failure criterion ΔR_f = 10 % (17.4 Ω in Fig. 2) is applied to the model, we can readily calculate the critical void volume (V_c) that must be depleted to cause failure [2]. In the case of Fig. 2, the critical void volume is 0.15 μm^3 since the electromigration failure occurs when the depletion length (L) reaches 0.9 μm. It is of interest to note how large this critical volume is, for example, compared to common slit-like voids that form along the line [4].

Clearly, the resistance model provides a simple and accurate way to simulate the electrical consequence of voiding and then to calculate the critical void volume that serves as a useful via-reliability parameter. But, it should be noted that the critical void volume varies as a function of both failure criterion and interconnect geometry. For example, when a different criterion ΔR_f = 20 % is used, V_c becomes 0.24 μm^3, which now becomes even larger (more than twice as large as the total volume of the W plug). In addition, at a given failure criterion, the longer the metal, the larger the critical void size. Simply, a failure criterion using the relative resistance increase is not a direct measure of the void size. One must be cautious about applying a failure criterion like the above to assess the interconnect reliability for a wide range of interconnect lengths [5]. Particularly, in studying failure mechanisms or kinetics, the absolute resistance increase ΔR should be used instead of the relative resistance increase $\Delta R/R_0$.

When the simulation results shown in Fig. 2 are combined with the resistance values experimentally measured with time, we can calculate the "drift velocity" of the void front moving away from the via. Since $d(\Delta R)/dL$ and $d(\Delta R)/dt$ are determined from the resistance simulation and the electromigration test, respectively, the drift velocity (v_d) can be obtained by the following simple relation

$$v_d = dL/dt = \frac{d(\Delta R)/dt}{d(\Delta R)/dL}, \tag{6}$$

which then determines the time for the void to grow to the point of failure (critical void size). From Eq. (6) finding v_d is simple for the linear drift region like that shown in Fig. 2. However, it is difficult to calculate v_d for the depletion stage $L < 0.36$ μm. Since the resistance variation is subtle, a typical experimental set-up to determine lifetimes rarely provides an accurate resistance profile. Moreover, when $L < 0.36$ μm, the resistance change is non-linear. As a consequence, v_d is not constant for the region. Further work is in progress, using a high-resolution resistance measurement technique, to characterize the initial depletion stage and then to calculate its drift velocity.

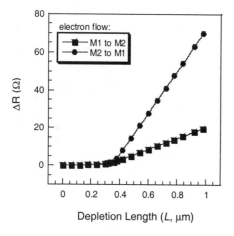

FIG. 3. Effect of electron-flow direction on the resistance variation as a function of Al(Cu) depletion length (L).

The resistance model is useful to understand the effect of refractory layer configuration on the via electromigration characteristics. The refractory layers such as TiN and Ti are incorporated to conduct electricity in the absence of Al(Cu) due to voiding. This well-known shunting effect, however, is limited depending on the arrangement of the refractory layers with respect to electron flow. An example is illustrated in Fig. 3. Note that there is a large difference in slope $d(\Delta R)/dL$ (110.4 Ω/μm vs. 29.2 Ω/μm) when the depletion length is larger than the W-plug size. As a consequence, the critical void volume (V_c) for the M1 failure is substantially smaller than that for the M2 failure; using the failure criterion $\Delta R_f = 10$ %, $V_c(M1) = 0.56V_c(M2)$.

If we suppose the void drift velocity is essentially the same for those two cases in Fig. 3, therefore, the lifetime of M2 could be almost twice as long as that of M2. This difference results from the fact that the shunting effect is substantially reduced for M1 since only the 250Å thick TiN layer (in contrast to the 300Å Ti / 600Å TiN bilayer) should carry electrons from the W via to the lower level Al(Cu). This result is useful since it can explain the experimental observations that the lower level metal is more susceptible to electromigration failure.

Effect of Contact Misalignment

Another subject that can be explored using the resistance model is the effect of contact misalignments on its electromigration characteristics. Figure 4 schematically illustrates the geometrical difference between a properly aligned contact and an intentionally misaligned contact studied here. Except that Fig. 4(b) represents a significant contact misalignment as large as 0.12 μm, both structures have the identical configuration like the W-plug to M2 connection studied above (Fig. 1). The initial resistance of the contacts is 87 Ω without considering the Si-tub resistance. The simulation results as a function of depletion length are shown in Fig. 5. Note that, for the misaligned contact, a sudden resistance increase is observed at a smaller depletion length (L = 0.24 μm). The critical void volume is then smaller for the misaligned contact, causing it is more susceptible to failure. This result is expected in light of the results discussed above. For the misaligned case there is no Al(Cu) above the W contact as soon as the depletion length reaches only 0.24 μm, in contrast to 0.36 μm for the aligned case. The impact of misalignment on the contact reliability may not be serious if its magnitude is relatively small with respect to the contact size. But, the results suggest that the reliability loss can become a serious concern as the interconnect feature size continuously decreases.

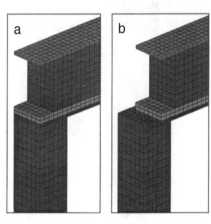

FIG. 4. A finite element model showing (a) a precisely aligned and (b) a misaligned W contact (depletion length: 0.18 μm).

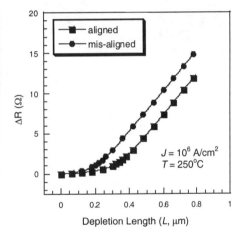

FIG. 5. Effect of contact misalignment on the resistance increase caused by Al(Cu) depletion.

Effects of Joule Heating

Up to this point the model has not considered the effect of Joule heating since the simulation is conduced at a relatively small current density (10^6 A/cm^2). Joule heating elevates the temperature and then the electrical resistivities of the metallic films. Furthermore, a non-uniform current density distribution near the vias/contacts, particularly, in the presence of voids, may lead to resistance values significantly larger than those at nominal testing temperatures. For these cases the resistance model requires a thermal simulation. Figure 6 shows the result from the resistance simulation conducted in conjunction with the thermal analysis when the depletion length (L) is 0.42 μm. Even with substantial current crowding near the void, the temperature increase is less than 4°C. It follows that the resistance increase (ΔR) after revising the resistivities is 9.4 Ω, which is only slightly larger that the value (8.1 Ω) at the identical depletion length in Fig. 3. Unless current densities are larger, therefore, reasonably accurate resistance values can still be obtained without conducting a time-consuming Joule-heat simulation.

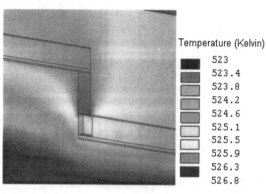

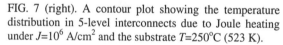

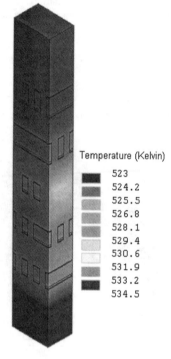

FIG. 6 (top). A contour plot showing the temperature distribution due to Joule heating under $J=10^6$ A/cm^2 and the substrate $T=250$°C (523 K). The depletion length is 0.42 μm and the consequent ΔR is 9.4 Ω.

FIG. 7 (right). A contour plot showing the temperature distribution in 5-level interconnects due to Joule heating under $J=10^6$ A/cm^2 and the substrate $T=250$°C (523 K).

However, there is an increasing concern on the impact of Joule heating since more and more metal layers are added to integrated circuits [6,7]. It is now common to have five or six layers of metals. Consequently, the distance between the top-level metal and the Si substrate becomes large so that the Si substrate cannot effectively dissipate the Joule heat generated at the upper level. An example illustrating a large temperature gradient through the metal levels due to Joule heating is shown in Fig. 7. This plot results from a thermo-electric coupled simulation conducted at the identical conditions (10^6 A/cm^2, 250°C) used for Fig. 6. Despite the fact that the material depletion and current crowding effects are not even considered for Fig. 7, the temperature increase is far more significant: $\Delta T = 11.5$°C at the top level. This result shows that Joule heating may become a serious threat to the reliability of multilevel interconnects. In fact, Joule heating is now one of the primary factors considered for interconnect current-density design rules.

SUMMARY

We have presented a 3-D thermo-electric interconnect resistance model developed using finite element analysis. The model calculates electrical resistance changes quickly and reliably in complex multilevel interconnects as a function of electromigration-induced Al(Cu) depletion. The model also evaluates the effect of Joule heating by calculating the local current distributions through the layers. The model provides a useful methodology for finding the critical void volume (V_c) and the void drift velocity (v_d) at given conditions (interconnect configuration and failure criterion). The model is useful also in understanding the effects of refractory layer (TiN, Ti) arrangements, contact misalignments, and current crowding on the electromigration characteristics of deep submicron contacts and vias.

ACKNOWLEDGMENTS

We would like to thank Dr. S. Rzepka at Dresden University of Technology, Germany. We also acknowledge S. Groothuis and F.S. Kelley at ANSYS, INC. for their valuable technical support.

REFERENCES

1. C.-K. Hu, M.B. Small, and P.S. Ho, J. Appl. Phys. **74**, 969 (1993).
2. A.S. Oates, Microelectron. Relib. **36**, 925 (1996).
3. ANSYS® Version 5.5. ANSYS® is a trademark of ANSYS, INC.
4. S.H. Kang, J.W. Morris, Jr., and A.S. Oates, JOM **51**, No.3, 16 (1999).
5. We limit our discussion to the failure-criterion issue only. For the stress-related short-length effect, refer to R.G. Filippi, R.A. Wachnik, H. Aochi, J.R. Lloyd, and M.A. Korhonen, Appl. Phys. Lett. **69**, 2350 (1996).
6. S. Rzepka, K. Banerjee, E. Meusel, and C. Hu, IEEE Trans. on Component Packaging and Manufacturing Tech. **21**, 406 (1998).
7. W.R. Hunter, IEDM Tech. Dig., 483 (1995).

INFLUENCE OF TEST STRUCTURE SHAPE AND TEST CONDITIONS IN HIGH ACCELERATED CU-ELECTROMIGRATION TESTS

J. ULLMANN, T. KÖTTER, W. HASSE, S.E. SCHULZ*
Institut für Halbleitertechnologie und Werkstoffe der Elektrotechnik, Universität Hannover, Appelstrasse 11a, D-30167 Hannover, Germany
* Zentrum für Mikrotechnologien, TU Chemnitz, Reichenhainerstrasse. 70, D-09107 Chemnitz

ABSTRACT

High accelerated electromigration tests were performed on 2 μm wide and 0.5 μm thick copper lines at ambient temperatures between 200 and 350°C and current densities between 2 and 10 MA/cm². Copper was deposited by sputtering and 800 μm long NIST test structures were fabricated by a reactive ion etching process. The activation energies for high current densities (5-10 MA/cm²) and 2 MA/cm² were found to be 0.54eV and 0.72eV respectively. Most of the failures were located at the transition region between the wide and the small part of the structure, even for 2 MA/cm². To minimize current crowding and temperature gradients, modified test structures with a stepped transition region were investigated experimentally and by finite element simulations using the FEM-simulator ANSYS. In the experiment no significant change of the MTTF and the failure location was observed. Also the simulation shows only a 20% decrease of the temperature gradients and no significant change of the mass flow divergence for a triple step modification. Further simulations show, that a decrease of the mass flow divergence can be achieved with a modified shape at the transition.

INTRODUCTION

As the dimensions of interconnects are scaled down a new metallization material, such as copper, with a higher electromigration resistance is required [1]. A wide range of activation energies, from 0.5 to 2 eV, for electromigration in copper lines have been reported [2,3,4]. Different transport mechanisms like bulk, grain boundary, surface and interface diffusion have been proposed to explain the results[2,5,6]. Due to the high electromigration resistance of copper, extreme test conditions must be applied to achieve acceptable times to failure. However, high accelerated eletromigration test should be designed carefully because temperature gradients generated by high current densities could lead to mass flow divergences which are not present under operating conditions. If surface or interface diffusion is the dominant transport mechanism this point may be more important since divergences generated by the micro structure (triple points) don't exist.

We have investigated the electromigration in copper lines using the well known NIST test structure (ASTM standard F 1259M-96) and some modifications of this structure. The transition region between the wide and the small part of test structure was modified to minimize current crowding and temperature gradients. The electromigration behaviour of the original and the modified test structures have been compared experimentally and by finite element simulation. The FEM-program ANSYS and a user routine[7] which calculates the mass flow and mass flow divergences due to electro- and thermomigration was used for simulation.

EXPERIMENTAL

A TiN(20nm)/W(22nm)/Cu(500nm)/W(66nm) sandwich was deposited by sputtering on top of 500nm field oxide. 1 - 12µm wide test structures were fabricated using a reactive ion etching process. After patterning a 50nm SiN barrier and a passivation (SiO(300nm)/SiN(550nm)) was deposited by PECVD. Finally the bond pads were opened and covered by a TiN/Al-Metallization.

Fig. 1 shows a cross-section of an unstressed interconnect. A small crack between the field oxide and the 50nm SiN-Barrier can be seen in the SEM micrograph indicating an adhesion problem between these layers. The resistivity of the copper films was found to be 2 µΩcm and a TCR of 3.33 10^{-3} 1/K was measured.

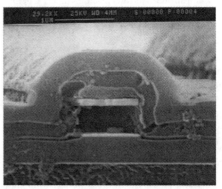

Fig. 1: Cross-section of an unstressed interconnect (Cu removed by etching)

Fig. 2 shows scematic draws of the investigated modifications of the test structure. Structure MOD 3 is equivalent to the NIST-Structure, for structure MOD 3/3 the transition region is divided into two steps and for structure MOD 3/0 into three steps. The aim of these modifications was to minimize the temperature gradients at the transition region.

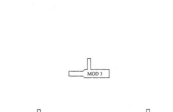

Fig. 2: Schematic draws of used modifications

2µm wide lines were tested in a Qualitau system at ambient temperature ranging from 200 - 350°C and current densities of 2, 5 and 10 MA/cm². A resistance change of ±20% was used as failure criteria.

After testing, the passivation and the tungsten on top of the copper was removed using a RIE process and the samples were inspected in a SEM.

EXPERIMENTAL RESULTS

The mean time to failure (MTTF) versus the invers temperature for 2µm wide NIST test structures (MOD 3) are shown in Fig. 3 for current densities of 2, 5 and 10 MA/cm². Also the 90% confidence intervals are ploted in Fig. 3. The activation energies for high current densities (5-10 MA/cm²) and 2 MA/cm² were found to be 0.54±0.07eV and 0.72±0.23eV respectively.

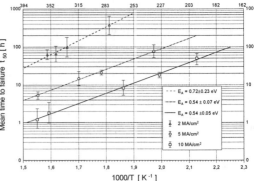

Fig. 3: Mean time to failure versus invers temperature

In Fig. 4 the current density exponent n is plotted versus interconnect temperature. A slight decrease of n with increasing temperature was observed. The average value of n was found to be 2.6±0.3.

The modified test structures MOD 3/0 and MOD 3/3 were stressed at an ambient temperature of 300°C and a current density of 5 MA/cm². Due to the joule heating at this test condition a temperature increase of the copper line of 19 K was observed. No significant difference was found between the modifications of the test structure. These results are supported by FEM simulations (see section FEM SIMULATION).

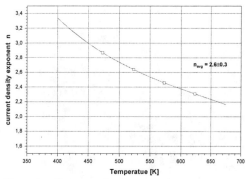

Fig. 4: Current density exponent n versus interconnect temperature

After removing of the passivation the stressed samples were inspected by SEM and the failure locations were determined. SEM micrographs of typical electromigration damage at the transition region of a MOD 3/0 test structure are shown in Fig. 5 and Fig. 6. At the cathode side holes and at the anode side hillocks have been observed. The location of the damage clearly shows that the mass flow divergences leading to the damage are caused by temperature gradients. Some structures have shown very flat copper extrusions at the anode side (Fig. 7). This failure mode is due to adhesion problems between the 50nm SiN barrier and the field oxide (see also Fig. 1). Damage at the small part of a test structure is shown in Fig. 8. It should be noticed that some copper has been removed from the side walls of the line. The distribution of the failure location along the test structure for 2μm wide lines and current densities of 2 and 10 MA/cm² is shown in Fig. 9. Most of the failures are located near the transition region at the anode side of the structure even for a current density of 2 MA/cm². About 25% of the failure occur in the wide part of the structure between the transition region and the bond pad. The location of the maximum of the distribution shows that temperature gradients are not negligible for the used current densities. The observed bad adhesion of the SiN barrier may lead to an easy diffusion path at the side walls of the lines. Also the low activation energy of 0.54eV (5-10MA/cm²), its dependance on the current density and the location and shape of the failures suggest that grain boundary and interface transport are contributing to the mass flow.

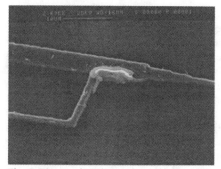

Fig. 5: Electromigration damage at the anode side (MOD 3/0)

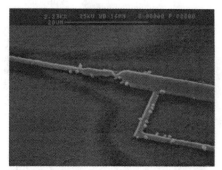

Fig. 6: Elelectromigration damage at the cathode side (MOD 3/0)

Fig. 7: Flat copper extrusion at the anode side (W top layer not removed)

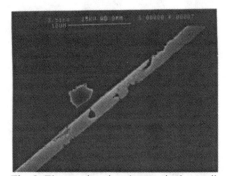

Fig. 8: Electromigration damage in the small part of the test structure

FEM SIMULATIONS

The main reasons of theses failures at the transition region are the current crowding, which leads to a high electromigration driven mass flow and on the other hand there is a high temperature gradient at these test conditions, which leads to a high mass flow divergence. Under operating conditions the temperature gradient can be neglected, but at high acclerated tests it can affect the measured activation energy, which leads to a wrong life time estimation. To minimize the influence of the temperature gradient, different variations of the test structure were investigated with finite element (FEM) simulator ANSYS and a user routine which calculates the resulting electro- and thermomigration.

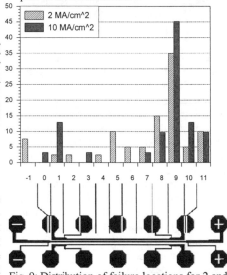

Fig. 9: Distribution of failure locations for 2 and 10 MA/cm²

Fig. 10: Single step

With the FEM-simulator, the current density, the temperature and the temperatur gradient can be calculated. With the additional user routine, the mass flow divergence and the location of the maximum mass flow divergence can be estimated with these informations[7]. In figure 10, the original single step and in figure 11, a triple step modification is shown. The simulations show slightly reduced temperature gradients (~25%) and the calculated mass flow divergences are nearly identical for all modifications. Also the points of the calculated failure location is corresponding (pointed with the arrow). This indicates, that the desired improvement of the test structure can not be achieved by the used modifications.

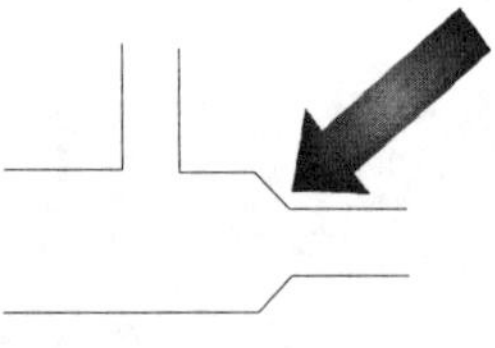

Fig. 11: Triple step

function	max. MFD [μm^{-3}]
f(x)=x	0,659 (Fig. 12)
f(x)=0,31· (e^x-1)	0,539
f(x)=0,5 · x ^{2}	0,555
f(x)=0,25 · x ^{3}	0,562 (Fig. 13)
f(x)=0,35 · x ^{2,5}	0,55
optimized	0,508 (Fig. 14)

Table 1: Max. mass flow divergence of different shapes

At this time the investigation was focused on the shape of the transition. The first step was to decrease the 45° angle at the first edge. This improves the situation, but only a zero degree slope leads to the best results. To find the optimal form, the transition region was divided into twenty segment and each segment was changed in its width until the maximum massflow divergence was at its minimum. The resulting maximum massflow divergence (MFD) is presented in table 1.

The figures 12,13 and 14 show the distribution and the maximum value of the massflow divergence. It can be seen that the value is decreasing about ~23% and that the maximum value stays in the region of the transition.

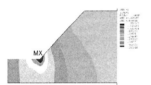 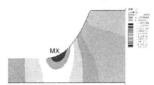 

Fig. 12: f(x)=x   Fig. 13: f(x)=0,25 · x ^{3} Fig. 14: Optimized shape

CONCLUSIONS

High accelerated electromigration tests were performed on 2 μm wide and 0.5 μm thick copper lines at ambient temperatures between 200 and 350°C and current densities between 2 and 10 MA/cm². Copper was deposited by sputtering and different modifications of the NIST test structure were fabricated by a reactive ion etching process. No significant differences in the measured MTTF's and calculated mass flow divergences were found between the modifications of the test structure. The desired improvement of the test structure can not be achieved by the used modifications. Further simulations show, that a decrease of the mass flow divergence can be achieved with a modified shape at the transition. The low activation energy of 0.54eV (5-10MA/cm²), its dependance on the current density and the location and shape of the failures suggest that grain boundary and interface transport are contributing to the mass flow.

ACKNOWLEDGMENTS

Financial support for this work was provided by the German Department of Education and Research under contract No. FKZ 01M 2972 D (FOKUM).

REFERENCES

1.   The National Technology Roadmap for Semiconductors, SIA, USA 1997

2.   C.-K. Hu, R. Rosenberg, K.Y. Lee, Applied Physics Letters, **74**, p. 2945 (1999)

3.   J. R. Lloyd, J. J. Clement, Thin Solid Films, **262**, p. 135 (1995)

4.   D. Save, F. Braud, J. Torres, F. Binder, C. Müller, J.-O. Weidner, W. Hasse, Microelectronic Engineering, **33**, 75 (1997)

5.   L. Arnaud, G. Tartavel, T. Berger, D. Mariolle, Y. Gobil, I. Touet, Proceedings 37th International Reliability Physics Symposium, San Diego, 1999

6.   N.D. McCusker, H.S. Gamble, B.M. Armstrong, Proceedings 37th International Reliability Physics Symposium, San Diego, 1999

7.   K. Weide, Fortschr.-Ber., VDI Reihe 9, Nr. 184, VDI Verlag 1994, Düsseldorf

ELECTROMIGRATION INDUCED EDGE   DRIFT VELOCITY MEASUREMENT BY BLECH PATTERN WITH MULTIPLE VOLTAGE PROBES

S.Shingubara, T. Osaka, H. Sakaue, T.Takahagi, and *A.H.Verbruggen
Dept.of Electrical Engineering, Hiroshima University
Kagamiyama 1-4-1, Higashi-hiroshima, 739-8527, Japan
*DIMES, Delft University of Technology, Lorentzweg1,2628 CJ Delft, The Netherlands
e-mail:shingu@ipc.hiroshima-u.ac.jp

ABSTRACT

Electromigration induced drift velocity measurement as well as failure mechanisms in Blech patterns are investigated by multiple probe resistance monitoring experiments.   When current density   J is higher than the critical value Jc ,   the present method turned out to be very effective to measure the edge drift velocity.   Investigation on a modified Blech pattern with a reservoir reveals detail information of mechanisms of type B void   which is formed at the intersection of current stressed area and a reservoir at a rather high current density stress.   Experimental results suggest an existence of compressive stress close to void nucleation site, which agrees well with proposed thermo-migration model.

INTRODUCTION

Electromigration induced edge displacement drift velocity measurement as well as failure mechanisms investigation is carried out by multiple probe resistance monitoring of Blech pattern. Since the idea of the critical current and length product was introduced by Blech [1],   the understandings of the mechanism of electromigration induced failures have been   significantly developed.   However, the dynamics of void nucleation and growth, and their relationship to hillock growth have not been understood clearly even though 23 years has passed.   The present study aims to understand dynamics of electromigration induced failure mechanisms by monitoring change of resistance distribution in the Blech pattern by multiple voltage probes. Usage of pure Al enables us to estimate evolution of stress distribution during DC stressing test as long as reversible (i.e. elastic) change of resistance has taken place [2].

EXPERIMENTS

Experimental Methods

We have used pure aluminum interconnects on TiN under-layer.   Both pure Al (purity: 6N, thickness: $0.3\,\mu$ m) and TiN (resistivity: $50\,\mu\,\Omega$cm, thichkness:$0.1\,\mu$ m) were sputter-

Conference Proceedings ULSI XV © 2000 Materials Research Society

deposited on SiO_2. Lift-off was carried out for pattern delineation of Al on TiN in order to avoid corrosion which occur due to RIE. We have prepared different samples with and without SiO_2 passivation glass. Voltage probes were placed every 12.5 μ m in 50 μ m length Blech pattern of 3.0 μ m width. Multiple probe resistance monitoring experiments during DC stressing test (current density: 0.05-0.2MA/cm^2, 210 °C) , have been carried out by standard scanner techniques [3].

Schematic illustration of the test pattern is shown in Fig.1. Five voltage probes are attached to the Blech pattern, and its segments which has the same length of 12.5 μ m are named from ch10 to 13 in the order from anode to cathode. In order to observe the reservoir effect, which is void formation at the end of no current stressed area, 5 voltage probes were attached to the reservoir , and the reservoir is devided to 5 segments of channel 5 to 9. In order to detect voltage change between channel 5 to channel 9, a small DC current I_2 which magnitude is 1/10 of I_1 was stressed. Ambient temperature during electromigration test was 210°C. Multiple probe resistance monitoring experiment was carried out by standard scanner techniques.

Voiding Phenomena Induced by Electromigration

Current density dependence of electromigration induced voids are summarized in Table 1 for both cases of Blech pattern and modified Blech pattern. In the case of Blech pattern, significant void formation (edge drift) was observed within 10 hours stressing test at 210°C when the current density was 0.2 MA/cm^2. However, no void was observed when current density was lower than 0.11 MA/cm^2 after 100 hours test. Thus a critical current density is between 0.11 and 0.2 MA/cm^2. On the other hand, three types of void were observed in the case of modified Blech pattern. This result is qualitatively consistent with our previous report (4). At a low current density of 0.093 MA/cm^2 , a void is formed at the edge of no current stressed area. We call this mode of void as type A which corresponds to the reservoir effect. When current density is increased to 0.20 MA/cm^2, Type B void is formed at the intersection between the current stressed area and the reservoir. Type AB void is a mixture of these two types. It appeared at an intermediate current density of 0.11 MA/cm^2. Critical current density for the modified Blech pattern is between 0.75 and 0.093 MA/cm^2, which is lower than that of Blech pattern, although effective length of the Al pattern where current passes are the same.

Edge Drift Velocity Measurements

A typical example of eletromigration induced edge drift in Blech pattern in the case without passivation glass is shown in Fig.2-(a). A huge void is formed at the cathode end of Al interconnect, and hillock is formed at the anode end concomitantly. Many studies had been carried out to measure EM edge drift velocity by dividing the Al depletion length by time. Resistance change behaviors of this case is shown in Fig.2-(b). Resistance of ch 1 increased almost linearly after some incubation time , and saturated after 250 min. Just after saturation of ch13, resistance of ch12 started to increase. These results indicate that an Al edge drifted at a constant velocity and it crossed the second voltage probe at 250min. We can estimate edge drift velocity very precisely from a linear extrapolation of resistance increase behavior. We have carried out similar experiment in the case with thin SiO_2 (20nm) passivation glass as shown in Fig.2-(c). The drift velocity of passivated case (Vd=0.89x10^{-9} m/s) was a little retarded from that of without passivation (Vd=1.04x10^{-9} m/s). Incubation time was prolonged by passivation from 40 to 60 min. These results suggest that the electromigration induced drift of atoms is suppressed by the compressive stress, since the passivation glass was deposited at room temperature by sputtering.

Mechansims of Type B Voids

Experimental result of the modified Blech pattern in the case of J=0.20 MA/cm^2 is shown in Fig.3. SEM photograph (Fig.3-(a)) indicates a huge void which extends from channel 4 to 9. Black and white contrast of void suggests that the void has reached to Al / TiN interface only between channel 4 and 5. Resistance change behavior at the initial stage is shown in Fig.3-(b). There is no incubation time for resistance increase in channel 5. Channel 1 exhibits gradual decrease in resistance without any incubation time. It should be noted that behavior of channel 6 is very complicated. Resistance decreased in the initial stage until 10 minutes, then increased. The amount of resistance decrease is about 2% in ratio. This suggests that compressive stress was built up in the initial stage at channel 6. Figure 4 shows dynamics of type B void inferred from observation. Since Type B void is formed at rather high current density conditions, Joule heating is considered to be dominant mechanism for void formation. Temperature gradient induced by Joule heating effect is schematically shown in Fig.4-(a). Temperature decreases monotonously from the intersection between current stressed area and reservoir to the edge of reservoir. Temperature gradient produces additional atomic flux called thermo-migration. Atomic flux ψ due to thermo-migration is expressed as, $- D (Q^*/T)(dT/dx)$, where D is diffusion coefficient, Q^* is effective heat of transport, and T is absolute temperature [5]. Rough estimation of atomic flux Ψ by thermo-migration and its divergence $d\Psi/dx$ are shown in Fig.4-(b), and (c) respectively. Negative divergence as well as positive divergence appear adjacently at a location slightly outside the intersection between current-stressed area and reservoir. This causes tensile stress and compressive stress simultaneously, and the former one produces nucleation of void. The compressive stress may lead to hillock formation, however, depletion of Al on TiN due to void growth bring about further enhancement of temperature gradient, and the void grow further and further.

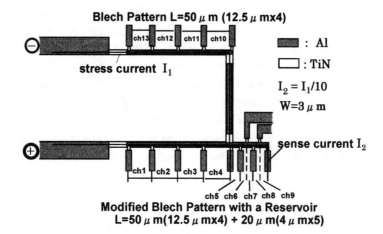

Blech Pattern L=50 μ m (12.5 μ mx4)

ch13 ch12 ch11 ch10

⊖

stress current I_1

▬ : Al

☐ : TiN

$I_2 = I_1/10$

W=3 μ m

sense current I_2

⊕

ch1 ch2 ch3 ch4

ch5 ch6 ch7 ch8 ch9

Modified Blech Pattern with a Reservoir
L=50 μ m(12.5 μ mx4) + 20 μ m(4 μ mx5)

FiIGURE 1 . Schematic diagram of experimental pattern. Blech pattern which has 5 voltage probes is connected to modified Blech pattern with a reservoir which has additional 5 probes. Modified Blech pattern is composed of 9 segments assigned to channel 1 to 9, and Blech pattern is composed of 4 segments assigned to channel 10 to 13.

TABLE 1. Void type of modified Blech pattern with a reservoir at various current densities. T=210°C

type A	type AB	type B

current density (MA/cm²)	0.056	0.075	0.093	0.11	0.20
Blech Pattern	no void	no void	no void	no void	void (edge drift)
Modified Blech Pattern	no void	no void	type A	type AB	type B

(a)

(b) unpassivated

$V_d = 1.04 \times 10^{-9}$ m/s

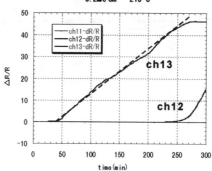

(c) passivated by SiO_2 (t=20nm)

$V_d = 0.89 \times 10^{-9}$ m/s

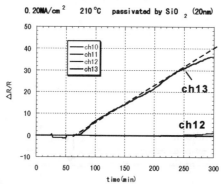

Fig.2 Comparison between passivated and unpassivated cases.
(a) SEM photograph of the test pattern of the unpassivated case after
      the current stressing test .
(b),(c) Resistance change behaviors of the Blech pattern of unpassivated (b)
      and passivated (c) cases.

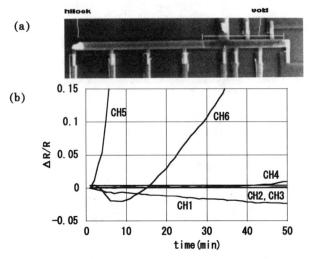

(a)

(b)

FIGURE 3. (a) SEM photograph of modified Blech pattern after DC stress test of 0.20 MA/cm^2, 210℃. (b) Resistance change ratio profile at the initial stage.

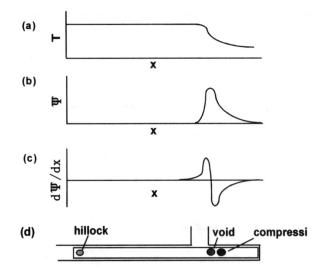

FIGURE 4 . Dynamics of type B void formation in a modified Blech pattern. (a) Temperature distribution along the Al interconnect when Joule heating is not negligible. (b) Distribution of atomic flux Ψ produced by thermomigration. (c) Distribution of atomic flux divergence $d\Psi/dx$. (d) A positive atomic flux divergence corresponding to atom depletion appear at a location slightly outside of the intersection, and a compressive stress appears due to a negative atomic flux divergence .

CONCLUDING REMARKS

We have investigated multiple-probe resistance change monitoring for electromigration stresse test of the Blech pattern as well as the modified Blech pattern with a reservoir. Electromigration induced drift velocity of Al pattern edge was accurately measured by linear extrapolation of the resistance increase and subsequent saturation of the first segment of Al interconnect at anode side.   This method is also very sensitive to detect incubation time of void nucleation and determining position of it.   This method is also very powerful to examine electromigration induced failure mechanisms.   Mechanism of type B void, which is formed at the intersection between the current stressed area and the no current stressed area at rather high current density conditions,   identified to be caused by thermomigration. Further investigation would be needed to clarify the reservoir effect.

REFERENCES

1.   I.A.Blech,   J.Appl.Phys. 42   (1976)   5671.
2.   A.H.Verbruggen, M.J.C.van den Homberg, L.C.Jacobs, A.J.Kalkman, J.R.Kraayveld, and S.Radelaar ,   *STRESS INDUCED PHENOMENA IN METALLIZATION*, the Fourth Workshop,   Edited by H.Okabayashi,S.Shingubara, and P.S.Ho,   AIP Conf. Proc. 418, (1997) 135.
3.   S.Shingubara, T.Osaka, H.Sakaue, T.Takahagi, and A.H.Verbruggen,   *STRESS INDUCED   PHENOMENA IN METALLIZATION*, the Fifth Workshop,   Edited by E. Arzt, O.Kraft, and P.S.Ho,   AIP Conf. Proc. (1999) to be published.
4.   S.Shingubara,   T. Osaka,   S.Abdeslam,   H. Sakaue,   and   T. Takahagi,   AIP Symp.Proc. vol.418, *Fourth Int.Workshop on Stress Induced   Phenomena in Metallization* (1997) p.159.
5.   A.S. Sorbello,   in *Electro- and Thermo- Transport in Metals and Alloys*, edited by R.E.Hummel and H.B.Huntington, AIME, New York (1977).

A NOVEL INTERCONNECT STACK FOR IMPROVED RESISTANCE TO ELECTROMIGRATION AND STRESS-INDUCED VOIDING

R. JAISWAL[1], T.S. LEE[1], K.K. LOOI[1], I. LIM[1], A. VELAGA[1], S. M. MERCHANT[2], J.S. HUANG[2], J. LI[2], S. KARTHIKEYAN[2], J. ZHANG[3], S. LAI[3] AND G. YAO[3].
[1]Silicon Manufacturing Partners Pte. Ltd., Singapore 738406.
[2]Lucent Technologies - Bell Laboratories, Orlando, FL 32819, USA.
[3]Applied Materials, Santa Clara, CA 95054, USA.
jaiswalrajneesh@smp.st.com.sg

Abstract

Stress-induced voids and early electromigration failures were observed in Al-Cu interconnect structures when an existing metallization scheme was extended for use in semiconductor devices for 0.25µm technology. Stress-induced voids were reduced by modifying the interlevel dielectric deposition conditions, but electromigration data showed a bimodal distribution. A series of experiments were conducted to study the impact of Al-Cu and cladding layer (Ti/TiN) deposition conditions on interconnect resistance to stress-induced voiding and electromigration. The impact of Ti/TiN underlayer thickness on electromigration behavior was also studied. The robustness of different schemes was judged based on their electromigration performance. Using ionized metal plasma (IMP) deposited Ti as an alternative to standard Ti in the underlayer, and a thinner TiN, a superior crystallographic Al-texture was observed. Based on these results, a new metallization scheme was proposed which showed no electromigration-induced failures and practically no increase in the line sheet resistance. The new IMP Ti-based metallization scheme had no impact on the device performance and was compatible with dielectric deposition conditions. In this paper we discuss the various interconnect schemes evaluated, their deposition parameters studied, and their electromigration performance.

Introduction

In MOS based technologies, linear scaling results in a linear increase in interconnect current densities [1]. Based on an exponential decrease in feature size, scaling rules predict sub-half micron line widths capable of supporting current densities approaching $10^6 A/cm^2$. In view of such high current densities, a major concern in device reliability is the electromigration resistance of metal lines [1]. Cu metallization has been proposed [2], as an alternative to Al-based alloys, because of lower electrical resistance and superior resistance to electromigration, to meet the demands of 0.25µm technology and beyond. So far, however, Cu-related integration issues and the introduction of new hardware have prevented its use in mass production [3].

This work aims to provide a robust metal stack (IMP Ti/TiN/Al-Cu/TiN) using Al-Cu, in combination with conventional contact and via W-plugs, with a view to extend present interconnects to future generations of technology. In this paper, we report various interconnect schemes evaluated and their electromigration performance. X-ray diffraction data is presented to show the impact of IMP Ti as an underlayer on Al-Cu orientation and the data is correlated with the stack electromigration performance [4].

Experiment

Lots were processed using different interconnect schemes. All metal layers were deposited in an Applied Materials Endura HP sputter system. All the metal layer processes were clamped except for IMP Ti. The Al-Cu thickness used in these experiments was varied between 4000A to 5000A.

In order to understand Al-Cu layer stress relief under different Al-Cu and Ti/TiN underlayer deposition conditions, wafers were processed with different Al-Cu and underlayer deposition temperatures, different underlayer TiN thickness and, different Ti as underlayers (see Table I).

TABLE I

SPLIT TYPE	SPLIT CONDITION
Al-Cu deposition temperature	Temperature 200°C to 450°C
Ti and TiN cladding layer temperatures	100°C to 250°C
TiN thickness	300A to 600A
Ti underlayer deposition	Standard Ti and IMP Ti

The effort was to minimize deviation from an existing metal stack used in the manufacturing line for other products. All wafers were inspected under a microscope after fabricating the interconnect stack (Ti/TiN/Al-Cu/TiN), subsequent dielectric deposition of high-density plasma (HDP) oxide, and after a furnace anneal for stress-induced voids in metal comb structures. HDP oxide was deposited in an AMAT HDP system after metal patterning at 450°C. The pedestal cooling backside He pressure was varied between 3 and 7 Torr for the inner zone and between 6 and 9.5 Torr for the outer zone. The bias and RF conditions were kept constant during HDP deposition. Devices were evaluated for electromigration integrity in package-level stress tests at an ambient temperature of 250°C and a current density of 2 MA/cm^2. Interconnect lines were considered to fail after a 20% increase in line sheet resistance.

Results and Discussion

Visual inspection results showed stress-induced metal voids on thin metal lines but the broad lines remained unaffected. Such a line width dependence of stress-induced voids has been previously observed [5]. Failure analysis showed missing Al in the void structures. Fig. 1 shows a light optical micrograph of a voided metal line structure. Fig. 2 shows a SEM cross-section of the metal comb structure after furnace anneal and after electromigration test. Voiding in the metal lines can be attributed to stress relief of thin metal lines during oxide deposition and subsequent furnace anneal [5]. Voids in the metal lines were eliminated using a high Al deposition temperature (≥400C) but early electromigration failures still occurred and the metal had a very grainy appearance. Fig. 3 shows the electromigration performance at different Al deposition temperatures. Electromigration resistance also improved with higher cladding (Ti/TiN) deposition temperatures compared to standard conditions. Thinner TiN underlayers also showed better electromigration resistance.

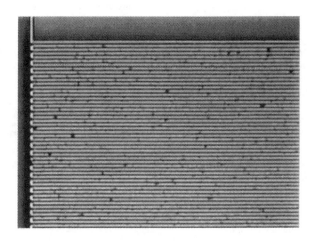

Fig. 1: Light optical micrograph of stress-induced voids in Al lines in dense comb structures observed after final furnace heat treatment.

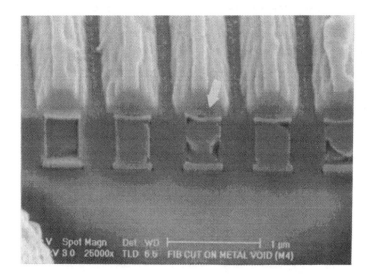

Fig. 2: SEM of a stress-induced void observed in the Al interconnect.

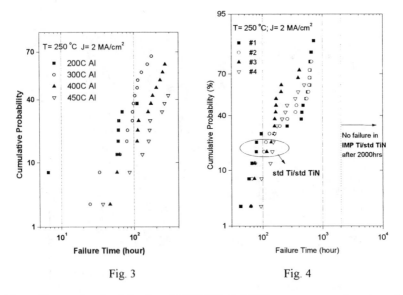

Fig. 3

Fig. 4

Fig. 3: Electromigration behavior of Ti/TiN/Al-Cu/TiN interconnect stacks, where the Al-Cu was deposited at various temperatures from 200°C to 450°C.

Fig. 4: Electromigration behavior of Ti/TiN/Al-Cu/TiN interconnect stacks for 0.25 μm technology. The std Ti/std TiN standard underlayer stack (labeled samples #1 to #4) shows failure at <1000 hrs; early failures are evident in samples #1, #2 and #3. In comparison, the IMP Ti/std TiN underlayers show no failure at >2000 hrs.

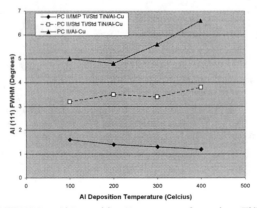

Fig. 5: Al (111) FWHM vs Al Deposition Temperature for various Ti/TiN underlayers. Texture data for Al-Cu deposited on IMP Ti/Std TiN vs Std Ti/Std TiN is shown. PC II is the pre-clean step prior to metals deposition.

A remarkable improvement in electromigration resistance was observed using IMP Ti instead of standard Ti as part of the interconnect stack, as shown in Fig. 4. In that figure, the standard interconnect stack underlayers are designated as std Ti/std TiN. The improved interconnect stack underlayers are designated as IMP Ti/std TiN. Post HDP-oxide deposition and post-furnace anneal inspections showed a smooth metal surface with no abnormal grains or hillocks. Test chips showed no failure even after 2000 hours of electromigration testing, with little or no change in line sheet resistance. Electrical tests showed that the new stack made no impact on the chip performance. The line sheet resistance was comparable to the original metal stack

X-ray diffraction rocking curve studies on the Al-Cu film showed that an IMP Ti underlayer results in a highly textured Al film with a narrow (<1°) Al <111> peak full-width at half-maximum (FWHM), as shown in Fig. 5. Unlike the Al-Cu film on top of a standard Ti underlayer where the Al <111> FWHM increases with Al deposition temperature, the Al <111> FWHM is independent of the Al-Cu deposition temperature and, therefore, more thermally stable when an IMP Ti underlayer is used. It was also found that a thinner TiN layer between the Ti and Al-Cu layers results in a narrower Al <111> FWHM, although this trend is not very pronounced when the Ti layer is deposited with IMP technique.

Conclusion

A robust interconnect stack with improved resistance to electromigration and stress-induced voiding has been developed using IMP Ti and thin TiN underlayers for a 0.25μm CMOS IC technology. Metal voids in thin metal lines have been eliminated and metal line structures have gone through electromigration tests for >2000 hours without any failure. This high resistance to both stress-induced voiding and electromigration can be attributed to an improved Al <111> structure resulting from an IMP Ti underlayer. The scheme has been successfully extended to 0.20 μm technology and is currently under test for a 0.18 μm process.

Acknowledgements

The authors wish to thank Gary Randel of SMP for useful discussions.

References

1. M. T. Bohr, Solid State Technology, v. 39, p. 105, (1996).
2. J. Slattery, S. Luce, T. McDevitt, T. Stamper, R. Goldblatt and G. Biery, Future Fab International, issue 6, p. 155.
3. A. E. Braun, Semiconductor International, v. 22, p. 58, (1999).
4. M. Kageyama, Y. Tatara, and H. Onoda, Jpn. J. Appl. Phys., v. 32, p. 4694, (1993).
5. A. Sidhwa, S. Melosky, S. Saunders, A. Ravaglia, J. Ling, S. Guisinger and H. Hoang, Proc. VMIC, p. 485, (1997).

CONTACT RESISTANCE MODELLING AS A FUNCTION OF SILICIDE TYPE FOR ULSI DEVICES

G. K. Reeves*, A. S. Holland* and P.W. Leech**
*RMIT University, Melbourne, Vic. 3001, Australia.
**CSIRO Div. of Manufacturing Science and Technology, Clayton Vic. 3169, Australia.

ABSTRACT

The development of low resistance ohmic contacts for ULSI devices has resulted in the widespread use of various silicide materials as part of the contact structure. The use of silicide materials combined with continuing developments in semiconductor process technology have enabled significant reductions to be made in device contact resistance Rc. This reduction is assessed by experimentally deriving the specific contact resistance $\rho_c(\Omega.cm^2)$ of the contact interface. The Cross Kelvin Resistor (CKR) test structure is commonly used to determine ρ_c by combining computer modelling with the experimentally measured Kelvin resistance of the test structure. This paper models the dependence of the Kelvin (contact) resistance on the type of silicide used in the contact, and shows how this dependence varies with some of the important geometrical dimensions of the silicide.

INTRODUCTION

Assumptions (such as two-dimensional (2D) modelling of CKR structures, single interface contacts) on which many computer models of the CKR test structure are based [1], are not suitable for the comprehensive analysis and error correction of silicide contacts. The multilayer nature of the metal-silicide-silicon contact introduces additional complexity for modelling and characterisation, as it contains *two* interfaces each with its own specific contact resistance value [2,3], namely the metal-silicide interface (ρ_{ca}) and the silicide-silicon interface (ρ_{cu}) as shown in fig.1. This differs from the single overall ρ_c quoted for such contacts [4]. As low overall ρ_c values in the mid $10^{-9}\Omega.cm^2$ range have been reported [5], it is essential to develop accurate CKR test structure models to better understand the *origin* and *influence* of the various contributions to contact resistance (and ρ_c). The electrical model for a silicide contact should thus take into account the individual interfaces, the contact geometry and the resistivities of the silicide and the doped silicon. This paper focuses on modelling the contact resistance rather than attempting to derive an overall ρ_c for the contact. Thus error correction procedures to derive ρ_c from CKR resistance measurements are not described.

Many factors influence the choice of silicide. These include their processing, electrical and material properties. To date there appears to be no work reported on modelling the influence of the type of silicide on the electrical properties of the contacts. In this paper we use Finite Element modelling of reacted silicide contact structures to compare the dependency of the contact resistance on the type of silicide in the contact structure (we make the reasonable approximation of Kelvin resistance Rk ≈ Rc as discussed later in this paper). The type of silicide is accounted for in our model by varying (i) the barrier height Φ_b of the silicide to silicon interface and (ii) the resistivity of the silicide. The effects of (i) are modelled through a variation of the silicide-silicon specific contact resistance (ρ_{cu}), while (ii) can be directly incorporated in the model. By using the three-dimensional (3D) model of the CKR as shown in fig.1, we account for the effects of the thickness of the semiconductor junction (t), the silicide depth (h) and the metal-silicide interface ρ_{ca}. In addition, the influence of varying the silicide overlap around the contact via (δ) can also be modelled.

741

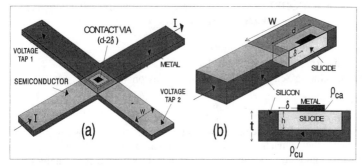

Fig.1 (a). CKR test structure for a silicide-well contact and **(b)** cross-section of the contact showing metal-silicide (ρ_{ca}) and silicide-silicon (ρ_{cu}) interfaces.

MODELLING PARAMETERS

The vertical doping profile for the heavily doped n-type silicon region of the contact was simulated by using a phosphorus implant of 25KeV at $2 \times 10^{15} \text{cm}^{-2}$ followed by a rapid thermal anneal. This gave a junction depth of approximately 150nm. The magnitude of ρ_{cu} is a function of the doping density and barrier height Φ_b[6]. Additional calculations were performed from the data in [6] for Φ_b values of 0.55 and 0.65eV, (the most commonly used contact silicides have a Φ_b covering 0.55-0.65eV). Thus values of ρ_{cu} as a function of doping density were obtained for silicides with $\Phi_b = 0.55$, 0.60 and 0.65eV. Table I gives the doping density and silicon resistivity vs depth profile. The corresponding value of ρ_{cu} for $\Phi_b = 0.55eV$ is given in the last column. The ρ_{cu} data is shown in fig.2 for the three values of Φ_b; e.g. that part of a silicide-silicon contact interface located at h=75nm for $\Phi_b = 0.60eV$ will have a ρ_{cu} of $6 \times 10^{-6} \Omega.\text{cm}^2$. Table II lists some of the common silicides, their Φ_b values and resistivities ρ_b. Three pairs of parameters (Φ_b, ρ_b) corresponding to NiSi, TiSi$_2$ and MoSi$_2$ (shaded region in Table II) are used in our calculations. Similar values of Φ_b and ρ_b

TABLE I. Doping profile, resistivity and ρ_{cu} values for $\Phi_b = 0.55eV$.

Depth (nm)	Doping (cm^{-3})	Resistivity ($\Omega.\text{cm}$)	ρ_{cu} ($\Omega.\text{cm}^2$)
0	2.8×10^{20}	2.9×10^{-4}	1.1×10^{-8}
15	3.2×10^{20}	2.6×10^{-4}	8×10^{-9}
30	2.7×10^{20}	3×10^{-4}	1.2×10^{-8}
45	1.6×10^{20}	4.8×10^{-4}	3.8×10^{-8}
60	8×10^{19}	9×10^{-4}	2.2×10^{-7}
75	3.2×10^{19}	1.8×10^{-3}	2.2×10^{-6}
90	1×10^{19}	6×10^{-3}	1×10^{-4}
105	2.1×10^{18}	1.5×10^{-2}	1.2×10^{-2}
120	3.5×10^{17}	3.4×10^{-2}	0.2
135	5.2×10^{16}	1.1×10^{-1}	-
150	5.2×10^{15}	8×10^{-1}	-

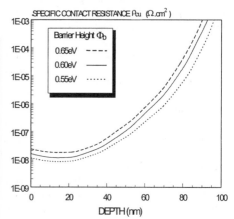

Fig.2. Specific contact resistance for the silicide-silicon interface (ρ_{cu}) as a function of the depth of the interface in the doped silicon (values of Φ_b as shown).

TABLE II. Silicide type, barrier height Φ_b and resistivity for n-Si. Shaded region indicates three sets of data used in the model calculations and in fig.2.

Silicide Type	Barrier height Φ_b (eV)	Resistivity ρ_b ($\mu\Omega.cm$)
TaSi$_2$	0.56-0.59	50
NiSi	0.65	14
TiSi$_2$	0.60	15
MoSi$_2$	0.55	22
CoSi$_2$	0.64	15

suggest that the CoSi$_2$ results will be similar to NiSi. The model takes the depth of the doped silicon layer as $t=150nm$, the width of the semiconductor tap $W=1.0\mu m$ (fig.1), the silicide well depth is h, the silicided area is a square d x d (d is fixed at $0.5\mu m$ in our model) and the contact overlap of the silicide around the metal via is δ. A fixed value for ρ_{ca} of $4\times10^{-9}\Omega.cm^2$ is used. This corresponds with our experimentally measured value for the Al-TiSi$_2$ interface [7]. In the absence of experimental data for ρ_{ca} for other metal-silicide interfaces we use this value in our model for the metal-silicide interfaces (except where noted for data in later figures when the influence of ρ_{ca} on Rc is being calculated).

MODELLING RESULTS AND DISCUSSION

The modelling results are obtained from the CKR structure of fig.1. While this structure gives the Kelvin resistance Rk ((V2-V1)/I), it has been shown [7,8] that the physical region of Rk is a reasonable approximation to the physical region that would be associated with the device contact resistance Rc. The contact metallization is a common physical boundary for both Rc and Rk, while the potential V2 on the semiconductor tap defines the second physical boundary for Rk. This second equipotential boundary can thus be conveniently used for identifying the approximate physical region associated with Rc (although the boundary's exact location will vary slightly between contacts as the individual electrical and geometrical parameters of the

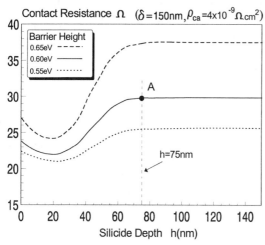

Fig.3. Dependence of Rc on the silicide depth for the three sets of Φ_b and ρ_b values highlighted in Table II.

contact change). Thus references to Rc in the text are the values of Rk obtained from the model of fig.1. The appearance of the three curves in fig.2 might suggest the modelling results for the three silicides will be near identical. However the logarithmic scale for ρ_{cu} ensures that even small changes in Φ_b translate into significant changes in ρ_{cu} which give quite different values of Rc. The dependence of Rc on silicide depth h is illustrated in fig.3 for the three silicides chosen from Table II. The individual combinations of (Φ_b, ρ_b) clearly have a significant effect on Rc for all silicide depths. For this model the minimum in Rc occurs when the silicide depth is approximately the same as the peak in the doping profile with the least dependence being shown by the silicide with the lowest barrier height. The plateau in Rc for $h > 80nm$ indicates that the majority of the device current is already entering the silicide at depths $< 80nm$. This is due to the higher Si resistivity and the higher values of ρ_{cu} for $h > 80nm$. Point A in fig.3 and in subsequent figures is used as a reference to indicate a common data point where $h = 75nm$, $\Phi_b = 0.6eV$, $\rho_b = 15\mu\Omega.cm$, $\delta = 150nm$ and $\rho_{ca} = 4 \times 10^{-9}\Omega.cm^2$.

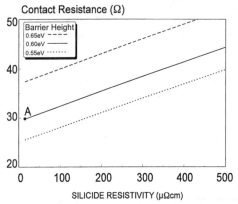

Contact Resistance (Ω)

SILICIDE RESISTIVITY (μΩcm)

Fig.4. Dependence of Rc on the silicide resistivity for a three silicide barrier heights.

Since the results in fig.3 incorporate both Φ_b and ρ_b variations, further modelling was undertaken by keeping Φ_b fixed and varying only ρ_b in order to determine the significance of just the silicide resistivity. An example of these results is shown in fig.4 where the influence of the silicide resistivity was determined by remodelling and varying the resistivity from 15 to $500\mu\Omega.cm$ (for $h = 75nm$, and $\delta = 150nm$). The graph shows a near linear dependence on resistivity. For the three barrier heights, only very large increases in ρ_b have a significant effect on Rc. (It was also observed from the modelling that when a doping profile which was constant with depth was used, a similar dependence to that of fig.4 was obtained). Increasing ρ_b from 15 to $50\mu\Omega.cm$ caused only a minor increase in Rc ($\sim 3\%$), indicating that for this range of resistivity range, ρ_b has only a second order effect. This is in marked contrast to the influence of Φ_b. In fig.5 the restriction of $\rho_{ca} = 4 \times 10^{-9}\Omega.cm^2$ is removed in order to examine the contribution ρ_{ca} makes to Rc. This figure shows the influence of the metal-silicide interface (ρ_{ca}) on Rc for various δ values

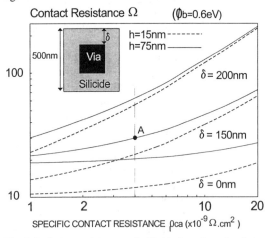

Contact Resistance Ω   ($\Phi b = 0.6eV$)

SPECIFIC CONTACT RESISTANCE ρ_{ca} ($\times 10^{-9}\Omega.cm^2$)

Fig.5. Influence of the metal-silicide interface (ρ_{ca}) on Rc for various contact via sizes δ (d=500nm).

using the TiSi$_2$ data and with the silicide depth=15nm and 75nm (solid line). Clearly the ρ_{ca} interface can significantly increase Rc particularly when the contact has (i) values of ρ_{ca} ($\geq 2 \times 10^{-9}\Omega.cm^2$) and (ii) a decreasing contact via size (δ increasing). These observations are important for silicide contacts due to the increasing use of various barrier materials at this interface and the decreasing contact dimensions. The vertical dashed line indicates where $\rho_{ca} = 4 \times 10^{-9}\Omega.cm^2$ (experimental values for the Al-TiSi$_2$ interface of $3-5 \times 10^{-9}\Omega.cm^2$ have been measured [6]). The curves for h = 15nm show a more rapid increase than for h=75nm. Thus the influence of an increasing value of ρ_{ca} is more pronounced when the silicide depth is near where the minimum in Rc is observed ($\sim$15-20nm in fig.3). In an actual device, the value of ρ_{ca} may not be as low as the value assumed for figs.3 and 4. The experimental value of $\sim$4x10$^{-9}\Omega$.cm^2 was derived for a contact with no barrier metal (such as Ti-W) between the silicide and aluminium. Thus consideration may need to be given to the contribution ρ_{ca} makes to Rc. Fig.6 is similar to fig.5. Here the

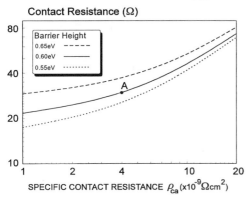

Φ_b=0.6eV, δ=150nm and h=75nm result of fig.5 is compared to the results computed for Φ_b=0.55 and 0.65eV. It can be seen from fig.6 that the importance of silicide type (i.e. Φ_b) diminishes as the contribution of ρ_{ca} to Rc becomes more dominant (as ρ_{ca} increases). The results of figs. 5 and 6 give a clear indication of an acceptable range for ρ_{ca} for contacts with various geometries and barrier heights. However additional experimental data on ρ_{ca} for various metal-barrier-silicide interfaces is required in order to precisely determine the conditions under which these experimental values will make a significant impact on Rc.

Fig.6. Influence of the metal-silicide interface (ρ_{ca}) on Rc for the three types of silicide in Table II (δ=150nm and h=75nm).

CONCLUSION

It has been shown from the modelling of a silicide well contact that small changes in the silicide-silicon barrier height Φ_b have a strong influence on ρ_{cu} which in turn strongly influence the contact resistance Rc. The influence of the silicide's resistivity on Rc is much weaker than Φ_b. Thus in choosing a silicide for a low resistance contact, Φ_b is of more significance than ρ_b. It is also shown that the metal-silicide interface ρ_{ca} may make an important contribution to Rc. For the sub-micron sized contacts modelled here, values of ρ_{ca} typically $\geq$2x10$^{-9}\Omega$.cm^2 will make a noticeable contribution to Rc. This contribution is of increasing importance for the decreasing contact via openings found in ULSI devices.

ACKNOWLEDGMENTS

Two of the authors (GKR and ASH) would like to acknowledge the support of an Australian Research Council Grant in undertaking this work.

REFERENCES

[1] W. M. Loh, et al., IEEE Trans. Electron Dev., vol. ED-34, no.3, 1987, pp.512-523.
[2] G.K. Reeves and H.B. Harrison, IEEE Trans. Electron Dev., vol. ED-42, 1995, pp.1536-1547.
[3] G.K. Reeves, A.S. Holland, and P.W. Leech, MRS Proc., vol. 514, Spring 1998, pp.363-368.
[4] P. Revva, A. Kastanas and A. Nassiopoulou, J. Electrochem. Soc. vol. 144, 1997. pp.4072-4076.
[5] T. Ohmi, Microelectronics Jnl. vol. 26, no.6, 1995, pp. 565-619.
[6] C.Y. Chang, Y.K Fang and F.M. Sze, Solid State Electronics vol.14, 1971, pp.541-550.
[7] G.K. Reeves, A.S. Holland, and P.W. Leech, MRS Proc., vol. 564, Spring 1999 (in-press).
[8] A.S. Holland, G.K. Reeves and P.W. Leech, Proc. of ESSDERC '98, Sept. 1998. pp.128-131.

THE GROWTH OF EXTRUSIONS AT W-TERMINATED ALSICU LINES AND AN APPROACH FOR AN EXTRAPOLATION TO USE CONDITIONS

F.UNGAR, A.V.GLASOW
Infineon Technologies, Reliability Methodology, Otto-Hahn-Ring 6, D-81739 Munich, Germany

ABSTRACT

Bipolar transistors were stressed with high current densities at elevated temperatures in order to characterize the reliability of their interconnection. The main interest was focused on the emitter lead. Latter mostly has to be designed as small as possible to obtain the highest possible performance, but concurrently has to carry high current densities.

It was found that not voiding within these lines is the main concern for the operational use, but the growth of extrusions, resulting e.g. in shorts.

Additionally, a strong increase of the copper concentration within the AlSiCu-line was observed, which could lead to a contamination of the active regions of the transistors.

In this work we show an approach to extrapolate the results obtained from highly accelerated tests to use conditions. A good correlation to Black's Law was found.

INTRODUCTION

With decreasing geometry of metal lines and contacts, the demands on the reliability of the metallization of integrated circuits increase rapidly. Besides voiding due to electromigration additional reliability concerns become more and more severe as a result of high current densities.

The purpose of these investigations was to determine if the current density in short lines (<50 µm) can be increased in conformity with the Blech-effect [1]. Blech claimed that in short lines electromigration is suppressed by a counteracting force (backflux).

For bipolar transistors the interconnects are more and more the current limiting factor. Blech's effect could therefore be used to adapt the electrical design rules e.g. for the emitter lead and to allow a higher current density.

However, using higher current densities raises the probability of the growth of extrusions which are responsible for shorts between interconnect lines.

In order to extrapolate the results obtained from accelerated life-tests to operating conditions a valid model has to be found. In this work it was considered, that the development and growth of extrusions are the result of the accumulation of metal atoms due to electromigration at a flux divergence (e.g. a W stud). In this case Black's [2] equation can be used as the base of the extrapolation model.

EXPERIMENT

Tests were performed on npn-transistors arising from a standard 25 GHz BICMOS process [s. figure 1 and 2]. Emitter leads (width = 2µm) with three different lengths l_E (26µm, 42µm and 62µm) were stressed in order to detect the influence of the interconnect geometry. The collector is connected with stacked vias to a wide metal 2 line. The inter-metal dielectric consists of silane-oxide. The metal stack of the metal lines was 44/100/400/20/33nm Ti/TiN/AlSiCu/Ti/TiN for metal 1 and 44/100/800/22/33nm for metal 2 respectively.

Electrons are moving from both sides of the transistor via contacts into the emitter. The electron flux causes metal atoms to migrate out of the wide connecting lines (reservoir) towards the emitter. As the tungsten contacts do inhibit a further migration, metal atoms are accumulated and so a compressive stress is induced.

747

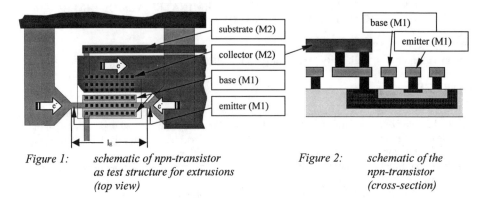

Figure 1:   schematic of npn-transistor   Figure 2:   schematic of the
             as test structure for extrusions   npn-transistor
             (top view)   (cross-section)

Electrical tests were performed on wafer level at varying chuck temperatures (T = 200 ...
300°C) and emitter line current densities (j = 3 ... 7 MA/cm²). The base of the transistor was
connected via a 100 Ohm resistor to the collector in order to turn it on, but still force the main current through the collector.

The time to failure was defined as the time until the first extrusion becomes visible through a standard optical microscope (s. figure 5). This time correlates very well with a sudden decrease of the resistance of the structure (s. figure 3).

The generated extrusions increase both the cross-section of the line and the conductivity. The sample size was at least 10 structures per stress condition. The temperature increase in the test structure due to Joule heating was determined with fluorescence-micro-thermography (FMT).

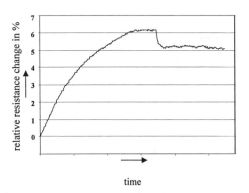

Figure 3:   typical resistance drop
             during the test after
             formation of an extrusion

RESULTS

After electrical stress test structures were analysed by FIB (Focussed Ion Beam). Figure 6
and 7 show typical failure scenarios of the test structures after extrusions have been detected
between emitter and base leads.

A migration of material from both sides of the emitter towards the tungsten plugs leads to an
accumulation of material at these plugs. Because of this accumulation a compressive stress is
induced which finally leads to a crack in the encapsulating IMOX. After the oxide is cracked,
metal atoms are moving into these cracks, reducing the compressive stress (figure 6). In some
cases the thickness of the emitter line increases up to 30% (figure 6 and 7). Both figures show

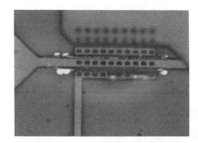

Figure 4: test structure with extrusions between emitter and base (SEM)

Figure 5: test structure after 200h of electrical stress showing extrusions (optical microscope)

the typical development of a crack in the oxide. The direction of the crack tends towards the surface of the chip until an interface between two layers is reached. Due to reduced adhesion between these layers the crack now will follow this interface.

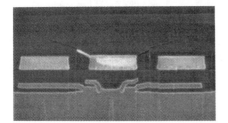

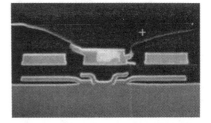

Figure 6: test structure after an extrusion is detected (SEM after FIB)

Figure 7: test structure after 200h of electrical stress (SEM after FIB)

For the electrical stress the emitter current density has been varied depending on the test temperature in order to achieve similar times to failure. The temperature of the emitter lead was determined with FMT measurements to take the additional JOULE heating into account.

In table 1 the MTFs for tests at a current density of $j=5$ MA/cm² and different temperatures are summarized.

line length	line temperature			
	210°C	260°C	285°C	310°C
26μm	28,05 h	3,67 h	1,72 h	0,64 h
42μm	14,60 h	2,93 h	0,85 h	0,32 h
62μm	9,03 h	1,86 h	0,74 h	0,22 h

Table 1: MTFs for varying line lengths and stress temperatures at $j=5MA/cm^2$

With increasing temperature the failure time was found to decrease very fast caused by the increased migration of the metal atoms. As shown in figure 8 for the different types of emitter lead, the time to the first appearance of extrusions depends nearly logarithmically on the test temperature. A logarithmic dependence of the MTF was also found for measurements with varying current densities.

Figure 9 displays the results for three line lengths and three stress current densities. From figures 8 and 9 it is obvious, that time to failure not only depends on current density and test temperature, but is strongly correlated to the geometry of the structure. The line-length-dependence is shown in figure 10.

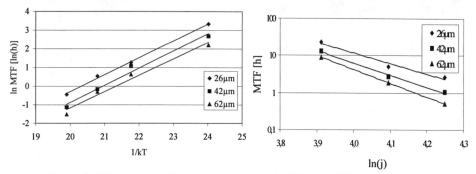

Figure 8: failure time vs. line temperature at j=5 MA/cm²

Figure 9: failure time vs. current density *at T=200°C*

According to Black`s equation the activation energy and the current density exponent have been determined for the process of the development of extrusions in the test structure. In table 2 the activation energies for different current densities and all three lengths are summarized.

From the data in table 2 the activation energy has been found to be independent of the length of the test structure. The Ea is slightly higher than values obtained from conventional EM-test structures for Al-grain boundary migration. The current density exponent has been determined with values of about n=5 at these tests.

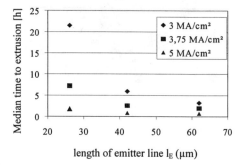

Figure 10: length-dependence of the formation of extrusions at $T_{chuck}=275°C$

line length	Current density	
	j=3,75MA/cm²	j=5MA/cm²
26µm	*1,15 eV*	*0,90 eV*
42µm	*1,08 eV*	*0,92 eV*
62µm	*1,12 eV*	*0,87 eV*

Table 2: Activation energies for varying lengths of the emitter lead

Using Black's equation, values have been extrapolated to operating conditions. At use conditions ($T_{use}=125°C$; $j_{use}=0,275MA/cm²$) even the most critical values show a minimum time of more than 10^3 years until the first appearance of an extrusion. With these exceptional high MTFs a formation of extrusions is no reliability concern at use conditions. This is also reflected from the experience of previous EM tests and field returns.

Additional quantitative EDX analysis performed after electrical stress revealed an increased concentration of copper as high as up to 40wt% within the emitter lead and the extrusion (see figure 11 and 12). Supposing that mainly copper-diffusion is activated by the electron flow, these high concentrations are caused by accumulation of copper at Al grain boundaries.

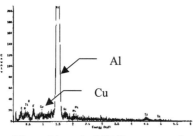

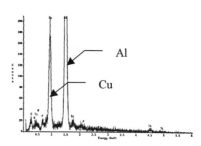

Figure 11: typical EDX spectrum of the line before electrical stress

Figure 12: typical EDX spectrum of the line after electrical stress of 200h

CONCLUSION

As expected by Blech the nucleation of voids within the line is suppressed by the counteracting backflux force. However, the dramatic increase of copper in the emitter line leads to the conclusion, that not aluminum migration but copper diffusion (interstitial or along Al grain boundaries) is the main cause for the evolution of the high compressive stress which finally leads to the cracking of the passivation. This correlates well with the high activation energy obtained during the experiments [3].

Significant is the point that such high concentrations of non-encapsulated copper in the region of high frequency transistors can deteriorate their performance and contaminate the integrated circuit. Further analysis will be done in order to characterize the amount of copper in the surrounding oxide.

The growth of extrusions are and will be a severe reliability concern especially for lines embedded in the mechanically weak low k dielectrics and for copper-damascene structures with poor adhesion between the SiN cap-layer and the main inter-metal dielectric. This work shows a method to determine the parameters needed for extrapolation to use conditions. Standard extrusion monitors like parallel lines do not necessarily detect existing extrusions. Sudden decreases of the structure's resistance are a good sign for this event and need to be taken seriously. The most critical geometric environment needs to be determined in order to be able to generate a "worst-case" test-structure.

LITERATURE

[1] : I.A.Blech, Electromigration in thin aluminum films on TiN, J.Appl.Phys., Vol. 47 (1976)
[2] : J.R.Black, Electromigration - a brief survey ..., IEEE Trans. On Electron Devices, ED16 (1969)
[3] : Kaur et al, Handbook of Grain and Interphase Boundary Diffusion Data, Ziegler Press (1989)

AUTHOR INDEX

Achen, A., 357
Adolphi, B., 257
Ahn, Dong-Joon, 507
Akasaka, Yasushi, 521
Anoshkina, E., 313
Aoi, Nobuo, 649
Aoyama, J., 545
Armstrong, N., 109
Asano, T., 219
Assous, M., 357

Babu, S.V., 597
Baklanov, M.R., 577, 615
Bao, T.I., 431
Barth, Hans-Joachim, 319
Bartha, J.W., 257
Basak, S., 677
Beaudoin, Stephen P., 467
Bender, H., 149, 167
Bendjus, B., 257
Bertz, A., 395
Beyer, G.P., 149, 167
Blanc, R., 357
Bloomfield, M.O., 663
Boey, J.Y., 201
Booms, L.M., 473
Broadbent, E.K., 117
Brongersma, S.H., 149, 167
Brown, I.G., 245
Brown, W.L., 85
Bufaroosha, M.S., 201
Buonanno, M., 349

Cale, T.S., 207, 663
Case, C.B., 349
Castro, L., 173
Cerullo, M., 633
Chang, C.L., 431
Chang, W., 431
Chao, L.C., 47
Chao, Y.C., 455
Charneski, Lawrence J., 213
Chen, C.C., 47
Chen, Linlin, 101
Chen, S.C., 461
Cheung, David, 379
Cheung, R., 173
Childs, K.D., 557
Chiu, Y.H., 455
Cho, Sung-Lae, 239
Chopra, Nasreen, 379
Chow, L., 313
Chung, H.D., 605

Chung, Ju-Hyuk, 583
Chung, U-In, 583
Clarke, D., 513
Clevenger, L.A., 17, 501
Clough, S.P., 557
Conard, T., 615
Contolini, R.J., 117
Curtis, Gary, 637
Cygan, Z.T., 201

DeHaven, P., 77
Demolliens, O., 357
Dew, S.K., 669
Dixit, G., 539
Doi, H., 295
Domae, Shinichi, 271
Drese, R., 85
Dubin, Valery M., 63
Dundas, C., 637

Educato, J., 705
Egashira, Y., 545
Engelhardt, Manfred, 417
Erjavec, James, 467
Etoh, Ryuji, 565
Evans, David R., 213

Fan, H., 201
Ferguson, J., 437
Fischer, A.H., 137
Fischer, D., 257
Fluegel, J., 77
Forsythe, G., 349
Francois-Saint-Cyr, H., 313
Fronczek, F.R., 201
Froyen, L., 577
Fujikawa, Takao, 155, 187
Fujiwara, Norito, 487
Fukasawa, Masanaga, 41, 365
Fukuda, Takuya, 649
Fulton, Dakin, 101
Furusawa, Takeshi, 609
Furuta, P., 327
Furuya, Akira, 27

Galewski, C.J., 513
Gambino, J., 501
Gao, T., 33
Gao, Xiaoping, 449
Gebhart, T., 69
Geenen, L., 409
Gessner, T., 125, 195, 335, 395
Gilet, J.M., 251

Golovin, D., 85
Gomi, H., 589
Gonella, R., 251
Goto, Hidekazu, 289
Gray, W.D., 33
Grief, M., 677
Griffin, G.L., 201
Grivna, G.M., 627
Gross, M.E., 85
Guinn, K.V., 633

Hah, S.R., 605
Hamamura, H., 283
Harada, Takeshi, 271
Hariharaputhiran, M., 597
Harris, K.K., 307
Hasegawa, Toshiaki, 41, 365
Hasse, W., 721
Hasunuma, M., 699
Hawker, C.J., 327
Hay, J., 327
Hayashi, Y., 425
Hayashi, Yoshihiro, 27
Hedrick, J.L., 327
Helneder, J., 69, 125, 417
Hemmes, D.G., 77
Herner, S.B., 539
Hey, P., 167, 705
Holland, A.S., 741
Holmes, D., 615
Hong, J., 207
Hong, Jeongeui, 527
Hong, S.J., 605
Horiike, Y., 545
Hosokawa, N., 295
Hsieh, K., 133
Hsieh, W.Y., 133
Hsiung, C.S., 133
Hsu, Sheng Teng, 213, 487
Hu, C-K., 691
Huang, J., 437
Huang, J.S., 643, 735
Huang, K., 437
Huang, M.H., 455
Huang, Tzufang, 379
Huang, Y.M., 403
Huot, A., 137
Hussain, A., 313
Hutagalung, F.R., 443

Iggulden, R.C., 17, 501
Ikeda, Koichi, 41, 365
Im, Se-Joon, 227, 239
Israel, Barak, 301
Ito, N., 93
Ito, S., 699
Itoh, H., 545

Itoh, Hitoshi, 649
Ivanov, I.C., 245

Jackson, R.L., 77, 117
Jaiswal, R., 735
James, A.M., 201
Jang, S.M., 431
Janssen, G.C.A.M., 161
Jeng, S.M., 571
Jin, Won-Hwa, 527
Joshi, R.V., 669

Kadomura, Shingo, 41, 365
Kanamori, J., 461
Kanazawa, Masato, 565
Kaneko, H., 699
Kang, Geung-Won, 583
Kang, Ho-Kyu, 387
Kang, S.H., 643, 713
Kanno, I., 533
Karthikeyan, S., 735
Kawai, M., 699
Kawazoe, T., 277
Kelley, D.R., 109
Kersch, Alfred, 655
Kim, Chang-Young, 507
Kim, Choon-Hwan, 507
Kim, Chul-Sung, 583
Kim, Chung-Tae, 343
Kim, Dae-Yong, 63
Kim, Hong-Seok, 527
Kim, Hyun-Mi, 495
Kim, Ki-Bum, 227, 239, 495
Kim, Kyoung-Ho, 227
Kim, Kyu-Hyun, 527
Kim, Pilsung, 527
Kim, Sam-Dong, 343
Kim, Si-Bum, 343
Kim, Soo-Hyun, 227, 239
Kim, Soo-Jin, 507
Kim, Woo-Hyun, 507
Kitabjian, P., 167
Kito, Hideyuki, 41
Klaasen, W.A., 9, 691
Koerner, Heinrich, 319, 417
Komai, Naoki, 41
Komiyama, H., 283, 443, 545
Kondo, Tomoyasu, 155
Kondoh, E., 219
Kötter, T., 721
Ku, Ja-Hum, 583
Kumihashi, Takao, 609
Kwon, Dong-Chul, 387
Kwon, Hyug-Jin, 507
Kwon, Tae-Seok, 527

Lacour, C., 357

Lagrange, S., 149
Lai, S., 735
Lanckmans, F., 409
Lang, Chi-I, 379
Langsam, Michael, 449
Lankmans, F., 615
Lardon, N., 357
Laursen, T., 677
Lee, Eung-Joon, 583
Lee, Haebum, 63
Lee, Hyeon-Deok, 387
Lee, In-Haeng, 507
Lee, Jang-Eun, 583
Lee, Ju-Hyung, 379
Lee, Kyung-Bok, 507
Lee, Moon-Yong, 583
Lee, Peter, 379
Lee, Sang-In, 387, 605
Lee, T.S., 735
Lee, Won-Jun, 527
Lee, Young-Hyun, 583
Lee, Youngjun, 527
Leech, P.W., 741
Li, Hua, 577
Li, J., 735
Li, Y., 597
Liang, M.S., 47, 143, 265, 431, 455,
    571, 621
Lim, I., 735
Lin, X., 643
Lingk, C., 85
Littau, K.A., 539
Liu, C.C., 403
Liu, C.S., 265, 621
Liu, J.C., 47
Liu, Kuowei, 379
Loboda, M.J., 371, 437
Locke, P.S., 77
Loh, E.G., 443
Londergan, A.R., 513
Looi, K.K., 735
Lopatin, Sergey D., 63, 181
Lou, I-Shing, 379
Low, Kia Seng, 319
Lu, Yung-Cheng, 379
Lui, M.H., 47
Lur, W., 133, 403

Ma, Jim, 379
Ma, Wen, 479
Maa, Jer-shen, 487
Machida, Shuntaro, 609
Maddalon, C., 357
Maeda, Keiichi, 41
Maekawa, Atsushi, 609
Maex, Karen, 33, 149, 167, 409, 577, 615
Maisonobe, J.C., 357

Malhotra, S.G., 17, 77
Mandal, Robert, 379
Markley, Thomas J., 449
Masui, Takuya, 155, 187
Matsunaga, Hironori, 649
Matsuo, Kouji, 521
Matthews, Paul, 379
Matz, W., 257
Maverick, A.W., 201
Mayer, S.T., 53, 117
Maynard, H., 349
McGahan, W.A., 473
McKinley, J.M., 307
McNevin, S.C., 633
Meindl, James D., 3
Melov, V., 257
Meng, Michael, 683
Merchant, S.M., 85, 307, 735
Meyer, H., 69
Meynen, H., 33
Miller, R.D., 327
Miner, J., 349
Miyano, Kiyotaka, 521
Miyata, Koji, 41, 365
Moffat, T.P., 109
Mohler, C.E., 473
Monteiro, O.R., 245
Moon, J.T., 605
Moriyama, M., 277
Morohashi, T., 557
Motakef, Shahrnaz, 449
M'saad, Hichem, 479
Murakami, M., 277
Muranaka, S., 533
Murella, K., 677

Naik, Mehul, 379, 705
Nakajima, Kazuaki, 521
Nakajima, T., 545, 551
Nakamura, T., 233
Nakamura, Yoshitaka, 289
Ngai, C., 437
Nguyen, C.V., 327
Nishino, S., 425
Nojiri, Kazuo, 609
Nomura, E., 93

Oates, A.S., 643, 713
Obeng, Y.S., 643, 713
Ogawa, H., 545
Ogawa, Shinichi, 565
Oh, M., 85, 307
Ohba, T., 161, 557
Ohnishi, Shigeo, 487
Ohnishi, T., 187
Oiwa, R., 557
Okabayashi, H., 93

Okada, O., 295
O'Neill, Anthony, 319
O'Neill, Mark L., 449
Osaka, T., 727

Palmans, R., 149
Pamler, Werner, 319
Pan, T., 705
Pang, Ben, 379
Papapanayioutou, D., 245
Parikh, S., 705
Park, Heung-Lak, 507
Park, Hyung-Soon, 507
Park, Jong-Wang, 583
Park, Ki-Chul, 239
Park, Suh-Hu, 583
Park, Sung-Eon, 495
Park, Yun-Ho, 387
Parks, C., 77
Passemard, G., 357
Patel, A., 437
Patton, E., 77
Paul, D.F., 557
Peace, Steve, 637
Peikert, M., 257
Peng, S.C., 683
Petrov, N., 173
Pierce, D., 705
Poon, Tze, 379
Proost, J., 167

Ramarajan, S., 597
Rana, V.V.S., 705
Rathi, S., 437
Raupp, Gregory B., 467
Reeves, G.K., 741
Reid, J.D., 53, 117
Remenar, J.F., 327
Rennau, M., 395
Rha, Sa-Kyun, 527
Rice, D.P., 327
Richard, E., 149, 167
Richards, D.F., 663
Richardson, K., 313
Riedel, S., 125, 195
Ritzdorf, Tom, 101, 637
Roberts, David A., 449
Robeson, Lloyd M., 449
Rosenberg, R., 691
Rossman, Kent, 479
Runnels, S.R., 677
Ryu, Choon Kun, 343
Ryu, Inn-Cheol, 507

Saitoh, S., 93
Sakaue, H., 727
Sans, C.A., 513

Sasai, H., 533
Sato, Haruhiko, 271
Schaekers, M., 577
Schmid, Guenter, 417
Schnabel, R.F., 501
Schreiber, J., 257
Schrenk, Michael, 417
Schulz, S.E., 125, 195, 395, 721
Schwarzer, R.A., 137
Schwerd, Markus, 69, 125, 319, 417
Seidel, T.E., 513
Seidel, Uwe, 417
Sekiguchi, A., 295
Sekiguchi, Mitsuru, 271
Sezi, Recai, 417
Shacham-Diamand, Yosi, 173, 301
Shimizu, N., 233
Shimogaki, Y., 283, 443, 545
Shingubara, S., 727
Shirafuji, T., 425
Shue, S.L., 143, 265, 571, 621
Sierocki, Paul R., 449
Sikita, John, 467
Simon, A.H., 17, 77
Sleeckx, E., 577
Smy, T., 669
Sohn, Hyunchul, 507
Stafford, C.R., 109
Stamper, A.K., 9, 691
Stevie, F.A., 307, 313
Streiter, R., 335
Struyf, H., 33
Sugiarto, Dian, 379
Sugiyama, M., 545
Sugiyama, Tatsuo, 565
Suguro, Kyoichi, 521
Sun, B., 539
Suzuki, K., 295
Suzuki, Kohei, 155, 187
Sverdlov, E., 173
Sverdlov, Yelena, 301
Swaanen, M., 251

Tago, Kazutami, 609
Taguchi, Mitsuru, 41, 365
Taguchi, Yoji, 155
Taguwa, T., 589
Tai, W.W., 349
Takahagi, T., 727
Tamaru, Tsuyoshi, 289
Tanaka, M., 277
Tanaka, T., 545, 551
Tanaka, Y., 539
Tang, B., 437
Tao, H.J., 455
Tatsumi, Tetsuya, 41, 365
Taylor, J. Ashley, 633

Thies, A., 69
Ting, C., 245
Tobe, R., 295
Tokashiki, K., 633
Tokunaga, Kazuhiko, 41
Tokunaga, Takafumi, 609
Toney, M., 327
Torres, J., 251
Tsai, C.S., 47, 455
Tsai, C.Y., 403
Tsai, M.H., 143
Tsai, W.J., 143
Tseng, Eric, 683
Tsutsumi, Kikuko, 565

Uchibori, C.J., 233
Uchida, Yoko, 289
Ueno, K., 93
Uhlig, M., 395
Ullmann, J., 721
Ungar, F., 747
Urabe, K., 589
Usui, T., 699

Vandervorst, W., 409
Vanhaelemeersch, S., 33, 615
Van Hove, M., 33
Vaudin, M.D., 109
Velaga, A., 735
Vellaikal, Manoj, 479
Verbruggen, A.H., 727
Vereecke, G., 577
Vervoort, I., 149, 167
Volksen, W., 327
von Glasow, A., 137, 747

Wachnik, R.A., 9
Wang, M.Y., 571
Watanabe, T., 699
Weber, S.J., 17, 501
Wee, Yong-Jin, 387

Weiss, K., 125, 195
Weiss, U., 335
Wenzel, C., 257
Werner, Christoph, 655
Werner, T., 395
Wieser, E., 257
Willecke, Ralf, 379
Wolf, H., 335
Wong, S. Simon, 63
Wu, J.Y., 403

Xiao, S.Q., 295
Xiao, X., 335
Xu, P., 437
Xu, X.B., 295

Yamamoto, R., 283
Yamamoto, T., 589
Yamamura, Ikuhiro, 41
Yamashita, K., 545, 551
Yang, D., 207
Yang, F., 473
Yang, J.J., 349
Yang, S.C., 455
Yao, G., 735
Yau, Wai-Fan, 379
Yoon, B.U., 605
Yoon, D., 327
Yoshie, T., 461
Yost, Dennis, 379, 705
Young, B.R., 455
Yu, C.H., 143, 265, 431, 571, 621
Yunogami, Takashi, 609

Zhang, Fengyan, 487
Zhang, H-M., 539
Zhang, J., 735
Zhang, P., 167
Zhou, D., 313
Zhu, Q., 513
Zhuang, Weiwei, 213

SUBJECT INDEX

ab initio calculation, 545
additives - electrodeposition
  analysis, 117
  Cu electrodeposition, 53, 63, 69, 85, 101,
    109, 149
adhesion
  Cu, 155, 195
  layer
    CF polymer, 395
    Zr, 233
aerogel, 335, 349
Ag, 219
agglomeration - Cu on barrier, 271
Al (aluminum), 501
  CVD, 17, 27, 545, 551
  damascene, 27
  dual damascene, 699
  electromigration, 727
  interlayer, 227
  oxidized surface, 551
  reflow, 27, 699
  texture, 27
ammonia post-treatment, 283, 289
amorphization - Ta, 257
amorphous-C:F, 425, 443
amorphous-SiC:H, 437
annealing - Cu, 133, 155
ASET, 649
ashing, 533
aspect ratio, 507
Au - adsorption on Al, 551
Auger electron spectroscopy, 557
automated crystal orientation measurement,
  137

barrier
  Ag, 219
  BLOκ , 437
  Ta, 257, 265
  TaN, 239, 265
  TaSi, 257
  TiN, 295
  W, 539
bias thermal stressing (BTS), 407
bipolar transistors, 747
Blech pattern, 727
bottom-up fill - electrodeposited Cu, 53, 101,
  149
brighteners, 53

capacitance(/)
  Cu/CVD SiOC, 33
  Cu/FLARE, 41

Cu/FSG, 403
  voltage measurement, 409
CF polymer, 395
C_4F_8, 609
C_5F_8, 425
chemical mechanical polishing (CMP)
  Cu, 327
  FSG, 431, 479
  low-k, 605
  model,
    Cu, 677
    W, 683
  Ta, 327
chemical vapor deposition (CVD)
  Al, 17, 27, 545, 551
  Cu, 17, 125, 201, 207, 213
  low-k, 379
  Ru, RuO_2, 495
  SiOCH, 343
  TaN, 227, 239
  TiN, 17, 227, 271, 283, 289, 295, 589
  W, 539
chlorine (Cl) removal - CVD TiN, 283, 289
cleaning
  Cu contamination/removal, 637
  HSQ, 455
  post
    Al etch, 533
    damascene etch, 501, 615, 621
    window etch, 643
cobalt
  silicide, 301
  tungsten phosphide (CoWP), 301
conformality - CVD TiN, 257
contact(s), 501
  etching, 633
  resistance, 507
  silicides, 741
contamination - Cu, 637
corrosion - W, 533
critical void volume, 713
Cu (copper)
  adhesion, 233
  adsorption on Al, 551
  CVD, MOCVD, 125, 195, 201, 207, 213
  damascene, 85, 137, 149
  deposition equipment, 655
  diffusion, 227, 277, 307, 313
    barrier, 301
  dissolution, 327
  drift dissolution, 407
  dual damascene, 41, 47

Cu (copper) (*continued*)
electrodeposition, 9, 17, 53, 63, 69, 77, 85, 93, 101, 109, 117, 125, 133, 143, 155, 161, 167, 187, 245
additives, 53, 63, 69, 85, 101, 109, 117, 149
electroless deposition, 173
electromigration, 137, 691, 705, 721
high pressure anneal, 187
integration, 9, 417
interconnect, 137
ionized magnetron sputtering, 17, 663
oxide, 615, 621
post etch clean, 615, 621
recrystallization, 63, 77, 85, 93, 109, 167
reliability, 9, 137, 691, 705, 721
removal, 637
resistivity, 69, 77, 85, 109, 143
seed layer, 9, 245
silicide, 319
silicon nitride, 319
sputtered, 187
stress, 63, 69, 143
migration, 9
surface oxide, 615, 621
texture, 63, 69, 77, 85, 93, 109, 137, 143, 219, 227, 265
Cu(hfac)$_2$(allylamine), 201
Cu(hfac)$_2$(propylene glycol), 201
Cu(hfac)(tmvs), 195
current density - Cu electrodeposition, 53, 63, 133
Cu(tmvs)(hfac), 207
CVD - *see* chemical vapor deposition
cyclic voltammetric stripping (CVS), 117

damascene, 479
Al, 27
BLOĸ, 437
Cu, 9, 17, 85, 125, 137, 149, 271, 349, 621
DARPA, 3
defect analysis, 557
density functional theory, 551
diamond - Cu diffusion, 313
dielectric constant, 437, 443, 449
diffusion
Cu, 307
barrier for
Cu
Al, 227
CoWP, 301
Ta, 257, 265
TaN, 227, 239, 265
TaSi, 257
TiN, 227, 277

W, 539
dimethylaluminum hydride (DMAH), 545, 551
dissolution - Cu, Ta, 327
DRAM, 17, 501, 527
contact, 507
electrode, 487
drift
diffusion - Cu, 409
velocity measurement, 727
dry etching - SOG, 607
dual damascene
Al, 17, 501, 699
cleans, 251, 621
Cu, 9, 17, 479, 691, 705
etch process, 47, 425
low-k, 41, 365, 371, 379, 403, 417

ECD - *see* electrodeposition
electrical resistance model, 713
electrochemical deposition - *see* electrodeposition
electrode, 487
electrodeposition - Cu, 9, 17, 53, 63, 69, 77, 85, 93, 101, 109, 117, 125, 133, 143, 155, 161, 167, 187, 245
acceleration, 53
additives, 53, 63, 69, 85, 101, 109, 117, 149
bath age, 69
current density, 53, 63, 133, 143
forcefill, 161
impurities, 85, 143
induction delay, 133
on-line analysis, 117
radial effect, 77
electroless deposition
CoWP, 301
Cu, 173, 181
electromigration
Al, 387, 643, 699, 727, 735, 747
Cu, 9, 137, 251, 691, 705, 721
electron backscatter detection (EBSD), 93, 137
electroplating - *see* electrodeposition
embedded DRAM, 589
endpoint detection - plasma etch, 633
energy dispersive x-ray spectroscopy (EDX), 557
epitaxy - CoSi$_2$, 565
etch stop, 437, 633
etching
aromatic hydrocarbon low-k, 357
HSQ, 455
PFC elimination, 649
porous silica, 349
residues, 615, 621

extrusions, 747
  Cu, 705

Fe, 69
feature-scale planarization, 677
FeRAM (ferroelectric RAM) electrode, 487
filtered cathodic arc deposition - Cu, 245
finite element
  model - Cu electromigration, 721
  modeling, 741
FLARE (fluorinated aryl ether), 41, 365
flow modulation, 283
fluorine barrier, 539
forcefill - Cu, 161
formaldehyde-free solution, 173
Fourier transform infrared spectroscopy
  (FTIR) - low-k, 343, 349
FSG (fluorinated silicate glass), 403, 431, 479

gap fill, 357
gate
  electrode, 521
  poly-Si, 583
global warming potential, 425
glyoxylic acid, 173
gold - see Au, 551
grain
  boundary diffusion, 277, 705
  growth - Cu, 77, 109
  size
    electrolessly deposited Cu, 181
    electrodeposited Cu, 63, 85, 167

hard mask, 357
hardness testing, 85
heat sink, 335
high
  density plasma (HDP) SiO$_2$, 577
  hydrogen silsesquioxane (HSQ), 455, 461
  performance liquid chromatography
    (HPLC), 117
  pressure anneal, 155, 187

implantation
  Ge, 571
  N, 257, 571
  O, 257
integration
  Al, 17, 387, 461, 589, 699
  Cu, 9, 17, 33, 41, 251, 349, 371, 379, 403,
    417, 479
  W, 527
Interconnect Focus Center, 3
interconnects
  Al, 17, 27, 431, 699, 735
  Cu, 9, 17, 137, 431, 479, 691, 705

interlayer - Al, 227
ion beam induced CVD - TaN, 239
ionized sputtering
  Cu, 17, 663
  Ta, 17, 507
  Ti, 17
iron - see Fe
IR-RAS, 545
Ir-Ta-O, 487

Joule heating, 713
junction leakage, 565

Kikuchi pattern, 137

levelers, 53, 101
low-k dielectric, 327
  aerogel, 335, 349
  aromatic hydrocarbon, 409
  BLOκ, 437
  C:F, 395, 425, 443
  CMP, 605
  Cu drift, 409
  dual damascene, 365
  etch, 349, 357, 455, 649
  FLARE, 41, 365
  fluorinated silicon oxide (FOx, FSG,
    SiOF), 387, 431, 479
  hydrogen silsesquioxane (HSQ), 455, 461
  integration, 371, 379
  methyl hydrosilsesquioxane (MHSQ),
    461
  methyl siloxane SOG, 609
  nanoglass, 349
  nanoporous
    polymer, 327, 449
    SiOCH, 343
  organic SOG, 699
  oxazole, 417
  polyarylene ether, 409
  porous
    organosilicates, 597
    parylene-N, 467
    SiOCH, 343
  post etch clean, 615
  SiLK, 357, 473
  SiC:H, 371, 379
  SiOCH, 33, 371
  trimethyl silane (3MS), 33

MARCO, 3
mechanical, 85
MESA model, 677
methylhydrosilsesquioxane (MHSQ), 461
methylsilsesquioxane, 597
MOCVD - see chemical vapor deposition

modeling
  CMP
    Cu, 683
    W, 677
  ionized magnetron sputtering, 663
  porous dielectric, 597
  silicide contact resistance, 741

nanoglass, 349
nanoporous
  polymer, 449
  SiOCH, 343
Nb reflow liner, 699
NH_3 - *see* ammonia
nucleation - Cu CVD, 207

optical
  emission
    diagnostic, 633
    spectroscopy, 425
  monitoring - SiLK curing, 473
organic spin-on-glass (SOG), 461, 609, 699
organosilanes, 371
outgassing - low-k, 349
overfill - electrodeposited Cu, 149
oxazole - low-k dielectric, 417
oxide resputtering, 583
oxidized Al, 551
oxygen (O_2) plasma treatment, 461, 495

pad conditioning, 683
particle
  analysis, 557
  formation, 319
parylene-N, 467
passivation - via etch, 455
PDEAT - $Ta(NEt_2)_5$, 239
PECVD - *see* plasma-enhanced CVD
peroxide-based slurries, 327
PFC-free etching, 649
physical vapor deposition (PVD)
  Cu, 85, 187
  Ti/TiN, 735
  W, 539
planarization, 627
  CMP, 327
plasma
  damage, 589
  enhanced CVD (PECVD)
    a-SiC:H, 437
    C:F, 395, 425, 443
    SiC:H, SiOCH, 343, 371
    SiN_x, 319
    SiO_2, 577
    SiOC, 33
    SiOF (FSG), 387, 431, 479
    Ti, 589

  tolerance - Ag, 219
  treatment - low-k, 417, 605
pole figures, 85
polymer foam, 467
porogens, 597
porous dielectrics
  organosilicates, 597
  parylene-N, 467
  silica, 349
post etch clean, 501
  Al, 277
  contacts, 589
  dry, 621
  HSQ, 455
  wet, 615
pre-amorphization implant, 571

rapid thermal annealing (RTA)
  electrodeposited Cu, 167
  WN_x, 517
recrystallization
  electrodeposited Cu, 63, 77, 85, 93, 109, 167
  PVD Cu, 85, 187
reflow Al, 27, 699
reliability - *see* electromigration model, 713
removal rate model - CMP, 683
resistivity
  electrodeposited Cu, 69, 77, 85, 109
    annealing, 167
  electrolessly deposited Cu, 181
  sputtered Cu/Zr, 233
  W, WN_x, 513
Ru
  CVD, 495
  oxide CVD, 495
ruthenocene, 495

seed layer - Cu, 125, 245
SiCH (black diamond), 379
silane, 319
silicide, 527, 571, 577, 583, 741
silicon nitride (SiN_x), 307, 319
SiLK, 357, 473
silver - *see* Ag
SIMS - Cu diffusion, 307, 313
simulation
  Cu/aerogel, 335
  Cu CMP, 655
  CVD W, 669
  electrodeposited Cu, 655
  PVD
    Cu, 655, 663
    Ta, 655
    TaN, 665
    Ti, 665

SiO$_2$ - Cu diffusion, 307
SiOC, SiOCH, 33, 343
SiOF, 387
spin-on low-k
    nanoporous polymer, 449, 461
    oxazole, 417
    SiLK, 473
sputter etch - dielectric planarization, 627
step coverage, 627
stress
    electrodeposited Cu, 17, 69, 143
        annealing, 167
    low-k, 371
    migration
        Al, 735
        Cu, 9
sulfuric acid, 521
suppressor, 53
surface(/)
    interface diffusion, 691
    reaction model, 545

Ta (tantalum)
    dissolution, 327
    IPVD, 17
TaN (tantalum nitride)
    MOCVD, 227, 239
    PVD simulation, 669
Ta(NEt$_2$)$_5$ - PDEAT, 239
TDEAT - Ti(NEt$_2$)$_4$, 295
tetramethylsilane, 343
texture
    Ag, 219
    Al, 27, 735
    electrodeposited Cu, 63, 69, 77, 85, 93,
        109, 137, 143, 227, 265, 271
    electrolessly deposited Cu, 181
    Ta, 257
    TaN, 239, 257
    TiN, 277
    W, 539
thermal conductivity
    desorption, 577
        spectroscopy, 479
    low-k, 387
    optimization, 335
thin film, 85
Ti (titanium), 271, 565
    adsorption on Al, 551

ionized metal plasma deposition, 507
IPVD, 17
PVD simulation, 669
silicide, 527
TiCl$_4$, 283, 289, 589
TiN (titanium nitride), 271, 277, 565
    CVD, 17, 227, 283, 289
Ti(NEt$_2$)$_4$ (TDEAT), 295
Ti(NMe$_2$)$_4$ (TDMAT), 271
trenching, 609
trimethylsilane (TMS), 371, 437
tungsten - see W
twinning - Cu, 93

vanadium - adsorption on Al, 551
vapor deposition polymerization -
    parylene-N, 467
via
    chain resistance - dual damascene Cu, 621
    cleans, 501
    damascene etch, 47
    filling - Cu, 173, 187
void
    Al electromigration, 699, 727
    Cu, 155, 161
    drift velocity, 713
voltammetry, 109

W (tungsten)
    bit lines, 527
    CVD simulation, 669
    gate, 513
    nitride (WN$_x$), 513
    oxide, 521
    plugs, 533
water vapor - effect on Cu CVD, 207
WF$_6$ encroachment, 461
whisker formation, 521

x-ray
    diffraction - see texture, pole figures
    photoelectron spectroscopy (XPS) -
        Cu surface, 615

Zr (zirconium), 233